Ergebnisse der Mathematik und ihrer Grenzgebiete

3. Folge · Band 18

A Series of Modern Surveys in Mathematics

A. E. Brouwer A. M. Cohen A. Neumaier

Distance-Regular Graphs

Springer-Verlag
Berlin Heidelberg New York
London Paris Tokyo

Andries E. Brouwer Arjeh M. Cohen
Stichting Mathematisch Centrum
Centrum voor Wiskunde en Informatica
Kruislaan 413, NL-1098 SJ Amsterdam

Arnold Neumaier
Institut für Angewandte Mathematik, Universität Freiburg
Hermann-Herder-Straße 10, D-7800 Freiburg im Breisgau

Mathematics Subject Classification (1980): 05-XX, 05-02, 05B25, 05B05, 05B30, 05CXX, 05C25, 05C35, 05C50, 05C75, 20F32, 20B25, 51Exx, 51B25, 94Bxx

ISBN-13: 978-3-642-74343-6 e-ISBN-13: 978-3-642-74341-2
DOI: 10.1007/978-3-642-74341-2

Softcover reprint of the hardcover 1st edition 1989
2141/3140-543210

To Agnete
Peggy
Ilse

Preface

Graphs are studied in a wide variety of contexts. This is due to the fact that the concept of a graph is both general and convenient. It is convenient because mathematical as well as intuitive notions can often be formulated very easily in terms of symmetric relations; it is general because there are so many symmetric relations on a set. In particular, there is no hope for the classification of all finite graphs in the vein of the classification of, say, all finite fields. However, the situation may change if regularity or symmetry is assumed to exist.

Since the ancient determination of the five platonic solids, the study of symmetry and regularity has always been one of the most fascinating aspects of mathematics, and even now there are many challenging problems in this area. One intriguing phenomenon is the fact that quite often arithmetical regularity properties of an object imply the uniqueness of the object, and (often as a consequence of the uniqueness proof) imply the existence of many symmetries, i.e., large automorphism groups. This interplay between regularity and symmetry properties of graphs is the theme of this book.

Many regularity properties are naturally expressible in terms of an association scheme, which is by far the most important unifying concept in algebraic combinatorics. Of special interest are two classes of association schemes, those which are P-polynomial or Q-polynomial. The P-polynomial association schemes (i.e., those with a linear distribution diagram with respect to some relation) are essentially the same objects as distance-regular graphs. As we shall see, the theory of distance-regular graphs has connections to many parts of graph theory, design theory, coding theory, geometry (both finite and Euclidean), and group theory. Indeed, most finite objects of sufficient regularity are closely related to certain distance-regular graphs.

Starting from very elementary regularity properties, the concept of a distance-regular graph arises naturally as a common setting for regular graphs which are extremal in one sense or another. Moreover, several other important regular combinatorial structures are equivalent to special families of distance-regular graphs: symmetric 2-designs, regular two-graphs, symmetric nets and Hadamard matrices. These connections are developed in Chapter 1. The systematic study of distance-regular graphs is postponed to Chapter 4. Parameter restrictions are discussed in Chapter 5, which allow us to compute a list of parameters of putative distance-regular graphs on not too many vertices in Chapter 14.

Chapter 2 gives a short treatment of general association schemes. Association schemes were first studied by statisticians in the guise of partially balanced designs, cf. BOSE & NAIR [110] and BOSE & SHIMAMOTO [111]. It was soon noticed that powerful existence conditions could be obtained by the representation theory of the Bose-Mesner algebra of an association scheme (BOSE & MESNER [109]). Partially balanced designs for the classical groups, closely related to distance-regular graphs, were found by RAY-CHAUDHURI [606]. Independently, group theorists studied the centralizer algebra of a permutation group and its representations (cf. WIELANDT [782]). Later it became clear that both approaches are essentially the same (cf. DELSARTE [251], HIGMAN [383]). Distance-regular graphs were introduced by Biggs around 1970 (see e.g. BIGGS [68]), and their basic theory was developed by him and others (Damerell, Gardiner, Meredith, Smith).

Connections with coding theory, made prominent by DELSARTE [251] (and discussed in Chapter 11 in this book), led to a thorough investigation of two particular classes of distance-regular graphs, the Hamming and Johnson graphs. Their characterization by parameters attracted many researchers (see Chapter 9.1.1) and was finally settled by TERWILLIGER [719] and NEUMAIER [561]. The crucial observation is that these graphs have a so-called root representation in a suitable eigenspace of the Bose-Mesner algebra; therefore, root system techniques developed by CAMERON, GOETHALS, SEIDEL & SHULT [180] can be applied to provide the geometric information required to determine the graphs. A treatment is given in Chapter 3.

Distance-regular graphs of diameter 2 are just the connected strongly regular graphs, and these exist in great numbers. The situation seems to be somewhat different for larger diameter, and we attempt to give (in Chapter 6) an exhaustive list of the known primitive distance-regular graphs of diameter > 2, and of many imprimitive ones. Earlier, more specialized collections have been compiled by BIGGS [77] (for diameter 3), and by Bannai (for large diameter). Bannai's examples are closely related to the geometry of projective and polar spaces, and constructions and properties for these graphs are collected in Chapter 9. Several mathematicians including Askey, Bannai and Stanton observed that all known primitive distance-regular graphs of large diameter have the special property that they are Q-polynomial, and Bannai directed the attention to the problem of deciding whether all Q-polynomial distance-regular graphs of large diameter are already known. This led to the result of LEONARD [487] that Q-polynomial distance-regular graphs satisfy rather restrictive parameter conditions (cf. Chapter 8), and that, for large diameter, their eigenvalues must be integral (BANNAI & ITO [33]). Results by TERWILLIGER [718, 721, 727] indicate that the known techniques are sufficient to determine the possible parameter sets for Q-polynomial graphs. Recent progress in the characterization of distance-regular graphs of large diameter by parameters or geometric properties, documented e.g. in the survey paper [34] by BANNAI & ITO, nourishes the hope that we are not very far from a complete solution of Bannai's problem.

A glance through the list of known distance-regular graphs of diameter $d > 3$ reveals the fact that most of them are in fact distance-transitive; i.e., their automorphism group G is so large that the ordered pairs of vertices at each given distance form an orbit under the natural action of G on the set of ordered pairs. The automorphism groups of most distance-transitive graphs are closely related to finite simple groups, and it is a remarkable fact that almost all nonabelian finite simple groups occur in this context; the only possible exceptions are the Chevalley groups of type E_8, F_4 and ${}^2E_6$ and some sporadic simple groups.

Infinite families of distance-transitive graphs belonging to the Lie groups can be described in a simple and uniform way by means of Tits systems, i.e., BN-pairs (see Chapter 10); the graphs obtained for the classical groups are isomorphic to those in Bannai's list. The distance-transitive graphs which are not of Lie type have a more or less sporadic nature; thus for each case, special constructions are required, relying on an associated design, code, geometry, or the group involved, see Chapters 11-13. The sporadic nature seems to be related to arithmetical coincidences, and indeed most of the sporadic graphs (discussed in Chapter 13) are uniquely determined by their parameters. Recent results of SAXL, PRAEGER & YOKOYAMA [597] on primitive permutation groups (based on the classification of finite simple groups) indicate that the complete classification of all distance-transitive graphs is in sight. The distance-transitive graphs of small valency have been determined by BIGGS & SMITH [84] (for valency 3) SMITH [681, 682, 683] (for valency 4) and recently (up to valency 13) by IVANOV, IVANOV & FARADJEV [282] and IVANOV & IVANOV [434]; see Chapter 7.

Concerning the nomenclature for graphs, we often found it convenient to attach a name to a graph in order to be able to refer to it. The name chosen is not necessarily that of the author who first discovered the graph or recognised it as being distance-regular (indeed, otherwise many of our graphs would be called 'Biggs graph', defeating the purpose of naming) - we hope that the authors involved will forgive us.

About the history of this book: Inspired by Bannai's lectures Fall 1979 at Ohio State University and Spring 1980 in Oberwolfach, the first two authors together with H.A. Wilbrink spent a few weeks working out the intersection arrays and geometric descriptions of the known families of distance-regular graphs with unbounded diameter. (The result of this formed the basis for the present Chapter 9.) Seeing these intersection arrays, the third author found that they all could be described in terms of four parameters, and invented the term *classical* for graphs with such intersection arrays. While Bannai had mainly been interested in the case of large diameter, the third author collected all distance-regular graphs of diameter at least three (all strongly regular graphs are classical), in order to find out what other graphs were classical, or at least Q-polynomial. Now, having these lists, the last two authors decided to publish this information in some form, and the first author joined the project at a time where it became clear that the result would be a book rather than a paper.

Thus, the main emphasis of this book is on describing the known distance-regular graphs, on classifying and, if possible, characterizing them. The structure of these graphs is touched upon insofar as necessary to describe the graphs, determine their intersection arrays, or to characterize them.

Therefore, a number of topics relating to distance-regular graphs have not been treated. Although many distance-regular graphs can be constructed from square 2-designs and symmetric nets, we do not treat these objects, since a proper treatment of their theory and construction would fill a separate book (cf. LANDER [474] for square 2-designs). Another book would be required to cover the present knowledge about strongly regular graphs (no such book is available at present); thus we concentrate on diameter > 2. Relations to orthogonal polynomials are hardly mentioned; a thorough discussion of this point is contained in the recent book by BANNAI & ITO [33]. Neither do we say much about topics like perfect codes and tight designs in graphs.

We felt free to cite, by reference only, all results which are rather technical or do not fit naturally into the line of development of the book; this applies, e.g., to nonexistence results for Moore geometries, a treatment of locally t-arc-transitive graphs, the integrality of eigenvalues of Q-polynomial graphs of large diameter, and to a number of geometric characterization theorems.

On the other hand, we have not hesitated to enlarge the scope of the exposition whenever that seemed natural. Thus, we treat graphs of Coxeter and Lie type in some generality, not restricting ourselves to the finite distance-regular case. Similarly, more of the theory of Euclidean representations has been given than was necessary for the characterization of certain families of classical graphs.

The different chapters of the book are fairly self-contained and in principle not much more is required from the reader than some standard linear algebra and elementary properties of groups. (To help the reader, an appendix contains a summary of other concepts and notation used in this book.) But it might be that certain chapters will be somewhat heavy-going if it is here for the first time that the reader encounters concepts like linear codes, root systems, polar spaces, or Chevalley groups. The book is not recommended as a first introduction to such subjects, or, e.g., to algebraic graph theory. (Instead, the reader might wish to consult ARTIN [9], BATTEN [46], BIGGS [71], CAMERON & VAN LINT [181], CARTER [187], CONWAY & SLOANE [215], GROVE & BENSON [339], HAEMERS [340], LANDER [474], VAN LINT [502].)

An early draft of this book was read by J.T.M. van Bon, who saved us from many mistakes and contributed numerous improvements. Several other people have contributed some remarks or corrections.

A very early draft of this book was produced by the second author with help of the typing office at the CWI in Amsterdam. The current version was typeset by the first author using TROFF at the CWI in Amsterdam.

Amsterdam, Eindhoven and Freiburg, January 1989

A.E. Brouwer
A.M. Cohen
A. Neumaier

Contents

3. REPRESENTATION THEORY

4. THEORY OF DISTANCE-REGULAR GRAPHS

5. PARAMETER RESTRICTIONS FOR DISTANCE-REGULAR GRAPHS

6. CLASSIFICATION OF THE KNOWN DISTANCE-REGULAR GRAPHS

7. DISTANCE-TRANSITIVE GRAPHS

8. Q-POLYNOMIAL DISTANCE-REGULAR GRAPHS

A. APPENDIX

Chapter 1
Special Regular Graphs

A connected graph Γ is called *distance regular* if there are integers b_i, c_i ($i \geqslant 0$) such that for any two points γ, $\delta \in \Gamma$ at distance $i = d(\gamma,\delta)$, there are precisely c_i neighbours of δ in $\Gamma_{i-1}(\gamma)$ and b_i neighbours of δ in $\Gamma_{i+1}(\gamma)$. In particular, Γ is regular of valency $k = b_0$. The sequence

$$\iota(\Gamma) := \{b_0, b_1, \cdots, b_{d-1}; c_1, c_2, \cdots, c_d\} ,$$

where d is the diameter of Γ, is called the *intersection array* of Γ (cf. BIGGS [71]); the numbers c_i, b_i, and a_i, where

$$a_i = k - b_i - c_i \quad (i = 0, \cdots, d) \tag{1}$$

is the number of neighbours of δ in $\Gamma_i(\gamma)$ for $d(\gamma,\delta) = i$, are called the *intersection numbers* of Γ. Clearly

$$b_0 = k \ , \ \ b_d = c_0 = 0 \ , \ \ c_1 = 1. \tag{2}$$

By counting edges $\delta\epsilon$ with $d(\gamma,\delta) = i$, $d(\gamma,\epsilon) = i+1$ we see that $\Gamma_i(\gamma)$ contains k_i points, where

$$k_0 = 1 \ , \ \ k_1 = k \ , \ \ k_{i+1} = k_i b_i / c_{i+1} \quad (i = 0, \cdots, d-1). \tag{3}$$

Trivial examples are the polygons; they have intersection array $\{2,1, \cdots, 1; 1, \cdots, 1, c_d\}$, where $c_d = 2$ for the $2d$-gon and $c_d = 1$ for the $(2d+1)$-gon. Other simple examples arise from the five Platonic solids: their vertices and edges form distance-regular graphs with intersection arrays $\{3; 1\}$ (tetrahedron), $\{4,1; 1,4\}$ (octahedron), $\{3,2,1; 1,2,3\}$ (cube), $\{5,2,1; 1,2,5\}$ (icosahedron), $\{3,2,1,1,1; 1,1,1,2,3\}$ (dodecahedron).

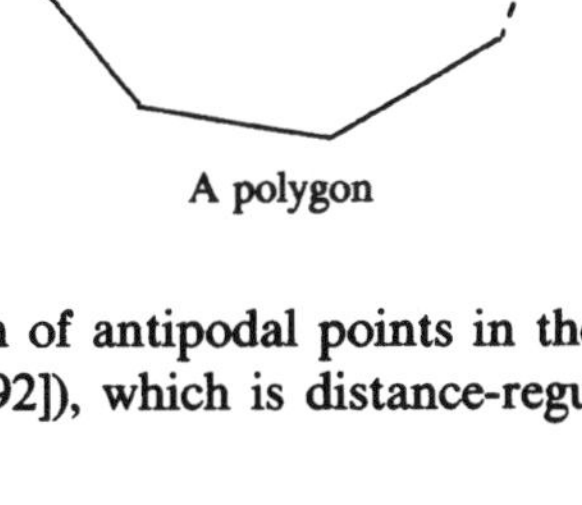

A polygon

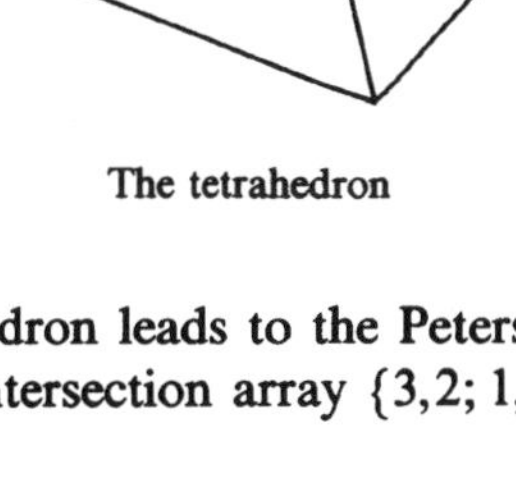

The tetrahedron

Identification of antipodal points in the dodecahedron leads to the Petersen graph (cf. PETERSEN [592]), which is distance-regular with intersection array $\{3,2; 1,1\}$. The line

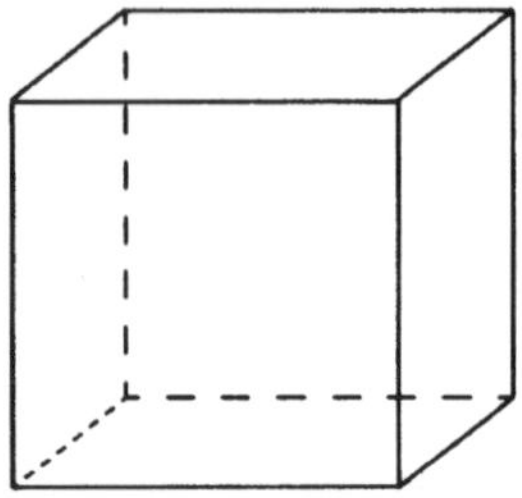

The cube

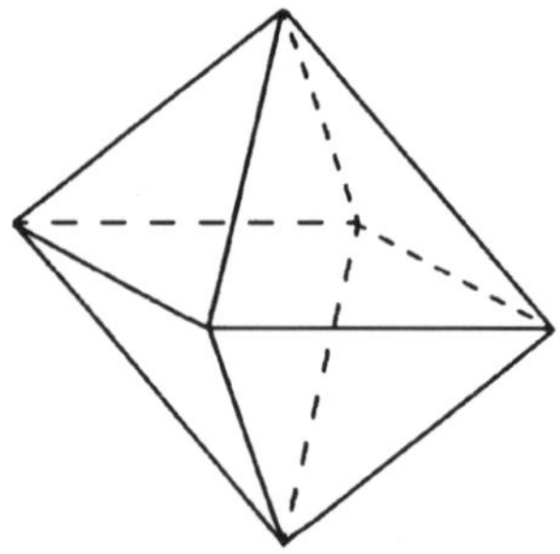

The octahedron

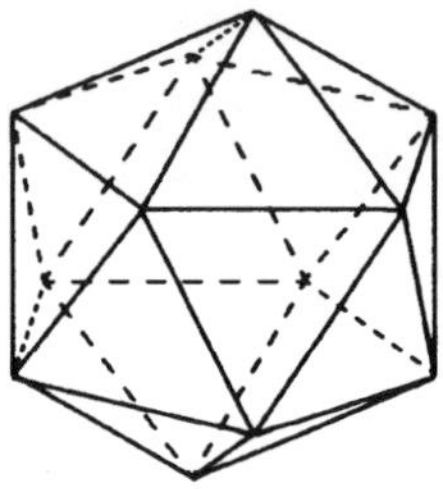

The icosahedron

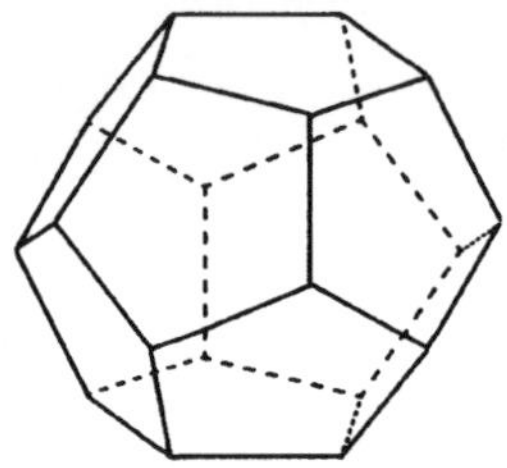

The dodecahedron

graph of the Petersen graph is also distance-regular; its intersection array is $\{4,2,1;1,1,4\}$. It is not difficult to verify by drawing that in each case the intersection array determines the graph uniquely (up to isomorphism, of course).

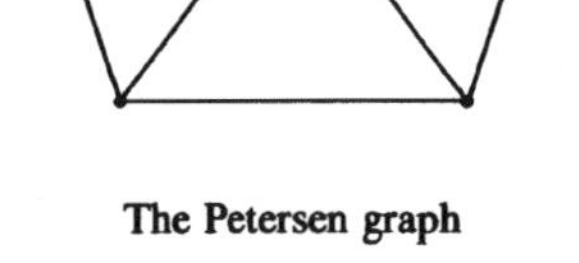

The Petersen graph

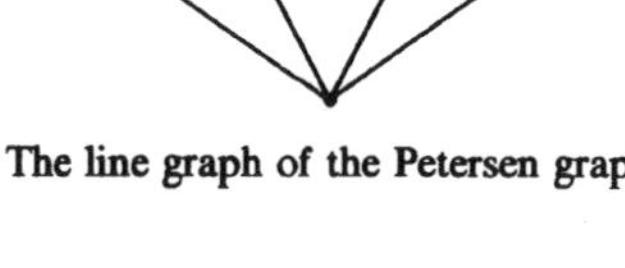

The line graph of the Petersen graph

Most distance-regular graphs, however, are more difficult to describe or to characterize by arithmetical or geometrical properties. The second part of the book (Chapters 9, 10, 11, 12 and 13) is entirely devoted to such questions.

In this chapter we start with the study of certain rather general regularity properties of graphs which hold for various classes of distance-regular graphs. Much more can be said; the material included here has been selected in view of later applications. Distance-regular graphs can be viewed as association schemes which are P-polynomial. Chapter 2 is devoted to that part of the theory of association schemes that is relevant to the study of distance-regular graphs. Chapter 3 considers representations of graphs in Euclidean space, an important tool for the characterization of a large family of distance-regular graphs by parameters. The systematic treatment of distance-regular graphs starts in Chapter 4 with elementary considerations, and is continued in Chapter 5 with conditions on their parameters. An overview of the known families of graphs and some related theory is given in Chapter 6. Distance-transitive graphs are studied in Chapter 7, and the theory of Q-polynomial graphs is treated in Chapter 8. For unexplained definitions and notation, we refer to the Appendix.

1.1 Edge regular and co-edge-regular graphs

Since many distance-regular graphs occur naturally as the extremal cases of graphs with weaker regularity properties, we shall first consider regular graphs having one or more of the following properties:

R1. Any two adjacent vertices have precisely $\lambda = \lambda(\Gamma)$ common neighbours.

R2. Any two vertices at distance 2 have precisely $\mu = \mu(\Gamma)$ common neighbours.

R3. Any two nonadjacent vertices have precisely $\mu = \mu(\Gamma)$ common neighbours.

Clearly R2 is a consequence of R3, and these two properties are equivalent if Γ is connected of diameter at most 2.

A regular graph with v points and valency k is called *edge-regular* with parameters (v,k,λ) if R1 holds, *amply regular* with parameters (v,k,λ,μ) if R1 and R2 hold, *co-edge-regular* with parameters (v,k,μ) if R3 holds, and *strongly regular* with parameters (v,k,λ,μ) if R1 and R3 hold.
Clearly, every distance-regular graph is amply regular ($\lambda = a_1$, $\mu = c_2$), and it is strongly regular if its diameter is at most 2.
A graph satisfying R3 (or only R2) with $\mu = 0$ is a disjoint union of cliques. A graph satisfying R3 with $\mu > 0$ is connected and has diameter at most two. Graphs satisfying R3 with $\mu = 1$ are studied in detail in §1.17. Clearly, for a non-complete co-edge-regular graph we have $\mu \leqslant k$, and a co-edge-regular graph with $\mu = k$ is a complete multipartite graph $K_{s \times t}$. We call a co-edge-regular graph *nontrivial* if it is not complete and $0 < \mu < k$. This definition also applies to strongly regular graphs.

It is easy to see that the complement $\overline{\Gamma}$ of a regular graph Γ is again regular (with valency $v-1-k$), and that the complement of an edge-regular graph Γ is co-edge-regular (with $\mu(\overline{\Gamma}) = v-2k+\lambda$), and, conversely, that the complement of a co-edge-regular graph is edge-regular (with $\lambda(\overline{\Gamma}) = v-2k+\mu-2$). In particular, the complement of a strongly regular graph is again strongly regular.

An edge-regular graph has $vk/2$ edges, $vk\lambda/6$ triangles, and $k\lambda/2$ triangles containing a fixed point. This shows that $vk\lambda$ is divisible by 3 and that, if k is odd, v and λ are even.

If γ is a vertex of the graph Γ, then we shall write $\Gamma(\gamma)$ for the set of neighbours of γ in Γ, and, more generally, $\Gamma_i(\gamma)$ for the set of vertices at distance i from γ in Γ. Also, $\gamma \sim \delta$ for vertices γ, δ of Γ means that γ and δ are adjacent in Γ, while $\gamma^{\perp}$ is the set $\{\gamma\} \cup \Gamma(\gamma)$.

1.1.1. Lemma. *In an amply regular graph, the number k_2 of points at distance 2 from a vertex γ is independent of γ and satisfies the relation*

$$k_2\mu = k(k-1-\lambda)\,. \tag{4}$$

Proof. Count the number of edges between $\Gamma(\gamma)$ and $\Gamma_2(\gamma)$ in two ways. □

1.1.2. Proposition (cf. Bose & Dowling [106]). *Let Γ be a connected graph satisfying* R1 *and* R2 *with $\mu > 1$. Then Γ is regular (and hence amply regular).*

Proof. Let $\gamma\delta$ be an edge and put $A := \Gamma(\gamma)\setminus\delta^{\perp}$, $B := \Gamma(\delta)\setminus\gamma^{\perp}$, and $C := \Gamma(\gamma)\cap\Gamma(\delta)$. Obviously, $|C| = \lambda$. Let p, q, r be the number of edges between vertices of A and C, B and C, A and B, respectively, and let t be the number of edges in C. There are $\lambda^2 - t = p+t+\lambda$ triangles containing γ and a point of C, and $\lambda^2 - t = q+t+\lambda$ triangles containing δ and a point of C. Therefore, $p = q$. Moreover, there are $|A|(\mu-1) = p+r$ paths of length two in $\Gamma\setminus\{\gamma\}$ containing δ and a point of A, and $|B|(\mu-1) = q+r$ paths of length two in $\Gamma\setminus\{\delta\}$ containing γ and a point of B. Consequently, $|A|(\mu-1) = |B|(\mu-1)$, and, since $\mu > 1$, we have $|A| = |B|$. Hence γ and δ have the same valency $|A|+|C| = |B|+|C|$. We have shown that any two adjacent vertices have the same valency. Since Γ is connected, this implies that Γ is regular. □

This same observation was made by Enomoto [275], who also remarked that if Γ is a connected graph satisfying R2 with $\mu > 1$ without induced $K_{1,1,2}$ subgraphs (so that each edge lies in a unique maximal clique), then Γ is regular.

The following result will have a useful application later (cf. Theorem 1.4.3); it is an immediate corollary of Theorem 1.17.1.

1.1.3. Lemma. *A co-edge-regular graph with $\mu(\Gamma) = 1$ is strongly regular.*

Proof. Let Γ have parameters (v,k,μ). Let $d(\gamma,\delta) = 2$ with $\Gamma(\gamma)\cap\Gamma(\delta) = \{\epsilon\}$. Then $\{\gamma,\epsilon\}\subseteq C$ and $\{\epsilon,\delta\}\subseteq C'$ where C, C' are maximal cliques of Γ. Since $\mu = 1$, $\Gamma(\gamma)\setminus C\subseteq\Gamma_2(\delta)$, and for each $\gamma'\in\Gamma(\gamma)\setminus C$ we get a unique vertex $\delta'\in\Gamma(\delta)\cap\Gamma(\gamma')$. This sets up a one-to-one correspondence between $\Gamma(\gamma)\setminus C$ and $\Gamma(\delta)\setminus C'$. (Indeed, if $\delta'\in C'$, then either $\delta' = \epsilon$ and $\{\gamma,\gamma',\epsilon\}$ is a clique meeting the maximal clique C in two points, which is impossible, or $\delta'\sim\epsilon$ and $|\Gamma(\gamma')\cap\Gamma(\epsilon)| > 1$, contradiction again.) Since γ, δ both have valency k, it follows that $|C| = |C'|$. Since Γ is connected and C, C' were arbitrary maximal cliques having a common vertex, it follows that Γ is strongly regular with $\lambda = |C|-2$. □

A graph Γ is said to be *locally* $\mathfrak{X}$ where $\mathfrak{X}$ is a graph or a class of graphs, when for each $\gamma\in\Gamma$ the subgraph induced by $\Gamma(\gamma)$ is isomorphic to (respectively a member of) $\mathfrak{X}$. Let us give three results that we shall need later.

1.1.4. Proposition. *Let Γ be a connected graph that is locally a pentagon. Then Γ is isomorphic to the icosahedron.*

Proof. Straightforward. □

1.1.5. Proposition. *Let Γ be a connected graph that is locally complete multipartite. Then Γ is either triangle-free or complete multipartite. In particular, if Γ is locally $K_{m_1,\cdots,m_t}$ then all m_i are equal (to m, say), and $\Gamma \cong K_{(t+1)\times m}$.*

Proof. If Γ is not triangle-free, then Γ contains a maximal induced complete multipartite subgraph K with $s \geqslant 3$ classes. If $\gamma \in \Gamma \setminus K$ has distance one from K then consideration of $\Gamma(\delta)$ for the neighbours δ of γ in K shows that K has neighbours of γ in $s-1$ or s classes of K, and that if γ is adjacent to one point of a class, it is adjacent to all points of the class. But this implies that $K \cup \{\gamma\}$ is complete multipartite. Since K is maximal and Γ is connected, $\Gamma = K$. □

1.1.6. Corollary. *A connected graph Γ with $b_1(\Gamma) = 1$ is a polygon or $K_{m\times 2}$ for some $m \geqslant 2$.*

Proof. If $\gamma \in \Gamma$, then $\Gamma(\gamma) \cong K_{a\times 2}$ for some a, so the above proposition applies. □

A graph Γ is called *coconnected* when the complementary graph $\overline{\Gamma}$ is connected; the *cocomponents* of Γ are the components of $\overline{\Gamma}$.

1.1.7. Lemma. *A connected amply regular graph that is not locally coconnected, is complete multipartite.*

Proof. Let Γ be amply regular with parameters (k,λ,μ), and let $\gamma \in \Gamma$ be a vertex such that $\Gamma(\gamma)$ is not coconnected. If some cocomponent of $\Gamma(\gamma)$ is complete, then it is a singleton, and, since $\Gamma(\gamma)$ is regular, $\Gamma(\gamma)$ is complete, but then also Γ is complete. Now suppose that Γ is not complete. Then any cocomponent A of $\Gamma(\gamma)$ has size at least $k-\mu+1$, for any two nonadjacent vertices of it have μ common neighbours among which the points in $\{\gamma\} \cup \Gamma(\gamma)\setminus A$. Also, if $|A| = a$, then A is regular of valency $\lambda-(k-a)$; it follows that A is a cocomponent of $\Gamma(\beta)$ for any $\beta \in \Gamma(\gamma)\setminus A$, and then also a cocomponent of $\Gamma(\alpha)$ for any $\alpha \in \Gamma(\beta)\setminus\Gamma(\gamma)$. Thus, for $\alpha \in \Gamma_2(\gamma)$ the set $\Gamma(\alpha) \cap \Gamma(\gamma)$ is a union of cocomponents of $\Gamma(\gamma)$, and since $\Gamma(\alpha) \cap \Gamma(\gamma)$ has size μ it cannot miss any cocomponent, so that $\Gamma(\alpha) = \Gamma(\gamma)$ (since Γ is regular). Now $\{\gamma\} \cup \Gamma_2(\gamma)$ is a coclique, a cocomponent of Γ; consequently, for any $\beta \in \Gamma(\gamma)$, $\Gamma(\beta)$ is not coconnected and our analysis applies with an arbitrary vertex of Γ instead of γ. It follows that Γ is complete multipartite. □

For an extensive study of graphs Γ that are locally Δ, where either Γ or Δ is a small graph, see Hall [348]. Graphs that are locally a tree were studied in Blass, Harary & Miller [87]. For locally Petersen graphs, see Theorem 1.16.5. For locally icosahedral graphs, see Blokhuis et al. [94]. For locally dodecahedral graphs, see Coxeter [228] and Vanden Cruyce [761]. For locally Kneser graphs, see Hall [349]. See also, e.g., Hall & Shult [351] and Johnson & Shult [445, 446].

1.1.8. Lemma. *In an edge-regular graph with parameters (v,k,λ) where $\lambda > 0$ we have $v \geqslant 3(k-\lambda)$, with equality if and only if for each triangle T and vertex $\gamma \notin T$ we have $|\gamma^\perp \cap T| \in \{1,2\}$.*

Proof. Let $a = |T^\perp \setminus T|$. Now

$$v \geqslant a+3(1+(\lambda-a-1)+(k-2(\lambda-a-1)-a-2)) = a+3(k-\lambda).$$ □

1.2 Line graphs

A graph without a subgraph of the shape $K_{2,1,1}$:

has the special property that each edge is in a unique maximal clique; such a graph Γ is the collinearity graph of the partial linear space formed by the vertices of Γ as points and the maximal cliques of Γ as lines. In particular this is the case when Γ is edge-regular with $\lambda \leqslant 1$ or co-edge-regular with $\mu = 1$. The amply regular graphs with $\lambda = 0$ and $\mu = 1$ are just the regular graphs of girth at least 5. The line graphs of regular graphs of girth at least 5 are also amply regular, and characterized by $k = 2\lambda + 2$, $\mu = 1$:

1.2.1. Proposition (BOSE & DOWLING [106]). *Let Γ be an amply regular graph with parameters (v,k,λ,μ) where $\lambda \leqslant 1$ or $\mu = 1$. Then all maximal cliques of Γ have size $\lambda+2$, and every vertex of Γ is in precisely $k / (\lambda + 1)$ maximal cliques. In particular,*

$$(\lambda+1) \mid k \ , \ (\lambda+1)(\lambda+2) \mid vk. \tag{5}$$

Moreover, $k = 2(\lambda+1)$ if and only if Γ is the line graph of a regular graph Δ of valency $\lambda+2$. If $\mu = 1$, then Δ has girth at least 5.

Proof. Since $\lambda \leqslant 1$ or $\mu = 1$, the λ common neighbours of an edge $\gamma\delta$ are mutually adjacent. Hence every maximal clique is a $(\lambda+2)$-clique. Since the sets $L \setminus \{\gamma\}$, where L is a maximal clique containing γ, partition $\Gamma(\gamma)$, there are precisely $k / (\lambda+1)$ such cliques. The total number of maximal cliques is $vk / ((\lambda+1)(\lambda+2))$, whence (5).
If $k = 2(\lambda+1)$, then every point is in precisely two maximal cliques, whence Γ is the line graph of the graph Δ whose vertices are the maximal cliques of Γ and in which two vertices are adjacent whenever they have a point of Γ in common. As each vertex on a maximal clique is on a unique other line, the valency of Δ is $\lambda+2$, and when $\mu = 1$, the girth of Δ is at least 5. The converse is immediate. □

Remark. By more complicated counting arguments, BROUWER & NEUMAIER [136] showed that an amply regular graph with parameters $(v,k,\lambda,2)$ such that $k < \frac{1}{2}\lambda(\lambda+3)$ contains no $K_{2,1,1}$. Thus,

$$\textit{If} \quad \mu = 2 \ \textit{and} \ k < \tfrac{1}{2}\lambda(\lambda+3) \ \textit{then} \ (\lambda+1) \mid k \ \textit{and} \ (\lambda+1)(\lambda+2) \mid vk. \tag{6}$$

Below we find line graphs of graphs of girth at least 5 as an extremal case of amply regular graphs. First a result on the global structure. Recall that a *(co)clique extension* of a graph Γ is the graph obtained from Γ by replacing each vertex γ by some (co)clique C_γ and letting $\alpha \sim \beta$ (for $\alpha \in C_\gamma$, $\beta \in C_\delta$, $\gamma \neq \delta$) if and only if $\gamma \sim \delta$.

1.2.2. Lemma. *Let Γ be a regular graph without triangles and without induced $2K_2$. Then Γ is a coclique extension of a point (i.e., a coclique), of an edge (i.e., a complete bipartite graph), or of a pentagon.*

Proof. Let Γ be regular of degree k. If $k = 0$ we have a coclique, so assume $k > 0$. Then Γ is connected, since $2K_2$ does not occur.
Suppose α and β are vertices of Γ at distance 3. Then $\Gamma(\alpha) \cup \Gamma(\beta)$ must induce a

complete bipartite graph $K_{k,k}$, otherwise we find an induced K_3 or $2K_2$. But if $\gamma \in \Gamma(\alpha)$, we find $\Gamma(\gamma) \supset \Gamma(\beta) \cup \{\alpha\}$ and Γ is not regular, contradiction.
Suppose Γ is bipartite. Since Γ has diameter at most two, Γ is complete bipartite.
For the remainder of the proof we may assume that Γ contains a pentagon $\gamma_0 \sim \gamma_1 \sim \gamma_2 \sim \gamma_3 \sim \gamma_4 \sim \gamma_0$. If $\alpha \sim \gamma_0$, then α is adjacent to precisely one of γ_2, γ_3. Let $C_i = \{\alpha \mid \alpha \sim \gamma_{i-1}, \gamma_{i+1}\}$ (indices mod 5). If $\alpha_1 \in C_1$ and $\alpha_2 \in C_2$, then the set $\{\gamma_0, \alpha_1, \alpha_2, \gamma_3\}$ induces $2K_2$, unless $\alpha_1 \sim \alpha_2$. Thus $C_i \cup C_{i+1}$ induces a complete bipartite graph. Looking at the degrees of the vertices γ_i, and putting $|C_0| = a$, we find $|C_1| = k-2-a$, $|C_2| = a$, etc., so that $a = k-2-a$, $a = \frac{1}{2}k-1$. This means that we have seen the entire graph, and since γ_i and each vertex in C_i have the same neighbours, we have a coclique extension of a pentagon. □

1.2.3. Theorem (cf. Taylor & Levingston [711], Neumaier [555]). *Let Γ be a connected noncomplete amply regular graph with parameters (v,k,λ,μ).*

(i) *We have*

$$k \geqslant 2\lambda + 3 - \mu, \tag{7}$$

and, with $b_2 = \min\{b_2(\alpha,\beta) \mid d(\alpha,\beta) = 2\}$, every 2-claw is in at most $\max\{0, k-2\lambda-3+\mu-b_2\}$ quadrangles.

(ii) *Equality holds in (7) if and only if Γ is the icosahedron or the line graph of a regular graph of girth at least 5. In case Γ has diameter 2 this happens only when Γ is the pentagon.*

Proof. (i) For a fixed 2-claw $\gamma \sim \beta \sim \delta$, denote by q the number of quadrangles containing γ, β and δ, and by r the number of 3-claws containing γ, β and δ. Counting in two ways the number of vertices in $\Gamma(\beta) \setminus \{\gamma, \delta\}$ adjacent to γ but not to δ, we obtain $k-2-(\lambda+r) = \lambda-(\mu-1-q)$. This leads to $k = 2\lambda+3-\mu+q+r$, whence $k \geqslant 2\lambda+3-\mu$. If $q > 0$ then there is a vertex $\alpha \in \Gamma_2(\beta)$ adjacent to γ and δ, and therefore $r \geqslant b_2(\alpha,\beta) \geqslant b_2$. This implies $q \leqslant k-2\lambda-3+\mu-b_2$ and completes the proof of (i).
(ii) Now assume that $k = 2\lambda+3-\mu$. Then $q = r = 0$, i.e., Γ has neither induced quadrangles nor 3-claws. In particular, for each vertex $\gamma \in \Gamma$, the graph $\overline{\Gamma(\gamma)}$ does not contain an induced $2K_2$ or K_3. By the above lemma it follows that this graph is an a-coclique extension of a point, an edge or a pentagon. In the first case Γ is locally a clique and hence a union of cliques, contradiction. In the second case $\Gamma(\gamma)$ is the union of two cliques, $k = 2(\lambda+1)$, $\mu = 1$, and, by the above proposition, Γ is the line graph of a regular graph of girth at least 5. Finally, if $\overline{\Gamma(\gamma)}$ is an a-coclique extension of the pentagon, then $\Gamma(\gamma)$ is an a-clique extension of the pentagon and $k = 5a$, $\lambda = 3a-1$, $\mu = a+1$, $k-\lambda-1 = 2a$. (Note that the three cases yield different parameter sets, so that the case is independent of the vertex γ chosen.) In this last case, let $\Gamma(\gamma)$ be the union of five a-cliques C_j (uniquely determined by: all vertices of C_j have the same neighbours in $\Gamma(\gamma)$), and, for $\alpha \in C_0$, let $\Gamma(\alpha)$ be the union of five a-cliques A_j. Then $A_0 = \{\gamma\} \cup C_0 \setminus \{\alpha\}$, $A_1 = C_1$, $A_4 = C_4$. Now considering $\Gamma(\beta)$ for $\beta \in C_1$, we find a contradiction unless $a = 1$. But if $a = 1$, then Γ is locally a pentagon and by Proposition 1.1.4 Γ is the icosahedron.
Finally, if Γ is a line graph of diameter 2, then Γ does not contain an induced $2K_2$ and hence all its induced cycles are pentagons; one immediately concludes that Γ itself is a pentagon. □

1.2.4. Corollary. *With the hypothesis of the theorem the following hold.*

(i) *If* Γ *contains a quadrangle, then*

$$k \geqslant 2\lambda + 4 - \mu + b_2 . \tag{7a}$$

(ii) *If* Γ *has diameter at least three and does not contain 3-claws, then equality holds in (7), and the conclusion of 1.2.3 (ii) holds.*

(iii) *If* $k < 3(\lambda - \mu + 2)$ *then* Γ *does not contain 3-claws.*

Proof. Part (i) follows immediately from 1.2.3 (i).
(ii) If $\alpha \sim \beta \sim \gamma \sim \delta$, where $d(\alpha,\delta) = 3$, then we have $\beta^\perp \cap \gamma^\perp = (\alpha^\perp \cap \gamma^\perp) \cup (\beta^\perp \setminus \alpha^\perp)$ so that $\lambda + 2 = \mu + (k - \lambda - 1)$ and $k = 2\lambda - \mu + 3$.
(iii) If $\Gamma(\gamma)$ contains a 3-coclique, then $k = |\Gamma(\gamma)| \geqslant 3 + 3\lambda - 3(\mu - 1)$. □

1.3 Strongly regular graphs

A graph of diameter 1 is a clique and hence distance-regular with intersection array $\{k;1\}$. A graph of diameter 2 is distance-regular if and only if it is a strongly regular graph with $\mu > 0$. In terms of the parameters (v,k,λ,μ), the intersection array is given by $\{k, k-1-\lambda;\ 1, \mu\}$. If $\mu = k$, then the graph is a complete multipartite graph $K_{s\times t}$ $(t > 1)$; otherwise, we have $0 < \mu < k$, i.e., a nontrivial strongly regular graph. The parameters of such a graph Γ can be naturally expressed in terms of its eigenvalues. We show that the eigenvalues are integers with the exception of the case where Γ is a *conference graph*, i.e., a strongly regular graph whose parameters satisfy $v = 4\mu + 1$, $k = 2\mu$, $\lambda = \mu - 1$ (equivalently, a distance-regular graph with intersection array $\{2\mu, \mu;\ 1, \mu\}$).

1.3.1. Theorem. *Let* Γ *be a strongly regular graph with parameters* (v,k,λ,μ) *that is neither a clique nor a coclique. Then the following hold.*

(i) *The eigenvalues of* Γ *are* k, r, s *with* $r \geqslant 0$, $s \leqslant -1$, *where* r, s *are the roots of the quadratic equation* $\theta^2 + (\mu - \lambda)\theta + (\mu - k) = 0$.

(ii) Γ *is a conference graph precisely when* r *and* s *have the same multiplicity, and then* $r, s = (-1 \pm \sqrt{v})/2$; *otherwise,* r *and* s *are integers.*

(iii) *If* $\mu > 0$, *then the parameters can be expressed in terms of* r, s *and* μ *as*

$$k = \mu - rs, \quad v = (k-r)(k-s)/\mu, \quad \lambda = \mu + r + s. \tag{8}$$

(iv) $k = r$ *if and only if* $s = -1$ *if and only if* $\mu = 0$ *if and only if* Γ *is a union of cliques; in this case, the multiplicity* m *of the eigenvalue* k *equals the number of components of* Γ, *and that of* -1 *is* $v - m$.

(v) $r = 0$ *if and only if* $\mu = k$ *if and only if* Γ *is complete multipartite; in this case* $s = -m$, *where* m *is the size of the classes of* Γ.

(vi) *When the eigenvalues* k, r *and* s *are distinct, their multiplicities are* 1, f *and* $g = v - 1 - f$, *respectively, where*

$$f = \frac{(s+1)k(k-s)}{\mu(s-r)} . \tag{9}$$

The complement of Γ *is strongly regular with parameters* $(v,\bar{k},\bar{\lambda},\bar{\mu})$*, where (for* $\mu > 0$*)*

$$\begin{aligned}\bar{k} &= v-k-1 = \bar{\mu}-(r+1)(s+1)\,,\\ \bar{\lambda} &= v-2k+\mu-2 = \bar{\mu}-r-s-2\,,\\ \bar{\mu} &= v-2k+\lambda = rs(r+1)(s+1)/\mu\,,\end{aligned} \tag{10}$$

and has eigenvalues $\bar{k}$, $-s-1$, $-r-1$.

Proof. For $\gamma,\delta\in\Gamma$, the (γ,δ)-entry of the square of the adjacency matrix A of Γ is equal to k, λ, μ in the respective cases where $\gamma=\delta$, $\gamma\delta$ is an edge, and $\gamma\delta$ is a non-edge. Thus, $A^2 = kI+\lambda A+\mu(J-I-A)$, where I is the identity matrix and J is the matrix all of whose entries are 1. Equivalently

$$A^2+(\mu-\lambda)A+(\mu-k)I = \mu J. \tag{11}$$

Now suppose θ is an eigenvalue of A with eigenvector $\xi = (\xi_1,\cdots,\xi_v)$. Then ξ is also an eigenvector of μJ. Applying both sides of (11) to the eigenvector, we get

$$\xi_j(\theta^2+(\mu-\lambda)\theta+(\mu-k)) = \mu\sum_{i=1}^{v}\xi_i \quad (j=1,\cdots,v)$$

If the right hand side of this expression is nonzero, then all ξ_j $(j=1,\cdots,v)$ are equal and $\theta = k$ (as $A\mathbf{1} = k\mathbf{1}$, where $\mathbf{1}$ denotes the vector all of whose entries are 1); in particular, we have

$$k^2+(\mu-\lambda)k+\mu-k = \mu v.$$

Otherwise, θ satisfies the quadratic equation of the theorem, and, since A has negative eigenvalues (for $k>0$ and $\operatorname{tr} A = 0$) and the constant term of the equation is $\mu-k\leqslant 0$, there are two solutions $\theta = r$ and $\theta = s$ with $r\geqslant 0$ and $s<0$. Hence $rs = \mu-k$, $r+s = \lambda-\mu$ and $(k-r)(k-s) = \mu v$. In particular, we have $k = r \Leftrightarrow \mu = 0 \Leftrightarrow \Gamma$ is a union of cliques $\Leftrightarrow k = \lambda+1 \Leftrightarrow s = -1$, and we have $r = 0 \Leftrightarrow \mu = k \Leftrightarrow \Gamma$ is complete multipartite.

Now, assume $\mu > 0$. Then (8) follows from the identities $rs = \mu-k$ and $r+s = \lambda-\mu$. Since $v>0$, we have $k>r>s$; in particular, the eigenvalue k has multiplicity 1. Therefore, the multiplicities f and g of r and s satisfy $v = 1+f+g$, and $1\cdot k+f\cdot r+g\cdot s = \operatorname{tr} A = 0$. Solving these equations for f and g and using (8) we obtain (9).

If $f\neq g$, then $r = (-k-(\lambda-\mu)g)/(f-g)$, so r is rational; since r is an algebraic integer, it is an integer, and hence $s = \lambda-\mu-r$ is an integer as well. (Alternatively: if r is not integral then its algebraic conjugates are also eigenvalues with the same multiplicity, so that r, s are quadratic irrationalities, and $f = g$.)

If $f = g$, then $f = g = \frac{1}{2}(v-1)$, and $k = -gs-fr = -f(s+r) = \frac{1}{2}(v-1)(\mu-\lambda)$. But $0<k<v-1$, so $\mu-\lambda = 1$ and $k = \frac{1}{2}(v-1)$. Finally, using $(v-k-1)\mu = k(k-\lambda-1)$, we get $\mu+\lambda = k-1$, so $\mu = \frac{1}{2}k = \frac{1}{4}(v-1)$, and $\lambda = \frac{1}{4}(v-1)-1$, i.e., Γ is a conference graph.

The other statements are easily shown to hold. □

Conference matrices and Paley graphs

Conference graphs are thus named because of their relation to conference matrices: a matrix C of order n with zero diagonal and entries ± 1 elsewhere is called a *conference matrix* when $CC^{\mathrm{T}} = (n-1)I$ (cf. WALLIS, STREET & WALLIS [768], p. 293). Given a symmetric conference matrix C, we may (after multiplying rows and columns by ± 1 if necessary) assume that the first row and first column of $C+I$ contains ones only. If S is the matrix obtained by deleting the first row and first column of C, then $A = \frac{1}{2}(S+J-I)$ is the adjacency matrix of a conference graph, and conversely, the adjacency matrix A of a conference graph yields a symmetric conference matrix

$$C = \begin{bmatrix} 0 & \mathbf{1}^{\mathrm{T}} \\ \mathbf{1} & A \end{bmatrix}.$$

If a conference matrix of order n exists, then $n \equiv 2 \pmod 4$, and $n-1$ is the sum of two squares (BELEVITCH [48], cf. VAN LINT & SEIDEL [504]); see also the remarks following Proposition 1.10.5.

Examples of conference graphs are the *Paley graphs* $QR(q)$, defined for prime powers $q \equiv 1 \pmod 4$, which have $\mathbf{F}_q$ as vertex set and where two vertices are adjacent whenever their difference is a square, cf. PALEY [582]. (QR is a mnemonic for 'quadratic residue'.)

The Paley graphs $QR(p)$ with prime p are the only nontrivial strongly regular graphs that admit a regular cyclic automorphism group (see Corollary 2.10.6).

The Hoffman bound

In strongly regular graphs, those cliques and cocliques with the property that every vertex outside is adjacent with the same number of vertices inside are characterized by the fact that they satisfy with equality the so-called *Hoffman bound*:

1.3.2. Proposition (HOFFMAN [398]). *Let Γ be a regular graph with v vertices, valency k, and with smallest eigenvalue $-m$.*

(i) *If C is a coclique of Γ, then $|C| \leqslant (1+k/m)^{-1}v$, with equality if and only if every vertex $\gamma \notin C$ has the same number of neighbours (namely m) in C.*

(ii) *If Γ is strongly regular and C is a clique of Γ, then $|C| \leqslant 1+k/m$, with equality if and only if every vertex $\gamma \notin C$ has the same number of neighbours (namely μ/m) in C.*

Proof. In §3.7, a simple representation theoretic proof of the Hoffman bound will be given. Here, we confine ourselves to the case where Γ is strongly regular, and give a proof by 'quadratic counting'.
(i) Let C be a coclique of size $c > 0$, and write $e_\gamma := |\Gamma(\gamma) \cap C|$ for $\gamma \notin C$. By counting in two ways the number of edges $\alpha\gamma$ with $\alpha \in C, \gamma \notin C$ and the number of paths $\alpha \sim \gamma \sim \beta$ with $\alpha, \beta \in C$, $\gamma \notin C$, we find

$$\sum e_\gamma = ck, \quad \sum \binom{e_\gamma}{2} = \binom{c}{2}\mu,$$

where the sum extends over the $v-c$ vertices γ outside C. Thus, $e = ck/(v-c)$ is the average of the e_γ over all $\gamma \notin C$, and we have

$$0 \leqslant \sum (e_\gamma - e)^2 = \sum (2\tbinom{e_\gamma}{2} + (1-2e)e_\gamma + e^2)$$
$$= c(c-1)\mu + (1-2e)ck + e^2(v-c)$$
$$= ((c-1)(v-c)\mu + (v-c-2ck)k + ck^2)c \,/\, (v-c)$$
$$= (v(k-\mu) + (v\mu + \mu - k - k^2)c - \mu c^2)c \,/\, (v-c)$$
$$= (vm \,/\, (k+m) - c)(r(k+m) + c\mu)c \,/\, (v-c)$$

by (8). Hence $c \leqslant vm/(k+m) = v/(1+k/m)$, proving the inequality in (i). If equality holds, we must have $e_\gamma = e$ for all $\gamma \notin C$, and, since then $e = ck/(v-c) = m$, statement (i) follows (for strongly regular graphs Γ). Now (ii) follows from (i) by taking complements and using (10). □

There is an extensive literature on strongly regular graphs. The interested reader is referred to Brouwer & van Lint [134], Cameron [167], Hubaut [413] and Seidel [656].

1.4 Strongly regular graphs as extremal graphs

1.4.1. Proposition (cf. Seidel [656]). *Let Γ be a nontrivial regular graph with v vertices and valency k. Suppose that any two adjacent vertices have at least λ common neighbours, and that any two nonadjacent vertices have at least μ common neighbours. Then*

$$v \leqslant k + 1 + k(k-1-\lambda)/\mu, \tag{12}$$

with equality if and only if Γ is strongly regular with parameters (v,k,λ,μ).

Proof. The number of induced paths of length 2 is at most $vk(k-1-\lambda)$ and at least $v(v-1-k)\mu$, hence (12). Clearly, equality in (12) forces $\lambda(\gamma,\delta) = \lambda$ for all edges $\gamma\delta$ and $\mu(\gamma,\delta) = \mu$ for all non-edges $\{\gamma,\delta\}$, and therefore implies strong regularity. □

Thus, the number v of vertices of a co-edge-regular graph is bounded by k and μ. For edge-regular graphs, a bound on v in terms of k and λ is more difficult to obtain, and sometimes even impossible. E.g., the line graph Γ of an infinite tree of valency t is edge-regular with valency $k(\Gamma) = 2t-2$, $\lambda(\Gamma) = t-2 = \frac{1}{2}k(\Gamma)-1$, and from Proposition 1.2.1 one can find finite edge-regular graphs with $\lambda = \frac{1}{2}k-1$ and arbitrarily large number of vertices. It seems likely that v can be bounded in terms of k and λ if $\lambda > \frac{1}{2}k-1$; however we can prove it only for $\lambda \geqslant \frac{2}{3}k-1$.

1.4.2. Lemma. *Let Γ be a connected edge-regular graph with parameters (v,k,λ) and $\lambda \geqslant \frac{2}{3}k-1$. If Γ is not complete, then Γ has diameter $d = 2$ and $v \leqslant 2k-2$ vertices. Moreover, the numbers $k_2 = v-1-k$ and $b_1 = k-1-\lambda$ satisfy*

$$kb_1 > k_2(k+1-2b_1). \tag{13}$$

Proof. Let $\gamma,\delta \in \Gamma$ be at distance two. Among the k neighbours of a vertex $\alpha \in \Gamma(\gamma) \cap \Gamma(\delta)$ there are b_1 vertices nonadjacent with γ and b_1 nonadjacent with δ. Hence at least $k-2b_1$ neighbours of α are adjacent with γ and δ. Therefore

$$\mu(\gamma,\delta) \geq k+1-2b_1 \quad \text{if } d(\gamma,\delta) = 2.$$

Now suppose that $k \geq 3b_1$. This is equivalent with $k \geq 3(k-1-\lambda)$ and hence with $\lambda \geq \frac{2}{3}k-1$. If β, δ are at distance 3 and $\gamma \in \Gamma(\beta) \cap \Gamma_2(\delta)$, then $\mu(\gamma,\delta) \geq k+1-2b_1 \geq b_1+1$; but γ has only b_1 neighbours in $\Gamma_2(\beta)$, contradiction. Therefore Γ has diameter $d = 2$.

Counting edges between $\Gamma(\gamma)$ and $\Gamma_2(\gamma)$ we find $kb_1 \geq k_2(k+1-2b_1)$. If equality holds, then Γ is strongly regular with $\mu = k+1-2b_1 = 2\lambda+3-k$. By Theorem 1.2.3 this means that Γ is the pentagon, but that violates $\lambda \geq \frac{2}{3}k-1$. Consequently, equality does not hold, and we proved (13). We can now estimate

$$\begin{aligned} k_2 &< kb_1/(k+1-2b_1) = b_1+b_1(2b_1-1)/(k+1-2b_1) \\ &\leq b_1+b_1(2b_1-1)/(b_1+1) = 3b_1-3+3/(b_1+1) \leq k-3+3/(b_1+1) \end{aligned}$$

and hence $k_2 \leq k-3$ (since if $b_1 = 1$, then Γ is a complete graph minus a matching and $k_2 = 1$, k even). Consequently, $v = 1+k+k_2 \leq 2k-2$. □

The icosahedron has $k = 5$, $\lambda = 2$, $d = 3$, so the bound is 'sharp'. Similarly, the octahedron has $v = 6$, $k = 4$, $\lambda = 2$ and achieves equality in the bound on v.

An edge-regular graph for which λ is very close to k is already strongly regular:

1.4.3. Theorem. *Let Γ be a connected edge-regular graph with parameters (v,k,λ).*

(i) *If $\lambda \geq k-\frac{1}{2}-\sqrt{k}$, then Γ is complete multipartite.*

(ii) *If $\lambda \geq k+\frac{1}{2}-\sqrt{2k+2}$, then Γ is the complement of a strongly regular graph Δ with $\mu(\Delta) \leq 2$.*

Proof. We first note that the complement Δ of Γ is co-edge-regular with $\bar{\mu} := \mu(\Delta) = v-2k+\lambda$. If Γ has diameter $d \leq 2$ then $k_2 = v-1-k$ and $\bar{\mu} = k_2-b_1$ so that in this case (13) can be rewritten as

$$k\bar{\mu} < (b_1+\bar{\mu})(2b_1-1). \tag{13a}$$

Furthermore, (12) applied to Δ with the parameter $\bar{\lambda} = 0$, and

$$\bar{k} = k_2 = b_1+\bar{\mu}, \qquad v = 1+k+\bar{k}$$

gives the inequality

$$k\bar{\mu} \leq (b_1+\bar{\mu})(b_1+\bar{\mu}-1) \tag{14}$$

and equality in (14) implies that Δ (and hence Γ) is strongly regular, with $\lambda(\Delta) = 0$. Recall that if $\bar{\mu} = 0$, then Δ is a disjoint union of cliques, and Γ is complete multipartite.

(i) Suppose that $\lambda \geq k-\frac{1}{2}-\sqrt{k}$. Then $b_1 = k-1-\lambda \leq \sqrt{k}-\frac{1}{2}$, whence $k \geq (b_1+\frac{1}{2})^2 > b_1^2+b_1$, and therefore $k \geq b_1^2+b_1+1 \geq 3b_1$. Hence the hypothesis of Lemma 1.4.2 is satisfied, so, by the above, (13a) holds. But $k \geq b_1^2+b_1+1$; this contradicts (13a) for $\bar{\mu} > 2$, (13a) or (14) for $\bar{\mu} = 2$, and (14) for $\bar{\mu} = 1$. Hence $\bar{\mu} = 0$. This proves (i).

(ii) Suppose now that $\lambda \geq k+\frac{1}{2}-\sqrt{2k+2}$. Then

$$b_1 = k-1-\lambda \leqslant \sqrt{2k+2}-\tfrac{3}{2}\text{, so } k \geqslant -1+\tfrac{1}{2}(b_1+\tfrac{3}{2})^2 > \tfrac{1}{2}b_1(b_1+3).$$

Since this is an integer, $k \geqslant \frac{1}{2}b_1(b_1+3)+1 = \frac{1}{2}(b_1+1)(b_1+2) \geqslant 3b_1$. This contradicts (13a) for $\overline{\mu} > 4$, and (13a) or (14) for $\overline{\mu} \in \{3,4\}$. Finally, if $\overline{\mu} = 2$, then we have equality in (14) whence Δ is strongly regular, and, if $\overline{\mu} = 1$, then Δ is strongly regular by Lemma 1.1.3. This proves (ii). □

The *complete union* of a number of graphs is the complement of the disjoint union of their complements. By complementing the second part of the theorem we obtain:

1.4.4. Corollary. *Let Γ be a nontrivial co-edge-regular graph with parameters (v,k,μ). If*

$$v > k+\tfrac{1}{2}(k-\mu+1)(k-\mu+2),$$

then Γ is a complete union of strongly regular graphs Δ_i with $\mu(\Delta_i) \leqslant 2$ for each i.

Proof. Use $\overline{k} = v-1-k$ and $\overline{\lambda} = v-2k+\mu-2$ and simplify. □

Examples.

(i) The complement Γ of an $n\times(n+1)$-grid Δ is an edge-regular graph with $k(\Gamma) = n(n-1)$, $\lambda(\Gamma) = (n-1)(n-2)$; thus we have an infinite family of edge-regular graphs with $\lambda = k+1-\sqrt{4k+1}$ which are not strongly regular. This comes very close to the bounds of the theorem.

(ii) If Γ and Γ' are co-edge-regular graphs with parameters (v,k,μ) and (v',k',μ') where $v-v' = k-k' = \mu-\mu'$, then their complete union is co-edge-regular with parameters $(v+v',\ k+v',\ \mu+v')$. This explains the dependency of the inequality of the Corollary on $v-k$ and $k-\mu$ only.

Problems.

(i) Is v bounded in terms of k for an edge-regular graph with parameters (v,k,λ) if $\lambda > \frac{1}{2}k-1$? (Line graphs of graphs with girth at least 5 satisfy $\lambda = \frac{1}{2}k-1$.) Find a good diameter bound.

(ii) Find, for given λ, the smallest k such that infinitely many edge-regular graphs with parameters (v,k,λ) exist which are locally connected. (Note that if a graph does not contain subgraphs $K_{1,1,2}$, then it is locally a disjoint union of cliques.) Or weaken 'locally connected' to $\mu(\gamma,\delta) > 1$ for all γ, $\delta \in \Gamma$.

(iii) (cf. Example (i) above) Is every edge-regular graph with parameters (v,k,λ) satisfying $\lambda > k+1-\sqrt{4k+1}$ necessarily strongly regular?

1.5 Taylor graphs and regular two-graphs

A distance-regular graph with intersection array

$$\{k,\mu,1;\ 1,\mu,k\}$$

is called a *Taylor graph*. For a Taylor graph, $k_2 = k$, $k_3 = 1$, and $v = 2k+2$; hence Γ is an antipodal double cover of a graph with $k+1$ vertices and valency k, i.e., of a complete graph K_{k+1}. As observed by TAYLOR & LEVINGSTON [711], there is a one-to-one correspondence between antipodal double covers of complete graphs and certain designs, called two-graphs; in particular, the two-graphs corresponding to Taylor graphs are precisely the regular two-graphs introduced in TAYLOR [710].

A *two-graph* is a design $(X,\mathfrak{D})$ with block size 3 such that every 4-subset of X contains an even number of blocks. Instead of $(X,\mathfrak{D})$ we also write $\mathfrak{D}$ if X is clear from the context. A subset of X is called *coherent* if all its 3-subsets are blocks; in particular all 2-subsets are coherent, and blocks are just the coherent triples. A two-graph is called *regular* if it is a 2-design, i.e., if every 2-set is contained in a constant number $\lambda = \lambda(\mathfrak{D})$ of coherent triples. The *complement* of a two-graph $(X,\mathfrak{D})$ is the design on X whose blocks are the triples outside $\mathfrak{D}$. Clearly, this is again a two-graph. The complement $(X,\overline{\mathfrak{D}})$ of a regular two-graph $\mathfrak{D}$ is also regular with $\lambda(\overline{\mathfrak{D}}) = |X|-2-\lambda(\mathfrak{D})$.

If Δ is an arbitrary graph with vertex set X, then the set $\mathfrak{D}(\Delta)$ of all 3-subsets of X containing an odd number of edges of Δ defines a two-graph. Moreover, all two-graphs arise in this way. Indeed, if $(X,\mathfrak{D})$ is a two-graph, then the *derived design* $(X\setminus\{w\},\mathfrak{D}_w)$ at $w \in \mathfrak{D}$ (where $\mathfrak{D}_w = \{D\setminus\{w\} \mid w \in D \in \mathfrak{D}\}$, cf. §A.6) is a graph on $X\setminus\{w\}$, and $\Delta_w := \mathfrak{D}_w + \{w\}$ is a graph with $\mathfrak{D}(\Delta_w) \cong \mathfrak{D}$.

1.5.1. Proposition (TAYLOR [710]). *If $(X,\mathfrak{D})$ is a regular two-graph and $\infty \in X$, then $\mathfrak{D}_\infty$ is either a complete graph or a strongly regular graph satisfying $k = 2\mu$. Conversely, if Δ is either a complete graph or a strongly regular graph satisfying $k = 2\mu$, then $\mathfrak{D}(\Delta+\{\infty\})$ is a regular two-graph.*

Proof. Let $(X,\mathfrak{D})$ be a regular two-graph. Then $\Delta := \mathfrak{D}_\infty$ is regular of valency $k(\Delta) = \lambda(\mathfrak{D})$. Suppose Δ is not complete. In order to show that Δ is strongly regular it suffices to show that $\mu(\Delta) = \frac{1}{2}k(\Delta)$; indeed, since $(\overline{\mathfrak{D}})_\infty = \overline{\mathfrak{D}_\infty}$ we can then find $\lambda(\Delta)$ from $\mu(\overline{\Delta})$. Let $\{\gamma,\delta\}$ be a nonedge of Δ and put $\mu = \mu(\gamma,\delta)$. The number of incoherent triples on $\{\gamma,\delta\}$ in $\mathfrak{D}$ equals $1+\mu+(v(\Delta)-2-2k(\Delta)+\mu) = |X|-2-2k(\Delta)+2\mu$ but it also equals $|X|-2-\lambda(\mathfrak{D}) = |X|-2-k(\Delta)$, so that $\mu = \frac{1}{2}k(\Delta)$, independent of γ, δ.

Conversely, let Δ be a strongly regular graph with parameters (v,k,λ,μ). If $k = 2\mu$, then $\lambda = \frac{1}{2}(3k-1-v)$ by Lemma 1.1.1 (using $k_2 = v-k-1$), and we find for $\gamma,\delta \in X$ that the number N of coherent triples on $\{\gamma,\delta\}$ in $\mathfrak{D} = \mathfrak{D}(\Delta+\{\infty\})$ satisfies $N = k$ when $\infty \in \{\gamma,\delta\}$, $N = 1+\lambda+\mu(\overline{\Delta}) = 1+\lambda+(v-2k+\lambda) = k$ when $\{\gamma,\delta\}$ is an edge of Δ, and $N = 2(k-\mu) = k$ when $\{\gamma,\delta\}$ is a nonedge of Δ. Thus we have $N = k$ in all cases and $\mathfrak{D}$ is a regular two-graph. □

To each two-graph $(X,\mathfrak{D})$ and point $\infty \in X$ we may associate a graph $\Gamma = \Gamma_\infty(\mathfrak{D})$ whose vertices are the symbols x^+, x^- $(x \in X)$, and whose edges are the pairs $x^\sigma y^\tau$ where $x \neq y$ and either $\sigma = \tau$, $\{x,y,\infty\}$ coherent or $\sigma \neq \tau$, $\{x,y,\infty\}$ not coherent.

If $(X,\mathfrak{D})$ is *complete,* i.e., if all 3-subsets of X are coherent, then Γ consist of two disjoint copies of K_n $(n = |X|)$; otherwise Γ is connected, we have $d(x^+,x^-) = 3$ for all $x \in X$, and Γ is a regular graph with $v = 2n$, $k = k_2 = n-1$, $k_3 = 1$ and therefore an antipodal double cover of the complete graph on X.

1.5.2. Lemma. *Let $(X, \mathfrak{D})$ be a two-graph and $v, w \in X$. Then $\Gamma_v(\mathfrak{D}) \cong \Gamma_w(\mathfrak{D})$.*

Proof. One easily checks that the mapping $\phi: X \to X$ defined by $\phi(v^\sigma) = w^\sigma$, $\phi(w^\sigma) = v^\sigma$, and $\phi(x^\sigma) = x^\tau$ $(x \neq v, w)$ with $\sigma = \tau$ if and only if $\{v,w,x\}$ is coherent, is an isomorphism. □

The *switching class* of the two-graph $\mathfrak{D}$ is the set of graphs Δ such that $\mathfrak{D} = \mathfrak{D}(\Delta)$. Two graphs Δ' and Δ'' are *switching equivalent* when $\mathfrak{D}(\Delta') = \mathfrak{D}(\Delta'')$; this holds if and only if there is a subset S of their vertex set X such that Δ' and Δ'' induce the same graphs on S and on $X \setminus S$, while they induce complementary relations on $S \times (X \setminus S)$. If $\mathfrak{D} = \mathfrak{D}(\Delta)$, then we may construct a graph $\Gamma_\Delta(\mathfrak{D})$ with vertices x^+, x^- $(x \in X)$ and edges $x^\sigma y^\tau$ where $x \neq y$ and either $\sigma = \tau$, xy an edge of Δ, or $\sigma \neq \tau$ and xy not an edge of Δ. Now $\Gamma_u(\mathfrak{D}) = \Gamma_\Delta(\mathfrak{D})$ for $\Delta = \mathfrak{D}_u + \{u\}$. Generalizing the above lemma, we can say that the isomorphism type of $\Gamma_\Delta(\mathfrak{D})$ depends only on the switching class of Δ.

1.5.3. Theorem (TAYLOR & LEVINGSTON [711]). *Let X be a set of cardinality n and $\infty \in X$. There is a one-to-one correspondence between antipodal double covers Γ (with vertex set $\{x^+, x^- \mid x \in X\}$) of the complete graph K_n with vertex set X and non-complete two-graphs $(X, \mathfrak{D})$ given by*

$$\Gamma \mapsto \mathfrak{D} = \{\{x,y,z\} \mid \text{there are signs } \rho, \sigma, \tau \text{ such that } x^\rho y^\sigma z^\tau \text{ is a triangle in } \Gamma\}$$
$$= \mathfrak{D}(\gamma^\perp) \text{ for any vertex } \gamma \text{ of } \Gamma,$$

and

$$\mathfrak{D} \mapsto \Gamma = \Gamma_\infty(\mathfrak{D}).$$

Furthermore, the following conditions concerning a corresponding pair Γ, $\mathfrak{D}$ are equivalent:

(i) *Γ is a Taylor graph (i.e., is distance-regular);*

(ii) *$(X, \mathfrak{D})$ is a regular two-graph;*

(iii) *For some (and then for every) $\gamma \in \Gamma$, the graph $\Delta = \Gamma(\gamma)$ is strongly regular and satisfies $k(\Delta) = 2\mu(\Delta)$.*

Moreover, if (i)-(iii) *hold, then $\lambda(\Gamma) = \lambda(\mathfrak{D}) = k(\Delta) = 2\mu(\Delta)$ and (if $k(\Delta) > 0$ then) $\lambda(\Delta) = \frac{1}{2}(3k(\Delta) - n)$.*

Proof. If $\alpha \in \Gamma$ then write α^- for its antipode, i.e., the unique vertex in $\Gamma_3(\alpha)$. The maps $\Gamma \mapsto \mathfrak{D}$ and $\mathfrak{D} \mapsto \Gamma$ given above are each other's inverses and we have $\mathfrak{D} = \mathfrak{D}(\gamma^\perp)$ since two vertices of Γ are adjacent if and only if their antipodes are adjacent.

(i)⇔(ii). Let $\gamma \in \Gamma$ and $\delta \in \Gamma_2(\gamma)$. Then $\delta = \alpha^-$ for some $\alpha \in \Gamma(\gamma)$ and $\mu(\gamma,\delta)$ is just the number of incoherent triples containing α and γ. Therefore, α and γ are in $n-2-\mu(\gamma,\delta)$ coherent triples. In particular, $\mu(\Gamma) = \mu$ if and only if $\mathfrak{D}$ is a regular two-graph with $\lambda(\mathfrak{D}) = n-2-\mu$; and then, since always $b_2(\Gamma) = 1$ and $b_1(\Gamma) = \mu(\Gamma)$, we see that $\lambda(\Gamma) = k(\Gamma) - 1 - \mu(\Gamma) = n-2-\mu = \lambda(\mathfrak{D})$. Thus, Γ is a Taylor graph if and only if $\mathfrak{D}$ is a regular two graph.

(ii)⇔(iii). If $\Delta = \Gamma(\gamma)$ then $\Delta = \mathfrak{D}_\gamma$ and Δ is complete if and only if $\mathfrak{D}$ is. Also $\mathfrak{D} = \mathfrak{D}(\Delta + \{\gamma\})$ so that (ii)⇔(iii) is just the contents of the above proposition.

The stated equations for the various parameters have already been derived above and in the proof of the proposition. □

1.5.4. Corollary. *A graph Γ is distance-regular with intersection array $\{k, k-1, 1; 1, k-1, k\}$ if and only if Γ is a complete bipartite graph $K_{k+1,k+1}$ with a 1-factor removed, i.e., the complement of a $2\times(k+1)$-grid. Thus, there is a unique such graph Γ for each $k \geqslant 2$.* □

Remarks. (i) The 'graph extension construction' $\Delta \mapsto \Gamma_\infty(\mathfrak{D}(\Delta+\{\infty\}))$ was first used in SHULT [671].
(ii) If Γ is a Taylor graph with intersection array $\{k, \mu, 1; 1, \mu, k\}$ ($\mu < k-1$) then Γ_2 is a Taylor graph with intersection array $\{k, \mu', 1; 1, \mu', k\}$ ($\mu' = k-1-\mu$), and conversely. (Here Γ_2 is the graph with the same vertex set as Γ, where two vertices are adjacent whenever they have distance 2 in Γ.) The two Taylor graphs Γ and Γ_2 correspond to a complementary pair of regular two-graphs.

1.5.5. Theorem (NEUMAIER [555]). *Let Γ be a connected amply regular graph with parameters (v,k,λ,μ) and diameter $d \geqslant 3$. Then*

$$k_2 \geqslant k \geqslant \lambda+\mu+1. \tag{15}$$

Moreover, $k_2 = k$ (equivalently, $k = \lambda+\mu+1$), if and only if Γ is a polygon or a Taylor graph.

Proof. By Lemma 1.1.1, $k_2 \geqslant k$ is equivalent to $k \geqslant \lambda+\mu+1$, which is equivalent to $b_1 \geqslant c_2$, and this is obvious for $d \geqslant 3$. The equality case is characterized in Lemma 5.1.2. □

Sometimes, the switching class of a regular two-graph contains a strongly regular graph.

1.5.6. Theorem. *Let Δ be a strongly regular graph with parameters (v,k,λ,μ). Then the two-graph $\mathfrak{D} = \mathfrak{D}(\Delta)$ is regular if and only if $v = 2(2k-\lambda-\mu)$. If this is the case, then $\mathfrak{D}_\infty$ is strongly regular with parameters $(v-1, 2(k-\mu), k+\lambda-2\mu, k-\mu)$, and $\Gamma_\infty(\mathfrak{D})$ is distance-regular with intersection array $\{v-1, 2(k-\lambda-1), 1; 1, 2(k-\lambda-1), v-1\}$ on $2v$ vertices.*

Proof. $\mathfrak{D}(\Delta)$ is regular when $\lambda+(v-2k+\lambda) = 2(k-\mu)$. The parameters follow immediately. □

Relations of two-graphs with equiangular lines and switching classes of graphs are discussed in §3.8. For a recent account of further theory and the construction of two-graphs see SEIDEL [654, 655], SEIDEL & TAYLOR [658], BUSSEMAKER et al. [154], NEUMAIER [557], FRANKL & FÜREDI [299].

1.6 Square 2-designs

Any connected bipartite graph Γ is the incidence graph of a design $(X,\mathfrak{B})$, and this design is determined up to duality by Γ. The graph Γ is regular if and only if $(X,\mathfrak{B})$ is a *square* 1-*design*, i.e., a 1-(v,k,r) design with $k = r$, or, equivalently, with $|X| = |\mathfrak{B}|$. The graph Γ is amply regular if and only if $(X,\mathfrak{B})$ is a square 1-design with the properties that, for any two points, there are either 0 or μ blocks containing both and that any two blocks have precisely 0 or μ points in common (cf. Condition SS_μ of §1.7). Below, we observe that Γ is (bipartite) distance-regular of diameter 3 if and only if $(X, \mathcal{B})$ is a square 2-design. In the next section we describe the designs (partial λ-geometries) that correspond to distance-regular graphs of diameter 4.

When the design has a polarity without absolute points, we can write its incidence matrix as a symmetric matrix with zero diagonal and regard it as the adjacency matrix of a graph Δ; then Γ is the *bipartite double* of Δ; this situation is studied in §1.11.

1.6.1. Theorem. *A bipartite graph with v vertices is distance-regular of diameter* $d = 3$ *(with intersection array* $\{k, k-1, k-\mu; 1, \mu, k\}$*) if and only if it is the incidence graph of a square 2-design (with parameters* 2-(w,k,μ)*, where* $k > 1$ *and* $w = \frac{1}{2}v = 1 + k(k-1)/\mu$*).*

Proof. Let X_1 and X_2 be the parts of a bipartite distance-regular graph Γ of diameter 3. Interpret the adjacency relation of Γ as incidence on $X_1 \times X_2$. It is obvious that this defines a square 2-design with the stated parameters. The converse also holds as any two blocks of a square 2-(w,k,μ) design meet in exactly μ points of the design. □

Remarks. (i) If Γ is a bipartite distance-regular graph with intersection array $\{k, k-1, k-\mu; 1, \mu, k\}$ $(k-\mu > 1)$ then Γ_3 is a bipartite distance-regular graph with intersection array $\{k_3, k_3-1, k-\mu; 1, \bar{\mu}, k_3\}$, and conversely, where $k_3 = w-k = (k-1)(k-\mu)/\mu$ and $\bar{\mu} = w-2k+\mu = (k-\mu)(k-\mu-1)/\mu$. The graphs Γ and Γ_3 correspond to a pair of complementary square 2-designs.
(ii) For $k-\mu = 1$, corresponding to the trivial square 2-design formed by the k-subsets of a $(k+1)$-set, we get again the complements of $2\times(k+1)$-grids, cf. Corollary 1.5.4.

There is an extensive theory on square 2-designs. The interested reader should consult the book by Lander [474] for details. A list of the known square 2-designs is contained in Beth, Jungnickel & Lenz [62], Appendix, §2. For the most important nonexistence criterion, the Bruck-Chowla-Ryser Theorem, see Theorem 1.10.4.

1.7 Partial λ-geometries

We generalize the results of the previous section and show that bipartite amply regular graphs of girth 4 (i.e., with $\mu > 1$) are equivalent to semisymmetric designs. A design $(X, \mathcal{D})$ is called *semisymmetric* if, for some $\lambda = \lambda(\mathcal{D}) > 1$, the following property holds:

SS_λ. There is a pair of disjoint blocks; any two distinct blocks meet in exactly 0 or λ points, and any two distinct points are in 0 or λ blocks.

The incidence graph Γ of a semisymmetric design is bipartite, whence $\lambda(\Gamma) = 0$, and has $\mu(\Gamma) = \lambda(\mathcal{D}) > 1$. Thus, by Proposition 1.1.2, Γ is amply regular. In particular, the regularity of Γ implies that every block contains the same number k of points and every point is in k blocks. Moreover, there are equally many points and blocks. Therefore, semisymmetric designs are 1-designs. Conversely, if Γ is amply regular with $\mu > 1$ and bipartite with parts X_1 and X_2, we get a semisymmetric design by interpreting the adjacency relation between X_1 and X_2 as incidence. Semisymmetric designs with $\lambda = 2$ are known as *semibiplanes*. For examples and further properties of semisymmetric designs, we refer to Hughes & Piper [415] (Chapter 7.4), Hughes

[414], WILD [785], and JANKO & VAN TRUNG [443]. Here, we are interested in a special class of semisymmetric designs related to bipartite distance-regular graphs of diameter 4, introduced by DRAKE [271]; cf. also CAMERON & DRAKE [177], and NEUMAIER [556].

A (square) *partial λ-geometry with nexus e* is a 1-(v,k,k) design satisfying SS_λ for some $\lambda < k$ and

PG_e. For any block B and any point $\gamma \notin B$ there are precisely e blocks $C \ni \gamma$ intersecting B, and precisely e points in B adjacent to γ in the point graph.

(As Cameron and Drake defined them, non-square partial λ-geometries have $\lambda = 1$ and are just partial geometries (cf. §A.6); we shall not be interested in those, so for us 'partial λ-geometry' will imply 'square'.) It is not difficult to show that the second condition in PG_e is a consequence of the first. A partial λ-geometry with $\lambda > 1$ is a semisymmetric design; a partial 1-geometry is nothing but a partial geometry with equally many points and blocks. Clearly, the conditions SS_λ and PG_e (for $e > 0$), translated into conditions on the incidence graph Γ, say $c_2(\Gamma) = \lambda$ (and $\operatorname{diam}(\Gamma) \geqslant 4$) and $c_3(\Gamma) = e$ (and $\operatorname{diam}(\Gamma) \leqslant 4$). Thus we have

1.7.1. Theorem (Cameron & Drake, cf. DRAKE [271]). *A bipartite graph Γ is distance-regular with diameter $d = 4$ if and only if Γ is the incidence graph of a partial λ-geometry with nexus e for certain integers λ and e. If this is the case, then $c_2(\Gamma) = \lambda$ and $c_3(\Gamma) = e$.* □

A simple restriction on the parameters is implied by the following:

1.7.2. Lemma. *Let Γ be a distance-regular bipartite graph of even diameter. Then its parameters satisfy $\mu \mid k$.*

Proof. Let $\Delta = \Gamma_2$. If θ is an eigenvalue of Γ then $(\theta^2 - k)/\mu$ is an eigenvalue of Δ, since the adjacency matrices A_Γ and A_Δ of Γ and Δ satisfy $A_\Gamma^2 = kI + \mu A_\Delta$. But Γ has eigenvalue 0 (indeed, Γ has an odd number $d+1$ of distinct eigenvalues, and with θ, $-\theta$ is also an eigenvalue of Γ) and hence $-k/\mu$ is an eigenvalue of Δ. Now $-k/\mu$ is a rational algebraic integer and therefore integral. It follows that $\mu \mid k$. □

In particular we find (for $d = 4$) that the parameters of a partial λ-geometry satisfy $\lambda \mid k$, a result due to NEUMAIER [556].

In §6.11 (Table 6.9) we shall list all known partial λ-geometries with nexus e and block size $k > e+1$ in terms of their incidence graphs. By definition (of SS_λ) we have $k > e$. The partial λ-geometries with $k = e+1$ (corresponding to incidence graphs with an intersection array $\{m\lambda, m\lambda-1, (m-1)\lambda, 1;\ 1, \lambda, m\lambda-1, m\lambda\}$) are too abundant to be described here; they are called *symmetric (m,λ)-nets*; these are designs satisfying SS_λ whose point graph and block graph are complete multipartite graphs $K_{m\lambda\times m}$. Thus, these correspond precisely to the bipartite antipodal distance-regular graphs of diameter four.

1.7.3. Proposition (DRAKE [271]). *Let $\mathfrak{D}$ be a partial λ-geometry with $k = e+1$. Then $\mathfrak{D}$ is a symmetric net. Conversely, every symmetric net is such a partial λ-geometry.*

Proof. The point graph Δ of $\mathfrak{D}$ is strongly regular and satisfies $k(\Delta) = k(k-1)/\lambda$ and $\mu(\Delta) = ke/\lambda$. (Indeed, it is a halved graph of a distance-regular graph of diameter 4 - cf. §1.11.) If $k = e+1$, then $\mu(\Delta) = k(\Delta)$ whence Δ is a complete multipartite graph $K_{m\lambda\times m}$, where $m = k/\lambda$. The same argument applies to the block graph. □

We refer to BETH, JUNGNICKEL & LENZ [62] (I §7.18/19, II §8, VIII §3, XII §6), DRAKE [271], and JUNGNICKEL [448] for theory and constructions of symmetric nets and the closely related concept of generalized Hadamard matrices; see also BERMAN [56], BUTSON [157], DAWSON [245], HINE & MAVRON [391], JUNGNICKEL [447,449], DE LAUNEY [478,479], MAVRON [524], SEBERRY [649], and SHAD [662]. Some families of symmetric nets are constructed in Remark 9.3.13; see also below.

A connection with affine resolvable designs

A 2-design is called *affine resolvable* when being disjoint is an equivalence relation on the blocks, and any two non-disjoint blocks meet in μ points, for some constant μ. (Now $\mu = k^2/v$.) Examples of affine resolvable designs are the designs of points and hyperplanes in an affine space. For a survey of affine resolvable designs, see SHRIKHANDE [669].

Affine resolvable designs yield symmetric nets (and hence bipartite antipodal distance-regular graphs of diameter four):

1.7.4. Proposition (MAVRON [523], Theorem 1.4; see also MCFEAT & NEUMAIER [527]).

(i) *Given an affine resolvable 2-(v,k,λ) design, where $v = mk$, $\lambda = (k-1)/(m-1)$ (and hence $r = (mk-1)/(m-1)$ and $b = mr$) for some prime power m, we can construct a symmetric (m,k)-net.*

(ii) *Given a symmetric (m,k)-net, its derived design is a resolvable 2-(v,k,λ) design, where $v = mk$ and $\lambda = k-1$ (and hence $r = mk-1$ and $b = mr$).*

Proof. (i) Given the affine resolvable 2-design with the stated parameters, arbitrarily number the blocks from each parallel class from 1 to m. Let A_i be the $v\times r$ incidence matrix with rows indexed by the points, columns by the parallel classes, and entries 1 if the point is on the i-th block of the parallel class, and 0 otherwise. Since m is a prime power, there exists an $m\times m(m-1)$ array B with entries from $I_m = \{1,2,\cdots,m\}$ such that all columns are permutations of I_m, and such that given any two rows, each ordered pair of distinct symbols from I_m occurs in exactly one column. (Such an array B can be obtained from an orthogonal array $OA(m,m)$ (which is essentially just an affine plane $AG(2,m)$) by deleting m (constant) columns.) Replace in this array each symbol i by the matrix A_i. This yields the $m^2k\times m(mk-1)$ matrix D. Next, take an identity matrix of order m and replace each symbol by a (constant) column of mk such symbols. This yields the $m^2k\times m$ matrix E. It is straightforward to check that the $m^2k\times m^2k$ matrix (DE) is the incidence matrix of a symmetric (m,k)-net. Part (ii) is obvious. □

In the special case $m = 2$ the resolvable designs found under (ii) are affine resolvable (e.g., because $b = v+r-1$), and we find a one-to-one correspondence between these so-called 'Hadamard 3-designs' and symmetric $(2,k)$-nets, a result due to DRAKE [271]. In the next section we look at this situation in some more detail.

1.8 Hadamard matrices

A distance-regular graph with intersection array

$$\{2\mu, 2\mu-1, \mu, 1;\ 1, \mu, 2\mu-1, 2\mu\} \tag{16}$$

is called a *Hadamard graph of order* 2μ. By §1.7, Hadamard graphs are incidence graphs of certain symmetric nets; however, there is a neater construction in terms of Hadamard matrices.

A *Hadamard matrix of order* n is an $n\times n$-matrix H with $H_{i,j}\in\{1,-1\}$ $(i,j = 1,\cdots,n)$ such that $HH^{\mathsf T} = nI$. The definition implies that two distinct rows of H have inner product zero whence they agree in precisely $n/2$ places. In particular, if $n > 1$, then n is even. If $n > 2$, then one finds that any three distinct rows agree in precisely $n/4$ places. Thus, either $n \leq 2$ or n is divisible by 4. The case $n \leq 2$ is trivial. It is a long-standing conjecture that Hadamard matrices exist for all orders $n \equiv 0 \pmod 4$. Hadamard matrices are known to exist for many orders, e.g., for $n = 4m < 428$ and for $n = 2^s m$ with $m^2 \leq 2^s$; see WALLIS, STREET & WALLIS [768], SEBERRY [767], GERAMITA & SEBERRY [322] and the survey article of HEDAYAT & WALLIS [367] for the many known construction techniques and SAWADE [637] for the current state of knowledge. (See also AGAIAN [2].)

1.8.1. Theorem (DRAKE [271], SHAD [662], DELORME [246]). *Let H be a Hadamard matrix of order k. Then the graph $\Gamma(H)$ whose vertices are the symbols $r_i^+, r_i^-, c_i^+, c_i^-$, $(i = 1, \cdots, k)$, and whose edges are the 2-sets $\{r_i^\epsilon, c_j^\eta\}$ with $\epsilon = \eta$ if $H_{i,j} = 1$ and $\epsilon \neq \eta$ if $H_{i,j} = -1$, is a Hadamard graph of order k. Conversely, every Hadamard graph arises in this way.*

Proof. Let $\gamma = r_i^+$. It is easily checked that $\Gamma(\gamma)$ consists of the vertices c_j^ϵ where ϵ has the sign of $H_{i,j}$, that $\Gamma_2(\gamma)$ consists of all vertices r_j^ϵ with $j \neq i$, that $\Gamma_3(\gamma)$ consists of the vertices c_j^ϵ where ϵ has the opposite sign of $H_{i,j}$ and that $\Gamma_4(\gamma)$ consists of r_i^- only. Hence Γ has diameter $d = 4$. Also, it readily follows that $c_i(\Gamma) = 1, \frac{1}{2}k, k-1, k$ for $i = 1, 2, 3, 4$, respectively. Since Γ is obviously bipartite and has valency $k = 2\mu$, this forces $b_i(\Gamma) = k - c_i(\Gamma) = k, k-1, \frac{1}{2}k, 1$ for $i = 1, 2, 3, 4$. Hence Γ is Hadamard graph of order k.
Conversely, let Γ be a Hadamard graph. Then $k_1 = 2\mu$, $k_2 = 4\mu - 2$, $k_3 = 2\mu$ and $k_4 = 1$. Hence, to every $\gamma \in \Gamma$ there corresponds a unique opposite vertex in $\Gamma_4(\gamma)$. Now fix $\gamma \in \Gamma$. Clearly, the opposites of vertices from $\Gamma(\gamma)$ are in $\Gamma_3(\gamma)$, and those of vertices in $\Gamma_2(\gamma)$ are in $\Gamma_2(\gamma)$. Hence the vertices of Γ can be labelled (in many ways) such that opposite vertices get labels with different signs, the 4μ vertices in $\Gamma(\gamma) \cup \Gamma_3(\gamma)$ get the labels r_i^ϵ ($\epsilon \in \{+,-\}$, $i = 1, \cdots, k$), and the remaining 4μ vertices get the labels c_i^ϵ ($\epsilon \in \{+,-\}$, $i = 1, \cdots, k$). It is immediate that the $k \times k$-matrix H defined by $H_{i,j} = 1$ if $r_i^+ \sim c_j^+$, and $H_{i,j} = -1$ otherwise, is a Hadamard matrix with $\Gamma(H) = \Gamma$. □

1.8.2. Corollary. *If $\mu > 2$ is odd, then there is no distance-regular graph with intersection array $\{2\mu, 2\mu-1, \mu, 1; 1, \mu, 2\mu-1, 2\mu\}$.* □

[This also follows from inequality (12) (the 'partial λ-geometry bound') in §5.8.]

Remark. There are unique Hadamard matrices of orders 1, 2, 4, 8 and 12 (up to H-equivalence, that is, up to permutation of rows and columns and multiplication of rows and columns by ± 1). The associated Hadamard graphs are $2K_2$, C_8, 2^4 and unique graphs with intersection arrays $\{8,7,4,1;1,4,7,8\}$ and $\{12,11,6,1;1,6,11,12\}$. On the other hand, there are precisely 5 inequivalent Hadamard matrices of order 16 and 3 of order 20 (HALL [353, 355]). See also §7.6B.

1.9 Hadamard graphs as extremal graphs

The results in this section are inspired by a paper of MULDER [547] concerning $(0,\mu)$-graphs, i.e., graphs in which any two vertices have 0 or μ common neighbours (cf. Theorem 1.11.1 (ii)).

1.9.1. Proposition. *Let Γ be a connected amply regular graph of diameter d with parameters (v,k,λ,μ), where $\lambda \leqslant \mu \neq 1$. Then*

$$c_i(\gamma,\delta) \geqslant \mu-2+i \tag{17}$$

for all $\gamma, \delta \in \Gamma$ at distance i $(i = 2, \cdots, d)$.

Proof. Let β, γ be vertices at distance i, and let $\delta \in \Gamma(\beta) \cap \Gamma_{i-1}(\gamma)$. Let N be the number of edges $\delta'\epsilon$ with $\delta' \in \Gamma(\beta)\setminus\{\delta\}$, $\epsilon \in \Gamma(\delta) \cap \Gamma_{i-2}(\gamma)$. Bounding N in two ways gives

$$(c_i(\beta,\gamma)-1)(\max(\lambda,\mu)-1) \geqslant N \geqslant c_{i-1}(\gamma,\delta)(\mu-1).$$

Since $\lambda \leqslant \mu \neq 1$, this implies $c_i(\beta,\gamma) \geqslant c_{i-1}(\gamma,\delta)+1$. Now, (17) follows by induction as $c_2(\gamma,\delta) = \mu$ for all γ, δ at distance 2. □

1.9.2. Corollary. *Let Γ be a connected amply regular graph of diameter $d \geqslant 4$ with parameters (v,k,λ,μ), where $\lambda \leqslant \mu \neq 1$. Then*

$$d \leqslant k+4-2\mu \leqslant k. \tag{18}$$

Proof. Let α, δ be two vertices at distance d, and take $\gamma \in \Gamma_2(\alpha) \cap \Gamma_{d-2}(\delta)$. Since $d \geqslant 4$, we have $k = k(\Gamma) \geqslant \mu(\alpha,\gamma)+c_{d-2}(\gamma,\delta)$. Hence, using (17) we find $k \geqslant \mu+(\mu-2+d-2)$, which implies (18). □

Remark. TERWILLIGER [715] calls an amply regular graph with $\mu > 1$ a $(2,\mu,\lambda,k)$-graph. From the proof of his Theorem 3.4 one can deduce that for $\mu > 2$, $\lambda \leqslant \mu$ and $2 \leqslant i \leqslant \frac{1}{2}d-1$, there are γ, δ at distance i with $k \geqslant (3-\frac{2}{\mu})c_i(\gamma,\delta)$. Thus (17) implies

$$k \geqslant (3-\frac{2}{\mu})(\mu-3+[\frac{d}{2}]) \quad \text{if } \mu \geqslant \max(3,\lambda),\ d \geqslant 6 \tag{18a}$$

which, for large d, is stronger than (18).

1.9.3. Theorem. *Let Γ be a connected amply regular graph of diameter at least 4 with parameters (v,k,λ,μ). Then $k \geqslant 2\mu$, with equality if and only if Γ is a polygon or a Hadamard graph.*

Proof. Obviously, polygons and Hadamard graphs satisfy $k = 2\mu$. Let $\alpha, \epsilon \in \Gamma$ be vertices at distance 4. If $\gamma \in \Gamma_2(\alpha) \cap \Gamma_2(\epsilon)$, then γ has μ neighbours in $\Gamma(\alpha) \cap \Gamma(\gamma)$ and μ neighbours in $\Gamma(\epsilon) \cap \Gamma(\gamma) \subseteq \Gamma_3(\alpha)$. Therefore, $k \geqslant 2\mu$.
Now suppose that $k = 2\mu > 2$. Then, by the above,

$$\Gamma(\gamma) \subseteq \Gamma(\alpha) \cup \Gamma(\epsilon) \quad \text{for all } \gamma \in \Gamma_2(\alpha) \cap \Gamma_2(\epsilon). \tag{19}$$

Fix $\beta \in \Gamma(\alpha) \cap \Gamma_3(\epsilon)$, $\delta \in \Gamma_2(\beta) \cap \Gamma(\epsilon) \subset \Gamma_3(\alpha)$. If $\lambda > 0$, then β has a neighbour $\beta' \in \Gamma(\alpha)$. If every common neighbour of β and δ is adjacent to β', then β and β' have at least μ common neighbours, whence $\lambda \geqslant \mu$, and we get (from Theorem 1.5.5) $k \geqslant \lambda+\mu+1 \geqslant 2\mu+1 = k+1$, a contradiction. Hence there is a point $\gamma \in \Gamma(\beta) \cap \Gamma(\delta)\setminus\Gamma(\beta')$. But $\gamma \in \Gamma_2(\alpha) \cap \Gamma_2(\epsilon)$, and, by (19), the λ common neighbours of β and γ are in $\Gamma(\alpha)$, so α and β have $\lambda+1$ common neighbours, which is absurd. Therefore, $\lambda = 0$. Take $\beta' \in \Gamma(\alpha)\setminus\{\beta\}$. Then β and β' have μ common neighbours. By the proposition, however, $\Gamma(\beta) \cap \Gamma_2(\epsilon)$ contains $c_3(\epsilon,\beta) \geqslant \mu+1$ points. Since $k = 2\mu$, this implies that β' has a neighbour in $\Gamma_2(\epsilon)$. In particular, $\beta' \in \Gamma_3(\epsilon)$, showing that

$$\Gamma(\alpha) \subseteq \Gamma_3(\epsilon). \tag{20}$$

Now let $\gamma' \in \Gamma(\beta)\setminus\{\alpha\}$. Then $\gamma' \in \Gamma_2(\alpha)$, and, by (19), γ' has no neighbour in $\Gamma_2(\alpha) \cap \Gamma_2(\epsilon)$. Since $\mu > 1$, there is a point $\beta' \in \Gamma(\alpha) \cap \Gamma(\gamma')\setminus\{\beta\}$. Since β and β' have less than $\mu = \mu(\beta,\delta)$ neighbours $\notin \alpha^\perp$, there is a vertex $\gamma \in \Gamma(\beta) \cap \Gamma(\delta)\setminus\Gamma(\beta')$. Clearly, $\gamma \in \Gamma_2(\gamma') \cap \Gamma_2(\epsilon)$, and $\Gamma(\gamma) \cap \Gamma(\alpha) \cap \Gamma(\gamma') \subseteq \Gamma(\alpha) \cap \Gamma(\gamma')\setminus\{\beta'\}$. Since $\mu(\gamma,\gamma') = \mu(\alpha,\gamma') = \mu$, this implies that $\Gamma(\gamma) \cap \Gamma(\gamma')\setminus\Gamma(\alpha) \neq \emptyset$, so, by (19), $\Gamma(\gamma) \cap \Gamma(\gamma') \cap \Gamma(\epsilon) \neq \emptyset$. In particular, $\gamma' \in \Gamma_2(\epsilon)$. Since $\gamma' \in \Gamma(\beta)\setminus\{\alpha\}$ and $\beta \in \Gamma(\alpha) \cap \Gamma_3(\epsilon) = \Gamma(\alpha)$ were arbitrary, we get $\Gamma_2(\alpha) \subseteq \Gamma_2(\epsilon)$, and by symmetry, $\Gamma_2(\alpha) = \Gamma_2(\epsilon)$. Now (19) and (20) yield $\Gamma_3(\epsilon) = \Gamma(\alpha)$, and by symmetry we also have $\Gamma_3(\alpha) = \Gamma(\epsilon)$. Therefore, $\alpha^\perp \cup \epsilon^\perp \cup \Gamma_2(\alpha)$ is the set of all vertices, and Γ is bipartite. Now one finds $c_4(\alpha,\epsilon) = 2\mu$, and $c_3(\alpha,\delta) = c_3(\delta,\alpha) = 2\mu-1$ for $\delta \in \Gamma_3(\alpha)$, $c_3(\beta,\epsilon) = c_3(\epsilon,\beta) = 2\mu-1$ for $\beta \in \Gamma_3(\epsilon)$. Therefore, for every $\beta \in \Gamma(\alpha)$, there is a unique vertex $\bar{\beta} \in \Gamma(\epsilon)\setminus\Gamma_2(\beta)$, and clearly $\Gamma_4(\beta) = \{\bar{\beta}\}$. Since Γ is connected, it readily follows that Γ is distance-regular with intersection array $\{2\mu, 2\mu-1, \mu, 1; 1, \mu, 2\mu-1, 2\mu\}$. Thus Γ is a Hadamard graph. □

1.10 Square divisible designs

In §§1.5-8 we exhibited connections between certain classical combinatorial configurations and certain imprimitive distance-regular graphs of diameter 3 or 4. In this section we shall describe another such connection, pointed out by Godsil and Mathon.

A *divisible design* (also known as 'group-divisible design') $GD(k,g,n,\lambda_1,\lambda_2)$ is a design $(X,\mathcal{B})$ with blocks of size k provided with a partition $\mathcal{G}$ of X into n subsets ('classes' or 'parts' or 'groups') of size g, such that any two points from the same class occur together in λ_1 blocks, while any two points from different classes occur together in λ_2 blocks. Clearly, $v := |X| = gn$, and each point is in $r = ((n-1)g\lambda_2 + (g-1)\lambda_1)/(k-1)$ blocks.

When the number of blocks of a design equals the number of points then the design is called *square*. Thus, a divisible design is square when $r = k$. (And then $rk - v\lambda_2 = k - \lambda_1 - g(\lambda_2 - \lambda_1)$.)

Let N be the point-block incidence matrix of the design. Then

$$\det NN^\top = rk(rk - v\lambda_2)^{n-1}(r-\lambda_1)^{n(g-1)}.$$

The design is called *regular* when $\det NN^\top \neq 0$.

1.10.1. Theorem (CONNOR [210]; a more general result is given by HOFFMAN [396]). *If the parameters of a regular square divisible design $GD(k,g,n,\lambda_1,\lambda_2)$ satisfy $(k-\lambda_1, \lambda_2-\lambda_1) = 1$, then the dual of this design is also a divisible design, with the same parameters.* □

Let $(a,b)_p$ denote the Hilbert norm residue symbol for $\mathbf{Q}_p$, defined by $(a,b)_p = +1$ if $aX^2 + bY^2 = Z^2$ has a solution $(X,Y,Z) \neq (0,0,0)$ in $\mathbf{Q}_p$, and $(a,b)_p = -1$ otherwise (cf. SERRE [661], Chap. III, or RAGHAVARAO [601], Appendix A6). By SERRE [661], Ch. IV, §3.2, Theorem 8, the equation $aX^2 + bY^2 = Z^2$ has a solution $(X,Y,Z) \neq (0,0,0)$ in rational integers if and only if $a > 0$ or $b > 0$, and $(a,b)_p = 1$ for all odd primes p. We have the following analogue of the Bruck-Chowla-Ryser Theorem for square divisible designs:

1.10.2. Theorem (BOSE & CONNOR [105], cf. RAGHAVARAO [601], Theorem 12.4.4 and RAJKUNDLIA [603], Theorem 1.6).
Suppose a regular square divisible design $GD(k,g,n,\lambda_1,\lambda_2)$ exists. Then

(i) $(k^2 - v\lambda_2)^{n-1}(k-\lambda_1)^{n(g-1)}$ *is a square, and*

(ii) $(k^2 - v\lambda_2, (-1)^{n(n-1)/2} g^n \lambda_2)_p (k-\lambda_1, (-1)^{n(g-1)(g+n-1)/2} g^n)_p = 1$ *for all odd primes p.* □

In the special case $\lambda_1 = 0$ one usually writes $GD[k,\lambda,g;gn]$ instead of $GD(k,g,n,0,\lambda)$; then this theorem reduces to

1.10.3. Corollary. *Suppose a square divisible design $GD[k,\lambda,g;gn]$ exists, where $k < n$.*

(i) *If n is even, then $k - g\lambda$ is a square. If moreover $n \equiv 2 \pmod 4$ and g is even, then $(k,-1)_p = 1$ for all odd primes p, i.e., k is a sum of two squares.*

(ii) *If n is odd and g is even, then k is a square and $(k-g\lambda, (-1)^{(n-1)/2} g\lambda)_p = 1$ for all odd primes p.*

(iii) *If n and g are both odd, then $(k-g\lambda, (-1)^{(n-1)/2} g\lambda)_p (k, (-1)^{(g-1)/2} g)_p = 1$ for all odd primes p.* □

Note that part (iii) is repeatedly stated incorrectly in the literature.

In the special case $g = 1$ a divisible design is just a 2-design, and the Bose-Connor Theorem reduces to the Bruck-Chowla-Ryser Theorem.

1.10.4. Theorem (BRUCK & RYSER [140], CHOWLA & RYSER [197], SHRIKHANDE [666]).
Suppose a square 2-(v,k,λ) design exists, where $0 < \lambda < k$. If v is even, then $k-\lambda$ is a square. If v is odd, then $(k-\lambda, (-1)^{(v-1)/2}\lambda)_p = 1$ for all odd primes p, i.e., the equation

$$(k-\lambda)X^2 + (-1)^{(v-1)/2}\lambda Y^2 = Z^2$$

has a solution in integers X, Y, Z, not all zero. □

Now suppose that Γ is an antipodal distance-regular graph of diameter 3 with $\lambda = \mu$. Then the adjacency matrix A of Γ is the point-block incidence matrix of a square divisible design; in fact, if Γ is an antipodal r-cover, then the design will have parameters $GD[k,\lambda,g;ng]$ with $g = r$, $n = k+1$, $k = r\lambda + 1$. (And conversely, any square $GD[k,\lambda,g;ng]$ possessing a polarity without absolute points yields such a distance-regular graph Γ.) As Godsil and Mathon point out, this means that the Bose-Connor Theorem implies parameter restrictions for antipodal distance-regular graphs. Let us make this more explicit.

1.10.5. Proposition. *Let Γ be a distance-regular graph with intersection array $\{r\mu+1, (r-1)\mu, 1; 1, \mu, r\mu+1\}$. Then*

(i) *If r is even, then $k = r\mu + 1$ is a sum of two squares.*

(ii) *If $r\mu$ is odd, then the equation*

$$kX^2 + (-1)^{(r-1)/2} rY^2 = Z^2$$

has a solution in integers X, Y, Z, not all zero.

Proof. Corollary 1.10.3 simplifies, since $k - g\lambda = 1$. In case (i), μ is even, cf. Corollary 4.2.6. □

For example, the intersection array $\{3\mu+1, 2\mu, 1; 1, \mu, 3\mu+1\}$ is feasible for all μ, but ruled out by the above theorem for $\mu = 3, 7, 11, 13, 15, \cdots$. Also $\{6,4,1; 1,1,6\}$ ($v = 35$) is ruled out.
For $r = 2$ these conditions are equivalent to the restriction that the number of vertices of a conference graph must be the sum of two squares (cf. §1.3).
When $\lambda = \mu - 2$, then $A + I$ is the point-block incidence matrix of a square divisible design $GD[k+1, \lambda+2, r; (k+1)r]$ with $k+1 = (\lambda+2)r$, but the design is not regular, and we find no restrictions.

Next, when Γ is a bipartite antipodal distance-regular graph of diameter 4, then Γ is the point-block incidence graph of a symmetric $(k/\mu, \mu)$-net, that is, of a square divisible design $GD[k, \mu, k/\mu; k^2/\mu]$. Again this design is not regular, and we find no restrictions.

Note, however, that this equivalence of bipartite antipodal distance-regular graphs of diameter four and square divisible designs $GD[k, \mu, k/\mu; k^2/\mu]$ means that the existence of an antipodal distance-regular graph Δ of diameter 3 with $\lambda_\Delta = \mu_\Delta - 2$ implies the existence of an antipodal bipartite distance-regular graph Γ of diameter 4 with $k_\Gamma = k_\Delta + 1$, $\mu_\Gamma = \mu_\Delta$ and $v_\Gamma = 2v_\Delta$. (For example, starting with the 3-cube, we find the 4-cube; more generally, starting with the graphs of Table 6.10, (A1) we find graphs equivalent to symmetric (p^i, p^{i+2j})-nets.) This is a special instance of the more general construction of the 'extended bipartite double', cf. the next section.

In general, a square divisible design $GD[k, \lambda, g; gn]$ will have point-block incidence graph Γ with distribution diagram

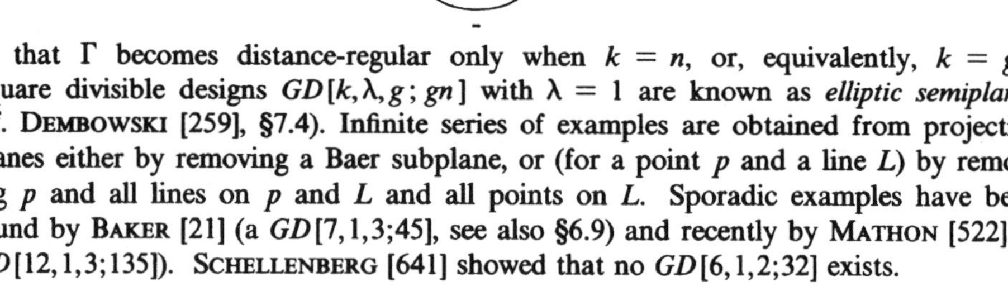

so that Γ becomes distance-regular only when $k = n$, or, equivalently, $k = g\lambda$. Square divisible designs $GD[k, \lambda, g; gn]$ with $\lambda = 1$ are known as *elliptic semiplanes* (cf. Dembowski [259], §7.4). Infinite series of examples are obtained from projective planes either by removing a Baer subplane, or (for a point p and a line L) by removing p and all lines on p and L and all points on L. Sporadic examples have been found by Baker [21] (a $GD[7,1,3;45]$, see also §6.9) and recently by Mathon [522] (a $GD[12,1,3;135]$). Schellenberg [641] showed that no $GD[6,1,2;32]$ exists.

1.11 The bipartite double of a graph

Let Γ be a graph. Denote by $\tilde{\Gamma}$ the graph whose vertices are the symbols γ^+, γ^- ($\gamma \in \Gamma$) and whose edges are the 2-sets $\{\gamma^\sigma, \delta^\tau\}$, with $\gamma \sim \delta$ and $\sigma \neq \tau$. Then clearly the map $\gamma^\sigma \mapsto \gamma$ ($\gamma^\sigma \in \tilde{\Gamma}$) is a surjective morphism from $\tilde{\Gamma}$ onto Γ. The graph $\tilde{\Gamma}$ is called the *bipartite double* of Γ.

For example, if Γ is a $(2e+1)$-gon, then $\tilde{\Gamma}$ is a $(4e+2)$-gon, and if Γ is a complete graph K_n, then $\tilde{\Gamma}$ is isomorphic to $K_{n,n}$ minus a 1-factor (i.e., the complement of the $2\times n$-grid).

Early occurrences of the concept of bipartite double of a graph can be found in BIGGS [71] and BIGGS & GARDINER [81].

The graph Γ is not determined by $\tilde{\Gamma}$. For example, there are just two nonisomorphic strongly regular graphs Γ with parameters $(v,k,\lambda,\mu) = (16,6,2,2)$ (cf. SHRIKHANDE [668] and Theorem 3.12.4 below), and both yield the folded 6-cube (with intersection array $\{6,5,4;1,2,6\}$) for $\tilde{\Gamma}$. Part (iii) of the proposition below shows how much of Γ one can easily retrieve from $\tilde{\Gamma}$. In fact, as noted in §1.6, a connected bipartite graph $\tilde{\Gamma}$ determines (up to duality) a design $(X,\mathfrak{B})$, where in our case X is the vertex set of Γ and $\mathfrak{B} = \{\Gamma(\gamma) \mid \gamma \in \Gamma\}$, and from this design we find Γ by fixing a polarity without absolute points.

If Γ is a bipartite graph, then the connected components of Γ_2 are called the *halved graphs* of Γ. If Γ is connected then it has precisely two halved graphs. If Γ is regular with v vertices, then its halved graphs have $\frac{1}{2}v$ vertices, explaining the 'halved'.

The *spectrum* of a graph is the multiset containing its eigenvalues, each with its multiplicity. Union of multisets is union of sets and addition of multiplicities. Below we shall use the cardinality of a multiset to mean the number of distinct elements in it. When A and B are multisets then $A+B$ denotes the multiset $(a+b \mid a \in A,\ b \in B)$, and $-A$ denotes $(-a \mid a \in A)$. When writing down multisets explicitly we shall usually write multiplicities as exponents.

1.11.1. Theorem. *Let Γ be connected and let $\tilde{\Gamma}$ be its bipartite double. Then:*

(i) *$\tilde{\Gamma}$ is bipartite; it is connected if and only if Γ is not bipartite.*

(ii) *$\tilde{\Gamma}$ is amply regular with parameters $(2v,k,0,\mu)$ if and only if Γ has v vertices, is regular of valency k and any two vertices of Γ have either 0 or μ common neighbours.*

Now assume moreover that Γ is not bipartite and has diameter $d \geqslant 2$. Then:

(iii) *If Γ has no triangles, then $\tilde{\Gamma}$ has halved graphs isomorphic to Γ_2; if each edge of Γ is in at least one triangle, then $\tilde{\Gamma}$ has halved graphs isomorphic to $\Gamma_1 \cup \Gamma_2$.*

(iv) *$\tilde{\Gamma}$ has diameter $\tilde{d} \leqslant 2d+1$; we have $\tilde{d} = 2d+1$ if and only if for some $\gamma \in \Gamma$ the subgraph $\Gamma_{\leqslant d-1}(\gamma)$ is bipartite.*

(v) *If Γ has spectrum Θ, then $\tilde{\Gamma}$ has spectrum $\Theta \cup -\Theta$; consequently $\tilde{\Gamma}$ can be distance-regular only if $|\Theta \cup -\Theta| = \tilde{d}+1$.*

(vi) *$\tilde{\Gamma}$ is distance-regular of diameter $\tilde{d} = 2d+1$ if and only if Γ is distance-regular with $a_i = 0$ $(i < d)$ and $a_d > 0$. In this case $\tilde{\Gamma}$ is an antipodal 2-cover of Γ.*

(vii) *If Γ is distance-regular and $j = \min\{i \mid a_i \neq 0\} < d$ then $\tilde{\Gamma}$ will be distance-regular if and only if $d = 2j$, $a_j = c_{j+1}$, $b_{j-i} = c_{j+i+1} = c_{j+i} + a_{j+i}$ $(i = 1, \cdots, j-1)$; if this is the case, then $\tilde{d} = 2j+1 = d+1$. In particular, $\tilde{\Gamma}$ will be distance-regular (with $\tilde{d} = 3$) when Γ is strongly regular with $\lambda = \mu > 0$.*

Proof. Straightforward. □

An example of a graph Γ with the property required in (ii) which is not edge-regular is

depicted in §1.16. If Γ is the grid 2×4 (which has spectrum $4^1 2^1 0^3 (-2)^3$), then $\tilde{\Gamma}$ is the 4-cube 2^4 (with spectrum $4^1 2^4 0^6 (-2)^4 (-4)^1$). In particular, $\tilde{\Gamma}$ can be distance-regular while Γ is not even edge-regular.
The condition $|\Theta| = d+1$ for the spectrum of a graph Γ of diameter d is necessary but not sufficient for being distance-regular; in Example 9.2.6 we exhibit a graph that is not distance-regular with the same diameter and spectrum as the 4-cube. The situation (vi) will be investigated further in §4.2D.

The extended bipartite double of a graph

For a graph Γ, denote by $\hat{\Gamma}$ the graph whose vertices are the symbols γ^+, γ^- $(\gamma \in \Gamma)$ and whose edges are the 2-sets $\{\gamma^\sigma, \delta^\tau\}$, with $\gamma = \delta$ or $\gamma \sim \delta$, and $\sigma \neq \tau$. The graph $\hat{\Gamma}$ is called the *extended bipartite double* of Γ.

1.11.2. Theorem. *Let Γ be connected and let $\hat{\Gamma}$ be its extended bipartite double. Then:*

(i) *$\hat{\Gamma}$ is bipartite and connected.*

(ii) *$\hat{\Gamma}$ is amply regular with parameters $(2v, k+1, 0, \mu)$ if and only if Γ is amply regular with parameters $(v, k, \mu-2, \mu)$.*

(iii) *$\hat{\Gamma}$ has halved graphs isomorphic to $\Gamma_1 \cup \Gamma_2$.*

(iv) *$\hat{\Gamma}$ has diameter $\hat{d} = d+1$.*

(v) *If Γ has spectrum Θ, then $\hat{\Gamma}$ has spectrum $(-\Theta-1) \cup (\Theta+1)$.*

(vi) *$\hat{\Gamma}$ is distance-regular if and only if Γ is distance-regular with $b_i + c_{i+1} = k+1$ $(1 \leqslant i \leqslant d-1)$. If this is the case, then $\hat{\Gamma}$ has parameters $\hat{k} = \hat{c}_{d+1} = k+1$, $\hat{c}_i = c_i$, $\hat{b}_{i+1} = b_i$ $(0 \leqslant i \leqslant d)$.* □

For examples and further discussion, see §4.2F (and §1.10, §9.4C, §11.3H). Note that if Γ is bipartite, then $\hat{\Gamma}$ is the direct product $\Gamma \times K_2$; in particular, if Γ is the m-cube, then $\hat{\Gamma}$ is the $(m+1)$-cube (see the next section).

1.12 Direct products and Hamming graphs

The *direct product* of t graphs $\Gamma^1, \cdots, \Gamma^t$ is the graph with vertex set $\Gamma^1 \times \cdots \times \Gamma^t$ whose edges are the 2-sets $\{\gamma, \delta\}$ for which there is an index m such that $\gamma_m \sim \delta_m$ and $\gamma_l = \delta_l$ for all $l \neq m$. An easy induction argument shows that we have $d(\gamma,\delta) = d(\gamma_1,\delta_1) + \cdots + d(\gamma_t,\delta_t)$.

1.12.1. Proposition (cf. Doob [266], Weichsel [774], Song [694]). *Let Γ be the direct product of $t \geqslant 2$ nonempty graphs Γ^l with v_l vertices, diameter d_l and spectrum Θ_l $(l = 1, \cdots, t)$. Then:*

(i) *The graph Γ has $v = v_1 \cdots v_t$ vertices, has diameter $d = d_1 + \cdots + d_t$ and has spectrum $\Theta_1 + \cdots + \Theta_t$.*

(ii) *Γ is regular (of valency k) if and only if each Γ^l is regular (of valency k_l, say), and then $k = k_1 + \cdots + k_t$.*

(iii) *Γ is amply regular with parameters (v, k, λ, μ) if and only if $\mu = 2$ and each Γ^l is amply regular with parameters (v_l, k_l, λ, μ).*

(iv) *The graph* Γ *is distance-regular if and only if there is a number s such that each* Γ^l *is distance-regular with* $b_i(\Gamma^l) = k_l - si$, $c_i(\Gamma^l) = i$ $(i = 1, \cdots, d_l)$*; in this case,* $b_i(\Gamma) = k - si$ *and* $c_i(\Gamma) = i$ $(i = 1, \cdots, d)$.

Proof. Straightforward. □

The direct products of cliques are called *Hamming graphs*. The Hamming graph $H(d,n)$ is the direct product of d cliques of size n. By the proposition, $H(d,n)$ is distance-regular of diameter d with intersection array

$$\{d(n-1), (d-1)(n-1), (d-2)(n-1), \cdots, n-1;\ 1, 2, 3, \cdots, d\}.$$

The Hamming graph $H(d, 2)$ is also called the *d-cube*. It is often denoted by 2^d or Q_d.

Since a direct product of Hamming graphs $H(2,4)$ and complete graphs K_4 is again a Hamming graph $H(n, 4)$, and hence distance-regular, and since the Shrikhande graph (cf. §3.11) has the same parameters as the Hamming graph $H(2,4)$, it follows from the above proposition that the direct product of a number of Shrikhande graphs and some complete graphs K_4 is also distance-regular. These graphs were first constructed by Doob [266].

Let N be a set of size n containing 0. The vertices of $H(d,n)$ can be identified with the elements of N^d, called the *words* of length d over the alphabet N. The distance between two words x, y in $H(d,n)$ is called the *Hamming distance* of x and y; it is the number of entries in which x and y differ. The *weight* of a word is its distance from the zero word $(0, \cdots, 0)$, i.e., the number of nonzero entries. A subset of N^d is called a *code of length d over the alphabet N*; in case $n = 2$ a *binary code*, and in case $n = 3$ a *ternary code*. The *minimum weight* of a code C is the smallest weight among all weights of words in C. A code is *linear* if the alphabet is a finite field and $ax + by := (ax_1 + by_1, \cdots, ax_d + by_d) \in C$ for all $x, y \in C$, $a, b \in N$. The *coset graph of a linear code C* is the graph whose vertices are the cosets of C, adjacent if and only if their union contains vectors at Hamming distance 1. If C has minimum weight at least five, then clearly its coset graph is amply regular with parameters $(v,k,\lambda,\mu) = (|C|, d(n-1), n-2, 2)$. A number of interesting distance-regular graphs arise in this way, see Chapter 11.

1.13 *d*-cubes as extremal graphs

A *rectagraph* is a connected triangle-free graph with the property that every path of length two is in a unique quadrangle. By Proposition 1.1.2, a rectagraph is regular; thus the rectagraphs are precisely the connected amply regular graphs with $\lambda = 0$ and $\mu = 2$.

If Γ is rectagraph of valency k with $a_2(\Gamma) = 0$ and $c_3(\Gamma) = 3$, then each component of Γ_2 is locally a triangular graph $T(k) = \binom{k}{2}$; if Γ is bipartite, then Γ_2 has two components (the two halved graphs of Γ (cf. §1.11)), otherwise only one. Conversely, by Neumaier [561] (Propositions 2 (ii) and 3), every connected locally triangular graph is the halved graph of a bipartite rectagraph; indeed, for $k > 4$ the missing points are uniquely found as the maximal cliques. Note also that the bipartite rectagraphs are precisely the incidence graphs of *semibiplanes*, i.e. semisymmetric designs with $\mu = 2$ (cf. §1.7).

Examples of rectagraphs are the coset graphs of binary linear codes with minimum weight at least 5, in particular the k-cubes (coset graphs of the zero code of length k). For relations between k-cubes and a large class of rectagraphs of valency k, see §4.3B. Here we consider only the extremal case.

1.13.1. Proposition (MULDER [547], NEUMAIER [558]). *Let Γ be a rectagraph with valency k. Then Γ has at most 2^k vertices, with equality if and only if Γ is the k-cube.*

Proof. Fix $\gamma \in \Gamma$, and let N_i be the number of edges $\alpha\beta$ with $\alpha \in \Gamma_i(\gamma)$ and $\beta \in \Gamma_{i+1}(\gamma)$. By Proposition 1.9.1, $c_i(\gamma,\alpha) \geqslant i$, and so $b_i(\gamma,\alpha) \leqslant k - c_i(\gamma,\alpha) \leqslant k - i$. Therefore, $k_{i+1}(\gamma)(i+1) \leqslant N_i \leqslant k_i(\gamma)(k-i)$; by induction we find $k_i(\gamma) \leqslant \binom{k}{i}$. Thus, $v = \sum k_i(\gamma) \leqslant \sum \binom{k}{i} = 2^k$. Now suppose that $v = 2^k$. Then equality holds throughout; in particular, Γ is distance-regular with parameters $c_i = i$, $b_i = k - i$, $k_i = \binom{k}{i}$, and for each $\gamma \in \Gamma$, there is a unique vertex $\bar{\gamma} \in \Gamma_k(\gamma)$. Now, fix $\alpha \in \Gamma$ and map each $\gamma \in \Gamma$ to the characteristic vector $c(\gamma)$ of the set of all $\beta \in \Gamma(\alpha) \cap \Gamma_{i-1}(\gamma)$, where $i = d(\alpha,\gamma)$. Then $c(\gamma)$ has weight $d(\alpha,\gamma)$, and it is easy to show that c is an isomorphism from Γ onto the k-cube. □

Remark. A similar characterization of the Hamming graph $H(d,q)$ was given by Enomoto (see Theorem 9.2.2).

1.13.2. Theorem. *Let Γ be a connected amply regular graph with parameters (v,k,λ,μ) where $\lambda \leqslant \mu \neq 1$. Then the diameter of Γ is at most k, with equality if and only if Γ is the d-cube.*

Proof. By Corollary 1.9.2, Γ has diameter $d \leqslant k$. Assume $d = k$. By the same corollary, we must have $\mu = 2$. Now choose two points α,ω at distance k, and define Γ_g to be the set of vertices for which the distances to α and ω sum to k. If $\gamma \in \Gamma_g$ has distance i from α, then by Proposition 1.9.1, $\Gamma(\gamma)$ contains at least i vertices of $\Gamma_{i-1}(\alpha)$ and at least $k-i$ vertices of $\Gamma_{k-i-1}(\omega) \cap \Gamma(\gamma) \subseteq \Gamma_{i+1}(\alpha)$. But the valency is k, whence $c_i(\alpha,\gamma) = i$, $b_i(\alpha,\gamma) = k-i$. By symmetry, also $c_{k-i}(\omega,\gamma) = k-i$, $b_{k-i}(\omega,\gamma) = i$. Consequently, $\Gamma(\gamma) \subseteq \Gamma_g$. Since $\alpha \in \Gamma_g$ and Γ is connected, this shows that $\Gamma_g = \Gamma$. But then the argument of the previous proof applies with equality throughout, establishing that Γ is the d-cube. □

1.14 Gamma spaces and singular lines

A partial linear space (X,L) with the property that if l is a line and x is a point not on l then x is collinear with no, one or all points of l, is called a *gamma space*. (Cf. D.G. HIGMAN [387], J.I. HALL [350].) A subspace of (X,L) with the property that any two of its points are collinear is called a *singular subspace*. Let us describe circumstances under which it is possible to retrieve the set of lines L when only the collinearity graph Γ of (X,L) is given.

Recall that for a vertex γ of a graph Γ, we write $\gamma^\perp$ for the set of vertices consisting of γ and its neighbours; and if A is a set of vertices, we let $A^\perp = \bigcap_{\gamma \in A} \gamma^\perp$ be the set of all vertices at distance at most one from each vertex of A. *Singular lines* of Γ are sets l

of the form $\{\gamma,\delta\}^{\perp\perp}$ where γ and δ are adjacent vertices. Clearly, singular lines are cliques, and Γ is the point graph (collinearity graph) of the design (X,L), where X is the vertex set of Γ and L the collection of singular lines of Γ. Any maximal clique that contains two vertices of a singular line l contains l entirely.

We are interested in the situation where singular lines of a graph Γ behave as lines, i.e., where two singular lines have at most one point in common. If that is the case, then the design (X,L) is a gamma space with the property that each $l \in L$ is in the intersection of all maximal cliques containing l. Conversely, if (X,L) is a gamma space with this property, then L is the collection of singular lines of its collinearity graph Γ. (D.G. HIGMAN [385] calls such graphs Γ *isotropic*.)

In the finite case, we have an easy criterion.

1.14.1. Lemma. *Let Γ be an edge-regular graph and l a singular line of Γ. Then $l = \{\gamma,\delta\}^{\perp\perp}$ for any two points γ, $\delta \in l$. In particular, two singular lines of Γ have at most one point in common.*

Proof. Suppose $\beta \in \{\gamma,\delta\}^{\perp\perp}$. Then $\{\gamma,\delta\}^{\perp} \subseteq \{\beta,\gamma\}^{\perp}$, and hence, since Γ is edge-regular, $\{\gamma,\delta\}^{\perp} = \{\beta,\gamma\}^{\perp}$. Consequently, $\{\gamma,\delta\}^{\perp\perp} = \{\beta,\gamma\}^{\perp\perp}$. □

1.15 Generalized quadrangles with line size three

In this section, we are concerned with nondegenerate generalized quadrangles all of whose lines have size 3. According to a result of Cameron they are finite. This result and the well-known classification of all finite generalized quadrangles with line size 3 is the content of the theorem below.

Recall that a *generalized quadrangle* is a partial linear space such that if x is a point not on a line l, then x is collinear with a unique point of l. It is called *degenerate* if some point is collinear with all other points. Thus, in a degenerate generalized quadrangle the set of lines forms a pencil. Write L_x for the set of lines on x. One easily sees that if x and y are noncollinear, then $|L_x| = |L_y|$. It follows that $|L_x|$ is independent of x unless all lines have size two and the point graph (collinearity graph) is a complete bipartite graph with halves of different cardinality. Dually, the size of the lines is a constant, unless the point graph is a non-square grid. (Note that the concept of generalized quadrangle is almost self dual, but we insist that lines have at least two points, while a point need not be on any line.) A generalized quadrangle is determined by its point graph (since the lines are the maximal cliques).

1.15.1. Lemma. *Let Q be a finite generalized quadrangle with order (s,t), that is, with line size $s+1$ and $t+1$ lines on each point. Then the point graph of Q is strongly regular with parameters*

$$v = (s+1)(st+1)\,,\quad k = s(t+1)\,,\quad \lambda = s-1\,,\quad \mu = t+1 \tag{21}$$

and eigenvalues k, $s-1$, $-t-1$ of multiplicities 1, f and g, where

$$f = \frac{st(s+1)(t+1)}{s+t}\,,\quad g = \frac{s^2(st+1)}{s+t}\,.$$

Conversely, if Γ is a strongly regular graph with parameters (21) such that all its singular *lines have size $s+1$, then Γ is the point graph of a generalized quadrangle of order (s,t). In particular, if $s = 2$, then every strongly regular graph with parameters (21) is the point graph of a generalized quadrangle with line size 3.*

Proof. Straightforward. □

Examples. All generalized quadrangles Q whose lines have size 3 are known (as has been discovered, independently, by - in chronological order - SEIDEL [653], SHULT [670], THAS [729], FREUDENTHAL [301], DIXMIER & ZARA [265]; cf. PAYNE & THAS [588]). The three nondegenerate examples are found by taking the points and lines on a nondegenerate quadric with Witt index 2 in the projective space $PG(d,2)$, $d = 3,4,5$. Let us look at these one by one.

(i) The 3×3-grid Q_1, a generalized quadrangle of order (2,1). The point graph has parameters $(v,k,\lambda,\mu) = (9,4,1,2)$.

(ii) The geometry Q_2, a generalized quadrangle of order (2,2) whose 15 points are the 2-subsets of a 6-set X and whose 15 lines are the partitions of X into 2-subsets. (This is known as Sylvester's duad-syntheme geometry, cf. SYLVESTER [705]). Equivalently, points are the transpositions and lines are the remaining odd involutions of the symmetric group S_6. Another geometric description is found by taking the points and totally isotropic lines in $PG(3,2)$ provided with a nondegenerate symplectic form. The point graph has parameters $(v,k,\lambda,\mu) = (15,6,1,3)$.

(iii) The geometry Q_4, a generalized quadrangle of order (2,4) with 27 points and 45 lines obtained from Q_2 by adding 12 further points i^-, i^+ $(i \in X)$ and 30 further lines $\{i^-, j^+, (ij)\}$ $(i,j \in X, i \neq j)$. This generalized quadrangle is famous for its connections with the 27 lines on a cubic surface (see SCHLÄFLI [642], BAKER [20] and HIRSCHFELD [393], Chapter 20; one easily recognizes a double six in the present construction), and indeed, another geometric construction is found by taking the lines and the points on a nondegenerate Hermitean quadric in $PG(3,4)$, which is a cubic surface. The point graph has parameters $(v,k,\lambda,\mu) = (27,10,1,5)$.

1.15.2. Theorem. *Let Q be a generalized quadrangle all of whose lines have size 3. Then Q is isomorphic to a coclique, a set of $2n+1$ points and n lines intersecting in a common point, or one of the above generalized quadrangles Q_i $(i = 1,2,4)$.*

First proof. Clearly one gets one of the described examples if Q contains no quadrangle. From now assume that Q contains a quadrangle.

STEP 1. $Q_1 \subset Q$.
If Q contains a quadrangle then two further lines must join the third points on the lines through the quadrangle; their intersection gives one further point. Hence Q contains the 3×3-grid Q_1.

STEP 2. *If $Q \neq Q_1$, then $Q_2 \subset Q$.*
Label the points of the grid Q_1 inside Q (obtained in step 1) by transpositions as follows:

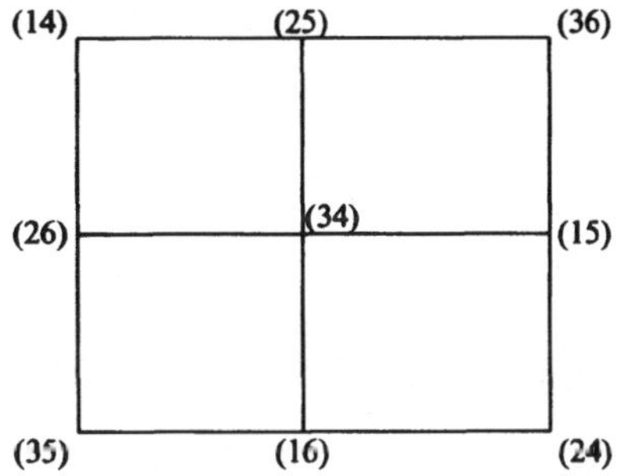

Any further point is adjacent to a 3-coclique in Q_1, and by symmetry we may assume that there is a point in $Q \setminus Q_1$ labelled (56) adjacent with (14),(34), and (24). This defines three lines which we may take to be (14) (23) (56), (24) (13) (56), and (34) (12) (56), thereby giving names to three further points (23), (13), and (12), outside Q_1. Then one sees that (12) must be adjacent with three uniquely determined points of Q_1, one of which is (34); label the other two by (35) and (36); they provide two further lines (12) (35) (46) and (12) (36) (45). Similarly for (23) and (13), so that names are given to all points of Q_1. Now four further lines through the 15 points labelled so far can be found showing that Q contains Q_2.

STEP 3. *If $Q \neq Q_1, Q_2$, then Q is isomorphic to Q_4.*
Any further point is adjacent to a 5-coclique in Q_2. The only 5-cocliques are the sets $\{(ij) \mid j \neq i\}$ for fixed i. By symmetry we may assume that Q contains a point 1^- adjacent to all $(1i)$, $(i \neq 1)$. This defines 5 lines which we may take to be $\{1^-, i^+, (1i)\}$, $i \neq 1$. Then i^+ is forced to be adjacent with the points (ij), $j \neq i$, and we get 20 more lines which have to be $\{j^-, i^+, (ij)\}$, $1 \neq i \neq j \neq 1$, with 5 further points j^-. Finally, j^- is forced to be adjacent with $(1j)$, giving 5 more lines $\{j^-, 1^+, (1j)\}$ and one new point 1^+. So we have found Q_4 inside Q. If $Q \neq Q_4$, then any further point is adjacent to a 9-coclique in Q_4. But if such a 9-coclique intersects the double six $\{i^-, i^+ \mid i \in \{1, \cdots, 6\}\}$ in m points, it is impossible to extend such an m-coclique to a 9-coclique by adding points of Q_2. Hence no further case arises. □

Second proof. As in the first proof, we may restrict attention to the case where Q contains Q_1. Suppose that Q strictly contains Q_1. Define a graph $\Delta = (Q, \approx)$ on $Q \setminus Q_1$ by letting $\gamma \approx \delta$ if and only if γ is collinear with δ and the line $\gamma\delta$ meets Q_1.

Each component of Δ is isomorphic to $K_{3,3}$. Indeed, each $\gamma \in Q \setminus Q_1$ determines a 3-coclique (coline) $\gamma^\perp \cap Q_1$ in Q_1, so Δ has valency 3, and if γ, δ are adjacent in Δ, then the corresponding colines meet in a single point. But Q_1 contains two parallel classes of colines, so Δ is bipartite. Finally, if $\gamma \sim \delta \sim \epsilon \sim \zeta$ is a path in Δ and $\beta \in Q_1$ is the third point of the line $\gamma\delta$, then ζ cannot be collinear with β (since δ and ζ determine distinct and parallel colines) and not with δ (since $\delta\epsilon\zeta$ would be a triangle) so $\gamma \sim \zeta$. If the colines determined by γ and ζ meet in α, then $\alpha\gamma\zeta$ is a line, so γ is adjacent to ζ in Δ.

Let the components of Δ be the sets D_j $(1 \leqslant j \leqslant r)$. Each $D_j \cup Q_1$ induces a generalized quadrangle with 15 points and 15 lines so that a point outside is adjacent to 5 points inside. It follows that points from distinct D_j are adjacent if and only if they belong to disjoint colines in Q_1, and the lines of Q disjoint from Q_1 have as point graph $2(\overline{3\times r})$. But if $r \geqslant 4$, then $\overline{3\times r}$ contains two triangles with a common edge, impossible; and if $r = 2$, then $\overline{3\times 2}$ is a hexagon and does not contain triangles, impossible again. So $r \in \{0,1,3\}$ and $|Q| \in \{9,15,27\}$, while in each case the structure is uniquely determined. □

Third proof (cf. CAMERON [171], Theorem 5.3). Fix a point $\infty \in Q$ and label the two points on each line through ∞ arbitrarily with 0 and 1. Write $R = Q\setminus\infty^{\perp}$. Let $L = L_{\infty}$ be the set of lines on ∞. We may assume that $L \neq \emptyset$ and that $R \neq \emptyset$. Now we can define a labelling $\lambda : R \to 2^L$ of the points nonadjacent to ∞ by $\lambda(\gamma) = \lambda_\gamma$, where $\lambda_\gamma(l)$ is the label of the neighbour of γ on the line l ($l \in L$). If $\alpha,\beta \in R$, $\alpha \sim \beta$ then the vectors λ_α and λ_β have precisely one coordinate in common. Conversely, for any $\alpha \in R$ and $l \in L$ there is precisely one $\beta \in R$ with $\beta \sim \alpha$ and $\lambda_\alpha(l) = \lambda_\beta(l)$. If γ and δ have distance two in R then λ_γ and λ_δ agree in all but two coordinates, l_1 and l_2. If now $m \in L\setminus\{l_1,l_2\}$ and $\epsilon \in R$ with $\epsilon \sim \delta$ and $\lambda_\epsilon(m) = \lambda_\delta(m)$ then λ_γ and λ_ϵ differ in all but three coordinates (and in particular $\epsilon \notin \gamma^{\perp}$). If there are at least 4 lines on γ then γ and ϵ must have a common neighbour in R and λ_γ and λ_ϵ differ in three and agree in two coordinates, so that $|L| = 5$.
Thus, in any case $|L| = t+1$ is finite and $|R| = (s+1)(st+1)-1-s(t+1) = s^2t = 4t$. The labels we see in one component of R are all vectors in case $t+1$ is even, and all vectors of a given parity in case $t+1$ is odd. If $t+1 = 5$ then $|R| = 16 = \frac{1}{2}\cdot 2^5$ and if $t+1 = 2$, then $|R| = 4 = 2^2$; in both cases each label occurs precisely once and Q is completely determined. If $t+1 = 3$, then $|R| = 8 = 2^3$ and each label occurs twice; R is a triangle-free double cover of K_4 and hence $K_{4\times 2}$, and again Q is determined. □

Fourth proof. (This proof uses more background knowledge than the previous ones; it first shows how to embed Q in a projective space; once this is done, classical geometry will do the rest.) As before we may suppose that Q contains a quadrangle. Let $\overline{V}$ be the vector space over $\mathbf{F}_2$ with basis Q. There is a unique quadratic form $\overline{q}$ on $\overline{V}$ with associated bilinear form $\overline{b}$ (related by $\overline{q}(u+v) = \overline{q}(u)+\overline{q}(v)+\overline{b}(u,v)$) defined by $\overline{q}(x) = 0$, $\overline{b}(x,y) = 0$ if $x \sim y$ and $\overline{b}(x,y) = 1$ if $x \not\sim y$ for x, $y \in Q$. Let $R = \operatorname{Rad}\overline{b}$, $V = \overline{V}/R$ and let $\pi : \overline{V} \to V$ be the quotient map. Since Q is nondegenerate, we have $x^{\perp} \neq Q$ and $x^{\perp} \neq y^{\perp}$ for $x \neq y$, $x,y \in Q$. Thus $\pi x \neq 0$ for $x \in Q$, and $\pi|_Q$ is injective, so that we may identify Q with the subset πQ of $V\setminus\{0\}$. If $\{x,y,z\}$ is a line of Q, then $x+y+z \in R$ so that $\{x,y,z\}$ is a line in the projective space PV (identified with $V\setminus\{0\}$). The forms $\overline{q}$ and $\overline{b}$ induce forms q and b on V (defined by $q(v+R) = \overline{q}(v)$ and $b(u+R,v+R) = \overline{b}(u,v)$ for u, $v \in \overline{V}$) and for these forms the points and lines of Q are totally singular.
CLAIM. *Q is the geometry of totally singular points and lines in the orthogonal space (V,q).*
Indeed, let $v \neq 0$ be a singular point of V. Since Q spans V, we can write $v = \sum_{i=1}^{s} x_i$ with $x_i \in Q$ ($1 \leqslant i \leqslant s$); suppose this sum is chosen such that s is minimal. If $x_i \sim x_j$ then $x_i + x_j \in Q$ and the number of summands is not minimal, contradiction. Hence $\{x_1, \cdots, x_s\}$ is a coclique. Since v is singular, $0 = q(v) = \binom{s}{2}$, so that $s \equiv 0$ or 1 (mod 4). Suppose $s \geqslant 4$. If $y \in Q$, $y \sim x_1, x_2$, $y \not\sim x_j$ then $x_1+x_2+x_j = (x_1+y)+((x_2+y)+x_j)$ is a sum of two elements in Q, contradiction. Thus, if y, $z \sim x_1, x_2$, then y, $z \sim x_j$ for all j ($1 \leqslant j \leqslant s$), but then $x_1+x_2+x_3+x_4 = ((x_1+y)+(x_2+z))+((x_3+y)+(x_4+z))$ is a sum of two

elements of Q, contradiction. Hence $s < 4$, so $s = 1$ and $v \in Q$.
If $\{x,y,x+y\}$ is a totally singular line, then $b(x,y) = 0$, so that this is a line of Q. This proves our claim.

Now the classification is immediate from the known theory of orthogonal geometry: Since Q must contain projective lines but not projective planes, we have dim $V = 3$, 4 or 5, where we have a hyperbolic quadric if dim $V = 3$ and an elliptic quadric in case dim $V = 5$. □

Remark. The advantage of the third proof is that one already finds that Q is finite before having detailed information on the possible finite examples. Unfortunately, none of the proofs seems to generalize to arbitrary $s > 2$ and in fact it is unknown whether all nondegenerate generalized quadrangles whose lines have finite length $s+1$ are finite. For $s = 3$, Kantor has a group-theoretic proof answering the question in the affirmative; a simple geometric argument in the style of the second proof was given by Brouwer (both unpublished; Kantor's result is referred to in CAMERON [175]).

1.15.3. Corollary. *Let Γ be a connected amply regular graph with parameters (v,k,λ,μ), where $k = 3\mu-3$ and $\lambda = 2\mu-4$. Then either $\mu = 3$ and Γ is a conference graph, i.e., distance-regular with intersection array $\{6,3;1,3\}$, or $\mu \in \{2,4,6,10\}$, and Γ is a Taylor graph, i.e., distance-regular with intersection array $\{3\mu-3, \mu, 1; 1, \mu, 3\mu-3\}$. In each case there is a unique graph.*

Proof. Suppose that Γ has diameter d. If $d \leqslant 2$, then Γ is strongly regular, and Theorem 1.3.1 gives irrational eigenvalues $\frac{1}{2}(\mu-4\pm\sqrt{\mu^2+4})$. Hence Γ is a conference graph, and $\lambda = \mu-1$, so $\mu = 3$. But it is easy to check that the only such graph is the Paley graph $QR(13)$. If $d \geqslant 3$, then Γ is a Taylor graph by Theorem 1.5.5; hence Theorem 1.5.3 applies and shows that $\Delta = \Gamma(\gamma)$ is strongly regular with parameters $v(\Delta) = 3\mu-3$, $k(\Delta) = 2\mu-4$, $\mu(\Delta) = \mu-2$. Consequently, its complementary graph $\overline{\Delta}$ is strongly regular with parameters $\overline{v} = 3\mu-3$, $\overline{k} = \mu$, $\overline{\lambda} = 1$. By the lemma, $\overline{\Delta}$ is the point graph of a generalized quadrangle with line size 3, whence known, and Γ can be uniquely reconstructed from Δ. □

Remark. In view of the equivalence of Taylor graphs and regular two-graphs, this also follows from TAYLOR [710]. The Taylor graphs arising are the 3-cube ($\mu = 2$), the Johnson graph $J(6,3)$ ($\mu = 4$), the halved 6-cube $D_{6,6}(1)$ ($\mu = 6$), and the Gosset graph $E_7(1)$ ($\mu = 10$); see Chapters 3, 9 and 10 for definitions.

1.16 Regular graphs without quadrangles

A regular graph of girth at least 5 has no quadrangles. On the other hand, for regular graphs of girth at most 4, the requirement that there be no quadrangles is very restrictive; it is equivalent to the requirement that for any two vertices γ, δ at distance 2, the subgraph $\Gamma(\gamma) \cap \Gamma(\delta)$ is a clique. Recently, TERWILLIGER [717] proved some beautiful results on such graphs Γ assuming the extra property that all these cliques have constant size, i.e., $\mu(\Gamma)$ is constant. The known examples with $\mu(\Gamma) > 1$ are all distance-regular.

Let $x \equiv y$ if $x^\perp = y^\perp$, then $\equiv$ is an equivalence relation, and we shall write $\overline{\Gamma}$ for the quotient $\Gamma/\equiv$ and $\overline{x}$ for the equivalence class of the vertex x. (I.e., $\overline{\Gamma}$ has vertices

$\bar{x}$ for $x \in \Gamma$ and $\bar{x} \sim \bar{y}$ when $x \sim y$, $\bar{x} \neq \bar{y}$.) $\bar{\Gamma}$ is called the *reduced graph* of Γ, and Γ is called *reduced* when all equivalence classes have size one (cf. §A.8).

A *Terwilliger graph* is a non-complete graph Γ such that, for any two vertices γ, δ at distance two, $\Gamma(\gamma) \cap \Gamma(\delta)$ is a clique of size μ (for some fixed $\mu \geqslant 0$).
Terwilliger's main result on such graphs is that if Γ is a regular Terwilliger graph, then its reduced graph $\bar{\Gamma}$ is strongly regular.

1.16.1. Proposition. *Let Γ be a regular connected Terwilliger graph, and suppose that $b_2(\alpha,\beta) = b_2(\beta,\alpha)$ for any two vertices $\alpha, \beta \in \Gamma$, $d(\alpha,\beta) = 2$. (This holds in particular when Γ has diameter 2.) Then, for each $\gamma \in \Gamma$, the graph $\Delta := \Delta_\gamma := \gamma^\perp \setminus \bar{\gamma}$ is a coconnected regular Terwilliger graph, and $\mu(\Delta) = \mu(\Gamma) - |\bar{\gamma}|$.*

Proof. Since Γ is regular and not a clique, $\gamma^\perp$ is not a clique, and hence Δ is nonempty and not a clique. Obviously, Δ is a Terwilliger graph with $\mu(\Delta) = \mu(\Gamma) - |\bar{\gamma}|$. Let Δ_0 be a cocomponent of Δ containing some nonedge of Δ. Then each point of Δ_0 is adjacent to each point of $\Delta \setminus \Delta_0$. Since Δ_0 contains a nonedge and Δ is a Terwilliger graph, $\Delta \setminus \Delta_0$ is a clique contained in the radical of Δ and hence is empty. It follows that $\Delta = \Delta_0$ is coconnected.
Now suppose that Δ is not regular. Since Δ is coconnected, we can find nonadjacent vertices $\alpha, \beta \in \Delta$ with different valencies in Δ. For any set $E \subseteq \Gamma(\alpha) \cap \Gamma(\beta)$ define $E_\alpha := \{\xi \in \Gamma(\alpha) \setminus \Gamma(\beta) \mid \{\alpha,\beta,\xi\}^\perp = E\}$ and $E_\beta := \{\xi \in \Gamma(\beta) \setminus \Gamma(\alpha) \mid \{\alpha,\beta,\xi\}^\perp = E\}$. Then, for $\omega = \alpha$ and $\omega = \beta$, we have

$$k_\Gamma(\omega) = \mu + \sum_E |E_\omega|, \quad k_\Delta(\omega) = \mu(\Delta) + \sum_{E \ni \gamma} |E_\omega|.$$

Since $k_\Gamma(\alpha) = k_\Gamma(\beta)$ but $k_\Delta(\alpha) \neq k_\Delta(\beta)$, there must be a proper subset E of $\Gamma(\alpha) \cap \Gamma(\beta)$ such that $|E_\alpha| \neq |E_\beta|$. By hypothesis, $E \neq \varnothing$. Now, if $\xi \in E_\alpha$, then $\Gamma(\xi) \cap \Gamma(\beta)$ has $|E|$ points in E and $\mu - |E|$ points in E_β. (Indeed, if $\delta \in \Gamma(\xi) \cap \Gamma(\beta) \setminus E$, then since $\Gamma(\xi) \cap \Gamma(\beta)$ is a clique, we have $E \subseteq \Gamma(\delta)$; but also $\Gamma(\alpha) \cap \Gamma(\delta)$ is a clique, so $\Gamma(\delta) \cap \Gamma(\alpha) \cap \Gamma(\beta) \subseteq E$ and $\delta \in E_\beta$.) It follows that the edges between E_α and E_β form a regular bipartite graph of valency $\mu - |E| > 0$, which is impossible since $|E_\alpha| \neq |E_\beta|$. Hence Δ must be regular. □

Remark. The Terwilliger graph depicted below is regular of valency 3 and has $\mu = 1$. Its diameter is 3. However, some local graphs are $K_1 + K_2$ - these are not regular, and the equivalence classes of these local graphs do not have constant size.

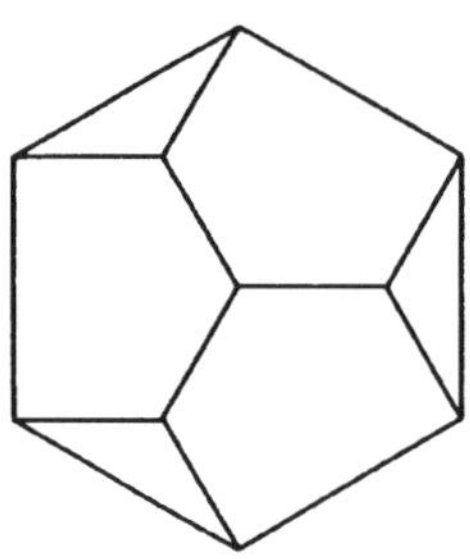

The hypothesis on b_2 in the above proposition may be replaced by the hypothesis $\mu = 2$, provided that Γ is not the result of doubling all points in a regular Terwilliger graph with $\mu = 1$.

Problem. Is it true that if Γ is a regular Terwilliger graph with $|\bar{\delta}| < \mu$ for at least one vertex $\delta \in \Gamma$, then $\Delta := \gamma^\perp \setminus \bar{\gamma}$ is regular for each vertex $\gamma \in \Gamma$?

1.16.2. Proposition. *Let Γ be a connected regular Terwilliger graph of diameter 2. Then all equivalence classes $\bar{\gamma}$ ($\gamma \in \Gamma$) have the same cardinality, and Γ is a strongly regular Terwilliger graph.*

Proof. Write k_γ for the valency of $\Delta_\gamma = \gamma^\perp \setminus \bar{\gamma}$ where $\gamma \in \Gamma$. If $\alpha \sim \beta$, $\bar{\alpha} \neq \bar{\beta}$, then $k_\alpha = |\{\alpha,\beta\}^\perp| - |\bar{\alpha}| - 1$, so that $\kappa := k_\alpha + |\bar{\alpha}| = k_\beta + |\bar{\beta}|$ is independent of the vertex chosen. Counting edges between Δ_γ and $\Gamma_2(\gamma)$ we find

$$|\Delta_\gamma| \cdot (k - k_\gamma - |\bar{\gamma}|) = (v - |\Delta_\gamma| - |\bar{\gamma}|)\mu ,$$

whence, using $|\Delta_\gamma| = k + 1 - |\bar{\gamma}|$,

$$(k + 1 - |\bar{\gamma}|)(k - \kappa + \mu) = (v - |\bar{\gamma}|)\mu ,$$

and this linear equation determines $|\bar{\gamma}|$ uniquely since $k > \kappa$. Thus, $|\bar{\alpha}|$ is constant, and the proposition follows. □

1.16.3. Theorem. *Let Γ be an amply regular Terwilliger graph with parameters (v,k,λ,μ), $\mu > 1$. Then, for any $\gamma \in \Gamma$, the reduced graph $\overline{\Gamma(\gamma)}$ of $\Gamma(\gamma)$ is a strongly regular Terwilliger graph with parameters*

$$\bar{v} = \frac{k}{s}, \quad \bar{k} = \frac{\lambda - s + 1}{s}, \quad \bar{\mu} = \frac{\mu - 1}{s},$$

$$\bar{\lambda} = \frac{(\lambda - s + 1)(\lambda - 2s + 1) - (\mu - 1)(k - \lambda - 1)}{s(\lambda - s + 1)},$$

where s is the size of the equivalence classes of $\Gamma(\gamma)$. Moreover, every singular line has size $s + 1$, and every vertex is in k / s singular lines. In particular,

$$\mu = s + 1 \text{ or } \mu \geqslant s^2 + s + 1,$$

$$s \,|\, \gcd(k, \lambda + 1, \mu - 1), \quad \lambda - s + 1 \,|\, (\mu - 1)(k - s), \quad s \leqslant \bar{\lambda} + 1 .$$

Proof. $\delta \equiv \epsilon$ in $\Gamma(\gamma)$ means $\{\gamma,\delta\}^\perp = \{\gamma,\epsilon\}^\perp = \{\gamma,\delta,\epsilon\}^\perp$, so the equivalence class in $\Gamma(\gamma)$ containing δ has the same size as the equivalence class in $\Gamma(\delta)$ containing γ, so s is constant. Now the formulae for $\bar{v}$, $\bar{k}$, $\bar{\lambda}$ and $\bar{\mu}$ follow by simple counting.
Γ with its singular lines is a Gamma space, and clearly, all μ-graphs in Γ are singular subspaces. Since there are subspaces of size μ, we have either $\mu = s + 1$ or $\mu \geqslant s^2 + s + 1$. Also, the singular lines of $\overline{\Gamma(\gamma)}$ have size at least $s + 1$ (cf. §A.8). Taking, if necessary, repeatedly local graphs, we eventually arrive at a graph with $\mu = 1$ (and the next stage is a union of cliques); if $\bar{\mu} = 1$, then $\mu = s + 1$. Now if $\bar{\mu} = 1$, then $\overline{\Gamma(\gamma)}$ has lines of size $\bar{\lambda} + 2$ so that $s \leqslant \bar{\lambda} + 1$. The rest is obvious. □

Examples. Starting with an amply regular Terwilliger graph Γ with $\mu > 1$ and taking the reduced local graph a number of times we obtain a strongly regular graph with $\mu = 1$. The only known such graphs are the pentagon, the Petersen graph and the Hoffman-Singleton graph. These graphs have lines of size 2 so if they occur as reduced local graph, then in fact they occur as local graph. The only graph that is locally a pentagon is the icosahedron (which has diameter 3). There are three locally Petersen graphs (HALL [345]), see below. All three are distance-regular, but only two of these are Terwilliger graphs (with diameters 3 and 4, respectively). No locally Hoffman-Singleton graph is known; but the intersection arrays $\{50,42,9;1,2,42\}$ or $\{50,42,1;1,2,50\}$ might belong to such a graph which is distance-regular. Feasible parameter sets for strongly regular graphs with $\mu = 1$ are for example $(v,k,\lambda) = (400,21,2)$, $(3250,57,0)$, $(3905,64,3)$. In conclusion, we know of three examples of distance-regular Terwilliger graphs with $\mu > 1$, and these might well be all such graphs.

Problems.

(i) Settle the existence problem for locally Hoffman-Singleton graphs, in particular for distance-regular graphs with array $\{50,42,9;1,2,42\}$ or $\{50,42,1;1,2,50\}$. VAN BON [pers. comm.] showed that a locally Hoffman-Singleton graph cannot be distance-transitive.

(ii) Show (or disprove) that all Terwilliger graphs with $\mu > 1$ have diameter $\leqslant 4$. If possible, classify them all.

Remarks. (i) Taking the singular spaces of a regular Terwilliger graph of rank i to be the objects of type i, $0 \leqslant i \leqslant m-1$ (where $m-1$ is the rank of a maximal clique) we obtain a geometry with Buekenhout diagram

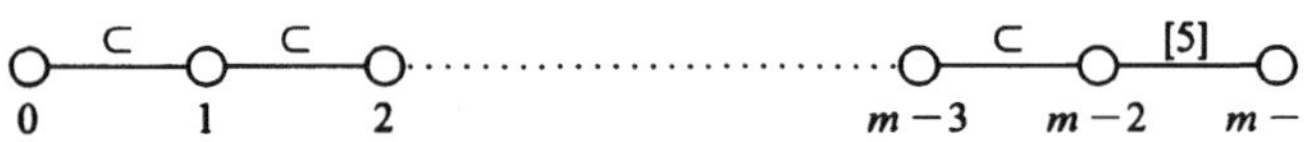

where [5] denotes a partial linear space of girth 5, and incidence is symmetrized containment. See BUEKENHOUT [143, 144] for definitions and properties of geometries with diagrams.
(ii) There are three graphs that are locally an icosahedron, cf. BLOKHUIS et al. [94], but of course these cannot be Terwilliger graphs; none of them is distance-regular.
(iii) Some locally locally Petersen graphs are constructed in IVANOV & SHPECTOROV [438]; in particular they construct examples for He over $3.S_7$ and for $PSp_4(5)$ over $P\Sigma L_2(25)$; the former graph is not distance-regular; the latter is strongly regular (but not rank 3). Of course, the latter example is part of an infinite tower

$$O_{2n+1}(5) \supseteq O^-_{2n}(5) \supseteq O_{2n-1}(5) \supseteq O^-_{2n-2}(5) \supseteq \cdots$$

that gives rise to a flag-transitive geometry with the above Buekenhout diagram; the corresponding graphs were first described by Wilbrink, cf. BROUWER & VAN LINT [134], §7D.

As a simple application of the above theorem we have:

1.16.4. Corollary (cf. HALL [345]). *Let Γ be an amply regular graph with parameters (v,k,λ,μ), $k = 10$, $\lambda = 3$. Then $\mu \geqslant 2$, and one of the following holds:*

(i) *Γ is locally Petersen.*

(ii) *There are vertices α, β at distance 2 with $b_2(\alpha,\beta) \leqslant \mu$.*

Proof. Suppose that $b_2(\alpha,\beta) > \mu$ for all α, β at distance 2. Then Theorem 1.2.3 applies with $b_2 \geqslant \mu+1$, showing that every 2-claw is in at most $\max\{0, k-2\lambda-3+\mu-b_2\} = 0$ quadrangles. Thus Γ contains no quadrangles and is a Terwilliger graph. Proposition 1.2.1 shows that $\mu \neq 1$. Hence $\mu \geqslant 2$, so that by the above theorem, each $\Gamma(\gamma)$ is strongly regular. But a strongly regular graph with 10 vertices and valency 3 must be the Petersen graph. Hence (i) holds. □

Remark. If one looks for possible amply regular graphs with $k = 10$, $\lambda = 3$ which are not locally Petersen, one finds a feasible intersection array $\{10,6,1;1,3,10\}$. But an exhaustive search by Hoare (quoted in BIGGS [77]) showed that there is no distance-regular graph with this intersection array. (As Godsil and Mathon remarked, this graph is also excluded by the analogue of the Bruck-Chowla-Ryser theorem for square divisible designs, see Corollary 1.10.3.) Also, in BLOKHUIS & BROUWER [92] it is shown that if $\mu = 2$, then Γ is locally Petersen, and that $\mu \geqslant 4$ implies that Γ is strongly regular and either is the triangular graph $T(7)$ with $(v,k,\lambda,\mu) = (21,10,3,6)$, or is one of the ten (cf. PAULUS [584], BUSSEMAKER, MATHON & SEIDEL [154]) strongly regular graphs with $(v,k,\lambda,\mu) = (26,10,3,4)$. Thus, it may well be that case (ii) only holds for the strongly regular graphs mentioned.

1.16.5. Theorem (HALL [345]). *There are up to isomorphism precisely three connected locally Petersen graphs, namely:*

(i) *The unique distance-regular graph with* 21 *vertices and intersection array* $\{10,6;1,6\}$*: the complement of the triangular graph* $T(7)$.

(ii) *The unique distance-regular graph with* 63 *vertices and intersection array* $\{10,6,4,1;1,2,6,10\}$*: the Conway-Smith graph* (A8) *described in §13.2B.*

(iii) *The unique distance-regular graph with* 65 *vertices and intersection array* $\{10,6,4;1,2,5\}$*: the Doro graph* (D4) *described in §12.1.* □

1.16.6. Corollary. *Let Γ be an amply regular Terwilliger graph of diameter d.*

(i) *If $d = 2$ and $v < 50\mu$, then Γ is the pentagon or the Petersen graph.*

(ii) *If $k < 50(\mu-1)$, then Γ is the icosahedron ($v = 12, d = 3$) or a locally Petersen graph (and then $v = 63, d = 4$ or $v = 65, d = 3$).*

Proof. (i) Suppose $v < 50\mu$ and $d = 2$. Then Γ is strongly regular. If $\mu = 1$, then from Theorem 1.3.1 we find after a short calculation that Γ must be the pentagon or the Petersen graph. Thus, a further example with a minimal number of vertices has $\mu > 1$, and the residue $\overline{\Gamma}$ of a vertex is again a Terwilliger graph with $d = 2$. Since it cannot be a polygon or a Petersen graph, we have $\overline{v} \geqslant 50\overline{\mu}$ by minimality of v, hence $k = \overline{v}s \geqslant 50\overline{\mu}s = 50(\mu-1) \geqslant 25\mu$. Hence $b_1 = (v-k-1)\mu/k < 49\mu/25\mu < 2$, and Corollary 1.1.6 gives a contradiction.
(ii) Suppose $k < 50(\mu-1)$. Then $\mu > 1$ and $\overline{v} = k/s < 50(\mu-1)/s = 50\overline{\mu}$, so that the residues are pentagons or Petersen graphs. Since $s \leqslant \overline{\lambda}+1 = 1$, i.e., $\overline{v} = k$, we have locally a pentagon if $k = 5$, giving the icosahedron, or locally a Petersen graph if $k = 10$, giving one of the two locally Petersen graphs with $v = 63$ or 65, since the complement of $T(7)$ contains quadrangles. □

1.17 Geodetic graphs of diameter two

A graph is called *geodetic* (cf. ORE [580], p. 105) if any two vertices are joined by a unique shortest path. In this section we study finite geodetic graphs of diameter two, that is, finite graphs with the property that any two nonadjacent vertices have a unique common neighbour. The results are mostly due to STEMPLE [702], KANTOR [456] and BRIDGES, FINUCANE & MENA [115]. See also LEE [480] and SCAPELLATO [639]. This section is a condensed version of BLOKHUIS & BROUWER [93], where full proofs of all results given here and some additional information may be found.

Examples of geodetic graphs of diameter two:

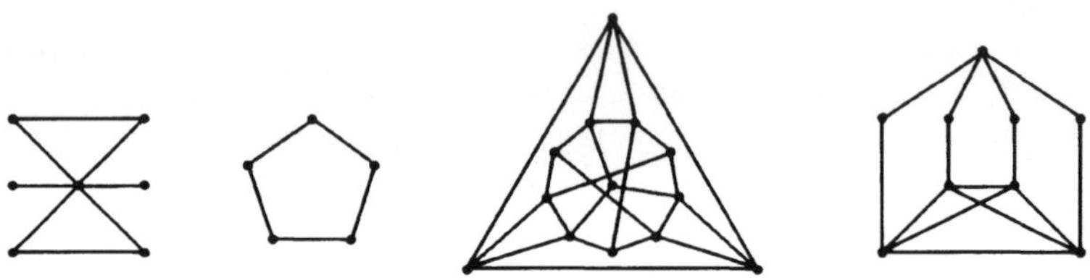

Let us call maximal cliques *lines*. Clearly in a geodetic graph each edge is in a unique line. Lines with more than two points will be called *thick*.

1.17.1. Theorem. *Let Γ be a geodetic graph of diameter two. Then we have one of three cases. Either*

(i) *Γ contains a vertex adjacent to all other vertices; in this case Γ is a degenerate generalized quadrangle (a collection of lines with one common vertex),*

or

(ii) *Γ is regular, and then strongly regular (with $\mu = 1$),*

or

(iii) *precisely two valencies a, b occur, $a < b$. If we let A and B denote the sets of vertices with valency a and b, respectively, then A is a coclique, lines meeting A and B have size 2, and lines contained in B have size l, where $l = b - a + 2 > 2$. Also, $v = ab + 1$.*

Proof. STEP 1. *Let L, M be two lines meeting in the point ϵ. Then for $\gamma \in L \setminus \{\epsilon\}$ and $\delta \in M \setminus \{\epsilon\}$ we have $|\gamma^\perp \setminus L| = |\delta^\perp \setminus M|$. In particular, all points of $L \setminus \{\epsilon\}$ have the same valency.*
Each $\alpha \in \gamma^\perp \setminus L$ has a unique neighbour $\beta \in \delta^\perp \setminus M$.

STEP 2. *If for some vertex γ and line L we have $\gamma^\perp \subseteq L$ then* (i) *holds.*
In this situation no line M is disjoint from L (for if $\alpha, \beta \in M$ and $\alpha \sim \alpha' \in L$, $\beta \sim \beta' \in L$ then $\alpha\beta\beta'\alpha'$ is a 4-gon, contradicting $\mu = 1$) and if M, N are lines distinct from L, then M and N meet L in the same point (otherwise Γ has diameter > 2).

Now assume that we are not in case (i). Then by Step 1 if L is a line with at least 3 points, then all points of L have the same valency.

STEP 3. *Let L be a line with l points, each of valency b. Then $v = b(b-l+2)+1$.* Choose γ, $\delta \in L$. There is a one-to-one correspondence between vertices at distance two from L and pairs of vertices in $(\gamma^{\perp} \setminus L) \times (\delta^{\perp} \setminus L)$. Thus $v = l + l(b+1-l) + (b+1-l)^2$.

If Γ is regular, then Step 3 implies that all lines have the same size so that Γ is edge-regular and hence strongly regular. This is case (ii). If $L = \{\gamma,\delta\}$ is a line where $k(\gamma) = a$ and $k(\delta) = b$, then by the same counting argument as in Step 3 we find $v = 2+(a-1)+(b-1)+(a-1)(b-1) = ab+1$. Since Γ is connected, this proves everything. □

Let us describe all known examples not of type (i).

a) Three strongly regular graphs with $\mu = 1$ are known: the pentagon, with $(v,k,\lambda,\mu) = (5,2,0,1)$, the Petersen graph, with $(v,k,\lambda,\mu) = (10,3,0,1)$ and the Hoffman-Singleton graph, with $(v,k,\lambda,\mu) = (50,7,0,1)$. Infinitely many feasible parameter sets exist; the smallest unsettled case is $(v,k,\lambda,\mu) = (400,21,2,1)$.

b) For each natural number l we can construct a graph Γ on $2l+1$ vertices by taking a clique B_0 and a coclique A, both of size l, and an extra point ∞, where $\Gamma(\infty) = A$ and the points of A are joined to those of B_0 by a matching. These graphs are of type (iii) (provided that $l \geqslant 3$; for $l = 2$ we get the pentagon, for $l = 1$ the graph $K_{1,2}$) and $a = 2$, $b = l$, $B = B_0 \cup \{\infty\}$.

c) For each natural number n such that an affine plane $AG(2,n) = (X, \mathfrak{L})$ exists, we can construct a graph Γ by setting $A = X$, the set of points, and $B = \mathfrak{L}$, the set of lines of the affine plane, where a point is adjacent to a line when it is on the line, and two lines are adjacent when they are parallel. For $n = 2$ this is the Petersen graph; for $n \geqslant 3$ these graphs are of type (iii), and we have $v = 2n^2+n$, $|A| = n^2$, $|B| = n^2+n$, $a = n+1$, $b = 2n-1$, $l = n$.

d) For each natural number n such that an affine plane $AG(2,n) = (X, \mathfrak{L})$ exists, let Γ_0 be the graph constructed under c) and enlarge the set B with $n+1$ points corresponding to the $n+1$ parallel classes of lines; join each line to the point representing its direction, and join all $n+1$ new points mutually. This yields a graph of type (iii) with $v = 2n^2+2n+1$, $|A| = n^2$, $|B| = (n+1)^2$, $a = n+1$, $b = 2n$, $l = n+1$.

e) For each polarity π of a projective plane $PG(2,n) = (X, \mathfrak{L})$, let Γ be the graph with point set X, where the points x, $y \in X$ are joined when the point x is on the line y^{π}. Now Γ is a graph of type (iii) with $v = n^2+n+1$, $a = n$, $b = n+1$, $l = 3$ and A is the set of absolute points of π.

Problem. Prove that no other examples exist.

Remark. These families of examples overlap for small parameters: $(v,a,l) = (7,2,3)$ is found under both b) and e), $(v,a,l) = (13,3,3)$ under both d) and e) and $(v,a,l) = (21,4,3)$ under both c) and e).

From now on, assume that Γ is a graph of type (iii), so that in particular Γ has empty radical. It is easy to characterize the above examples by the fact that they have small a or l.

1.17.2. Proposition (W.G. Bridges, cf. KANTOR [456]). *If $l = 3$ then Γ is the graph of a polarity in $PG(2,n)$, as described in Example* e). *In particular, $v = n^2 + n + 1$ and $a = n$.*

Proof. Add loops at the vertices in A, then any two vertices have a unique common neighbour; the resulting adjacency matrix is the incidence matrix of a projective plane with polarity. If the plane is not $PG(2,n)$, i.e., if some line contains all but one of the points, then Γ has a nonempty radical, contradiction. □

1.17.3. Proposition. *Suppose that some vertex $\beta \in B$ is not on a thick line, so that β is an isolated point in the graph B. Then $\Gamma(\beta) = A$ so that $|A| = b$ and each vertex in $B \setminus \{\beta\}$ has precisely one neighbour in A and is on c thick lines, where $c = (b-1)/(l-1)$.*
If $c = 1$ then $a = 2$ and we have example b).
Otherwise, the graph $\Delta = \Gamma_2(\beta)$ is distance-regular with intersection array $\{b-1, a-2, 1; 1, 1, b-1\}$ and eigenvalues $b-1$, -1, $\frac{1}{2}(l-3) \pm \sqrt{D}$ (where $D = a - 1 + \frac{1}{4}(l-1)^2$) with multiplicities 1, $b-1$, $\frac{1}{2}b(a-2)(1 \mp \frac{l-3}{2\sqrt{D}})$, respectively.
If $c = 2$, then $l = 3$ and Γ is uniquely determined (with Δ being the line graph of the Petersen graph); otherwise $c \geqslant 3$.
Conversely, if Δ is a distance-regular graph with intersection array $\{k, k-1-\lambda, 1; 1, 1, k\}$, i.e., if Δ is antipodal with $d = 3$ and $\mu = 1$, then Δ can be embedded uniquely as $\Gamma_2(\beta)$ in a graph Γ of type (ii) *when $\lambda = 0$ and of type* (iii) *when $\lambda > 0$.*

Proof. We have $B \setminus \{\beta\} = \bigcup_{\alpha \in A} B_\alpha$, where $B_\alpha = B \cap \alpha^\perp \setminus \{\beta\}$, and each point of B_α has a unique neighbour in each set B_ϵ, for $\epsilon \neq \alpha$. This shows that $\Gamma_2(\beta) = B \setminus \{\beta\}$ carries a distance-regular graph with parameters as claimed. The spectrum follows by straightforward calculation. The case $c = 1$ is trivial. That $c \neq 2$ for $l > 3$ can be seen by ad hoc arguments or by invoking the known classification of distance-regular graphs that are line graphs (see §4.2E). If $c = 2$, $l = 3$ then we have the graph belonging to a polarity in $PG(2,4)$ where the set of absolute points is a line. Conversely, if Δ is an antipodal graph as described, then adding a new vertex for each equivalence class of vertices at maximal distance 3 (with these vertices as neighbours) and a last vertex β with these vertices as neighbours, we find a graph Γ as required. □

Remarks. (i) In the case $c = 2$, instead of adding new vertices, we can modify the graph Δ (with intersection array $\{4,2,1; 1,1,4\}$) by making points at distance 3 adjacent. Our new graph $\Delta \cup \Delta_3$ is now the generalized quadrangle Q_2 on 15 points. If Γ is the Hoffman-Singleton graph (cf. §13.1) and $\gamma \in \Gamma$ then $\Gamma_2(\gamma)$ is distance-regular with intersection array $\{6,5,1; 1,1,6\}$.
(ii) The situation of this Proposition occurs in Example e) when q is even (and A is a line). Now Δ has intersection array $\{q, q-2, 1; 1, 1, q\}$, cf. Table 6.10, (A3).

1.17.4. Theorem. *If $a \leqslant 2l-2$, then $a \in \{2, l, l+1\}$, and Γ is one of the graphs in Example* b), c) *or* d).

Proof. By our standing assumption $l > 2$. If $l = 3$ then $a \leqslant 4$ and by Proposition 1.17.2 and the remark after the examples we have b), c), or d). Now let $l > 3$. If a point of B is on c thick lines then it has $b - c(l-1)$ neighbours in A. For $a \leqslant 2l-2$ this means that a point of B has b or $b-(l-1) = a-1$ or $b-2(l-1) = a-l$ neighbours in A. If some point of B has b neighbours in A, then $|A| = b$ and all other points of B have precisely one neighbour in A and are on $c = (b-1)/(l-1)$ thick lines. If $c = 1$, then $a = 2$ and we have example b). If $c = 2$, then $l = 3$, and if $c > 2$, then $a \geqslant 2l$. Thus, we may assume that each point of B is on either one or two thick lines. If each point of B is on one thick line, then $|B| = al$ and $|B|(a-1) = |A|a$ and $v = ab+1 = a(a+l-2)+1$ imply $l = a-1$ and we have example c). Thus we may assume that $C := \{\beta \in B \mid \beta \text{ is on two thick lines}\}$ is nonempty. If L is a thick line with m points in $B \setminus C$, $m > 0$, then each point of $L \setminus C$ has distance one to $a-1$ and distance two to $(m-1)(a-1)+(l-m)(a-l)$ points of A, so that $|A| = l(a-l)+m(l-1)$ and in particular m is constant, independent of L. If $m = l$ then the two lines on a point of C are entirely contained within C so that a point of C has $(a-l)(2l-1)$ points at distance at most two in A, and $l(a-l)+l(l-1) = (a-l)(2l-1)$ implies $a = 2l$, contrary to the assumption. Thus $m < l$, and each point of $L \cap C$ is on a unique thick line entirely contained within C, and we find too many neighbours in A unless $a = l$. But now $A \cup (B \setminus C)$ induces itself a geodetic graph of diameter two which must be an affine plane example; it follows that $m = l-1$ and that C consists of one thick line, and we have example d). □

1.17.5. Theorem. *Suppose that D is not a square. Then $a \leqslant 2l-2$ or $l = 3$ (so that the graph Γ is of a known type).*

Proof (sketch). Consider the matrix $M = (m_{\gamma\delta})$ with $m_{\gamma\delta} = 0$ if $\gamma \notin \delta^\perp$, 1 if $\gamma \sim \delta$, $\frac{1}{2}(l-1)$ if $\gamma = \delta \in A$, and $\frac{1}{2}(3-l)$ if $\gamma = \delta \in B$. This matrix satisfies

$$M^2 = J + (a-1+\tfrac{1}{4}(l-1)^2)I,$$

and has eigenvalues $a+\frac{1}{2}(l-1)$, $\sqrt{D}$ and $-\sqrt{D}$ (where $D = a-1+\frac{1}{4}(l-1)^2$), with multiplicities 1, m_1 and m_2. We have $m_1+m_2 = v-1 = ab$ and

$$a+\tfrac{1}{2}(l-1)+(m_1-m_2)\sqrt{D} = \operatorname{tr} M = |A|\frac{l-1}{2} + |B|\frac{3-l}{2}. \tag{22}$$

If $\sqrt{D}$ is not rational, then, since tr$M$ is rational, we must have $m_1 = m_2$ and we can solve the equations (22) and $|A|+|B| = v = ab+1$ for $|A|$ and $|B|$. We find

$$|A| = \tfrac{1}{2}ab+1-\tfrac{1}{2}a\frac{b-2}{l-2} \quad\text{and}\quad |B| = \tfrac{1}{2}ab+\tfrac{1}{2}a\frac{b-2}{l-2}. \tag{23}$$

Applying ‘interlacing’ to M and the 2×2 matrix obtained from M by taking average row sums in the four submatrices induced by the partition of the vertex set into A and B, we find (similar to Haemers [340], 2.1.2)

$$\left|a\frac{|A|}{|B|} - \frac{l-1}{2}\right| \leqslant \sqrt{D}. \tag{24}$$

From this the theorem follows easily for large l, and by a detailed case argument for $l = 4, 5$. □

We now have the main result: either the graph is known, or a is large and D is a square. Let us mention three further results.

1.17.6. Proposition. $|A| < |B|$.

Proof. Regard A as the set of points and B as the set of lines of a linear space. A linear space has at least as many lines as points, with equality if and only if it is a projective plane. In our case we cannot have a projective plane: a pair of adjacent points in B corresponds to a pair of disjoint lines. □

Another way of constructing a partial linear space is due to LEE [480]. Let L be a thick line, and take as Points the $l(a-1)$ vertices at distance one from L, and as Lines the $(a-1)^2$ vertices at distance two from L, with adjacency as Point-Line incidence. The resulting structure is a transversal design $TD[l;a-1]$.

1.17.7. Proposition. *If B is disconnected, then either it contains an isolated point (and we have the situation of Proposition 1.17.3) or it is a disjoint union of lines (and we have the affine plane example c). Otherwise B has diameter* 3 *or* 4. □

Assuming some regularity, we find a distance-regular graph on B (and a Steiner system on A).

1.17.8. Proposition. *If each point of B has a constant number i of neighbours in A, then B is distance-regular of diameter 3 with intersection array $\{b-i, b-i-l+1, i; 1, 1, b-2i+1\}$ and i satisfies $|B|i = |A|a$ and $|B|i(i-1) = |A|(|A|-1)$, i.e., $i(i+1-l) = a-1$ and $|A| = i(b-i+1)$.* □

As an example of this situation, take example e) with a unitary polarity ($n = q^2$). Then $b = q^2+1$, $i = q+1$, $|A| = q^3+1$ and we find the unitary nonisotropics graph (cf. §12.4).
For any graph Δ with the above intersection array, Δ_3 is strongly regular (by Proposition 4.2.17); in our case it is the block graph of the Steiner system $S(2,i,i(b-i+1))$ on A.

Chapter 2
Association Schemes

The first part of this chapter contains a short account of the basic theory of (symmetric) association schemes. Such schemes are essentially partitions of a complete graph into regular subgraphs which are interrelated in a specific way. For a more extensive treatment, see BANNAI & ITO [33]. The last part of this chapter treats some special topics. Although we shall develop large parts of the theory of distance-regular graphs independently of the results of this chapter, we shall use concepts and results about association schemes for more specialized topics such as, e.g., Q-polynomial orderings (Chapter 8) and codes in graphs (Chapter 11). Multiplicity formulas (2.2.2) and bounds (2.3.3) as well as the Krein conditions (2.3.2) developed here in general context will recur for distance-regular graphs in Chapter 4.

2.1 Association schemes and coherent configurations

Let X be a finite set. An *association scheme with d classes* is a pair $(X,\mathcal{R})$ such that

(i) $\mathcal{R} = \{R_0,R_1, \cdots ,R_d\}$ is a partition of $X \times X$;

(ii) $R_0 = \Delta := \{(x,x) \mid x \in X\}$;

(iii) $R_i = R_i^{\mathsf{T}}$ (i.e., $(x,y) \in R_i \Rightarrow (y,x) \in R_i$) for all $i \in \{0, \cdots ,d\}$;

(iv) there are numbers p_{ij}^k (the *intersection numbers* of the scheme) such that for any pair $(x,y) \in R_k$ the number of $z \in X$ with $(x,z) \in R_i$ and $(z,y) \in R_j$ equals p_{ij}^k.

The number $n_i := p_{ii}^0$ of $z \in X$ with $(x,z) \in R_i$ (which is independent of $x \in X$) is called the *valency* of R_i, and the total number of points of X is

$$n := |X| = \sum_{i=0}^{d} n_i .$$

Observe that $n_0 = 1$. Occasionally we shall use v instead of n and k_i or v_i instead of n_i.

Many authors write $\square_{ijk}$ for p_{ij}^k, where $\square \in \{a,s,p,x\}$. The concept defined here is the association scheme as originally defined by BOSE & SHIMAMOTO [111]; it is what DELSARTE [251] (cf. BANNAI & ITO [33]) calls a *symmetric* association scheme; in his more general definition, (iii) is replaced by the weaker statement:

(iii′) for each i ($0 \leqslant i \leqslant d$) there is an index j ($0 \leqslant j \leqslant d$) such that $R_i = R_j^{\mathsf{T}}$, i.e., if $(x,y) \in R_i$, then $(y,x) \in R_j$.

But he also requires

(v) $p_{ij}^k = p_{ji}^k$ for all $i,j,k \in \{0, \cdots ,d\}$

in order to ensure that the Bose-Mesner algebra (see §2.2) will be commutative. D. G. HIGMAN [382, 383, 384, 386] studied the even more general concept of a *coherent configuration*, that is, a pair $(X, \mathfrak{R})$ satisfying (i),

(ii′) Δ is a union of some R_j $(0 \leqslant j \leqslant d)$,

(iii′), and (iv). In this book, we shall need non-symmetric association schemes only in Chapter 10, where we use the terminology of association schemes for pairs $(X, \mathfrak{R})$ satisfying only (i), (ii), (iii′), and (iv). Higman calls these *homogeneous coherent configurations.*

2.1.1. Lemma. *The parameters n_i, p_{ij}^k of an association scheme with d classes satisfy the following relations:*

(i) $p_{0j}^k = \delta_{jk}$,

(ii) $p_{ij}^0 = \delta_{ij} n_j$,

(iii) $p_{ij}^k = p_{ji}^k$,

(iv) $p_{ij}^k n_k = p_{ik}^j n_j$,

(v) $\sum_j p_{ij}^k = n_i$,

(vi) $\sum_l p_{ij}^l p_{lk}^m = \sum_r p_{ir}^m p_{jk}^r$.

Proof. The statements (i) - (v) are obvious. The expressions at both sides of (vi) count quadruples (z,u,x,y) with $(z,u) \in R_i$, $(u,x) \in R_j$ and $(x,y) \in R_k$ for a fixed pair $(z,y) \in R_m$. □

Remark. Various inequalities involving the p_{ij}^k can be given. For example, one has

$$\textit{If } p_{st}^l \neq 0 \textit{ then } p_{ij}^l \leqslant \sum_{r=0}^{d} \min(p_{ir}^s, p_{jr}^t). \tag{1}$$

Proof. For a given triple $x, y, z \in X$ with $(x,y) \in R_l$, $(x,z) \in R_s$, $(y,z) \in R_t$, count in two ways the number of pairs (w,r) with $(w,x) \in R_i$, $(w,y) \in R_j$, $(w,z) \in R_r$. □

MATHON [520] (p. 127-128) has a similar inequality, and an example ruled out both by his and by the above inequality. In this chapter such inequalities will not play any rôle, but in Chapter 4 we shall see that the basic inequalities on the parameters of a distance-regular graph are special cases of (1) above.

2.2 The Bose-Mesner algebra

The relations R_i of a (symmetric) association scheme are described by their $\{0,1\}$-adjacency matrices A_i defined by

$$(A_i)_{xy} = \begin{cases} 1 & \text{if } (x,y) \in R_i, \\ 0 & \text{otherwise.} \end{cases}$$

The A_i are determined as symmetric $n \times n$-matrices up to ordering of the elements of X. The axioms (i)-(iv) of an association scheme are now expressed by

$$\text{(i)} \ \sum_{i=0}^{d} A_i = J, \ \text{(ii)} \ A_0 = I, \ \text{(iii)} \ A_i = A_i^{\mathsf{T}}, \ \text{(iv)} \ A_i A_j = \sum_{k=0}^{d} p_{ij}^k A_k.$$

From (i) we see that the A_i are linearly independent, and from (iii) and (iv) that they generate a $(d+1)$-dimensional commutative algebra $\mathfrak{A}$ of symmetric matrices. This algebra was first studied by BOSE & MESNER [109] and is called the *Bose-Mesner algebra* of the scheme.

Since the A_i commute, they can be diagonalized simultaneously (cf. GANTMACHER [308], Ch. IX, §15); we find a decomposition of $\mathbf{C}^n$ as a direct sum of $d+1$ eigenspaces of dimension f_j $(0 \leqslant j \leqslant d)$. One of the eigenspaces has dimension 1 (since $J \in \mathcal{Q}$ and J has n as an eigenvalue of multiplicity 1), so we may suppose that $f_0 = 1$. The f_j are called the *multiplicities* of the scheme; if $\{E_j \mid 0 \leqslant j \leqslant d\}$ is the basis of minimal idempotents of $\mathcal{Q}$ (cf. §2.6), then

$$f_j = \text{rk } E_j = \text{tr } E_j, \quad \sum_{j=0}^{d} E_j = I, \quad \text{and} \quad E_0 = n^{-1}J.$$

Define matrices P and Q (called the *eigenmatrices* of the association scheme) by

$$A_j = \sum_{i=0}^{d} P_{ij}E_i \text{ and } E_j = \frac{1}{n}\sum_{i=0}^{d} Q_{ij}A_i. \tag{2}$$

Then $PQ = QP = nI$. Also $A_jE_i = P_{ij}E_i$, which shows that the numbers P_{ij} are the eigenvalues of A_j (occurring with multiplicity f_i). In particular, since the A_j are integral matrices, the P_{ij} are algebraic integers. Note that conjugate algebraic integers occur with the same multiplicity; in particular, if all multiplicities f_i are different, then the P_{ij} are integers.

2.2.1. Lemma. *The matrices P and Q of an association scheme satisfy:*

(i) $P_{i0} = 1, \quad Q_{i0} = 1,$

(ii) $P_{0j} = n_j, \quad Q_{0j} = f_j,$

(iii) $P_{ij}P_{ik} = \sum_l p^l_{jk}P_{il},$

(iv) $f_iP_{ij} = n_jQ_{ji}, \quad \sum_i f_iP_{ij}P_{ik} = nn_j\delta_{jk}, \quad \sum_i n_iQ_{ij}Q_{ik} = nf_j\delta_{jk},$

(v) $P_{ij}Q_{ki} = \sum_l p^k_{jl}Q_{li},$

(vi) $|P_{ij}| \leqslant n_j, \quad |Q_{ij}| \leqslant f_j,$

(vii) $\sum_j P_{jk} = \sum_i p^i_{ik}.$

Proof. (i) follows from $A_0 = I$ and from $E_0 = n^{-1}J$, (ii) from $A_jJ = n_jJ$ and from $\text{tr } E_j = f_j$, (iii) from $A_jA_k = \sum_l p^l_{jk}A_l$. The three parts of (iv) are equivalent since $PQ = nI$, and follow from $\sum_i f_iP_{ij}P_{ik} = \text{tr }(A_jA_k) = \delta_{jk}n_jn$. (v) follows from (iii) and (iv), or by evaluating A_jE_i in two ways. The first part of (vi) is obvious, and the second part follows from the first part and (iv). Finally, (vii) follows from

$$n\sum_j P_{jk} = \sum_{i,j} Q_{ij}P_{ji}P_{jk} = \sum_{i,j,l} p^l_{ik}Q_{ij}P_{jl} = n\sum_i p^i_{ik}. \qquad \square$$

If we let L_i be the matrix whose (k,j)-entry is $L_i(k,j) = p^k_{ij}$, then (iii) may be expressed by

$$PL_jP^{-1} = \text{diag }(P_{0j}, P_{1j}, \cdots, P_{dj}).$$

Consequently the P_{hi} are eigenvalues of L_i, and it follows that $A_i \mapsto L_i$ defines an isomorphism between the Bose-Mesner algebra of the scheme and the algebra generated by the L_i. The L_i are called the *intersection matrices.* (This is Delsarte's notation.

Many authors use L_i^{T} instead.)

It is possible to compute the matrices P and Q, and, in particular, the multiplicities, from the L_i. Let Δ_n and Δ_f denote the diagonal matrices diag$(n_0, \cdots, n_d)$ and diag$(f_0, \cdots, f_d)$, respectively, and observe that Lemma 2.2.1 (iv) may be expressed as

$$\Delta_n Q = (\Delta_f P)^{\mathrm{T}}. \tag{3}$$

From Lemma 2.1.1 (iv) we see that $\Delta_n L_i$ is symmetric so that if u is a right eigenvector of L_i, then $(\Delta_n u)^{\mathrm{T}}$ is a left eigenvector of L_i. Let e_h be the h-th unit vector, then $L_i Q e_h = P_{hi} Q e_h$, i.e., common right eigenvectors of the L_i are the Qe_h, and common left eigenvectors are the $(\Delta_n Q e_h)^{\mathrm{T}} = e_h^{\mathrm{T}} \Delta_f P$. Noting that $(Qe_h)_0 = (\Delta_n Q e_h)_0 = f_h$ and that the inner product $(\Delta_n Q e_h, Qe_i)$ equals $\delta_{ih} f_h n$, we have:

2.2.2. Proposition. *Let u_i and $v_i^{\mathrm{T}} = (\Delta_n u_i)^{\mathrm{T}}$ $(i = 0, \cdots, d)$ be the common right and left eigenvectors of the matrices L_i, respectively, normalized such that $(u_i)_0 = (v_i)_0 = 1$. Then $f_i = n / (u_i, v_i)$ $(i = 0, \cdots, d)$, the rows of P are the v_i^{T} $(i = 0, \cdots, d)$, and the columns of Q are the $f_i u_i$ $(i = 0, \cdots, d)$.* □

2.2.3. Corollary. $(u_i, v_i) = \sum_j n_j (u_i)_j^2 = \sum_j n_j^{-1} (v_i)_j^2$ *is rational.* □

If for some j the matrix L_j has $d+1$ distinct eigenvalues (as is almost always the case), then it suffices to consider this L_j in the above proposition. In particular if the scheme is P-polynomial (cf. §2.7), then A_1 generates the Bose-Mesner algebra so that L_1 has $d+1$ distinct eigenvalues, and L_1 alone suffices to compute the multiplicities, P and Q. In case L_j has $d+1$ distinct *integral* eigenvalues, the eigenvalues of all L_i are rational and therefore also integral. Note also that since the P_{ij} are eigenvalues of L_j, the requirement that the P_{ij} must be algebraic integers is met as soon as the p_{ij}^k are integral.

A. The Frame quotient

The algebraic integrality of the P_{ij} implies a simple and useful divisibility condition relating valencies and multiplicities of an association scheme.

2.2.4. Theorem (FRAME [297], cf. WIELANDT [782] §30). *The valencies n_i and the multiplicities f_i $(0 \leqslant i \leqslant d)$ of a d-class association scheme with n points have the property that the number*

$$q = n^{d-1} \prod_{i=0}^{d} \frac{n_i}{f_i}$$

is integral, and a square in the field obtained by adjoining to $\mathbf{Q}$ the entries of the matrix P defined in (2).

Proof. Since $QP = nI$, (3) implies $n\Delta_n = P^{\mathrm{T}} \Delta_f P$, so that $(\det P)^2 \cdot \prod f_i = n^{d+1} \prod n_i$. But the entries of P are algebraic integers and the sum of the columns of P is the vector $(n, 0, \cdots, 0)^{\mathrm{T}}$, so $n^{-1} \det P$ is an algebraic integer and $(n^{-1} \det P)^2 = n^{d-1} \prod \frac{n_i}{f_i}$ is rational, hence an integer. □

The number q in Theorem 2.2.4 is called the *Frame quotient*. Note that Theorem 30.1 (B) of WIELANDT [782] is false; for example, in the association scheme of the Perkel graph (cf. §13.3) we find $q = 57^2 \cdot \dfrac{6\cdot 20\cdot 30}{18^2\cdot 20} = 5\cdot 19^2$, not a square. In the special case of a metric scheme, another expression for q is given in Proposition 2.7.2.

Another consequence of the identity $n\Delta_n = P^{\mathsf T}\Delta_f P$ is obtained by comparing their elementary divisors:

2.2.5. Proposition (BANNAI & ITO [33], p. 74; cf. WIELANDT [782], Theorem 30.4). *If l of the multiplicities f_i are divisible by a given prime power p^a, then at least l of the numbers nn_i are divisible by p^a.* □

The identity $n\Delta_n = P^{\mathsf T}\Delta_f P$ yields another way of computing the multiplicities f_i, as will be sketched below.

2.2.6. Lemma. *The $(d+1)\times(d+1)$ matrix G with (integral) entries*

$$G_{jk} := \sum_{l=0}^{d} p^l_{jk}\Big(\sum_{i=0}^{d} p^i_{li}\Big) \quad (j, k = 0, \cdots, d)$$

satisfies $G = P^{\mathsf T}P$, and hence

$$\det(xG - n\Delta_n) = n^2 q \prod_{i=0}^{d}(x - f_i).$$

Proof. Since $\sum_i p^i_{li} = \operatorname{tr} L_l = \sum_i P_{il}$, we have $G_{jk} = \sum_{i,l} p^l_{jk}P_{il} = \sum_i P_{ij}P_{ik} = (P^{\mathsf T}P)_{jk}$; thus $G = P^{\mathsf T}P$ and $xG - n\Delta_n = P^{\mathsf T}(xI - \Delta_f)P$ so that $\det(xG - n\Delta_n) = (\det P)^2\prod(x - f_i) = n^2 q\prod(x - f_i)$. □

Thus, the multiplicities may be found as the zeros of the polynomial $p(x) = \det(xG - n\Delta_n)$. In particular, all zeros of $p(x)$ must be integral, and the highest coefficient $\det G = n^2 q$ of $p(x)$ divides all coefficients and all function values $p(s)$, s integral. Since $p(x)$ has integral coefficients, we can decide in error-free arithmetic whether or not all zeros of $p(x)$ are integral (see OBRESCHKOFF [574], §37 for efficient methods; be aware of overflow problems). This has been used by BIGGS [75] to compile the first list of feasible parameter sets for distance-regular graphs with small diameter and valency.

Remarks. (i) Noting that $p(1) = 0$ since $(G - n\Delta_n)\mathbf{1} = 0$, one can deflate the zero $f_0 = 1$, and finds for the $d\times d$ matrices $\Delta'_n = \operatorname{diag}(n_1, \cdots, n_d)$ and G' with entries $G'_{jk} = G_{jk} - n_j G_{0k} = \sum_l (p^l_{jk} - n_j\delta_{kl})(\sum_i p^i_{li})$ the relation

$$\begin{bmatrix} 1 & 0 \\ -\Delta'_n\mathbf{1} & I \end{bmatrix}(xG - n\Delta_n)\begin{bmatrix} 1 & 0 \\ \mathbf{1} & I \end{bmatrix} = \begin{bmatrix} n(x-1) & \cdots \\ 0 & xG' - n\Delta'_n \end{bmatrix},$$

so that, with smaller coefficients,

$$\det(xG' - n\Delta'_n) = nq\prod_{i=1}^{d}(x - f_i).$$

(ii) We can also compute, for selected integers s, the signature of $sG - n\Delta_n$ (or $sG' - n\Delta'_n$), and find thus the number of multiplicities f_i smaller than, equal to, and greater than s. This can be turned into a method for finding the multiplicities using a bisection method, cf. WILKINSON & REINSCH [786].

B. Pseudocyclic association schemes

A d-class association scheme is called *pseudocyclic* when $f_1 = f_2 = \cdots = f_d$. As we shall see in §12.7, one may construct strongly regular graphs and distance-regular graphs of diameter 3 from such schemes.

2.2.7. Proposition (MATHON [520]; HOLLMANN [403], p. 84). *A d-class association scheme is pseudocyclic if and only if for some constant f we have $n_i = f$ $(1 \leqslant i \leqslant d)$ and $\sum_i p_{ii}^k = f-1$ $(1 \leqslant k \leqslant d)$.*

Proof. If all f_i $(i \neq 0)$ are the same, then $f_i = (n-1)/d$ $(1 \leqslant i \leqslant d)$; in particular $\gcd(f_i,n) = 1$ for all i. Therefore, using Theorem 2.2.4, $\prod(n_i/f_i)$ is an integer, and hence $\prod n_i \geqslant (\frac{n-1}{d})^d$. By the arithmetic-geometric mean inequality it follows that $n_i = (n-1)/d$ $(1 \leqslant i \leqslant d)$.

Now the scheme is pseudocyclic if and only if the first row of Q is $(1\, f \cdots f)$, i.e., if and only if for $1 \leqslant k \leqslant d$ we have $\sum_i P_{ik} = \frac{f-1}{f} n_k = f-1$. By Lemmas 2.2.1 (vii) and 2.1.1 (iv) we are done. □

Note that in order to be pseudocyclic it does not suffice to have $n_1 = n_2 = \cdots = n_d$; a counterexample is given by the Johnson scheme $J(7,2)$ (cf. §2.7, §9.1).

The above result may be expressed combinatorially. Let $R_i(x) = \{y \mid (x,y) \in R_i\}$.

2.2.8. Corollary. *Let $(X, \mathfrak{R})$ be a d-class association scheme. Then the following are equivalent:*

(i) *$(X, \mathfrak{R})$ is pseudocyclic (with $f_1 = \cdots = f_d = f$);*

(ii) *$(X, \mathfrak{B})$ is a 2-$(v,f,f-1)$ design, where $\mathfrak{B} = \{R_i(x) \mid x \in X,\ 1 \leqslant i \leqslant d\}$;*

(iii) *$(X, \mathfrak{B}')$ is a 2-$(v,f+1,f+1)$ design, where*

$$\mathfrak{B}' = \{\{x\} \cup R_i(x) \mid x \in X,\ 1 \leqslant i \leqslant d\}.$$ □

For more information on (mostly 3-class) pseudocyclic association schemes, see §12.7. For the case $f \leqslant 2$, see the remarks in §2.4.

2.3 The Krein parameters

The Bose-Mesner algebra $\mathfrak{A}$ is not only closed under matrix multiplication but also under componentwise (Hadamard, Schur) multiplication (denoted $\circ$). The A_i form a basis of minimal idempotents with respect to this multiplication. Consequently, we may write

$$E_i \circ E_j = \frac{1}{n} \sum_{k=0}^{d} q_{ij}^k E_k$$

for certain real numbers q_{ij}^k, the *Krein parameters* or *dual intersection numbers*. Note that our q_{ij}^k are those of DELSARTE [251] but differ from those of SEIDEL [656] by a factor n. The Krein parameters have properties 'dual' to those of the p_{ij}^k.

More generally, there is a formal duality between ordinary multiplication, the numbers p_{ij}^k and the matrices A_i and P on the one hand, and Schur multiplication, the numbers q_{ij}^k and the matrices E_i and Q on the other hand. If two association schemes have the property that the intersection numbers of one are the Krein parameters of the other, then the converse is also true. Two such schemes are said to be *formally dual* to each other. A scheme is called *formally self-dual* if $P = Q$ for some ordering of the primitive idempotents; it follows that $f_i = n_i$ and $q_{ij}^k = p_{ij}^k$. One scheme may have several formal duals, or none at all - formal duality is a matter of parameters, not of structure. But when an association scheme is invariant under a regular abelian group of automorphisms there is a natural way to define a dual scheme. Such a scheme is called *self-dual* when it is isomorphic to its dual. For a further discussion of duality, see §2.10B.

The following lemma is the exact analogue of Lemma 2.1.1 and Lemma 2.2.1 (iii), (v).

2.3.1. Lemma. *The Krein parameters q_{ij}^k of an association scheme satisfy:*

(i) $q_{0j}^k = \delta_{jk}$,

(ii) $q_{ij}^0 = \delta_{ij} f_j$,

(iii) $q_{ij}^k = q_{ji}^k$,

(iv) $q_{ij}^k f_k = q_{ik}^j f_j$,

(v) $\sum_j q_{ij}^k = f_i$,

(vi) $\sum_l q_{ij}^l q_{lk}^m = \sum_l q_{il}^m q_{jk}^l$,

(vii) $Q_{ij} Q_{ik} = \sum_l q_{jk}^l Q_{il}$,

(viii) $P_{ki} Q_{ij} = \sum_l q_{jl}^k P_{li}$.

Proof. Following SEIDEL [656], let us define $\Sigma(A) := \sum_{i,j} A_{ij}$, the sum of all entries of a matrix A. Then $\Sigma(M \circ N) = \mathrm{tr}\,(MN^{\mathrm{T}})$ and $JAJ = \Sigma(A)J$. Now $\Sigma(E_i) = \delta_{0i} n$ since $\Sigma(E_i)J = JE_iJ = nE_0E_iJ = \delta_{0i} nJ$. Now, (i) follows from $E_0 = n^{-1}J$, and (ii) from $q_{ij}^0 = \Sigma(E_i \circ E_j) = \mathrm{tr}\,(E_iE_j) = \delta_{ij} f_j$; (iii) is obvious; (vi), (vii) and (viii) follow by evaluating $E_i \circ E_j \circ E_k$, $A_i \circ E_j \circ E_k$, and $(A_i \circ E_j)E_k$ in two ways; (iv) follows from (vi) by taking $m = 0$, and (v) from (viii) by taking $i = 0$. □

The q_{ij}^k are not necessarily integral, but they are nonnegative; these important restrictions, discovered by SCOTT [647] (see also [648]), are known as the *Krein conditions*. As before, let $\Sigma(A)$ denote the sum of all entries of the matrix A.

2.3.2. Theorem. *For all $i,j,k \in \{0,1, \cdots ,d\}$,*

$$n f_k q_{ij}^k = n^2 \Sigma(E_i \circ E_j \circ E_k) = \sum_{l=0}^{d} n_l Q_{li} Q_{lj} Q_{lk} \geqslant 0, \tag{4}$$

with equality if and only if

$$\sum_{x \in X} E_i(u,x) E_j(v,x) E_k(w,x) = 0 \quad \text{for all } u,v,w \in X. \tag{5}$$

Proof. Since the E_i are symmetric idempotent matrices,

$$\sum_{u \in X} E_i(u,x)E_i(u,y) = E_i(x,y). \tag{6}$$

Hence, if we denote the left hand side of (5) by $q(u,v,w)$, we have

$$\begin{aligned}\Sigma(E_i \circ E_j \circ E_k) &= \sum_{x,y \in X} E_i(x,y)E_j(x,y)E_k(x,y) \\ &= \sum_{u,v,w \in X} \Big(\sum_{x \in X} E_i(u,x)E_j(v,x)E_k(w,x)\Big)\Big(\sum_{y \in X} E_i(u,y)E_j(v,y)E_k(w,y)\Big) \\ &= \sum_{u,v,w \in X} q(u,v,w)^2 \geqslant 0\end{aligned}$$

by (6). Since both

$$n^2\Sigma(E_i \circ E_j \circ E_k) = n^2 \mathrm{tr}\,((E_i \circ E_j)E_k) = n\,\mathrm{tr}\,(\sum_l q_{ij}^l E_l E_k) = n\,\mathrm{tr}\,(q_{ij}^k E_k) = n f_k q_{ij}^k$$

and (by (2))

$$n^2\Sigma(E_i \circ E_j \circ E_k) = \frac{1}{n}\sum_l Q_{li}Q_{lj}Q_{lk}\Sigma(A_l) = \sum_l n_l Q_{li}Q_{lj}Q_{lk},$$

inequality (4) holds, and equality holds in (4) if and only if $\sum q(\alpha,\beta,\gamma)^2 = 0$, which is equivalent to (5). □

Remarks. (i) Another proof of the Krein conditions can be given as follows. The number $\frac{1}{n}q_{ij}^k$ is an eigenvalue of $E_i \circ E_j$ which is a principal submatrix of the Kronecker product $E_i \otimes E_j$. As all eigenvalues of the E_l and hence of $E_i \otimes E_j$ are 0 or 1, it follows by interlacing (cf. §3.3) that $0 \leqslant \frac{1}{n}q_{ij}^k \leqslant 1$. (Since $q_{ij}^k \geqslant 0$ and $\sum_j q_{ij}^k = f_i \leqslant n$ imply that $\frac{1}{n}q_{ij}^k \leqslant 1$, the latter inequality does not yield new information.)

(ii) For practical calculations the inequalities are best written $q_{ijk} \geqslant 0$, where $q_{ijk} = \sum_l n_l (u_i)_l (u_j)_l (u_k)_l$. Note that $q_{ij}^k = n^{-1} f_i f_j q_{ijk}$ by the theorem and Proposition 2.2.2, and that q_{ijk} is symmetric in i, j and k.

(iii) The argument of the proof can be applied to

$$\sum_x E_{i_1}(u_1,x) \cdots E_{i_s}(u_s,x)$$

and

$$\sum_{x_1,x_2} E_{i_1}(u_1,x_1) \cdots E_{i_s}(u_s,x_1)(E_{j_1}(v_1,x_1) \cdots E_{j_t}(v_t,x_1) - E_{j_1}(v_1,x_2) \cdots E_{j_t}(v_t,x_2))$$

in place of $q(u,v,w)$, and then gives the inequalities

$$\sigma_{i_1 \cdots i_s} := \sum_{l=0}^{d} n_l Q_{li_1} \cdots Q_{li_s} = n^{s-1}\Sigma(E_{i_1} \circ \cdots \circ E_{i_s}) \geqslant 0 \tag{4a}$$

and

$$\sigma_{i_1 \cdots i_s j_1 \cdots j_t} \geqslant \frac{1}{n}\sigma_{i_1 \cdots i_s}\sigma_{j_1 \cdots j_t}. \tag{4b}$$

As was remarked by Cameron and Delsarte (according to BIGGS [75]), (4a) follows from (4) and hence imposes no new restrictions on the parameters. (Indeed, this follows immediately from Lemma 2.3.1 (vii) and induction on s.)

Another restriction (called the *absolute bound*), due to SCOTT [648] for the case of rank 3 graphs, to SEIDEL [656] for the case of strongly regular graphs, and to NEUMAIER

[554] for the general case, is found by considering the rank of $E_i \circ E_j$.

2.3.3. Theorem. *The multiplicities f_i $(0 \leqslant i \leqslant d)$ of a d-class association scheme satisfy:*

$$\sum_{q_{ij}^k \neq 0} f_k \leqslant \begin{cases} f_i f_j & \text{if } i \neq j, \\ \frac{1}{2} f_i(f_i+1) & \text{if } i = j. \end{cases} \tag{7}$$

Proof. The left hand side is rk $(E_i \circ E_j)$. But rk $(E_i \circ E_j) \leqslant$ rk $(E_i \otimes E_j) = f_i f_j$. If $i = j$, the entries of $E_i \circ E_i$ are the squares of the entries of E_i. But if the rows of E_i are linear combinations of f_i rows, then those of $E_i \circ E_i$ are linear combinations of the $f_i + \binom{f_i}{2}$ rows that are the elementwise products of any two of those f_i rows. □

Problem. Is there a geometric condition which is satisfied precisely when equality holds in (7)?

2.4 Imprimitivity

In this section we discuss the concepts of imprimitivity, subscheme and quotient scheme for an association scheme. Most of the material here can also be found in RAO, RAY-CHAUDHURI & SINGHI [604]; see also CAMERON, GOETHALS & SEIDEL [178].

An association scheme $(X, \mathfrak{R})$ is called *imprimitive* if some union of relations is an equivalence relation distinct from Δ and $X \times X$.

Suppose that $\tilde{0}$ is a set of indices such that $\bigcup_{i \in \tilde{0}} R_i$ is an equivalence relation with equivalence classes $X_1, \cdots, X_{\tilde{n}}$. Then for each j $(1 \leqslant j \leqslant \tilde{n})$ we have $|X_j| = \sum_{i \in \tilde{0}} n_i = n'$, say, and $n = \tilde{n} \cdot n'$.

On each X_j we find a *subscheme* $(X_j, \mathfrak{R}')$, where $\mathfrak{R}' = \{R'_i \mid i \in \tilde{0}\}$ and $R'_i = R_i \cap (X_j \times X_j)$. Each of these subschemes has the same parameters

$$n' = |X_j|, \quad n'_i = n_i \ (i \in \tilde{0}), \quad p'^k_{ij} = p^k_{ij} \ (i,j,k \in \tilde{0}).$$

(Of course these subschemes need not be pairwise isomorphic.)

On the other hand, let us define a relation $\sim$ on the set of indices of the relations of $(X, \mathfrak{R})$ as follows: let $a \sim b$ if $p^i_{ab} \neq 0$ for some $i \in \tilde{0}$. It is straightforward that $\sim$ is an equivalence relation: if we suppose $a \sim b \sim c$, then we can find points $w, x, y, z \in X$ with relations as shown in the diagram, where $i, j \in \tilde{0}$.

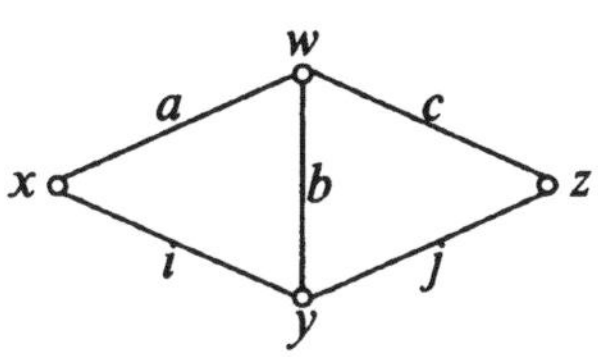

But now $(x,z) \in R_k$ for some $k \in \tilde{0}$ and $p^k_{ac} \neq 0$ so that $a \sim c$. The equivalence relation on X is uniquely determined by $\sim$ as it is determined by the class $\tilde{0}$. As suggested by the notation, $\tilde{0}$ is the equivalence class of 0, and we write $\tilde{a}$ for the equivalence class containing a.

(Conversely, when $\tilde{0}$ is a proper subset of the set of indices such that $\sim$ is an equivalence relation, and $\tilde{0} \neq \{0\}$, then $(X, \mathfrak{R})$ is imprimitive. Thus, imprimitivity of a scheme can be recognized from its parameters.)

Now we can define a *quotient scheme* $(\tilde{X}, \tilde{\mathfrak{R}})$ of $(X, \mathfrak{R})$ with respect to $\tilde{0}$ whose point set is the set $\tilde{X}$ of equivalence classes on X, and whose relations are $\tilde{R}_{\tilde{i}}$, with

$$\tilde{R}_{\tilde{i}} = \{(\tilde{x}, \tilde{y}) \mid \text{ for } x \in \tilde{x} \text{ and } y \in \tilde{y} \text{ we have } (x,y) \in R_i \text{ with } i \in \tilde{i}\}.$$

This quotient scheme is an association scheme with parameters

$$\tilde{n} = |\tilde{X}|, \quad \tilde{n}_{\tilde{i}} = \frac{1}{n'} \sum_{i \in \tilde{i}} n_i = n_i \Big/ \sum_{j \in \tilde{0}} p^i_{ij},$$

$$\tilde{p}^{\tilde{k}}_{\tilde{i}\tilde{j}} = \frac{1}{n'} \sum_{i \in \tilde{i},\, j \in \tilde{j}} p^k_{ij} \text{ for any } k \in \tilde{k}.$$

Remarks. (i) Suppose that in an association scheme $(X, \mathfrak{R})$ we have $n_a = 1$ for some $a \neq 0$. Then the involution $x \mapsto \bar{x}$ defined by $(x, \bar{x}) \in R_a$ is an automorphism of $(X, \mathfrak{R})$ and the scheme is imprimitive since $R_0 \cup R_a$ is an equivalence relation. More generally, let $A = \{a \mid n_a = 1\}$. Then for each $a \in A \setminus \{0\}$ we find an involution σ_a, and if we set $\sigma_0 = 1$, then $\sigma_a \sigma_b = \sigma_b \sigma_a = \sigma_c$, where c is determined by $p^c_{ab} \neq 0$. Thus, A has in a natural way the structure of an elementary abelian 2-group (in particular: $|A|$ is a power of two), and A acts on $(X, \mathfrak{R})$.

(ii) (Rao, Ray-Chaudhuri & Singhi [604]) Suppose that $p^k_{ij} \leq 1$ for all $i, j, k \neq 0$. Then by Lemma 2.1.1 (iv) we have $n_i = f$ for some f and all $i \neq 0$. If $f = 1$ then we are in the case studied above, so suppose $f > 1$. We cannot have $p^j_{ii} = p^j_{mm} = 1$ for distinct nonzero i, m, for

$$f + \sum_{j \neq 0} p^j_{ii} p^j_{mm} = \sum_j p^j_{ii} p^m_{jm} = \sum_j p^m_{ij} p^j_{im} = \sum_j (p^m_{ij})^2 = \sum_j p^m_{ij} = n_i = f$$

by Lemma 2.1.1 (iv), (vi), (v). But now $\sum_i p^j_{ii} = 1$, $\sum_j p^j_{ii} = f - 1$, $d = \sum_{i,j} p^j_{ii} = (f-1)d$, where i, j denote nonzero indices, so that $f = 2$. Thus the scheme is pseudocyclic by Proposition 2.2.7. Now one may verify that X can be given the structure of an abelian group G such that the relations become $R_g = \{(x,y) \mid x - y = \pm g\}$ $(g \in G)$. Of course, conversely, given an abelian group G of odd order, one finds in this way an association scheme with $p^k_{ij} \leq 1$ (and $q^k_{ij} \leq 1$) for all $i, j, k \neq 0$. In fact this is the finest translation association scheme for the group G; all other translation association schemes for G can be derived from this one by merging classes. (Cf. §2.10.)
Rao, Ray-Chaudhuri & Singhi [604] continue from here and set up a Jordan-Hölder theory entirely analogous to the usual one in group theory.

Dual imprimitivity

Dualizing the above, we find that subscheme and quotient scheme are dual concepts.

First consider a quotient scheme. The Bose-Mesner algebra $\tilde{\mathfrak{A}}$ of the quotient scheme consists precisely of those matrices $\tilde{A}$ for which $\tilde{A} \otimes J_{n'} \in \mathfrak{A}$. We have

$\sum_{j\in\bar{j}} A_j = \tilde{A}_{\bar{j}}\otimes J_{n'}$; now $\sum_{j\in\bar{j}} A_j$ has eigenvalue $\sum_{j\in\bar{j}} P_{ij}$ in the eigenspace corresponding to E_i and hence $\tilde{A}_{\bar{j}}$ has eigenvalue $n'^{-1}\sum_{j\in\bar{j}} P_{ij}$ in the eigenspace corresponding to E'_i whenever $E_i = n'^{-1}E'_i\otimes J_{n'}$; for all other i we find $\sum_{j\in\bar{j}} P_{ij} = 0$. (On the other hand, if $\tilde{u}$ is an eigenvector of $\tilde{A}_{\bar{j}}$, then $\tilde{u}\otimes\mathbf{1}_{n'}$ is an eigenvector of A_j, and we find that this eigenvalue of $\tilde{A}_{\bar{j}}$ also equals $\frac{\tilde{n}_{\bar{j}}}{n_j}P_{ij}$.) Thus, the quotient scheme inherits a subcollection of the set of eigenspaces.

Let $\bar{0}$ be the set of indices of eigenspaces E_i of the form $E_i = \frac{1}{n'}E'_i\otimes J_{n'}$. Then the E'_i $(i\in\bar{0})$ are the idempotents of $\tilde{\mathfrak{A}}$ and we see that $q_{ij}^k = 0$ for $i,j\in\bar{0}$, $k\notin\bar{0}$.

Now consider a subscheme. The minimal idempotents are here $\sum_{i\in\bar{i}} E_i$ (restricted to $X_j\times X_j$) for certain subsets $\bar{i}$ of the set of indices of eigenspaces; let $\bar{0}$ be the index set containing 0, then

$$\sum_{i\in\bar{0}} E_i = \frac{1}{n'}I_{\tilde{n}}\otimes J_{n'} \text{ and } q_{ab}^i \neq 0 \text{ for some } i\in\bar{0} \text{ if and only if } \bar{a} = \bar{b}.$$

Thus, to a subscheme corresponds an equivalence relation on the set of eigenspaces. Here $\overline{f_{\bar{i}}} = \frac{1}{\tilde{n}}\sum_{i\in\bar{i}} f_i$ and $\overline{q_{\bar{i}\bar{j}}^{\bar{k}}} = \frac{1}{\tilde{n}}\sum_{i\in\bar{i},\,j\in\bar{j}} q_{ij}^k$ for any $k\in\bar{k}$.

Remarks. (iii) Suppose $(X,\mathfrak{R})$ is dually imprimitive, that is, that there is a subset $\bar{0}$ of the set of indices of eigenspaces distinct from $\{0\}$ and from $\{0,\cdots,d\}$ such that if $q_{ij}^k \neq 0$ and $i,j\in\bar{0}$ then $k\in\bar{0}$. Then $\frac{n}{\tilde{n}}\sum_{i\in\bar{0}} E_i$ is an equivalence relation, where $\tilde{n} = \sum_{i\in\bar{0}} f_i$. (Indeed, let $E = \frac{n}{\tilde{n}}\sum_{i\in\bar{0}} E_i$, then $E\circ E = E$ since for $k\in\bar{0}$ we have $\sum_{i\in\bar{0},j\in\bar{0}} q_{ij}^k = \sum_{i\in\bar{0},j} q_{ij}^k = \sum_{i\in\bar{0}} f_i = \tilde{n}$; in particular all entries in E are 0 or 1. Also $E^2 = \frac{n}{\tilde{n}}E$ and $EJ = \frac{n}{\tilde{n}}J$ and $E = E^\top$ so that rows in E are either orthogonal or identical and E is an equivalence relation.) It follows that $(X,\mathfrak{R})$ is imprimitive, and the concepts of imprimitivity and dual imprimitivity coincide.

(iv) Suppose that in some association scheme $(X,\mathfrak{R})$ we have $f_a = 1$ for some $a\neq 0$. Then $E_a\circ E_a = \frac{1}{n}E_0$ (since rk $E_a\circ E_a = 1$) so that $n^2E_a\circ E_a\circ E_j = E_j$ for all j, and, since rk $E_a\circ E_j \leqslant$ rk E_j, we see that for each j there is a j' such that $E_a\circ E_j = \frac{1}{n}E_{j'}$. (Indeed, $q_{aj}^{j'} = 1$ since $f_j = f_{j'}$ and $q_{aj}^{j'}q_{aj'}^{j} = 1$.) All entries of E_a are $\pm\frac{1}{n}$. Thus, the scheme is imprimitive, and $\{\{j,j'\}\,|\,0\leqslant j\leqslant d\}$ defines an equivalence relation on the set of eigenspaces with $\bar{0} = \{0,a\}$. More generally, let $A = \{a\,|\,f_a = 1\}$. Then for $a,b\in A$ define $c = a+b$ by $q_{ab}^c\neq 0$. Then $(A,+)$ is an elementary abelian 2-group and in particular $|A|$ is a power of 2. The 2-group structure of A was also found by Chang [189].

(v) Suppose that $q_{ij}^k \leqslant 1$ for all $i,j,k\neq 0$. Then just as in Remark (ii) above (but using Lemma 2.3.1 instead of Lemma 2.1.1) we find that $(X,\mathfrak{R})$ is pseudocyclic with $n_i = f_i = f$ for $i\neq 0$, and $f\leqslant 2$. Now it is easy to see that $p_{ij}^k\leqslant 1$ for all nonzero i,j,k, and we have the situation of Remark (ii), i.e., the scheme is derived from an abelian group.

(vi) Clearly, if $\tilde{n} = 2$ ('bipartite scheme') then if $p_{ij}^k\neq 0$ and $i,j\notin\tilde{0}$, then $k\in\tilde{0}$. Dually, if $n' = 2$ ('antipodal 2-cover') then if $q_{ij}^k\neq 0$ and $i,j\notin\bar{0}$, then $k\in\bar{0}$. Indeed, if we write $J_2 = \begin{bmatrix}1&1\\1&1\end{bmatrix}$ and $K_2 = \begin{bmatrix}1&-1\\-1&1\end{bmatrix}$, then the idempotents E_i with

$i \in \bar{0}$ are precisely those of the form $F \otimes J_2$ while the others are of the form $F \otimes K_2$ by Remark (i) above; but $(F \otimes K_2) \circ (F' \otimes K_2) = (F \circ F') \otimes J_2$ is a linear combination of E_i with $i \in \bar{0}$.

(vii) DELSARTE [254] shows that $f_i = 1$ if and only if $|P_{ij}| = n_j$ for all j (i.e., if and only if $|Q_{ji}| = 1$ for all j). And indeed, this is easy to see directly: If $P_{ij} = \epsilon_j n_j$ ($\epsilon_j = \pm 1$) then each eigenvector u in the i-th eigenspace satisfies $u_x = \epsilon_j u_y$ when $(x,y) \in R_j$. Thus, if $|P_{ij}| = n_j$ for all j, then u is determined up to a constant, and $f_i = 1$. Conversely, if $f_i = 1$, then we saw above in Remark (iv) that all entries of E_i are $\pm \frac{1}{n}$, i.e., $Q_{ji} = \pm 1$ for all j. But $P_{ij} = f_i^{-1} n_j Q_{ji}$ and $f_i = Q_{0i}$, so $|P_{ij}| = n_j$. Dually, one has $n_i = 1$ if and only if $|Q_{ij}| = f_j$ for all j (i.e., if and only if $|P_{ji}| = 1$ for all j).

2.5 Subsets in association schemes

The results in this section are mostly due to DELSARTE [251].

Let Y be a nonempty subset of X. Define a vector $a = (a_i)_{0 \le i \le d}$ (the *inner distribution* of Y) by

$$a_i = \frac{1}{|Y|} \chi^T A_i \chi,$$

where $\chi = \chi_Y$ is the characteristic vector of Y. In other words,

$$a_i = \frac{1}{|Y|} \cdot \# \{(y,y') \in Y \times Y \,|\, (y,y') \in R_i\}.$$

Clearly, $a_0 = 1$ and $\sum_{i=0}^d a_i = |Y|$.

Also define an $n \times (d+1)$ matrix B (the *outer distribution* of Y) by

$$B_{xi} := \# \{y \in Y \,|\, (x,y) \in R_i\} = (A_i \chi)_x.$$

2.5.1. Lemma. *The inner distribution a and outer distribution B of a nonempty subset Y of X satisfy*

(i) $a = |Y|^{-1} \chi^T B$,

(ii) $B\mathbf{1} = |Y| \mathbf{1}$,

(iii) $(BQ)_{xj} = n(E_j \chi)_x$,

(iv) $(|Y| aQ)_j = n \chi^T E_j \chi$,

(v) $B^T B = n^{-1} |Y| P^T \Delta_{aQ} P$.

Proof. (i) and (ii) are obvious; (iii) follows from $(BQ)_{xj} = (\sum_i Q_{ij} A_i \chi)_x = n(E_j \chi)_x$; (iv) follows from $(|Y| aQ)_j = (\chi^T BQ)_j = n \chi^T E_j \chi$; but (iii) and (iv) imply $Q^T B^T B Q = n |Y| \Delta_{aQ}$, hence (v). □

The vector aQ is called the *MacWilliams transform* of the vector a.

2.5.2. Proposition (DELSARTE [251], p. 26). *The inner distribution a of any nonempty subset Y of X satisfies $aQ \geqslant 0$. Moreover, if $(aQ)_j = 0$, then the outer distribution B of Y satisfies $(BQ)_{xj} = 0$ for all $x \in X$.*

Proof. We have $(|Y|aQ)_j = n\chi^T E_j \chi \geqslant 0$ since E_j is positive semi-definite. But if $\chi^T E_j \chi = 0$, then already $E_j \chi = 0$, and the second assertion follows from 2.5.1 (iii). □

It turns out that this proposition is a powerful tool for obtaining bounds on the parameters of subconfigurations of a given configuration, and structural information when the bound is met with equality. For example, if $(X, \mathfrak{R})$ is the *Hamming scheme* (X the set of vectors of length m, say, over an alphabet of size q (so that $n = q^m$) and $(u,v) \in R_i$ when the vectors u and v differ in precisely i coordinates, cf. §9.2), then subsets Y of X are called *codes,* and one is usually interested in codes with a large minimum distance δ; for the inner distribution a of such a code this means that $a_1 = \cdots = a_{\delta-1} = 0$ and we find parameter restrictions (as was first remarked in DELSARTE [249]). In many cases it has been possible to prove uniqueness of the extremal codes using this proposition. (Indeed, this is how one shows uniqueness of the ternary Golay code, the Nordstrom-Robinson code and the Nadler/van Lint codes; see SLOANE [678], DELSARTE & GOETHALS [257], SNOVER [690], and GOETHALS [328].) Also, the important asymptotic bounds given in MCELIECE, RODEMICH, RUMSEY & WELCH [526] have been obtained in this way; cf. also BEST & BROUWER [60], BEST et al. [61], MACWILLIAMS & SLOANE [515], Ch. 17. Similarly, one may find bounds on the center density and 'kissing number' of sphere packings (ODLYZKO & SLOANE [575]), parameters of t-designs (DELSARTE, HAEMERS & WEUG [343]) and orthogonal arrays, sizes of cliques and cocliques in graphs and so on. See also CONWAY & SLOANE [215], Chapters 9, 13-14.

Remark. One can generalize and strengthen the above result (DELSARTE [254]).
Let Y and Z be two subsets of X, with inner distributions a and b, and characteristic vectors χ and ψ, respectively. Then

$$(|Y|aQ)_i(|Z|bQ)_i = n^2(\chi^T E_i \chi)(\psi^T E_i \psi) \geqslant n^2(\chi^T E_i \psi)^2,$$

with equality if and only if $E_i\chi$ and $E_i\psi$ are linearly dependent. In particular, for $Z = \{z\}$, we find

$$|Y|(aQ)_i f_i \geqslant ((BQ)_{zi})^2,$$

a strengthening of Proposition 2.5.2.

A very useful application is the following result (DELSARTE [251], Theorem 3.9, p. 32).

2.5.3. Proposition. *Let $\{\{0\}, I_1, I_2\}$ be a partition of $\{0, \cdots, d\}$, and assume that Y and Z are nonempty subsets of X such that the inner distribution b of Y satisfies $b_i = 0$ for $i \in I_1$, and the inner distribution c of Z satisfies $c_i = 0$ for $i \in I_2$. Then $|Y| \cdot |Z| \leqslant |X|$, and equality holds if and only if for all $i \neq 0$ we have $(bQ)_i = 0$ or $(cQ)_i = 0$.*

Proof. Define $\beta_i = f_i^{-1}|Z|^{-1}(cQ)_i$. Then $\beta_0 = 1$, $\beta_i \geqslant 0$ for all i, and $\sum_i \beta_i Q_{ki} = n_k^{-1}|Z|^{-1} n c_k$. Now

$$|Y| = \sum_k b_k = (bQ)_0 \leqslant \sum_{i,k} b_k Q_{ki} \beta_i = \sum_i Q_{0i} \beta_i = \frac{n}{|Z|}. \qquad \square$$

Let us investigate a special case of this situation somewhat closer. Let, for $I \subseteq \{0, \cdots, d\}$, the *I-sphere* around the point $x \in X$ be the set $\{y \in X \mid (x,y) \in R_i \text{ for some } i \in I\}$. A nonempty subset Y of X is called *perfect* (more precisely, *I-perfect*) when the I-spheres around its points form a partition of X.

2.5.4. Corollary ('Lloyd's theorem', cf. LLOYD [508], DELSARTE [251], p. 63). *Let Y be I-perfect, with inner distribution a. Then $\sum_{i \in I} P_{ji} = 0$ for all $j \neq 0$ such that $(aQ)_j \neq 0$.*

Proof. Apply the previous theorem, with Z an I-sphere. If c is the inner distribution of Z, then

$$|Z| f_j^{-1}(cQ)_j = f_j^{-1} \sum_i \sum_{g,h \in I} n_g p_{ih}^g Q_{ij} = \sum_i \sum_{g,h \in I} p_{gh}^i P_{ji} = \sum_{g,h \in I} P_{jg} P_{jh} = (\sum_{i \in I} P_{ji})^2.$$

□

As another application we find an upper bound for the minimum distance and covering radius of a code.

2.5.5. Proposition (MACWILLIAMS & SLOANE [515], p. 172, 175; DELSARTE [251], p. 60, 67). *Let Y be a nonempty subset of X with inner distribution a and outer distribution B. Let $r+1$ be the number of j such that $(aQ)_j \neq 0$. Then $r+1 = \text{rk } B$. Let $I \subseteq \{0, \cdots, d\}$.*

(i) ('MacWilliams inequality') *If the I-spheres around the points $y \in Y$ are pairwise disjoint, then $|I| \leqslant r+1$.*

(ii) *If the I-spheres around the points $y \in Y$ cover X, and I is chosen minimal with this property, then $|I| \leqslant r+1$.*

Proof. We saw in Lemma 2.5.1 (v) that $B^{\mathsf{T}} B = n |Y| P^{\mathsf{T}} \Delta_{aQ} P$, and since $\text{rk } B = \text{rk } B^{\mathsf{T}} B$ and P is invertible, this proves the statement on rk B.

(i) Fix $y \in Y$ and for each $i \in I$ choose an x_i such that $(x_i, y) \in R_i$. Then the rows of B corresponding to the x_i are linearly independent, so $|I| \leqslant \text{rk } B$.

(ii) For each $i \in I$ choose an x_i such that $(x_i, y) \in R_i$ for some $y \in Y$, while not $(x_i, z) \in R_j$ for any $z \in Y$, $j \in I \setminus \{i\}$. Again the corresponding rows of B are linearly independent. □

Remarks. (i) If for some Y equality holds in MacWilliams' inequality, then Y is perfect, but the converse need not hold. However, in the special case where the scheme is metric (see §2.7) and $I = \{0, \cdots, e\}$ we find that if some I-perfect set exists, then there is a polynomial ψ of degree e with $\psi(P_{j1}) = \sum_{i \in I} P_{ji}$ that has at least r zeros (because of Lloyd's theorem), so that $|I| = e+1 = r+1$.

(ii) We shall need a slightly strengthened version of this proposition in Theorem 11.1.1 (ii): Let $I = \{i_0, i_1, \cdots, i_t\}$ and put $I_j = \{i_0, i_1, \cdots, i_j\}$. Suppose that for each j, $1 \leqslant j \leqslant t$, the union of the I_j-spheres around the points of Y strictly contains the union of the I_{j-1}-spheres around them. Then $t \leqslant r$.

2.6 Characterization of the Bose-Mesner algebra

Define (as before) the Hadamard product of two $n \times n$-matrices A and B by $(A \circ B)_{xy} = A_{xy} B_{xy}$. Let us call two (0,1)-matrices A, A' *disjoint* if $A \circ A' = 0$, and two symmetric idempotent matrices E, E' *orthogonal* if $EE' = 0$. Then we can characterize the Bose-Mesner algebra as follows.

2.6.1. Theorem. *Let $\mathcal{C}$ be a vector space of symmetric $n \times n$ matrices.*

(i) *$\mathcal{C}$ has a basis of mutually disjoint* (0,1)*-matrices if and only if $\mathcal{C}$ is closed under Hadamard multiplication.*

(ii) *$\mathcal{C}$ has a basis of mutually orthogonal idempotents if and only if $\mathcal{C}$ is closed under ordinary multiplication.*

(iii) *$\mathcal{C}$ is the Bose-Mesner algebra of an association scheme if and only if $\mathcal{C}$ is closed under both ordinary and Hadamard multiplication, and $I, J \in \mathcal{C}$.*

Proof. In each case, we show the nontrivial direction only.
(i) Suppose that $\mathcal{C}$ is closed under Hadamard multiplication. Call a (0,1)-matrix $D \in \mathcal{C}$ *primitive* if $D \neq 0$ and for every $A \in \mathcal{C}$ there is a number $\alpha = \alpha_D(A)$ such that $A \circ D = \alpha D$. Clearly, distinct primitive (0,1)-matrices are disjoint. We show that the set $\mathcal{D}$ of primitive (0,1)-matrices of $\mathcal{C}$ is a basis of $\mathcal{C}$. Indeed, for every $A \in \mathcal{C}$, the matrix $A - \sum_{D \in \mathcal{D}} \alpha_D(A) D$ belongs to $\mathcal{D}^\perp = \{B \in \mathcal{C} \mid B \circ D = 0 \text{ for all } D \in \mathcal{D}\}$; hence it suffices to show that $\mathcal{D}^\perp = \{0\}$. But if this is not the case, then there is a matrix $B \in \mathcal{D}^\perp \setminus \{0\}$ with a minimal number of nonzero entries, and some entry equal to 1. Since $B \circ B - B$ is in $\mathcal{D}^\perp$ and has fewer nonzero entries than B, we have $B \circ B - B = 0$, so that B is a (0,1)-matrix. Similarly, for any $A \in \mathcal{C}$, the matrix $A \circ B - \alpha B$ is in $\mathcal{D}^\perp$ and for some α has fewer nonzero entries than B, so that $A \circ B - \alpha B = 0$. Hence $B \in \mathcal{D}$, contradiction.

(ii) Suppose that $\mathcal{C}$ is closed under ordinary multiplication. Then $(\mathcal{C}, \cdot)$ is commutative since for $A, B \in \mathcal{C}$ we have $BA = B^\top A^\top = (AB)^\top = AB$. We proceed 'dually' to (i). Call an idempotent $E \in \mathcal{C}$ *primitive* if $E \neq 0$ and for every $A \in \mathcal{C}$ there is a number $\alpha = \alpha_E(A)$ such that $AE = \alpha E$. Clearly, distinct primitive idempotents are orthogonal. To show that the set $\mathcal{E}$ of primitive idempotents of $\mathcal{C}$ is a basis of $\mathcal{C}$, it suffices to show that the set $\mathcal{E}^\perp = \{B \in \mathcal{C} \mid BE = 0 \text{ for all } E \in \mathcal{E}\}$ consists of 0 only. But if this is not the case, then there is a matrix $B \in \mathcal{E}^\perp \setminus \{0\}$ of minimal rank having some eigenvalue equal to 1. As in (i) one now finds that B is an idempotent, and $B \in \mathcal{E}$, contradiction.

(iii) Suppose that $\mathcal{C}$ is closed under both ordinary and Hadamard multiplication, and that $I, J \in \mathcal{C}$. There is a primitive (0,1)-matrix A_0 with $I \circ A_0 = A_0$, and since JA_0 must be symmetric, we have $A_0 = I$. If $A_1, \cdots, A_d$ are the remaining primitive (0,1)-matrices, then $J \circ A_i = A_i$ for all i, whence $J = A_0 + A_1 + \cdots + A_d$. Thus, the A_i satisfy 2.2 (i)-(iii), and since $A_i A_j$ must be expressible as a linear combination $\sum p_{ij}^k A_k$, 2.2 (iv) holds. Hence we have an association scheme whose Bose-Mesner algebra is $\mathcal{C}$. □

Remark. Similarly, coherent configurations can be characterized in terms of *coherent algebras*, i.e., algebras of square matrices containing I, J, closed under (conjugate) transposition, ordinary multiplication and Hadamard multiplication. See HIGMAN [386].

2.7 Metric and cometric schemes

An association scheme together with an ordering of its relations (resp. minimal idempotents) is called *metric* (resp. *cometric*) if $p_{ij}^k \neq 0$ implies $k \leqslant i+j$ and moreover $p_{ij}^{i+j} \neq 0$ for all i, j, k (resp. the same condition with q_{ij}^k instead of p_{ij}^k). Since $n_k p_{ij}^k$ is symmetric in i, j, k we in fact have $|i-j| \leqslant k \leqslant i+j$ when $p_{ij}^k \neq 0$ and similarly in the dual case. Clearly, the ordering is determined by the first relation R_1 (resp. the first minimal idempotent E_1).

If $(X, \mathfrak{R})$ is a metric scheme with ordering $R_0, \cdots, R_d$, then (X, R_1) is a distance-regular graph, and $(x,y) \in R_j$ if and only if x and y have distance j in the graph (X, R_1). Conversely, if Γ is a distance-regular graph of diameter d and $(x,y) \in R_i$ if and only if x and y have distance i in Γ, then $(X, \{R_0, \cdots, R_d\})$ is a metric scheme.

Note that an association scheme may well be metric (cometric) for several distinct orderings of its relations (eigenspaces). For example, the *Johnson scheme* $\binom{S}{m}$ which has as points the m-subsets of a fixed n-set S and relations R_j defined by $(x,y) \in R_j$ if $|x \setminus y| = j$ is metric. In the special case $n = 2m+1$ also (X, R_m) is a distance-regular graph (known as the Odd graph) and in this case $R_0, R_m, R_1, R_{m-1}, \cdots$ is another metric ordering of the relations, cf. §9.1D. The case of several metric structures has been fairly well studied, cf. Bannai & Bannai [27], Gardiner [313] and §3.4 of Bannai & Ito [33]; see also §4.2D. Not much is known about the case of several coexisting cometric structures, but Bannai and Ito [33], §3.7, show that there are at most two distinct cometric structures when the scheme is metric and $d \geqslant 34$ (cf. Theorem 8.3.5). Association schemes that are both metric and cometric are studied in Chapter 8; almost all the important results are due to Leonard [485, 486, 487]. For the non-symmetric case, some results can be found in Damerell [241] and Leonard [488, 489].

Let P, Q be the eigenmatrices of an association scheme. The scheme is called *P-polynomial* (resp. *Q-polynomial*) for a given ordering of its relations (eigenspaces) if there exist real numbers $z_0, \cdots, z_d$ and polynomials p_k (resp. q_k) of degree k with real coefficients such that $P_{ik} = p_k(z_i)$ for $0 \leqslant i, k \leqslant d$ (resp. $Q_{ik} = q_k(z_i)$). Often the polynomials are normalized by a linear transformation such that $p_1(z) = z$ (respectively, $q_1(z) = z$). Then, for a P-polynomial scheme, $P_{ik} = p_k(\theta_i)$ where the $\theta_i = P_{i1}$ are the eigenvalues of L_1 (and hence of A_1), and, for a Q-polynomial scheme, $Q_{ik} = q_k(f_1 u_{1i})$, where $u_1 = (u_{10}, \cdots, u_{1d})^{\mathsf{T}}$ is the right eigenvector of L_1 with $u_{10} = 1$ corresponding to the first idempotent E_1 (cf. Proposition 2.2.2). Sometimes the $f_1 u_{1i} = Q_{i1}$ are called the *dual eigenvalues* of the scheme. According to Lemma 2.2.1 (iv) we have $\sum_i f_i P_{ij} P_{ik} = n n_j \delta_{jk}$ and $\sum_j n_j Q_{ji} Q_{jl} = n f_i \delta_{il}$ so that the p_k (resp. q_k) will be orthogonal polynomials.

2.7.1. Proposition (Delsarte [251], Theorems 5.6 and 5.16). *Let $(X, \mathfrak{R})$ be an association scheme with ordering $R_0, \cdots, R_d$ of its relations and $E_0, \cdots, E_d$ of its eigenspaces. The following assertions are equivalent:*

(i) *$(X, \mathfrak{R})$ is a metric association scheme (that is, (X, R_1) is a distance-regular graph);*

(i′) $p_{1i}^{i+1} \neq 0$, $p_{1i}^k = 0$ *for* $k > i+1$ $(i = 0, \cdots, d-1)$;

(ii) *there are polynomials p_i of degree i such that $A_i = p_i(A_1)$;*

(iii) $(X, \mathfrak{R})$ *is P-polynomial.*

Dually, the following assertions are equivalent.

(i) $(X, \mathfrak{R})$ *is a cometric association scheme;*

(i′) $q_{1i}^{i+1} \neq 0$, $q_{1i}^{k} = 0$ *for* $k > i+1$ $(i = 0, \cdots, d-1)$;

(ii) *there are polynomials* q_i *of degree i such that* $E_i = q_i(E_1)$, *where multiplication is* $\circ$, *the Hadamard multiplication;*

(iii) $(X, \mathfrak{R})$ *is Q-polynomial.*

Proof. Clearly, (i) implies (i′), giving

$$A_1 A_i = \sum_{k=0}^{i+1} p_{1i}^{k} A_k.$$

Since $p_{1i}^{i+1} \neq 0$, this allows us to express A_{i+1} as a linear combination of $A_0, \cdots, A_i$ and $A_1 A_i$, and hence, inductively, as a polynomial of degree $i+1$ in A_1. Hence (i′) implies (ii). If (ii) holds, then

$$p_i(z) p_j(z) = \sum_{k=0}^{d} p_{ij}^{k} p_k(z),$$

so that $p_{ij}^{k} = 0$ for $k > i+j$, and $p_{ij}^{i+j} \neq 0$. Hence (ii) implies (i).
Since we can express any polynomial in A_1 in terms of the eigenvalues $\theta_i = P_{1i}$ of A_1 as

$$p(A_1) = \sum_i p(\theta_i) E_i, \tag{8}$$

(ii) implies $P_{ik} = p_k(\theta_i)$, hence (iii). Conversely, if (iii) holds, then z_i is linear in θ_i so that we may assume w.l.o.g. that $z_i = \theta_i$. Then (8) shows that (ii) holds.
In the dual case the arguments run entirely similar. □

A 2-class association scheme is metric for both orderings 012 and 021 of its relations, unless it is imprimitive, say with (X,R_1) complete multipartite and (X,R_2) a union of cliques, in which case we only have the P-polynomial ordering 012. Similarly, a 2-class association scheme is cometric for both orderings 012 and 021 of its eigenspaces, unless it is imprimitive, in which case we only have the Q-polynomial ordering 012, if E_1 is the eigenspace with $A_1 E_1 = 0$.

A. The Frame quotient in a metric scheme

Using the 3-term recurrence relation found above we can give another expression for the Frame quotient $q = (n^{-1}\det P)^2$ (see §2.2A) in the case of a metric scheme:

2.7.2. Proposition. *For a P-polynomial d-class association scheme, let* $\theta_0 > \cdots > \theta_d$ *be the distinct eigenvalues of* A_1, *and let* $c_i = p_{1,i-1}^{i}$. *Then*

$$n \prod_{i=1}^{d} c_i = \prod_{i=1}^{d} (\theta_0 - \theta_i)$$

and the Frame quotient can be written as $q = r^2$, *where*

$$r = \frac{\prod_{0<i<j} (\theta_i - \theta_j)}{\prod_{i=1}^{d-1} c_i^{d-i}}.$$

In particular, if all eigenvalues of A_1 *are rational integers, then r is an integer.*

Proof. From the proof of 2.2.4, we may take $r = n^{-1}\det P$. By assumption there are polynomials

$p_k(z) = \sum_{j=0}^{k} \beta_{jk} z^j$ such that $P_{ik} = p_k(\theta_i)$. This implies

$$P = \begin{bmatrix} 1 & \theta_0 & \theta_0^2 & \cdots & \theta_0^d \\ 1 & \theta_1 & \theta_1^2 & \cdots & \theta_1^d \\ \vdots & & & & \vdots \\ 1 & \theta_d & \theta_d^2 & \cdots & \theta_d^d \end{bmatrix} \cdot \begin{bmatrix} \beta_{00} & \beta_{01} & \cdots & \beta_{0d} \\ & \beta_{11} & \cdots & \beta_{1d} \\ & & \ddots & \vdots \\ 0 & & & \beta_{dd} \end{bmatrix},$$

whence

$$\det P = (-1)^{d(d+1)/2} \prod_{i<j} (\theta_i - \theta_j) \cdot \beta_{00}\beta_{11} \cdots \beta_{dd}.$$

Now the relations $z p_i(z) = p_{1i}^{i-1} p_{i-1}(z) + p_{1i}^{i} p_i(z) + p_{1i}^{i+1} p_{i+1}(z)$ (cf. (8) above) show that $\beta_{ii} = c_{i+1}\beta_{i+1,i+1}$, and since $\beta_{00} = 1$, this gives

$$\beta_{00}\beta_{11} \cdots \beta_{d-1,d-1} = \left[\prod_{i=1}^{d-1} c_i^{d-i} \right]^{-1}.$$

Moreover, since $\sum_k p_k(\theta_i) = n\delta_{i0}$, we have

$$\sum_k p_k(z) = \beta_{dd} \prod_{i=1}^{d} (z - \theta_i).$$

Hence

$$n = \beta_{dd} \prod_{i=1}^{d} (\theta_0 - \theta_i).$$

□

2.8 Subsets of cometric schemes; the Assmus-Mattson theorem

Let $T \subseteq \{0, \cdots, d\}$. A subset Y of a d-class association scheme $(X, \mathfrak{R})$ is called a *T-design* if $(aQ)_i = 0$ for all $i \in T$, where a is the inner distribution of Y. This definition is motivated by the fact that for $T = \{1, \cdots, t\}$, a T-design in the Johnson scheme $J(v,k)$ is a classical t-(v,k,λ) design. For the identification of T-designs in the classical association schemes, see Delsarte [252], Munemasa [548] and Stanton [700].

Roughly speaking, the Assmus-Mattson Theorem (Assmus & Mattson [17], Theorem 4.2) says that if a code C in the Hamming scheme $H(v,q)$ is sufficiently nice, then the (supports of the) code words of a given weight k form a t-design (in $J(v,k)$). In a general association scheme, the spheres around a point are not necessarily association schemes themselves, and we need the concept of a relative t-design with respect to a point.

The subset Y of $(X, \mathfrak{R})$ is called a *relative T-design with respect to x* (for $x \in X$) when $E_i\chi$ and $E_i\psi$ are linearly dependent for all $i \in T$, where $\chi = \chi_Y$ and $\psi = \chi_{\{x\}}$; by the remark before Proposition 2.5.3 this is true if and only if $|Y|(aQ)_i f_i = ((BQ)_{xi})^2$ for all $i \in T$.

When $(X, \mathfrak{R})$ is Q-polynomial, we have a natural ordering on the eigenspaces; in this case we call Y a (relative) *t-design* when it is a (relative) T-design for $T = \{1, \cdots, t\}$. (Note that $(aQ)_0 = |Y|$ and that $E_0\chi$ is a multiple of $\mathbf{1}$ for any χ, so that it makes no difference whether $0 \in T$ in case of a relative T-design, while it is impossible to have $0 \in T$ for a (nonempty) T-design.)

2.8.1. Theorem (DELSARTE [254]). *Let Y be a t-design in the Q-polynomial association scheme $(X, \mathfrak{R})$, and assume that $Y_i := \{y \in Y \mid (x,y) \in R_i\}$ is nonempty for s nonzero values of i. Then each Y_i is a relative $(t+1-s)$-design with respect to x.*

Proof. For an arbitrary subset Y of an arbitrary association scheme $(X, \mathfrak{R})$ we have $\chi_Y^\top E_j \chi_Y = n^{-1} |Y| (aQ)_j$. From this it follows by straightforward computation (using $\|M\|^2 = \operatorname{tr} M^\top M$ and $(E_k)_{xx} = n^{-1} f_k$ and $q_{kl}^0 = \delta_{k,l} f_k$) that

$$\|nE_k \Delta_Y E_l - |Y| \delta_{k,l} E_k\|^2 = |Y| \sum_{j>0} q_{kl}^j (aQ)_j, \tag{9}$$

where Δ_Y is the diagonal matrix with (x,x)-entry 1 if $x \in Y$ and 0 otherwise. It follows that if $(X, \mathfrak{R})$ is a Q-polynomial association scheme, then Y is a t-design if and only if

$$nE_k \Delta_Y E_l = |Y| \delta_{k,l} E_k \quad \text{when } k+l \leqslant t. \tag{10}$$

Let $M = \{i > 0 \mid Y_i \neq \varnothing\}$, then $|M| = s$. By Q-polynomiality of $(X, \mathfrak{R})$ we have $Q_{jk} = q_k(z_j)$ for certain numbers z_j and polynomials $q_k(z)$ of degree k. The polynomial

$$f_i(z) = \prod_{j \in M \setminus \{i\}} \frac{z - z_j}{z_i - z_j}$$

has degree $s-1$ in z (for $i \in M$), and $f_i(z_j) = \delta_{ij}$ for $i, j \in M$. Write

$$f_i(z) = \sum_{k=0}^{s-1} c_{ik} q_k(z)$$

with certain coefficients c_{ik}. Then we have (for $i \in M$):

$$A_i + \sum_{j \notin M} f_i(z_j) A_j = \sum_{j=0}^{d} f_i(z_j) A_j = \sum_{k=0}^{s-1} c_{ik} \sum_{j=0}^{d} Q_{jk} A_j = n \sum_{k=0}^{s-1} c_{ik} E_k$$

hence (with $\chi = \chi_{Y_i}$ and $\psi = \chi_{\{x\}}$):

$$\chi = \Delta_Y A_i \psi = n \sum_{k=0}^{s-1} c_{ik} \Delta_Y E_k \psi - \Delta_Y f_i(z_0) \psi$$

and (for $l \leqslant t+1-s$ and some constant c):

$$E_l \chi = n \sum_{k=0}^{s-1} c_{ik} E_l \Delta_Y E_k \psi - E_l c \psi = \Big(|Y| \sum_{k=0}^{s-1} \delta_{k,l} c_{ik} - c\Big) E_l \psi. \qquad \square$$

2.8.2. Corollary. *Under the assumptions of the above theorem, if not more than $t+1$ nonidentity relations occur between the elements of Y, then Y is regular, i.e., for each i the value $|Y_i|$ does not depend on the choice of $x \in Y$.*

Proof. For $s \leqslant t+1$ we find $|Y_i| \mathbf{1} = nE_0 \chi = n(c_{i0}|Y| - c)\mathbf{1}$, and c_{i0} and c do not depend on x but only on M and on whether $x \in Y$ or not. $\square$

This theorem (and corollary) can be improved a little bit in several ways. Let $R_g(x) = \{z \in X \mid (x,z) \in R_g\}$ be the g-sphere around x. Suppose that Y contains $R_g(x)$ for $g \in G$. Then we can take $M = \{i \notin G \mid Y_i \neq \varnothing\}$; the additional terms $E_l \Delta_Y A_g \psi$ in the expression for $E_l \chi$ above do not harm since $\Delta_Y A_g \psi = A_g \psi$ so that $E_l \Delta_Y A_g \psi = P_{lg} E_l \psi$. Thus, in the above theorem we may take for s the number of indices i such that $\varnothing \neq Y_i \neq R_i(x)$. Noting that $R_i(x)$ is a relative l-design with respect to x for any l, we can also take for s the number of i such that Y_i is neither empty nor a relative l-design with respect to x. Then the conclusion is that each Y_j is a relative $\min(l, t+1-s)$-design with respect to x.

Under stronger assumptions on t, we find that Y is not only regular, but even carries a subscheme. Let R_Y denote the restriction of a relation R to $Y \times Y$, and put $\mathfrak{R}_Y = \{R_Y \mid R \in \mathfrak{R} \text{ and } R_Y \neq \emptyset\}$. If $(Y, \mathfrak{R}_Y)$ is again an association scheme, we call it a *subscheme* of $(X, \mathfrak{R})$. Let $\mathcal{Q}$ be the Bose-Mesner algebra of $(X, \mathfrak{R})$, and put $\mathcal{Q}_Y = \{A_Y \mid A \in \mathcal{Q}\}$, where A_Y is the submatrix of A consisting of the rows and columns indexed by Y. Now Y carries a subscheme if and only if $\mathcal{Q}_Y$ is closed under multiplication.

2.8.3. Proposition. *Let Y be a t-design in the Q-polynomial association scheme $(X, \mathfrak{R})$, and assume that precisely s nonidentity relations occur between the elements of Y, where $t \geqslant 2s-2$. Then $(Y, \mathfrak{R}_Y)$ is a Q-polynomial subscheme of $(X, \mathfrak{R})$, with idempotents $(cE_i)_Y$ $(i = 0, \cdots, s-1)$ and $(I - \sum_{i=0}^{s-1} cE_i)_Y$, where $c = |X| / |Y|$.*

Proof. By (10) we find that the $(cE_i)_Y$ $(i = 0, \cdots, s-1)$ and $(I - \sum_i cE_i)_Y$ are mutually orthogonal idempotents. But $\mathcal{Q}_Y$ is spanned by $s+1$ matrices $(A_j)_Y$, so $\dim \mathcal{Q}_Y = s+1$, and since $\mathcal{Q}_Y$ has a basis of mutually orthogonal idempotents, it is closed under multiplication. Now Q-polynomiality of $(Y, \mathfrak{R}_Y)$ follows from that of $(X, \mathfrak{R})$. □

2.9 Distribution diagrams and the group case

Given $i \in \{0, 1, \cdots, d\}$, the intersection numbers p_{ij}^k, $0 \leqslant j, k \leqslant d$ of an association scheme (or, more generally, of a coherent configuration) with d classes can be conveniently represented in a directed graph whose vertex set is $\{0, \cdots, d\}$ in which a directed edge with label p_{ij}^k is drawn whenever the latter number is nonzero (note that $n_k p_{ij}^k = n_j p_{ik}^j$ so that p_{ij}^k is nonzero if and only if p_{ik}^j is nonzero); furthermore, the number p_{jj}^0 is often written inside the node representing the vertex j. Such a labelled graph is called the *distribution diagram* of the association scheme with respect to the class with number i. Instead of using arrows to depict directed edges, we usually write the label p_{ij}^k at the starting point of an edge representing the (virtual) arrow; loops are usually not drawn at all, even if the label is present. For example, the Petersen graph becomes a (symmetric) association scheme, when we define $(x,y) \in R_i$ if and only if $d(x,y) = i$ $(i = 0,1,2)$, where d is the distance function. We find (for R_1) the diagram

As another example,

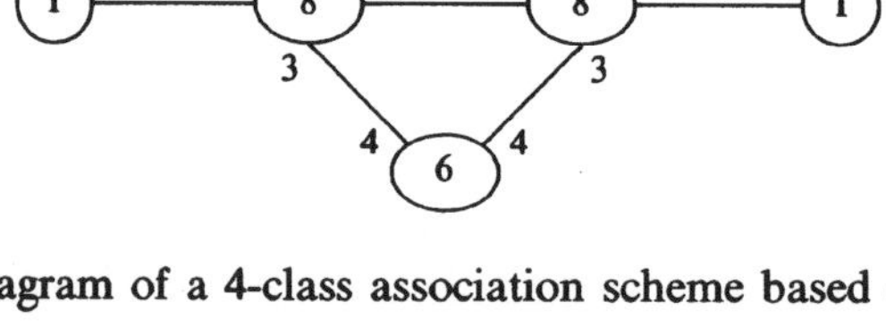

is the distribution diagram of a 4-class association scheme based on a finite set of size

1+8+6+8+1=24 with respect to a class, say R_1, of size $8\times 24 = 192$ in which $p_{11}^1 = 3$. (In fact this is the scheme on the vertices of the 24-cell in $\mathbb{R}^4$, where R_1 is the relation of being joined by an edge of this polytope. The graph (X,R_1) is called $F_{4,1}(1)$ and will be encountered in §10.3.7.)

2.9.1. Proposition.

(i) *A distance-regular graph* Γ *of diameter* d *gives rise to a* d*-class association scheme* $(\Gamma,\{R_0,\cdots,R_d\})$*, where* $R_i = \{(\gamma,\delta)\in\Gamma\times\Gamma \mid d(\gamma,\delta) = i\}$.

(ii) *An association scheme has a linear distribution diagram if and only if it is isomorphic to the association scheme of a distance-regular graph.*

Proof. Straightforward. □

Now, let us consider the group case. (For details, cf. WIELANDT [782].) Let G be a permutation group on X (with action written on the left). Then G has a natural action on $X\times X$, the orbits of which are called *orbitals*. To each $g\in G$ there corresponds a (permutation) matrix g^* with

$$g^*_{xy} = \begin{cases} 1 & \text{if } y = gx, \\ 0 & \text{otherwise.} \end{cases}$$

The *centralizer algebra* of this permutation representation is the algebra $\mathcal{C}$ of all complex matrices A commuting with g^* for all $g\in G$. We have $A\in\mathcal{C}$ if and only if $A_{xy} = A_{gx,gy}$ for all $x,y\in X$, $g\in G$, in other words, if and only if A, viewed as a function on $X\times X$, is constant on the orbitals. Thus, dim $\mathcal{C}$ is the number of orbitals, and the characteristic functions of the orbitals form a basis of $\mathcal{C}$. In fact $\mathcal{C}$ is closed under componentwise multiplication $\circ$, and this basis is the basis of minimal idempotents for the multiplication $\circ$.

Let π be the permutation character of G on X. The identity matrix I is a minimal idempotent if and only if G is transitive on X, that is, if and only if the number $e_1 = (\pi,1)$ of orbits equals one.

Let $\pi = \sum_i e_i\chi_i$ be the decomposition of π into irreducible characters. Then $\mathcal{C}$ is the direct sum of full matrix algebras of $e_i\times e_i$ matrices, so that dim $\mathcal{C} = \sum_i e_i^2$ and $\mathcal{C}$ is commutative if and only if $e_i = 1$ for all i, i.e., if and only if π is *multiplicity-free*.

If $A\in\mathcal{C}$ then also $A^\top\in\mathcal{C}$. We have $A = A^\top$ for all $A\in\mathcal{C}$ if and only if all orbitals are *self-paired*, that is, if and only if for all distinct x, $y\in X$ there is a $g\in G$ that interchanges x and y. This condition is known as *generous transitivity* (cf. NEUMANN [566]). Clearly, if all matrices in $\mathcal{C}$ are symmetric, then $AB = (AB)^\top = B^\top A^\top = BA$ for A, $B\in\mathcal{C}$, so that $\mathcal{C}$ is commutative. Thus, we have the implications

generously transitive ⇒ multiplicity-free ⇒ transitive.

In the terminology of this chapter: The orbitals turn X into a coherent configuration. When G is transitive, we get a homogeneous coherent configuration. If G is multiplicity-free then (v) of §2.1 holds. If G is generously transitive then also (iii) of §2.1 holds, and we find a (symmetric) association scheme.

Let us call an irreducible character of G *real* when it is the character of a real representation, and *symplectic* when it is real-valued but not real. (The remaining possibility is that of a *complex* character.)

2.9.2. Proposition (CAMERON [162]). *Let G act multiplicity-freely with permutation character π. Then the number of self-paired orbitals equals the number of real irreducible constituents of π. It follows that the action is generously transitive if and only if it is multiplicity-free and all irreducible constituents of π are real.*

Proof. Let G be an arbitrary permutation group with permutation character π. We mimic the proof of 'Burnside's Lemma' (which says that $(\pi,1)$ is the number of orbits of G, and is not Burnside's, see NEUMANN [568]). Let us write $g(x,y) := (gx,gy)$ and $G(x,y) := \{(gx,gy) \mid g \in G\}$. The number N of self-paired orbitals of G equals

$$N = \sum_{\substack{x,y \\ (y,x) \in G(x,y)}} \frac{1}{|G(x,y)|} = \frac{1}{|G|} \sum_{\substack{x,y \\ (y,x) \in G(x,y)}} |G_{xy}|$$

$$= \frac{1}{|G|} \sum_{\substack{x,y,g \\ (y,x) = g(x,y)}} 1 = \frac{1}{|G|} \sum_g \pi(g^2).$$

But if we put

$$\nu_\chi = \frac{1}{|G|} \sum_g \chi(g^2),$$

then it is known (see, e.g., ISAACS [422], p. 50) that if χ is irreducible, then ν_χ equals $+1$, -1, or 0 according to whether χ is real, symplectic, or complex.
Thus, N equals the number of real irreducible constituents of π minus the number of symplectic irreducible constituents of π, counted with multiplicity. Since the representation is real, symplectic irreducible constituents occur with even multiplicity, and hence are absent in case π is multiplicity-free. □

2.9.3. Proposition. *Suppose G acts transitively on X. Consider the homogeneous coherent configuration $(X, \mathcal{R})$ whose relations are the G-orbitals on X. Then $(X, \mathcal{R})$ is imprimitive if and only if G acts imprimitively on X.*

Proof. The 'only if' part is immediate, as the equivalence classes of an equivalence relation $\bigcup_{i \in \tilde{0}} R_i$ (for some set $\tilde{0}$ of relation indices) form a system of imprimitivity for G on X. Conversely, if $X_1, \cdots, X_t$ are proper subsets of X of size ≥ 2 partitioning X and permuted by G (i.e., a system of imprimitivity for G), then we let $\tilde{0}$ be the set of indices i with $R_i \subseteq \bigcup_j X_j \times X_j$. Observe that $0 \in \tilde{0} \neq \{0\}$ since $|X_1| \geq 2$ implies the existence of an index i with $(x,y) \in R_i \cap (X_1 \times X_1)$, leading to $R_i = \{(gx,gy) \mid g \in G\} \subseteq \bigcup_j X_j \times X_j$. Then $\bigcup_{i \in \tilde{0}} R_i = \bigcup_j X_j \times X_j$ is a nontrivial equivalence relation, and so $(X, \mathcal{R})$ is imprimitive. □

Remark. A quotient of G acts primitively on a proper quotient of $(X, \mathcal{R})$.

The distribution diagram with respect to a given orbital will also be called the *double coset diagram* with respect to this orbital. The reason for this name is the fact that if $H = G_x$ is the stabilizer of some point $x \in X$, then X can be identified (as a set on which G acts) with G/H, and the map $(g_1H, g_2H) \mapsto Hg_2^{-1}g_1H$ defined on $X \times X = (G/H) \times (G/H)$ induces a bijective correspondence between the orbitals and the double cosets of G with respect to H. If, for some $g \in G$, we have $HgH = Hg^{-1}H$, then the double coset diagram with respect to the corresponding orbital coincides with the double coset graph $DC(G,H,g)$ (cf. §A.4).

2.10 Translation association schemes

In this section we consider association schemes for which the underlying set X has naturally the structure of an abelian group. We write the group operation as addition, and regard a finite abelian group as a $\mathbf{Z}$-module, with the outer multiplication of $s \in \mathbf{Z}$ and $x \in X$ defined as $sx := x + \cdots + x$ (s terms) if $s > 0$, $sx := 0$ if $s = 0$, and $sx := -(-s)x$ if $s < 0$. We note that $sx = 0$ for all x if the group order $|X|$ divides s, and if s is coprime to $|X|$ then the mapping $x \mapsto sx$ is a bijection of X.

We call an association scheme $(X, \mathcal{R})$, where the underlying set X has the structure of an abelian group, a *translation association scheme* if, for all classes $R \in \mathcal{R}$,

$$(x,y) \in R \Rightarrow (x+z, y+z) \in R, \tag{11}$$

i.e., if the class containing $(x,y) \in X \times X$ depends only on $x - y$.

We recall some elementary facts about the (complex) centralizer algebra $\mathcal{C}(X)$ which are basic to our investigations. Let $n = |X|$. A basis of $\mathcal{C}(X)$ is given by the $n \times n$ permutation matrices P_x ($x \in X$) indexed by X, whose nonzero entries are ones in the positions (a,b) with $b - a = x$. These matrices commute, and satisfy

$$P_x P_y = P_{x+y}, \quad (P_x)^s = P_{sx}, \quad P_x \circ P_y = \delta_{xy} P_x, \quad P_x^{\mathrm{T}} = P_{-x} \tag{12}$$

for all $x, y \in X$, $s \in \mathbf{Z}$. This implies that $\mathcal{C}(X)$ is a commutative algebra over $\mathbf{C}$ closed under transposition and under Hadamard multiplication. The *restricted centralizer algebra* $\mathcal{C}_r(X)$, consisting of all real, symmetric matrices in $\mathcal{C}(X)$, is a real subalgebra of $\mathcal{C}(X)$, also closed under Hadamard multiplication, and spanned by the matrices $P_x + P_{-x}$ ($x \in X$).

The Bose-Mesner algebra of a translation association scheme on X is a subalgebra of $\mathcal{C}_r(X)$, since the (0,1)-adjacency matrices of a translation association scheme on X can be expressed in terms of the P_x as

$$A_i = \sum_{x \in N_i} P_x, \tag{13}$$

where

$$N_i = \{x \in X \,|\, (0,x) \in R_i\}. \tag{14}$$

Note that $N_0 = \{0\}$.

The Bose-Mesner algebra of a translation association scheme becomes a commutative *Schur ring* when the P_x are identified with the basis elements of the group ring of X. See Wielandt [782], Bercov [52] and Tamaschke [709] for an extended theory of Schur rings (which are essentially the same things as coherent configurations with a regular automorphism group). In the following, we discuss those aspects of the theory of Schur rings that have close connections to the main topic of this book.

A. Multiplier theorems and cyclotomic schemes

2.10.1. Theorem (SCHUR [644], cf. WIELANDT [782], Theorem 23.9). *Let $(X,\mathfrak{R})$ be a (not necessarily symmetric) translation association scheme, and let s be an integer coprime to $|X|$. Then*

$$sR := \{(sx,sy) \mid (x,y) \in R\} \in \mathfrak{R} \text{ for all } R \in \mathfrak{R}.$$

Proof. Suppose first that s is a prime. Then, for $i = 0, \cdots, d$,

$$(A_i)^s \equiv \sum_{x \in N_i} P_{sx} \pmod{s} \tag{15}$$

and since $(A_i)^s$ belongs to the Bose-Mesner algebra of this scheme, we also have

$$(A_i)^s = \sum_{k=0}^{d} d_{ik} A_k \tag{16}$$

for suitable integers d_{ik}. If s and $|X|$ are coprime, then the right hand side of (15) is a (0,1)-matrix, and a comparison of (15) and (16) shows that

$$\sum_{x \in N_i} P_{sx} = \sum_{k \in I_i} A_k, \tag{17}$$

where $I_i = \{k \mid d_{ik} \equiv 1 \pmod{s}\}$. Clearly, I_i must be nonempty. Summing (17) over $i = 0, \cdots, d$, we find

$$J = \sum_{x \in X} P_{sx} = \sum_{i=0}^{d} \sum_{k \in I_i} A_k,$$

whence the I_i form a partition of $\{0, \cdots, d\}$. Therefore, each I_i contains a unique number i', and (17) implies that $sR_i = R_{i'}$ for $i = 0, \cdots, d$. This proves the theorem for all primes s not dividing $|X|$, and hence for all positive s coprime to $|X|$. Since $sR = s'R$ when $s \equiv s' \bmod |X|$, the general case follows. □

Let q be a prime power, d a divisor of $q-1$, and a be a primitive root in $\mathbb{F}_q$. Let $K \subset \mathbb{F}_q^*$ be the subgroup of the d-th powers. Then we can define the *cyclotomic scheme* on $X = \mathbb{F}_q$ by $R_i = \{(x,y) \mid y - x \in a^i K\}$ $(1 \leq i \leq d)$ (DELSARTE [251]). This association scheme is symmetric when $(q-1)/d$ is even, or when q is a power of two. For $d = 2$, the cyclotomic schemes are just the association schemes of the Paley graphs (cf. §1.3). The restricted centralizer algebra of $(\mathbb{F}_q, +)$ is just the Bose-Mesner algebra of the cyclotomic scheme with $K = \{-1,1\}$.

2.10.2. Corollary. *A translation association scheme with a prime number of points is a cyclotomic scheme.* □

2.10.3. Corollary (BURNSIDE [151], p. 341). *A transitive permutation group of prime degree p is either a subgroup of $AGL(1,p)$ (and hence solvable), or 2-transitive.*

Proof. By the above, the corresponding association scheme must be cyclotomic, and if the group is not 2-transitive, then the number of classes is > 1, in which case the automorphism group is a subgroup of $AGL(1,p)$ (cf. Proposition 12.7.5). □

2.10.4. Proposition. *Let $(X, \mathfrak{R})$ be a primitive translation association scheme with $d \geqslant 2$ classes, and let N_i be as in (14). Then:*

(i) *Each of the sets N_i ($i \neq 0$) generates X.*

(ii) *If N_i is a union of cosets of a subgroup Y of X, then $i = 0$ or $Y = \{0\}$.*

(iii) *If s is a prime dividing $|X|$ then, for every $i = 0, \cdots, d$, and every nonzero $z \in X$, the number of solutions of $sx = z$ with $x \in N_i$ is divisible by s.*

Proof. (i) The connected components of the graph of the relation R_i are the cosets $x + Y_i$ of the subgroups Y_i of X generated by N_i. Thus, if $Y_i \neq \{0\}, X$, then the partition into cosets of Y_i yields an imprimitivity relation, contradiction. Since $Y_i \neq \{0\}$ for $i \neq 0$, we must have $Y_i = X$.

(ii) This follows similarly, since the partitions into cosets of $Z = \{x \in X \mid x + N_i = N_i\}$ (which contains Y) yields an imprimitivity relation when $i \neq 0$ and $Y \neq \{0\}$.

(iii) With the notation of the proof of Theorem 2.10.1, we interpret the equations (15) and (16) in the case where s is a prime dividing $|X|$. If, for some $i \in \{0, \cdots, d\}$ and some $z \in X$, the number of solutions of $sx = z$ with $x \in N_i$ is not divisible by s, then $d_{ik} \not\equiv 0 \pmod{s}$ for the index k with $(z, 0) \in R_k$. Hence A_k must be the sum of certain P_{sx} ($x \in X$), i.e., N_k is contained in the proper subgroup $Y = \{sx \mid x \in X\}$ of X. Now (i) implies $k = 0$, and hence $z = 0$. □

2.10.5. Theorem (cf. WIELANDT [781]). *Let $(X, \mathfrak{R})$ be a primitive translation association scheme with $d \geqslant 2$ classes. If X has some cyclic Sylow subgroup (and in particular if X is cyclic), then X has prime order and $(X, \mathfrak{R})$ is cyclotomic.*

Proof. By assumption, $X = Y \oplus Z$ with a cyclic group Y of prime power order p^e, and a group Z of order coprime to p. In particular,

$$U = \{x \in X \mid px = 0\} = \{y \in Y \mid py = 0\}$$

is the unique subgroup of X of order p. Hence for each $z \neq 0$ the equation $px = z$ has either 0 or p solutions $x \in X$, and, by Proposition 2.10.4 (iii), all these solutions must lie in the same set N_i. Therefore, each set $N_i \setminus U$ is a union of cosets of U. We now suppose that $X \neq U$ (else Corollary 2.10.2 applies), and derive a contradiction. Choose $i \neq 0$ such that $\alpha = |N_i \cap U|$ is minimal. Then $\alpha \leqslant (p-1)/d$, and in particular $\alpha + 1, 2\alpha < p$. We write

$$K = \sum_{x \in U} P_x, \quad B = \sum_{x \in N_i \cap U} P_x, \quad C = \sum_{x \in M} P_x,$$

where M is a set of representatives of the cosets of U in $N_i \setminus U$. Then

$$K^2 = pK, \ KB = \alpha K, \ A_i = B + KC,$$

and since the P_x commute, this implies

$$A_i^2 - 2\alpha A_i = B^2 - 2\alpha B + pKC^2. \tag{18}$$

Now

$$A_i^2 - 2\alpha A_i = \sum \alpha_k A_k$$

with integers $\alpha_k = p_{ii}^k - 2\alpha\delta_{ik}$. By Proposition 2.10.4 (i), each set N_k ($k \neq 0$) contains an element in $X \setminus U$. Since the $(a - x, a)$-entries of $B^2 - 2\alpha B$ are zero when $x \notin U$, (18) implies that all α_k ($k \neq 0$) are divisible by p. Therefore (18) shows that all nondiagonal entries of $B^2 - 2\alpha B$ are divisible by p, and since these entries lie between $-2\alpha > -p$ and $\alpha < p$, we see that $B^2 - 2\alpha B$ is a multiple of I. Now one finds that $(N_i \cap U) \cup \{0\}$ is a subgroup of U of order $\alpha + 1 < p$, forcing $N_i \cap U = \varnothing$. This implies that N_i is a union of cosets of U, contradicting 2.10.4(ii). □

2.10.6. Corollary (KELLY [464], BRIDGES & MENA [116]). *A strongly regular graph with a regular cyclic automorphism group is a Paley graph with a prime number of vertices.*

Proof. The corresponding 2-class association scheme is a cyclic translation association scheme. □

For further results in this direction, see WIELANDT [782], §25, BERCOV [52] and WIELANDT [783] (the latter for the case $|X| = p^2$). Their results suggest the following

Conjecture. *Let $(X, \mathfrak{R})$ be a primitive translation association scheme. Then one of the following holds:* (i) *X is elementary abelian,* (ii) *$(X, \mathfrak{R})$ is a Hamming scheme,* (iii) *$(X, \mathfrak{R})$ is of Latin square type, or of negative Latin square type.*
Here an association scheme is called of (negative) Latin square type when every (0,1)-matrix with zero diagonal in its Bose-Mesner algebra is the adjacency matrix of a strongly regular graph with $\mu = s(s+1)$ (respectively $\mu = r(r+1)$).

Remarks. (i) The Hamming schemes are translation association schemes with $X = X_1 \oplus X_2 \oplus \cdots \oplus X_m$ for arbitrary abelian groups X_i of the same order. Examples of translation association schemes with elementary abelian X are the cyclotomic schemes; many others will be constructed in Chapters 9 and 11.

(ii) If one were to allow nonabelian groups in the definition of a translation association scheme, then the conjecture would become false, since, e.g., the association scheme of the generalized quadrangle $GQ(2,4)$ has a regular nonabelian automorphism group (of prime power order) $3 \cdot 3^2$.

B. Duality

Translation association schemes arise in dual pairs, thus accounting for the formal duality observed in §2.3. The duality results as a consequence of the duality theorems for finite abelian groups. We first review the facts about these (see, e.g., HUPPERT [416], §V.6 or HASSE [365], §1.5.1).

Let X be a finite abelian group (written additively). The *character group* X^* of X is the group of homomorphisms from X into the multiplicative group of nonzero complex numbers, i.e., the set of mappings (*characters*) $\chi : X \to \mathbf{C} \setminus \{0\}$ with $\chi(x+y) = \chi(x)\chi(y)$ for all $x, y \in X$ with product defined by $(\chi_1\chi_2)(x) := \chi_1(x)\chi_2(x)$ $(x \in X)$. Clearly $\chi(0) = 1$. We shall denote the identity of X^* by 1; it is the character defined by $1(x) = 1$ for all $x \in X$.

2.10.7. Proposition. *The character group X^* of a finite abelian group X is isomorphic to X. Moreover, there exists an isomorphism $f : X \to X^*$ such that $f(x)(y) = f(y)(x)$ for all $x, y \in X$.*

Proof. Since X is a direct product of cyclic groups, we can find a basis $\{x_1, \cdots, x_r\}$ of X (i.e., $X = \langle x_1, \cdots, x_r \rangle$, and $\sum m_i x_i = 0$ implies $m_i x_i = 0$ for all i). Then a basis $\{\xi_1, \cdots, \xi_r\}$ of X^* may be defined by $\xi_i(x_j) = 1$ for $i \neq j$, and $\xi_i(x_i) = \zeta_{m_i}$, some primitive m_i-th root of unity, where m_i is the order of x_i. (Note that any element of X^* is determined by its values on the x_i.) Now $x_i \mapsto \xi_i$ defines an isomorphism between X and X^*, as required. □

Each $x \in X$ defines a character x^{**} on X^* by $x^{**}(\chi) = \chi(x)$. Thus, X is naturally embedded in X^{**}, and by the above proposition $|X| = |X^{**}|$, so that X^{**} can be canonically identified with X. Below we shall not distinguish between X and X^{**}.

2.10.8. Proposition. *Let Y be a subgroup of X. Then for each $\chi \in X^*$ we have*

$$\sum_{y \in Y} \chi(y) = \begin{cases} |Y| & \text{if } \chi(y) = 1 \text{ for all } y \in Y, \\ 0 & \text{otherwise.} \end{cases}$$

In particular, for each $\chi \in X^*$

$$\sum_{x \in X} \chi(x) = \begin{cases} |X| & \text{if } \chi = 1, \\ 0 & \text{otherwise,} \end{cases}$$

and for each $x \in X$

$$\sum_{\chi \in X^*} \chi(x) = \begin{cases} |X| & \text{if } x = 0, \\ 0 & \text{otherwise.} \end{cases}$$

Proof. Suppose $\chi(y_0) \neq 1$ for some $y_0 \in Y$. Then

$$\chi(y_0) \sum_{y \in Y} \chi(y) = \sum_{y \in Y} \chi(y + y_0) = \sum_{y \in Y} \chi(y)$$

and it follows that $\sum_y \chi(y) = 0$. The second statement follows by taking $Y = X$, and the third by applying the second to X^*. □

Taking $Y = X$, it follows that the matrix Ω with entries $\Omega_{x\chi} = \chi(x)$ is nonsingular: $\Omega\overline{\Omega}^{\mathsf{T}} = |X|I$. (Note that Proposition 2.10.7 implies that Ω may be taken to be symmetric.)

Now let $(X, \mathfrak{R})$ be a d-class translation association scheme (with relations R_j described by $(x,y) \in R_j \Leftrightarrow y - x \in N_j$). For each character χ of X (considered as a vector indexed by X) we have

$$A_j\chi = \Big(\sum_{x \in N_j} \chi(x)\Big)\chi.$$

Thus, χ is an eigenvector; varying χ we find all eigenvalues of A_j. We can define the *dual scheme* $(X^*, \mathfrak{R}^*)$ by $\mathfrak{R}^* = \{R_0^*, \cdots, R_d^*\}$, where $(\chi, \psi) \in R_i^*$ if and only if $\eta = \chi^{-1}\psi$ is an eigenvector belonging to the i-th eigenspace, i.e., if and only if $E_i\eta = \eta$. Let $N_i^* = \{\eta \in X^* \mid E_i\eta = \eta\}$. Then the above observation can be expressed as

$$P_{ij} = \sum_{x \in N_j} \eta(x) \quad \text{for } \eta \in N_i^*. \tag{19}$$

We shall see that $(X^*, \mathfrak{R}^*)$ is again a translation association scheme, and that its dual scheme $(X^{**}, \mathfrak{R}^{**})$ is canonically isomorphic to $(X, \mathfrak{R})$.

We now introduce the *Fourier transform* $\tau: \mathcal{C}(X) \to \mathcal{C}(X^*)$ by

$$(P_x)^\tau := \sum_{\chi \in X^*} \chi(x) P_\chi \quad (x \in X),$$

$$(\sum_{x \in X} \alpha_x P_x)^\tau := \sum_{x \in X} \bar{\alpha}_x (P_x)^\tau \quad (\alpha_x \in \mathbf{C}),$$

where $\bar{\alpha}$ denotes the complex conjugate of $\alpha \in \mathbf{C}$. Note that τ is semilinear, and that its restriction to $\mathcal{C}_r(X)$ is linear. We shall use the same symbol τ for the Fourier transform $\tau : \mathcal{C}(X^*) \to \mathcal{C}(X^{**}) = \mathcal{C}(X)$.

2.10.9. Proposition. *For all $A, B \in \mathcal{C}(X)$ and all $\alpha \in \mathbf{C}$ we have*

(i) $(A+B)^\tau = A^\tau + B^\tau, \quad (\alpha A)^\tau = \bar{\alpha} A^\tau,$

(ii) $(AB)^\tau = A^\tau \circ B^\tau, \quad (A \circ B)^\tau = |X|^{-1} A^\tau B^\tau,$

(iii) $A^{\tau\tau} = |X| A,$

(iv) $I^\tau = J, \quad J^\tau = |X| I.$

Moreover, if $A \in \mathcal{C}_r(X)$, then $A^\tau \in \mathcal{C}_r(X^*)$.

Proof. Everything is immediate from the definition of τ, its semilinearity, and Proposition 2.10.8. E.g., part (iii) follows from

$$(P_y)^{\tau\tau} = (\sum_\chi \chi(y) P_\chi)^\tau = \sum_\chi \sum_x \overline{\chi(y)} \chi(x) P_x = \sum_x \sum_\chi \chi(x-y) P_x = |X| P_y.$$

Concerning the last part, since $\mathcal{C}_r(X)$ is spanned by the $P_x + P_{-x}$ $(x \in X)$, it suffices to consider the case $A = P_x + P_{-x}$. But

$$(P_x + P_{-x})^\tau = \sum_\chi (\chi(x) + \chi(-x)) P_\chi = \sum_\chi \operatorname{Re} \chi(x) (P_\chi + P_{\chi^{-1}}) \in \mathcal{C}_r(X^*). \qquad \square$$

2.10.10. Theorem (TAMASCHKE [708], DELSARTE [251], Theorem 2.9). *For any d-class translation association scheme $(X, \mathfrak{R})$ with valencies n_i, multiplicities f_i, and eigenmatrices P, Q, the dual scheme $(X^*, \mathfrak{R}^*)$ is a d-class translation association scheme with valencies $n_i^* = f_i$, multiplicities $f_i^* = n_i$, and eigenmatrices $P^* = Q$, $Q^* = P$.*

Proof. For $\eta = \chi^{-1}\psi \in N_k^*$ we have

$$(E_i^\tau)_{\chi\psi} = |X|^{-1} \sum_j Q_{ji} \sum_{z \in N_j} \eta(z) = |X|^{-1} \sum_j P_{kj} Q_{ji} = \delta_{ik} = (A_i^*)_{\chi\psi},$$

so that $A_i^* = E_i^\tau$. Since the image under τ of the Bose-Mesner algebra of $(X, \mathfrak{R})$ is an algebra spanned by the A_i^* $(i = 0, \cdots, d)$, and $A_i^* \in \mathcal{C}_r(X)$ for all i, $(X, \mathfrak{R}^*)$ is a translation association scheme. The assertions about the parameters now follow easily from §2.2. $\square$

It follows that

$$Q_{ij} = \sum_{\chi \in N_j^*} \chi(x) \quad \text{for } x \in N_i \tag{20}$$

and that

$$E_j = |X|^{-1} \sum_{x \in X} \sum_{\chi \in N_j^*} \chi(x) P_x. \tag{21}$$

2.10.11. Corollary. *Let $(X, \mathfrak{R})$ be a translation association scheme with $|X| = n$. Then*

(i) *The entries of the eigenmatrix Q are algebraic integers, and the numbers q_{ij}^k are integers.*

(ii) *The dual Frame quotient*

$$q^* := n^{d-1} \prod_{i=0}^{d} \frac{f_i}{n_i} = n^{2d-2} / q$$

is integral.

(iii) *If l of the valencies n_i are divisible by a given prime power p^a, then at least l of the numbers nf_i are divisible by p^a.* □

Remarks. (i) $(X, \mathfrak{R})$ is P-polynomial if and only if $(X, \mathfrak{R}^*)$ is Q-polynomial. In particular, if $(X, \mathfrak{R})$ is both P- and Q-polynomial, then so is $(X, \mathfrak{R}^*)$. Many examples of P- and Q-polynomial translation association schemes are known; see (C8)-(C12) in Table 6.1, and (P3) in Table 6.3. For further results on P-polynomial translation association schemes over an elementary abelian group, see §11.1C.

(ii) The results of this section do not essentially depend on the regular abelian subgroup, but rather on the fact that $\mathcal{C}(X)$ is a commutative algebra of matrices, with identity, closed under Hadamard multiplication (i.e., a commutative coherent algebra), which possesses a semilinear duality operator τ satisfying $I^\tau = J$ and $(AB)^\tau = A^\tau \circ B^\tau$. Such algebras are in fact equivalent to (not necessarily symmetric) formally self-dual association schemes, and the association schemes of Latin square type or negative Latin square type provide other examples. This observation and further dual properties of objects in a formally self-dual scheme are discussed in NEUMAIER [563].

C. Additive codes

An *additive code* in a translation association scheme $(X, \mathfrak{R})$ is a subgroup of X. The *weight* of an element $x \in X$ is the number $w(x) = k$ such that $x \in N_k$. The *weight distribution* of a code $Y \subseteq X$ is the vector $(a_0, \cdots, a_d)$ defined by

$$a_k = |Y \cap N_k| \quad (k = 0, \cdots, d);$$

for an additive code, the weight distribution equals the inner distribution. The numbers $k \neq 0$ for which $a_k \neq 0$ are called the *weights* of the code. The *degree* of an additive code is the number of its distinct nonzero weights. The *dual* Y' of an additive code Y is the subgroup $\{\chi \in X^* \mid \chi(y) = 1 \text{ for all } y \in Y\}$ of X^*.

The following result of DELSARTE [251] (Theorem 6.3) relates the weight distribution of an additive code with that of its dual, thus generalizing corresponding results by MACWILLIAMS [514] about linear codes over finite fields.

2.10.12. Theorem. *The weight distributions a and a' of an additive code Y and its dual code Y' are related by the formulae*

$$a' = \frac{1}{|Y|} aQ, \quad a = \frac{|Y|}{|X|} a'P.$$

Proof. Let χ_Y be the characteristic vector of Y. Using Proposition 2.10.8, equation (21) and Lemma 2.5.1 (iv) we find

$$|Y|^2 a_k' = |Y|^2 \cdot |Y' \cap N_k^*| = |Y| \sum_{\eta \in N_k^*} \sum_{y \in Y} \eta(y) =$$

$$= \sum_{\eta \in N_k^*} \sum_{y,y' \in Y} \eta(y'-y) = |X| \chi_Y^T E_k \chi_Y = |Y| (aQ)_k.$$

Multiply by P and use $QP = |X|\, I$ to get the second formula. □

Note that the codes Y and Y' live in different association schemes. In order to exhibit the relation with the classical MacWilliams relations, let us digress for a moment and look at inner products on abelian groups.

Let X be a finite abelian group. The *exponent* of X is the smallest integer $e > 0$ satisfying $ex = 0$ for all $x \in X$; by Lagrange's theorem, the exponent is a divisor of $|X|$, and every prime factor of $|X|$ divides e.

A mapping $(\cdot,\cdot)$ from $X \times X$ into the group $\mathbf{Z}/e\mathbf{Z}$ of integers modulo the exponent e of X is called an *inner product* of X if, for all $x, y, z \in X$,

$$(x,y) = (y,x),$$
$$(x,y+z) = (x,y) + (x,z), \text{ and}$$
$$(x,w) = 0 \text{ for all } w \in X \Leftrightarrow x = 0.$$

Every finite abelian group X has an inner product. Indeed, X is a direct product of cyclic groups, hence isomorphic to a group $(\mathbf{Z}/e_1\mathbf{Z}) \oplus \cdots \oplus (\mathbf{Z}/e_m\mathbf{Z})$. The exponent e is the lowest common multiple of $e_1, \cdots, e_m$, and the formula

$$(x,y) := \sum \frac{e}{e_l} x_l y_l \quad \text{for } x = (x_1, \cdots, x_m),\ y = (y_1, \cdots, y_m)$$

defines an inner product of this group. In particular, if X is elementary abelian of order p^m (p prime), then $(x,y) := \sum x_l y_l$ is an inner product on X.

Let ω be a fixed primitive e-th root of unity. Then there is a one-to-one correspondence between isomorphisms $f: X \to X^*$ and nondegenerate biadditive forms $(\cdot,\cdot): X \times X \to \mathbf{Z}/e\mathbf{Z}$ given by

$$\omega^{(x,y)} = (f(y))(x).$$

Proposition 2.10.7 says that for suitable f the obtained form will be symmetric, i.e., an inner product.

For a fixed inner product $(\cdot,\cdot)$ on X, let, for $Y \subseteq X$,

$$Y^\perp := \{x \in X \mid (x,y) = 0 \text{ for all } y \in Y\}.$$

Clearly, $Y^\perp$ is a subgroup of X, $X^\perp = \{0\}$, and $Y^{\perp\perp} \supseteq Y$. When Y is a subgroup then $|Y| \cdot |Y^\perp| = |X|$ and it follows that $Y^{\perp\perp} = Y$. It is possible to identify Y' with $Y^\perp$ using the isomorphism $f: X \to X^*$ corresponding to the inner product, but this identification in general only preserves the group structure, not the association scheme structure. However, in the case of the Hamming scheme $H(n,q)$ with the usual inner product, we find $x \in N_i \Leftrightarrow f(x) \in N_i^*$, and Theorem 2.10.12 above specializes to the classical MacWilliams relations. (Note that the classical 'inner product' on $H(n,q)$ (identified with $\mathbf{F}_q^n$), given by $[x,y] = \sum x_i y_i$, is a map into $\mathbf{F}_q$; from this map we obtain an inner product in our present sense (into $\mathbf{F}_p$, where q is a power of the prime p) by putting $(x,y) = \operatorname{tr}[x,y]$. For linear codes Y, both inner products yield the same code $Y^\perp$.)

In Proposition 2.8.3 above we saw a sufficient condition for a subset of a Q-polynomial association scheme to carry a subscheme. In the case of an additive code in a translation association scheme, we can be more precise:

2.10.13. Theorem (Delsarte [251], Theorem 6.10). *Let Y be an additive code in a translation association scheme $(X, \mathcal{R})$, Y' the dual code, and B' the outer distribution matrix of Y' in $(X^*, \mathcal{R}^*)$. Then $(Y, \mathcal{R}_Y)$ is a subscheme of $(X, \mathcal{R})$ if and only if the matrix B' has precisely $s+1$ distinct rows, where $s+1 = \operatorname{rk} B'$ is the number of weights of Y.*

Proof. Since $Y'' = Y$, the equality $s+1 = \operatorname{rk} B'$ follows from Proposition 2.5.5. Let us denote the adjacency matrix of the relation $(R_i)_Y$ by $(A_i)_Y$. The dimension of the linear span of the matrices $(A_i)_Y$ ($0 \leqslant i \leqslant d$) is the number of weights of Y, i.e., $s+1$. We have an association scheme if and only if this equals the dimension of the algebra generated by the $(A_i)_Y$. But the dimension of this algebra is the number of distinct rows of B'; indeed, these matrices are diagonalized simultaneously by the characters on Y, and the dimension of this algebra is the number of eigenspaces. Now the eigenvalue of $(A_i)_Y$ belonging to $\eta \in X^*$ (or, rather, to $\eta\,|\,_Y$) equals:

$$\sum_{y \in N_j \cap Y} \eta(y) = \frac{|Y|}{|X|} \sum_{y \in N_j} \eta(y) \sum_{\chi \in Y'} \chi(y)$$
$$= \frac{|Y|}{|X|}\Big(\sum_{y \in N_j} \sum_{\psi \in X^*} \psi(y) P_\psi \chi_{Y'} \Big)_{\eta^{-1}}$$
$$= |Y|(E_j{}^*\chi_{Y'})_{\eta^{-1}} = \frac{|Y|}{|X|}(B'Q^*)_{\eta^{-1}j},$$

where $\chi_{Y'}$ is the characteristic vector of Y', and since Q^* is nonsingular, our claim follows. □

2.11 Representation diagrams, Krein modules and spherical designs

In this section we give more intrinsic interpretations of the Krein parameters, based on work of TERWILLIGER [721], DELSARTE, GOETHALS & SEIDEL [258] and GOETHALS & SEIDEL [330].

Let $(X, \mathfrak{R})$ be an association scheme, fixed throughout this section. The representation diagram of a minimal idempotent $E = E_k$ is the undirected graph Δ_E with vertices $0, 1, \cdots, d$, where two vertices i, j are joined whenever $i \neq j$ and $q_{ij}^k = q_{ji}^k \neq 0$. By relabelling the idempotents we may restrict ourselves to the case $k = 1$. (This excludes $k = 0$, but the representation diagram of E_0 is a coclique, and the theory becomes trivial.) Thus we consider a fixed minimal idempotent $E = E_1$. It has rank $f = f_1$, and if we denote the x-column of E by $\overline{x}$, then $(\overline{x}, \overline{y}) = (E^\top E)_{xy} = E_{xy}$ so that

$$(\overline{x}, \overline{y}) = E_{xy} = \frac{f}{n} u_i \quad \text{if } (x,y) \in R_i, \tag{22}$$

where the $u_i = f^{-1} Q_{i1}$ are the components of the normalized right eigenvector u corresponding to E (cf. Proposition 2.2.2).

2.11.1. Proposition. *The number of components of the representation diagram Δ_E of E equals the number of $i \in \{0, \cdots, d\}$ with $u_i = 1$.*

Proof. By Lemma 2.3.1 (iv) and (vii), the matrix B with entries $B_{ij} = q_{ij}^1 / f_i = q_{j1}^i / f$ satisfies

$$(QB)_{ij} = \sum Q_{il} B_{lj} = \sum Q_{il} q_{j1}^l / f = Q_{ij} Q_{i1} / f = Q_{ij} u_i$$

so that $QB = \Delta_u Q$, i.e., $u_0, \cdots, u_d$ are the eigenvalues of B. Note that B has constant row sums 1 by Lemma 2.3.1 (v). Since B is nonnegative and $B_{ij} \neq 0$ if and only if ij is an edge of the representation diagram, the result follows from Perron-Frobenius theory (cf. Proposition 3.2.2). □

The relation $QB = \Delta_u Q$ has first been observed by MCKAY [530] in the context of explaining relations between binary polyhedral groups and Dynkin diagrams.

We now introduce the Krein module as a tool for the formulation of Terwilliger's characterization of acyclic representation diagrams. The $n^2 f$-dimensional vector space $\mathfrak{M}$ of $n \times n$-matrices with rows and columns indexed by X and with entries in the column space of E can be given the structure of a Euclidean space by introducing the positive definite inner product

$$(M,N) := \sum_{x,y \in X} (M_{xy}, N_{xy}) \quad \text{for } M, N \in \mathfrak{M},$$

and also the structure of an $\mathcal{A}$-module by defining for every matrix A in the Bose-Mesner algebra $\mathcal{A}$ and every matrix $M \in \mathfrak{M}$ a product AM by

$$(AM)_{xy} = \sum_{u \in X} A_{xu} M_{uy}.$$

The *Krein module* of E is then defined as the $\mathcal{A}$-submodule $\hat{\mathcal{A}}$ of $\mathfrak{M}$ generated by $\{\hat{A} \mid A \in \mathcal{A}\}$, where the linear mapping $\hat{}: \mathcal{A} \to \mathfrak{M}$ is defined by

$$\hat{A}_{xy} := A_{xy}\bar{x} \quad (x, y \in X).$$

Since $\mathcal{A}$ is spanned by $A_0, \cdots, A_d$ and also by $E_0, \cdots, E_d$, the Krein module is spanned by the matrices $A_i\hat{A}_j$ and also by the $E_i\hat{E}_j$. The following lemma relates the Krein module with the Krein parameters. Recall that $\Sigma(A)$ denotes the sum of the elements of the matrix A.

2.11.2. Lemma.

(i) $(A\hat{B}, C\hat{D}) = \Sigma(AC \circ BD \circ E)$ *for all* $A, B, C, D \in \mathcal{A}$. *In particular,*

$$(\hat{A}, \hat{B}) = \Sigma(I \circ AB \circ E) = \operatorname{tr}(AB \circ E) \text{ for } A, B \in \mathcal{A}.$$

(ii) *The matrices* $E_i\hat{E}_j$ *are mutually orthogonal matrices with norm*

$$(E_i\hat{E}_j, E_i\hat{E}_j) = \frac{f}{n} q_{ij}^1 \geqslant 0.$$

Proof. Note that

$$(A\hat{B})_{xy} = \sum_{u \in X} A_{xu} B_{uy} \bar{u}, \tag{23}$$

so that by (22)

$$\begin{aligned}(A\hat{B}, C\hat{D}) &= \sum_{x,y \in X} \Big(\sum_u A_{xu} B_{uy} \bar{u}, \sum_v C_{xv} D_{vy} \bar{v}\Big) \\ &= \sum_{u,v \in X} \Big(\sum_x A_{xu} C_{xv}\Big)\Big(\sum_y B_{uy} D_{vy}\Big)(\bar{u}, \bar{v}) \\ &= \sum_{u,v \in X} (AC)_{uv} (BD)_{uv} E_{uv} = \Sigma(AC \circ BD \circ E),\end{aligned}$$

giving (i).

Since $E_iE_k = 0$ for $i \neq k$, this implies that $(E_i\hat{E}_j, E_k\hat{E}_l) = \Sigma(E_iE_k \circ E_jE_l \circ E) = 0$ if $(i,j) \neq (k,l)$, so the $E_i\hat{E}_j$ are orthogonal; thus (ii) follows, since by (i) and Theorem 2.3.2 we have

$$(E_i\hat{E}_j, E_i\hat{E}_j) = \Sigma(E_i^2 \circ E_j^2 \circ E) = \Sigma(E_i \circ E_j \circ E) = \frac{f}{n} q_{ij}^1. \qquad \square$$

Of particular interest are the skew-symmetric matrices in the Krein module. If we define the linear mapping $\delta: \hat{\mathcal{A}} \to \hat{\mathcal{A}}$ by $\delta(M) := M - M^{\mathsf{T}}$, then $\delta\hat{\mathcal{A}}$ consists precisely of the skew-symmetric matrices in $\hat{\mathcal{A}}$.

2.11.3. Lemma. *The map δ satisfies:*

(i) $\delta(A\hat{B}) = A\hat{B} - B\hat{A}$,

(ii) $(\delta(A\hat{B}),\delta(C\hat{D})) = 2(\Sigma(AC\circ BD\circ E) - \Sigma(AD\circ BC\circ E))$,

(iii) $(\delta(\hat{A}_i),\delta(\hat{A}_j)) = 2fn_i(1-u_i)\delta_{ij}$.

Proof. (i) holds since $(A\hat{B})^{\mathrm{T}} = B\hat{A}$ by (23), (ii) is an obvious consequence of the previous lemma, and (iii) follows since

$$I\circ A_iA_j\circ E = n_i\delta_{ij}I\circ E = n_i\delta_{ij}\frac{f}{n}I,$$

$$A_i\circ A_j\circ E = \delta_{ij}A_i\circ E = \delta_{ij}\frac{f}{n}u_iA_i. \qquad \square$$

2.11.4. Theorem (TERWILLIGER [721]). *Let E be a minimal idempotent of an association scheme with associated representation diagram Δ_E and Krein module $\hat{\mathcal{Q}}$. Then:*

(i) *The dimension of $\delta\hat{\mathcal{Q}}$ equals the number of edges in Δ_E.*

(ii) *$\delta\hat{\mathcal{Q}} = \{\delta\hat{A} \mid A \in \mathcal{Q}\}$ if and only if Δ_E is a forest (and a tree if $u_i \neq 1$ for $i \neq 0$).*

Proof. Let v, e, and c denote the number of vertices, edges, and components of Δ_E.
(i) The subspace $\delta\hat{\mathcal{Q}}$ is spanned by the matrices $\Delta_{ij} := \delta(E_i\hat{E}_j) = E_i\hat{E}_j - E_j\hat{E}_i$. By Lemma 2.11.2, we have $\Delta_{ij} \neq 0$ precisely when ij is an edge of Δ_E, and since Δ_{ij} is orthogonal to all $\Delta_{i'j'}$ except to $\Delta_{ji} = -\Delta_{ij}$, we have $\dim \delta\hat{\mathcal{Q}} = e$.

(ii) By Lemma 2.11.3 (iii) and Proposition 2.11.1, the $\delta(\hat{A}_i)$ with $u_i = 1$ vanish, and the $\delta(\hat{A}_i)$ with $u_i \neq 1$ form a set of $v-c$ linearly independent matrices. Thus by (i) we have $e \geqslant v-c$, with equality if and only if $\delta\hat{\mathcal{Q}}$ is spanned by this set, i.e., $\delta\hat{\mathcal{Q}} = \{\delta\hat{A} \mid A \in \mathcal{Q}\}$. Now a classical theorem on graphs (see e.g. BOLLOBÁS [96], p. 8) implies that this holds precisely when Δ_E is a forest (and a tree if $c = 1$, i.e., if $u_i \neq 1$ for $i \neq 0$). $\square$

Remark. By dualizing the arguments we get a characterization of association schemes whose distribution diagram is a forest.

2.11.5. Corollary. *For all $i,j \in \{1,\cdots,d\}$ we have*

$$\sum_{k=0}^{d} n_ku_k(p^k_{ii}p^k_{jj} - (p^k_{ij})^2) \geqslant \sum_{u_k \neq 1} n_k\frac{(u_i-u_j)^2}{1-u_k}(p^k_{ij})^2,$$

and equality holds for all i,j precisely when Δ_E is a forest.

Proof. Using (22), we get from Lemma 2.11.3 (ii) and Lemma 2.1.1 (iv) that

$$(\delta(A_i\hat{A}_j),\delta(A_k\hat{A}_l)) = 2f\sum_{h=0}^{d} n_hu_h(p^h_{ik}p^h_{jl} - p^h_{jk}p^h_{il}),$$

and, in particular, that

$$(\delta(A_i\hat{A}_j),\delta(\hat{A}_k)) = 2fn_k(u_i-u_j)p^k_{ij}.$$

This last equation implies that the orthogonal projection of $\delta(A_i\hat{A}_j)$ to $\{\delta\hat{A} \mid A \in \mathcal{Q}\}$ is given by

$$\sum_{u_k \neq 1} p_{ij}^k \frac{u_i - u_j}{1 - u_k} \delta(\hat{A}_k).$$

By comparing its norm with the norm of $\delta(A_i\hat{A}_j)$, the assertion follows. □

Note that these inequalities are consequences of the Krein conditions, as can be seen by expressing $\delta(A_i\hat{A}_j)$ and its projection in terms of the $E_k\hat{E}_l$.

2.11.6. Theorem (Terwilliger [721]). *Let $(X, \mathcal{R})$ be a metric association scheme, and E a minimal idempotent with associated representation diagram Δ_E and Krein module $\hat{\mathcal{Q}}$. Then the following are equivalent:*

(i) *$(X, \mathcal{R})$ is cometric with first idempotent $E_1 = E$,*

(ii) *Δ_E is a path,*

(iii) *Δ_E is a tree,*

(iv) *$u_i \neq 1$ for $i \neq 0$, and $\delta\hat{\mathcal{Q}} = \{\delta\hat{A} \mid A \in \mathcal{Q}\}$.*

Proof. (i) ⇔ (ii) ⇒ (iii) are obvious, and (iii) ⇔ (iv) follows from Theorem 2.11.4. In order to prove (iii) ⇒ (ii), let s be a leaf of Δ_E, and t its neighbour. The $\theta_i = P_{i1}$ are the eigenvalues of A_1, and since the powers of A_1 span the $(d+1)$-dimensional Bose-Mesner algebra, the θ_i must be distinct. We now show that θ_s satisfies a quadratic equation (independent of s) with nonzero constant coefficient; this shows that there are at most two leaves s, so that Δ_E must be a path.

Since $Q_{i1} = fu_i$, Lemma 2.3.1 (vii) gives for $j = 1$, $k = s$ the equation $fu_iQ_{is} = q_{1s}^tQ_{it} + q_{1s}^sQ_{is}$, which implies in view of Lemma 2.2.1 (iv) a relation $u_iP_{si} = \alpha P_{si} + \beta P_{ti}$ with certain constants α, β. Taking $i = 0$, we find $u_0 = \alpha + \beta$, so that

$$(u_0 - u_i)P_{si} = \beta(P_{si} - P_{ti}). \tag{24}$$

Now Lemma 2.2.1 (iii) gives

$$P_{s1}P_{sk} = \sum_l p_{1k}^l P_{sl} \tag{25}$$

and

$$\sum_l p_{1k}^l(u_0 - u_l)P_{sl} = \beta \sum_l p_{1k}^l(P_{sl} - P_{tl}) = \beta(P_{s1}P_{sk} - P_{t1}P_{tk})$$
$$= \beta(P_{s1} - P_{t1})P_{sk} + \beta P_{t1}(P_{sk} - P_{tk}),$$

so that

$$\sum_l p_{1k}^l(u_0 - u_l)P_{sl} = ((u_0 - u_1)P_{s1} + (u_0 - u_k)P_{t1})P_{sk}. \tag{26}$$

Since $p_{1k}^l = 0$ for $l \neq k-1, k, k+1$ and since $p_{1k}^{k-1} \neq 0$, (25) shows that P_{sd} and $P_{s,d-1}$ are not both zero (otherwise all P_{si} would be zero, but $P_{s0} = 1$); therefore the determinant of the homogeneous linear system for P_{sd} and $p_{1d}^{d-1}P_{s,d-1}$ obtained from (25) and (26) for $k = d$, vanishes. This gives

$$(u_0 - u_1)\theta_s + (u_0 - u_d)(\theta_t - a_d) = (\theta_s - a_d)(u_0 - u_{d-1}), \tag{27}$$

where $a_d = p_{1d}^d$. And by eliminating P_{s2} from (25) and (26) for $k = 1$, we get

$$(u_2-u_1)\theta_s(\theta_s-\lambda)+(u_0-u_1)\theta_s\theta_t+(u_0-u_2)n_1 = 0, \tag{28}$$

where $\lambda = p^1_{11}$. Since $u_i \neq 1 = u_0$ for $i \neq 0$, we can solve (27) for θ_t and substitute it into (28); this gives the required quadratic equation for θ_s, and its constant term is $(u_0-u_2)n_1 \neq 0$. □

Remark. By dualizing the above argument we see that if the distribution diagram of the relation R_1 of a cometric association scheme is a tree, then it is a path, the scheme is metric, and (X,R_1) is a distance-regular graph. For examples of association schemes with a tree that is not a path as distribution diagram, cf. Remark 9.3.12 (ii), §13.5 and Remark 13.6.2 (iii).

Problem. Are there association schemes having a distribution diagram and a representation diagram which form nonisomorphic trees?

The Krein parameter q^1_{11} can also be interpreted in the context of spherical designs. The connection between association schemes and spherical designs is given by (22), which implies that E is the Gram matrix of the set $\overline{X}$ of columns of E in the f-dimensional column space of E (the *eigenspace* of the association scheme corresponding to E). All vectors $\overline{x}$ ($x \in X$) have the same norm $(\overline{x},\overline{x}) = f$, and the cosine of the angle between $\overline{x}$ and $\overline{y}$ is $(\overline{x},\overline{y})/\sqrt{(\overline{x},\overline{x})(\overline{y},\overline{y})} = u_i$ if $(x,y) \in R_i$. Thus, in the terminology of DELSARTE, GOETHALS & SEIDEL [258], $\overline{X}$ is a *spherical code* with set of angles $\{u_0,u_1,\cdots,u_d\}$, and we can apply their results and those of GOETHALS & SEIDEL [330] to association schemes. In particular, we get certain linear inequalities for the numbers

$$p_s = \frac{1}{n}\sum_{l=0}^{d} n_l u_l^s;$$

(note that, in the notation of Remark 2.3 (iii), $p_s = n^{-1}f^{-s}\sigma_{1\cdots 1}$ (s indices)), so that in particular

$$p_0 = 1,\ \ p_1 = 0,\ \ p_2 = \frac{1}{f},\ \ p_3 = \frac{q^1_{11}}{f^2}. \tag{29}$$

For example, we have the *Sidel'nikov inequalities* (SIDEL'NIKOV [674]) which assert that $p_s \geqslant 0$ if s is odd, and

$$p_s \geqslant \frac{1{\cdot}3\cdots(s-1)}{f(f+2)\cdots(f+s-1)} \quad \text{if } s \text{ is even.} \tag{30}$$

A stronger set of inequalities involve so-called *Gegenbauer polynomials* $Q_k(x)$, a family of orthogonal polynomials defined by

$$Q_0(x) = 1,\ \ Q_1(x) = fx,\ \ Q_{k+1}(x) = \frac{f+2k}{k+1}\left(xQ_k(x) - \frac{f+k-3}{f+2k-4}Q_{k-1}(x)\right)$$

($k \geqslant 1$). In terms of these polynomials we have for all $i \geqslant 0$ the inequality

$$\sum_l n_l Q_i(u_l) \geqslant 0 \tag{31}$$

which leads to linear inequalities for the p_s. These inequalities imply those of Sidel'nikov since x^s is a nonnegative linear combination of $Q_0(x),\cdots,Q_s(x)$ (GOETHALS & SEIDEL [330]). A spherical code satisfying (31) with equality for

$i = 1, \cdots, t$ is a *spherical t-design* and has a particularly homogeneous structure, see DELSARTE et al. [258]. Using explicit expressions for $Q_i(x)$, $i \leqslant 5$ (see, e.g., [258]), we find that the spherical code $\overline{X}$ constructed from E is a spherical 2-design, and the inequalities (31) reduce to

$$p_3 \geqslant 0, \quad p_4 \geqslant \frac{3}{f(f+2)}, \quad p_5 \geqslant \frac{10}{f+6} p_3$$

for $i = 3, 4, 5$. In particular, $\overline{X}$ is a spherical 3-design precisely when $q_{11}^1 = 0$, and the Krein condition $q_{11}^1 \geqslant 0$ is a special case of the inequalities (31).

If S is a subset of a Euclidean space V, the definition

$$u*v := \sum_{x \in S} (u,x)(v,x)x \quad \text{for } u, v \in S$$

turns V into the *Norton algebra* $N(S)$, a commutative nonassociative algebra satisfying the identity

$$(u*v, w) = (v*w, u) = (w*u, v)$$

$(= \sum_{x \in S} (u,x)(v,x)(w,x))$; cf. GOETHALS & SEIDEL [330], CAMERON, GOETHALS & SEIDEL [178] and SMITH [688]. The product $*$ is identically zero precisely when the Gram matrix $E = \operatorname{Gram} S$ of S (cf. §3.4) satisfies $\Sigma(E^2 \circ E^2 \circ E^2) = 0$. In particular, if $S = \overline{X}$ is the set of columns of the minimal idempotent E_1 of an association scheme on X, then $E = E_1^2 = E_1$; thus $N(\overline{X})$ is nontrivial precisely when $q_{11}^1 > 0$. Since $x*x = q_{11}^1 x$ for all $x \in X$, we find in this case that $N(\overline{X})$ is generated by idempotents, namely by the vectors $(q_{11}^1)^{-1/2} x$ $(x \in X)$.

If G is a group of isometries of V stabilizing S, then G acts naturally as a group of automorphisms on the Norton algebra $N(S)$; in particular, this applies to the Norton algebra $N(\overline{X})$ belonging to idempotents of association schemes $(X, \mathfrak{R})$ for generously transitive permutation groups G, the case studied in SMITH [688].

Chapter 3
Representation Theory

Motivated by applications to the classification of certain distance-regular graphs we consider representations of graphs by sets of vectors in a Euclidean space.
After reviewing basic facts on nonnegative and positive semidefinite matrices we introduce in Section 3.5 spherical and affine representations of a graph. We show that these representations are related to the eigenvalues of the graph and we prove Hoffman's coclique bound and other consequences for special subgraphs.

Representations with even integral distance squares lead naturally to lattices. In particular, if, in addition, adjacent vertices have images at squared distance 2, we get a root lattice, i.e., an integral lattice generated by norm 2 vectors (roots). Root lattices are introduced and classified in Section 3.10. The fundamental fact that the only root lattices are the lattices A_n, D_n, E_6, E_7, and E_8 leads to the line graph theorem of Cameron, Goethals, Seidel and Shult, a beautiful characterization of the graphs with smallest eigenvalue $\geqslant -2$ in terms of generalized line graphs and certain exceptional graphs represented in E_8. In particular, Seidel's classification of strongly regular graphs with smallest eigenvalue -2 is obtained and extended to edge-regular and co-edge-regular graphs.

Section 3.15 uses root lattices to determine the amply regular graphs with $\mu > 1$ having a root representation (an affine representation of a special kind). Among the graphs obtained we find several well-known families of distance-regular graphs: Johnson graphs, Hamming graphs, half-cubes, and also some graphs discovered by Doob.

Applications to the classification of distance-regular graphs will be given in the subsequent chapter in Section 4.4. Each eigenspace determines a graph representation and in the particular case when the second largest eigenvalue is $b_1 - 1$ and a quadrangle exists, we get a root representation. As a consequence, a large class of distance-regular graphs can be characterized by their intersection array (Theorem 4.4.11).

We shall work in the space $\mathbb{R}^n$ of real column vectors $x = (x_1, \cdots, x_n)^\mathsf{T}$, equipped with the usual inner product $(x,y) = x^\mathsf{T} y = x_1 y_1 + \cdots + x_n y_n$. The number $\|x\|^2 = (x,x)$ will be called the *norm* of $x \in \mathbb{R}^n$.

3.1 Nonnegative matrices

In this section we state and prove the Perron-Frobenius theorem on nonnegative matrices. For a matrix (or vector) A, we write $A > 0$ ($\geqslant 0$) when all its entries are positive (nonnegative).

Let T be a real $n \times n$ matrix with nonnegative entries. T is called *primitive* if for some k we have $T^k > 0$; T is called *irreducible* if for all i, j there is a k such that $(T^k)_{ij} > 0$. Note that the matrix $T = (t_{ij})$ is irreducible if and only if the directed graph G_T with vertices $\{1, \cdots, n\}$ and edges (i,j) whenever $t_{ij} > 0$ is strongly connected. When T is irreducible, $I + T$ is primitive. The *period* d of an irreducible matrix T is the greatest common divisor of the integers k for which $(T^k)_{ii} > 0$. It is independent of the i chosen.

3.1.1. Theorem (PERRON [591], FROBENIUS [303, 304]). *The following holds for a real irreducible $n \times n$ matrix T with $T \geqslant 0$.*

(i) *There is a real number θ_0 and a real vector x_0 with $Tx_0 = \theta_0 x_0$, $\theta_0 > 0$, $x_0 > 0$. If $x \geqslant 0$, $x \neq 0$ and $Tx \geqslant \theta x$, then $\theta_0 \geqslant \theta$.*

(ii) *The eigenvalue θ_0 of T has geometric and algebraic multiplicity one.*

(iii) *For each eigenvalue θ of T we have $|\theta| \leqslant \theta_0$. If T is primitive then $|\theta| = \theta_0$ implies $\theta = \theta_0$. In general, if T has period d, then T has precisely d eigenvalues θ with $|\theta| = \theta_0$, namely $\theta = \theta_0 e^{2\pi i j/d}$ for $j = 0, 1, \cdots, d-1$. In fact the entire spectrum of T is invariant under rotation of the complex plane over an angle $2\pi / d$ about the origin.*

(iv) *Any (nonzero) nonnegative left or right eigenvector of T has eigenvalue θ_0. More generally, if $x \geqslant 0$, $x \neq 0$ and $Tx \leqslant \theta x$, then $x > 0$ and $\theta \geqslant \theta_0$; moreover, $\theta = \theta_0$ if and only if $Tx = \theta x$.*

(v) *If $0 \leqslant S \leqslant T$ and $S \neq T$, then every eigenvalue σ of S satisfies $|\sigma| < \theta_0$.*

Proof. (i) Let $P = (I+T)^{n-1}$. Then $P > 0$ and $PT = TP$. Let $B = \{x \in \mathbf{R}^n \mid x \geqslant 0 \text{ and } x \neq 0\}$. For $x \in B$ define:

$$\Theta(x) = \max\{\theta \in \mathbf{R} \mid \theta x \leqslant Tx\} = \min_{\substack{1 \leqslant i \leqslant n \\ x_i \neq 0}} \frac{(Tx)_i}{x_i}.$$

Now $\Theta(\alpha x) = \Theta(x)$ for $\alpha \in \mathbf{R}$, $\alpha > 0$, and ($x \leqslant y$, $x \neq y$ implies $Px < Py$, so) $\Theta(Px) \geqslant \Theta(x)$; moreover, $\Theta(Px) > \Theta(x)$ unless x is an eigenvector of T. Put $C = \{x \mid x \geqslant 0 \text{ and } \|x\| = 1\}$. Since C is compact and $\Theta(\ )$ is continuous on $P[C]$ (but not in general on C !), there is an $x_0 \in P[C]$ such that

$$\theta_0 := \sup_{x \in B} \Theta(x) = \sup_{x \in C} \Theta(x) = \sup_{x \in P[C]} \Theta(x) = \Theta(x_0).$$

Now $x_0 > 0$ and x_0 is an eigenvector of T, so $Tx_0 = \theta_0 x_0$, and $\theta_0 > 0$.

(iii) For a vector $x = (x_1, \cdots, x_n)^{\mathsf{T}}$, write $|x| = (|x_1|, \cdots, |x_n|)^{\mathsf{T}}$. If $Tx = \theta x$, then by the triangle inequality we have $T|x| \geqslant |\theta|\,|x|$. For nonzero x this means $|\theta| \leqslant \Theta(|x|) \leqslant \theta_0$. If $|\theta| = \theta_0$, then $T|x| = \theta_0 |x|$ and we have equality in the triangle inequality $|\sum_j t_{ij} x_j| \leqslant \sum_j t_{ij} |x_j|$; this means that all numbers $t_{ij} x_j$ $(1 \leqslant j \leqslant n)$ have the same angular part (argument). If T is primitive then we can apply this reasoning with T^k instead of T, where $T^k > 0$, and conclude that all x_j

have the same angular part. Consequently, in this case x is a multiple of a real vector and may be taken real, nonnegative. Now $Tx = \theta x$ shows that θ is real, and $|\theta| = \theta_0$ that $\theta = \theta_0$. In the general case T^d is a direct sum of primitive matrices $T^{(0)}, \cdots, T^{(d-1)}$, and if $x = (x^{(0)}, \cdots, x^{(d-1)})$ is the corresponding decomposition of an eigenvector of T (with eigenvalue θ), then also $(x^{(0)}, \zeta x^{(1)}, \cdots, \zeta^{d-1} x^{(d-1)})$ is an eigenvector of T, with eigenvalue $\zeta\theta$, for any d-th root of unity ζ. (Here we assume that the $T^{(i)}$ are ordered in such a way that in G_T the arrows point from the subset corresponding to $T^{(i)}$ to the subset corresponding to $T^{(i+1)}$.) Now T^d has a unique eigenvalue of maximum modulus; indeed, let $T_{(i)}^{(i+1)}$ be the (nonsquare) submatrix of T describing the arrows in G_T between the subset corresponding to $T^{(i)}$ to the subset corresponding to $T^{(i+1)}$, then $T^{(i)} = \prod_{j=0}^{d-1} T_{(i+j)}^{(i+j+1)}$ and, if $T^{(i)}z = \gamma z$, $z > 0$, then $T^{(i-1)}z' = \gamma z'$ where $z' = T_{(i-1)}^{(i)} z \neq 0$, so that all $T^{(i)}$ have the same eigenvalue of maximum modulus. It follows that T has precisely d eigenvalues of maximum modulus.

(iv) Doing the above for left eigenvectors instead of right ones, we find $y_0 > 0$ with $y_0^\top T = \eta_0 y_0^\top$. If $Tx = \theta x$ and $y^\top T = \eta y^\top$, then $\eta y^\top x = y^\top T x = \theta y^\top x$. It follows that either $\theta = \eta$ or $y^\top x = 0$. Taking $y \in B$, $x = x_0$ or $x \in B$, $y = y_0$ we see that $\theta = \eta$ ($= \theta_0 = \eta_0$). Similarly, if $Tx \leqslant \theta x$, $x \in B$ then $\theta_0 y_0^\top x = y_0^\top T x \leqslant \theta y_0^\top x$ so that $\theta_0 \leqslant \theta$; also $0 < Px \leqslant (1+\theta)^{n-1} x$, so $x > 0$. If $\theta = \theta_0$, then $y_0^\top (Tx - \theta x) = 0$ so $Tx = \theta x$.

(v) If $s \neq 0$, $Ss = \sigma s$, then $T|s| \geqslant S|s| \geqslant |\sigma||s|$, so $|\sigma| \leqslant \theta_0$. But, if $|\sigma| = \theta_0$, then $|s|$ is eigenvector of T and $|s| > 0$ and $(T-S)|s| = 0$, so $S = T$.

(ii) (Recall that the geometric multiplicity of an eigenvalue is the dimension of the corresponding eigenspace, while the algebraic multiplicity is its multiplicity as a root of the characteristic polynomial. The geometric multiplicity is not more than the algebraic multiplicity.)

In order to prove that θ_0 is a simple root of χ_T, the characteristic polynomial of T, we have to show that $\frac{d}{d\theta}\chi_T(\theta)$ is nonzero for $\theta = \theta_0$. But $\chi_T(\theta) = \det(\theta I - T)$ and $\frac{d}{d\theta}\chi_T(\theta) = \sum_i \det(\theta I - T_{ii})$ (where T_{ii} is the principal minor of T obtained by deleting the i-th row and column), and by (v) we have $\det(\theta I - T_{ii}) > 0$ for $\theta = \theta_0$. □

Remark. For a complex matrix S, let $|S|$ be the matrix with entries $|S|_{ij} = |S_{ij}|$. The *spectral radius* of a matrix is the maximum of the absolute values of its eigenvalues.

In case $T \geqslant 0$ but T not necessarily irreducible, we can say the following.

(i) *The spectral radius θ_0 of T is an eigenvalue, and there are nonnegative left and right eigenvectors corresponding to it.*

(ii) *If $|S| \leqslant T$ and S has eigenvalue σ, then $|\sigma| \leqslant \theta_0$.*

Proof. (i) Use continuity arguments; (ii) the old proof still applies. □

For more details, see the exposition of the Perron-Frobenius Theory in GANTMACHER [308] (Vol. 2, Chapter XIII); cf. also VARGA [763], MARCUS & MINC [516], BERMAN & PLEMMONS [55], SENETA [660] (Chapter 1).

3.2 Adjacency matrices and eigenvalues of graphs

In this section we apply the Perron-Frobenius Theory to get information on graph spectra (mainly on the largest and smallest eigenvalue of a graph). For more information on the spectra of graphs, cf. BIGGS [71], SCHWENK & WILSON [646], and the monographs by CVETKOVIĆ, DOOB & SACHS [235] and CVETKOVIĆ, DOOB, GUTMAN & TORGAŠEV [234]. For tables with the spectra of small graphs, see also CVETKOVIĆ & PETRIĆ [237] and BUSSEMAKER et al. [156].

Given a (finite) graph Γ, its *adjacency matrix* A is the 0-1 matrix with rows and columns indexed by the vertex set of Γ, and where $A_{\gamma\delta} = 1$ if and only if $\gamma \sim \delta$. By definition the *eigenvalues* of Γ are those of A. Since A is nonnegative, the results of the previous section apply. But we know more: since Γ is undirected, the matrix A is symmetric (with zero diagonal), and this means that all eigenvalues are real and that $\mathbb{R}^\Gamma$ has a basis consisting of eigenvectors of A.

Let Γ have $v > 0$ vertices and average valency $\overline{k} = v^{-1} \sum_{\gamma \in \Gamma} k(\gamma)$. Let A have eigenvalues $\theta_1 \geqslant \theta_2 \geqslant \cdots \geqslant \theta_v$.

3.2.1. Lemma. *We have $\theta_1 \geqslant \overline{k}$, with equality if and only if Γ is regular.*

Proof. Let $\{u_i \mid 1 \leqslant i \leqslant v\}$ be an orthonormal set of eigenvectors of A, where $Au_i = \theta_i u_i$. We can express the all-1 vector $\mathbf{1}$ in terms of this basis: $\mathbf{1} = \sum a_i u_i$. Now $v = \mathbf{1}^\top \mathbf{1} = \sum a_i^2$ and $v\overline{k} = \mathbf{1}^\top A \mathbf{1} = \sum a_i^2 \theta_i$, so that $\overline{k} \leqslant \theta_1$ with equality if and only if $a_i(\theta_1 - \theta_i) = 0$ for all i, i.e., if and only if $A\mathbf{1} = \theta_1 \mathbf{1}$. □

3.2.2. Proposition. *Γ is regular if and only if $\sum_{i=1}^{v} \theta_i^2 = v\theta_1$. If this is the case, then the degree of Γ is $k = \theta_1$, and the multiplicity $m(\theta_1)$ of θ_1 is the number of connected components of Γ.*

Proof. We have $\sum_{i=1}^{v} \theta_i^2 = \operatorname{tr} A^2 = \sum_{\gamma \in \Gamma} k(\gamma) = v\overline{k}$, and the first part of the statement follows from the above lemma. Concerning the second part, the spectrum of a disconnected graph is the union of the spectra of its components (counting multiplicities), while the adjacency matrix of a connected graph is irreducible, so that by the Perron-Frobenius Theorem 3.1.1 its largest eigenvalue has multiplicity one. □

3.2.3. Proposition. *Let Γ be connected with largest eigenvalue θ_1. Then $-\theta_1$ is an eigenvalue if and only if Γ is bipartite. If this is the case then for each eigenvalue θ of Γ, $-\theta$ is also an eigenvalue, with the same multiplicity.*

Proof. This follows immediately from part (iii) of the Perron-Frobenius Theorem 3.1.1 if we note that the adjacency matrix of a connected graph is primitive unless the graph is bipartite, in which case this matrix has period 2. □

3.2.4. Proposition. *Let Γ be a connected graph, and let Δ be a nonempty proper subgraph of Γ. Then the largest eigenvalue of Δ is strictly less than the largest eigenvalue of Γ.*

Proof. This follows immediately from part (v) of Theorem 3.1.1. □

This proposition easily leads to the determination of all graphs with largest eigenvalue at most 2. The names given to these graphs come from the theory of Lie groups, where they appear as extended Dynkin diagrams of type A, D, E; see also §3.10 and §10.1.

3.2.5. Theorem (SMITH [686], cf. LEMMENS & SEIDEL [483]). *The only connected graphs having largest eigenvalue 2 are the following graphs (the number of vertices is one more than the index given).*

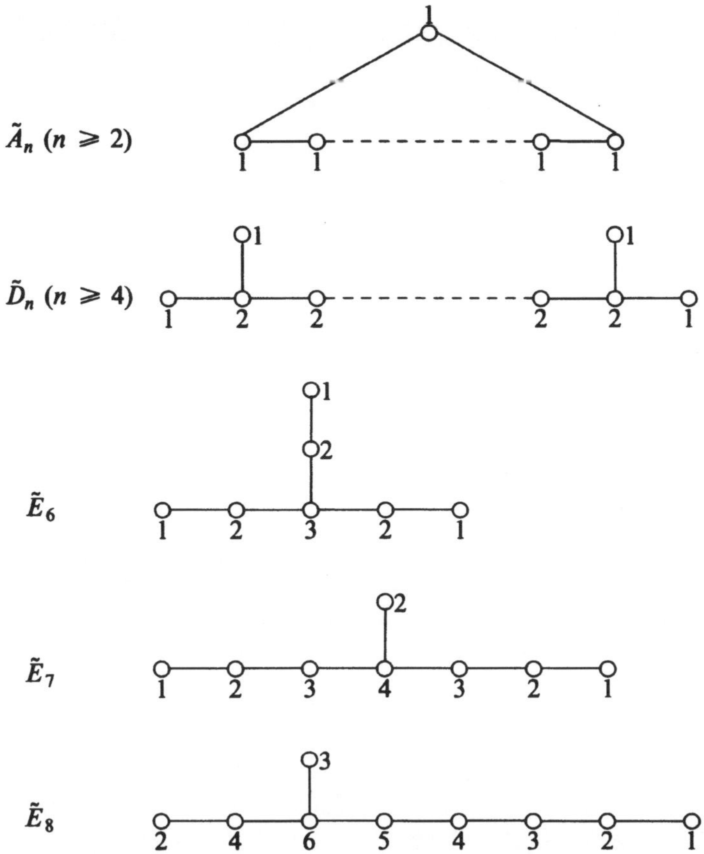

For each graph, the corresponding eigenvector is indicated by the integers at the vertices. Moreover, each connected graph with largest eigenvalue < 2 *is a subgraph of one of the above graphs, and each connected graph with largest eigenvalue* > 2 *contains one of these graphs.*

Proof. First we remark that the graphs in the theorem have largest eigenvalue 2. Indeed, twice the integer at any vertex equals the sum of the integers at the adjacent vertices. Next we observe that a 4-claw is of type $\tilde{D}_4$. Now, given any connected graph Γ, by considering the possible valencies of the vertices of Γ it is easy to show that either Γ is a subgraph of one of the graphs in the theorem, or Γ contains one of those graphs as a subgraph. However if the largest eigenvalue of Γ is $\leqslant 2$, then it follows from Proposition 3.2.4 that Γ cannot contain one of the graphs in the theorem as a proper subgraph. This proves the theorem. □

Remarks. (i) We can give the spectra of the graphs with largest eigenvalue ≤ 2 explicitly. The graphs A_n ($n \geq 0$), D_n ($n \geq 4$) and E_n ($n = 6, 7, 8$) (for pictures, see Table 10.1) have eigenvalues $2\cos(\pi m_j / h)$ ($1 \leq j \leq n$), where h is the Coxeter number and the m_j are the *exponents* of the corresponding Coxeter system ($m_j = d_j - 1$), see Table 10.3. In particular, their largest eigenvalues are $2\cos(\pi / h)$.

[Indeed, for a Coxeter element $c = w'w''$, with w' and w'' as in the remark following 10.4.4, we find that $c + c^{-1}$ has matrix $A^2 - 2I$, where A is the adjacency matrix of our graph. But it is known (cf. BOURBAKI [113], Ch. V, §6.2) that c has eigenvalues $\exp(2\pi i m_j / h)$ ($1 \leq j \leq n$). On the other hand, it is straightforward to determine the spectrum of A directly (cf. CVETKOVIĆ & GUTMAN [236]), so this is a way of determining the exponents for these Coxeter systems. See also GOODMAN, DE LA HARPE & JONES [335].]

The graphs $\tilde{A}_n$ ($n \geq 2$) have eigenvalues $2\cos(2\pi j / (n+1))$ ($0 \leq j \leq n$); the graphs $\tilde{D}_n$ ($n \geq 4$) have eigenvalues 2, 0, 0, -2, $2\cos(\pi j / (n-2))$ ($1 \leq j \leq n-3$); $\tilde{E}_6$ has spectrum 2, 1^2, 0, $(-1)^2$, -2; $\tilde{E}_7$ has spectrum 2, $\sqrt{2}$, 1, 0^2, -1, $-\sqrt{2}$, -2 and $\tilde{E}_8$ has spectrum 2, τ, 1, τ^{-1}, 0, $-\tau^{-1}$, -1, $-\tau$, -2, where $\tau = (1+\sqrt{5})/2 = 2\cos(\pi / 5)$.

(ii) It is possible to extend this result a bit, and classify graphs with largest eigenvalue at most $\tau^{3/2} = \sqrt{2+\sqrt{5}} \approx 2.058171$. Indeed, CVETKOVIĆ, DOOB & GUTMAN [233] show that if Γ has largest eigenvalue θ and $2 < \theta \leq \tau^{3/2}$, then Γ is a tree without vertices of degree ≥ 4 and with at most two vertices of degree 3, and a precise classification of such graphs is derived in BROUWER & NEUMAIER [135]. On the other hand, SHEARER [665] shows that any real number $r \geq \tau^{3/2}$ is a limit point of a set of spectral radii of graphs (in fact of trees), and hence classification beyond $\tau^{3/2}$ seems hopeless.

We give one other result on the behaviour of largest eigenvalues. A *pendant path* π of a graph Γ is an induced path one of whose end points has degree one in Γ, and such that its other end point is the only vertex of π adjacent to some vertex in $\Gamma \setminus \pi$. Given an edge $\gamma\delta$ in a graph $\Gamma = (V,E)$, the graph obtained from Γ by *subdividing* the edge $\gamma\delta$ is the graph $\Delta = (V \cup \{\infty\}, (E \setminus \{\gamma\delta\}) \cup \{\gamma\infty, \infty\delta\})$, where ∞ is some new symbol.

3.2.6. Proposition (HOFFMAN & SMITH [402]). *Let Γ be a connected graph, and assume that Δ is obtained from Γ by subdividing an edge $\gamma\delta$ of Γ. Let $\theta_1(\Gamma)$ and $\theta_1(\Delta)$ be the largest eigenvalues of Γ and Δ, respectively. Then*

(i) $\theta_1(\Gamma) < \theta_1(\Delta)$ *if $\gamma\delta$ is an edge of a pendant path of Γ;*

(ii) $\theta_1(\Gamma) = \theta_1(\Delta) = 2$ *if $\gamma\delta$ is not in a pendant path, and Γ is isomorphic to $\tilde{A}_n$ or $\tilde{D}_n$ for some n;*

(iii) $\theta_1(\Gamma) > \theta_1(\Delta)$ *otherwise.*

Proof. (i) follows from Proposition 3.2.4, and (ii) is obvious. For (iii), first observe that by Theorem 3.2.5 we have $\theta_1 := \theta_1(\Gamma) > 2$. Let $Ax = \theta_1 x$, where A is the adjacency matrix of Γ, and $x > 0$. Put $a = x_\gamma$ and $b = x_\delta$. We may suppose that $a \leq b$, and, in case γ is a vertex of degree 2 adjacent to δ and ϵ, that $a \leq x_\epsilon$. (For: when $\Gamma(\gamma) = \{\delta,\epsilon\}$, then the graphs obtained from Γ by subdividing $\gamma\delta$ and $\gamma\epsilon$, respectively, are isomorphic.) Let B be the adjacency matrix of Δ, and y a vector with $y_\infty = a$, $y_\gamma = c$, and $y_\alpha = x_\alpha$ for $\alpha \in \Gamma$, $\alpha \neq \gamma$, where c is still to be determined. Then $y \geq 0$ and $By \leq \theta_1 y$ when $0 \leq c \leq a$, $b + c \leq \theta_1 a$ and $a + \theta_1 a - b \leq \theta_1 c$. We can satisfy these inequalities for c precisely when $(a + \theta_1 a - b)/\theta_1 \leq \theta_1 a - b$, i.e., when $\theta_1 a - b \geq a/(\theta_1 - 1)$. But if γ has degree 2, then $\theta_1 a - b = x_\epsilon \geq a$

$> a / (\theta_1 - 1)$, and if γ has degree at least 3, then $\theta_1 a - b \geqslant 2a / \theta_1 > a / (\theta_1 - 1)$. Thus, we can always find a vector y with $y \geqslant 0$, $By \leqslant \theta_1 y$, $By \neq \theta_1 y$, and by Theorem 3.1.1 (iv) it follows that $\theta_1(\Delta) < \theta_1$. □

3.3 Interlacing

The following theorem is a classical result, cf. FISCHER [286] and COURANT & HILBERT [220] (Vol. 1, Ch. I, §4). Our treatment follows HAEMERS [340] (Ch. I).

3.3.1. Theorem. *Let S be a real $n \times m$ matrix such that $S^\top S = I$. Let A be a real symmetric matrix of order n, and define $B = S^\top AS$. Let the eigenvalues of A be $\theta_1 \geqslant \theta_2 \geqslant \cdots \geqslant \theta_n$ and let those of B be $\eta_1 \geqslant \eta_2 \geqslant \cdots \geqslant \eta_m$.*

(i) 'Interlacing': *We have $\theta_j \geqslant \eta_j \geqslant \theta_{n-m+j}$ $(1 \leqslant j \leqslant m)$.*

(ii) *If for some j equality holds in either inequality, then there is a vector v such that $Bv = \eta_j v$ and $ASv = \eta_j Sv$.*

(iii) *If, for some integer $l \geqslant 1$, we have $\eta_j = \theta_j$ for $1 \leqslant j \leqslant l$, then for each vector v such that $Bv = \eta_l v$ we have $ASv = \eta_l Sv$.*

(iv) *If, for some integer l, we have $\eta_j = \theta_j$ for $1 \leqslant j \leqslant l$ and $\eta_j = \theta_{n-m+j}$ for $l+1 \leqslant j \leqslant m$, then $SB = AS$.*

Proof. Let $u_1, \cdots, u_n$ be an orthonormal basis of eigenvectors of A such that $Au_i = \theta_i u_i$ for all i, and let $v_1, \cdots, v_m$ be such a basis for B, with $Bv_j = \eta_j v_j$ for all j. Write $U_i = \langle u_1, \cdots, u_i \rangle$ and $V_j = \langle v_1, \cdots, v_j \rangle$.
By expanding the vector v on the basis $\{v_1, \cdots, v_m\}$ one sees that if $v \in V_j$, then $v^\top Bv \geqslant \eta_j v^\top v$. Similarly it follows that if $u \in U_{j-1}{}^\perp$, then $u^\top Au \leqslant \theta_j u^\top u$. In both cases, equality implies that we have an eigenvector.
(i) Let v be a nonzero vector in $V_j \cap (S^\top U_{j-1})^\perp$ (this is a subspace of dimension at least one). Then $Sv \in U_{j-1}{}^\perp$. Now

$$\eta_j \leqslant \frac{v^\top Bv}{v^\top v} = \frac{v^\top S^\top ASv}{v^\top S^\top Sv} \leqslant \theta_j.$$

Applying the same argument to $-A$ and $-B$ instead of A and B, one finds $\eta_j \geqslant \theta_{n-m+j}$. This proves (i), and (ii) follows immediately from the above.
(iii) Induction on l. We may suppose that $\eta_l \neq \eta_{l-1}$. Now $U_{l-1} = SV_{l-1}$ so that $v \in V_l \cap V_{l-1}{}^\perp = V_l \cap (S^\top U_{l-1})^\perp$, and the above applies.
(iv) We have $SBv_j = \eta_j Sv_j = ASv_j$ for all j. Since the v_j form a basis, it follows that $SB = AS$. □

Remark. This theorem generalizes directly to complex Hermitean matrices instead of real symmetric matrices (with conjugate transpose instead of transpose) with virtually the same proof.

If the assumptions of (iv) hold, we say that the interlacing is *tight*. Below we formulate some corollaries in terms of graphs, since that is what we are mainly interested in. Of course analogous statements hold for arbitrary real symmetric matrices (indeed, in §2.3 we saw how to derive the Krein conditions from such an interlacing result).

3.3.2. Corollary. *Let Γ be a graph and Δ an induced subgraph. Then the eigenvalues of Δ interlace those of Γ.*

Proof. Apply the above theorem, where A and B are the adjacency matrices of Γ and Δ, and $S_{\alpha\beta}$ equals 1 when $\alpha = \beta \in \Delta$, and is 0 otherwise. □

Remark (personal communication by W.H. Haemers). As an application we can show that if Γ is a strongly regular graph that is not complete multipartite, then for each vertex $\gamma \in \Gamma$ the subgraph $\Gamma_2(\gamma)$ is connected. Indeed, if not, then its largest eigenvalue $k - \mu$ would have multiplicity at least two, and hence would be not larger than the second largest eigenvalue r of Γ. Thus, $x^2 + (\mu - \lambda)x + \mu - k \leqslant 0$ for $x = k - \mu$, i.e., $(k - \mu)(k - \lambda - 1) \leqslant 0$, contradiction.

3.3.3. Corollary. *Let Γ be a graph, and $\Pi = \{X_1, \cdots, X_m\}$ be a partition of its vertex set into nonempty parts. Let B_{ij} be the average number of neighbours in X_j of a vertex in X_i. Then the eigenvalues of B interlace those of Γ. If the interlacing is tight, then each vertex in X_i has precisely B_{ij} neighbours in X_j, i.e., the partition Π is regular in the sense of §A.4.*

Proof. Apply the above theorem to the adjacency matrix A of Γ and the matrix S such that $S_{\gamma i}$ equals $|X_i|^{-1/2}$ if $\gamma \in X_i$ and is 0 otherwise. (Then $S^{\mathsf{T}}AS = D^{-1}BD$ if D is the diagonal matrix with $D_{ii} = |X_i|^{1/2}$, so that B has the same eigenvalues as $S^{\mathsf{T}}AS$.) □

For more detailed eigenvalue inequalities, see Haemers [340].

3.4 Gram matrices

Real symmetric $n \times n$-matrices G are in bijective correspondence with quadratic forms q on $\mathbf{R}^n$ via the relation

$$q(x) = x^{\mathsf{T}}Gx \quad (x \in \mathbf{R}^n).$$

Two quadratic forms q and q' on $\mathbf{R}^n$ are *congruent*, i.e., there is a nonsingular $n \times n$-matrix S such that $q(x) = q'(Sx)$ for all $x \in \mathbf{R}^n$, if and only if their corresponding matrices G and G' satisfy $G = S^{\mathsf{T}}G'S$. Moreover, this occurs for some S if and only if G and G' have the same rank and the same number of nonnegative eigenvalues (Sylvester [706]'s 'law of inertia for quadratic forms', cf. Gantmacher [308], Vol. 1, Chapter X, §2); recall the basic fact that every real symmetric square matrix has real eigenvalues only, and an orthonormal basis of eigenvectors. We shall now be concerned with matrices having nonnegative eigenvalues only.

3.4.1. Lemma. *Let G be a real symmetric $n \times n$-matrix. Then* (i) *and* (ii) *are equivalent:*

(i) *For all $x \in \mathbf{R}^n$, $x^{\mathsf{T}}Gx \geqslant 0$.*

(ii) *All eigenvalues of G are nonnegative.*

Proof. There is an orthogonal matrix Q and a diagonal matrix D whose nonzero entries are the eigenvalues of G such that $G = Q^{\mathsf{T}}DQ$. If (ii) holds, then $x^{\mathsf{T}}Gx = (Qx)^{\mathsf{T}}D(Qx) \geqslant 0$ implies (i). Conversely, (ii) follows from (i) by choosing x to be an eigenvector. □

A symmetric $n \times n$-matrix G satisfying (i) and (ii) is called *positive semidefinite.* It is called *positive definite* when $x^{\mathsf T}Gx = 0$ implies $x = 0$, or, equivalently, when all its eigenvalues are positive. For any collection X of vectors of $\mathbf{R}^n$, we define its *Gram matrix* as the square matrix G indexed by X (or by some index set for X) whose (x,y)-entry G_{xy} is the inner product $(x,y) = x^{\mathsf T}y$. This matrix is always positive semidefinite, and it is definite if and only if the vectors in X are linearly independent. (Indeed, if we use H to denote the $n \times |X|$-matrix whose columns are the vectors of X, then $G = H^{\mathsf T}H$, and $x^{\mathsf T}Gx = (Hx,Hx) \geqslant 0$.) Conversely, if G is an arbitrary symmetric positive semidefinite matrix of rank n then G is congruent to a diagonal matrix consisting of zeros and n ones,

$$G = (S_{11}^{\mathsf T} \;\; S_{12}^{\mathsf T})\begin{pmatrix} I & 0 \\ 0 & 0 \end{pmatrix}\begin{pmatrix} S_{11} \\ S_{12} \end{pmatrix} = S_{11}^{\mathsf T}\, S_{11},$$

so that G is the Gram matrix of the set of columns of S_{11} in $\mathbf{R}^n$.

3.5 Graph representations

A (*Euclidean*) *representation* of a graph Γ is a map $\rho : \Gamma \to \mathbf{R}^n$ such that the images of adjacent vertices $\gamma, \delta \in \Gamma$ have constant squared distance.

In the sequel, we usually write $\overline{\gamma}$ for the image $\rho(\gamma)$ of a vertex γ under ρ, and

$$[\gamma,\delta] := (\overline{\gamma} - \overline{\delta}, \overline{\gamma} - \overline{\delta})$$

for the squared distance of the images of $\gamma, \delta \in \Gamma$. Note that 'squared distance' refers to the Euclidean distance in $\mathbf{R}^n$, whereas otherwise 'distance' means graph distance. The representation is called *faithful* when ρ is injective.

3.5.1. Lemma. *Let* $\gamma \mapsto \overline{\gamma}$ $(\gamma \in \Gamma)$ *be a representation of a graph* Γ. *Then, for* $\gamma, \gamma_1, \gamma_2, \delta, \delta_1, \delta_2 \in \Gamma$, *we have:*

$$\begin{aligned}
(\overline{\gamma},\overline{\delta}) &= \tfrac{1}{2}((\overline{\gamma},\overline{\gamma}) + (\overline{\delta},\overline{\delta}) - [\gamma,\delta]),\\
(\overline{\gamma}_1 - \overline{\gamma}_2,\overline{\delta}) &= \tfrac{1}{2}((\overline{\gamma}_1,\overline{\gamma}_1) - (\overline{\gamma}_2,\overline{\gamma}_2) - [\gamma_1,\delta] + [\gamma_2,\delta]),\\
(\overline{\gamma}_1 - \overline{\gamma}_2,\overline{\delta}_1 - \overline{\delta}_2) &= \tfrac{1}{2}([\gamma_1,\delta_2] + [\gamma_2,\delta_1] - [\gamma_1,\delta_1] - [\gamma_2,\delta_2]).
\end{aligned}$$

Proof. Straightforward. □

A representation is called *spherical* if all image vectors have the same norm $(\overline{\gamma},\overline{\gamma}) =: p$; then the images of adjacent vertices γ,δ have constant inner product $(\overline{\gamma},\overline{\delta}) = p - a/2 =: q$, where a is the constant squared distance. A (p,q,r)-*representation* of Γ is a representation with the property

$$(\overline{\gamma},\overline{\delta}) = \begin{cases} p & \text{if } \gamma = \delta, \\ q & \text{if } \gamma \sim \delta, \\ r & \text{if } \gamma \not\sim \delta. \end{cases}$$

If only the weaker requirement

$$(\bar{\gamma},\bar{\delta}) = \begin{cases} p & \text{if } d(\gamma,\delta) = 0, \\ q & \text{if } d(\gamma,\delta) = 1, \\ r & \text{if } d(\gamma,\delta) = 2 \end{cases}$$

holds, then we call the representation a *weak (p,q,r)-representation*. Thus, (weak) (p,q,r)-representations are spherical. Note that the Cauchy-Schwarz inequality implies $|q| \leqslant p$ if Γ contains adjacent points, and $|r| \leqslant p$ if Γ contains nonadjacent points (or, in the weak case, points at distance 2).

Example. The cross-polytope, described by an orthonormal basis and their negatives, is a $(1,0,-1)$-representation of the graph $K_{n\times 2}$, and the standard representation of an n-dimensional cube by $(1,-1)$-vectors is a weak $(n,n-2,n-4)$-representation.

From a given representation one can easily construct other representations by scaling and shifting. Indeed, if t_1,t_2 are nonnegative numbers, one may choose a vector c of norm t_2 orthogonal to all $\bar{\gamma}$ ($\gamma \in \Gamma$) - if necessary by extending $\mathbf{R}^n$ to $\mathbf{R}^{n+1}$ - and consider the representation ρ' with $\rho'(\gamma) = c + \sqrt{t_1}\,\rho(\gamma)$. Thus, the shift is by c, and the scaling by $\sqrt{t_1}$. If ρ is a (weak) (p,q,r)-representation then ρ' is a (weak) $(pt_1+t_2, qt_1+t_2, rt_1+t_2)$-representation. If $q \neq r$ then the number

$$b = (p-q)/(q-r),$$

called the *base* of the representation, is invariant under scaling and shifting.

Spherical representations of a graph Γ are closely related to the extremal eigenvalues of Γ, i.e., of its adjacency matrix A.

3.5.2. Proposition. *Let Γ be a nonempty graph. Then Γ has a $(p, 1,0)$-representation if and only if its smallest eigenvalue is $\geqslant -p$, and a $(p, -1,0)$-representation if and only if its largest eigenvalue is $\leqslant p$.*

Proof. If Γ has smallest eigenvalue $\geqslant -p$, then the matrix $pI+A$ is symmetric and positive semidefinite and hence can be interpreted as the Gram matrix of a set of vectors in $\mathbf{R}^f$ for some f. Clearly, this gives a $(p, 1,0)$-representation of Γ. Conversely, the Gram matrix of the image of Γ under a $(p, 1,0)$-representation is $G = pI+A$, and since this is positive semidefinite, its eigenvalues are nonnegative, and hence those of A are $\geqslant -p$. The second statement follows in the same way, considering $G = pI-A$. □

3.5.3. Proposition. *Let Γ be a regular graph with $v > 1$ points, valency k, second largest eigenvalue θ_2, and smallest eigenvalue θ_v.*

(i) *If $M \geqslant \theta_2$ and $a \geqslant (k-M)/v$, then Γ has an $(M+a,a-1,a)$-representation; in particular, Γ always has a $(\theta_2+1,0,1)$-representation.*

(ii) *If $-m \leqslant \theta_v$ and $a \leqslant (k+m)/v$, then Γ has an $(m-a,1-a,-a)$-representation.*

(iii) *If Γ has a (p,q,r)-representation in $\mathbf{R}^n$ with $p > q \neq r$, then (putting $b = (p-q)/(q-r)$) we have: if $b > 0$ then $\theta_v \geqslant -b-1$, and if $b < 0$ then $\theta_2 \leqslant -b-1$.*

Proof. Since $AJ = JA = kJ$, the eigenvalues of $G = MI-A+aJ$ are $M-k+av$ and $M-\theta_i$ ($2 \leqslant i \leqslant v$). Hence, if $M \geqslant \theta_2$ and $a \geqslant (k-M)/v$, then G is positive semidefinite and defines an $(M+a,a-1,a)$-representation. The second statement of

(i) follows by taking $M = \theta_2$, $a = 1$. Hence (i) holds, and (ii) follows similarly by considering $G' = mI + A - aJ$. Finally, if Γ has a (p,q,r)-representation, then the Gram matrix G of the image of Γ can be expressed in terms of the adjacency matrix as

$$G = pI + qA + r(J - I - A) = (p-r)I + (q-r)A + rJ.$$

If $b > 0$, then $(b+1)I + A + \frac{r}{q-r}J$ is positive semidefinite, and its eigenvalue $b+1+\theta_v$ must be nonnegative. If $b < 0$, then $-(b+1)I - A + \frac{r}{r-q}J$ is positive semidefinite, and its eigenvalue $-b-1-\theta_2$ is nonnegative. □

Remark. (i) We can be slightly more precise. If $v > n$, then G has eigenvalue 0 with multiplicity at least $v-n$. But $G\mathbf{1} = 0$ if and only if $\sum_{\gamma\in\Gamma} \bar{\gamma} = 0$, or, equivalently, if and only if $p + qk + r(v-k-1) = 0$. Putting $\epsilon = 1$ if this is the case, and $\epsilon = 0$ otherwise, we see that the eigenvalue $-b-1$ has multiplicity at least $v-n-\epsilon$. In particular, if $v-n-\epsilon > 0$, then $-b-1$ equals θ_v (if $b > 0$) or θ_2 (if $b < 0$). In case $-b-1 = k$ and $\epsilon = 1$ (i.e., in case $r = 0$) we even find that this eigenvalue has multiplicity at least $v-n$, so that Γ has at least $v-n$ components.
(ii) If $r = 0$ then we need not require Γ to be regular in part (iii). If moreover $b < 0$, the conclusion even becomes $\theta_1 \leqslant -b-1$.

3.5.4. Corollary.

(i) *A graph with smallest eigenvalue $\geqslant -1$ is a disjoint union of complete graphs.*

(ii) *A regular graph with second largest eigenvalue $\leqslant 0$ is a complete multipartite graph.*

Proof. If a graph Γ has a $(1,1,0)$-representation (or a $(1,0,1)$-representation), then its vertices are adjacent (or nonadjacent) if and only if they have the same image. Now, (i) follows from Proposition 3.5.2, and (ii) from Proposition 3.5.3. □

For a more general result see Corollary 3.12.3.

An *affine (a,b)-representation* of a graph Γ is a representation $\gamma \mapsto \bar{\gamma}$ $(\gamma \in \Gamma)$ whose squared distances satisfy:

$$[\gamma,\delta] = \begin{cases} a \text{ if } d(\gamma,\delta) = 1, \\ b \text{ if } d(\gamma,\delta) = 2. \end{cases}$$

An affine (a,b)-representation is called *strict* if $[\gamma,\delta] \in \{0,a,b\}$ implies $d(\gamma,\delta) \leqslant 2$, and *complete* if $[\gamma,\delta] \in \{0,a,b\}$ for all γ, δ. In particular, any affine representation of a graph of diameter at most two is strict and complete.

The image of a graph in a complete affine (a,b)-representation, $a \neq b$, is a *2-distance set*, i.e., a set X of vectors such that distinct points from X have two distinct distances only (cf. Larman, Rogers & Seidel [475], Seidel [657], Blokhuis [88] and Neumaier [550, 553, 564]). Conversely, if X is an arbitrary 2-distance set in $\mathbb{R}^n$ with distances a and b, then the graph Γ whose vertices are the vectors in X, adjacent if they have distance a, has the identity as a complete affine (a,b)-representation. We also note that every weak (p,q,r)-representation is an affine $(2p-2q, 2p-2r)$-representation.

3.5.5. Lemma. *Let Γ be a graph with an affine (a,b)-representation. If $0 < a < b$, then $\overline{\gamma} = \overline{\delta}$ implies $\gamma = \delta$ or $d(\gamma,\delta) \geqslant 4$.*

Proof. Suppose $\overline{\gamma} = \overline{\delta}$ and $\gamma \neq \delta$. If $d(\gamma,\delta) = 1$, then $a = (\overline{\gamma}-\overline{\delta},\overline{\gamma}-\overline{\delta}) = 0$. If $d(\gamma,\delta) = 2$, then $b = (\overline{\gamma}-\overline{\delta},\overline{\gamma}-\overline{\delta}) = 0$. If $d(\gamma,\delta) = 3$, then with $\alpha \in \Gamma(\gamma) \cap \Gamma_2(\delta)$, we have $a = (\overline{\gamma}-\overline{\alpha},\overline{\gamma}-\overline{\alpha}) = (\overline{\alpha}-\overline{\delta},\overline{\alpha}-\overline{\delta}) = b$. In each case, we have a contradiction. Hence the result. □

3.5.6. Proposition. *Let Γ have an affine (a,b)-representation. Then:*

(i) *For all $\gamma \in \Gamma$, the neighbourhood $\Gamma(\gamma)$ has an $(a, a/2, a-b/2)$-representation.*

(ii) *If Γ is regular and the images of nonadjacent points of Γ have squared distance b, then there is a positive number p such that Γ has a $(p, p-a/2, p-b/2)$-representation.*

Proof. As usual, $\overline{\gamma}$ denotes the image of $\gamma \in \Gamma$ under the representation.
(i) If $\alpha,\beta \in \Gamma(\gamma)$ then by Lemma 3.5.1, $(\overline{\alpha}-\overline{\gamma},\overline{\beta}-\overline{\gamma}) = \frac{1}{2}([\alpha,\gamma]+[\beta,\gamma]-[\alpha,\beta]-[\gamma,\gamma])$ $= a-\frac{1}{2}[\alpha,\beta]$. Hence the representation of $\alpha \in \Gamma(\gamma)$ by $\overline{\alpha}-\overline{\gamma}$ defines an $(a, a/2, a-b/2)$-representation of $\Gamma(\gamma)$.
(ii) Suppose that Γ has v points and valency k. By the same lemma, the center of mass $c = v^{-1}\sum_{\gamma\in\Gamma}\overline{\gamma}$ satisfies

$$(\overline{\alpha},c) = \frac{1}{2v}\left[\sum_{\gamma\in\Gamma}((\overline{\alpha},\overline{\alpha})+(\overline{\gamma},\overline{\gamma})-[\alpha,\gamma])\right] = \tfrac{1}{2}(\overline{\alpha},\overline{\alpha})+\frac{1}{2v}h,$$

where

$$h = \sum_{\gamma\in\Gamma}(\overline{\gamma},\overline{\gamma})-ka-(v-1-k)b$$

is independent of $\alpha \in \Gamma$. Therefore, with $p = (c,c)-h/v$, using the lemma once again, we find

$$(\overline{\alpha}-c,\overline{\beta}-c) = (\overline{\alpha},\overline{\beta})-(\overline{\alpha},c)-(\overline{\beta},c)+(c,c) = p-\tfrac{1}{2}[\alpha,\beta].$$

This implies that the representation $\alpha \mapsto \overline{\alpha}-c$ $(\alpha \in \Gamma)$ is a $(p, p-a/2, p-b/2)$-representation of Γ. □

3.6 The absolute bound

By a theorem of DELSARTE, GOETHALS & SEIDEL [258], the number v of vertices of a spherical s-distance set in $\mathbb{R}^n$ satisfies the so-called *absolute bound*

$$v \leqslant \binom{n+s-1}{n-1}+\binom{n+s-2}{n-1}.$$

Since (p,q,r)-representations of graphs provide examples of spherical 2-distance sets, we can apply this for $s = 2$ to graph representations. Here we give another proof of this bound for spherical 2-distance sets, and sharpen it in a special case.

3.6.1. Theorem. *Let Γ be a graph on v vertices with a (p,q,r)-representation in $\mathbb{R}^n$, $p \neq q, r$. Then Γ has at most $\frac{1}{2}n(n+3)$ vertices. Moreover, if Γ is a disjoint union of cliques, then $v \leqslant n+1$, except when $s = (p-q)/(r-q)$ is an integer > 1; in this case $v \leqslant n+t-1 \leqslant 2n-2$, where t is the number of maximal cliques of size s.*

Proof. Let P be the space of polynomials of degree at most 2 in n variables with constant coefficient zero. P contains the v polynomials

$$p_\gamma(z) = p((\bar{\gamma},z)-q)((\bar{\gamma},z)-r)+qr((z,z)-p),$$

and since $p_\gamma(\bar{\gamma}) = p(p-q)(p-r) \neq 0$ and $p_\gamma(\bar{\delta}) = 0$ for $\delta \neq \gamma$, these polynomials are linearly independent. Hence $v \leqslant \dim P = \binom{n+1}{2}+n = \frac{1}{2}n(n+3)$.
Now suppose that Γ is a disjoint union of r cliques, and let G be the Gram matrix of the image of Γ. The matrix G has two distinguished invariant subspaces X and Y, namely the $(v-r)$-dimensional space X consisting of the vectors x with $\sum_{\gamma\in C} x_\gamma = 0$ for each maximal clique C of Γ, and the r-dimensional space Y spanned by the characteristic vectors y_C of the maximal cliques. The Gram matrix of these characteristic vectors is $\bar{G} = (p-q)D+(q-r)D^2+rDJD$, where D is the diagonal matrix whose diagonal entries are the clique sizes of Γ. Since $p \neq q$, a straightforward calculation shows that $\operatorname{rk} \bar{G} \geqslant r-1$ unless $s = (p-q)/(r-q)$ is an integer > 1; in this case $\operatorname{rk} \bar{G} = r+1-t$, where t is the number of maximal cliques of size s. Now $y^\mathsf{T}Gy = \eta^\mathsf{T}\bar{G}\eta$ for $y = \sum \eta_C y_C \in Y$, and $x^\mathsf{T}Gx = (p-q)x^\mathsf{T}x \neq 0$ for $x \in X\setminus\{0\}$. Since $\mathbf{R}^n$ is the orthogonal direct sum of the invariant subspaces X and Y, this implies $n \geqslant \operatorname{rk} G = \dim X + \operatorname{rk} \bar{G}$. Hence $v = \dim X + r \leqslant n+r-\operatorname{rk}\bar{G}$, i.e., $v \leqslant n+1$ unless s is an integer > 0, and then $v \leqslant n+t-1$. In the latter case $t \leqslant v/2$ implies $t \leqslant n-1$ and hence $v \leqslant n+t-1 \leqslant 2n-2$. □

3.7 Representations of subgraphs

In this section we derive results about representations of subgraphs of Γ given a representation of the graph Γ. First a technical result about submatrices of a Gram matrix. The statements of the proposition below are weak forms of so-called 'interlacing' results, see §3.3.

3.7.1. Proposition. *Let G be a positive semidefinite $n\times n$-matrix. Then:*

(i) $\det G \geqslant 0$,

(ii) *If* $G = \begin{bmatrix} G_{11} & G_{12} \\ G_{21} & G_{22} \end{bmatrix}$, *where G_{11} is a square submatrix, then for every matrix C of the same size as G_{12}, the matrix $H = G_{22} - C^\mathsf{T}G_{12} - G_{21}C + C^\mathsf{T}G_{11}C$ is positive semidefinite. In particular, if G_{11} is nonsingular, then the matrix*

$$G/G_{11} := G_{22} - G_{21}G_{11}^{-1}G_{12}$$

is positive semidefinite.

(iii) *For every symmetric partition of G into $n_i\times n_j$-matrices $G_{ij} = G_{ji}^\mathsf{T}$, i.e.,*

$$G = \begin{bmatrix} G_{11} & G_{12} & \cdots & G_{1l} \\ \vdots & \vdots & & \vdots \\ G_{l1} & G_{l2} & \cdots & G_{ll} \end{bmatrix},$$

where $\sum n_i = n$, the matrix $\bar{G}$ of dimension $l\times l$, whose (i,j)-entry is the arithmetic mean $\mathbf{1}^\mathsf{T}G_{ij}\mathbf{1}/n_in_j$ of the entries of G_{ij}, is positive semidefinite.

Proof. (i) holds since the determinant is the product of the eigenvalues.

To get (ii) we apply Lemma 3.4.1 (i) with $x = \begin{bmatrix} -Cy \\ y \end{bmatrix}$ and find $y^{\mathsf{T}}Hy \geqslant 0$ for all y, so that H is positive semidefinite; by putting $C = G_{11}^{-1}G_{12}$ (so that $C^{\mathsf{T}} = G_{21}G_{11}^{-1}$) we see that G / G_{11} is positive semidefinite.
Similarly, (iii) follows by applying, for given $z \in \mathbf{R}^{l}$, Lemma 3.4.1 (i) to the matrix $\overline{G}$ and taking for x the vector whose entries corresponding to the i-th block of the partition are $n_i^{-1}z_i$; for then $z^{\mathsf{T}}\overline{G}z \geqslant 0$ yields (iii). □

The matrix G / G_{11}, defined in (ii) above is called the *Schur complement* of G_{11} in G. The matrix $\overline{G}$ of (iii) is said to be the *condensed matrix* of G with respect to the given partition.

As a first application, let us prove the *Hoffman bound* on the size of a coclique (cf. Section 1.3):

3.7.2. Proposition (HOFFMAN [398]). *Let Γ be a connected graph with v vertices, regular with valency $k > 0$, and smallest eigenvalue $-m$. If Γ contains a coclique C, then $|C| \leqslant v / (1 + k / m)$. Moreover, equality implies that every point of $\Gamma \setminus C$ is adjacent to exactly m points of C.*

Proof. By Proposition 3.5.3 (ii), Γ has an $(m-a,\ 1-a,\ -a)$-representation, where $a = (k+m) / v$. Put $c = |C|$. The condensed Gram matrix of the image of C and a point $\gamma \in \Gamma \setminus C$ adjacent with e points of C is the 2×2-matrix

$$\overline{G} = \begin{bmatrix} mc^{-1} - a & ec^{-1} - a \\ ec^{-1} - a & m - a \end{bmatrix} .$$

Therefore $(1,0)\overline{G}(1,0)^{\mathsf{T}} = mc^{-1} - a \geqslant 0$ which gives $c \leqslant m / a = v / (1 + k / m)$. If equality holds, then $\det \overline{G} = -(ec^{-1} - a)^2 \geqslant 0$ implies that $e = ac = m$. □

Remark. This result also follows immediately from Corollary 3.3.3.

Clearly, if a graph Γ has a weak (p,q,r)-representation, then the same holds for every subgraph of Γ. For particular subgraphs, however, a sharper statement is possible.

3.7.3. Proposition. *Let Γ be a graph and $\gamma \in \Gamma$.*

(i) *If Γ has a weak (p,q,r)-representation, then $\Gamma(\gamma)$ has a $(p-s,\ q-s,\ r-s)$-representation, where $s = q^2 / p$.*

(ii) *If Γ has a (p,q,r)-representation, then $\Gamma \setminus \gamma^{\perp}$ has a $(p-t,\ q-t,\ r-t)$-representation, where $t = r^2 / p$.*

(iii) *If Γ has a weak (p,q,r)-representation, then $\Gamma_2(\gamma)$ has a weak $(p-t,\ q-t,\ r-t)$-representation, where $t = r^2 / p$.*

Proof. Let G be the Gram matrix of the image of $\Gamma(\gamma)$. Then the Gram matrix of the image of $\gamma^{\perp}$ is the matrix

$$\begin{bmatrix} p & q\mathbf{1}^{\mathsf{T}} \\ q\mathbf{1} & G \end{bmatrix}$$

with Schur complement $G' = G - sJ$, where $s = q^2 / p$. By Proposition 3.7.1 (ii), G' is positive semidefinite; hence it defines a representation of $\Gamma(\gamma)$. As points of $\Gamma(\gamma)$

have distance $\leqslant 2$ in Γ, this is a $(p-s, q-s, r-s)$-representation. Hence (i) holds. Parts (ii) and (iii) are proved in the same way. □

In Proposition 4.4.2 we shall show that many distance-regular graphs have a weak $(b^2, b, 1)$-representation with $b > 1$. Here we prove some geometric properties of graphs with such a representation.
Recall that if C is a singular subspace (i.e., a clique such that every singular line not contained in it intersects it in at most one vertex), then res (C) denotes the quotient graph obtained from $C^\perp \setminus C$ by identifying vertices α, β in the latter graph whenever $\alpha^\perp = \beta^\perp$ (cf. §A.8).

3.7.4. Proposition. *Let Γ be a graph with a weak $(b^2, b, 1)$-representation, $b > 1$, and suppose that every singular line of Γ has size $\geqslant s+1$. Let C be a singular subspace of Γ of size $c > 0$. Then:*

(i) res (C) *has a* $\left(\dfrac{b+sc}{1+(s-1)c}, 1, \dfrac{1-c}{b}\right)$*-representation.*

(ii) res (C) *has smallest eigenvalue* $\geqslant -\dfrac{b}{1+(s-1)c} - 1$; *in particular, the residue of a point has smallest eigenvalue* $\geqslant -1-b/s$.

(iii) *If* res (C) *contains more than one element, then*

$$sc(c-1-b) \leqslant (c-1)^2 + b^2.$$

Proof. An element $\gamma^* \in$ res (C) is an equivalence class of $\geqslant 1+(s-1)c$ points adjacent to all points of C, since with γ, all points on singular lines joining γ with C belong to γ^*. From each equivalence class we select a set of $1+(s-1)c$ points and consider the condensed Gram matrix $\overline{G}$ of the matrix of the representation restricted to the union of C with all of these sets. In terms of the adjacency matrix A_{res} of res (C) we find

$$\overline{G} = \begin{bmatrix} b + \dfrac{b^2-b}{c} & b\mathbf{1}^{\mathrm{T}} \\ b\mathbf{1} & J + (b-1)A_{\text{res}} + (b-1)\left(1 + \dfrac{b}{1+(s-1)c}\right)I \end{bmatrix}.$$

The Schur complement of $\overline{G}$ is the matrix

$$\overline{H} = (b-1)\left[A_{\text{res}} + \left(1 + \frac{b}{1+(s-1)c}\right)I - \frac{c-1}{c-1+b}J\right]$$

$$= \frac{(b-1)b}{c+b-1}\left[\frac{b+sc}{1+(s-1)c}I + A_{\text{res}} + \frac{1-c}{b}(J - I - A_{\text{res}})\right].$$

Scaling by $\dfrac{c+b-1}{(b-1)b}$ now gives (i), and the positive semidefiniteness of

$$A_{\text{res}} + \left[1 + \frac{b}{1+(s-1)c}\right]I = \frac{1}{b-1}\overline{H} + \frac{c-1}{c+b-1}J$$

yields (ii). Finally, if res (C) contains more than one point then, by definition, it contains two nonadjacent points. The Cauchy-Schwarz inequality for the representation (i) therefore implies that $\frac{c-1}{b} \leqslant \frac{b+sc}{1+(s-1)c}$ which gives the inequality in (iii). □

3.7.5. Theorem. *Let Γ be a graph with a weak $(b^2,b,1)$-representation, $b > 1$, and suppose that every singular line of Γ has size $\geqslant s+1$. Then:*

(i) *If $s \geqslant 3$ and $b \leqslant s^2-s+1$, or $s = 2$ and $b \leqslant 2$, then distinct maximal cliques intersect in a singular line, a point, or the empty set.*

(ii) *If $s \geqslant 4$ and $b \leqslant s+1$ then every edge is in at most two maximal cliques.*

Proof. If (i) is not valid then there are maximal cliques intersecting in a singular subspace C containing a proper triangle $\alpha\beta\gamma$. But then $c = |C| \geqslant s^2+s+1$, since all points on singular lines joining α with a point of the line through β and γ belong to C. Clearly, res (C) contains more than one point, so that by (iii) of the above proposition, $sc(c-1-b) \leqslant (c-1)^2+b^2$. For $s \geqslant 3$ we eliminate b using $b \leqslant s^2-s+1$, and obtain $(s-1)(c+s-1)(c-s^2-s-1) + (s-1)^3-2s \leqslant 0$, contradiction. For $s = 2$ we get a similar contradiction from $b \leqslant 2$. This proves (i).
Under the assumption of (ii), the conclusion of (i) holds, whence the residue of a singular line l is a coclique whose size r is just the number of maximal cliques containing l. In the $\left[\frac{b+s^2+s}{s^2}, 1, \frac{-s}{b}\right]$-representation of res (l) obtained from the above proposition with $c = s+1$, the condensed Gram matrix of the r-coclique is the 1×1-matrix with entry $\frac{1}{r}\left[\frac{b+s^2+s}{s^2}-(r-1)\frac{s}{b}\right]$. Since this must be a nonnegative number, $(r-1)s^3 \leqslant b(b+s^2+s) \leqslant (s+1)^3$, whence $r \leqslant 1+(\frac{s+1}{s})^3 \leqslant 1+(\frac{5}{4})^3 < 3$. Therefore $r \leqslant 2$, i.e., every singular line is in at most two maximal cliques. Since every edge is in a unique singular line, (ii) follows. □

In the same way one can prove that $b \leqslant s-2$ implies that every edge is in a unique maximal clique. Assuming that Γ is edge-regular, one can deduce from Corollary 3.12.3 below that this already holds under the weaker arithmetical assumption $b < s$.

3.8 Graph switching, equiangular lines, and representations of two-graphs

Spherical $(m,-1,1)$-representations of graphs are closely related to equiangular lines in Euclidean spaces. Indeed, the images of distinct vertices of a graph Γ under an $(m,-1,1)$-representation are vectors x,y of length m with inner product -1 or 1; hence their angle ϕ satisfies $\cos^2\phi = (x,y)^2/((x,x)(y,y)) = m^{-2}$. This implies that the lines along these vectors form a set $\mathfrak{L}$ of equiangular lines at mutual angle arccos $(1/m)$. Conversely, for any set $\mathfrak{L}$ of equiangular lines in $\mathbf{R}^n$ at mutual angle arccos $(1/m)$ we may choose a vector of norm m on each line; then for any two such vectors x, y $(x \neq y)$ we have $(x,y)^2 = (x,x)(y,y)\cos^2\phi = 1$ whence $(x,y) \in \{1,-1\}$. Therefore these vectors form an $(m,-1,1)$-representation of a suitable graph Γ.

Since each line contains two vectors of norm m, the above correspondence is not one-to-one. This ambiguity can be described in terms of graph switching. Let Γ be a graph, let S be a set of vertices of Γ, and let S^c be the complement $\Gamma\setminus S$. *Switching* Γ with respect to S (or S^c) is the operation of replacing all edges of Γ between S and S^c by the new edges $\{\gamma_1,\gamma_2\}$ $(\gamma_1 \in S,\ \gamma_2 \in S^c,\ \gamma_1 \not\sim \gamma_2)$, leaving the edges inside S and inside S^c (cf. SEIDEL [655]). If Γ^1 is obtained from Γ by switching with respect to S_1

and Γ^2 is obtained from Γ^1 by switching with respect to S_2, then Γ^2 can be directly obtained from Γ by switching with respect to the symmetric difference $(S_1 \cap S_2) \cup (S_1^c \cap S_2^c)$. Therefore switching defines an equivalence relation on the set of graphs with a given vertex set.

One easily sees that, if S is a set of vertices of a graph Γ with an $(m,-1,1)$-representation then the replacement of $\bar{\gamma}$ by $-\bar{\gamma}$ for all $\gamma \in S$ gives an $(m,-1,1)$-representation of the graph Γ' obtained from Γ by switching with respect to S. Conversely, it is clear that $(m,-1,1)$-representations of two graphs Γ and Γ' give rise to the same set of equiangular lines if and only if they are switching equivalent. Thus sets of equiangular lines at mutual angle arccos $(1/m)$ are in one-to-one correspondence with classes of switching equivalent graphs with an $(m,-1,1)$-representation.

Every graph Γ has $(m,-1,1)$-representations for sufficiently large m. Indeed, denote by B the $(0,-1,1)$-*adjacency matrix* of Γ (introduced by VAN LINT & SEIDEL [504]), i.e., the symmetric matrix indexed by the vertices of Γ with entries

$$B_{\gamma\delta} = 0 \text{ if } \gamma = \delta,\ B_{\gamma\delta} = -1 \text{ if } \gamma \sim \delta,\ B_{\gamma\delta} = 1 \text{ if } \gamma \not\sim \delta.$$

If Γ has an $(m,-1,1)$-representation, then the Gram matrix of its image is $G = B + mI$; hence the smallest eigenvalue of B is $\geqslant -m$. Conversely, if the smallest eigenvalue of B is $\geqslant -m$, then $G = B + mI$ is positive semidefinite, hence the Gram matrix of a set of vectors giving an $(m,-1,1)$-representation of Γ.

3.8.1. Proposition. *Let Γ be a graph. If Γ is switching equivalent to a graph with largest eigenvalue $\leqslant (m-1)/2$ then Γ has an $(m,-1,1)$-representation.*

Proof. Without loss of generality, we may assume that the largest eigenvalue of Γ itself is $\leqslant (m-1)/2$. Then Γ has a spherical $((m-1)/2,-1,0)$-representation. Scaling by $\sqrt{2}$ and shifting by a norm 1 vector orthogonal to all image vectors turns such a representation into an $(m,-1,1)$-representation of Γ. □

Remark. The line graphs Γ of partial linear spaces $\mathfrak{L}$ with n points and constant line size $s+1$ have an $(s+1,1,0)$-representation in $\mathbf{R}^n$, where a vertex of Γ, i.e., a line of $\mathfrak{L}$, is represented by the characteristic vector of the line.
Suppose that Γ has a (p,q,r)-representation. If there is a vector c such that $\frac{1}{2}(c,c) = (c,\bar{\gamma}) = q+r$ for all $\gamma \in \Gamma$, then $(c-\bar{\gamma},c-\bar{\delta}) = (\bar{\gamma},\bar{\delta})$, $(c-\bar{\gamma},\bar{\delta}) = q+r-(\bar{\gamma},\bar{\delta})$ for $\gamma, \delta \in \Gamma$; so the replacement of $\bar{\gamma}$ by $c-\bar{\gamma}$ for $\gamma \in S$ yields a (p,q,r)-representation for the graph obtained from Γ by switching with respect to S. In particular, if $\mathfrak{L}$ is a partial linear space with $2m^2$ points and constant line size m, then this applies with $c = m^{-1}(1,\cdots,1)^T$ to the $(m,1,0)$-representation constructed above for the line graph of $\mathfrak{L}$.
In particular, this applies with $m=2$ to line graphs of graphs; cf. Lemma 3.11.1.

Equiangular lines and hence $(m,1,-1)$-representations are closely related to two-graphs and the associated antipodal double covers of complete graphs; cf. TAYLOR [710], SEIDEL & TAYLOR [658], BUSSEMAKER et al. [154]. Indeed, for any set Λ of equiangular lines with angles arccos $(1/m)$, the graph Γ whose vertices are the $2|\Lambda|$ norm m vectors on lines of Λ, adjacency being defined by inner product -1 is an antipodal double cover of a complete graph (apart from a degenerate case where it may be the disjoint union of two complete graphs). The two-graph $\mathfrak{D}$ associated with Γ by Theorem 1.5.3 can also be described as the two-graph whose points are the lines of Λ and whose coherent triples are the triples of lines at obtuse angles, i.e., those triples which can be represented by three norm m vectors with mutual inner product -1.

In terms of an arbitrary set X of norm m vectors, one along each line, the two-graph consists of X as point set, and three vectors in X form a coherent triple if and only if their three mutual inner products have product -1. If such a set X is considered as a $(m, -1, 1)$-representation of a graph Δ, this amounts to taking as points the vertices of Δ and as coherent triples the 3-sets on which Δ induces a graph with an odd number of edges. This gives a geometric interpretation to the two-graph $\mathfrak{D}(\Delta)$ introduced in §1.5; in particular it shows that this two-graph only depends on the switching class of Δ.

We shall say that a mapping $x \mapsto \hat{x}$ which associates to each point x of a two-graph $(X, \mathfrak{D})$ a line $\hat{x}$ in $\mathbf{R}^n$ is a *representation* of the two-graph if

(i) the set $\Lambda = \{\hat{x} \mid x \in X\}$ is equiangular, and

(ii) $\{x,y,z\} \in \mathfrak{D}$ if and only if $\hat{x}$, $\hat{y}$, $\hat{z}$ are at obtuse angles.

The *norm* of the two-graph is the minimal possible m for the angle arccos $(1/m)$ of its representations. We can compute the norm of a two-graph $(X, \mathfrak{D})$ from all three constructions given in §1.5:

a) If Δ is a graph such that $\mathfrak{D} = \mathfrak{D}(\Delta)$, then Δ has a $(m, -1, 1)$-representation if and only if the $(0, -1, 1)$-adjacency matrix B of Δ (i.e., the matrix $B = J - I - 2A$, where A is the $(0,1)$-adjacency matrix of Δ) has smallest eigenvalue $\geqslant -m$. Hence the norm of $\mathfrak{D}$ is minus the smallest eigenvalue of B.

b) If Δ' is a derived graph of $\mathfrak{D}$, with adjacency matrix A', then $\mathfrak{D} = \mathfrak{D}(\Delta' + \{\infty\})$ so that the norm of $\mathfrak{D}$ is minus the smallest eigenvalue of the matrix

$$B = \begin{bmatrix} 0 & \mathbf{1}^\mathsf{T} \\ \mathbf{1} & J - I - 2A' \end{bmatrix}.$$

c) If Γ is the antipodal double cover of a complete graph K_{k+1} associated with $\mathfrak{D}$, then (for a suitable numbering of the vertices) the adjacency matrix A'' of Γ can be written as

$$A'' = \frac{1}{2}\begin{bmatrix} J - I - B & J - I + B \\ J - I + B & J - I - B \end{bmatrix},$$

where B is the $(0, -1, 1)$-adjacency matrix of $\gamma^\perp$ (for any $\gamma \in \Gamma$). One easily verifies that $A''\binom{x}{y} = \lambda\binom{x}{y}$ if and only if $(J - I)(x + y) = \lambda(x + y)$ and $-B(x - y) = \lambda(x - y)$; therefore the eigenvalues of A'' are those of $J - I$ (namely k and -1) and those of $-B$. Since $\mathfrak{D} = \mathfrak{D}(\gamma^\perp)$, this implies that the norm of $\mathfrak{D}$ is given by the second largest eigenvalue of Γ.

From the equivalence of Taylor graphs and regular two-graphs we may deduce the fact that $\mathfrak{D}$ is a regular two-graph if and only if the matrix B in (c) (and hence in (a), (b)) has two eigenvalues only (cf. TAYLOR [710]).

3.9 Lattices and integral representations

A *lattice* in $\mathbf{R}^n$ is a discrete set of vectors in $\mathbf{R}^n$ which is closed under addition and subtraction (and hence under multiplication by integers). An *integral lattice* is a lattice in which the inner product of any two vectors is integral. An integral lattice is called *even* if it contains only vectors of even norm. Note that an integral lattice is even when it has a basis consisting of vectors of even norm. The set of vectors of norm m in a lattice $\mathbf{L}$ is denoted by $\mathbf{L}^{(m)}$. The *direct sum* of two lattices $\mathbf{L}_1$ and $\mathbf{L}_2$ is defined if $\mathbf{L}_1$ and $\mathbf{L}_2$ are orthogonal, i.e., if $(x_1,x_2) = 0$ for $x_1 \in \mathbf{L}_1$, $x_2 \in \mathbf{L}_2$, as

$$\mathbf{L}_1 \oplus \mathbf{L}_2 = \{x_1 + x_2 \mid x_1 \in \mathbf{L}_1, x_2 \in \mathbf{L}_2\}.$$

$\mathbf{L}$ is called *irreducible* if $\mathbf{L} = \mathbf{L}_1 \oplus \mathbf{L}_2$ implies $\mathbf{L}_1 = \{0\}$ or $\mathbf{L}_2 = \{0\}$, and *reducible* otherwise. If X is a set of vectors in $\mathbf{R}^n$ such that $(x,y) \in \mathbf{Z}$ for all $x,y \in X$, then

$$\mathbf{L} = \{\sum_{x \in X} a_x x \mid a_x \in \mathbf{Z}\}$$

is an integral lattice, the lattice *generated* (or *spanned*) by X. Every integral lattice $\mathbf{L}$ has an *integral basis*, i.e., a set of linearly independent vectors spanning $\mathbf{L}$; the determinant of the Gram matrix of an integral basis is an invariant of $\mathbf{L}$, called the *discriminant*. If $\mathbf{L}$ is an integral lattice, the *dual lattice* is defined as

$$\mathbf{L}^* = \{z \in \mathbf{R} \otimes \mathbf{L} \mid (z,x) \in \mathbf{Z} \text{ for all } x \in \mathbf{L}\},$$

where $\mathbf{R}\otimes\mathbf{L}$ denotes the subspace of $\mathbf{R}^n$ spanned by $\mathbf{L}$. The dual lattice $\mathbf{L}^*$ is an integral lattice if and only if $\mathbf{L}$ is *unimodular*, i.e., has discriminant ± 1; in this case $\mathbf{L}^* = \mathbf{L}$. Otherwise we only have $\mathbf{L} \subseteq \mathbf{L}^* \subseteq d^{-1}\mathbf{L}$, where d is the discriminant of $\mathbf{L}$, and the factor group $\mathbf{L}^*/\mathbf{L}$ is an abelian group of order $|d|$ (cf. Cohn [209], Chapters 4, 5).

We say that a representation $\gamma \mapsto \bar{\gamma}$ $(\gamma \in \Gamma)$ of a graph Γ is *integral* when the squared distance $[\gamma,\delta]$ of the images of any two points $\gamma, \delta \in \Gamma$ is an *even* integer. Associated with an integral representation is the lattice $\mathbf{L}(\Gamma)$ generated by all vectors $\bar{\gamma} - \bar{\delta}$, $\gamma, \delta \in \Gamma$; equivalently,

$$\mathbf{L}(\Gamma) = \{\sum_{\gamma \in \Gamma} a_\gamma \bar{\gamma} \mid a_\gamma \in \mathbf{Z}, \sum_{\gamma \in \Gamma} a_\gamma = 0\}.$$

Of course, $\mathbf{L}(\Gamma)$ depends on the representation of Γ, but it will be clear from the context which representation is meant.

By Lemma 3.5.1, the inner product of any two generating vectors is an integer so that $\mathbf{L}(\Gamma)$ is an even integral lattice. If Γ is connected and $\gamma = \gamma_0, \gamma_1, \cdots, \gamma_l = \delta$ is a path, then $\bar{\gamma} - \bar{\delta} = \sum_{i=1}^{l} (\bar{\gamma}_{i-1} - \bar{\gamma}_i)$, so that $\mathbf{L}(\Gamma)$ is already generated by the vectors $\bar{\gamma} - \bar{\delta}$ where $\gamma, \delta \in \Gamma$ are adjacent. In particular, by definition of a representation, $\mathbf{L}(\Gamma)$ is, then, generated by vectors of constant norm a.

If an integral representation of a graph Γ is spherical, say $(\bar{\gamma},\bar{\gamma}) = p$ for all $\gamma \in \Gamma$, then we may assume that p is an even integer (otherwise shift the representation; this does not affect squared distances). Now the lattice $\mathbf{L}^+(\Gamma)$ generated by the vectors $\bar{\gamma}$, $\gamma \in \Gamma$, i.e.,

$$\mathbf{L}^+(\Gamma) = \{\sum_{\gamma \in \Gamma} a_\gamma \bar{\gamma} \mid a_\gamma \in \mathbf{Z}\},$$

is also an even integral lattice. $\mathbf{L}(\Gamma)$ is a sublattice of $\mathbf{L}^+(\Gamma)$, possibly of one dimension lower (where $\dim \mathbf{L} := \dim \mathbf{R}\otimes\mathbf{L}$).

3.10 Root systems and root lattices

A *root lattice* is an integral lattice $\mathbf{L}$ in $\mathbf{R}^n$ generated by vectors of norm 2; the vectors of norm 2 in $\mathbf{L}$ are called the *roots* of $\mathbf{L}$. One can easily see that the set $\Phi = \mathbf{L}^{(2)}$ of all roots of $\mathbf{L}$ forms a *root system* in $V := \mathbf{R} \otimes \mathbf{L}$, i.e., Φ is a finite set of nonzero vectors spanning V such that

$$\text{if } \lambda \in \mathbf{R} \text{ and } r, \lambda r \in \Phi \text{ then } \lambda = \pm 1, \tag{1}$$

$$\text{if } r, s \in \Phi \text{ then } s - 2\frac{(r,s)}{(r,r)} r \in \Phi, \tag{2}$$

and

$$2\frac{(r,s)}{(r,r)} \in \mathbf{Z} \text{ for all } r, s \in \Phi. \tag{3}$$

For a systematic account of root systems, see Chapter VI of Bourbaki [113]. Here, we restrict ourselves to giving properties which are relevant for later applications. Also, by studying root lattices we only obtain root systems in which all roots have equal length. Bourbaki relaxes condition (1), and calls a root system *reduced* if (1) holds. In Chapter 10 we shall come back to root systems.

For $r \in \mathbf{L}^{(2)}$, the *reflection* w_r at the hyperplane orthogonal to r through the origin maps $x \in \mathbf{R}^n$ to

$$w_r(x) := x - 2\frac{(x,r)}{(r,r)} r = x - (x,r)r. \tag{4}$$

It leaves the lattice invariant and preserves inner products. Therefore, the same holds for the group $W = W(\mathbf{L})$ generated by all reflections w_r, $r \in \mathbf{L}^{(2)}$, the *Weyl group* of $\mathbf{L}$ (or, more precisely, of the root system $\mathbf{L}^{(2)}$). Since $\mathbf{L}^{(2)}$ is finite (and spans $\mathbf{R}^n$), W is finite, too.

3.10.1. Proposition. *If a root lattice $\mathbf{L}$ is irreducible, then its Weyl group W is transitive on the roots.*

Proof. If Φ_0 is a W-orbit of $\mathbf{L}^{(2)}$ and $\Phi_0 \neq \mathbf{L}^{(2)}$, then there are $r, s \in \mathbf{L}^{(2)}$ with $r \in \Phi_0$, $s \notin \Phi_0$, and $(r,s) \neq 0$. But then $(r,s) = \pm 1$, so $r \mp s$ is a root and either w_{r-s} or $w_{r+s} w_s$ belongs to W and moves r to s, a contradiction. □

Remark. Much more is true, and in the case of E_8, we shall need more in §3.12. The *root system graph* of a root system in which all roots have equal length is the graph whose vertices are the roots and in which two vertices are joined whenever their angle is $\pi/3$. In §10.3 we identify the root system graph for E_8 as the Coxeter graph $E_{8,8}$. But the j-cliques in this graph are the points of $E_{8,9-j}$ for $j = 1, 2, 3, 4, 5$, so the Weyl group W is transitive on j-cliques. Similarly, the root system graphs for D_n, E_6, E_7 are the Coxeter graphs $D_{n,2}$ ($n \geqslant 4$), $E_{6,2}$ and $E_{7,1}$, respectively, and W acts flag-transitively on these. The root system graph for A_n ($n \geqslant 2$) is a double cover of a triangular graph $T(n+1)$; in this case the Weyl group W does not act edge-transitively, but $\langle W, -1 \rangle$ does, where -1 is the map sending r to $-r$ for all vectors r.

Fundamental systems and classification

A *fundamental system* (of roots) is a set Π of roots of $\mathbb{L}$ such that $\mathbb{L}$ is generated by Π and such that

$$(r,s) \leqslant 0 \text{ for } r, s \in \Pi, r \neq s. \tag{5}$$

A fundamental system Π is called *reduced* if the roots in Π are linearly independent and *extended* otherwise. The *Dynkin diagram* Δ of a fundamental system Π not containing two opposite roots is the graph on Π obtained by joining two distinct roots $r, s \in \Pi$ by an edge if $(r,s) = -1$. If a fundamental system Π contains two opposite roots $r, -r$ then clearly $(r,s) = 0$ for all other $s \in \Pi$ so that $\mathbb{L}$ contains a one-dimensional direct summand (A_1 in the classification below). We shall not define the Dynkin diagram in this situation.

We shall show that every root lattice $\mathbb{L}$ has a reduced fundamental system Π, and that the Dynkin diagram Δ of Π has largest eigenvalue less than 2. Now Theorem 3.2.5 provides a determination of all Dynkin diagrams, and, since Δ determines Π, and Π determines $\mathbb{L}$, this yields a classification of all root lattices $\mathbb{L}$.

Let Π be a fundamental system of the root lattice $\mathbb{L}$, and let $G = ((r,s))_{r,s \in \Pi}$ be its Gram matrix. Then $G = 2I - A$, where A is nonnegative and integral; in case Δ is defined, A is the adjacency matrix of Δ. Since G is positive semidefinite, and positive definite if and only if Π is linearly dependent, A has largest eigenvalue at most 2, and largest eigenvalue strictly less than 2 when Π is reduced.

3.10.2. Proposition. *Let u be a vector in $\mathbb{R}^n$ not orthogonal to any root of the root lattice $\mathbb{L}$. Put $\Phi^+(u) := \{r \in \mathbb{L}^{(2)} \mid (r,u) > 0\}$ and*

$$\Pi(u) := \{r \in \Phi^+(u) \mid \text{ for all } s, t \in \Phi^+(u) \textit{ we have } r \neq s+t\}.$$

Then $\Pi(u)$ is a reduced fundamental system of roots, and written on this basis each root has either only positive or only negative coefficients.

Proof. First of all, $\mathbb{L}$ is generated by $\Pi := \Pi(u)$. (Indeed, $\mathbb{L}$ is generated by $\Phi^+ := \Phi^+(u)$, but if there are vectors $r \in \Phi^+$ not in the $\mathbb{Z}$-span of Π, then choose such an r with minimal (r,u) and derive a contradiction.) If $r, s \in \Pi$, $(r,s) > 0$, $r \neq s$, then $(r,s) = 1$ and $r-s \in \mathbb{L}^{(2)}$, so that $\pm(r-s) \in \Phi^+$, say $r-s \in \Phi^+$, and we find the contradiction $r = (r-s)+s \notin \Pi$. Thus, Π is a fundamental system, and each root in Φ^+ is a nonnegative integral linear combination of the roots in Π. If Π is not reduced, then we have a relation $\sum_{r \in \Pi} \gamma_r r = 0$ for certain γ_r, not all zero. Putting $x := \sum\{\gamma_r r \mid \gamma_r > 0\}$ and $y := -\sum\{\gamma_r r \mid \gamma_r < 0\}$, we find $x = y \neq 0$ hence $(x,y) > 0$, but also $(x,y) \leqslant 0$ since any two members of Π have nonpositive inner product, contradiction. Thus Π is reduced. □

3.10.3. Proposition. *Let Π be a reduced fundamental system of the root lattice $\mathbb{L}$. Then:*

(i) *For all $x \in \mathbb{R}^n$ there is an element $w \in W(\mathbb{L})$ such that $(w(x),r) \geqslant 0$ for all $r \in \Pi$.*

(ii) *$\Pi = \Pi(u)$ for some vector $u \in \mathbb{R}^n$. In particular, $W(\mathbb{L})$ acts transitively on the set of reduced fundamental systems of $\mathbb{L}$.*

(iii) *If $\mathbb{L}$ is irreducible, then there is a unique root $\tilde{r} \in \mathbb{L}^{(2)}$ such that $\tilde{\Pi} := \Pi \cup \{\tilde{r}\}$ is an extended fundamental system of roots.*

Proof. (i) Let $G = 2I - A$ be the Gram matrix of Π. Since G is positive definite, A

has largest eigenvalue less than 2. Let $(\gamma_r)_{r\in\Pi}$ be a positive eigenvector of A (cf. §3.1). If $x \in \mathbf{L}$, $s \in \Pi$, $(x,s) < 0$, then put $x' = w_s(x) = x-(x,s)s$. We have

$$\sum_{r\in\Pi} \gamma_r(x',r) = \sum_{r\in\Pi} \gamma_r(x,r)+(-2\gamma_s+\sum_{r\sim s}\gamma_r)(x,s) > \sum_{r\in\Pi} \gamma_r(x,r).$$

Since W is finite, we reach our goal after finitely many steps.
(ii) Use induction on $|\Pi|$. Choose $x \neq 0$ with $(x,r) \geqslant 0$ for all $r \in \Pi$, and let $\Pi_0 := \Pi \cap x^{\perp}$. Now Π_0 is a reduced fundamental system of a root lattice $\mathbf{L}_0$, and by induction $\Pi_0 = \Pi_0(u_0)$ for some vector u_0. It follows that $\Pi = \Pi(u)$ where $u = x+\epsilon u_0$ when $\epsilon > 0$ is sufficiently small. Now transitivity of W follows from (i).
(iii) Existence: choose $r_1 \in \Phi^+(u)$ with (r_1,u) maximal. Then $\Pi \cup \{-r_1\}$ is a fundamental system of roots.
Uniqueness: We may assume that $\mathbf{L} \neq A_1$, the unique root lattice of dimension one. Since linear dependencies of $\tilde{\Pi}$ correspond to eigenvectors with eigenvalue 2 of its Dynkin diagram $\tilde{\Delta}$, we find such an eigenvector (γ_r) with $\gamma_{\tilde{r}} = 1$. Inspection of the eigenvectors listed in Theorem 3.2.5 shows that for each Dynkin diagram Δ only one diagram $\tilde{\Delta}$ and eigenvector (γ_r) occurs. □

Let us write 'Dynkin diagram of $\mathbf{L}$' for 'Dynkin diagram of some reduced fundamental system of $\mathbf{L}$'; by Proposition 3.10.3 (ii), the Dynkin diagram of $\mathbf{L}$ is determined up to isomorphism. Summing up what we have proved in this section:

3.10.4. Theorem (WITT [790]). *There are unique irreducible root lattices with Dynkin diagram isomorphic to A_n $(n \geqslant 0)$, D_n $(n \geqslant 4)$ or E_m $(m \in \{6,7,8\})$, and no other irreducible root lattices exist.* □

In the following section we give explicit descriptions of the irreducible root lattices.

The irreducible root lattices

We now describe the irreducible root lattices. Together with each example we give some useful information on the dual lattice. We write e_i for the unit vector in $\mathbf{R}^n$ having a 1 in the i-th coordinate and zeros elsewhere, and note that a reflection w_r at $r = e_i - e_j$, $i \neq j$, has just the effect of interchanging the i-th and j-th coordinates. (Note that our first example A_n is presented on a hyperplane in $\mathbf{R}^{n+1}$ rather than in $\mathbf{R}^n$ in order to get a nicer presentation, and in order to make the inclusion $A_n \subset D_{n+1}$ obvious.)

(i) Define

$$A_n := \{x \in \mathbf{R}^{n+1} \mid x_i \in \mathbf{Z}, \sum_{i=1}^{n+1} x_i = 0\}.$$

The roots of A_n are the $n(n+1)$ vectors $e_i - e_j$ $(i \neq j)$. An extended fundamental system is given by the vectors $r_i = e_i - e_{i-1}$ $(i = 1, \cdots, n+1)$, where $e_0 = e_{n+1}$, with Dynkin diagram:

$\tilde{A}_n$ $(n \geqslant 2)$

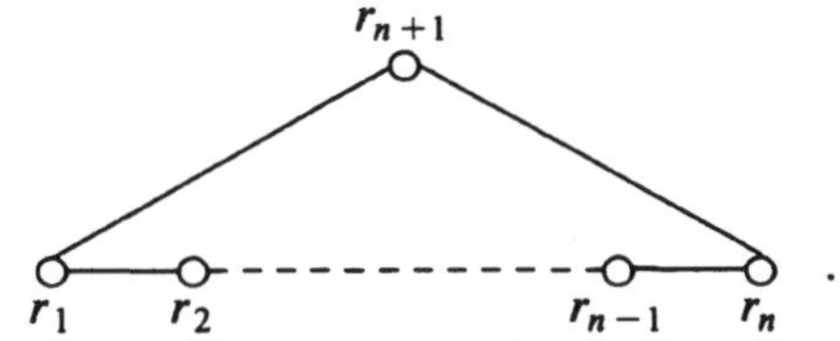

Removal of r_{n+1} leaves a reduced fundamental system. The dual lattice A_n^* is generated by A_n and $z = \frac{1}{n+1}(e_1 + \cdots + e_n - ne_{n+1})$. The discriminant of A_n is $n+1$, and A_n^* / A_n is a cyclic group of order $n+1$; the vectors of A_n^* in the coset $iz + A_n$ have norm $i(n+1-i)/(n+1) + 2k$ for some $k \in \mathbf{Z}$.

(ii) For $n \geqslant 4$, define

$$D_n := \{x \in \mathbf{R}^n \mid x_i \in \mathbf{Z}, \sum_{i=1}^{n} x_i \in 2\mathbf{Z}\}.$$

The roots of D_n are the $2n(n-1)$ vectors $\pm e_i \pm e_j$ $(i \neq j)$; an extended fundamental system is given by the vectors $r_1 = e_2 + e_1$, $r_i = e_i - e_{i-1}$ $(i = 2, \cdots, n)$, $r_{n+1} = -e_n - e_{n-1}$, with Dynkin diagram:

$\tilde{D}_n$ $(n \geqslant 4)$

Removal of r_{n+1} leaves a reduced fundamental system. The dual lattice D_n^* is generated by D_n and the vectors e_1, $\frac{1}{2}(e_1 + e_2 + \cdots + e_n)$. The discriminant of D_n is 4 and D_n^* / D_n is a Klein's four group if n is even and cyclic otherwise. The vectors of D_n^* have norm $m + k$, where $m \in \{0, \frac{n}{4}\}$ and $k \in \mathbf{Z}$.

(iii) Define E_8 as the lattice spanned by D_8 and $\frac{1}{2}(e_1 + \cdots + e_8)$. The roots of E_8 are the $240 = 112 + 128$ vectors $\pm e_i \pm e_j$ and $\frac{1}{2}(\pm e_1 \pm e_2 \cdots \pm e_8)$, where in a root of the second kind the number of minus signs is even. An extended fundamental system is given by the vectors

$$\begin{aligned} r_1 &= \tfrac{1}{2}(e_1 + e_2 + e_3 - e_4 - e_5 - e_6 - e_7 + e_8), \\ r_i &= e_i - e_{i-1} \quad (i = 2, \cdots, 7), \\ r_8 &= -e_8 - e_7, \\ r_9 &= \tfrac{1}{2}(e_1 + e_2 + e_3 + e_4 + e_5 + e_6 + e_7 + e_8), \end{aligned}$$

with Dynkin diagram:

$\tilde{E}_8$

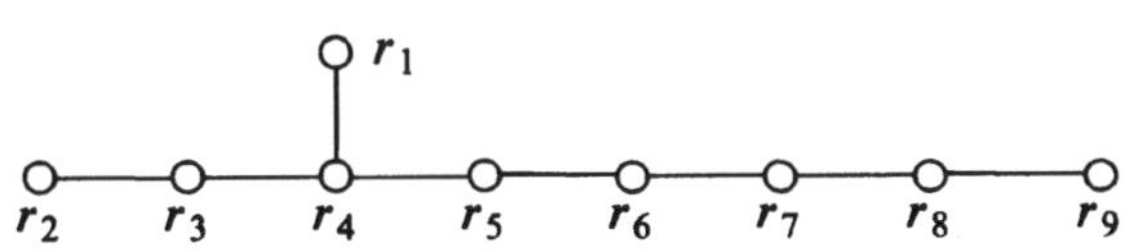

Removal of r_9 leaves a reduced fundamental system. E_8 is unimodular, i.e., $E_8^* = E_8$. Vectors in E_8 have norm $2k$, $k \in \mathbf{Z}$.

(iv) Define

$$E_7 := E_8 \cap r_9^{\perp} = \{x \in E_8 \mid x_1 + \cdots + x_8 = 0\}.$$

E_7 contains $126 = 56+70$ of the roots of E_8. An extended fundamental system is given by $r_1, \cdots, r_7$ as in (iii) and $r_8' = e_1 - e_8$, with Dynkin diagram:

$\tilde{E}_7$

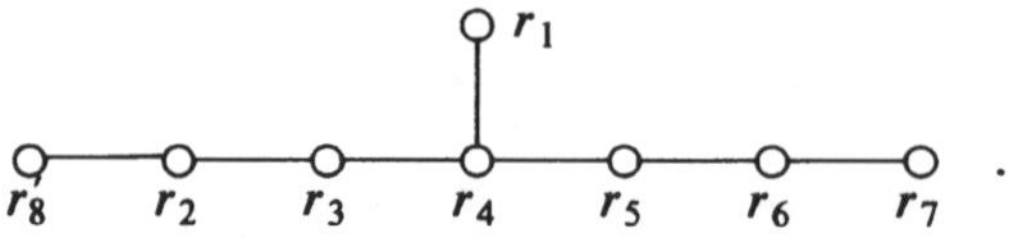

Removal of r_8' leaves a reduced fundamental system. The dual lattice E_7^* is generated by E_7 and $\frac{1}{4}(e_1+e_2+e_3+e_4+e_5+e_6-3e_7-3e_8)$. The discriminant of E_7 is 2, and hence E_7^* / E_7 is a cyclic group of order 2. Vectors in E_7^* have norm $m+2k$ where $m \in \{0, \frac{3}{2}\}$ and $k \in \mathbf{Z}$.

(v) Define

$$E_6 := E_8 \cap \{r_8, r_9\}^{\perp} = \{x \in E_8 \mid x_1 + \cdots + x_6 = x_7 + x_8 = 0\}.$$

E_6 contains $72 = 32+40$ roots of E_8. An extended fundamental system is given by $r_1, \cdots, r_6$ as for E_8 and $r_7' = e_7 - e_8$, with Dynkin diagram:

$\tilde{E}_6$

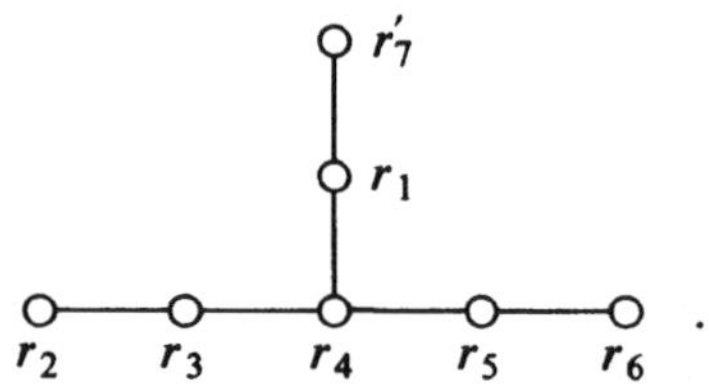

Removal of r_7' leaves a reduced fundamental system. The dual lattice E_6^* is generated by E_6 and $\frac{1}{3}(e_1+e_2+e_3+e_4-2e_5-2e_6)$. The discriminant of E_6 is 3, and E_6^* / E_6 is a cyclic group of order 3. Vectors in E_6^* have norm $m+2k$ where $m \in \{0, \frac{4}{3}\}$ and $k \in \mathbf{Z}$.

Another proof of the classification

Above (Theorem 3.10.4) we classified root lattices using the existence of fundamental systems and the classification of graphs with largest eigenvalue less than 2. Here we give an entirely different proof, reducing the classification of root lattices to that of generalized quadrangles with lines of size three. Our proof is inspired by a characterisation of the root systems of A_n, D_n, E_6, E_7 and E_8 (in terms of star-closed sets of lines) by Cameron et al. [180].

Theorem. *Every root lattice is the direct sum of lattices isomorphic to A_n, D_n, E_6, E_7, or E_8.*

Proof. Let $\mathbf{L}$ be a root lattice. Without loss of generality, $\mathbf{L}$ is irreducible and we may take roots $a, b \in \mathbf{L}^{(2)}$ with $(a,b) = 1$. (If no such roots exist then $\mathbf{L} \cong A_1$.) We define a design Q whose points are the roots $x \in \mathbf{L}^{(2)}$ satisfying $(x,a) = (x,b) = 1$, and whose lines are the mutually orthogonal triples of points. If $\{x,y,z\}$ is a line, then the norm of $x+y+z-a-b$ is zero, so that $z = a+b-x-y$; conversely, if x, y are orthogonal points of Q then $z = a+b-x-y$ has norm 2 and is orthogonal to x, y so that $\{x,y,z\}$ is a line. Therefore, every pair of orthogonal points of Q is on a unique line. Now let u be a point of Q not on the line $l = \{x,y,z\}$. Since $x+y+z = a+b$, we have $(u,x)+(u,y)+(u,z) = 2$; but (u,x), (u,y), $(u,z) \in \{-1,0,1\}$ so that u is orthogonal to precisely one point of l. This implies that Q is a (possibly degenerate) generalized quadrangle whose lines have size three. We now observe that the combinatorial structure of Q determines the inner products of points of Q: If x and y are on a line then $(x,y) = 0$; otherwise $(x,y) = \pm 1$, but -1 is impossible since then $x+y-\frac{2}{3}(a+b)$ would have norm $-\frac{2}{3}$. Therefore, the lattice $\mathbf{L}_0$ generated by a, b and the points of Q is determined up to isomorphism by the isomorphism type of Q.

We now show that $\mathbb{L} = \mathbb{L}_0$. Indeed, suppose that $\mathbb{L} \neq \mathbb{L}_0$. Since $\mathbb{L}$ is irreducible and generated by roots there are roots $r, s \in \mathbb{L}^{(2)}$ such that $r \notin \mathbb{L}_0$, $s \in \{a,b\} \cup Q$ and $(r,s) = 1$. If $(r,a) = (r,b) = \pm 1$, then $\pm r \in Q$. If $(r,a) = -(r,b) = \pm 1$, then $\pm r + b - a$ has norm zero so that $\pm r = a - b$. If $(r,a) = \pm 1$, $(r,b) = 0$, then $a \mp r \in Q$, and if $(r,a) = 0$, $(r,b) = \pm 1$ then $b \mp r \in Q$. Finally, if $(r,a) = (r,b) = 0$ then $s, s - r \in Q$. In each case it follows that $r \in \mathbb{L}_0$, contradiction. Thus, $\mathbb{L}$ is completely determined by Q. Now, for any a, b, the graph on Q is an $(n-2)$-coclique if $\mathbb{L} = A_n$, a set of $n-3$ lines having a common point if $\mathbb{L} = D_n$, and a nondegenerate generalized quadrangle with 9, 15, and 27 points for E_6, E_7, and E_8, respectively. But by Theorem 1.15.2 there are no other generalized quadrangles. Therefore, $\mathbb{L}$ is isomorphic to one of A_n, D_n, E_6, E_7, and E_8. □

3.11 Graphs represented by roots of E_8

A number of important graphs are represented by roots of E_8, and in the next section we shall see that it is important to know all the graphs represented by roots of E_8. Let us define the graph $E_8(1)$ by taking the 240 roots of E_8 as vertices, joining them by an edge whenever their inner product is one. The identity is a representation of $E_8(1)$, but not a weak (2,1,0)-representation since roots at distance 2 may have inner product 0 or -1. (Cf. the following double coset diagram of the action of the reflection group on the set of roots.)

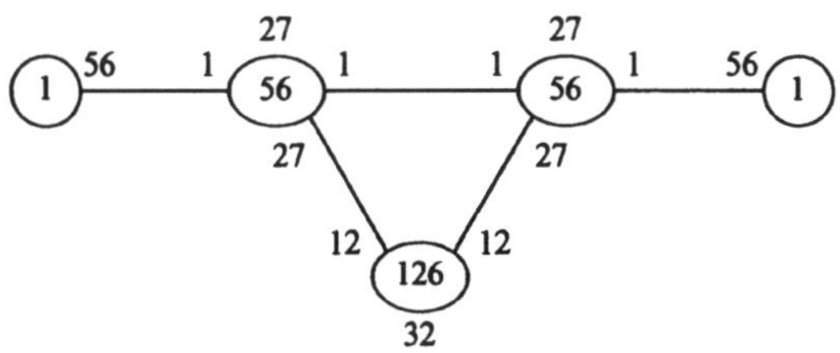

However, many subgraphs of $E_8(1)$ give interesting representations. In particular, let

$$c_i := e_i + e_8 \quad (i = 1, \cdots, 7),$$
$$c_8 := \tfrac{1}{2}(e_1 + e_2 + e_3 + e_4 + e_5 + e_6 + e_7 + e_8),$$

and denote by $E_n(1)$, $n \leqslant 7$, the induced subgraph on the set of common neighbours of $c_{n+1}, \cdots, c_8$. We get:

(i) The *Gosset graph* $E_7(1)$ with 56 vertices $e_i + e_j$, $c_8 - e_i - e_j$ $(1 \leqslant i < j \leqslant 8)$ and valency 27. This graph is the skeleton of the Gosset polytope 3_{21} (cf. COXETER [222]), and is related to the 28 bitangents of a quartic surface (cf. DICKSON [535]). A modern description of the structure of $E_7(1)$ is given by TAYLOR [710] in terms of a regular two-graph with 28 points. The Gosset graph $E_7(1)$ is represented by the identity as a weak (2,1,0)-representation.

$E_7(1)$ contains precisely 244 nonisomorphic regular subgraphs in a (2,1,0)-representation. Of these subgraphs, 53 are line graphs, contained in the triangular graph T(8), and 4 others are of the form $K_{m\times 2}$ $(3 \leqslant m \leqslant 6)$, represented by $e_i + e_8$, $c_8 - e_7 - e_i$ $(i = 1, \cdots, m)$. The 187 remaining subgraphs have been found by an exhaustive computer search of BUSSEMAKER et al. [153] (cf. Theorem 3.12.2). Some of these graphs are described below.

(ii) The *Schläfli graph* $E_6(1)$ with 27 vertices $e_i + e_7$, $e_i + e_8$ $(i \leqslant 6)$, $c_8 - e_i - e_j$ $(i < j \leqslant 6)$ and valency 16. From the description in §1.15 it is clear that $E_6(1)$ is the complement of the point graph of the generalized quadrangle of order (2,4) with 27 points. The graph $E_6(1)$ and its subgraphs, in particular $E_5(1)$ and $E_4(1)$, are represented as $(2,1,0)$-representations. $E_6(1)$ contains 24 regular subgraphs which are not line graphs (no. 164-187 in BUSSEMAKER et al. [153]).

(iii) The *Clebsch graph* $E_5(1)$ with 16 vertices e_6+e_7, e_i+e_8 $(i \leqslant 5)$, $c_8-e_i-e_j$ $(i<j\leqslant 5)$ and valency 10. It is strongly regular with parameters $(v,k,\lambda,\mu)=(16,10,6,6)$. The Clebsch graph is isomorphic to the halved 5-cube, and its complement (with parameters $(v,k,\lambda,\mu)=(16,5,0,1)$) to the folded 5-cube. The Clebsch graph is switching equivalent to the 4×4-grid and to the Shrikhande graph (see below). $E_5(1)$ contains 3 regular (and in fact co-edge-regular) subgraphs which are not line graphs (no. 185-187 in [153]); the two proper subgraphs are a graph with the 12 vertices e_i+e_8 $(i=2,3,4)$, $c_8-e_i-e_j$ $(i<j\leqslant 5,\ (i,j)\neq(1,5))$, and valency 7 and $\mu=4$ (the Clebsch graph minus $2K_2$); and a graph with the 8 vertices e_6+e_7, e_i+e_8 $(i=2,3,4)$, $c_8-e_i-e_{i+1}$ $(i\leqslant 4)$ and valency 4 and $\mu=2$:

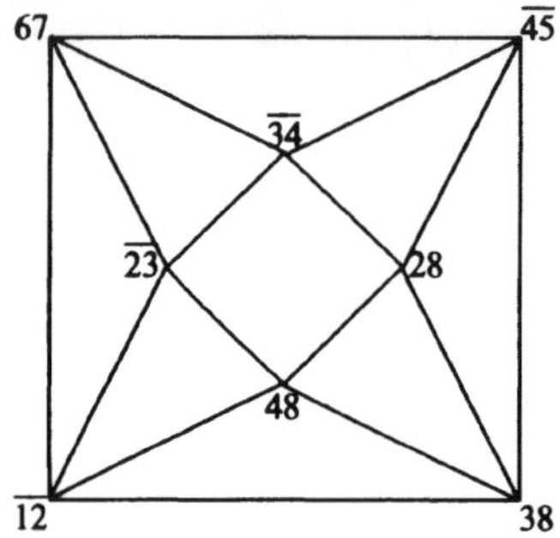

(iv) The graph $E_4(1)$ is the triangular graph $T(5)$ with 10 vertices and valency 6. This graph is strongly regular with parameters $(v,k,\lambda,\mu)=(10,6,3,4)$, the complement of the Petersen graph.

(v) The *Petersen graph* with 10 vertices and valency 3 is the complement of the triangular graph $T(5)$. It can also be obtained from the triangular graph $T(5)$ with vertices e_i+e_j, $(i<j\leqslant 5)$ by switching with respect to $\{e_i+e_{i+1} \mid 1\leqslant i\leqslant 4\} \cup \{e_5+e_1\}$. This graph is strongly regular with parameters $(v,k,\lambda,\mu)=(10,3,0,1)$.

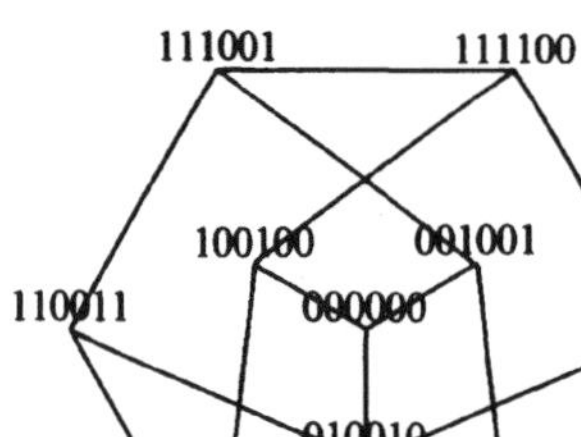

(The labels in the picture show that the Petersen graph is a subgraph of the halved 6-cube.)

(vi) The *Shrikhande graph* $L_2'(4)$ with 16 vertices and valency 6, obtained from the

4×4-grid $L_2(4)$ with vertices e_i+e_j $(i \leqslant 4 < j)$ by switching with respect to $\{e_i+e_{i+4} \mid i \leqslant 4\}$. This graph is strongly regular with the same parameters $(v,k,\lambda,\mu) = (16,6,2,2)$ as $L_2(4)$ (cf. SHRIKHANDE [668]). Since it is locally a hexagon, the Shrikhande graph is a quotient of the triangular lattice in $\mathbf{R}^2$.

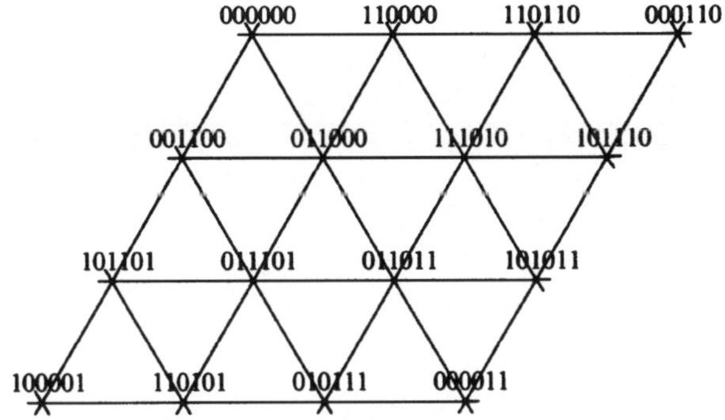

(The labels in the picture show that the Shrikhande graph is a subgraph of the halved 6-cube.) Up to isomorphism, the Shrikhande graph contains 3 regular proper subgraphs which are not line graphs, namely the 8-vertex graph drawn under (iii) (take the left half of the picture above), the 10-vertex graph obtained by deleting $3K_2$:

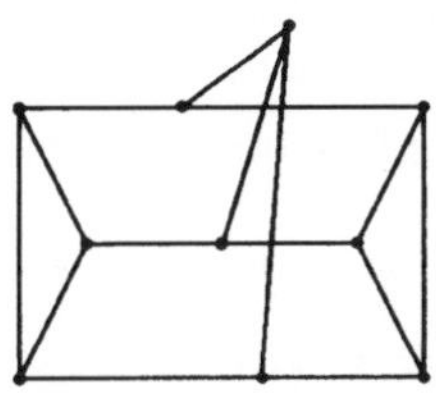

(diameter 3, $k = 3$, $\mu = 1$), and the 12-vertex graph obtained by deleting a 4-coclique:

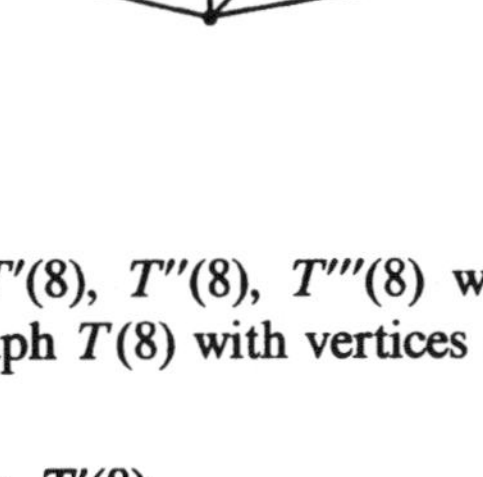

(diameter 3, $k = 4$).

(vii) The three *Chang graphs* $T'(8)$, $T''(8)$, $T'''(8)$ with 28 vertices and valency 12, obtained from the triangular graph $T(8)$ with vertices e_i+e_j $(i < j \leqslant 8)$ by switching with respect to one of the sets

$\{e_i+e_{i+4} \mid 1 \leqslant i \leqslant 4\}$ for $T'(8)$,

$\{e_i+e_{i+1} \mid 1 \leqslant i \leqslant 7\} \cup \{e_8+e_1\}$ for $T''(8)$, and

$\{e_1+e_2,\ e_2+e_3,\ e_3+e_1,\ e_4+e_5,\ e_5+e_6,\ e_6+e_7,\ e_7+e_8,\ e_8+e_4\}$ for $T'''(8)$.

These graphs are all strongly regular with the same parameters $(v,k,\lambda,\mu) = (28,12,6,4)$ as $T(8)$ (cf. CHANG [190, 191], and SEIDEL [652, 654] for equivalence under switching).

3.11.1. Lemma. *Let Γ be a subgraph of $E_7(1)$ such that Γ does not contain a pair of vertices at distance 3 in $E_7(1)$, i.e., such that the identity is a $(2,1,0)$-representation of Γ. Then Γ is switching equivalent to the line graph of a graph on 8 vertices. (And conversely, graphs switching equivalent to the line graph of a graph on 8 vertices have a $(2,1,0)$-representation in $E_7(1)$, cf. the remark following 3.8.1.)*

Proof. The only pairs of vectors at distance 3 in $E_7(1)$ are the pairs x, y with $(x,y) = -1$, i.e., the pairs $e_i + e_j$, $c - e_i - e_j$, where $c = \frac{1}{2}(e_1 + \cdots + e_8)$. If $S \subseteq \Gamma$ then the graph Γ' obtained from Γ by switching with respect to S has a $(2,1,0)$-representation ρ given by $\rho(x) = c - x$ for $x \in S$ and $\rho(x) = x$ for $x \in \Gamma \setminus S$. But any graph with a representation in the set of vectors $\{e_i + e_j \mid 1 \leqslant i < j \leqslant 8\}$ is the line graph of a graph on 8 vertices. □

3.11.2. Proposition (BUSSEMAKER & NEUMAIER [155]). *Let Γ be a regular subgraph of $E_7(1)$ in a $(2,1,0)$-representation, and suppose that it is not a line graph.*

(i) *If Γ is strongly regular, then Γ is either $K_{m\times 2}$ $(3 \leqslant m \leqslant 6)$ or one of the graphs of Schläfli, Clebsch, Petersen, Shrikhande, or Chang.*

(ii) *If Γ is edge-regular, then Γ is strongly regular.*

(iii) *If Γ is co-edge-regular, then Γ is strongly regular or one of the two regular subgraphs of the Clebsch graph with 8 and 12 vertices, respectively.*

Proof. By inspection of the tables in BUSSEMAKER et al. [153]. Graph no. 5 is the Petersen graph, no. 69 is the Shrikhande graph, no. 161-163 are the Chang graphs, no. 184 is the Schläfli graph, and no. 187 is the Clebsch graph. □

It would be interesting to have a geometric, computer free proof of this proposition.

Remark. Precisely 68 of the 187 exceptional regular graphs in $E_7(1)$ are cospectral with line graphs; these 68 are constructed explicitly (by switching line graphs) in CVETKOVIĆ & RADOSAVLJEVIĆ [238]. See also CVETKOVIĆ & DOOB [232].

3.12 Graphs with smallest eigenvalue at least -2

Let Δ be a graph with m vertices and let $a_1, \cdots, a_m$ be a set of nonnegative integers. A *generalized line graph* $L(\Delta;a_1, \cdots, a_m)$ (cf. HOFFMAN [398]) is a graph obtained from the line graph $L(\Delta)$ of a graph Δ with vertex set $\{1, \cdots, m\}$ by adding the vertices $(i,\pm l)$, where $i = 1, \cdots, m$, $l = 1, \cdots, a_i$, and joining $(i,\pm l)$ with all $\{i,j\}$ $(j \in \Delta(i))$ and all (i,l'), $l' \neq \pm l$. We get a $(2,1,0)$-representation of a generalized line graph by representing the vertices $\{i,j\}$ of $L(\Delta)$ by $e_i + e_j$ and the vertices $(i,\pm l)$ by $e_i \pm e_{(i,l)}$, $l = 1, \cdots, a_i$, where the $e_i, e_{(i,l)}$ form a set of orthonormal vectors; in particular, all generalized line graphs have (by Proposition 3.5.2) smallest eigenvalue $\geqslant -2$.

The knowledge of all root lattices leads to a proof of the following important theorem by CAMERON, GOETHALS, SEIDEL & SHULT [180] characterizing graphs with smallest eigenvalue -2 by using their $(2,1,0)$-representation.

3.12.1. Theorem. *Let Γ be a connected graph with smallest eigenvalue ≥ -2. Then one of the following holds:*

(i) *Γ is a generalized line graph.*

(ii) *Γ has a (2,1,0)-representation by roots of E_8; in particular, Γ is a subgraph of $E_8(1)$. The number v of vertices, and the average valency $\bar{k}$ are restricted by $v \leq \min(36, 2\bar{k}+8)$. Moreover, every vertex has valency at most 28.*

Proof. According to Proposition 3.5.2, the graph Γ has a (2,1,0)-representation $\gamma \mapsto \bar{\gamma}$ ($\gamma \in \Gamma$), and the lattice $\mathbf{L}^+(\Gamma)$ generated by the images $\bar{\gamma}$ ($\gamma \in \Gamma$) is an irreducible root lattice. Hence $\mathbf{L}^+(\Gamma)$ is one of A_n, D_n, E_6, E_7, E_8, and since $A_n, D_n \subset \mathbf{Z}^{n+1}$, $E_6 \subset E_7 \subset E_8$, the image Σ of Γ is a set of roots of some $\mathbf{Z}^{n+1}$ or of E_8, with non-negative inner products. Suppose first that $\Sigma \subseteq \mathbf{Z}^{n+1}$. We may sign and label the basis vectors e_i of $\mathbf{Z}^{n+1}$ in such a way that for some $m \geq 1$ we have $(x,e_i) \geq 0$ for all $x \in \Sigma$ and $i = 1, \cdots, m$, but for $i > m$, (x,e_i) takes both signs for $x \in \Sigma$. Denote by Δ the graph with vertex set $\{1, \cdots, m\}$ and edge set $\{ij \mid e_i + e_j \in \Sigma\}$. The restriction on inner products now shows that for each $i > m$ there is a unique $i^* \leq m$ such that $e_{i^*} \pm e_i \in \Sigma$, and that the remaining roots of Σ have the form $e_i + e_j$ ($i < j \leq m$). Hence Γ is isomorphic to the generalized line graph $\Gamma(\Delta, a_1, \cdots, a_m)$, where a_l denotes the number of $i > m$ with $i^* = l$.

Suppose now that $\Sigma \subseteq E_8$. Then clearly $\Gamma \subseteq E_8(1)$. Consider the 36-dimensional space of symmetric 8×8-matrices, equipped with the positive definite trace inner product $(P,Q) := \mathrm{tr}\,(PQ)$. Associated with the 240 roots r of E_8 are 120 rank one matrices $P_r = rr^{\mathsf{T}}$ with mutual inner product $(P_r,P_s) = \mathrm{tr}\,(rr^{\mathsf{T}}ss^{\mathsf{T}}) = (r,s)^2$. The Gram matrix of the v-set $X = \{P_r \mid r \in \Sigma\}$ is $G = 4I + A$, where A is the adjacency matrix of Γ. Hence the smallest eigenvalue of G is ≥ 2, G is positive definite, X is linearly independent, and $v = |X| \leq 36$. Now let P be the sum of all 120 matrices P_r and let Q be the sum of the P_r ($r \in \Sigma$). The Gram matrix of P and Q,

$$\begin{bmatrix} 120(4+56) & v(4+56) \\ v(4+56) & v(4+\bar{k}) \end{bmatrix}$$

then has determinant $3600v(8+2\bar{k}-v) \geq 0$ which implies $v \leq 2\bar{k}+8$.
To bound the maximal valency, let $\gamma \in \Gamma$. By transitivity of $E_8(1)$ we may assume that $\bar{\gamma} = c_8$. Now $\bar{\gamma}$ has 56 neighbours in $E_8(1)$, namely the vertices of $E_7(1)$. But since $(e_i + e_j, c_8 - e_i - e_j) = -1$, at most half of the 56 neighbours can occur in Γ. Hence the valency of Γ is at most 28. □

A graph with 36 vertices, maximal valency 28, and smallest eigenvalue -2 which is not a generalized line graph can, e.g., be obtained by adding to $K_8 + L(K_8)$ edges joining $i \in K_8$ with $jk \in L(K_8)$ whenever $i \notin \{j,k\}$; a (2,1,0)-representation in E_8 is given by the vectors $\frac{1}{2}(f_1 + \cdots + f_8) - f_i$ ($i \leq 8$) and $f_i + f_j$ ($i < j \leq 8$), where $f_1, \cdots, f_8$ are obtained from $e_1, \cdots, e_8$ by reversing the sign of one e_i. Thus, examples satisfying equality in (ii) of the theorem exist. It would be interesting to have a description of all maximal graphs occurring under (ii).

For regular graphs sharper results can be obtained. The following theorem was given in a slightly weaker form in CAMERON et al. [180]. The more precise statements (ii)-(iv) about graphs represented in $E_8(1)$ were obtained by BUSSEMAKER et al. [153] (with the help of a computer). The proof given here is computer free.

3.12.2. Theorem. *Let Γ be a connected regular graph with v points, valency k, and smallest eigenvalue $\geqslant -2$. Then one of the following holds:*

(i) *Γ is the line graph of a regular or a bipartite semiregular connected graph Δ.*

(ii) *$v = 2(k+2) \leqslant 28$ and Γ is a subgraph of $E_7(1)$, switching equivalent to the line graph of a graph Δ on 8 vertices, where the valencies of all vertices of Δ have the same parity.*

(iii) *$v = \frac{3}{2}(k+2) \leqslant 27$ and Γ is a subgraph of the Schläfli graph.*

(iv) *$v = \frac{4}{3}(k+2) \leqslant 16$ and Γ is a subgraph of the Clebsch graph.*

(v) *$v = k+2$ and $\Gamma \cong K_{m\times 2}$ for some $m \geqslant 3$.*

Moreover, $\mathbb{L}^+(\Gamma) \cong A_n$ if and only if (i) *holds with a bipartite graph Δ with $n+1$ vertices, and $\mathbb{L}^+(\Gamma) \cong D_n$ if and only if either* (i) *holds with a graph Δ with n vertices which is not bipartite or* (v) *holds with $m = n-1$.*

Proof. Suppose first that Γ is a generalized line graph $L(\Delta; a_1, \cdots, a_m)$. Since Γ is connected, Δ may be taken to be connected, too. If the vertex $i \in \Delta$ has valency k_i then $ij \in \Gamma$ has valency $2(a_i + a_j) + k_i + k_j - 2$ whereas $(i,l) \in \Gamma$ has valency $2(a_i - 1) + k_i$. Since Γ is regular but $k_j > 0$, Γ can contain only vertices of one kind; hence either $\Gamma = K_{m\times 2}$ or $\Gamma = L(\Delta)$. In the latter case we may assume w.l.o.g. that the vertices 1 and 2 of Δ are adjacent. From the regularity of Γ one finds now that vertices at an even distance from 1 have valency k_1, and vertices at an odd distance from 1 have valency k_2. In particular, if Δ is not regular then $k_1 \neq k_2$, Δ contains no odd circuit, and hence Δ is a bipartite semiregular graph.

Now suppose that Γ is not a generalized line graph. Then the lattice $\mathbb{L} = \mathbb{L}^+(\Gamma)$ is one of E_6, E_7, E_8. The vector $c := (k+2)^{-1}\sum_{\gamma\in\Gamma}\overline{\gamma}$ satisfies

$$(c,\overline{\gamma}) = 1 \text{ for all } \gamma \in \Gamma,\ (c,c) = \frac{v}{k+2} =: N.$$

If $N \geqslant 3$ then by Theorem 3.12.1 (ii), $2k+8 \geqslant v \geqslant 3(k+2)$, giving $k \leqslant 2$, but since regular graphs with valencies $k \leqslant 2$ are line graphs, $N < 3$. Since $\mathbb{L}$ is generated by the $\overline{\gamma}$ ($\gamma \in \Gamma$), c belongs to the dual lattice $\mathbb{L}^*$. If $\mathbb{L} = E_8$, then nonzero vectors of $\mathbb{L}^* = \mathbb{L}$ have norm $2, 4, 6, \cdots$ so that $N = 2$, $v = 2(k+2)$. By transitivity of $W(E_8)$ on roots we may assume that $c = \frac{1}{2}(e_1 + \cdots + e_8)$, and we see that Γ is a subgraph of $E_7(1)$, hence (by Lemma 3.11.1) switching equivalent to the line graph of a graph Δ with 8 vertices; in particular, $v \leqslant 28$. Moreover, $\sum_{\gamma\in\Gamma}\overline{\gamma} = (k+2)c$ implies, by choice of c, that the valency of each vertex of Δ has the same parity. Hence (ii) holds.

If $\mathbb{L} = E_7$, then nonzero vectors of $\mathbb{L}^*$ have norm $\frac{3}{2}, 2, \frac{7}{2}, \cdots$ so that $N \in \{\frac{3}{2}, 2\}$. If $N = 2$, then the lattice generated by $\mathbb{L}$ and c is a root lattice, hence contained in some E_8 and we get case (ii) as before. And if $N = \frac{3}{2}$, then $v = \frac{3}{2}(k+2)$. The graph $\Gamma' = \Gamma \cup \{\alpha_1, \alpha_2\}$, where α_1 and α_2 are adjacent and adjacent to all points of Γ, has the (2,1,0)-representation $\Sigma \cup \{c+a, c-a\}$, where a is a vector of norm $\frac{1}{2}$ orthogonal to E_7. As Γ' is not a generalized line graph, it is a subgraph of $E_8(1)$, by Theorem 3.12.1. By transitivity of $W(E_8)$ on 2-cliques we now find that $\Gamma \subseteq E_6(1)$. In particular, $v \leqslant 27$, and we have case (iii).

If $\mathbb{L} = E_6$, then nonzero vectors of $\mathbb{L}^*$ have norm $\frac{4}{3}, 2, \frac{10}{3}, \cdots$ so that $N \in \{\frac{4}{3}, 2\}$. For $N = 2$ we get case (ii) as before. And if $N = \frac{4}{3}$, then $v = \frac{4}{3}(k+2)$. Now the graph $\Gamma'' = \Gamma \cup \{\alpha_1,\alpha_2,\alpha_3\}$, where $\alpha_1, \alpha_2, \alpha_3$ form a triangle adjacent with all points of Γ, has the (2,1,0)-representation $\Sigma \cup \{c+a_1-a_2, c+a_2-a_3, c+a_3-a_1\}$, where a_1, a_2, a_3 are mutually orthogonal vectors of norm $\frac{1}{3}$ orthogonal to E_6. As before, we find $\Gamma \subseteq E_5(1)$, $v \leqslant 16$, hence case (iv).

The final statement about $\mathbb{L}^+(\Gamma)$ is easily seen to hold. □

Remark. There is a unique regular graph Γ with $\mathbb{L}^+(\Gamma) \cong E_6$ and $v = 2(k+2)$, namely the Petersen graph. There are precisely three graphs Γ with $\mathbb{L}^+(\Gamma) \cong E_7$ and $v = 2(k+2)$, namely the Shrikhande graph, and the two graphs obtained by deleting from the Shrikhande graph a 4-coclique and $3K_2$, respectively, cf. 3.11 (vi).
(Indeed, using the notation of the above proof, if $(c,c) = 2$, then $c \in \mathbb{L}$ and Γ is represented in $\{x \in \mathbb{L} \mid (x,c) = 1\}$. It follows that Γ is switching equivalent to a subgraph of the Petersen graph or the Shrikhande graph, respectively.)

3.12.3. Corollary (cf. DOOB & CVETKOVIĆ [268]). *Let Γ be a regular graph with v vertices and valency k.*

(i) *If Γ is connected and its smallest eigenvalue is > -2, then either Γ is complete or v is odd and $k = 2$ (i.e., Γ is an odd polygon).*

(ii) *If the second largest eigenvalue of Γ is < 1 then Γ is a coclique, a complete multipartite graph, or the complement of a union of odd polygons (and in particular $v = k+3$).*

Proof. (i) In this case $2I+A$ is positive definite; hence the (2,1,0)-representation maps Γ to a set Σ of linearly independent vectors. If $\mathbb{L}^+(\Gamma) = A_n$ then Σ contains precisely n points $e_i - e_j$ spanning A_n, so that Γ is the line graph of a tree on $n+1$ vertices, and the only possible regular graph is $L(K_{1,n})$, the complete graph K_n. If $\mathbb{L}^+(\Gamma) = D_n$ then Σ contains precisely n points $\pm e_i \pm e_j$ spanning D_n, and the only possible regular graph is a polygon (if n is odd). If $\mathbb{L}^+(\Gamma) \subseteq E_8$ no further graph is found.
(ii) Here, $I-A+J = 2I+(J-I-A)$ is positive definite, and we can apply the analysis of (i) to each component of the complement. □

As a consequence of Theorem 3.12.2 we are able to prove Seidel's characterization of strongly regular graphs with smallest eigenvalue -2 and to extend it to edge-regular and co-edge-regular graphs.

3.12.4. Theorem. *Let Γ be a connected regular graph with smallest eigenvalue -2.*

(i) (SEIDEL [653]) *If Γ is strongly regular then Γ is a triangular graph $T(n)$ ($n \geqslant 5$), a square grid $n \times n$ (also called a lattice graph $L_2(n)$) ($n \geqslant 3$), a complete multipartite graph $K_{n\times 2}$ ($n \geqslant 2$), or one of the graphs of Petersen, Clebsch, Schläfli, Shrikhande, or Chang.*

(ii) *If Γ is edge-regular, then Γ is strongly regular or the line graph of a regular triangle-free graph.*

(iii) *If Γ is amply regular, then Γ is strongly regular or the line graph of a regular graph of girth $\geqslant 5$.*

(iv) *If Γ is co-edge-regular, then Γ is strongly regular, an $m \times n$-grid, or one of the two regular subgraphs of the Clebsch graph with 8 and 12 vertices, respectively.*

Proof. Γ is not a complete graph and not a pentagon, since these graphs have smallest eigenvalue > -2. In view of Theorem 3.12.2 and Proposition 3.11.2 it is sufficient to

consider the case where $\Gamma = L(\Delta)$ is the line graph of a connected graph Δ, which is regular, or bipartite and semiregular.
Suppose first that Γ is edge-regular. If Δ is not regular then Δ is bipartite, and the valencies p, q of the two parts of Δ are distinct, say $p < q$. If $p = 1$, then $\Delta = K_{1,q}$ and Γ is complete, contradiction; hence $p > 1$. The common neighbours of two adjacent vertices $ab, ac \in L(\Delta)$ are the ad ($d \in \Delta(a) \setminus \{b,c\}$), and their number is $p-2$ or $q-2$ depending on the valency of a, contradiction. Hence Δ is regular, of valency h, say; clearly $h > 1$. If $a \in \Delta$ and $b, c \in \Delta(a)$, then ab and ac have $h-1$ or $h-2$ common neighbours in $L(\Delta)$ depending on whether bc is an edge or not. Since $\Gamma = L(\Delta)$ is edge-regular, this number must be constant. Hence either all $\Delta(a)$ are cliques and $\Delta = K_n$, in which case $\Gamma = T(n)$ is strongly regular, or all $\Delta(a)$ are cocliques, i.e., Δ is triangle-free. This proves (ii).
If Γ is amply regular then, in addition, disjoint edges of Δ have either μ transversals (edges meeting the given edges) or none. For a triangle-free Δ, this holds never if $\mu > 2$; for $\mu = 1$ if and only if Δ has no quadrangles; and for $\mu = 2$ if and only if $\Delta = K_{n,n}$ (one easily sees that a maximal $K_{m,n}$ subgraph of Δ agrees with Δ, and regularity implies $m = n$). In the latter case Γ is a square grid, which is strongly regular. This proves (iii).
Now suppose that Γ is co-edge-regular. Then any two disjoint edges of Δ have precisely μ transversals. Clearly $\mu \leqslant 4$, and equality forces $\Delta = K_n$ and $\Gamma = T(n)$. If $\mu = 3$ then Δ contains a $K_{2,1,1}$ and it is easily seen that Δ must be a $K_{m,1,1}$, contradicting the regularity of Δ. If $\mu = 2$ and Δ contains a triangle, then Δ must be a $K_{m,1}$ with an extra edge, contradicting the regularity of Δ. If $\mu = 2$ and Δ contains no triangle, then, as before, $\Delta = K_{m,n}$ and Γ is a grid $m \times n$. Finally, if $\mu = 1$ then Δ has girth $\geqslant 5$ and must be the pentagon; then Γ is the pentagon, contradiction. Thus a co-edge-regular line graph is a triangle graph or a grid, proving (iv).
Since every strongly regular graph is co-edge-regular, and the grids $m \times n$ are strongly regular only when $m = n$, this also implies (i). □

Line graphs, generalized line graphs, and graphs with smallest eigenvalue $\geqslant -2$ have been characterized by a finite list of forbidden subgraphs, see VAN ROOIJ & WILF [623], RAO, SINGHI & VIJAYAN [605], KUMAR, RAO & SINGHI [469] and the survey in BUSSEMAKER & NEUMAIER [155]. As a consequence, we have the following result about graphs with smallest eigenvalue slightly less than -2.

3.12.5. Theorem.

(i) (HOFFMAN [399]). *Let $\hat{\theta}_k$ be the supremum of the smallest eigenvalues of graphs with minimal valency k and smallest eigenvalue < -2. Then $(\hat{\theta}_k)_k$ forms a monotone decreasing sequence with limit $-1-\sqrt{2}$.*

(ii) (BUSSEMAKER & NEUMAIER [155]). *$\hat{\theta}_1$ is the smallest eigenvalue of the graph*

E_{10}

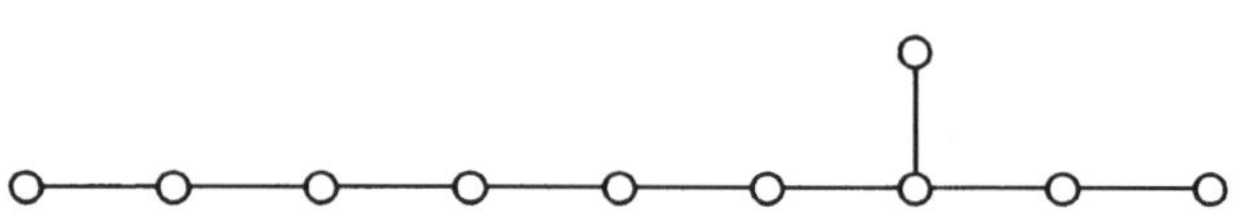

and this graph is the unique connected graph with this smallest eigenvalue (which is approximately -2.006594, the smallest root of the characteristic equation $\theta^2(\theta^2-1)^2(\theta^2-3)(\theta^2-4) = 1$).

Proof. See the given references. □

In connection with (ii) we mention that the complement of a disjoint union of at least two graphs with largest eigenvalue θ_0 has smallest eigenvalue $-1-\theta_0$, and taking for these graphs paths of length 2 gives graphs of arbitrarily large minimum valency and smallest eigenvalue $-1-\sqrt{2}$. Cf. also HOFFMAN [400]. Note that knowledge of non-trivial upper bounds for $\hat{\theta}_k$ would allow to weaken the hypothesis of Theorem 3.12.2.

Remarks. Of major interest would be a classification of the graphs Γ with smallest eigenvalue $\geqslant -3$, i.e., of special (3,1,0)-representations. There are many relevant examples of such graphs, e.g., all regular graphs of valency 3, the line graphs of partial linear spaces with line size 3, clique extensions of the graphs obtained from the graphs with smallest eigenvalue -2, where each vertex is replaced by a 2-clique, and a number of sporadic distance-regular graphs. The lattice $\mathbb{L}^+(\Gamma)$ is then generated by norm 3 vectors, and as a first step of classifying the graphs Γ one would try to classify the integral lattices generated by norm 3 vectors. This is a difficult problem, cf. NEUMAIER [559], PLESKEN & POHST [594] and CONWAY & SLOANE [214]. A slightly different view of the problem is obtained by looking instead at the (4,2,1)-representation obtained by adding to each image vector a fixed unit vector e orthogonal to $\mathbb{L}^+(\Gamma)$. Then the norm 4 vectors $2e$ and $e+\overline{\gamma}$ ($\gamma \in \Gamma$) generate an even lattice $\mathbb{L}$, and the points of the (4,2,1)-representation belong to

$$\mathbb{L}_u^{(4)} := \{x \in \mathbb{L}^{(4)} \,|\, (x,u) = 2\},$$

where $u = 2e$ is a particular norm 4 vector of $\mathbb{L}$. Now, in general, if $\mathbb{L}$ is an integral lattice generated by norm 4 vectors and $u \in \mathbb{L}^{(4)}$, then the vectors $x - \frac{1}{2}u$ $(x \in \mathbb{L}_u^{(4)})$ are norm 3 vectors with integral inner product; hence they again generate an integral lattice. In our situation, with $u = 2e$, we just recover the lattice $\mathbb{L}^+(\Gamma)$. So we may as well try to classify integral lattices $\mathbb{L}$ generated by norm 4 vectors and then study (4,2,1)-representations in $\mathbb{L}_u^{(4)}$, for $u \in \mathbb{L}^{(4)}$. In such a classification one has to account at least for root lattices, root lattices scaled by a factor $\sqrt{2}$, and their refinements by adding further norm 4 vectors. For example, the Leech lattice (LEECH [481], CONWAY [212]) arises as a refinement of $\sqrt{2}D_{24}$. A natural question is whether every integral lattice generated by norm 4 vectors can be embedded into a refinement of a root lattice; note that $\sqrt{2}\mathbb{L}$ is embeddable into $\mathbb{L} \oplus \mathbb{L}$. A few studies of general lattices of minimum norm 4 are available (CONWAY & SLOANE [213, 214], NEUMAIER [562]).

3.13 Equiangular lines

In their paper on equiangular lines, LEMMENS & SEIDEL [483] show that the converse of Proposition 3.8.1 holds in case $m = 3$ for graphs with more than 28 vertices. The following more precise result is implicitly contained in SHULT & YANUSHKA [673], Proposition 3.8 and Theorem 3.9; cf. also NEUMAIER [559]:

3.13.1. Theorem. *A graph has a* $(3,-1,1)$*-representation if and only if it is switching equivalent to a subgraph of some* nK_2 *(the disjoint union of n edges) or of the complement of the triangular graph* $T(8)$*.*

Proof. We first note that nK_2 is represented by $\{e_0 \pm \sqrt{2}e_i \,|\, i=1, \cdots, n\}$, and the complement of $T(8)$ is represented by the 28 vectors obtained by permuting the indices of the vector $2^{-3/2}(3e_1+3e_2-e_3-e_4-e_5-e_6-e_7-e_8)$, where the e_i $(i \geqslant 0)$ are mutually orthogonal unit vectors. It is easily checked that these are

$(3,-1,1)$-representations. Hence every subgraph of nK_2 or $\overline{T(8)}$, and every graph switching equivalent to it, has a $(3,-1,1)$-representation.
Conversely, suppose that Γ has a $(3,-1,1)$-representation. We may assume that Γ has an isolated vertex γ (otherwise we switch Γ with respect to $\Gamma(\gamma)$). Denote the images of γ and $\Gamma\setminus\{\gamma\}$ under the representation by w and Q_0, respectively, and let $\mathbb{L} = \mathbb{L}^+(\Gamma)$. Then $\mathbb{L}$ is an integral lattice, $Q_0 \cup \{w\} \subseteq \mathbb{L}^{(3)}$, and $(x,w) = 1$, $(x,y) \in \{1,-1\}$ for all $x,y \in Q_0$. Now let Q be a maximal subset of $\{z \in \mathbb{L}^{(3)} \mid (z,w) = 1\}$ containing Q_0 such that all inner products of vectors in Q are odd. Call vectors in Q points and 3-subsets xyz of Q satisfying $x+y+z = w$ lines. We show that Q is a generalized quadrangle.
If $x,y \in Q$, then $(x,y)^2 \leqslant (x,x)(y,y) = 9$; hence $(x,y) \in \{1,-1,3,-3\}$ since inner products are odd. But if $(x,y) = \pm 3$, then $y = \pm x$, and the negative sign is impossible since $(x+y,w) = 2$. Hence distinct points have inner product 1 or -1. If xyz is a line, then $(x,y)+(x,z) = (x,w)-(x,x) = -2$ whence $(x,y) = (x,z) = -1$, and, by symmetry $(y,z) = -1$. Conversely, if $(x,y) = -1$, then $z := w-x-y$ satisfies $(z,w) = 1$ and $(z,x) = (z,y) = -1$; hence $(z,z) = (z,w-x-y) = 3$, i.e., $z \in \{z \in \mathbb{L}^{(3)} \mid (z,w) = 1\}$. Moreover, for any $u \in Q$, $(z,u) = 1-(x,u)-(y,u)$ is odd since (x,u) and (y,u) are odd. Hence $z \in Q$ by maximality of Q. Thus we have shown that two points $x,y \in Q$ are adjacent (in the point graph of Q) if and only if $(x,y) = -1$. Now let xyz be a line and let u be a point of Q not on the line. Since $(u,x), (u,y), (u,z) \in \{1,-1\}$ and $(u,x)+(u,y)+(u,z) = (u,w) = 1$, precisely one of the three inner products is -1, i.e., u is adjacent to a unique point of the line. Hence Q is a generalized quadrangle with lines of size 3.
Now we may apply Theorem 1.15.2 which implies that either Q is contained in a set of $2n+1$ points and n lines intersecting in a common point z, say, or Q is contained in a generalized quadrangle Q_4 with 27 points. In the first case, switching Γ with respect to the vertex whose image is z (if such a vertex exists) results in a subgraph of $(n+1)K_2$. And in the second case, switching Γ with respect to the vertices whose images are in a fixed double six of Q_4 (cf. §1.15) results in a subgraph of $\overline{T(8)}$. □

Remarks. (i) Shifting a $(3,-1,1)$-representation of a graph Γ by a suitable norm 1 vector and scaling by $1/\sqrt{2}$ gives a (2,0,1)-representation of Γ and hence a (2,1,0)-representation of its complement $\overline{\Gamma}$. This relates the above result to Theorem 3.12.1 and Theorem 3.12.2; note that the subgraphs of $T(8)$ are just the line graphs of graphs Δ on 8 vertices.
(ii) It can be shown that given m with $3 < m < 1+2\sqrt{2}$, there is a number v_m such that every graph with an $(m,-1,1)$-representation and more than v_m vertices is switching equivalent to a subgraph of some nK_2 (Neumaier [564]). Note that in view of Proposition 3.2.4, a graph with largest eigenvalue $< \sqrt{2}$ is a subgraph of some nK_2 since a triangle has largest eigenvalue 2 and a path of length 2 has largest eigenvalue $\sqrt{2}$.
(iii) More generally, given $m > 1$, we may ask for the smallest number v_m such that every graph with an $(m,-1,1)$-representation and more than v_m vertices is switching equivalent to a graph with largest eigenvalue $\leqslant (m-1)/2$; if no such number exists we may put $v_m = \infty$. The above theorem implies that $v_3 = 28$ and Neumaier [564] shows that $v_5 < \infty$. In view of results of Lemmens & Seidel [483] and Goethals & Seidel [329], the exceptions for $m = 5$ seem to be related to the regular two-graph with 276 vertices (cf. §11.4H).

3.14 Root graphs

A *root representation* of a graph is an integral affine (2,4)-representation, i.e., a representation such that the squared distances $[\gamma,\delta] = (\bar{\gamma}-\bar{\delta},\bar{\gamma}-\bar{\delta})$ of the images of all $\gamma, \delta \in \Gamma$ are even integers and satisfy

$$[\gamma,\delta] = 2i \text{ if } d(\gamma,\delta) = i \leqslant 2.$$

A connected graph having a root representation is called a *root graph,* and a *strict root graph* if the representation is strict, i.e.,

$$[\gamma,\delta] = 2i \leqslant 4 \Leftrightarrow d(\gamma,\delta) = i \leqslant 2.$$

The name reflects the fact that the lattice $\mathbb{L}(\Gamma)$ of a root representation of Γ, defined in Section 3.9, is a root lattice:

3.14.1. Proposition. *Let Γ be a connected root graph. Then $\mathbb{L}(\Gamma)$ is a root lattice. Moreover, if Γ is locally connected, then $\mathbb{L}(\Gamma)$ is irreducible.*

Proof. Since Γ is connected, $\mathbb{L}(\Gamma)$ is generated by the norm 2 vectors $\bar{\gamma}-\bar{\delta}$ ($\gamma, \delta \in \Gamma$, $\gamma \sim \delta$), and is therefore a root lattice. Now suppose that $\mathbb{L}(\Gamma)$ is the direct sum of two lattices $\mathbb{L}_1$ and $\mathbb{L}_2$. If we call an edge $\gamma\delta$ of *type i* if the norm 2 vector $\bar{\gamma}-\bar{\delta}$ is in $\mathbb{L}_i$ then each edge is either of type 1 or of type 2. If $\alpha\beta\gamma$ is a triangle, then $(\bar{\alpha}-\bar{\beta},\bar{\alpha}-\bar{\gamma}) = 1$ so that edges in the same triangle have the same type. If Γ is locally connected, then this implies that all edges have the same type, whence $\mathbb{L}(\Gamma) = \mathbb{L}_1$ or $\mathbb{L}(\Gamma) = \mathbb{L}_2$, showing that $\mathbb{L}(\Gamma)$ is irreducible. □

Using the classification of root lattices we shall give a complete characterization of amply regular root graphs with $\mu > 1$ in the next section. This leads in the next chapter to an important classification theorem for distance-regular graphs (Theorem 4.4.11).

Remark. Related questions are considered by TERWILLIGER [724], who considers structure theorems for (not necessarily regular) graphs having a spherical representation such that $[\gamma,\delta]$ is an even integer for all $\gamma, \delta \in \Gamma$ and $[\gamma,\delta] = 2$ if and only if $d(\gamma,\delta) = 1$.

Root graphs generalize graphs with smallest eigenvalue $\geqslant -2$:

3.14.2. Theorem.

(i) *Every graph with smallest eigenvalue $\geqslant -2$ is a root graph.*

(ii) *A regular root graph of diameter $\leqslant 2$ has smallest eigenvalue $\geqslant -2$.*

Proof. (i) A graph with smallest eigenvalue $\geqslant -2$ has a spherical (2,1,0)-representation which is a root representation.
(ii) By Proposition 3.5.6 (ii), a regular root graph Γ of diameter $\leqslant 2$ has a spherical $(p,p-1,p-2)$-representation for some $p > 0$; thus by Proposition 3.5.3 (iii), Γ has smallest eigenvalue $\geqslant -2$. □

The graphs $K_{1,n}$ are root graphs of diameter 2 with smallest eigenvalue $-\sqrt{n}$, so regularity is required in part (ii) above.

Examples

The representation of $E_7(1)$ defined in 3.11 is a root representation, so that any induced subgraph of $E_7(1)$ is a root graph. (However, $E_8(1)$, as given in 3.10, contains vertices with squared distance 6 having a common neighbour, and hence defines no root representation). Further examples of root graphs are the *code graphs* $\Gamma(C)$, defined by the words of a binary code C of length n, words being adjacent if they differ in two entries (i.e., if their Hamming distance is 2); they have a natural representation by (0,1)-vectors in $\mathbf{R}^n$. We mention in particular the following amply regular code graphs.

(i) The *half-cube* $D_{n,n}(1)$ is the code graph of the binary code consisting of all even weight words of length n. $D_{n,n}(1)$ has 2^{n-1} vertices, valency $\binom{n}{2}$ and $\mu = 6$. The graph $D_{5,5}(1)$ is isomorphic to the Clebsch graph described in Section 3.11.

(ii) The *Johnson graph* $J(n,d)$ is the code graph of the binary code consisting of all words of length n and weight d. $J(n,d)$ has $\binom{n}{d}$ vertices, valency $d(n-d)$ and $\mu = 4$. In particular ($d = 1$), n-cliques are code graphs.

(iii) The *icosahedron* with 12 vertices, valency 5 and $\mu = 2$ can be viewed as the code graph of the binary code consisting of the words 000000, 110000, 001111, 111111, and those obtained by a cyclic permutation of the first five entries.

(iv) The *Shrikhande graph* with 16 vertices, valency 6 and $\mu = 2$ can be viewed as the code graph of the binary code consisting of the words 000000, 110000, 010111, 011011 and those obtained by a cyclic permutation of the six entries (cf. Example 3.11 (vi)).

(v) The *Petersen graph* with 10 vertices, valency 3 and $\mu = 1$ can be viewed as the code graph of the binary code consisting of the words 000000, 100100, 001111 and those obtained by a cyclic permutation of the six entries (cf. Example 3.11 (v)).

(vi) The direct product Γ of root graphs $\Gamma^1, \cdots, \Gamma^s$ is a root graph: simply represent $(\gamma_1, \cdots, \gamma_s)$ by the sum of the images of the γ_i. Clearly $\mathbf{L}(\Gamma)$ is the direct sum of $\mathbf{L}(\Gamma^1), \cdots, \mathbf{L}(\Gamma^s)$. In particular, the *Hamming graphs* (direct products of cliques) and *Doob graphs* (the direct product of 4-cliques and Shrikhande graphs, cf. §9.2B) are root graphs.

We first prove some easy results about the local structure of root graphs.

3.14.3. Lemma. *Let Γ be a root graph and $\alpha, \beta, \gamma \in \Gamma$. For the associated root representation we have*

$$(\overline{\alpha}-\overline{\gamma},\overline{\beta}-\overline{\gamma}) = 2-d(\alpha,\beta) \geqslant 0 \ \text{ if } \alpha,\beta \in \Gamma(\gamma) \tag{6}$$

$$= 3-d(\alpha,\beta) \geqslant 0 \ \text{ if } \alpha \in \Gamma(\gamma),\ \beta \in \Gamma_2(\gamma), \tag{7}$$

$$= 4-d(\alpha,\beta) \geqslant 0 \ \text{ if } \alpha,\beta \in \Gamma_2(\gamma), \tag{8}$$

provided that $d(\alpha,\beta) \leqslant 2$.
In particular, if $\overline{\gamma} = 0$, then points in $\Gamma(\gamma)$ are represented by norm 2 vectors, and points in $\Gamma_2(\gamma)$ are represented by norm 4 vectors.

Proof. By Lemma 3.5.1 we have

$$(\overline{\alpha}-\overline{\gamma},\overline{\beta}-\overline{\gamma}) = \tfrac{1}{2}([\alpha,\gamma]+[\beta,\gamma]-[\alpha,\beta]-[\gamma,\gamma])$$
$$= d(\alpha,\gamma)+d(\beta,\gamma)-d(\alpha,\beta),$$

which implies (6), (7) and (8). By putting $\alpha = \beta$ we obtain the second statement. □

3.14.4. Proposition. *Let Γ be a root graph and $\gamma \in \Gamma$. Then:*

(i) *The local subgraph $\Gamma(\gamma)$ has smallest eigenvalue $\geqslant -2$.*

(ii) *Two nonadjacent points $\alpha,\beta \in \Gamma(\gamma)$ have at most one common neighbour in $\Gamma_2(\gamma)$, and such a neighbour is represented by $\overline{\alpha}+\overline{\beta}-\overline{\gamma}$.*

(iii) *For $\alpha,\beta \in \Gamma$ at distance two, the graph $\Gamma(\alpha) \cap \Gamma(\beta)$ is complete multipartite with classes of size 1 or 2.*

Proof. (i) By Proposition 3.5.6 (i), $\Gamma(\gamma)$ has a (2,1,0)-representation. Thus, if A denotes the adjacency matrix of $\Gamma(\gamma)$, then the Gram matrix $G = 2I+A$ of the image of $\Gamma(\gamma)$ is positive semidefinite. This implies (i).
(ii) Now let α, $\beta \in \Gamma(\gamma)$ be nonadjacent. If $\delta \in \Gamma_2(\gamma)$ is a common neighbour of α and β then Lemma 3.5.1 implies

$$(\overline{\delta}-\overline{\alpha},\overline{\beta}-\overline{\gamma}) = \tfrac{1}{2}([\delta,\gamma]+[\alpha,\beta]-[\delta,\beta]-[\alpha,\gamma]) = \tfrac{1}{2}(4+4-2-2) = 2.$$

Thus the norm of $(\overline{\delta}-\overline{\alpha})-(\overline{\beta}-\overline{\gamma})$ is $[\delta,\alpha]-2(\overline{\delta}-\overline{\alpha},\overline{\beta}-\overline{\gamma})+[\beta,\gamma] = 2-4+2 = 0$. Therefore $(\overline{\delta}-\overline{\alpha})-(\overline{\beta}-\overline{\gamma}) = 0$, so $\overline{\delta} = \overline{\alpha}+\overline{\beta}-\overline{\gamma}$, and, by Lemma 3.5.5, δ is unique. This implies (ii).
(iii) Finally, let $d(\alpha,\beta) = 2$. If $\Gamma(\alpha) \cap \Gamma(\beta)$ is not as claimed in (iii), then it contains three distinct points γ, δ, δ' such that $\delta, \delta' \notin \Gamma(\gamma)$. But this contradicts part (ii). □

3.14.5. Corollary. *Let Γ be a root graph with $\mu(\Gamma) = \mu$. Then any two nonadjacent vertices $\alpha,\beta \in \Gamma(\gamma)$ have either $\mu-1$ or $\mu-2$ common neighbours in $\Gamma(\gamma)$.*

Proof. Among the μ common neighbours of α and β, one is γ, and at most one is in $\Gamma_2(\gamma)$. The remaining neighbours must be in $\Gamma(\gamma)$. □

3.14.6. Lemma. *Let Γ be a root graph with $\mathbf{L}(\Gamma) \subseteq \mathbf{Z}^{n+1}$ and γ, δ vertices at distance two with $\overline{\delta}-\overline{\gamma} = 2e_0$, where $e_0 = (1,0,\cdots,0)$. Then no vertex of $\Gamma(\gamma) \cap \Gamma(\delta)$ is adjacent to a vertex of $\Gamma(\gamma) \setminus \Gamma(\delta)$.*

Proof. Without loss of generality we may assume that $\overline{\gamma} = 0$ and $\overline{\delta} = 2e_0$. Lemma 3.14.3 shows that common neighbours of γ and δ are represented by suitable vectors $e_0 \pm e_l$, and, by permuting the e_i and changing their signs if necessary, we may assume that $M := \Gamma(\gamma) \cap \Gamma(\delta)$ is represented by the $\mu = 2s+t$ vectors

$$e_0 \pm e_l \ (l = 1,\cdots,s), \quad e_0+e_l \ (l = s+1,\cdots,s+t).$$

Let $\alpha \in M$ so that (for some i) $\overline{\alpha} = e_0 \pm e_i$, and suppose that γ has a neighbour $\beta \notin M$ adjacent with α. Then $d(\beta,\delta) = 2$; but either $\overline{\beta} = e_0 \pm e_j$, $(\overline{\beta},\overline{\delta}) = 2$, or $\overline{\beta} = \pm e_i \pm e_j$ $(0 \neq j \neq i \neq 0)$, $(\overline{\beta},\overline{\delta}) = 0$, both contradicting Lemma 3.14.3. Hence no point of $\Gamma(\gamma) \setminus \Gamma(\delta)$ is adjacent to a point of M. □

3.15 Classification of amply regular root graphs

3.15.1. Theorem. *Let Γ be a regular root graph with $\mu(\Gamma) = \mu > 2$. Then Γ is locally connected and one of the following holds.*

(i) *Γ is a complete multipartite graph with classes of size 1 or 2.*

(ii) *$\mu = 6$ and Γ is a half-cube.*

(iii) *$\mu = 4$ and Γ is a Johnson graph.*

(iv) *$\mathbf{L}(\Gamma) \cong E_6$, E_7, or E_8.*

Proof. By Corollary 3.14.5, Γ is locally connected, and by Proposition 3.14.1, $\mathbf{L}(\Gamma)$ is irreducible. Now we assume that we are not in case (iv). Then by Theorem 3.10.4, $\mathbf{L}(\Gamma)$ is isomorphic to A_n or D_n; in particular $\mathbf{L}(\Gamma) \subseteq \mathbf{Z}^{n+1} := \mathbf{Z}e_0 \oplus \mathbf{Z}e_1 \oplus \cdots \oplus \mathbf{Z}e_n$, and we may identify vectors in $\mathbf{L}(\Gamma)$ with $(n+1)$-tuples of integers.

Suppose that Γ contains points γ, δ at distance 2 such that the norm 4 vector $\overline{\gamma} - \overline{\delta}$ has the form $\pm 2e_j$ for some j. Since Γ is locally connected, it follows from Lemma 3.14.6 that $\Gamma(\gamma) \cap \Gamma(\delta) = \Gamma(\gamma)$. Therefore Γ has valency $k = \mu$, and Γ is complete multipartite with classes of constant size s. But $\Gamma(\gamma)$ has classes of size 1 or 2 only; therefore $s \leqslant 2$, and (i) holds.

We may now assume that Γ does not contain points γ, δ at distance 2 such that $\overline{\gamma} - \overline{\delta} = \pm 2e_j$ $(j = 1, \cdots, n)$. Let γ be fixed, and shift the representation such that $\overline{\gamma} = 0$. By Lemma 3.14.3, points of $\Gamma(\gamma)$ are represented by norm 2 vectors, i.e., vectors of the form $\pm e_i \pm e_j$. We now show that the basis of $\mathbf{Z}^{n+1}$ can be chosen such that all signs are positive. Without loss of generality, we may assume that $e_0, \cdots, e_n$ is a basis such that the set N of points of $\Gamma(\gamma)$ represented by vectors of the form $e_i + e_j$ has maximal size. Suppose that some $\delta \in \Gamma(\gamma) \setminus N$ satisfies $\overline{\delta} = \pm e_i \pm e_j$. If the sign of e_i in $\overline{\delta}$ is negative, then no point $\alpha \in N$ can have an image $e_i + e_k$ (otherwise $(\overline{\alpha}, \overline{\delta}) \geqslant 0$ would imply $\overline{\alpha} = e_i + e_j$, $\overline{\delta} = -e_i + e_j$, $d(\alpha,\delta) = 2$, $\overline{\alpha} - \overline{\delta} = 2e_i$, contradiction). Therefore all points of N have an image orthogonal to e_i, and we can replace e_i in the basis by $-e_i$ without removing points from N. Similarly, if the sign of e_j in $\overline{\delta}$ is negative we can replace e_j in the basis by $-e_j$ without removing points from N. Therefore we may assume that the signs of e_i and e_j in $\overline{\delta}$ are positive, $\overline{\delta} = e_i + e_j$. But then $\delta \in N$, contradiction. Hence $\Gamma(\gamma) \setminus N = \varnothing$, $\Gamma(\gamma) = N$, and all points of $\Gamma(\gamma)$ are represented by vectors of the form $e_i + e_j$ $(i \neq j)$. Now let Δ be the graph whose vertices are the $n+1$ indices $0, 1, \cdots, n$, and whose edges are the pairs ij such that some point of $\Gamma(\gamma)$ is represented by $e_i + e_j$. Since distinct points of $\Gamma(\gamma)$ are represented by distinct vectors (their difference must have norm 2 or 4) we may identify the point represented by $e_i + e_j$ with the edge ij of Δ.

Let us denote the projection of the image $\overline{\delta}$ of a point $\delta \in \Gamma$ to $\mathbf{Z}e_i$ by δ_i, so that $\overline{\delta} = \delta_0 e_0 + \cdots + \delta_n e_n$, and denote by $S(\delta)$ the set of indices i with $\delta_i \neq 0$. By Lemma 3.14.3, the points $\delta \in \Gamma_2(\gamma)$ are represented by norm 4 vectors $\overline{\delta} = \overline{\delta} - \overline{\gamma}$, and since they may not have the form $\pm 2e_i$ they must have the form $\pm e_i \pm e_j \pm e_k \pm e_l$, i.e., $S(\delta)$ is a 4-set and $\delta_i = \pm 1$ for $i \in S(\delta)$. The μ common neighbours of γ and δ are certain edges of the subgraph of Δ induced on $S(\delta)$. But Lemma 3.14.3(6) implies that with one edge the adjacent edges in $S(\delta)$ are also common neighbours of γ and δ, and since $\mu > 1$, the intersection $\Gamma(\gamma) \cap \Gamma(\delta)$ coincides with the set of edges of Δ contained in $S(\delta)$.

Now suppose that $\delta, \delta' \in \Gamma_2(\gamma)$ are distinct points with $S(\delta) = S(\delta')$. Then δ and δ' have μ common neighbours in $\Gamma(\gamma)$, namely the edges of $S(\delta)$; therefore $d(\delta,\delta') \leqslant 2$

and $\overline{\delta}-\overline{\delta}'$ has norm $\leqslant 4$. Lemma 3.5.5 implies $\overline{\delta} \neq \overline{\delta}'$, and then $S(\delta) = S(\delta')$ implies that $\overline{\delta}-\overline{\delta}'$ is of the form $\pm 2e_i$, contradiction. Hence $\delta \in \Gamma_2(\gamma)$ is uniquely determined by $S(\delta)$, which allows us to identify the point $\delta \in \Gamma_2(\gamma)$ with the 4-set $S(\delta)$. To simplify the notation we shall use the abbreviation *abcd* for a 4-set $\{a,b,c,d\}$.

Let us call a set $S \subseteq \{0,1, \cdots ,n\}$ *special* if there is a point $\delta \in \Gamma$ with

$$S(\delta) = S, \ d(\delta,\gamma) = \tfrac{1}{2}|S|, \ \overline{\delta} = \sum_{l \in S} e_l.$$

Clearly, the empty set is special. Application of Lemma 3.14.3 to pairs of nonadjacent points in $\Gamma_{\leqslant 2}(\gamma)$ yields the following facts (here distinct Latin letters denote distinct indices taken from $\{0, \cdots ,n\}$).

S1. If $\delta \in \Gamma_2(\gamma)$ then $S(\delta)$ is special unless $\overline{\delta} = e_i + e_j + e_k - e_l$ for suitable indices (the latter is possible only if $\mu \leqslant 3$).

S2. If *abcd* is special then precisely μ of the sets *ab, ac, ad, bc, bd, cd* are special.

S3. If *ab* and *cd* are special then precisely μ of the sets $\varnothing$, *ac, ad, bc, bd, abcd* are special.

S4. If *abcd* and *de* are special then none or μ of the sets *ad, bd, cd, abde, acde, bcde* are special.

(Indeed, if *abcd* and *de* represent vertices at distance two, then they have μ common neighbours, and these must be among the six sets listed. On the other hand, if one of these six sets is special, we must show that it represents a common neighbour of *abcd* and *de*. For *ad, bd, cd* this is clear, and the remaining three are certainly neighbours of *de* (when special). But if e.g. *abde* is special but not a neighbour of *abcd* then its distance to *abcd* is at least 3 and we find $\mu = 3$, *ae, be, de, ac, bc, cd* special; now *cd* is a common neighbour of *abcd* and *de*, so that these vertices have distance 2 and the other two common neighbours must be *acde, bcde*. Now $\varnothing$ and *acde* have more than 3 common neighbours: *ac, cd, ae, de*, contradiction.)

By definition, the edges of Δ are just the special 2-sets. Since $\mu \geqslant 3$, S3 implies that Δ has no subgraph consisting of two disjoint edges. Therefore Δ consists of a connected graph Δ_0 and a (possibly empty) set of isolated vertices. S3 implies

S5. Any two disjoint edges of Δ have $\mu-1$ or $\mu-2$ transversals (i.e. edges meeting both given edges).

Moreover, since Γ is regular and not complete, $\Gamma(\gamma)$ is not complete; therefore we have

S6. Δ_0 is not isomorphic to K_3 or $K_{s,1}$.

We now consider each value of μ separately. Clearly, S3 implies that $\mu \leqslant 6$.

Case 1: $\mu = 6$. By S1 and S2, a point of $\Gamma_2(\gamma)$ is a complete subgraph K_4 of Δ_0. Let K be a complete subgraph K_m of Δ_0 with maximal $m \geqslant 4$. If $\Delta_0 \neq K$ then there is a vertex $b \in \Delta_0$ at distance 1 from K, and a vertex $a \in K$ adjacent with b. By maximality of K, there is a vertex $c \in K$ with $c \not\sim b$. Let $d \in K \setminus \{a,c\}$. Then the edges *ab, cd* have at most 3 transversals, contradicting S5. Hence $\Delta_0 = K$ is complete and Γ has valency $\binom{m}{2}$. By S3, every 4-subset of K is special.

Thus, we find that Γ is locally triangular ($\Gamma(\gamma) \cong T(m)$ for all $\gamma \in \Gamma$) in a canonical way: the neighbours of any vertex δ are represented by the $\binom{m}{2}$ vectors $\overline{\delta} + \epsilon_i(\delta)e_i + \epsilon_j(\delta)e_j$ ($i,j \in K$, $i \neq j$), where the $\epsilon_i(\delta)$ are signs ± 1.
By induction on t we see that
(i) for $d(\gamma,\delta) < t$, the signs $\epsilon_i(\delta)$ are determined by $\epsilon_i(\delta) = -1$ if $i \in S(\delta)$ and $\epsilon_i(\delta) = 1$ otherwise;
(ii) for $d(\gamma,\zeta) \leqslant t$ we have $\overline{\zeta} = \sum_{i \in S} e_i$ for $S = S(\zeta)$;
(iii) the map $\delta \mapsto S(\delta)$ defines an isomorphism from $\Gamma_{\leqslant t}(\gamma)$ onto the subgraph of the halved m-cube (on the even-sized subsets of K) induced by all subsets of even size $\leqslant 2t$.
Indeed, we know this for $t \leqslant 1$. Let $t \geqslant 2$ and let $\delta \in \Gamma_{t-1}(\gamma)$. We have $|S(\delta)| = 2(t-1) \geqslant 2$, and considering the neighbour of δ mapping to $S(\delta) \setminus \{a,b\}$ for $a,b \in S(\delta)$ we find that $\epsilon_a(\delta) = -1$ for $a \in S(\delta)$; similarly, if $c \in K \setminus S(\delta)$, then considering the neighbour mapping to $S(\delta) \cup \{c\} \setminus \{a\}$ we find that $\epsilon_c(\delta) = 1$ for $c \notin S(\delta)$. This proves (i) and determines the representation $\overline{\zeta}$ of all neighbours ζ of δ. In particular we find (ii). In order to prove (iii) we note that if $\zeta \in \Gamma_t(\gamma)$, $\zeta \sim \delta \in \Gamma_{t-1}(\gamma)$ with $S(\delta) = S(\zeta) \setminus \{a,b\}$, and c is a third element of $S(\zeta)$, then the unique vertex $\delta' \in \Gamma_{t-1}(\gamma)$ mapped to $S(\zeta) \setminus \{a,c\}$ is a common neighbour of δ and ζ (as we see locally at δ), and repeating this argument we see that all vertices of $\Gamma_{t-1}(\gamma)$ mapped to $S(\zeta) \setminus \{p,q\}$ for two elements $p,q \in S(\zeta)$ are neighbours of ζ; now it follows immediately that our map is injective, and one easily checks that all adjacencies in $\Gamma_{t-1}(\gamma) \cup \Gamma_t(\gamma)$ can be seen inside $\delta^{\perp}$ for some $\delta \in \Gamma_{t-1}(\gamma)$, from which it follows that our map is an isomorphism.
Thus, Γ is isomorphic to the half m-cube.

[Comment: At an early stage in the proof we know that Γ is locally triangular, from which it is easy to view Γ as the halved graph of a quotient of the m-cube. The reason that quotients do not occur here is that the representation provides unique names for all points of the cube.]

Case 2: $\mu = 5$. By S1 and S2, a point of $\Gamma_2(\gamma)$ is a subgraph $K_{2,1,1}$ of Δ_0. Let K be a subgraph of Δ_0 isomorphic to $K_{m,1,\ldots,1}$ with a maximal number $m+s$ of vertices, $m \geqslant 2, s \geqslant 2$. Using S5 we find as before that $\Delta_0 = K_{m,1,\ldots,1}$. Let L be the m-coclique of Δ_0. If $s = 2$ and $K \setminus L = \{a,b\}$ then $abcd$ is special for all $c,d \in L$ (by S3), whence ab has the $1+2m+\binom{m}{2}$ neighbours γ, ac, bc, $abcd$ (c, $d \in M$). But γ has only $1+2m$ neighbours (ab and ac, bc for $c \in L$), contradicting the regularity of Γ. If $s > 2$, then we find that, for a, b, $e \in K \setminus L$ and c, $d \in L$, the set $abcd$ is special. Now S4 gives a contradiction. Therefore, the case $\mu = 5$ is impossible.

Case 3: $\mu = 4$. By S5, each pair of disjoint edges has 2 or 3 transversals. By S1 and S2, a point of $\Gamma_2(\gamma)$ is a special 4-set on which Δ_0 induces

or

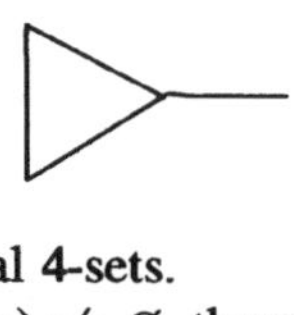

.

and by S3, all subgraphs of Δ_0 of these shapes are special 4-sets.
Suppose first that Δ_0 contains no quadrangle. Since $\Gamma_2(\gamma) \neq \emptyset$ there is a special 4-set $abcd$ with $a \sim b \sim c \sim a \sim d$. If $e \sim d$, then S4 implies that $abde$, $acde$, $bcde$ are

special. Since *bcde* is special, $e \sim b,c$, and since *ab* and *ce* cannot have 4 transversals, $e \not\sim a$. But then *aced* is a quadrangle, against the assumption. Therefore $e \not\sim d$. This implies that *bcde* is not special by S2. If $e \sim b$ then S4 with b and d interchanged shows that *abce* and *abde* are special. But *abce* forces $e \not\sim a$ and *abde* special forces $e \sim a$, contradiction. Therefore $e \not\sim b$ and by symmetry, $e \not\sim c$. Therefore any vertex of Δ_0 at distance 1 from *abcd* is adjacent to a only, the neighbours $\neq b, c$ of a form a coclique K of size t, say, and $\Delta_0 = \{a,b,c\} \cup K$. Now *ab* has the at least $t+4$ neighbours γ, *ac*, *bc*, *ae* ($e \in K$), and *abcd*, and γ has only $t+3$ neighbours *ab*, *ac*, *bc*, *ae* ($e \in K$). This contradicts the regularity of Γ.
Therefore Δ_0 contains a quadrangle *abcd* with $a \sim b \sim c \sim d$. Let e be a further vertex adjacent with d. If $e \not\sim b$ then $e \sim a$ (since otherwise *ab* and *de* have only one transversal) and by symmetry $e \sim c$. But then S4 is violated, contradiction. Therefore $e \sim b$. If in addition $e \sim a$ then the same argument with $\{a,c\}$ and $\{b,d\}$ interchanged shows that $e \sim c$; but then S4 is violated again. Therefore e is adjacent to precisely b and d. Hence a vertex of Δ_0 at distance 1 from a quadrangle Q is adjacent with precisely two vertices of Q, which are opposite. Now one sees that a maximal complete bipartite subgraph of Δ_0 containing Q is a component of Δ_0; hence Δ_0 is a complete bipartite graph $K_{s,t}$, and Γ has valency st. Let A and B be the parts of Δ_0. All 4-sets in $A \cup B$ with two points in A and two points in B are quadrangles and hence special. Similarly to what we did in the case $\mu = 6$, we can now construct an isomorphism from Γ onto the Johnson graph $J(s+t,s)$ of all subsets X of $A \cup B$ such that $|A \cap X| = |B \cap X|$; here vertices at distance i from γ will be mapped to subsets X with $|X \cap A| = |X \cap B| = i$.
[Thus, Γ is locally the $s \times t$-grid and has $\mu = 4$, so that by Theorem 9.1.3 we know that Γ is a quotient of $J(s+t,s)$; just as above, because of the unique labelling provided by the representation, quotients do not occur.]

CASE 4: $\mu = 3$. By S5, each pair of disjoint edges has 1 or 2 transversals. By S1 and S2, a point of $\Gamma_2(\gamma)$ is a special 4-set on which Δ_0 induces

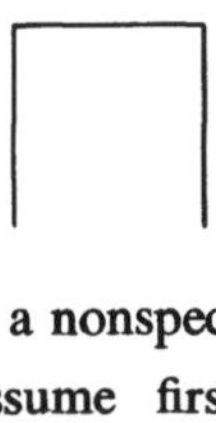 or  or 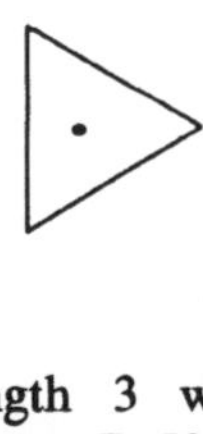 ,

or a nonspecial 4-set on which Δ_0 induces $K_1 + K_3$.
Assume first that Δ_0 contains an induced path P of length 3 with vertices $a \sim b \sim c \sim d$. By S3, *abcd* is special. Let e be at distance 1 from P. If $e \sim d$, then S4 implies that precisely two of the sets *abde*, *acde*, *bcde* are special. However, this is impossible: If $e \sim a$, then $e \sim b$ implies that *abde*, *bcde* are not special, $b, c \notin \Delta(e)$ implies that *abde*, *acde* and *bcde* are all special, and $e \not\sim b$, $e \sim c$ implies that *acde*, *bcde* are not special; contradiction. And if $e \not\sim a$ then $e \sim b$ since *ab* and *de* must have a transversal, and $e \not\sim c$ since *bc* and *de* cannot have 3 transversals. But then *acde* and *bcde* are not special, contradiction. Hence $e \not\sim d$, and by symmetry, $e \not\sim a$.
If $e \sim b$ and $e \sim c$, then S4 with (b,d) interchanged gives a contradiction. Therefore e is adjacent to just one of b and c (and if $e \sim b$ then *ebcd* is special). Hence the neighbours $\neq c$ of b form a coclique A, and the neighbours $\neq b$ of c form a coclique D. Now it is easy to see that $\Delta_0 = A \cup D \cup \{b,c\}$. Since γ has valency $|A| + |D| + 1$ and *bc* has $\geqslant |A| + |D| + 2$ neighbours γ, *ab* ($a \in A$), *cd* ($d \in D$) and P, this

contradicts the regularity of Γ. Therefore Δ_0 does not contain an induced path of length 3.

Now if $\Gamma_2(\gamma)$ contains a vertex *abcd* with $b \sim c \sim d \sim b$ and e is adjacent with d, then *abcd* and *de* have only two common neighbours, contradiction. Hence $e \not\sim d$ and by symmetry, $e \not\sim b,c$. Therefore *bcd* is an isolated triangle, $\Delta_0 = \{b,c,d\}$, contradicting S6. This shows that $\Gamma_2(\gamma)$ contains only special 4-sets forming a 3-claw.

Let *abcd* be such a special 4-set, $a \sim b, c, d$. If $e \sim d$ then $e \sim b, c$ since Δ_0 has no path of length 3; but then S4 is violated. Hence b, c, d are vertices of Δ_0 of valency 1, and since Δ_0 is connected and has no path of length 3, every vertex $\neq a$ of Δ_0 must be adjacent to a. By S6, Δ_0 contains at least one further edge *ef*. If there is another edge *eg* (or *fg*), then *ag*, *ef* have 3 transversals, contradicting S5, and if there is another edge *gh* then *ef*, *gh* have no transversal, contradicting S5. Therefore there is no further edge. Now *ef* must have a neighbour in $\Gamma_2(\gamma)$, but there is no special 4-set containing *ef*, contradiction. Therefore $\mu = 3$ is impossible.

This completes the proof of the theorem. □

Amply regular root graphs in E_8

For the case where $\mathbb{L}(\Gamma)$ is an exceptional root lattice the classification of regular root graphs Γ in part depends on a computer search by Bussemaker (personal communication). We only treat the amply regular case.

3.15.2. Proposition. *Let Γ be an amply regular root graph of diameter d with parameters (v,k,λ,μ). If $\mathbb{L}(\Gamma) \cong E_6$, E_7, or E_8, then either we have $d > 2$, $\mu = 1$, $k \leqslant 8$ and $(\lambda+1) \mid k$, or Γ is isomorphic to the Petersen graph, one of the Chang graphs, the Schläfli graph or the Gosset graph.*

Proof. If $d \leqslant 2$, then Theorem 3.14.2 implies that Γ is a strongly regular graph with smallest eigenvalue $\geqslant -2$, and by Corollary 3.12.3 and Theorem 3.12.4, we have the Petersen graph, a Chang graph, or the Schläfli graph, since the remaining graphs have $\mathbb{L}(\Gamma) \cong A_n$ or D_n. Hence we may suppose from now on that $d > 2$.

Fix $\gamma \in \Gamma$, and shift the representation such that $\bar{\gamma} = 0$. Then $\mathbb{L}^+(\Gamma(\gamma)) \subseteq \mathbb{L}(\Gamma) \subseteq E_8$, the restriction of the representation to $\Gamma(\gamma)$ is a spherical (2,1,0)-representation, and $\Gamma(\gamma)$ has smallest eigenvalue $\geqslant -2$. Suppose first that $\mu = 1$. Then $\Gamma(\gamma)$ is a disjoint union of r cliques of size $\lambda+1$. This implies that $\mathbb{L}^+(\Gamma(\gamma))$ is isomorphic to the direct sum of r copies of $A_{\lambda+1}$, whence $k = r(\lambda+1) \leqslant 8$, giving our first alternative. Now suppose that $\mu \geqslant 2$. By Theorem 3.15.1, Γ is locally connected, except possibly when $\mu = 2$. But in case $\mu = 2$ each component of $\Gamma(\gamma)$ is a regular graph of girth $g \geqslant 5$, hence contains a g-gon, and contributes at least 5 to the dimension of $\mathbb{L}^+(\Gamma(\gamma))$, so that there can be only one component. Therefore $\Gamma(\gamma)$ is a connected regular graph with k vertices and valency λ, and $k \geqslant \lambda+\mu+1$ by Theorem 1.5.5. We now use Theorem 3.12.2 to determine the possibilities for $\Gamma(\gamma)$.

Suppose first that $\Gamma(\gamma)$ is not a line graph. By Theorem 3.12.2, $\Gamma(\gamma)$ is one of the 187 exceptional regular subgraphs of $E_7(1)$. By Corollary 3.14.5, nonadjacent vertices of $\Gamma(\gamma)$ have $\mu-1$ or $\mu-2$ common neighbours in $\Gamma(\gamma)$. Inspection of a list of parameters of subgraphs of $E_7(1)$, given in Bussemaker & Neumaier [155], reveals the following possibilities with $k \geqslant \lambda+\mu+1$:

(k,λ,μ)	number of $\Gamma(\gamma)$	Γ ruled out by
(10,3,2)	1,2,5	computer
(10,3,3)	5	$k_2 = 20, m = 30, q = 0$
(12,4,3)	13	$k_2 = 28, m = 36, q = 2$
(14,5,3)	15,16	$k_2 = 14{\cdot}8/3$ nonintegral
(16,6,3)	56, 65-69	computer
(16,6,4)	69	$k_2 = 36, m = 72, q = 12$
(18,7,4)	105	$k_2 = 45, m = 72, q = 12$
(22,9,4)	135,148	count wheels
(24,10,5)	154	$k_2 = 24{\cdot}13/5$ nonintegral
(28,12,5)	161-163	count quadrangles
(28,12,6)	161-163	$k_2 = 70, m = 210, q \leqslant 114$
$(3\mu,2\mu-2,\mu)$, $3 \leqslant \mu \leqslant 9$	164-184	computer
(27,16,10)	184	- Gosset graph; exists -
(8,4,3)	185	Theorems 1.5.3-5

Here $k_2 = k(k-1-\lambda)/\mu$ is the number of vertices of $\Gamma_2(\gamma)$, m is the number of unordered pairs of nonadjacent points of $\Gamma(\gamma)$ having $\mu-2$ common neighbours in $\Gamma(\gamma)$, and q is the number of quadrangles in $\Gamma(\gamma)$. The entries marked 'computer' were disposed of by a computer search of Bussemaker. The case (27,16,10) is covered by Corollary 1.15.3, and the case (8,4,3) is ruled out by the Theorems 1.5.5 and 1.5.7 since $\Gamma(\gamma)$ is not strongly regular. In case $(k,\lambda,\mu) = (28,12,5)$, the three possibilities for the local graphs are the Chang graphs with $\mu = 4$; but this means that such graphs do not contain quadrangles, while the local graphs do contain quadrangles, contradiction. In case $(k,\lambda,\mu) = (22,9,4)$, the two possible local graphs have $(q,m) = (45,18)$ or $(60,12)$. Now count wheels on 5 vertices, i.e., subgraphs $K_{1,2,2}$. Let N_a be the number of vertices γ of Γ with local graph of type (number) a ($a = 135,148$). Choosing the center of the wheel first we find $45N_{135} + 60N_{148}$ such wheels; choosing a vertex on the boundary first, we find that this number also equals $(18N_{135} + 12N_{148}).2/4$, which is ridiculous. [The same argument applied to the case $(k,\lambda,\mu) = (16,6,3)$ shows that only local graphs of numbers 65 and 66 can occur. Both of these have $(q,m) = (3,12)$.] The remaining possibilities have a nonintegral k_2 or satisfy $m > k_2+q$. We now show that necessarily $m \leqslant k_2+q$, thereby ruling out these exceptions. Indeed, by Proposition 3.14.4(iii), the subgraphs $\Gamma(\gamma) \cap \Gamma(\delta)$ ($\delta \in \Gamma_2(\gamma)$) are complete multipartite with classes of size 1 or 2. Denote by s_i the number of $\delta \in \Gamma_2(\gamma)$ such that $\Gamma(\gamma) \cap \Gamma(\delta)$ has i classes of size 2. For each such δ, the intersection $\Gamma(\gamma) \cap \Gamma(\delta)$ contains precisely $\binom{i}{2}$ quadrangles. Since $\mu(\Gamma) = \mu$, two nonadjacent points in $\Gamma(\gamma)$ have a common neighbour $\delta \in \Gamma_2(\gamma)$ if and only if they have $\mu-2$ common neighbours in $\Gamma(\gamma)$, and then δ is unique. Therefore

$$\sum_{i \geqslant 0} s_i = k_2, \quad \sum_{i \geqslant 0} is_i = m, \quad \sum_{i \geqslant 0} \binom{i}{2} s_i \leqslant q,$$

and thus

$$0 \leqslant \sum_{i \geqslant 0} \tfrac{1}{2}(i-1)(i-2)s_i \leqslant k_2 - m + q,$$

as claimed. Thus, whenever some $\Gamma(\gamma)$ is not a line graph then the conclusion of the theorem holds.

Hence we may assume that $\Gamma(\gamma)$ is the line graph of a graph Δ with n vertices for each

$\gamma \in \Gamma$. By Theorem 3.12.2, Δ is regular, or bipartite and semiregular. Moreover, in the bipartite case $\mathbb{L}^+(\Gamma(\gamma)) \cong A_{n-1}$ and hence $n \leqslant 9$, and in the other case $\mathbb{L}^+(\Gamma(\gamma)) \cong D_n$ and hence $n \leqslant 8$. By Corollary 3.14.5, nonadjacent vertices of $\Gamma(\gamma)$ have $\mu - 1$ or $\mu - 2$ common neighbours in $\Gamma(\gamma)$; i.e., disjoint edges of Δ have $\mu - 1$ or $\mu - 2$ transversals (i.e. edges intersecting the given edges). In particular $\mu \leqslant 6$. A simple exercise (cf. Proposition 5 of NEUMAIER [561]) now shows that we have one of the following cases:

$$\begin{array}{ll}
\mu = 2, & \Delta = n\text{-gon } (n \leqslant 8), \\
\mu = 3, & \Delta = \text{ complement of } n\text{-gon } (n \leqslant 6), \\
\mu = 4, & \Delta = K_{3\times 2}, \\
\mu \in \{3,4\}, & \Delta = K_{p,q}\ (p+q \leqslant 9), \\
\mu \in \{5,6\}, & \Delta = K_n\ (n \leqslant 8).
\end{array}$$

The cases where Δ is the complement of the hexagon or $K_{3\times 2}$ were ruled out by computer; the remaining cases are handled as follows.

CASE 1: $\mu = 2$. Then $\Gamma(\gamma) = L(\Delta)$ is a k-gon, $5 \leqslant k \leqslant 8$. If $k \in \{7,8\}$, represent $\Gamma(\gamma)$ by the vectors $e_i + e_{i+1}$, $i \in \mathbb{Z}_k$. In particular, $\lambda = 2$. Let γ_i be the neighbour of γ represented by $e_i + e_{i+1}$, and let $\gamma_{ij} = \gamma_{ji}$ be the common neighbour of γ_i and γ_j distinct from γ, for $i - j \neq \pm 2$. Then γ_{ij} is represented by $e_i + e_{i+1} + e_j + e_{j+1}$, for $i - j \neq \pm 1$. The common neighbours of γ_{i+1} and $\gamma_{i,i-3}$ are γ_i and $\gamma_{i+1,i-3}$, so that $\gamma_{i,i-3} \not\sim \gamma_{i,i+1}$. It follows that $\gamma_{i,i-3} \sim \gamma_{i,i-1}$. Now $\gamma_{i,i-3}$ and $\gamma_{i+2,i-1}$ have distance at most two, since both are adjacent to $\gamma_{i,i-1}$, and Lemma 3.14.3 yields a contradiction.

For $k = 5,6$ one gets in the same way the icosahedron and the Shrikhande graph, which, however, have $\mathbb{L}(\Gamma) \cong D_6$.

CASE 2: $\mu = 3$. By Proposition 1.1.4, Γ is not locally a pentagon, so Δ cannot be the complement of a pentagon. Hence $\Gamma(\gamma) = L(K_{p,q})$ is a grid $p \times q$. Since nonadjacent points in $\Gamma(\gamma)$ have already 3 common neighbours in $\gamma^\perp$, the point γ is not in a quadrangle. Varying γ, we see that Γ does not contain quadrangles, but $\Gamma(\gamma)$ does, contradiction.

CASE 3: $\mu = 4$. Again $\Gamma(\gamma)$ is a grid $p \times q$ with $k = pq$ points and valency $\lambda = p + q - 2$, represented by the vectors $e_i + e_j$ $(i \in P, j \in Q)$, where P and Q are the parts of Δ. Since two nonadjacent points ab, $cd \in L(\Delta)$ have only 3 common neighbours in $\gamma^\perp$, they must have a common neighbour $\delta \in \Gamma_2(\gamma)$, and by Proposition 3.14.4(ii), δ is represented by $\overline{\alpha} + \overline{\beta} - \overline{\gamma} = e_a + e_b + e_c + e_d$. Since we may choose $a, c \in P$ and $b, d \in Q$ arbitrary but distinct, this yields $\binom{p}{2}\binom{q}{2}$ distinct norm 4 vectors, and, by Lemma 3.5.5, equally many points of $\Gamma_2(\gamma)$. Since the total number of points of $\Gamma_2(\gamma)$ is $k_2 = k(k-1-\lambda)/\mu = pq(pq-p-q+1)/4 = \binom{p}{2}\binom{q}{2}$, all points of $\Gamma_2(\gamma)$ are represented by vectors of shape $1^4 0^{p+q-4}$. An easy induction argument now establishes that Γ is represented by (0,1)-vectors of even weight, i.e., $\mathbb{L}(\Gamma) \subseteq D_{p+q}$, contradiction.

CASE 4: $\mu = 5$. Then $\Gamma(\gamma) = L(K_n)$ is a triangular graph $T(n)$, and $n > 3$ since $\Gamma(\gamma)$ is not complete. Since nonadjacent points in $\Gamma(\gamma)$ have already 5 common neighbours in $\gamma^\perp$, the neighbours of $\delta \in \Gamma_2(\gamma)$ in $\Gamma(\gamma)$ must form a clique. Since this must hold for every $\gamma \in \Gamma$, $\delta \in \Gamma_2(\gamma)$, it follows that Γ contains no quadrangle. But for $n > 3$, $T(n)$ contains a quadrangle, contradiction.

CASE 5: $\mu = 6$. Again $\Gamma(\gamma)$ is a triangular graph $T(n)$ with $K = \binom{k}{2}$ points and valency $\lambda = 2n-4$, represented by the vectors e_i+e_j $(1 \leq i < j \leq n)$. We now find that $\mathbb{L}(\Gamma) \subseteq D_n$ as in case 2, contradiction.

This completes the proof. □

Amply regular root graphs with $\mu = 2$

For $\mu \leq 2$ the situation is rather more complicated, due to the fact that Γ need not be locally connected. We have not attempted to settle the case $\mu = 1$, and restrict ourselves for $\mu = 2$ to the amply regular case with a strict root representation. (Note that we have met an example that is not amply regular in §3.11 (iii) - it was co-edge-regular with $(v,k,\mu) = (8,4,2)$.)

Recall (cf. §1.12) that the direct product $\Gamma \times \Delta$ of two graphs $\Gamma = (X,E)$ and $\Delta = (Y,F)$ is the graph with vertex set $X \times Y$ where $(x,y) \sim (x',y')$ if and only if $x = x'$ and $\{y,y'\} \in F$ or $\{x,x'\} \in E$ and $y = y'$. Clearly the direct product of strict root graphs Γ^i with $\mu(\Gamma^i) = 2$ again has this property.

3.15.3. Proposition. *Let Γ be a strict root graph with $\lambda(\Gamma) = \lambda$, $\mu(\Gamma) = 2$. Then any connected subgraph Δ of Γ with $\lambda(\Delta) = \lambda$, $\mu(\Delta) = 2$ (and any $(\lambda+2)$-clique Δ of Γ) is a direct factor of Γ, and $\Gamma \cong \Gamma' \times \Delta$ for a subgraph Γ' of Γ with $\lambda(\Gamma') = \lambda$, $\mu(\Gamma') = 2$.*

Proof. Fix $\infty \in \Delta$, and shift the representation such that $\overline{\infty} = 0$. Denote by Γ^0 the set of $\gamma \in \Gamma$ such that $(\overline{\gamma},\overline{\delta}) = 0$ for all $\delta \in \Delta$ (so that, in particular, $\infty \in \Gamma^0$), and let Γ' be the connected component of Γ^0 containing ∞. Let S be the set of all $(\gamma,\delta) \in \Gamma' \times \Delta$ such that there is a vertex $\epsilon \in \Gamma$ with $\overline{\epsilon} = \overline{\gamma}+\overline{\delta}$. Since the representation is strict and, in particular, injective, ϵ is uniquely determined by γ and δ; thus we have a canonical mapping $u: S \to \Gamma$ with $\overline{u(\gamma,\delta)} = \overline{\gamma}+\overline{\delta}$ for $(\gamma,\delta) \in S$. Since $\overline{\infty} = 0$, we have $(\gamma,\infty) \in S$, $u(\gamma,\infty) = \gamma$ for $\gamma \in \Gamma'$, and $(\infty,\delta) \in S$, $u(\infty,\delta) = \delta$ for $\delta \in \Delta$.

If $S \neq \Gamma' \times \Delta$, then we can choose $(\gamma,\delta) \in (\Gamma' \times \Delta) \setminus S$ at minimal distance i from (∞,∞). We must have $\gamma, \delta \neq \infty$; hence we can choose a vertex $\gamma' \sim \gamma$ on a geodesic in Γ' from ∞ to γ, and a vertex $\delta' \sim \delta$ on a geodesic in Δ from ∞ to δ. Then $\alpha = u(\gamma',\delta)$ and $\beta = u(\gamma,\delta')$ are both adjacent to $\epsilon' = u(\gamma',\delta')$, and since $(\overline{\alpha}-\overline{\epsilon'},\overline{\beta}-\overline{\epsilon'}) = (\overline{\delta}-\overline{\delta'},\overline{\gamma}-\overline{\gamma'}) = 0$, Lemma 3.14.3 gives $d(\alpha,\beta) = 2$. The second common neighbour ϵ of α and β cannot be adjacent to ϵ', since α and ϵ' already have λ common neighbours $u(\gamma',\delta'')$, where $\delta'' \in \Delta(\delta) \cap \Delta(\delta')$; therefore, by Proposition 3.14.4 (ii), $\overline{\epsilon} = \overline{\alpha}+\overline{\beta}-\overline{\epsilon'} = \overline{\gamma}+\overline{\delta}$, against the assumption that $(\gamma,\delta) \notin S$. Therefore $S = \Gamma' \times \Delta$.

Notice that the subgraph $u(S)$ of Γ is isomorphic to the direct product graph $\Gamma' \times \Delta$ since the representation is strict and $[\overline{u(\gamma,\delta)},\overline{u(\gamma',\delta')}] = [\overline{\gamma},\overline{\gamma'}]+[\overline{\delta},\overline{\delta'}]$.

Now suppose that $u(S) \neq \Gamma$, and choose $\beta \in \Delta$ at minimal distance (in Δ) from ∞ such that some vertex $\epsilon = u(\gamma,\beta)$ has a neighbour $\alpha \notin u(S)$. If $\beta \neq \infty$, then let $\beta' \sim \beta$ be on a geodesic in Δ from ∞ to β. By the choice of β, we have $\alpha \not\sim u(\gamma,\beta')$, and the second common neighbour ϵ' of α and $u(\gamma,\beta')$ is in $u(S)$; hence $\epsilon' = u(\gamma,\beta'')$ with $\beta'' \sim \beta'$ or $\epsilon' = u(\gamma',\beta')$ with $\gamma' \sim \gamma$. In both cases α is a common neighbour of ϵ and ϵ' not in $u(S)$, but ϵ and ϵ' already have λ or $\mu = 2$ common neighbours in $u(S)$, namely the $u(\gamma,\delta)$ with $\delta \in \Delta(\beta) \cap \Delta(\beta'')$ in the first case, and $u(\gamma,\beta')$ and $u(\gamma',\beta)$ in the second case, contradiction. Hence we must have $\beta = \infty$.

If $\alpha \in \Gamma^0$, then, since $\alpha \sim u(\gamma,\infty) = \gamma \in \Gamma'$, also $\alpha \in \Gamma'$, contradiction. Therefore there is a vertex $\delta \in \Delta$ such that $(\overline{\alpha},\overline{\delta}) \neq 0$. Let $\infty = \delta_0, \delta_1, \cdots, \delta_i = \delta$ be a geodetic path in Δ from ∞ to δ. We construct inductively vertices $\alpha = \alpha_0, \alpha_1, \cdots, \alpha_i$ by defining α_l as the second common neighbour of α_{l-1} and $u(\gamma,\delta_l)$; note that these are nonadjacent since $u(\gamma,\delta_l)$ and $u(\gamma,\delta_{l-1})$ already have λ common neighbours in $u(S)$. For the same reason we have $d(\alpha_l,u(\gamma,\delta_{l-1})) = 2$. Now Proposition 3.14.4(ii) gives $\overline{\alpha}_l = \overline{\alpha}_{l-1} + \overline{u(\gamma,\delta_l)} - \overline{u(\gamma,\delta_{l-1})}$, whence $\overline{\alpha}_l - \overline{\delta}_l = \overline{\alpha}_{l-1} - \overline{\delta}_{l-1} = \cdots = \overline{\alpha}_0 - \overline{\delta}_0 = \overline{\alpha}$, and Lemma 3.14.3 gives $0 = (\overline{\alpha}_l - \overline{u(\gamma,\delta_l)}, \overline{u(\gamma,\delta_{l-1})} - \overline{u(\gamma,\delta_l)}) = (\overline{\alpha} - \overline{\gamma}, \overline{\delta}_{l-1} - \overline{\delta}_l) = (\overline{\alpha}, \overline{\delta}_{l-1} - \overline{\delta}_l)$ since $\gamma \in \Gamma'$, whence $(\overline{\alpha},\overline{\delta}) = (\overline{\alpha},\overline{\delta}_i) = \cdots = (\overline{\alpha},\delta_0) = (\overline{\alpha},\overline{\infty}) = 0$, contradiction.

Therefore we must have $u(S) = \Gamma$, and since the representation is strict, u is an isomorphism. Thus $\Gamma \cong \Gamma' \times \Delta$, and a simple counting argument yields $\lambda(\Gamma') = \lambda$, $\mu(\Gamma') = 2$. □

3.15.4. Theorem. *Let Γ be an amply regular strict root graph with $\mu(\Gamma) = 2$. Then either Γ is a Hamming graph, or $\lambda = 2$ and Γ is a direct product of 4-cliques, icosahedra, and Shrikhande graphs.*

Proof. In view of the preceding proposition we may suppose w.l.o.g. that Γ contains no proper subgraph E with $\lambda(E) = \lambda$, $\mu(E) = 2 \leqslant |E|$ and no $(\lambda+2)$-clique. In particular, Γ is not a clique and we have $\lambda \geqslant 2$.

Suppose first that $\mathbb{L}(\Gamma) \cong \mathbb{L}_1 \oplus \mathbb{L}_2$. For every edge $\alpha\beta$ of Γ, $\overline{\alpha} - \overline{\beta}$ is a root of $\mathbb{L}(\Gamma)$, hence contained in $\mathbb{L}_1 \cup \mathbb{L}_2$; we call an edge $\alpha\beta$ of *type* j if $\overline{\alpha} - \overline{\beta} \in \mathbb{L}_j$ $(j \in \{1,2\})$. Since $(\overline{\alpha} - \overline{\beta}) + (\overline{\beta} - \overline{\gamma}) + (\overline{\gamma} - \overline{\alpha}) = 0$, all edges of a triangle have the same type. Now let E be a maximal subgraph such that all edges of E have the same type j. Clearly we may suppose that E contains an edge. Then $\mathbb{L}(E) \subseteq \mathbb{L}_j$. If $\alpha, \beta \in E$ have a common neighbour $\delta \notin E$, then $r = \overline{\alpha} - \overline{\delta}$ and $s = \overline{\beta} - \overline{\delta}$ are roots with $r - s \in \mathbb{L}_j$; since $r, s \in \mathbb{L}_1 \cup \mathbb{L}_2$, this forces $r, s \in \mathbb{L}_j$, so that $\overline{\delta} - \overline{\gamma} = \overline{\alpha} - \overline{\gamma} - r \in \mathbb{L}_j$ for every neighbour γ of δ in E. Therefore all edges of $E \cup \{\delta\}$ have type j, contradicting the maximality of E. It follows that E contains with two vertices all their common neighbours, so that $\lambda(E) = \lambda$, $\mu(E) = 2$. Our assumption now implies that $E = \Gamma$, so that all edges of Γ have type j and $\mathbb{L}(\Gamma) = \mathbb{L}_j$.

This proves that $\mathbb{L}(\Gamma)$ is irreducible, and hence one of the root lattices A_n, D_n, or E_6, E_7, E_8. But E_6, E_7, E_8 are ruled out by Proposition 3.15.2, whence $\mathbb{L}(\Gamma) \subseteq \mathbb{Z}^{n+1} = \mathbb{Z}e_0 \oplus \cdots \oplus \mathbb{Z}e_n$.

Suppose first that Γ contains points γ, δ at distance 2 such that the norm 4 vector $\overline{\gamma} - \overline{\delta}$ has shape $(\pm 2)0^n$. Now Lemma 3.14.6 shows that $\Gamma(\gamma)$ has a component of size at most 2, but this contradicts $\lambda \geqslant 2$. Thus, for $\gamma, \delta \in \Gamma$ at distance 2, the vector $\overline{\gamma} - \overline{\delta}$ must have shape $(\pm 1)^4 0^{n-3}$. We can now proceed as in the proof of Theorem 3.15.1. We fix $\gamma \in \Gamma$, shift the representation such that $\overline{\gamma} = 0$, and choose the basis of $\mathbb{Z}^{n+1}$ such that all points of $\Gamma(\gamma)$ are represented by vectors $e_i + e_j$. We define the graph Δ and the special sets as before and find that S1 - S6 remain valid; in particular, special 4-sets contain precisely two edges, and the union of two disjoint edges without transversal is special. Let Δ_0 be a connected component of Δ containing an edge. Since $\Gamma(\gamma)$ is regular of valency λ, every edge of Δ_0 intersects precisely λ other edges of Δ_0. If Δ_0 contains a triangle, then S5 implies that Δ_0 is a triangle, $\lambda = 2$, and Γ contains the $(\lambda+2)$-clique represented by $\{\emptyset\} \cup \Delta_0$, contradiction. Hence Δ_0 contains no triangle, and by S5, Δ_0 contains no quadrangle.

We need some further facts. Applying S4 to a path of length 4 we find:

S7. If $c \sim d \sim e \sim b \sim a$ is a path, then $acde$ is special.

S8. Δ_0 contains no subgraph of the form D_6

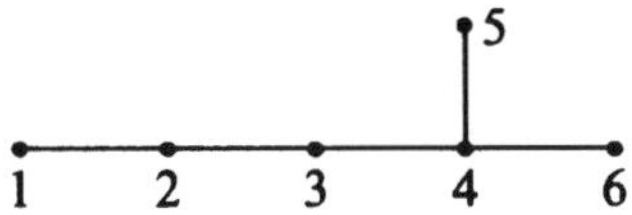

Indeed, S7 shows that 1345 and 1346 are special, so that S4 applied to 1345 and 46 gives a contradiction. We now distinguish three cases.

CASE 1. *Δ_0 is a tree.* Since $L(\Delta_0)$ must be regular of valency λ, we find $\Delta_0 = K_{n,1}$, contradicting S6.

CASE 2. *Δ_0 has girth $n \geq 6$.* Then S8 implies that no vertex of an n-gon has a further neighbour. Since Δ_0 is connected, it is an n-gon, with vertices labelled by $\mathbf{Z}_n$ and edges $\{i,i+1\}$ $(i \in \mathbf{Z}_n)$, say. In particular, $\lambda = 2$. By S3 and S7, the sets $\{i,i+1,k,k+1\}$ $(k-i \neq 0,\pm 1,\pm 2)$ and $\{i-1,i,i+1,i\pm 3\}$ are special. Now the points represented by the $n+2$ sets $\varnothing$, 23, 45, 0234, 1345, 2346, 3457, $\{3,4,k,k+1\}$ $(k \neq 1,2,3,4,5)$ are neighbours of the point represented by 34, so that two of these sets must coincide. This forces $n = 6$, and since the 16 special sets found so far represent the Shrikhande graph, a graph E with $\lambda(E) = 2 = \lambda$, $\mu(E) = 2$, there are no further special sets and Γ is the Shrikhande graph.

CASE 3. *Δ_0 has girth 5.* Then Δ_0 is not bipartite, and since $L(\Delta_0)$ must be regular of valency λ, Δ_0 is regular of valency $\frac{1}{2}\lambda + 1$. If $\lambda = 2$ then Δ_0 is the pentagon and Γ is locally a disjoint union of pentagons. Now the set of all $\delta \in \Gamma$ with $S(\delta) \subseteq \Delta_0$ is easily seen to be an icosahedron, and by our assumptions Γ itself is an icosahedron. If $\lambda > 2$, then for every vertex i of a pentagon $1 \sim 2 \sim 3 \sim 4 \sim 5 \sim 1$ of Δ_0, the set $\Delta_0(i)$ of neighbours of i is a $(\lambda-2)$-coclique, and by S8, $\Delta_0(i) \cup \Delta_0(i+2)$ (i mod 5) is a complete bipartite graph. Since there are no quadrangles, this forces $\lambda = 3$, and Δ_0 is the Petersen graph, so that Γ is locally a disjoint union of line graphs of the Petersen graph. In particular, Γ has valency $15t$, where t is the number of components of $\Gamma(\gamma)$. Now every $K_1 + K_{1,2}$ of the Petersen graph is in some path of length 4, and S7 shows that every 4-set containing two intersecting edges of Δ_0 is special. But there are 16 such 4-sets through an edge ab of Δ_0, and they determine 16 neighbours of the point δ of Γ represented by ab. Each of these 16 neighbours ϵ has $S(\epsilon) \subseteq \Delta_0$, but δ also has $15(t-1)$ neighbours with $S(\epsilon) \not\subseteq \Delta_0$, contradicting $k = 15t$. □

Problem. Is every regular root graph Γ with $\mu(\Gamma) = 2$ the direct product of $t \geq 1$ connected and locally connected subgraphs $\Gamma^1, \cdots, \Gamma^t$ of Γ? The line graphs of the graphs

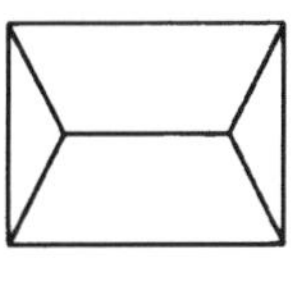

and

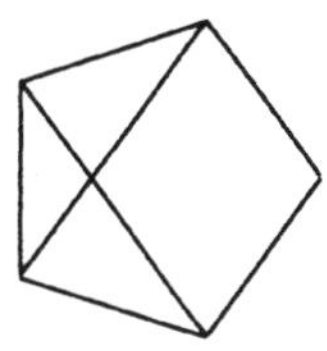

show that this is not true if '$\mu(\Gamma) = 2$' is replaced by '$\mu(\gamma,\delta) \in \{1,2\}$ for all $\gamma, \delta \in \Gamma$', or if the regularity assumption is dropped and '$\mu(\Gamma) = 2$' is replaced by '$\mu(\gamma,\delta) \geq 2$ for all $\gamma, \delta \in \Gamma$'. But it might be true that the answer to the above question is still affirmative if '$\mu(\Gamma) = 2$' is replaced by '$\mu(\gamma,\delta) \geq 2$ for all $\gamma, \delta \in \Gamma$'.

Chapter 4
Theory of Distance-Regular Graphs

We now come to the central topic of the book. The first section may be viewed as a short introduction to the subject. Although we shall develop large parts of the theory of distance-regular graphs independently of Chapter 2, we shall use concepts and results about association schemes for more specialized topics such as Q-polynomial orderings (Chapter 8) and codes in graphs (Chapter 11). In §4.2 we look at various constructions that, given a distance-regular graph, produce a new one. In §4.3 we show how certain conditions on the parameters force the presence of substructures, like lines or Petersen subgraphs. In §4.4 we use the results of Chapter 3 to obtain a characterization by parameters of the two most basic families of distance-regular graphs, the Johnson and Hamming graphs. Chapter 5 contains most of the known conditions on the parameters, Chapter 6 classifies the known distance-regular graphs in various families, Chapter 7 is concerned with distance-transitive graphs, Chapter 8 discusses the consequences of the Q-polynomial property, and the remaining chapters give all examples known to us.

4.1 Distance-regular graphs

In this section we introduce basic concepts and results concerning distance-regular graphs. Almost everything we say here will be a repetition of what was said in greater generality in Chapter 2.

A. Parameters

For convenience we repeat here the definition of a distance-regular graph. A connected graph Γ is called *distance-regular* if it is regular of valency k, and if for any two points $\gamma, \delta \in \Gamma$ at distance $i = d(\gamma,\delta)$, there are precisely c_i neighbours of δ in $\Gamma_{i-1}(\gamma)$ and b_i neighbours of δ in $\Gamma_{i+1}(\gamma)$. The sequence

$$\iota(\Gamma) := \{b_0, b_1, \cdots, b_{d-1}; c_1, c_2, \cdots, c_d\}, \qquad (1)$$

where d is the diameter of Γ, is called the *intersection array* of Γ; the numbers c_i, b_i, and a_i, where

$$a_i = k - b_i - c_i \quad (i = 0, \cdots, d) \qquad (1a)$$

is the number of neighbours of δ in $\Gamma_i(\gamma)$ for $d(\gamma,\delta) = i$, are called the *intersection numbers* of Γ. Clearly

$$b_0 = k\,,\ b_d = c_0 = 0\,,\ c_1 = 1. \qquad (1b)$$

By counting edges $\delta\epsilon$ with $d(\gamma,\delta) = i$, $d(\gamma,\epsilon) = i+1$ we see that $\Gamma_i(\gamma)$ contains k_i points, where

$$k_0 = 1\ ,\ k_1 = k\ ,\ k_{i+1} = k_i b_i / c_{i+1} \quad (i = 0, \cdots, d-1)\ ; \tag{1c}$$

therefore the total number of vertices is

$$v = 1 + k_1 + \cdots + k_d\ . \tag{1d}$$

As one easily checks, the matrices A_i with entries $(A_i)_{\gamma\delta} = 1$ if $d(\gamma,\delta) = i$, and $(A_i)_{\gamma\delta} = 0$ otherwise, satisfy the relations

$$\begin{gathered} A_0 = I\ ,\ A_1 = A\ , \\ AA_i = c_{i+1}A_{i+1} + a_iA_i + b_{i-1}A_{i-1} \quad (i = 0, \cdots, d)\ , \\ A_0 + A_1 + \cdots + A_d = J\ , \end{gathered} \tag{2}$$

where A is the adjacency matrix of Γ, $A_{-1} = A_{d+1} = 0$ and b_{-1} and c_{d+1} are unspecified. It follows that the matrices A_i can be written as polynomials in A of degree i,

$$A_i = v_i(A) \quad (i = 0, \cdots, d+1), \tag{3}$$

where the v_i are polynomials of degree i defined recursively by

$$\begin{gathered} v_{-1}(x) = 0, \quad v_0(x) = 1, \quad v_1(x) = x, \\ c_{i+1}v_{i+1}(x) = (x - a_i)v_i(x) - b_{i-1}v_{i-1}(x) \quad (i = 0, \cdots, d)\ . \end{gathered} \tag{4}$$

From (2) and (3) we find that

$$A_iA_j = \sum_{l=0}^{d} p_{ij}^l A_l \tag{5}$$

for certain numbers p_{ij}^l. By comparison of the (γ,δ)-entries we see that for γ, $\delta \in \Gamma$ at distance l, the number of $\epsilon \in \Gamma$ with $d(\gamma,\epsilon) = i$, $d(\delta,\epsilon) = j$ is just p_{ij}^l $(i, j, l = 0, \cdots, d)$. In particular, the p_{ij}^l are integers and we have $p_{0i}^l = \delta_{il}$, $p_{ij}^0 = \delta_{ij}k_i$ and

$$p_{ij}^l = 0 \text{ if } l > i+j \text{ or } l < |i-j|\ , \tag{6}$$

$$p_{1i}^{i-1} = b_{i-1}\ ,\ p_{1i}^i = a_i\ ,\ p_{1i}^{i+1} = c_{i+1}\ , \tag{7}$$

$$p_{ij}^l = p_{ji}^l\ . \tag{8}$$

By counting triples $\gamma,\delta,\epsilon \in \Gamma$ with $d(\gamma,\delta) = i$, $d(\gamma,\epsilon) = j$, $d(\delta,\epsilon) = l$ we find

$$p_{ij}^l k_l = p_{il}^j k_j = p_{lj}^i k_i\ . \tag{9}$$

Finally, if we compute $(AA_i)A_j = A(A_iA_j)$ using (2) and (5), and compare the (γ,δ)-entries for γ, $\delta \in \Gamma$ at distance l, we find

$$b_{i-1}p_{i-1j}^l + a_ip_{ij}^l + c_{i+1}p_{i+1j}^l = p_{ij}^{l-1}c_l + p_{ij}^l a_l + p_{ij}^{l+1}b_l\ , \tag{10}$$

which, solved for p_{i+1j}^l, can be used to compute the p_{ij}^l recursively, cf. §4.1D.

The preceding discussion implies that the vertex set of a distance-regular graph, together with the distance relations, forms a P-polynomial d-class association scheme; in this scheme, the distribution diagram with respect to the adjacency relation is linear:

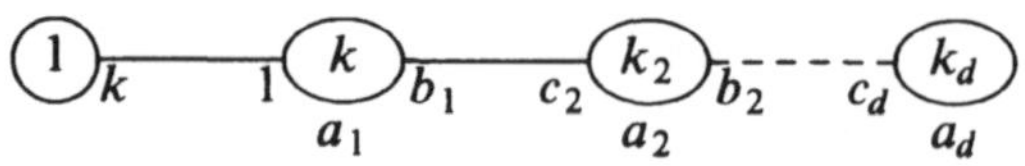

(cf. §2.9).

Conversely, it is obvious that a d-class association scheme having a connected and linear distribution diagram with respect to some relation R_l defines a distance-regular graph by calling points adjacent whenever they are in relation R_l.

Distance-regular graphs were introduced by BIGGS [68]. (The term 'distance-regular' only occurs much later; the early authors only talk about distance-transitive graphs. But the application to cages in section 5 of Biggs' paper shows that he was aware of the fact that no group is necessary for the basic feasibility analysis of intersection arrays.) They were also studied by DELSARTE [251] under the name *metrically regular graphs*. An elementary exposition of the basic theory is given in BIGGS [71]; cf. also MACWILLIAMS & SLOANE [515], Chapter 21, §4.

Note that in this book distance-regular graphs will always be finite. For results on the infinite case, see MACPHERSON [513] and MOHAR & OMLADIČ [536].

B. Eigenvalues

The adjacency matrix A of a distance-regular graph Γ has at least $d+1$ distinct eigenvalues since the matrices $A_i = v_i(A)$ $(i = 0, \cdots, d)$ are linearly independent. But $v_{d+1}(A) = A_{d+1} = 0$; therefore A has precisely $d+1$ distinct eigenvalues, namely the zeros of $v_{d+1}(x)$. Let these eigenvalues be $\theta_0 > \theta_1 > \cdots > \theta_d$. For every eigenvalue θ of A, the relation (2) implies the equation

$$\theta v_i(\theta) = c_{i+1}v_{i+1}(\theta) + a_i v_i(\theta) + b_{i-1}v_{i-1}(\theta) \quad (i = 0, \cdots, d) \tag{11}$$

(where $v_{-1}(\theta) = v_{d+1}(\theta) = 0$, and b_{-1} and c_{d+1} are unspecified). We also note that if θ is an integer then $v_i(\theta)$, being an eigenvalue of the integral matrix A_i, is an algebraic integer and hence an integer.

The valency k is a trivial eigenvalue of A, belonging to the eigenvector $\mathbf{1} = (1, \cdots, 1)^\top$. Using (1c) one easily verifies that $v_i(k) = k_i$; this motivates the introduction of the scaled quantities

$$u_i(\theta) := v_i(\theta) / k_i \quad (i = -1, \cdots, d+1). \tag{12}$$

The sequence $(u_0(\theta), \cdots, u_d(\theta))$ is fundamental for the study of distance-regular graphs; we call it the *standard sequence* of Γ corresponding to the eigenvalue θ. Using (1c) again, one finds from (11) and (12) the relations

$$u_{-1}(\theta) = 0\,, \quad u_0(\theta) = 1\,, \quad u_1(\theta) = \theta / k\,,$$

$$c_i u_{i-1}(\theta) + a_i u_i(\theta) + b_i u_{i+1}(\theta) = \theta u_i(\theta) \quad (i = 0, \cdots, d)\,, \tag{13}$$

which can be used to calculate the $u_i(\theta)$ when θ is known. The equations (13) imply that the eigenvalues of A are just the eigenvalues of the tridiagonal $(d+1)\times(d+1)$-matrix

$$L_1 := \begin{bmatrix} a_0 & b_0 & & & \\ c_1 & a_1 & b_1 & O & \\ & c_2 & \cdot & \cdot & \\ & & \cdot & \cdot & \cdot \\ & O & & \cdot & \cdot & b_{d-1} \\ & & & & c_d & a_d \end{bmatrix}, \tag{14}$$

and the vector $u := (u_0(\theta), \cdots, u_d(\theta))^T$ is a (right) eigenvector of L_1 corresponding to the eigenvalue θ, normalized such that its first component is one (cf. §2.2). [The vector $v = v(\theta)$ is a left eigenvector.] The standard sequence belonging to the trivial eigenvalue $\theta_0 = k$ consists entirely of ones. It is remarkable that, for most known distance-regular graphs, some other standard sequence (or even several), usually belonging to θ_1 or θ_d, satisfies a recurrence relation of the form

$$u_{i+1} + u_{i-1} = pu_i + r \quad (i = 1, \cdots, d-1)$$

or, even stronger,

$$u_{i-1} = bu_i + c \quad (i = 1, \cdots, d-1);$$

see Theorems 8.1.2 and 8.4.1 for a partial explanation.

The location of an eigenvalue in the spectrum can be determined from the signs of its standard sequence. In the following, a *sign change* of a sequence $w_0, \cdots, w_d$ is a pair (i,j) with $i < j$ and $w_i w_j < 0$, $w_l = 0$ for $i < l < j$.

4.1.1. Proposition. *Let* Γ *be a distance-regular graph with intersection array* (1). *Define the polynomials*

$$w_0(x) := 1, \quad w_1(x) := x, \tag{15a}$$

$$w_{i+1}(x) := (x - a_i)w_i(x) - c_i b_{i-1} w_{i-1}(x) \quad (i = 0, \cdots, d). \tag{15b}$$

(i) *The eigenvalues of* Γ *are precisely the zeros of* $w_{d+1}(x)$, *and for any eigenvalue* θ,

$$u_i(\theta) = \frac{w_i(\theta)}{b_0 \cdots b_{i-1}}, \quad v_i(\theta) = \frac{w_i(\theta)}{c_1 \cdots c_i} \quad (i = 0, \cdots, d).$$

(ii) *For* $\xi \in \mathbf{R}$, *the number of eigenvalues* $> \xi$ *of* Γ *equals the number of sign changes in the sequence* $w_0(\xi), \cdots, w_{d+1}(\xi)$.

Proof. Clearly, $w_i(x) := c_1 \cdots c_i v_i(x)$ satisfies (15), and (i) follows. And (ii) holds since the $w_i(x)$ form a Sturm sequence (see e.g. STOER & BULIRSCH [703], Ch. 5.6). □

Remark. $w_i(x)$ is the leading principal $i \times i$-minor of $xI - L_1$.

4.1.2. Corollary. *The standard sequence corresponding to the i-th eigenvalue* θ_i *has precisely i sign changes.* □

Since, for integral ξ, the $w_i(\xi)$ are integral, the above proposition allows a fast and error-free determination of all integral eigenvalues (and of the integral parts of nonintegral eigenvalues) and their standard sequences by repeated bisection of the interval $[-k,k]$ at integral points ξ, and checking for the signs of the $w_i(\xi)$. (See WILKINSON &

REINSCH [786] for a bisection algorithm that applies here with only minor modifications. Be aware of overflow problems!) If all but two eigenvalues have been found, we get the remaining two from

$$\mathrm{tr}\ L_1 = \sum_{i=0}^{d} a_i = \sum_{l=0}^{d} \theta_l$$

$$\mathrm{tr}\ {L_1}^2 = \sum_{i=0}^{d} (a_i^2 + 2b_i c_{i+1}) = \sum_{l=0}^{d} \theta_l^2,$$

as

$$\theta_{i,j} = \tfrac{1}{2}(\alpha \pm \sqrt{2\beta - \alpha^2}),$$

where

$$\alpha = \mathrm{tr}\ L_1 - \sum_{l \neq i,j} \theta_l, \quad \beta = \mathrm{tr}\ {L_1}^2 - \sum_{l \neq i,j} \theta_l^2\ ;$$

with very few exceptions, this accounts for all nonintegral eigenvalues of the known distance-regular graphs of valency $k > 2$. Note that a distance-regular graph of diameter 3 must have an integral eigenvalue $\theta \neq k$ unless it is the heptagon. (Indeed, this follows immediately from the Propositions 2.2.7 and 5.1.1; note that the proof given in BIGGS [77] would exclude the heptagon, and hence is incorrect. In fact, our argument shows that the only pseudocyclic distance-regular graphs of diameter at least 3 are the polygons; DOOB [267] gives this result in case $d \in \{3,4\}$ and ascribes it to Rama Kocherlakota.)

Problem. Show that any distance-regular graph with $d > 2$ and $k > 2$ has at least one integral eigenvalue $\theta \neq k$.

Especially for hand calculation (but see also §4.4.10-11), it is worthwhile to know that the eigenvalues of L_1 distinct from k are just the eigenvalues of the tridiagonal $d \times d$-matrix

$$T := \begin{bmatrix} -c_1 & b_1 & & & & \\ c_1 & k-b_1-c_2 & b_2 & & O & \\ & c_2 & \cdot & \cdot & & \\ & & \cdot & \cdot & \cdot & \\ & O & & \cdot & \cdot & b_{d-1} \\ & & & & c_{d-1} & k-b_{d-1}-c_d \end{bmatrix}; \tag{16}$$

indeed, one easily verifies that the numbers $t_i := u_{i-1}(\theta) - u_i(\theta)$ $(i = 1, \cdots, d)$ satisfy the relations

$$c_{i-1}t_{i-1} + (k - b_{i-1} - c_i - \theta)t_i + b_i t_{i+1} = 0 \quad (i = 1, \cdots, d)\ ; \tag{17}$$

hence the vector $t := (t_1, \cdots, t_d)^{\mathrm{T}}$ satisfies the equation $Tt = \theta t$. Since this accounts for d eigenvectors belonging to distinct eigenvalues, we get, up to scaling, all eigenvectors of T in that way. In particular, we can reconstruct the eigenvalues of A and the corresponding standard sequences from the eigenvalues and eigenvectors of T: If t is a nonzero vector satisfying (17) then θ is a nontrivial eigenvalue of A and $u_{i-1}(\theta) - u_i(\theta) = \gamma t_i$ for some constant γ, whence

$$u_0(\theta) = 1\ ,\ \ u_i(\theta) = u_{i-1}(\theta) - (1-\frac{\theta}{k})\frac{t_i}{t_1}$$

by (13). Note that t_1 cannot be zero. We shall call a sequence satisfying (17) with $t_1 > 0$ a *t-sequence* of Γ corresponding to θ.

C. Eigenspaces

The *θ-eigenspace* of a distance-regular graph Γ with adjacency matrix A is the space of all vectors $x \in \mathbf{R}^v$ with $Ax = \theta x$. The standard inner product of $\mathbf{R}^v$ induces on each eigenspace the structure of a Euclidean space. The θ-eigenspace is nontrivial only if θ is an eigenvalue of Γ. In this case it is closely related to the matrix

$$E(\theta) := \sum_{i=0}^{d} u_i(\theta)A_i\ ,$$

the *eigenmatrix* of Γ corresponding to the eigenvalue θ; here the $u_i(\theta)$ form the corresponding standard sequence. From (13) we derive the following

4.1.3. Lemma. *For every eigenvalue θ of Γ,*

$$AE(\theta) = \theta E(\theta), \tag{18}$$

and if θ, θ' are distinct eigenvalues then

$$E(\theta)E(\theta') = 0. \tag{19}$$

Proof. By (2) and (13),

$$\begin{aligned} AE(\theta) &= \sum_{i=1}^{d} u_i(\theta)AA_i = \sum_{i=0}^{d} u_i(\theta)(c_{i+1}A_{i+1} + a_iA_i + b_{i-1}A_{i-1}) \\ &= \sum_{i=0}^{d}(u_{i-1}(\theta)c_i + u_i(\theta)a_i + u_{i+1}(\theta)b_i)A_i = \sum_{i=0}^{d} \theta u_i(\theta)A_i \\ &= \theta E(\theta); \end{aligned}$$

hence (18) holds. Since $E(\theta)$ and $E(\theta')$ are polynomials in A they commute with A, and (18) implies $\theta E(\theta)E(\theta') = AE(\theta)E(\theta') = E(\theta)AE(\theta') = \theta' E(\theta)E(\theta')$. For $\theta \neq \theta'$ this implies (19). □

The precise relations between the θ-eigenspace, the standard sequence $u_i(\theta)$, and the eigenmatrix $E(\theta)$ can now be stated.

4.1.4. Theorem (cf. Biggs [68]). *The θ-eigenspace of a distance-regular graph Γ has dimension*

$$f(\theta) = v\,/\,(\sum_{i=0}^{d} k_i u_i(\theta)^2)\ ; \tag{20}$$

it is spanned by the set $S(\theta)$ of columns of the corresponding eigenmatrix $E(\theta)$. The vectors of $S(\theta)$ have length $\sqrt{v\,/\,f(\theta)}$, and their mutual angles take the d values

$$\arccos\, u_1(\theta), \cdots, \arccos\, u_d(\theta).$$

Proof. By (18), the columns of $E(\theta)$ belong to the eigenspace of Γ. By (3), $E(\theta) = \phi(A)$ where ϕ is the polynomial given by

$$\phi(x) = \sum_{i=0}^{d} v_i(\theta)v_i(x) / k_i .$$

Now (18) implies that $(x-\theta)\phi(x)$ is multiple of the minimal polynomial of A. Hence $\phi(\theta') = 0$ for all eigenvalues $\theta' \neq \theta$, and $\phi(\theta)$ is the only nonzero eigenvalue of $E(\theta)$. Its multiplicity is the multiplicity of θ as an eigenvalue of A, hence equals the dimension $f(\theta)$ of the eigenspace. Therefore the columns of $E(\theta)$ span the eigenspace. Calculation of the trace of $E(\theta)$ in two ways gives

$$f(\theta)\phi(\theta) = \operatorname{tr}(E(\theta)) = u_0(\theta)v = v, \tag{21}$$

and since $\phi(\theta) = \sum v_i(\theta)^2 / k_i = \sum k_i u_i(\theta)^2$, this implies (20). Finally, (18) implies $E(\theta)^2 = \phi(A)E(\theta) = \phi(\theta)E(\theta)$ whence by (21),

$$E(\theta)^2 = vf(\theta)^{-1}E(\theta). \tag{22a}$$

Since $E(\theta)$ is symmetric, $vf(\theta)^{-1}E(\theta)$ is the Gram matrix of the columns of $S(\theta)$; this implies the remaining assertions. □

Remarks. (i) The number $f_i = f(\theta_i)$ is the multiplicity of the eigenvalue θ_i of A. Formula (20) was first stated in Biggs [68]; a proof was given in Biggs [67]. Of course it is a special case of the formula in Proposition 2.2.2.
(ii) For the trivial eigenvalue $\theta = k$ we have $u_i(k) = 1$ for all i. Hence $E(k)$ is the all-one matrix J (of rank $f(k) = 1$). In particular, (19) implies

$$E(\theta)J = 0 \quad \text{for} \quad \theta \neq k . \tag{22b}$$

In the terminology of Delsarte, Goethals & Seidel [258], we may rephrase (22a) and (22b) by saying that the set $S(\theta)$ is a *spherical* 2-*design*.
(iii) Equations (19) and (22a) imply that the $d+1$ matrices

$$E_i = v^{-1}f_iE(\theta),$$

where $f_i = f(\theta_i)$, form the complete set of primitive idempotents of the $(d+1)$-dimensional algebra spanned by A, the Bose-Mesner algebra of Γ. Consequently,

$$A_i = \sum_{\theta} v_i(\theta)E(\theta).$$

The condition that $f(\theta_i)$ must be an integer severely restricts the possible intersection arrays. Further conditions on the parameters were already derived in §2.3: the Krein parameters must be nonnegative, and we have the 'absolute bound' on the multiplicities.

4.1.5. Proposition. *For $i, j, h = 0, \cdots, d$ we have*

$$0 \leqslant q_{ijh} = \sum_{l=0}^{d} k_l u_l(\theta_i)u_l(\theta_j)u_l(\theta_h)$$

with equality if and only if the entries $e_{\gamma\delta}^{(l)}$ of $E(\theta_l)$ satisfy

$$\sum_{\delta\in\Gamma} e_{\alpha\delta}^{(i)}e_{\beta\delta}^{(j)}e_{\gamma\delta}^{(h)} = 0 \quad \textit{for all } \alpha, \beta, \gamma \in \Gamma .$$

Moreover,

$$\sum_{q_{ijh} \neq 0} f_h \leqslant \begin{cases} f_i f_j & \text{if } i \neq j \\ \frac{1}{2} f_i (f_i + 1) & \text{if } i = j. \end{cases}$$

Proof. See Theorems 2.3.2 and 2.3.3. □

D. Feasible parameter sets

The following proposition lists the most obvious restrictions on putative parameter sets for distance-regular graphs. We shall call an intersection array *feasible* when it satisfies these conditions, even when it is ruled out by other considerations. Note that our usage of the word 'feasible' differs slightly from uses found in the literature (see below).

4.1.6. Proposition. *For a distance-regular graph, the following restrictions on the intersection array hold.*

(i) $k = b_0 > b_1 \geqslant b_2 \geqslant \cdots \geqslant b_{d-1} > b_d = 0$ *and* $1 = c_1 \leqslant c_2 \leqslant \cdots \leqslant c_d \leqslant k$.

(ii) *If* $i + j \leqslant d$, *then* $c_i \leqslant b_j$.

(iii) *All parameters* p^i_{jh} *(including the* k_i*) are nonnegative integers.*

(iv) *All multiplicities* f_j *are integers.*

Proof. For γ, δ at distance d, let $\gamma = \gamma_0, \gamma_1, \cdots, \gamma_d = \delta$ be a path of length d.
(i) (BIGGS [71], cf. QUIRIN [600], SMITH [680], ENOMOTO [275]) Counting neighbours of γ_i at distance $i-2$ from γ_1 gives $c_i \geqslant c_{i-1}$, and counting neighbours of γ_i at distance i from γ_1 gives $b_i \leqslant b_{i-1}$; hence (i) holds.
(ii) (TAYLOR & LEVINGSTON [711]) This follows similarly by counting neighbours of γ_i at distance $i-1$ from γ_0.
Finally, (iii) and (iv) are obvious. □

BIGGS [71], p. 144, calls the intersection array $\{b_0, \cdots, b_{d-1}; c_1, \cdots, c_d\}$ feasible if (i) and (iv) are satisfied, all k_i are integral, and vk and $k_i a_i$ are even ($1 \leqslant i \leqslant d$). The former condition clearly is part of (iii), but the latter restriction (our Lemma 4.3.1) is independent of (i)-(iv). BIGGS [75] lists three additional feasibility conditions, namely (iii), 4.1.5 and a condition like 5.5.5.

The p^i_{jh} required in (iii) can be computed recursively from

4.1.7. Lemma.

$$p^i_{0h} = \delta_{ih}, \quad p^i_{j0} = \delta_{ij}, \quad p^i_{jd+1} = 0,$$

$$p^i_{1h} = \begin{cases} c_i & \text{if } h = i-1, \\ a_i & \text{if } h = i, \\ b_i & \text{if } h = i+1, \\ 0 & \text{otherwise}. \end{cases}$$

$$p^i_{j+1h} = \frac{1}{c_{j+1}}(p^i_{jh-1}b_{h-1} + p^i_{jh}(a_h - a_j) + p^i_{jh+1}c_{h+1} - p^i_{j-1h}b_{j-1}).$$

In particular,

$$p_{ih}^{i+h} = \frac{c_{i+1} \cdots c_{i+h}}{c_1 \cdots c_h}, \qquad p_{ih}^{i-h} = \frac{b_{i-1} \cdots b_{i-h}}{c_1 \cdots c_h},$$

$$p_{i\,h+1}^{i+h} = p_{ih}^{i+h} \frac{a_i + \cdots + a_{i+h} - a_1 - \cdots - a_h}{c_{h+1}},$$

$$p_{i\,h+1}^{i-h} = p_{ih}^{i-h} \frac{a_i + \cdots + a_{i-h} - a_1 - \cdots - a_h}{c_{h+1}},$$

Proof. Use $\sum_s p_{1j}^s p_{sh}^i = \sum_t p_{jt}^i p_{1h}^t$ (cf. Lemma 2.1.1) and induction. □

For convenience of use we state some special cases of the above formulae:

$$p_{i\,i-1}^1 = \frac{c_i k_i}{k} = \frac{b_{i-1} k_{i-1}}{k}, \quad p_{ii}^1 = \frac{a_i k_i}{k}, \quad p_{i\,i+1}^1 = \frac{b_i k_i}{k},$$

$$p_{i-2\,2}^i = \frac{c_{i-1} c_i}{\mu}, \quad p_{i+1\,2}^{i-1} = \frac{b_{i-1} b_i}{\mu}, \quad p_{i-1\,i+1}^2 = \frac{k_i c_i b_i}{k b_1},$$

$$p_{i2}^{i-1} = \frac{b_{i-1}(a_i + a_{i-1} - a_1)}{\mu}, \quad p_{i2}^{i+1} = \frac{c_{i+1}(a_i + a_{i+1} - a_1)}{\mu}.$$

For computational purposes it is of interest to note that p_{ij}^l only depends on parameters a_s, b_s, c_s with $s < \frac{1}{2}(i+j+l)$ and possibly on c_s with $s = \frac{1}{2}(i+j+l)$. Note that the recursion found from (10) above is slightly different from that given in the lemma.

The f_j are computed using the formula

$$f_j = v / (\sum_i k_i u_i^2)$$

derived in Theorem 4.1.4, where $(u_i)_i$ is the standard sequence corresponding to the j-th eigenvalue θ_j, i.e., the corresponding right eigenvector of the $(d+1)\times(d+1)$ matrix

$$L_1 = \begin{bmatrix} 0 & k & & & \\ 1 & \lambda & b_1 & O & \\ & \mu & \cdot & \cdot & \\ & & \cdot & \cdot & \cdot \\ & O & & \cdot & \cdot & b_{d-1} \\ & & & & c_d & a_d \end{bmatrix}$$

normalized so that $u_0 = 1$.

Remarks. (i) If one uses finite precision arithmetic to calculate the f_j, then it is not always easy to decide whether f_j is integral or not. (See Lemma 2.2.6 and the discussion following for a way to proceed in exact arithmetic.) Sometimes one can show nonintegrality by observing that all symmetric functions in the eigenvalues with a given multiplicity are integral; in particular, if θ_j is nonintegral, then for some $i \neq j$ we must have $f_i = f_j$.

For example, for the intersection array $\{26,25,5,1;1,5,25,26\}$ and $v = 184$ we might find the spectrum 26^1 5.099^{46} 1.403^{79} -5.099^{46} -11.403^{12}; the multiplicity 79 occurs only once while 1.403 is nonintegral. Consequently this array does not belong to a distance-regular graph (and the 79 was in fact $\frac{13}{82}(287+33\sqrt{41}) = \frac{13}{2}(7+\frac{33}{\sqrt{41}}) \approx 78.9993$ with $\theta = -5+\sqrt{41} \approx 1.4031$).

(ii) It is possible to formulate general inequalities for association schemes of which (i) and (ii) of the above proposition are special cases. E.g.: *If* $p^l_{st} \neq 0$, *then* $p^l_{ij} \leqslant \sum_r \min(p^t_{rj}, p^s_{ir})$ (see the remark following Lemma 2.1.1).

(To see that this statement holds consider the picture

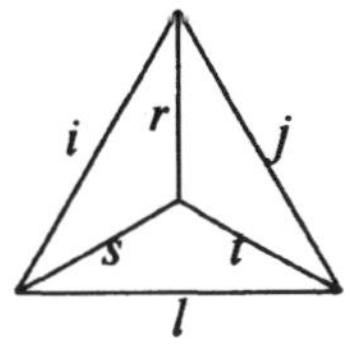

.)

For $i = l-1$, $t = l+1$, $j = s = 1$, this yields $c_l \leqslant c_{l+1}$; for $i = l+1$, $t = l-1$, $j = s = 1$, this yields $b_l \leqslant b_{l-1}$; and for $i = l-1$, $j = 1$, $s = l+t$, this yields $c_l \leqslant b_t$. Are there other useful specializations?

E. Imprimitivity and the Q-polynomial property

A graph Γ of diameter d is called *imprimitive* when for some i, $1 \leqslant i \leqslant d$, its distance-$i$ graph Γ_i (cf. §A.5) is disconnected. Note that when Γ is distance-regular this definition coincides with that given in §2.4 for association schemes.

Two obvious types of imprimitive graphs are the *bipartite* graphs of diameter $d \geqslant 2$ (here Γ_2 is disconnected) and the *antipodal* graphs (graphs such that Γ_d is an equivalence relation) of diameter $d \geqslant 2$. In Theorem 4.2.1 below we shall see that, conversely, any imprimitive distance-regular graph of valency > 2 is bipartite or antipodal.

Let $\theta_0 = k, \theta_1, \cdots, \theta_d$ be an ordering of the eigenvalues of a distance-regular graph Γ. We say that the ordering $\theta_0, \theta_1, \cdots, \theta_d$ is *Q-polynomial* (cf. §2.7) if there are a sequence $\sigma = (\sigma_0, \cdots, \sigma_d)$ and polynomials q_j of degree j ($j = 0, \cdots, d$) such that the eigenmatrices $E(\theta_j)$ can be written as

$$E(\theta_j) = \sum_{i=0}^{d} q_j(\sigma_i)A_i;$$

in this case, σ is called a *Q-sequence* of Γ. The graph Γ is called *Q-polynomial* if the eigenvalues of Γ allow a Q-polynomial ordering. Sometimes, a distance-regular graph has several Q-polynomial orderings.

Since $E(\theta_0), \cdots, E(\theta_d)$ are linearly independent, the $d+1$ numbers σ_i must be distinct. Clearly, the existence of the polynomials q_j is invariant under transformations of the sequence σ, i.e. under replacement of σ_i by $a\sigma_i + b$ ($i = 0, \cdots, d$) for any $a, b \in \mathbb{R}$ with $a \neq 0$. In particular we may normalize σ such that the linear polynomial q_1 becomes $q_1(x) = x$. Then $E(\theta_1) = \sum \sigma_i A_i$ whence $\sigma_i = u_i(\theta_1)$ is the standard sequence corresponding to θ_1. We call such a standard sequence a *standard Q-sequence* and summarize our discussion in the following

4.1.8. Proposition. *Every Q-sequence $\sigma = (\sigma_0, \cdots, \sigma_d)$ of a Q-polynomial distance-regular graph Γ consists of distinct elements σ_i and is a linear transform of some standard Q-sequence of Γ.* □

We shall discuss the properties of Q-polynomial orderings in some detail in Chapter 8; it turns out that the Q-sequence determines the Q-polynomial ordering and the Q-sequence and intersection array of Q-polynomial graphs are very restricted and determined by 5 parameters only (Proposition 8.1.5). Nevertheless, almost all *known* distance-regular graphs with large diameter satisfy this restrictive property. The importance of Q-polynomial orderings lies in the fact that they allow the investigation of certain very regular subsets of the points called *t*-designs. We do not discuss these matters and refer the interested reader to DELSARTE [251], MACWILLIAMS & SLOANE [515], NEUMAIER [550] and BANNAI & ITO [39].

F. Distance transitivity

A group of automorphisms of a connected graph Γ of diameter d is said to be *distance-transitive* on Γ if it is transitive (on the vertex set of Γ and) on each of the sets $\{(\gamma,\delta) \mid d(\gamma,\delta) = i\}$ for $0 \leqslant i \leqslant d$. A graph is called *distance-transitive* if it is connected and admits a distance-transitive group of automorphisms. Clearly, such a graph is flag-transitive, and hence can be represented as a coset graph $\Gamma(G,H,r)$, cf. §A.4. An easy criterion for distance-transitivity can be given in terms of the double coset graph (cf. §A.4):

4.1.9. Lemma. *The group G is distance-transitive on the coset graph $\Gamma(G,H,r)$ if and only if the associated double coset graph is linear (i.e., a path). In particular, a distance-transitive graph is distance-regular.*

Proof. Both conditions are equivalent to the requirement that the orbits of the stabilizer $G_\gamma = H$ of the vertex $\gamma = H$ of $\Gamma(G,H,r)$ are precisely the sets $\Gamma_i(\gamma)$ $(0 \leqslant i \leqslant d)$. □

The converse is not true, as was first shown by ADEL'SON-VEL'SKII et al. [1]. They gave the example of the block graph of the cyclic Steiner triple system on 13 points; this graph is strongly regular but its automorphism group has two orbits on the vertex set. The smallest distance-regular graph that is not distance-transitive is the Shrikhande graph. We shall meet numerous other examples (of arbitrarily large diameter) in subsequent chapters.

4.1.10. Theorem. *Suppose G is a distance-transitive group of automorphisms of a distance-regular graph Γ. Then the permutation representation of G on the vertex set of Γ is imprimitive if and only Γ is imprimitive. In this case, we have at least one of:*

(i) *Γ is bipartite; there is a normal subgroup G^0 of index 2 in G stabilizing the connected components of Γ_2, and distance-transitive on each;*

(ii) *Γ is an antipodal r-cover of a distance-transitive graph $\overline{\Gamma}$ (for some r, $1 < r \leqslant k$), there is a normal subgroup N stabilizing each class of mutually antipodal vertices and doubly transitive on each of these classes, such that the induced action of G on $\overline{\Gamma}$ has kernel N.*

Proof. Suppose that Γ is primitive and $\equiv$ is a G-invariant equivalence relation. If $\gamma \equiv \delta$ and $d(\gamma,\delta) = i > 0$, then, by distance transitivity of G, arbitrary points at distance i are equivalent, and, since Γ_i is connected, all points are equivalent. Therefore G is primitive. This yields the first part of the theorem. The second part easily follows from Theorem 4.2.1. □

Note that both (i) and (ii) may occur, as is shown by, e.g., the Foster graph and the $2m$-cubes.

4.1.11. Proposition. *Suppose G is a distance-transitive group of automorphisms of a distance-regular graph Γ of diameter d. Denote by π the permutation character of G on Γ. Then the following hold.*

(i) *The permutation rank of G on Γ is $d+1$ and hence equal to (π,π).*

(ii) *The characters χ_i $(i = 0, \cdots, d)$ of the representation of G on the θ_i-eigenspace of A in $\mathbf{R}^\Gamma$ are real, irreducible, and distinct. Furthermore,*

$$\pi = \sum_{i=0}^{d} \chi_i \quad \text{and} \quad \chi_i(1) = f(\theta_i),$$

where f is as in (20).

(iii) *The Bose-Mesner algebra of Γ is the full centralizer algebra of G.*

(iv) (Scott [648]) *If $q_{ij}^k \neq 0$, then χ_k occurs in $\chi_i\chi_j$, and if $q_{ii}^k \neq 0$, then χ_k occurs in χ_i^{2+}. In particular, if Γ is Q-polynomial, then the characters χ_i can be ordered in such a way that χ_{i+1} occurs in $\chi_1\chi_i$ $(0 \leqslant i < d)$.*

Proof (cf. §2.9). (i) Let H be the stabilizer in G of a vertex of Γ. By Frobenius reciprocity (cf. Gorenstein [338], Theorem 4.5) we have $(\pi,\pi) = (1_H,\pi|_H)$. But the right hand side is the number of H-orbits in the vertex set of Γ, which is by definition equal to the permutation rank of G, and by distance transitivity also equal to $d+1$.
(ii) Since the θ_i $(i = 0, \cdots, d)$ are distinct (cf. §4.1B), the $d+1$ eigenspaces of A in $\mathbf{R}^\Gamma$ are G-invariant. Thus, the χ_i are well defined, and obviously satisfy the equations in (ii). Since θ_i is real, the representation with character χ_i must be real. It remains to show that the χ_i are irreducible and mutually distinct. In view of (i), we have

$$d+1 = (\pi,\pi) = \sum_{i,j=0}^{d} (\chi_i,\chi_j) \geqslant \sum_{i=0}^{d} (\chi_i,\chi_i) \geqslant d+1,$$

whence $(\chi_i,\chi_j) = \delta_{i,j}$, proving (ii).
(iii) The Bose-Mesner algebra $\mathcal{Q}$ of Γ is generated by A. If $g \in G$ is viewed as a permutation matrix, then $g^\mathsf{T} A g = A$ (where $^\mathsf{T}$ is taken with respect to the standard basis Γ of $\mathbf{R}^\Gamma$). But $g^{-1} = g^\mathsf{T}$, so A and hence $\mathcal{Q}$ are contained in the centralizer algebra of G. By Schur's lemma (cf. Gorenstein [338], Theorem 5.3) and (ii), the dimension of the centralizer algebra is $d+1 = \dim \mathcal{Q}$. Therefore, it coincides with $\mathcal{Q}$.
(iv) We have $\chi_i(g) = \operatorname{tr}(P_g E_i) = \Sigma(P_g \circ E_i)$, where P_g denotes the permutation matrix representing the action of $g \in G$ on Γ. Consequently, $(\chi_i\chi_j)(g) = \operatorname{tr}(P_g E_i \otimes P_g E_j)$. The natural representation of G on $V \otimes V$, where $V = \mathbf{R}^\Gamma$, has an invariant subspace $\langle \gamma\otimes\gamma \mid \gamma \in \Gamma\rangle \cong V$, and if P is the natural projection onto this subspace, then $P(P_g E_i \otimes P_g E_j)P = P_g(E_i \circ E_j)$. Thus, if $q_{ij}^k \neq 0$ then we find that the representation $g \mapsto P_g E_k$ of G on V is a homomorphic image of the representation $g \mapsto P_g E_i \otimes P_g E_j$ on $V \otimes V$, so that χ_k is a constituent of $\chi_i\chi_j$. □

Problem. Can one recognize Q-polynomiality of a distance-transitive graph from the permutation character?

G. Distance-biregular graphs

Let Γ be a connected graph, and let $\gamma \in \Gamma$. The graph Γ is called *distance-regular around* γ when for $\delta \in \Gamma_i(\gamma)$ the numbers $a_i(\gamma) = |\Gamma_i(\gamma) \cap \Gamma(\delta)|$, $b_i(\gamma) = |\Gamma_{i+1}(\gamma) \cap \Gamma(\delta)|$ and $c_i(\gamma) = |\Gamma_{i-1}(\gamma) \cap \Gamma(\delta)|$ do not depend on the vertex δ chosen.

4.1.12. Theorem (Godsil & Shawe-Taylor [327]). *Suppose that Γ is distance-regular around each of its vertices. Then either Γ is distance-regular, or Γ is bipartite and the numbers $a_i(\gamma)$, $b_i(\gamma)$ and $c_i(\gamma)$ only depend on the part of the bipartition containing γ.* □

Graphs of the latter type (bipartite, with two sets of intersection numbers, say b_i', c_i' $(0 \leqslant i \leqslant d')$ and b_i'', c_i'' $(0 \leqslant i \leqslant d'')$) are called *distance-biregular*. These graphs were first introduced by Delorme [246] under the name 'graphes à régularité métrique forte'. He showed that the two components of Γ_2 are distance-regular, and derived various restrictions on the parameters. In particular, he found that the adjacency algebra of Γ (generated by the adjacency matrix A of Γ) has dimension $d+1$, where $d = \max(d',d'') = \operatorname{diam}(\Gamma)$, gave a bound on d in terms of the girth and valencies, and derived Krein inequalities for Γ (stronger than those for the two components of Γ_2).

About the spectra, we can say the following. (These remarks may be due to Gardiner.) Let Γ' and Γ'' be the two components of Γ_2, and let k' and k'' be the valencies in Γ of the vertices of Γ', Γ'', respectively. Then Γ' and Γ'' have eigenvalues $(\theta^2 - k')/\mu'$ and $(\theta^2 - k'')/\mu''$, respectively, where θ is an eigenvalue of Γ. (Of course, μ' stands for c_2', etc.) If $\theta \neq 0$ then these three eigenvalues have the same multiplicities in their respective graphs; if $\theta = 0$ then the multiplicity of the eigenvalue 0 in Γ is the sum of the multiplicities of $-k'/\mu'$ and $-k''/\mu''$ in Γ' and Γ'', respectively. (And one of these multiplicities will be zero when Γ' and Γ'' do not have the same diameter.)

Mohar & Shawe-Taylor [537] show that if one of the two valencies of Γ equals 2, then either Γ is $K_{2,q}$, or Γ arises by subdividing all edges of a distance-regular graph Γ'; in the latter case the two components of Γ_2 are Γ' and $L(\Gamma')$, and since $L(\Gamma')$ also must be distance-regular, we have one of the possibilities of the conclusion of Theorem 4.2.16. (And conversely, if both Γ' and $L(\Gamma')$ are distance-regular, then the subdivision of Γ', or, which amounts to the same, the point-edge incidence graph of Γ', is distance-biregular.)

Delorme showed that if $d' \neq d''$, then $\min(d',d'')$ is odd, and that if $d' = d''$ is odd, then Γ is distance-regular. But $d(\Gamma') = [\frac{1}{2}d']$ and $d(\Gamma'') = [\frac{1}{2}d'']$, so if Γ is distance-biregular but not distance-regular, then we have either $d' = 2e+1$, $d'' = 2e+2$, $d(\Gamma') = e$, $d(\Gamma'') = e+1$, or $d' = d'' = 2e$, $d(\Gamma') = d(\Gamma'') = e$ (where we have chosen Γ' such that $d(\Gamma') \leqslant d(\Gamma'')$).
One has 'Fisher's inequality': *if $d' = 2e+1$, $d'' = 2e+2$, then $k' > k''$*, and a 'Gardiner-Meredith inequality': *if $\mu' \geqslant 2$, then $c_{i+1}'' > c_i'$ and $c_{i+1}' > c_i''$.*
If $d = 2$, then $\Gamma \cong K_{p,q}$. If $d' = 3$, then we have a 2-(v,k,λ) design in which two

blocks meet in 0 or μ points (and both occur) - here $v = v(\Gamma')$, $k = k''$, $\lambda = \mu'$, $\mu = \mu''$. Examples are all 2-(v,k, 1) designs (Steiner 2-designs), and, e.g., the design of points and hyperplanes in an affine space. For $d = 4$, Delorme gives several geometric constructions, such as the incidence graphs of partial geometries; the incidence graphs of the points in $AG(3,q)$ and the planes from certain parallel classes, where the lines at infinity form a dual $(0,n)$-arc; and the incidence graphs of the points in $AG(6,q)$ and the 3-spaces from certain parallel classes, where the planes at infinity form one family in a $D_3(q)$ geometry. Finally, we have the incidence graphs of generalized polygons and the (inclusion) graphs of k-subsets and $(k+1)$-subsets of an n-set, or k-subspaces and $(k+1)$-subspaces of a vector space of dimension n. We know of no other examples with $d > 4$.

Remarks. (i) In Example 9.2.6 we shall meet a bipartite graph that is distance-regular (with constant a_i, b_i, c_i) around the points in one half of the bipartition, but not distance-biregular.
(ii) Distance-bitransitive graphs are studied in Shawe-Taylor [664].

H. Weakenings of distance-regularity

We have seen that a connected graph Γ is distance-regular if and only if $A_i = p_i(A)$ for certain polynomials p_i of degree i $(0 \leqslant i \leqslant d)$. If one drops the condition on the degree of p_i, one obtains the 'distance-polynomial graphs' studied by Weichsel [773]. For example, all regular connected graphs of diameter two are distance-polynomial, and so are the complements of distance-regular graphs (when connected). More generally, the complement of an arbitrary distance-polynomial graph is again distance-polynomial. Any distance-polynomial graph is regular (since $J = p(A)$ for some polynomial p).
A connected graph Γ is called *distance degree regular* when $|\Gamma_i(\gamma)|$ is independent of $\gamma \in \Gamma$ for all i, $0 \leqslant i \leqslant \operatorname{diam}\Gamma$. This holds if and only if $(A_iA_j)_{\gamma\delta} = (A_jA_i)_{\gamma\delta}$ for all i, j, and all pairs γ, δ of adjacent vertices of Γ. In particular, distance-polynomial graphs are distance degree regular, and so is any graph with a vertex-transitive group of automorphisms. The flag graph of K_4 (on 12 vertices) is distance degree regular, but not distance-polynomial. Hilano & Nomura [389] show that a distance degree regular graph satisfies $3k_j \geqslant 2(k+1)$ for all $j < d$, and determine the cases where equality holds.

4.2 Imprimitivity; new graphs from old

In this section we look at various relations between distance-regular graphs. If a distance-regular graph Γ is bipartite then its halved graphs are distance-regular, too; if Γ is antipodal then its antipodal quotient (folding) is again distance-regular. Sometimes the line graph of a distance-regular graph is distance-regular. Choosing another relation in the association scheme of Γ may yield a distance-regular graph again, either because the scheme has several P-polynomial structures, or because the new relation generates a coarser scheme which happens to be P-polynomial.

A. Imprimitivity

Recall that a graph Γ of diameter d is called *imprimitive* when for some i, $1 \leq i \leq d$, the graph Γ_i is disconnected. (Here Γ_i is the graph with the same vertices as Γ, where two vertices are adjacent when they have distance i in Γ.) Note that when Γ is distance-regular this definition coincides with that given in §2.4 for association schemes.

Two obvious types of imprimitive graphs are the *bipartite* graphs of diameter $d \geq 2$ (here Γ_2 is disconnected) and the *antipodal* graphs (graphs such that Γ_d is an equivalence relation) of diameter $d \geq 2$.

The following remarkable theorem was proved by Smith [680] (and later by Martinov [519]) for distance-transitive graphs; his proof is easily extended to arbitrary distance-regular graphs, see, e.g., Gardiner [313].

4.2.1. Theorem. *An imprimitive distance-regular graph of valency $k > 2$ is bipartite or antipodal (or both).*

Proof. Let Γ be an imprimitive distance-regular graph of valency $k > 2$ and diameter d. Let i be the smallest positive integer such that Γ_i is disconnected. If $\gamma,\delta,\epsilon \in \Gamma$ form a triple of type (j,i,i) (i.e. $d(\gamma,\delta) = j$, $d(\gamma,\epsilon) = d(\delta,\epsilon) = i$) then $p^j_{ii} > 0$ whence any path in Γ_j gives rise to a path in Γ_i. Hence the minimality of i implies $j \geq i$; therefore Γ contains no triples of type (j,i,i) with $j < i$.

We now distinguish several cases. If $i = 1$ then Γ is disconnected, contradiction. If $i = 2 < d$ then pick γ_0,γ_3 at distance three and consider a path $\gamma_0 \sim \gamma_1 \sim \gamma_2 \sim \gamma_3$. If $\delta \in \Gamma(\gamma_0) \cap \Gamma(\gamma_1)$ then either $\delta\gamma_0\gamma_2$ or $\delta\gamma_1\gamma_3$ is a triple of forbidden type (1,2,2); therefore $a_1 = 0$ and there are no triangles. If there is a circuit C of odd side > 3 then all vertices of C lie in the same connected component Δ of Γ_2; in particular, Δ contains edges of Γ. Now for any $\gamma,\delta \in \Delta$ with $\gamma \sim \delta$ (in Γ), all points adjacent in Γ to γ or δ are in Δ; since Γ is connected this implies $\Delta = \Gamma$, contradiction. Hence Γ contains no circuit of odd side and therefore Γ is bipartite. If $2 < i < d$, then pick γ_0,γ_d at distance d, and let $\gamma_0,\gamma_1, \cdots, \gamma_d$ be a path of length d. Since $k \geq 3$ there is $\delta \neq \gamma_i,\gamma_{i+2}$ with $\delta \sim \gamma_{i+1}$. But then $d(\delta,\gamma_0) = i+l$ with $l \in \{0,1,2\}$ and $\delta\gamma_{i+l}\gamma_l$ is a triple of forbidden type (j,i,i), $j \leq 2$; contradiction. Finally, if $i = d$ then all triples (j,d,d) with $j < d$ are forbidden whence Γ_d is a disjoint union of cliques, hence antipodal. □

When Γ is connected and bipartite of diameter $d \geq 2$, then Γ_2 has two components, and the graphs induced on these components by Γ_2 are called Γ^+ and Γ^- (or also $\frac{1}{2}\Gamma$ for an arbitrary one of these), the *halved* graphs of Γ. Clearly, the graphs $\frac{1}{2}\Gamma$ have diameter $[d/2]$.

When Γ is antipodal then we can define a new graph $\overline{\Gamma}$, the *folded* graph of Γ, which has the equivalence classes of Γ_d as vertices, and where two equivalence classes are adjacent whenever they contain adjacent vertices. Clearly, the graph $\overline{\Gamma}$ has diameter $[d/2]$.

The above theorem may be rephrased by saying that if $k > 2$ and I is a subset of $\{0, \cdots, d\}$ such that having distance in I is an equivalence relation, then $I = \{0\}$, or $I = \{0, d\}$, or $I = \{0, 2, 4, \cdots\}$, or $I = \{0, 1, \cdots, d\}$.

Given a distance-regular graph, we may obtain a primitive distance-regular graph after halving at most once and folding at most once (except in the case of $8m$-gons). More precisely, suppose that Γ is distance-regular of valency $k > 2$.

a) If Γ is antipodal with folded graph $\overline{\Gamma}$, then $\overline{\Gamma}$ is not antipodal, except when Γ has diameter $d \leqslant 3$, in which case $\overline{\Gamma}$ is complete, or when Γ is bipartite of diameter $d = 4$, in which case $\overline{\Gamma}$ is complete bipartite.
(For: if $\overline{\Gamma}$ has diameter e and is antipodal, then having distance in $\{0, e, d-e, d\}$ is an equivalence relation in Γ.)

b) If Γ is bipartite with halved graph $\frac{1}{2}\Gamma$, then $\frac{1}{2}\Gamma$ is not bipartite.
(For: if $\frac{1}{2}\Gamma$ is bipartite, then $0 = \lambda(\frac{1}{2}\Gamma) = p^2_{22}(\Gamma) \geqslant b_1 - 1$, so $b_1 = 1, k = 2$.)

c) If Γ is antipodal, and either has odd diameter d or is not bipartite, then $\overline{\Gamma}$ is primitive.
(For: if $\overline{\Gamma}$ is bipartite of diameter e, then having distance in

$$\{0, 2, \cdots, 2[\tfrac{1}{2}e], d-2[\tfrac{1}{2}e], \cdots, d-2, d\}$$

is an equivalence relation in Γ, so $d = 2e$ is even and Γ is bipartite.)
In fact, if Γ is antipodal and bipartite of odd diameter, then it is an antipodal 2-cover, and $\frac{1}{2}\Gamma \cong \overline{\Gamma}_2$ unless $d \leqslant 3$, in which case both $\frac{1}{2}\Gamma$ and $\overline{\Gamma}$ are complete.

d) If Γ is bipartite, and either has odd diameter d or is not antipodal, then $\frac{1}{2}\Gamma$ is primitive.
(For: if $\frac{1}{2}\Gamma$ is antipodal of diameter $e \geqslant 2$, then having distance in $\{0, 2e\}$ is an equivalence relation in Γ, so $d = 2e$ and Γ is antipodal.)

e) If Γ has even diameter $d = 2e$, and is both bipartite and antipodal, then the graphs $\frac{1}{2}\Gamma$ are antipodal, $\overline{\Gamma}$ is bipartite, and the graphs $\overline{\frac{1}{2}\Gamma} \cong \frac{1}{2}\overline{\Gamma}$ are primitive.

B. Parameters of halved graphs, folded graphs, and covers

The above discussion is essentially contained in the unpublished paper Biggs & Gardiner [81]. In this same paper we find the following useful result. (Part (ii) is also found in Gardiner [310].) Its proof is straightforward and will be left to the reader.

4.2.2. Proposition. *Let Γ be a distance-regular graph with intersection array $\iota(\Gamma) = \{b_0, b_1, \cdots, b_{d-1}; c_1, c_2, \cdots, c_d\}$ and diameter $d \in \{2m, 2m+1\}$. Then*

(i) *Γ is bipartite if and only if $b_i + c_i = k$ (i.e., $a_i = 0$) for $i = 0, \cdots, d$. In this case the halved graphs are distance-regular of diameter m with intersection array*

$$\iota(\Gamma^{\pm}) = \{\frac{b_0 b_1}{\mu}, \frac{b_2 b_3}{\mu}, \cdots, \frac{b_{2m-2} b_{2m-1}}{\mu}; \frac{c_1 c_2}{\mu}, \frac{c_3 c_4}{\mu}, \cdots, \frac{c_{2m-1} c_{2m}}{\mu}\}.$$

(ii) *Γ is antipodal if and only if $b_i = c_{d-i}$ for $i = 0, \cdots, d$, $i \neq m$. In this case Γ is an antipodal r-cover of its folded graph $\overline{\Gamma}$, where*

$$r = 1 + b_m / c_{d-m},$$

and if $d > 2$ then $\overline{\Gamma}$ is distance-regular of diameter m with intersection array

$$\iota(\overline{\Gamma}) = \{b_0, b_1, \cdots, b_{m-1}; c_1, c_2, \cdots, c_{m-1}, \gamma c_m\},$$

where

$$\gamma = r \ \textit{if}\ d = 2m, \ \ \gamma = 1 \ \textit{if}\ d = 2m+1. \qquad \square$$

Note that in case (i), although Γ^+ and Γ^- have the same parameters they need not be isomorphic; an example of this is Tutte's 12-cage (the generalized dodecagon of order (1,2), see §6.7 and §12.4). Of course, Γ^+ and Γ^- are isomorphic if Aut Γ is vertex transitive.

Reversing the point of view in (ii), we can ask for the parameters of an antipodal r-cover $\tilde{\Gamma}$ of a distance-regular graph Γ with intersection array $\{b_0, \cdots, b_{d-1}; c_1, \cdots, c_d\}$. If $\tilde{\Gamma}$ has diameter $2d > 2$, then $\iota(\tilde{\Gamma})$ has the form

$$r.\Gamma := \{b_0, b_1, \cdots, b_{d-1}, \frac{r-1}{r}c_d, c_{d-1}, \cdots, c_1; c_1, \cdots, c_{d-1}, \frac{1}{r}c_d, b_{d-1}, \cdots, b_0\},$$

(so that $r \leqslant c_d / \max(c_{d-1}, c_d - b_{d-1})$) and if $\tilde{\Gamma}$ has diameter $2d+1$, then $\iota(\tilde{\Gamma})$ has the form

$$(r.\Gamma)_t := \{b_0, b_1, \cdots, b_{d-1}, t(r-1), c_d, c_{d-1}, \cdots, c_1; c_1, \cdots, c_{d-1}, c_d, t, b_{d-1}, \cdots, b_0\}$$

for some integer t, where $t(r-1) \leqslant \min(b_{d-1}, a_d)$ and $c_d \leqslant t$. Clearly, given $\iota(\Gamma)$, there are only finitely many possibilities for r and t.

We can say the following about the spectra.

4.2.3. Proposition. *Let Γ be distance-regular with spectrum Φ, where $\theta \in \Phi$ has multiplicity $m(\theta)$.*

(i) *If Γ is bipartite, then Γ^+ and Γ^- have spectrum $\{(\theta^2 - k)/\mu \mid \theta \in \Phi\}$, where $(\theta^2 - k)/\mu$ has multiplicity $m(\theta) = m(-\theta)$ if $\theta \neq 0$ and $\frac{1}{2}m(0)$ if $\theta = 0$.*

(ii) *If Γ is antipodal of diameter $d \geqslant 3$, then $\overline{\Gamma}$ has a spectrum that is a subset $\overline{\Phi}$ of Φ, and for $\theta \in \overline{\Phi}$ the multiplicity of θ in the spectra of $\overline{\Gamma}$ and Γ agree.* ($\overline{\Phi}$ is obtained from Φ by taking every second eigenvalue, see below.) $\square$

(These statements depend on the form of the intersection array only, and not on the existence of a graph Γ. Thus, there is no feasible intersection array that can be ruled out because the halved graph would have eigenvalues with nonintegral multiplicity.)

Consider the case of a distance-regular graph Γ of diameter d with antipodal r-cover $\tilde{\Gamma}$ of diameter $\tilde{d} \geqslant 3$. For typographical reasons, let us write $e := \tilde{d}$. The eigenvectors of Γ yield eigenvectors of $\tilde{\Gamma}$ that are constant on the antipodal classes, so that the eigenvalues of Γ are also eigenvalues of $\tilde{\Gamma}$ with the same multiplicities. (Note that the multiplicity cannot increase, since if $(u_i)_i$ is the standard sequence of Γ corresponding to θ, then $(\tilde{u}_i)_i$, defined by

$$\tilde{u}_i = \begin{cases} u_i & \text{if } i \leqslant d \\ u_{e-i} & \text{otherwise} \end{cases}$$

is the standard sequence of $\tilde{\Gamma}$ corresponding to θ, and the formula $m(\theta) = v / \sum k_i u_i^2$ now shows that θ has the same multiplicity in Γ and $\tilde{\Gamma}$.)

It follows (e.g. by Proposition 4.1.1) that if $\tilde{\Gamma}$ has eigenvalues $\tilde{\theta}_0 > \tilde{\theta}_1 > \cdots > \tilde{\theta}_e$, then Γ has eigenvalues $\theta_0 = \tilde{\theta}_0$, $\theta_1 = \tilde{\theta}_2$, $\cdots$, $\theta_d = \tilde{\theta}_{2d}$, i.e. the eigenvalues of Γ interlace the 'new' eigenvalues of $\tilde{\Gamma}$.

The new eigenvalues arise from eigenvectors that are not constant on antipodal classes and hence have standard sequences $\tilde{u}$ with $\tilde{u}_e = -1/(r-1)$. For $i \leqslant d$ we have $\tilde{u}_{e-i} = -\tilde{u}_i/(r-1)$, and if $e = 2d$, this means that $u_d = 0$. Hence:

4.2.4. Proposition. *If $\tilde{d} = 2d$, then the d eigenvalues of $\tilde{\Gamma}$ that are not eigenvalues of Γ are the eigenvalues of the $d \times d$ matrix*

$$\begin{bmatrix} 0 & k & & & & \\ 1 & \lambda & b_1 & & O & \\ & \mu & \cdot & \cdot & & \\ & & \cdot & \cdot & \cdot & \\ & O & & \cdot & \cdot & b_{d-2} \\ & & & & c_{d-1} & a_{d-1} \end{bmatrix},$$

that is, are the roots of $u_d(\theta) = 0$. Thus, these eigenvalues do not depend on r, and their multiplicity is proportional to $r-1$.
If $\tilde{d} = 2d+1$, then the $d+1$ eigenvalues of $\tilde{\Gamma}$ (with parameters $(r.\Gamma)_t$) that are not eigenvalues of Γ are the eigenvalues of the $(d+1)\times(d+1)$ matrix

$$\begin{bmatrix} 0 & k & & & & \\ 1 & \lambda & b_1 & & O & \\ & \mu & \cdot & \cdot & & \\ & & \cdot & \cdot & \cdot & \\ & O & & \cdot & a_{d-1} & b_{d-1} \\ & & & & c_d & a_d - rt \end{bmatrix},$$

that is, the roots of $c_d u_{d-1}(\theta) + (a_d - rt - \theta)u_d(\theta) = 0$. Thus, these eigenvalues depend only on rt.

4.2.5. Corollary. *In particular, in case $d = 2$, $\tilde{d} = 4$, these eigenvalues θ are the two roots of $\theta^2 - \lambda\theta - k = 0$, and occur with multiplicity $m(\theta) = \dfrac{(r-1)v}{2+\lambda\theta/k}$. Consequently, either $\lambda = 0$, or $\lambda^2 + 4k$ is a square (and these eigenvalues are integral).* □

4.2.6. Corollary. *In particular, in case $d = 1$, $\tilde{d} = 3$, the two new eigenvalues θ are the two roots of $\theta^2 - (k-1-r\mu)\theta - k = 0$, (with $\mu = \mu(\tilde{\Gamma})$), and have multiplicity $m(\theta) = \dfrac{k(k+1)(r-1)}{k+\theta^2}$. Thus, either $k-1 = r\mu$ and $k\mu$ is even, or θ is an integer, and $(r-1)k\mu$ is even.*

Proof. The parity restrictions follow by counting edges in $\tilde{\Gamma}(\infty)$ for some $\infty \in \tilde{\Gamma}$. □

All of the above is already contained in BIGGS & GARDINER [81].

C. Structural conditions for the existence of covers

VAN BON [98] gives a simple geometric nonexistence criterion for distance-regular covers of even diameter of distance-regular graphs. For γ, $\delta \in \Gamma$ with $d(\gamma,\delta) = d$, let $C(\gamma,\delta)$ denote the union of all geodesics between γ and δ.

4.2.7. Proposition. *Suppose that Γ is distance-regular of diameter $d \geq 2$ and has a distance-regular antipodal r-cover of diameter 2d. Then for $\gamma, \delta \in \Gamma$, $d(\gamma,\delta) = d$, the subgraph $C(\gamma,\delta)\setminus\{\gamma,\delta\} = \bigcup_{j=1}^{d-1} \Gamma_j(\gamma) \cap \Gamma_{d-j}(\delta)$ is the disjoint union of r subgraphs of equal size.*

Proof. Let $\tilde{\Gamma}$ be a distance-regular antipodal r-cover of diameter $2d$ of Γ, with quotient map $\pi : \tilde{\Gamma} \to \Gamma$. Let $\gamma_1 \in \pi^{-1}(\gamma)$, and let $\pi^{-1}(\delta) = \{\delta_1, \cdots, \delta_r\}$. Let C_j be the union of all geodesics in $\tilde{\Gamma}$ between γ_1 and δ_j $(1 \leq j \leq r)$ and $C = \bigcup_j C_j \setminus \{\gamma_1,\delta_j\}$. Then $\pi|_C : C \to C(\gamma,\delta)\setminus\{\gamma,\delta\}$ is an isomorphism. □

For example, consider the graph $\overline{T(n)}$, the complement of the triangular graph $T(n)$. It is strongly regular, and connected when $n = 2$ or $n \geq 5$. If γ, δ are two nonadjacent vertices, then $\gamma^\perp \cap \delta^\perp$ is isomorphic to $\overline{T(n-3)}$, and in particular is connected for $n = 5$ or $n \geq 8$. Consequently, $\overline{T(n)}$ does not have antipodal covers of diameter 4 for $n = 5$ or $n \geq 8$. On the other hand $\overline{T(6)}$ and $\overline{T(7)}$ do have antipodal 3-covers of diameter 4, cf. §13.2.
Similarly, in the Hamming graph $H(d,n)$ we find $C(\gamma,\delta) \cong H(d,2)$ (if $n \geq 2$), and $C(\gamma,\delta)\setminus\{\gamma,\delta\}$ is connected for $d \geq 3$. Thus, $H(d,n)$ has no distance-regular antipodal covers of diameter $2d$ when $d \geq 3$, and in fact it is easy to check that if $d = 2$, the only cover occurring is the octagon, a 2-cover of the quadrangle $H(2,2)$.

A similar but slightly more complicated criterion can be given for covers with odd diameter.

4.2.8. Proposition. *Suppose that Γ is distance-regular of diameter $d \geq 2$, and has a distance-regular antipodal r-cover of diameter $2d+1$. Let γ, $\delta \in \Gamma$ with $d(\gamma,\delta) = d$, and put $E = \{\delta\} \cup (\Gamma(\delta) \cap \Gamma_d(\gamma))$. Then the collection of sets $C(\gamma,\epsilon)\setminus\{\gamma,\epsilon\}$ $(\epsilon \in E)$ can be partitioned into r nonempty parts such that sets from different parts are disjoint, and all edges joining vertices from sets in different parts are contained in $\Gamma(\gamma)$.* □

For example, consider $\Gamma = \overline{T(n)}$ again. If $\delta\epsilon$ is an edge in $\Gamma_2(\gamma)$, then $\{\gamma,\delta,\epsilon\}^\perp \cong \overline{T(n-4)} \neq \emptyset$ for $n \geq 6$. But Γ is connected of diameter 2 precisely when $n \geq 5$, so the only possibility for a cover arises when $n = 5$ (and then the cover must be a 2-cover, since $\overline{T(5)}$ has valency 3). And the Petersen graph $\overline{T(5)}$ does indeed have antipodal 2-covers of diameter 5, namely its bipartite double and the dodecahedron.

4.2.9. Corollary. *If $d \geq 2$, and any two adjacent vertices δ, ϵ in $\Gamma_d(\gamma)$ have a common neighbour in $\Gamma_{d-1}(\gamma)$, then Γ does not have distance-regular antipodal covers of diameter $2d+1$. In particular, this holds for the collinearity graph of a regular near 2d-gon.*

Proof. Clearly, if $\Gamma(\delta) \cap \Gamma(\epsilon) \cap \Gamma_{d-1}(\gamma) \neq \emptyset$, then $C(\gamma,\delta)\setminus\{\gamma,\delta\}$ and $C(\gamma,\epsilon)\setminus\{\gamma,\epsilon\}$ are not disjoint. But if Γ is a near $2d$-gon, and $\delta\epsilon$ is an edge in $\Gamma_d(\gamma)$, then by definition of regular near $2d$-gon, the line containing $\delta\epsilon$ has a (unique) point in $\Gamma_{d-1}(\gamma)$. □

For another criterion that applies both for odd and even diameter covers, see the remark after Corollary 5.2.4.

D. Generalized Odd graphs; several P-polynomial structures

A *generalized Odd graph* of diameter d, or regular thin near $(2d+1)$-gon (cf. §6.4), is a distance-regular graph Δ of diameter d such that $a_1(\Delta) = a_2(\Delta) = \cdots = a_{d-1}(\Delta) = 0$ and $a_d(\Delta) > 0$.
In this section we show that three families of distance-regular graphs coexist:
1. Graphs Γ of diameter d such that $\Gamma_d(\gamma)$ is a clique containing at least two points for each $\gamma \in \Gamma$, i.e., graphs Γ such that $k_d = a_d + 1 > 1$.
2. Regular thin near $(2d+1)$-gons Δ.
3. Bipartite antipodal 2-covers E of diameter $2d+1$.
These results are due to GARDINER [313] and BANNAI & BANNAI [27]; see also BANNAI & ITO [33], Theorem III.4.2.

4.2.10. Proposition. *If Γ is a distance-regular graph with $k_d = a_d + 1 > 1$, then $\Delta = \Gamma_d$ is a regular thin near $(2d+1)$-gon. Conversely, if Δ is a regular thin near $(2d+1)$-gon, and $d > 1$, then $\Gamma = \Delta_2$ is a distance-regular graph with $k_d = a_d + 1 > 1$.*
The correspondence between distances in Γ and Δ is as follows:

Γ	0	1	2	3	$\cdots$	$d-3$	$d-2$	$d-1$	d
Δ	0	2	4	6	$\cdots$	7	5	3	1.

Proof. (Let us use a bar to distinguish the parameters of Δ from those of Γ.)
(i) Since $a_d > 0$, we have $p^1_{dd} > 0$, so that for any two vertices γ, δ: $\bar{d}(\gamma,\delta) \leqslant 2d(\gamma,\delta)$ and $\bar{d}(\gamma,\delta) \leqslant 1 + 2(d - d(\gamma,\delta))$. Conversely, since $k_d = a_d + 1$, if $\bar{d}(\gamma,\delta)$ is even, then $d(\gamma,\delta) \leqslant \frac{1}{2}\bar{d}(\gamma,\delta)$, and if $\bar{d}(\gamma,\delta)$ is odd, then $d(\gamma,\delta) \geqslant d - \frac{1}{2}(\bar{d}(\gamma,\delta) - 1)$. From this we see that the correspondence $d \leftrightarrow \bar{d}$ is as claimed. Since the number of classes of Δ (viewed as association scheme) is d, the diameter of Δ, it follows that Δ is a distance-regular graph.
If $1 \leqslant j \leqslant d-1$, then $\bar{a}_j = \bar{p}^j_{1j} = p^i_{di}$ with $i = \frac{1}{2}j$ if j is even, and $i = d - \frac{1}{2}(j-1)$ otherwise. In the first case $p^i_{di} = 0$ by the triangle inequality; but since for each vertex γ the set $\Gamma_d(\gamma)$ is a clique, we have when $d(\gamma,\delta) = i$ that $\Gamma_d(\gamma) \subseteq \Gamma_{d-i}(\delta) \cup \Gamma_{d-i+1}(\delta)$ and it follows that $p^i_{di} = 0$ also in the second case. Finally $\bar{a}_d \neq 0$, for otherwise Δ would be bipartite, and $\Gamma = \Delta_2$ disconnected.
(ii) Now assume that Δ is a distance-regular graph with $\bar{a}_1 = \cdots = \bar{a}_{d-1} = 0$, $\bar{a}_d > 0$, and define Γ by $\Gamma = \Delta_2$. Again we easily verify the correspondence $d \leftrightarrow \bar{d}$, and from $\bar{a}_1 = 0$, $\bar{k} \geqslant 2$ one sees $k_d = a_d + 1 > 1$. □

4.2.11. Proposition. *If Δ is a regular thin near $(2d+1)$-gon, then its bipartite double E is distance-regular of diameter $2d+1$, an antipodal 2-cover of Δ.*
Conversely, if E is any distance-regular graph of odd diameter $2d+1$ that is bipartite and an antipodal 2-cover, then its antipodal quotient (folding) Δ is a regular thin near $(2d+1)$-gon, and E is the bipartite double of Δ. If $d > 1$ then the halved graph of E is $\Gamma := \Delta_2$ (so that $E_2 \cong 2\Gamma$ and $E_{2d} \cong 2\Delta$).

Proof. Straightforward, cf. Theorem 1.11.1. □

The graphs Γ are characterized by their property $k_d(k_d - 1) = k$, see Proposition 5.6.3.

Remark. The known regular thin near $(2d+1)$-gons are the Odd graphs O_{d+1} (R2), the folded $(2d+1)$-cubes (P1) and the two coset graphs (N3), (N4); their distance 2 graphs are the Johnson graphs $J(2d+1,d)$ (C1), the halved $(2d+1)$-cubes (C9) and the graphs (D7), (D8).

Let us investigate under what conditions the association scheme determined by Γ has two P-polynomial structures, i.e., when Γ_i (for some $i > 1$) is a distance-regular graph of the same diameter as Γ. Again, the results are mostly due to BANNAI & BANNAI [27] and GARDINER [313]. If Γ is an n-gon, then Γ_i is again an n-gon for any i such that $(n,i) = 1$. Also, if Γ is strongly regular, then Γ_2 is the complementary graph, and again strongly regular, and hence distance-regular, unless Γ was complete multipartite. Thus, we may suppose that $k > 2$ and $d > 2$.
Since the sequence $(k_j)_j$ is unimodal (Proposition 5.1.1), the valency k is either the smallest or the second smallest member of this sequence, and if $k \geqslant k_i$, then either $i = d-1$ and Γ is an antipodal 2-cover, or $i = d$. In particular, Proposition 5.1.1 implies immediately that any association scheme has at most two P-polynomial orderings, and that in our situation we have $i \in \{2, d-1, d\}$. Let us record this observation.

4.2.12. Theorem. *An association scheme different from that of a polygon has at most two P-polynomial structures.* □

4.2.13. Proposition. *Let Γ be a distance-regular graph of diameter $d \geqslant 3$ and not a polygon, such that Γ_{d-1} is also distance-regular of diameter d. Then Γ is an antipodal 2-cover, and either Γ is bipartite (and d is even) or $a_1 = \cdots = a_{e-1} = 0$, $a_e \neq 0$ for $e = [d/2]$ (and d is arbitrary). And conversely, for any such Γ the graph Γ_{d-1} is distance-regular, and the correspondence between distances is as follows:*

Γ	0	1	2	3	$\cdots$	$d-3$	$d-2$	$d-1$	d
Γ_{d-1}	0	$d-1$	2	$d-3$	$\cdots$	3	$d-2$	1	d.

Proof. We saw already that Γ must be an antipodal 2-cover, and that the neighbours of γ in Γ_{d-1} are the neighbours of its antipode in Γ. Now all statements follow straightforwardly. □

In the situation of this proposition, the folded graph $\overline{\Gamma}$ is either bipartite or a generalized Odd graph.
The first case occurs for $d = 4$ precisely when Γ is a Hadamard graph (see §1.8), and for $d = 6$ precisely when Γ is the 6-cube (by Corollary 5.8.3). For $d = 2e > 6$ the only known examples are the d-cubes.
The second case occurs for $d = 3$ precisely when Γ is a Taylor graph with $\mu < k-1$ (see §1.5); the other three known examples are the Wells graph (A12) ($d = 4$), the dodecahedron ($d = 5$) and the graph (A14) ($d = 6$).

4.2.14. Proposition. *Let Δ be a distance-regular graph of diameter $d \geqslant 3$ and not a polygon, such that $\Gamma = \Delta_2$ is also distance-regular of diameter d. Then either Δ is a Taylor graph with $\mu < k-1$ (and $d = 3$), or Δ is a generalized Odd graph and $\Delta = \Gamma_d$ (cf. 4.2.10).*

Proof. We must have $\Delta = \Gamma_i$ with $i \in \{2, d-1, d\}$. If $i = d-1$, then by the preceding proposition also $\Gamma = \Delta_{d-1}$, so that $d = 3$ and $i = 2$. If $i = 2$, the $d < 4$, so $d = 3$, and by the preceding proposition Δ is a Taylor graph. So let now $i = d$. If $\lambda_\Delta = 0$, then Proposition 4.2.10 tells us that Δ is a generalized Odd graph. But if Δ contains a triangle $\gamma\delta\epsilon$, then since $d \geqslant 3$ we can find vertices α, β with $d_\Delta(\alpha,\beta) = 3$,

$\alpha \sim \gamma$, $\beta \sim \delta$. We may assume $\alpha \not\sim \delta, \epsilon$ and find $d_\Gamma(\delta,\epsilon) = d \leqslant 2$, contradiction. □

4.2.15. Proposition. *Let Γ be a distance-regular graph of diameter $d \geqslant 3$, not a polygon, such that $\Delta = \Gamma_d$ is also distance-regular of diameter d. Then either $k_d = a_d + 1 > 1$ and Δ is a generalized Odd graph (cf. 4.2.10), or $a_d = 0$ and we have one of:*

(i) $d = 3$, *Γ_3 is connected, and the correspondence between distances is as follows:*

Γ	0	1	2	3
Δ	0	3	2	1;

(ii) $d = 4$, $b_3(\lambda + a_2 - a_3) = a_2$ *(i.e., $p^4_{34} = 0$), Γ_4 is connected, and the correspondence between distances is as follows:*

Γ	0	1	2	3	4
Δ	0	4	2	3	1;

(iii) $d \geqslant 5$ *and $a_i = 0$ for $2i < d$ and for $2i > d+2$, $p^d_{jd} = 0$ for $j \neq 0, 2$, and if d is even, then $a_i > 0$ for $i = \frac{1}{2}d, \frac{1}{2}d + 1$; the correspondence between distances is as follows:*

Γ	0	1	2	3	4	$\cdots$	$d-3$	$d-2$	$d-1$	d
Δ	0	d	2	$d-2$	4	$\cdots$	$d-3$	3	$d-1$	1.

Proof. If $a_d \neq 0$, then $\Gamma = \Delta_2$, and our result follows from the preceding proposition. So we may assume that $\Gamma = \Delta_d$ and $a_d = 0$. Since $p^{d-1}_{1d} \neq 0$, it follows that $\Gamma_{d-1} = \Delta_{d-1}$. If $d = 3$ then we are in case (i), so let $d \geqslant 4$. Let $\Gamma_2 = \Delta_j$ and $\Gamma_{d-2} = \Delta_h$, then $2 \leqslant j, h \leqslant d-2$, and the relations Δ_{j-1}, Δ_{j+1}, and possible Δ_j are the same as the relations Γ_{d-2} and possibly $\Gamma_{d-1} = \Delta_{d-1}$ and $\Gamma_d = \Delta_1$.
If $j \neq 2$, then Δ_1 is not among these, so $j = d-2$ and $\Gamma_{d-2} = \Delta_{d-3}$; by symmetry it follows that $\Delta_2 = \Gamma_{d-2}$, so $d = 5$ and we have

Γ	0	1	2	3	4	5
Δ	0	5	3	2	4	1.

It follows that $k_2 = k_3$ (by Lemma 5.1.1) and $p^5_{25} = 0$ (by the triangle inequality in Δ), so that $b_4 = 1$. From $k_2 = k_3 \geqslant k_4$ we find $k_5 = k_4 / k \leqslant k_2 / k < k$, but this holds for both Γ and Δ, contradiction.
Hence $j = 2$. Now one immediately sees that the correspondence between the distances is as claimed. Clearly, if $p^d_{jd} \neq 0$, then $j \in \{0,2,d\}$. This proves all our claims if $d = 4$, so let $d \geqslant 5$. By the triangle inequality we find $a_2 = 0$, and hence, by Lemma 5.5.1 (i), $\lambda = 0$, so that $p^d_{dd} = 0$. Since $p^{i+1}_{2,i} = c_{i+1}(a_{i+1} + a_i - \lambda)/\mu = 0$ $(0 \leqslant i \leqslant d-1,\ 2i \neq d-2, d-1, d, d+1, d+2)$ this proves our claims about the zero a_i. Finally, when $d = 2e$, then $p^e_{2,e-1} > 0$, so that $a_e > 0$, and $p^{e+2}_{2,e+1} > 0$, so that $a_{e+1} > 0$. □

Remarks. (i) Bannai & Ito [33], Theorem III.4.2, show also that, conversely, if the conditions mentioned in the above proposition are satisfied then Γ does have two P-polynomial structures. Brouwer [128] shows that in case (iii) of the above proposition either Γ and Δ are bipartite and d is odd, or $d \leqslant 6$; moreover, $k(b_{d-1} - 1) = (k_d - 1)\mu$, and if $d \geqslant 6$ or if $d = 5$, $a_3 > 0$, then also $\mu\mu(\Delta) = (b_{d-1} - 1)^2 + 1$ and $b_{d-1} = \frac{1}{2}(\mu + \mu(\Delta) - 3)$.

(ii) If Γ is the incidence graph of a square 2-design $\mathcal{B}$, then Γ_3 is also distance-regular, namely the incidence graph of the complement of $\mathcal{B}$ (cf. §1.7). This yields an example of case (i) in the above proposition. No other examples of the case $a_d = 0$ in the above proposition are known. The infinite family of intersection arrays mentioned in (48) of §8.3 would belong to case (ii).
(iii) More generally, it seems that in the primitive case $a_d = 0$ is very rare. The only known primitive distance-regular graph with $a_d = 0$ is the Livingstone graph (D12) which has intersection array $\{11,10,6,1;1,1,5,11\}$.

Problems. (i) Is there a regular thin near 7-gon with intersection array $\{7,6,6;1,1,2\}$? The distance 2 graph would have intersection array $\{42,30,2;1,10,36\}$ and $k = 42$, $k_3 = 7$. The distance distribution of the near 7-gon is $176 = 1+7+42+126$, and these numbers remind one of the Hoffman-Singleton graph (cf. §13.1), but there does not seem to be any connection.
(ii) Is there a primitive distance-regular graph Γ with diameter $d = 3$ and $a_3 = 0$? For any such graph, Γ_3 is distance-regular, again with $a_3 = 0$. An open case is the following pair of feasible intersection arrays,

$$v = 2002, \ \iota(\Gamma) = \{125,112,9;1,8,125\}, \ \iota(\Gamma_3) = \{126,125,9;1,9,126\}.$$

(iii) Are there primitive distance-regular graphs with diameter $d = 4$ and $a_4 = 0$, apart from the Livingstone graph? Two arrays with $k \leqslant 25$ pass all our existence tests: $\{21,18,12,4;1,1,6,21\}$ ($v = 1300$) and $\{22,21,21,3;1,1,3,22\}$ ($v = 4160$).

E. Distance-regular line graphs

We have seen (e.g. in §1.2) and will see (e.g. in §5.5) several inequalities on the parameters of a distance-regular graph that hold with equality only if the graph is a line graph. Thus, it is natural to investigate which distance-regular graphs are line graphs. The answer was given in the distance-transitive case by Biggs [73], and in the general case the answer turns out to be essentially the same. (A slightly stronger result can be found in Mohar & Shawe-Taylor [537].)

4.2.16. Theorem. *Let Δ be a connected graph such that its line graph $\Gamma = L(\Delta)$ is distance-regular. Then we have one of the following:*

(i) *$\Delta \cong K_{1,m}$ and $\Gamma \cong K_m$, a complete graph;*

(ii) *$\Delta \cong K_m$ and $\Gamma \cong T(m)$, a triangular graph;*

(iii) *$\Delta \cong K_{m,m}$ and $\Gamma \cong m \times m$, a lattice graph;*

(iv) *Δ is a polygon and $\Gamma \cong \Delta$;*

(v) *Δ is regular with diameter $d \geqslant 3$ and girth $g = 2d$, i.e., Δ is a regular generalized $2d$-gon of order $(1,s)$, say, where $d \in \{3,4,6\}$, and Γ is a generalized $2d$-gon of order $(s,1)$; in other words, Δ and Γ are the incidence graph and the flag graph of a regular generalized d-gon of order (s,s);*

(vi) *Δ is regular with diameter $d \geqslant 2$ and girth $g = 2d+1$, i.e., Δ is a Moore graph and $d = 2$; Δ has 10, 50 or perhaps 3250 vertices.*

Remark. Part (iii) is the case $d = 2$ of (v). Part (i) is the only case where Γ has smaller diameter than Δ.

Proof. (i) If Δ is not regular, then since Γ is regular, Δ must be bipartite and semiregular; but unless one of the valencies of Δ is one, Γ is not edge-regular. Thus

$\Delta \cong K_{1,m}$ for some m.
From now on we may assume that Δ is regular of valency k_0. If $k_0 \leq 2$, then we are in case (i) or (iv). So assume that $k_0 > 2$. Let d_Δ and d be the diameters of Δ and Γ, respectively, then clearly $|d_\Delta - d| \leq 1$, and if Δ is bipartite, then $d \leq d_\Delta$. Also, Δ has at least $d_\Delta + 1$ and Γ has precisely $d+1$ distinct eigenvalues. But the eigenvalues of Γ are the numbers $\theta + k_0 - 2$ where θ is an eigenvalue of Δ, and the number -2. Thus, if Δ is bipartite (i.e., if $-k_0$ is an eigenvalue of Δ) then Γ has at least $d_\Delta + 1$ distinct eigenvalues, and it follows that $d_\Delta = d$. Otherwise, Γ has at least $d_\Delta + 2$ distinct eigenvalues, and it follows that $d_\Delta = d - 1$.
(ii) Let Δ have girth g. First assume that g is even: $g = 2j$, and let e and f be opposite edges of a circuit of length g. Then $d_\Gamma(e,f) = j$, but both vertices of f have distance $j-1$ to some vertex of e, so that f has no neighbours in $\Gamma_{j+1}(e)$, i.e., $j = d$. By the above we see that Δ is bipartite and $d = d_\Delta = \frac{1}{2}g$, so that Δ is the incidence graph of a generalized polygon, and we have (iii), (iv) or (v). (That $d \in \{2,3,4,6\}$ follows from Theorem 6.5.1.)
Next assume that g is odd: $g = 2j+1$. Then, since Δ is not a circuit, we can find a g-circuit with edge e opposite to vertex α, and an edge f on α, but not in the circuit. Both vertices of e have distance j to α, and distance at least j to the other vertex of f, so that $d_\Gamma(e,f) = j+1$, and e has no neighbours in $\Gamma_{j+2}(f)$, i.e., $j+1 = d$, $j = d_\Delta$, and Δ is a Moore graph, and we have (ii) or (vi), cf. Theorem 6.7.1. □

Concerning parameters, if Δ is a Moore graph with valency $\kappa = k_0 > 2$ (and κ^2+1 vertices), then $\kappa \in \{3,7,57\}$ and its line graph $\Gamma = L(\Delta)$ has intersection array $\{2\kappa-2, \kappa-1, \kappa-2;\ 1, 1, 4\}$ and spectrum

4^1	2^5	$(-1)^4$	$(-2)^5$	$(\kappa = 3)$
12^1	7^{28}	2^{21}	$(-2)^{125}$	$(\kappa = 7)$
112^1	62^{1729}	47^{1520}	$(-2)^{89375}$	$(\kappa = 57)$

(Note that the sequence of multiplicities is not unimodal here.)

Concerning distance-transitive line-graphs, Γ is distance-transitive in cases (i)-(iv). In case (v) Γ is distance-transitive only when Δ is self-dual; examples are known for $g = 6$, $k = q+1$, q a prime power, for $g = 8$, $k = 2^{2m+1}+1$ and for $g = 12$, $k = 3^{2m+1}+1$ (cf. §10.7). In case (vi) we find the line graphs of the Petersen graph and the Hoffman-Singleton graph, and these are distance-transitive; however, as Aschbacher [10] showed, no distance-transitive Moore graph on 3250 vertices exists, and hence a corresponding line graph cannot be distance-transitive either. (Note that, if for some graph Δ its line graph $L(\Delta)$ is distance-transitive, then so is Δ itself, unless $\Delta \cong K_{1,m}$; in fact, except in very small cases, the vertices of Δ can be recognized as maximal cliques in $L(\Delta)$.)

F. Merging classes in distance-regular graphs

Sometimes, merging some classes of an association scheme yields again an association scheme. (For example, see §11.4F.) It is easy to give a necessary and sufficient condition for this to happen in the case of distance-regular graphs of diameter three, where this situation just amounts to the case when Γ_2 or Γ_3 are strongly regular.

4.2.17. Proposition (cf. BROUWER [125]). *Let* Γ *be a distance-regular graph of diameter* 3.
(i) Γ_2 *is strongly regular* $\Leftrightarrow c_3(a_3+a_2-a_1) = b_1a_2$.
(ii) Γ_3 *is strongly regular* $\Leftrightarrow \Gamma$ *has eigenvalue* $-1 \Leftrightarrow k = b_2+c_3-1$. □

(Concerning the significance of an eigenvalue -1, cf. §4.4.)

It is possible to generalize part (ii) to arbitrary diameter:

4.2.18. Proposition. *Let* Γ *be a distance-regular graph of diameter* d. *Then* $\Delta = \Gamma_1 \cup \Gamma_2$ *is distance-regular if and only if we have*

$$b_{j-1}+c_{j+1}-a_j = k+\mu-\lambda \tag{*}$$

for $2 \leqslant j \leqslant d-1$. *If this is the case, then* Δ *has diameter* $[(d+1)/2]$ *and parameters*

$$b_j(\Delta) = b_{2j-1}b_{2j}/\mu \quad (= k_1+k_2 \text{ if } j=0),$$
$$c_j(\Delta) = c_{2j-1}c_{2j}/\mu \quad (= c_d(k+\mu-\lambda+a_d-b_{d-1})/\mu \text{ if } j=(d+1)/2).$$

Proof. Straightforward. □

Note that (*) is equivalent to $b_j+c_{j+1} = k+1$ (j even, $0 \leqslant j \leqslant d-1$) and $b_j+c_{j+1} = b_1+\mu$ (j odd, $1 \leqslant j \leqslant d-1$).

In the situation of this proposition, we can say the following about the spectra. Δ has eigenvalues $(\theta^2+(\mu-\lambda)\theta-k)/\mu$, where θ runs through the eigenvalues of Γ. If d is even, it follows that for each eigenvalue $\theta \neq k$ of Γ also $\lambda-\mu-\theta$ is an eigenvalue. If d is odd, then all eigenvalues of Γ come in pairs θ_1, θ_2 with $\theta_1+\theta_2 = \lambda-\mu$, except for the valency k and one other eigenvalue θ_0. The standard sequence u corresponding to θ_0 must have $u_1 = u_2$, and it follows that $\theta_0 = -1$.

A number of near polygons, namely those with parameters

$$c_i = \frac{\mu}{s+1}\left(\left[{i \atop 1}\right]_s - \epsilon_i\right)+\epsilon_i \quad (i=1,\cdots,d)$$

where $s = \lambda+1$ and $\epsilon_i = 1$ if i is odd, $\epsilon_i = 0$ otherwise (i.e., $c_1 = 1$, $c_2 = \mu$, $c_3 = \mu s+1$, $c_4 = \mu(s^2+1)$, etc.), give examples of this situation.
In particular, we have:
(i) The distance 1-or-2 graph of the $(m-1)$-cube is the halved m-cube. The distance 1-or-2 graph of the folded $(2m-1)$-cube is the halved folded $2m$-cube.
(ii) The distance 1-or-2 graph of the bipartite double of the Odd graph O_m is the Johnson graph $J(2m,m)$. The distance 1-or-2 graph of the Odd graph O_m is the folded Johnson graph $\bar{J}(2m,m)$.

[More generally, one may remark the following: If $\Gamma_1 \cup \Gamma_2$ is distance-regular and Γ is antipodal, then Γ is an antipodal 2-cover. If Γ is an antipodal 2-cover of odd diameter with folded graph $\bar{\Gamma}$, then also $\bar{\Gamma}_1 \cup \bar{\Gamma}_2$ is distance-regular.]

(iii) The distance 1-or-2 graphs of the orthogonal dual polar graphs $B_{m,m}(q)$ are the half dual polar graphs $D_{m+1,m+1}(q)$. The distance 1-or-2 graphs of the symplectic dual polar graphs $C_{m,m}(q)$ are also distance-regular, but (for odd q) not isomorphic to previously known graphs (as was pointed out by V. A. Ustimenko, see §9.4C).
(iv) The distance 1-or-2 graph of a generalized hexagon of order (s,s) is strongly regular.
Several further examples of diameter three are known. With larger diameter we know

two more examples:
(v) The distance 1-or-2 graph of the graph with intersection array $\{22,21,20,3,2,1;1,2,3,20,21,22\}$ ((A14), see §11.3H) is the graph with intersection array $\{253,210,3;1,30,231\}$ ((D8), see §11.3E). The distance 1-or-2 graph of a graph with intersection array $\{21,20,16,6,2,1;1,2,6,16,20,21\}$ ((A16), see §11.3H) has intersection array $\{231,160,6;1,48,220\}$ ((D7), see §11.3F).

If the extended bipartite double $\hat{\Gamma}$ of a distance-regular graph Γ is distance-regular, then so is the distance 1-or-2 graph Δ of Γ, since Δ is a halved graph of $\hat{\Gamma}$. (And indeed, in Theorem 1.11.2 we found that $\hat{\Gamma}$ is distance-regular when $b_j + c_{j+1} = k+1$ holds for all j, i.e., when (*) holds and moreover $\mu = \lambda + 2$.)

Problem. For which graphs Γ is $\Gamma_1 \cup \Gamma_2$ distance-regular? Apart from the above examples also the quadratic forms graphs arise from such a construction.

4.3 Substructures

Sometimes it is possible to establish the existence of certain subconfigurations from the parameters of a graph. On the one hand this yields many new existence conditions: the total number of such subgraphs, the number on a vertex, on an edge, etc. must all be integers. On the other hand, substructures are a great help in uniqueness proofs. In this section we shall look at lines, cubes, Moore geometries and 7-point biplanes in a graph.

A. Lines

In the cases where it is known that the distance-regular graph Γ is the collinearity graph of a partial linear space with lines of size $s+1$, some additional restrictions on the parameters can be derived. For instance, the number of lines on a point and the total number of lines are integers, whence s divides k, and $s(s+1)$ divides vk. Even lines of size two yield some information: counting edges (and pairs of vertices at distance i) and triangles in various ways, we find

4.3.1. Lemma. *The parameters of a distance-regular graph satisfy the following divisibility conditions:*

$$vk_i \equiv k_i a_i \equiv 0 \pmod 2 \quad (1 \leqslant i \leqslant d) \quad \textit{and} \quad vk\lambda \equiv 0 \pmod 3\,. \qquad \square$$

For example, $\{7,5,5,2;1,1,1,5\}$ is feasible, but $k\lambda$ is odd in this case.

Often, one can conclude from the parameters that a graph has singular lines of size $\lambda+2$.

4.3.2. Proposition. *If, in a distance-regular graph, one of the conditions*

(i) $\mu = 1$ *or* $\lambda = 1$,

(ii) $\mu = 2$ *and* $k < \frac{1}{2}\lambda(\lambda+3)$,

(iii) $b_{d-1} = 1$ *and* $a_d = \lambda + 1$,

(iv) $b_i > 1$ *and* $c_i = b_1$

holds, then all singular lines have size $\lambda+2$. *In case* (iii) *we additionally have* $(\lambda+2) \mid k_d$.

Proof. In cases (i) and (ii) it is stated in §1.2 that all singular lines have size $\lambda+2$.
In case (iii), given some $\gamma \in \Gamma$, the common neighbours of two vertices α, β in $\Gamma_d(\gamma)$ must lie in $\Gamma_d(\gamma)$; in particular, if $\alpha \sim \beta$, then the $a_d - 1 = \lambda$ neighbours of α in $\Gamma_d(\gamma) \setminus \{\beta\}$ are all neighbours of β; thus $\Gamma_d(\gamma)$ is a union of cliques, and $(\lambda+2) \mid k_d$.
In case (iv) we choose an edge $\gamma\delta$ and one of the $p^1_{i,i+1} = k_i b_i / k > 0$ vertices $\alpha \in \Gamma_i(\gamma) \cap \Gamma_{i+1}(\delta)$. Put $D := \Gamma_{i+1}(\alpha) \cap \Gamma(\gamma)$, $C := \Gamma_i(\alpha) \cap D^{\perp}$. Since $b_1 = c_i$, we have $\gamma^{\perp} \setminus \Gamma_{i-1}(\alpha) \subseteq D^{\perp}$; in particular, D is a clique. Since $|D| = b_i > 1$, D contains two distinct vertices δ, δ'; consequently $|C| + |D| \leqslant \lambda(\delta,\delta') + 2 = \lambda + 2$. Since the λ common neighbours of $\gamma' \in C$ and $\delta' \in D$ must be in $C \cup D \setminus \{\gamma',\delta'\}$, we find that $C \cup D$ is a $(\lambda+2)$-clique. Since γ, $\delta \in C \cup D$, this is the singular line through $\gamma\delta$. □

Remark. We do not know any primitive distance-regular graphs Γ with $b_{d-1} = 1$, $\lambda > 0$.

4.3.3. Proposition. *If in a distance-regular graph* Γ *all singular lines have size* $\lambda+2$, *then*

$$(\lambda+1) \mid k, \quad (\lambda+1)(\lambda+2) \mid vk$$

and

$$a_2 \geqslant \lambda\mu.$$

If $\mu = 1$, *then also*

$$\frac{vk}{(\lambda+1)(\lambda+2)} \geqslant 1 + \frac{\lambda+2}{\lambda+1} b_1 + \frac{\lambda+2}{\lambda+1} b_1^2.$$

Proof. The divisibility conditions follow from §1.2; the first inequality follows by counting points in $\Gamma_2(\gamma) \cap \Gamma(\delta)$ where $\gamma, \delta \in \Gamma$ have distance 2, and the second inequality states that the total number of lines is at least the number of lines at distance at most two from a given line. □

Examples. The intersection arrays $\{8,4,1;1,1,8\}$ ($v = 45$), $\{10,5,4,2;1,2,2,10\}$ ($v = 96$), $\{23,16,14,2;1,2,8,14\}$ ($v = 576$), $\{42,28,5;1,2,30\}$ ($v = 729$) and $\{50,40,27;1,2,20\}$ ($v = 2401$) are feasible, but ruled out by this proposition. (The first array is also ruled out by Proposition 1.17.3.)

4.3.4. Proposition. *If all singular lines in* Γ *have size* $\lambda+2$, *and* $k = 2(\lambda+1)$, *then* Γ *is a line graph (and hence satisfies the conclusion of Theorem 4.2.16).*

Proof. $\Gamma = L(\Delta)$, where Δ has the singular lines as vertices, intersecting lines being adjacent. □

B. Cubes

Recall that a rectagraph is a connected graph with $\lambda = 0$ and $\mu = 2$. Every rectagraph is regular. Fairly often it is possible to show that a rectagraph must contain 3-cubes and to conclude that it is a (quotient of a) Hamming cube. (For $\lambda > 0$, see below.)

4.3.5. Lemma. *Let Γ be a rectagraph.*

(i) *If $a_2(\gamma,\delta) = 3$ (for $\gamma, \delta \in \Gamma$, $d(\gamma,\delta) = 2$) or $c_3(\gamma,\delta) = 3$ (for $\gamma, \delta \in \Gamma$, $d(\gamma,\delta) = 3$), then $\{\gamma\} \cup (\gamma^\perp \cap \Gamma_2(\delta)) \cup (\delta^\perp \cap \Gamma_2(\gamma)) \cup \{\delta\}$ is a 3-cube.*

(ii) *If $c_3 = 3$ and $a_2(\gamma,\delta) \leqslant 3$ for all $\gamma, \delta \in \Gamma$, $d(\gamma,\delta) = 2$, then any 3-claw determines a unique 3-cube. If $a_2(\gamma,\delta) \leqslant 2$ for all $\gamma, \delta \in \Gamma$, $d(\gamma,\delta) = 2$, then distances in the cube coincide with distances in Γ, and, in particular, $a_2 = 0$.*

Proof. Straightforward. □

This lemma shows that given a graph Γ with $\lambda = 0$, $\mu = 2$, $a_2 \in \{0,3\}$, $c_3 = 3$ we have a rank 4 Buekenhout geometry

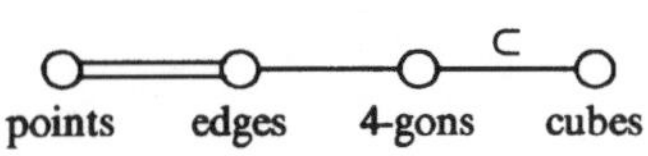

of truncated type C_k, and by results of Tits, Ronan and Brouwer & Cohen it follows that this geometry is a truncation of a quotient of the thin building of type C_k. This can be taken as starting point for characterizations of such Γ. See Tits [753], Ronan [620], Brouwer & Cohen [132], Chima & Shult [195], Shult [672]. Let us make this observation explicit.

4.3.6. Proposition (cf. Brouwer [123]). *Suppose Γ is a rectagraph of valency k such that any 3-claw determines a unique 3-cube. Then there is a map $\pi: 2^k \rightarrow \Gamma$ preserving distances $\leqslant 2$. If Γ does not contain pentagons, then π also preserves distance 3.*

Proof. Define $\pi(u)$ by induction on the weight $\mathrm{wt}(u)$ of u. Fix a vertex $\gamma_0 \in \Gamma$ and map $\mathbf{0}$ to it; map the set of vectors of weight one onto the set of neighbours of γ_0 in Γ. If $u \in 2^k$ with $\mathrm{wt}(u) \geqslant 2$ and e, e' are two unit vectors occurring in u (i.e., $\mathrm{wt}(u-e)$, $\mathrm{wt}(u-e') = \mathrm{wt}(u)-1$) then map u to the common neighbour of $\pi(u-e)$ and $\pi(u-e')$ distinct from $\pi(u-e-e')$. This does not depend on the choice of e and e', for if e'' is a third unit vector occurring in u then $\pi(u)$ is the common neighbour of $\pi(u-e)$ and $\pi(u-e'')$ distinct from $\pi(u-e-e'')$ in the cube on $\pi(u)$ and $\pi(u-e-e'-e'')$. For each v for which $\pi(v)$ has been defined already we have $d(\pi(u),\pi(v)) \leqslant d(u,v)$ with equality for $d(u,v) \leqslant 2$. (If $d(u,v) = 2$, $\pi(u) = \pi(v)$ then either $\mathrm{wt}(v) = \mathrm{wt}(u)-2$ and we find a contradiction with the definition of u, or $\mathrm{wt}(v) = \mathrm{wt}(u) \geqslant 2$, say $u = w+e_i+e_j$, $v = w+e_i+e_k$ for some w with $\mathrm{wt}(w) = \mathrm{wt}(u)-2$, and $\pi(u)$ and $\pi(v)$ have the three common neighbours $\pi(w+e_i)$, $\pi(w+e_j)$ and $\pi(w+e_k)$, again a contradiction.) If $d(\pi(u),\pi(v)) < d(u,v) = 3$ and Γ does not contain pentagons, then $d(\pi(u),\pi(v)) = 1$, $v = u-e-e'+f$ for three unit vectors e, e', f, where e, e' occur in u. Now $\pi(u)$ and $\pi(u-e-e')$ have common neighbours $\pi(v)$, $\pi(u-e)$, $\pi(u-e')$ and two of these must coincide, violating the induction hypothesis. □

We shall apply this proposition in §11.3D and §11.3E to show the uniqueness (given the parameters) of the coset graphs of the binary Golay code (extended or not). A related result can be found in Cameron [165], Theorem 5.10.

4.3.7. Corollary. *Let Γ, π be as above. If γ, δ are two adjacent vertices of Γ then each vector in $\pi^{-1}(\gamma)$ has distance one to a unique vector in $\pi^{-1}(\delta)$, and in particular $|\pi^{-1}(\gamma)| = |\pi^{-1}(\delta)|$. It follows that $v \,|\, 2^k$. If $c_i = i$, $a_i = 0$ $(i = 1, 2, \cdots, h-1)$, then for $\gamma \in \Gamma_h(\gamma_0)$ the set $\pi^{-1}(\gamma)$ consists of pairwise disjoint h-sets, so $h \,|\, c_h$.* □

This corollary rules out several otherwise feasible parameter sets for distance-regular graphs, such as, e.g., $\{6,5,4,3;1,2,3,6\}$ ($v = 52$) (but this also fails the Krein conditions) and $\{16,15,14,13;1,2,3,16\}$ with $v = 1152$. In fact, the only bipartite distance-regular graphs of diameter 4 with $\mu = 2$, $c_3 = 3$ have $c_4 \in \{4,8,24\}$.

4.3.8. Corollary. *Let Γ be a rectagraph of valency k such that any 3-claw determines a unique 3-cube. Then Γ has $v = 2^a$ vertices for some integer $a \leqslant k$.*

(i) *We have $v = 2^k$ if and only if Γ is the k-cube.*

(ii) *We have $v = 2^{k-1}$ if and only if there is a vector $u \in 2^k$ and a permutation $\sigma \in \mathrm{Sym}(k)$ where $\sigma^2 = 1$, σ fixes u and has at least 5 fix points on u, such that Γ is the quotient of 2^k obtained by identifying w with $w^\sigma + u$ for all $w \in 2^k$. (Now Γ will be bipartite when u has even weight; its halved graph will then be locally triangular when $\mathrm{wt}(u) \geqslant 8$.) The graph Γ is distance-regular only in case $u = \mathbf{1}$, $\sigma = 1$, where Γ is the halved k-cube.*

Proof. Straightforward. Note that the map described with respect to $\mathbf{0}$ by $v \mapsto v^\sigma + u$ is described with respect to the vector c by $c + v \mapsto c + v^\sigma + (c + c^\sigma + u)$. □

4.3.9. Proposition (NEUMAIER [561]). *Let Δ be a connected locally triangular graph. Then Δ is locally $T(m)$ for some $m \in \mathbb{N}$, and is the halved graph of a bipartite rectagraph Γ with $c_3 = 3$.*

Proof. The graph $T(m)$ is regular of valency $2(m-2)$ if $m \geqslant 2$, and is empty otherwise. It follows that if $\Delta(\delta) \cong T(m_\delta)$ for all $\delta \in \Delta$, and $\delta \sim \epsilon$, then $m_\delta = m_\epsilon$. By connectedness of Δ, $m_\delta = m$ for all $\delta \in \Delta$.
For $m \leqslant 4$ the statements are trivially fulfilled (for $m =$ 0-or-1, 2, 3, 4 the graph Δ is K_1, K_2, K_4, $K_{2,2,2,2}$, respectively, and in all cases Δ is the halved m-cube), so assume $m \geqslant 5$. Let Γ be the (bipartite) incidence graph of the points and the m-cliques of Δ. It is easily seen that Γ is a rectagraph. In $T(m)$, any point outside an $(m-1)$-clique is adjacent to two points of the $(m-1)$-clique. For Δ this means that any point outside an m-clique is adjacent to zero or three points of the m-clique, and we find that Γ has $c_3 = 3$. □

4.3.10. Lemma. *Let Δ be locally $T(m)$. Then the μ-graphs in Δ are isomorphic to the disjoint union of at most $[m/4]$ copies of $T(4)$.*

Proof. Each μ-graph in Δ is locally C_4, hence each component is isomorphic to $T(4)$. But if $aT(4)$ is a subgraph of $T(m)$, then $m \geqslant 4a$. □

The above enables us to classify locally $T(m)$ graphs Δ for small m. Indeed, by the above we find that Δ is the quotient of a halved cube (and in particular has a number of points that is a power of two), and by Proposition 1.13.1, Δ is isomorphic to the halved m-cube when it has 2^{m-1} vertices. By the lemma it then follows that, for $m \leqslant 7$, Δ must be the halved m-cube, while for $m = 8$ there is precisely one other possibility, namely the folded halved m-cube.

It is possible to generalize Proposition 4.3.6 to the case $\lambda > 0$. Again, our aim is to construct four types of objects: points, lines, grids, and 'cubes', i.e., subgraphs $H(i,\lambda+2)$ for $i = 0,1,2,3$. Now NOMURA [570] shows that if $\lambda \neq 2$, $a_2 = 2\lambda$, $\mu = 2$, $c_3 = 3$, then all singular lines have size $\lambda+2$. But when Γ has singular lines of size $\lambda+2$ and satisfies $a_2 = \lambda\mu$, $\mu > 1$, $\lambda > 0$, then Yanushka's Lemma

(Proposition 2.5 in Shult & Yanushka [673]) yields 'quads': geodetically closed subgraphs that are generalized quadrangles. If moreover $a_3 = \lambda c_3$, then Brouwer & Cohen [132] applies (cf. Remark (ii) following Theorem 9.4.4), and we find that Γ is a quotient of a dual polar space; if $\mu > 2$ then Γ must be a dual polar space.
Sometimes, a condition on a_3 is superfluous. E.g., when $\lambda = 1$, $a_2 = 2$, $\mu = 2$, $c_3 = 3$, one sees that any two vertices at distance three determine a subgraph $H(3,3)$ on 27 vertices, and it follows that Γ is a quotient of a Hamming graph $H(k, 3)$. In particular, v must be a power of 3. Thus, there is no distance-regular graph with intersection array $\{26,24,22,2; 1,2,3,22\}$ and $v = 2835$.

C. Moore geometries and Petersen graphs

4.3.11. Proposition (cf. Ivanov [432], Brouwer [127]). *Suppose Γ is a graph with $\mu = c_3 = 1$ and $a_2 > \lambda$. For γ, δ at distance two, put $\Delta(\gamma,\delta) := \{\gamma\} \cup A \cup C$, where C is the component of $\Gamma_2(\gamma)$ containing δ, and A is the union of all singular lines on γ containing a neighbour of a point of C. Then $\Delta = \Delta(\gamma,\delta)$ has diameter two, is a geodetically closed subspace, and is strongly regular with $k(\Delta) = a_2 + 1$, $\lambda(\Delta) = \lambda$, and $\mu(\Delta) = 1$, i.e., Δ is a Moore geometry, and the usual feasibility conditions apply. (In particular, if $\lambda = 0$, then $a_2 \in \{1,2,6,56\}$, and if $\lambda > 0$, then $a_2 > \lambda + 1$.) Furthermore, $\Delta(\gamma,\delta) = \Delta(\alpha,\beta)$ for any two nonadjacent vertices α, β in $\Delta(\gamma,\delta)$.*

Proof. Any line meeting Δ in two points is contained in Δ (since $c_3 = 1$), that is, Δ is a subspace. If $\gamma \sim \beta \sim \delta \sim \epsilon \sim \zeta \sim \eta$ with $\delta, \epsilon, \zeta, \eta \in C$, then $d(\beta,\zeta) \leq 2$ since ζ has at least two neighbours in $\Gamma_{\leq 2}(\beta)$, namely ϵ and the vertex in $\gamma^\perp \cap \zeta^\perp$; by induction we see that any vertex in C has distance at most two to β and hence to any point on the line $\gamma\beta$, and it follows that $d(\alpha,\theta) \leq 2$ for all $\alpha \in A$, $\theta \in C$. Also $d(\delta,\eta) \leq 2$ since δ has at least two neighbours in $\Gamma_{\leq 2}(\eta)$, namely ϵ and β, and by induction we see that Δ has diameter 2.
Suppose $\beta \notin \Delta$, $\beta \sim \alpha'$, $\alpha'' \in \Delta$. Since $c_3 = 1$ we have $\beta \notin \Gamma_3(\gamma)$. If $\beta \in \Gamma_2(\gamma)$ then α', $\alpha'' \in \Gamma(\gamma)$, contradicting $\mu = 1$. If $\beta \sim \gamma$, then α', $\alpha'' \in C$ and $\beta \in \Delta$, contradiction. Thus Δ is geodetically closed.
Clearly, $\lambda(\Delta) = \lambda$ and $\mu(\Delta) = 1$, and by Theorem 1.17.1 either Δ is a degenerate generalized quadrangle, or Δ is strongly regular. But in the former case $a_2 = \lambda$, contrary to our hypothesis. Thus Δ is strongly regular and its valency is its valency in δ, $a_2 + 1$.
Now suppose $\alpha \in A$, $d(\alpha,\delta) = 2$. Then $\Delta \cap \Gamma_2(\alpha)$ is regular of valency a_2 and hence must be the component of δ in $\Gamma_2(\alpha)$. This shows that $\Delta(\alpha,\delta) \subseteq \Delta(\gamma,\delta)$ and by symmetry $\Delta(\alpha,\delta) = \Delta(\gamma,\delta)$. Repeating this argument shows that $\Delta(\gamma,\delta) = \Delta(\alpha,\beta)$ whenever α, $\beta \in \Delta$, $\alpha \not\sim \beta$. □

4.3.12. Corollary. *Under the assumptions of the proposition we have*

(i) $(a_2 - \lambda) \mid (k - \lambda - 1)$,

(ii) $(a_2 + 1)(a_2 - \lambda) \mid k(k - \lambda - 1)$,

(iii) $(a_2 + 1)(a_2 - \lambda)(1 + (a_2 + 1)(a_2 - \lambda + 1)) \mid vk(k - \lambda - 1)$,

(iv) $b_2 \geq a_2 - \lambda$.

Proof. For (i)-(iii), count the number of subgraphs Δ of Γ on an edge, on a point, and their total number, respectively. For (iv), consider a subgraph Δ meeting both $\Gamma(\gamma)$ and $\Gamma_3(\gamma)$. □

As an application of this proposition, we can show that if Γ is distance-regular with $d \geqslant 5$ and $b_2 = 1$, then Γ is the dodecahedron. (For: $\mu = c_3 = 1$ by 4.1.6(ii), so $a_2 = k-2$ and (i) implies $\lambda = k-3$, so that $\lambda = 0$, $k = 3$ by (ii); now apply the classification of cubic distance-regular graphs (Theorem 7.5.1).)

The special case $\lambda = 0$, $a_2 = 2$ of this proposition produces Petersen subgraphs and was used in BROUWER [127] to show uniqueness of the distance-regular graphs with intersection arrays $\{7,6,4,4;1,1,1,6\}$ ($v = 330$) and $\{7,6,4,4,4,1,1,1;1,1,1,2,4,4,6,7\}$ ($v = 990$) and nonexistence of distance-regular graphs with intersection array $\{7,6,4,4,3,1,1,1;1,1,1,3,4,4,6,7\}$ ($v = 660$) (cf. §11.4D).

Let us give another sufficient condition for the presence of Petersen subgraphs.

4.3.13. Lemma. *Let Γ be a graph with $\lambda = 0$, $\mu = 1$, $a_2 = 2$. Then the following two conditions are equivalent.*

(i) *Each induced hexagon with at least one pair of vertices at distance 3 contains three such pairs.*

(ii) *Any two vertices of Γ at distance 2 are contained in a (unique) Petersen subgraph.*

When these conditions hold, then for each $\gamma \in \Gamma$ the subgraph $\Gamma_2(\gamma)$ is the disjoint union of $\frac{1}{6}k_2$ hexagons.

Proof. Suppose (i) holds. Fix $\infty \in \Gamma$ and, for $\alpha \in \Gamma_2(\infty)$, let $\bar{\alpha}$ be the unique common neighbour of ∞ and α. From $a_2 = 2$ it follows that $\Gamma_2(\infty)$ is a union of polygons; let $\alpha_0 \sim \cdots \sim \alpha_{g-1} \sim \alpha_0$ be such a polygon; clearly $g \geqslant 5$. Since $d(\alpha_i,\overline{\alpha_{i+1}}) = 2$ ($i \in \mathbf{Z}_g$), the vertex $\overline{\alpha_{i+1}}$ has two neighbours in $\Gamma_2(\alpha_i)$, say ∞ and β_i. Let γ_i be the common neighbour of α_i and β_i. Since $d(\alpha_i,\overline{\alpha_{i+1}}) = 2$, it follows from the hypothesis applied to the hexagon $\infty \sim \overline{\alpha_i} \sim \alpha_i \sim \gamma_i \sim \beta_i \sim \overline{\alpha_{i+1}} \sim \infty$ that also $d(\infty,\gamma_i) = 2$. But then $\gamma_i \in \{\alpha_{i-1},\alpha_{i+1}\}$ and $\beta_i \in \{\alpha_{i-2},\alpha_{i+2}\}$. If $\beta_i = \alpha_{i+2}$ then we have the triangle $\overline{\alpha_{i+1}}\alpha_{i+1}\alpha_{i+2}$, contradicting $\lambda = 0$. Thus $\beta_i = \alpha_{i-2}$, so that $\overline{\alpha_{i+1}} = \overline{\alpha_{i-2}}$ ($i \in \mathbf{Z}_g$). Now $\overline{\alpha_0} \sim \alpha_{3j} \sim \alpha_{3j+1} \sim \overline{\alpha_1}$ for all $j \in \mathbf{Z}_g$, so that there are only two distinct points α_{3j}, and $g = 6$. Using (i) again, we find (ii). The implication (ii) $\Rightarrow$ (i) is obvious, and so is the last statement. □

As examples of the use of this lemma we mention the cases of the intersection arrays $\{15,14,12;1,1,9\}$ ($v = 506$) and $\{9,8,6,3;1,1,3,8\}$ ($v = 280$), cf. §11.4B and §11.3G. Exploiting the fact that we have equality in the inequality of Proposition 5.5.4(ii) or (iii) (for $i = 3$), we find that the hypothesis about hexagons is satisfied, and hence find Petersen subgraphs. This leads to uniqueness proofs for distance-regular graphs with these intersection arrays.

D. 7-point biplanes

4.3.14. Proposition (cf. CHIMA [194]). *Let Γ be a graph with $\lambda = 0$, $\mu = 2$, $c_3 = 4$ and $a_2(\gamma,\delta) \leqslant 3$ for all γ, $\delta \in \Gamma$. Then any pair of vertices at mutual distance 3 determines a unique 'hex': a geodetically closed subgraph on 14 points with intersection array $\{4,3,2;1,2,4\}$.*

Proof. Let $d(\gamma,\delta) = 3$. Then $(\Gamma(\gamma) \cap \Gamma_2(\delta)) \cup (\Gamma_2(\gamma) \cap \Gamma(\delta))$ induces an 8-gon, say $(\alpha_0\,\alpha_1\,\alpha_2\,\alpha_3\,\alpha_4\,\alpha_5\,\alpha_6\,\alpha_7\,\alpha_0)$ with $\gamma \sim \alpha_{2j}$ $(0 \leqslant j \leqslant 3)$. Let β be the common neighbour of α_0 and α_4 distinct from γ and ϵ the common neighbour of α_2 and α_6 distinct from γ. The vertex α_6 has four neighbours γ, α_5, α_7, ϵ in $\Gamma_2(\alpha_1)$ so $d(\alpha_1,\alpha_6) = 3$ and α_1 and α_6 determine an 8-gon $(\alpha_5\,\delta\,\alpha_7\,\alpha_0\,\gamma\,\alpha_2\,\epsilon\,\zeta\,\alpha_5)$ for some vertex $\zeta \in \Gamma(\alpha_1)$. Similarly there is a vertex η adjacent to β, α_3 and α_7. We must have $d(\gamma,\zeta) = 3$ and γ and ζ determine an 8-gon $(\alpha_0\,\alpha_1\,\alpha_2\,\epsilon\,\alpha_6\,\alpha_5\,\alpha_4\,\beta\,\alpha_0)$ so $\zeta \sim \beta$ and $\eta \sim \epsilon$. Now we have found a closed configuration: a 14-point subgraph with distance distribution diagram

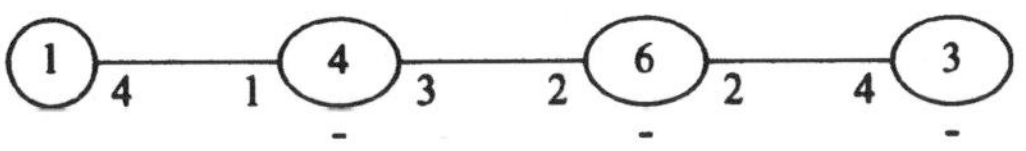

$v = 14,$

the incidence graph of the biplane 2-(7,4,2). □

4.3.15. Corollary. *The numbers* $\frac{1}{3}k_3, \frac{1}{2}b_2, \frac{1}{6}b_1b_2, \frac{1}{42}vk_3$ *are integral, and we have* $k \equiv 2$ *or* 4 (mod 6).

Proof. These are the number of hexes on a given point, on a pair of points at distance two, on an edge (respectively) and the total number of hexes. Since the edges and hexes on a point form a $S(3,4,k)$ design, we have $k \equiv 2$ or 4 (mod 6). □

For example, this rules out the existence of distance-regular graphs with intersection arrays $\{10,9,8,6;1,2,4,10\}$ ($v = 200$), $\{120,119,118,116;1,2,4,120\}$ ($v = 421500$), and $\{21,20,16,8,2,1;1,2,4,16,20,21\}$ ($v = 1536$). Also the array $\{8,7,6,4;1,2,4,8\}$ ($v = 100$) can be ruled out.

[One may proceed as follows: first, by the proposition, there are hexes; by counting, any two hexes that have an edge in common have a quadrangle in common. Let $H(\alpha,\beta)$ denote the hex on α and β, for $d(\alpha,\beta) = 3$. Fix a vertex $\infty \in \Gamma$, and call the elements of $\Gamma_i(\infty)$ Points, Edges, Quadrangles and Graphs, for $i = 1,2,3,4$, respectively, and define an incidence $*$ between these objects by $\gamma * \delta$ if $\gamma \in C(\infty,\delta)$ or $\delta \in C(\infty,\gamma)$. Now each Edge is incident with two Points, each Quadrangle with 4 Points and 4 Edges, and each Graph with all 8 Points, with 16 Edges (4 on each Point) and with 8 Quadrangles (4 on each Point, 2 on each Edge). Each Quadrangle γ induces a quadrangle in the graph formed by the Points and Edges incident with a given graph $\delta \sim \gamma$. (Indeed, if β is an Edge with two Points in γ, then $\beta \in H(\infty,\gamma)$ and if β is an Edge of δ but not of γ, then $\delta \in H(\beta,\gamma) = H(\infty,\gamma)$, contradiction.) It follows that the Points and Edges incident with a given Graph form a graph on 8 vertices, regular of valency 4, with the property that each edge is in two distinct quadrangles, and the only such graph is $K_{4,4}$. Thus, each Graph induces a partition of the set of Points into two 4-sets, and these 4-sets form a 2-(8,4,9) design. If $\infty \sim \alpha \sim \beta \sim \gamma \sim \delta$ is a geodesic, then there is a unique $\gamma' \in \Gamma_2(\alpha) \cap \delta^\perp$ such that $\{\alpha,\gamma,\gamma'\}^\perp = \varnothing$. Let $\{\gamma,\gamma'\}^\perp = \{\delta,\delta'\}$. Then $\delta' \in \Gamma_4(\infty)$ (otherwise $\infty \in H(\alpha,\delta)$), and δ and δ' determine the same partition of the set of Points. Indeed, all three vertices of $H(\alpha,\delta) \cap \Gamma_4(\infty) = \{\delta,\delta',\delta''\}$ determine this same partition. Varying α we see that in our design each block is repeated at least 5 times, so that no two distinct blocks can meet in more than one point, a contradiction.]

4.4 Representations of distance-regular graphs

This section is a continuation of Chapter 3; here we apply the theory developed there to the case of distance-regular graphs. The connection with graph representations is provided by the basic observation in the proposition below. The applications will culminate in a characterization of all distance-regular graphs with $\mu > 1$ and second largest eigenvalue $\theta_1 \geqslant b_1 - 1$. In particular, this settles the problem of characterizing Johnson graphs and half cubes by their intersection arrays. The proof of Theorem 6.3.2 also relies on the methods discussed in this section. For other applications, see Theorems 8.1.2, 8.4.5, 9.4.7, 9.5.8, and 11.3.6.

4.4.1. Proposition. *Let Γ be a distance-regular graph with v points, valency k, and intersection numbers a_i, b_i, c_i, and let $\theta \neq k$ be an eigenvalue of Γ with multiplicity f. Then Γ has a spherical representation in $\mathbb{R}^f$ such that*

$$\sum_{\gamma\in\Gamma} \bar{\gamma} = 0\,, \tag{23}$$

$$(\bar{\gamma},\bar{\delta}) = u_i \quad \textit{for all } \gamma,\delta \in \Gamma \textit{ with } d(\gamma,\delta) = i, \tag{24}$$

where $(u_0,u_1,\cdots,u_d)$ is the standard sequence corresponding to θ, i.e.,

$$u_0 = 1\,,\ u_1 = \theta/k\,,\ c_iu_{i-1}+a_iu_i+b_iu_{i+1} = \theta u_i \quad (i = 1,\cdots,d-1), \tag{25a}$$

$$c_du_{d-1}+a_du_d = \theta u_d, \tag{25b}$$

$$|u_i| \leqslant 1\,. \tag{26}$$

Proof. Map $\gamma \in \Gamma$ to the corresponding column of the eigenmatrix $E(\theta)$ scaled via multiplication by $\sqrt{f/v}$, and apply Theorem 4.1.4. Relation (23) follows from $E(\theta)J = 0$, and (26) holds since u_i is the cosine of the angle between $\bar{\gamma}$ and $\bar{\delta}$ for $d(\gamma,\delta) = i$. □

We call the representation defined in this proof the *standard representation* of Γ corresponding to θ.

4.4.2. Proposition. *Let Γ be a distance-regular graph, and let $\theta \neq k, -1$ be an eigenvalue of Γ. Put*

$$b := \frac{b_1}{\theta+1},\quad c := \frac{b(\lambda+1+b)}{k-\theta},\quad p := \frac{k}{\lambda+2}. \tag{27}$$

Then Γ has a weak $(c-b^2,c-b,c-1)$-representation if $0 < b < 1$, a weak $(p,p-1,p-2)$-representation if $b = 1$, and a weak $(b^2-c,b-c,1-c)$-representation if $b \notin [0,1]$. In particular, if $b \notin [-1-\lambda,1]$, then Γ has a weak $(b^2,b,1)$-representation.

Proof. For $i \leqslant 2$, the standard sequence $(u_i)_i$ corresponding to θ satisfies

$$u_i = \begin{cases} 1-\dfrac{i}{p} & \text{if } b = 1,\\[2mm] \dfrac{k-\theta}{kb(b-1)}(b^{2i}-c) & \text{otherwise.}\end{cases}$$

Hence scaling and shifting establishes the claim. □

Remark. Note that $u_1-u_2 = (k-\theta)(\theta+1)/(kb_1)$ so that the condition $\theta \neq k, -1$ is equivalent to the requirement that the corresponding representation distinguishes distance 1 and distance 2. The number b is the base of the representation since $b = (u_0-u_1)/(u_1-u_2)$. Note that (since $u_2 \leqslant u_0$ by (26))

$$b > 0 \text{ if } \theta > -1 \text{ and } b \leqslant -1 \text{ if } \theta < -1.$$

We shall be particularly interested in the representations corresponding to the second largest or the least eigenvalue of Γ, since these lead to strong consequences for subgraphs. For example, we have the following slightly strengthened version of a result of TERWILLIGER [720]:

4.4.3. Theorem. *Let Γ be a distance-regular graph of diameter at least 3 with eigenvalues $k = \theta_0 > \theta_1 > \cdots > \theta_d$, and put $b^+ = \dfrac{b_1}{\theta_1+1}$, $b^- = \dfrac{b_1}{\theta_d+1}$. Then $b^+ > 0$, $b^- < -1$, and we have:*

(i) *Each neighbourhood $\Gamma(\gamma)$ is a graph with smallest eigenvalue $\geqslant -1-b^+$ and second largest eigenvalue $\leqslant -1-b^-$ (here the second largest eigenvalue is taken to be the valency λ in case $\Gamma(\gamma)$ is disconnected).*

(ii) *If $b^+ < 1$ then either $\mu = 1$ or Γ is the icosahedron.*

(iii) *If $b^- > -2$ then either $\lambda = 0$ or Γ is the icosahedron.*

Proof. By Corollary 3.5.4, $\theta_1 > -1 > \theta_d$ so that $b^+ > 0$, $b^- < -1$. If we apply Proposition 3.7.3 to the representation of Proposition 4.4.1, we find that $\Gamma(\gamma)$ has a $(u_0-u_1^2, u_1-u_1^2, u_2-u_1^2)$-representation in $\mathbf{R}^{f-1}$. If $\theta \neq -1$, then the base of this representation is $b = b_1/(\theta+1)$, and an application of Proposition 3.5.3 shows that for $\theta > -1$, $\Gamma(\gamma)$ has smallest eigenvalue $\tau_{\min} \geqslant -1-b$, and for $\theta < -1$, $\Gamma(\gamma)$ has second largest eigenvalue $\tau_1 \leqslant -1-b$. For $\theta = \theta_1, \theta_d$, this implies (i).
Now suppose that $\theta = \theta_1 > b_1-1$ and $\mu > 1$. If Γ contains an induced quadrangle $\gamma_1 \sim \gamma_2 \sim \gamma_3 \sim \gamma_4 \sim \gamma_1$ then $\bar\gamma_1+\bar\gamma_3-\bar\gamma_2-\bar\gamma_4$ has norm $4(u_0-2u_1+u_2) = 4(k-\theta_1)(b_1-\theta_1-1)/kb_1 < 0$, contradiction. Therefore Γ contains no induced quadrangle, and each pair of nonadjacent points of $\Gamma(\gamma)$ has $\mu-1$ common neighbours in $\Gamma(\gamma)$; in particular $\Gamma(\gamma)$ is connected. But by (i), $\Gamma(\gamma)$ has smallest eigenvalue > -2, and by Corollary 3.12.3 (i), $\Gamma(\gamma)$ is a clique or a polygon with an odd number of vertices. If $\Gamma(\gamma)$ is a clique then Γ is complete, contradiction. If $\Gamma(\gamma)$ is a pentagon then $k = 5$, Γ is locally a pentagon and hence the icosahedron. If $\Gamma(\gamma)$ is a polygon with $k > 5$ points then it contains nonadjacent points without common neighbour in $\Gamma(\gamma)$, contradiction. This proves (ii).
Finally, suppose that $\theta = \theta_d < -b_1/2-1$ and $\lambda > 0$. By (i), $\Gamma(\gamma)$ has second largest eigenvalue < 1, and by Corollary 3.12.3 (ii), $\Gamma(\gamma)$ is a complete multipartite graph, or $k = \lambda+3$. In the latter case, Lemma 1.4.2 implies that $k = 5$, $\lambda = 2$, and we find the icosahedron. Otherwise, Γ is complete multipartite (cf. Proposition 1.1.5), against the assumption $d \geqslant 3$. This proves (iii). □

Remarks. (i) The icosahedron has $k = 5$, $\lambda = \mu = 2$, $\theta_1 = \sqrt{5}$, $\theta_d = -\sqrt{5}$ and hence $b^+ = \frac{1}{2}(-1+\sqrt{5}) < 1$, $b^- = \frac{1}{2}(-1-\sqrt{5}) > -2$.

(ii) A number of distance-regular graphs with $\mu = 1$ and $b^+ < 1$ are known; e.g., the generalized polygons with $t = 1$, the line graphs of the Petersen graph (A0) and of the Hoffman-Singleton graph (D5) and the Coxeter graph (D9).

(iii) If $b^+ = 1$, i.e., if $\theta_1 = b_1 - 1$, then $\Gamma(\gamma)$ has smallest eigenvalue $\geqslant -2$, and Theorem 3.12.2 can be applied; see Theorem 4.4.11 below.

(iv) If $b^- = -2$, i.e., if $\theta_d = -b_1/2-1$, then $\Gamma(\gamma)$ has second largest eigenvalue $\leqslant 1$, and its complement has smallest eigenvalue $\geqslant -2$. Again Theorem 3.12.2 can be applied; however, the corresponding distance-regular graphs have not yet been classified.

4.4.4. Theorem. *Let Γ be a distance-regular graph of diameter $d \geqslant 3$ with eigenvalues $k = \theta_0 > \theta_1 > \cdots > \theta_d$, and corresponding multiplicities $1 = f_0, f_1, \cdots, f_d$. If $f_i < k$ for some $i \neq 0$, then $i \in \{1, d\}$. Putting $b = b_1/(\theta_i+1)$ we have that each neighbourhood $\Gamma(\gamma)$ has eigenvalue $-1-b$ with* *multiplicity at least* $k-f_i$*; in case* $-1-b=\lambda$ *its multiplicity is at least* $k-f_i+1$. *Moreover, either* θ_i+1 *is an integer dividing* b_1*, or* θ_1+1 *and* θ_d+1 *are conjugate algebraic integers dividing* b_1.

Proof. Put $f=f_i$, $\theta=\theta_i$. The $(u_0-u_1^2, u_1-u_1^2, u_2-u_2^2)$-representation of $\Gamma(\gamma)$ in $\mathbb{R}^{f-1}$ found above has the property that $\sum_{\delta\in\Gamma(\gamma)}\bar{\delta}=0$. Also, $\theta\neq-1$ since otherwise we have $u_1=u_2$ so that the representation of $\Gamma(\gamma)$ is a regular k-simplex in $\mathbb{R}^{f-1}$, which is impossible for $k>f$. Now by the remark following Proposition 3.5.3, the eigenvalue $-1-b$ of $\Gamma(\gamma)$ has multiplicity at least $k-(f-1)-1=k-f$, and one more in case $-1-b$ is the valency of $\Gamma(\gamma)$.
With b^+ and b^- as above we have $-1-b^+\leqslant\tau_{\min}=-1-b$ (if $\theta>-1$) or $-1-b^-\geqslant\tau_1=-1-b$ (if $\theta<-1$). This forces $\theta=\theta_1$ or θ_d. Since the eigenvalues of Γ and $\Gamma(\gamma)$ are algebraic integers, $\theta+1$ and $b/(\theta+1)$ are algebraic integers. Since conjugate eigenvalues have the same multiplicity, θ can be irrational only when θ_1 and θ_d are conjugate quadratic algebraic integers. □

Remarks. For an application of this theorem, see Theorem 9.4.7. This theorem rules out intersection arrays like $\{36,25,8;1,4,20\}$ ($v=352$, $f_1=32$, $\theta_1=14$), $\{72,45,16;1,8,54\}$ ($v=598$, $f_1=45$, $\theta_1=26$), and $\{105,90,49,7,1;1,7,49,90,105\}$ ($v=2912$, $f_1=78$, $\theta_1=35$). Examples with $f_1<k$ are the Johnson graphs $J(n,m)$ ($k=m(n-m)$, $f_1=n-1$), the halved n-cubes ($k=n(n-1)/2$, $f_1=n$) and the graphs with $(v,k,f_1)=(378,45,42)$, $(1134,117,54)$, $(1024,231,77)$ discussed in §13.2C, §13.2D and §11.3F, respectively.
It is still possible to derive some information from the next case $f=k$; see Ivanov & Shpectorov [437] (cf. Theorems 11.3.6 and 9.5.8). Examples with $f=k$ are all forms graphs ($\theta=\theta_1$), the k-cubes ($\theta=\theta_1=k-2$), the folded halved k-cubes ($\theta=\theta_1=(k^2-9k+16)/2$ for even $k\geqslant 6$), and all antipodal graphs of diameter 3 ($\theta=\theta_2=-1$).

4.4.5. Theorem. *Let* Γ *be a distance-regular graph of diameter at least* 3 *with second largest eigenvalue* $\theta_1<b_1-1$*, and let* $b^+=b_1/(\theta_1+1)$. *Suppose that every singular line of* Γ *has size at least* $s+1$. *Then:*

(i) *If* $s\geqslant 3$ *and* $b^+\leqslant s^2-s+1$*, or if* $s=2$ *and* $b^+\leqslant 2$*, then distinct maximal cliques intersect in a singular line, a point, or the empty set.*

(ii) *If* $s\geqslant 4$ *and* $b^+\leqslant s+1$*, then every edge is in at most two maximal cliques.*

Proof. Apply Theorem 3.7.5. □

Maybe this result will help to prove uniqueness of the Grassmann graphs (C2).

4.4.6. Proposition. *Let* Γ *be a distance-regular graph of diameter* $d\geqslant 2$ *with eigenvalues* $k=\theta_0>\theta_1>\cdots>\theta_d$ *and put* $b^+=b_1/(\theta_1+1)$ *and* $b^-=b_1/(\theta_d+1)$. *Then:*

(i) (cf. Hoffman [398]) *The size of a clique* C *in* Γ *is bounded by*

$$|C|\leqslant 1-k/\theta_d\,. \tag{28}$$

If equality holds then every $\gamma\notin C$ *is adjacent to either* 0 *or* $e=b^-+1-k/\theta_d$ *points of* C. *No vertex of* Γ *has distance* d *to* C.

(ii) *If* Γ *contains a nonempty induced complete bipartite subgraph* $K_{s,t}$ *then*

$$\frac{2st}{s+t} \leqslant b^+ + 1 . \tag{29}$$

Proof. Both parts of this proposition are direct consequences of Proposition 2.5.2, and also follow easily using representations. Let us give one proof of each kind.
(i) We apply Proposition 2.5.2. A clique C of size c has inner distribution $a = (1, c-1, 0, \cdots, 0)$ and specializing $(aQ)_j \geqslant 0$ we find that $1 + (c-1)\theta_j / k \geqslant 0$ for all j. The strongest of these inequalities is equivalent with (28). When equality holds, we have $(BQ)_{\gamma d} = 0$ for all $\gamma \in \Gamma$, and this is equivalent to $\sum_i e_i P_{di} / k_i = 0$, where $e_i = |\Gamma_i(\gamma) \cap C|$ and P_{di} is the eigenvalue of A_i in the eigenspace corresponding to θ_d. In case $d(\gamma, C) = 1$ this becomes $e\theta / k + (c-e)(\theta^2 - \lambda\theta - k) / \mu k_2 = 0$, where $e = e_1$ and $\theta = \theta_d = P_{d1}$. Solving this for e, we find the stated result. We can compute $|C \cap \Gamma_i(\gamma)|$ in the same way for vertices γ with $d(\gamma, C) = i$. But if $d(\gamma, C) = d$ then it follows that $P_{dd} = 0$, and then with induction $P_{di} = 0$ for all i, contradicting $P_{d0} = 1$. Hence $d(\gamma, C) < d$ for all $\gamma \in \Gamma$.
(ii) In the standard representation corresponding to $\theta = \theta_1$, the condensed Gram matrix of the image of a $K_{s,t}$ is

$$\overline{G} = \begin{bmatrix} \dfrac{u_0 + (s-1)u_2}{s} & u_1 \\ u_1 & \dfrac{u_0 + (t-1)u_2}{t} \end{bmatrix};$$

hence $0 \leqslant (1, -1)\overline{G}(1, -1)^{\mathrm{T}} = (u_1 - u_2)((b^+ + 1)(s+t) - 2st) / st$ by (25), giving (29). □

Remark. In case (i) we can say slightly more. From the proof given it follows immediately that if C is a clique of size $1 - k / \theta_d$, then it is completely regular in the sense of §11.1A. See also Remark (iv) following Proposition 4.4.8.

Another interlacing result giving conditions for a regular graph to contain subgraphs $K_{s,t}$ can be found in HAEMERS [340], Theorem 2.1.8.

4.4.7. Proposition. *If $u_l \in \{1, -1\}$ for some $l > 0$ then one of the following holds:*
(i) $\theta = k$, $u_i = 1$ *for all* $i \geqslant 0$.
(ii) $\theta = -k$, $u_i = (-1)^i$ *for all* $i \geqslant 0$, *and* Γ *is bipartite.*
(iii) $l = d$, $u_{d-i} = u_d u_i$ *for all* $i \geqslant 0$, *and* Γ *is antipodal.*

Proof. In a standard representation, all image vectors have norm 1, and $(\overline{\gamma}, \overline{\delta}) = \pm 1$ implies $\overline{\gamma} = \pm\overline{\delta}$. Hence, for $\nu = 1, 2$, the relations $\equiv_\nu$ defined by $\gamma \equiv_\nu \delta$ if and only if $(\overline{\gamma}, \overline{\delta})^\nu = 1$, are equivalence relations, and $\gamma \equiv_\nu \delta$ if and only if $d(\gamma, \delta) \in I_\nu := \{l \mid u_l^\nu = 1\}$. Let $l_\nu \in I_\nu \setminus \{0\}$ be minimal. If $l_\nu = 1$ then $I_\nu = \{0, \cdots, d\}$ since Γ is connected, and one easily finds (i) or (ii). If $1 < l_\nu < d$ then Γ_{l_ν} is disconnected, and by Theorem 4.2.1, Γ is bipartite and $l_\nu = 2$; this gives (ii) again. Finally, if $l_\nu = d > 2$ then Γ_d is disconnected, and by Theorem 4.2.1, Γ is antipodal. For $\gamma, \delta \in \Gamma$ at distance d and $\alpha \in \Gamma_i(\gamma) \cap \Gamma_{d-i}(\delta)$ we now find $u_{d-i} = (\overline{\alpha}, \overline{\delta}) = (\overline{\alpha}, u_d \overline{\gamma}) = u_d u_i$, so that (iii) holds. □

As an application of the above proposition we derive a lower bound on the multiplicity of nontrivial eigenvalues of distance-regular graphs. The proof is based on ideas of

GODSIL [325] who used similar arguments to show that there are only finitely many coconnected distance-regular graphs with an eigenvalue of given multiplicity $f > 2$ (see Theorem 5.3.2).

4.4.8. Proposition. *Let Γ be a coconnected distance-regular graph of valency k, and let θ be an eigenvalue with multiplicity f. Then:*

(i) $f = 1 \Leftrightarrow |\theta| = k$,

(ii) $f = 2 \Leftrightarrow |\theta| \neq k = 2$,

(iii) *If $f > 2$ and $b_i > 1$, then $f > i+1$, and*

$$(f-i-1)(f-i+2) \geqslant 2b_i, \tag{30a}$$

$$f \geqslant i+b_i \text{ if } \lambda = 0, \tag{30b}$$

$$f \geqslant i+2+\tfrac{1}{2}b_i \text{ if } \lambda = 1 \text{ or } \mu = 1, \tag{30c}$$

$$f \geqslant i+2+\frac{\lambda}{\lambda+1}b_i \text{ if } \lambda > \mu = 1 \text{ and } i \leqslant 1. \tag{30d}$$

Proof. If $\theta = \pm k$, then $u_i = (\pm 1)^i$ and Theorem 4.1.4 gives $f = 1$. If $|\theta| \neq k = 2$, then Γ is a polygon, $\theta = 2\cos(j\pi/k)$ $(0 < j < k)$, and Theorem 4.1.4 gives $f = 2$. Hence we may suppose that $|\theta| \neq k > 2$. In particular, the standard sequence for θ satisfies u_0 $(= 1) > u_1, u_2$ by (26) and Proposition 4.4.7, since otherwise $u_0 = u_2 = 1$, $d = 2$, and Γ is not coconnected. We consider the standard representation of Γ in $\mathbb{R}^f$ corresponding to θ. The image of a geodetic path of length l has Gram matrix

$$G_l = \begin{bmatrix} u_0 & u_1 & \cdot & \cdot & \cdot & u_l \\ u_1 & u_0 & \cdot & \cdot & \cdot & u_{l-1} \\ \cdot & \cdot & \cdot & \cdot & & \cdot \\ \cdot & & \cdot & \cdot & \cdot & \cdot \\ \cdot & & & \cdot & \cdot & u_1 \\ u_l & u_{l-1} & \cdot & \cdot & u_1 & u_0 \end{bmatrix},$$

and there is a maximal number q such that G_q is nonsingular. If $q < d$, then G_{q+1} is a singular positive semidefinite matrix with 1-dimensional null space, and there is a unique null vector x of G_{q+1} with $x_{q+1} = 1$. For a geodetic path $\gamma_0 \sim \gamma_1 \sim \cdots \sim \gamma_{q+1}$ of length $q+1$ we then have $\sum_{j=0}^{q+1} x_j\overline{\gamma_j} = 0$, so that $\overline{\gamma_{q+1}} = -\sum_{j=0}^{q} x_j\overline{\gamma_j}$ is uniquely determined by $\gamma_0, \cdots, \gamma_q$. Thus, if $\delta \in \Gamma(\gamma_q) \cap \Gamma_{q+1}(\gamma_0)$, then $\overline{\delta} = \overline{\gamma_{q+1}}$, and since $l := d(\delta,\gamma_{q+1}) \leqslant 2$ and $u_l = (\overline{\delta},\overline{\gamma_{q+1}}) = (\overline{\delta},\overline{\delta}) = u_0$, we have $l = 0$, i.e., $\delta = \gamma_{q+1}$. Therefore $b_q = 1$ if $q < d$.

Now suppose that $b_i > 1$. Then $b_i > b_q$ so that $i < q$, i.e., every geodetic path of length $i+1$ is linearly independent. In particular, since such a path has $i+2$ vertices, we have $f \geqslant i+2$. In case $f \leqslant 2$ this forces $b_1 \leqslant 1$ so that, by Corollary 1.1.6, Γ is a clique, a polygon or $K_{m\times 2}$; but the latter is not coconnected. This proves (i) and (ii). It remains to prove (30a-d). Let $\Delta = \Gamma(\gamma_0) \cap \Gamma_{i+1}(\gamma_i)$, and consider the projection $z = \sum z_j\overline{\gamma_j}$ of the image of $\delta \in \Delta$ on the subspace spanned by $\gamma_0, \cdots, \gamma_i$. Then $u_{l+1} = (\overline{\gamma_l},\overline{\delta}) = (\overline{\gamma_l},z) = \sum u_{|j-l|}z_j$ $(l = 0, \cdots, i)$ and since the coefficient matrix G_i of this linear system is nonsingular, z is independent of $\delta \in \Delta$. Thus, the representation of Δ given by $\hat{\delta} = \overline{\delta} - z$ is orthogonal to $\gamma_0, \cdots, \gamma_i$ and hence represents Δ in

$\mathbb{R}^{f-1-i}$. Since any two vertices $\alpha, \beta \in \Delta$ have distance $\leqslant 2$ in Γ, we have $(\hat{\alpha},\hat{\beta}) = (\overline{\alpha}-z,\overline{\beta}-z) = (\overline{\alpha},\overline{\beta})-(z,z) = u_l - p$, where $l = d(\alpha,\beta) \leqslant 2$ and $p = (z,z)$. Hence Δ has a (u_0-p, u_1-p, u_2-p)-representation in $\mathbb{R}^{f-1-i}$, and an application of Theorem 3.6.1 completes the proof. □

Remarks. (i) Part (i) is due to DELSARTE [254], Remark 5.5. More precisely, ROOS [625] shows that if $f = 1$, $\theta \neq k$, then Γ is bipartite and $P_{ij} = (-1)^j k_j$, if θ belongs to the i-th eigenspace. See also Remark (iv) of §2.4.

(ii) If Γ is not coconnected then $d \leqslant 2$ and $\Gamma \cong \overline{aK_m}$ for suitable a, m. If $d = 2$ then $a \geqslant 2$, and the eigenvalue $-m$ of $\overline{aK_m}$ has multiplicity $f = a-1$ for all m.

(iii) If Γ has large geometric girth, then for each eigenvalue $\theta \neq \pm k$ its multiplicity $f = m(\theta)$ is large. Sometimes this leads to a bound on the (geometric) girth. The argument of TERWILLIGER [714] shows that if T is a tree in Γ such that for all $\gamma, \delta, \epsilon \in T$ we have $d_T(\gamma,\delta) = d_T(\gamma,\epsilon) \Rightarrow d_\Gamma(\gamma,\delta) = d_\Gamma(\gamma,\epsilon)$, then f is at least the number of leaves (vertices of degree 1) of T. (This bound can be sharp; for example, it shows that the multiplicities of a distance-regular graph with $\lambda = 0$ are not less than the valency, and the Higman-Sims graph has spectrum $22^1\, 2^{77}\, (-8)^{22}$.)
Now let Γ be distance-regular, and assume that $(c_1,a_1,b_1) = \cdots = (c_{r-1},a_{r-1},b_{r-1}) \neq (c_r,a_r,b_r)$. If $a_1 = 0$, then TERWILLIGER [714] shows that $f \geqslant k_{[r/2]} = k(k-1)^{[r/2]}$, and a similar result was found by GODSIL [325]. If $a_1 \neq 0$ and $r \geqslant 3$ then BANNAI & ITO [35] show $f \geqslant (k/2)^{[r/2]}$.
For graphs with small f, cf. the remarks following Theorem 4.4.4.

(iv) POWERS [595] shows that if Γ is not a clique but contains a c-clique, then $c \leqslant f_1$. Equality holds e.g. for the halved n-cubes and the triangular graphs.

4.4.9. Proposition. *Consider a standard representation of a distance-regular graph* Γ.

(i) *If* Γ *contains an induced quadrangle* Q *then*

$$u_0 - 2u_1 + u_2 \geqslant 0, \tag{31}$$

or, equivalently, $\theta \leqslant b_1 - 1$. *Equality implies that the images of opposite vertices of* Q *have equal sum, and*

$$u_i = 1 - i(\lambda+2)/k \quad \text{for } i \geqslant 0. \tag{32}$$

(ii) *If* Γ *contains a cube* Q *such that opposite vertices in the cube have distance 3 in* Γ *then*

$$u_0 - 3u_1 + 3u_2 - u_3 \geqslant 0. \tag{33}$$

Equality implies that the images of the two 4-cocliques of Q *have equal sum, and there are numbers* a,c *such that*

$$u_i = 1 + ia + i^2c \quad \text{for } i \geqslant 0. \tag{34}$$

(iii) *Let* $\gamma_1\gamma_2$ *and* $\delta_1\delta_2$ *be two vertex-disjoint edges, and suppose that both distances* l *and* $l+1$ *occur twice among* $d(\gamma,\delta)$, $\gamma \in \{\gamma_1,\gamma_2\}$, $\delta \in \{\delta_1,\delta_2\}$. *Then*

$$u_l + u_{l+1} \leqslant u_0 + u_1. \tag{35}$$

If equality holds then $\overline{\gamma}_1 + \overline{\gamma}_2 = \overline{\delta}_1 + \overline{\delta}_2$, *and*

$$u_{i+l} + u_{i+l+1} = u_i + u_{i+1} \quad \text{for } i = 0, \cdots, d-1-l. \tag{36}$$

Proof. (i) Let $x = \overline{\gamma}_1 + \overline{\gamma}_2$, $y = \overline{\delta}_1 + \overline{\delta}_2$, where γ_1, γ_2 and δ_1, δ_2 are the two pairs of opposite vertices of Q. Then

$$0 \leqslant (x-y, x-y) = (x,x) - 2(x,y) + (y,y) = 2(u_0 + u_2) - 8u_1 + 2(u_0 + u_2)$$
$$= 4(u_0 - 2u_1 + u_2) = 4(u_0 - 1)(b_1 - 1 - \theta)/b_1 ,$$

giving (31). Now suppose that equality holds in (31). Then $x = y$. Pick a point $\alpha \in \Gamma_{i-1}(\gamma_1) \cap \Gamma_{i+1}(\gamma_2)$. Then $u_{i-1} + u_{i+1} = (\overline{\alpha}, x) = (\overline{\alpha}, y) = 2u_i$, and (32) follows inductively. The proof of (ii) and (iii) is completely analogous and will be left to the reader. □

Equality in (31) implies $\theta = b_1 - 1$; we shall treat this case further in Theorem 4.4.11. The classification of Q-polynomial distance-regular graphs with a standard sequence (34) is discussed in TERWILLIGER [727]; together with the results of BUSSEMAKER & NEUMAIER [155] (cf. Theorem 6.3.2), his results imply that, for large diameter, the only possibilities are partition graphs and halved $(2d+1)$-cubes. The case of a standard sequence (36) is trivial for $l = 1$ (Proposition 4.4.6), and has been settled by TERWILLIGER [723] for $l = 2$; the graphs arising are the antipodal covers of complete graphs, the dodecahedron, d-cubes, folded $(2d+1)$-cubes and the doubled Odd graphs.

We now specialize to the situation where θ is an integral eigenvalue. In this case the eigenvector equation $Tx = \theta x$, where T is the reduced tridiagonal $d \times d$-matrix of §4.1B, has a rational, and hence integral solution; therefore there exists an integral t-sequence.

4.4.10. Proposition. *Let θ be an integral eigenvalue of Γ with corresponding integral t-sequence $(t_1, \cdots, t_d)$. Then the standard representation corresponding to θ can be scaled such that for all $\alpha, \gamma, \delta \in \Gamma$ with $\gamma \sim \delta$,*

$$(\overline{\alpha}, \overline{\gamma} - \overline{\delta}) = \begin{cases} 0 & \text{if } d(\alpha,\gamma) = d(\alpha,\delta), \\ t_i & \text{if } d(\alpha,\gamma) < d(\alpha,\delta) = i, \\ -t_i & \text{if } d(\alpha,\delta) < d(\alpha,\gamma) = i. \end{cases} \tag{37}$$

In particular, the scaled representation is an integral affine $(2t_1, 2t_1 + 2t_2)$-representation of Γ, the edges of Γ are represented by vectors of norm $2t_1$ in $\mathbb{L}(\Gamma)$, and the vertices of Γ are represented by vectors of norm $N := kt_1/(k-\theta)$ in the dual lattice $\mathbb{L}(\Gamma)^*$.

Proof. Scale the standard representation by multiplication of the inner products by N. Then the vertices are represented by vectors of norm N, and (37) follows since $N(u_{i-1} - u_i) = t_i$. If $d(\gamma,\delta) = i$ then (37) implies that $(\overline{\gamma} - \overline{\delta}, \overline{\gamma} - \overline{\delta}) = 2\Sigma_{j=1}^{i} t_j$; therefore we have an integral affine $(2t_1, 2t_1 + 2t_2)$-representation. Finally, the image $\overline{\alpha}$ of $\alpha \in \Gamma$ has integral inner products with all $\overline{\gamma} - \overline{\delta}$, and since $v\overline{\alpha} = \sum_{\delta \in \Gamma} (\overline{\alpha} - \overline{\delta}) \in \mathbb{L}(\Gamma)$ by (23), the vector $\overline{\alpha}$ belongs to $\mathbb{R} \otimes \mathbb{L}(\Gamma)$ and hence to $\mathbb{L}(\Gamma)^*$. □

4.4.11. Theorem. *Let Γ be a distance-regular graph with (second largest) eigenvalue $\theta_1 = b_1 - 1$. Then at least one of the following holds:*

(i) *Γ is a strongly regular graph with smallest eigenvalue -2.*

(ii) *$\mu = 1$, i.e., Γ has numerical girth at least 5.*

(iii) $\mu = 2$, *and* Γ *is a Hamming graph* (C8), *a Doob graph* (C8a), *or a locally Petersen graph* ((D4) *or* (A8)).

(iv) $\mu = 4$, *and* Γ *is a Johnson graph* (C1).

(v) $\mu = 6$, *and* Γ *is a half cube* (C9).

(vi) $\mu = 10$, *and* Γ *is the Gosset graph* $E_7(1)$ ((C5), $q = 1$).

Remark. The possibilities in (i) are described in Theorem 3.12.2, and those in (ii) are not (yet) classified. Examples are the Odd graphs, the generalized hexagon of order (2,2) and the generalized octagon of order (4,2).

Proof. If $d = 1$, then Γ is complete, i.e., a Hamming graph. If $d = 2$, then Γ is strongly regular, and by Theorem 1.3 the eigenvalues $\theta_1 = r$ and $\theta_2 = s$ of Γ satisfy the equation $x^2 + (\mu - \lambda)x + \mu - k = 0$. Therefore, $r + s = \lambda - \mu$, $rs = \mu - k$, and by the hypothesis, $b_1(s+1) = (r+1)(s+1) = -k + \lambda + 1 = -b_1$. Since $b_1 \neq 0$, we find $s = -2$, and (i) holds.
Now assume that $d \geqslant 3$, and, without loss of generality, let $\mu > 1$. We first note that by Theorem 4.4.3, all neighbourhoods $\Gamma(\gamma)$ are graphs with smallest eigenvalue at least -2. Suppose first that Γ contains no induced quadrangle. Then each $\Gamma(\gamma)$ is strongly regular by Theorem 1.16.3 (note that an m-clique extension of a graph with smallest eigenvalue θ has smallest eigenvalue $m(\theta+1) - 1$ so that clique extensions of strongly regular graphs cannot occur here), and the possibilities for $\Gamma(\gamma)$ are completely described by Corollary 3.12.3 and Theorem 3.12.4 (i). The only cases without a quadrangle are cliques, the pentagon, and the Petersen graph. If $\Gamma(\gamma)$ is a clique for some $\gamma \in \Gamma$, then Γ itself is a clique, and if Γ is locally a pentagon, then Γ is the icosahedron, and $\theta_1 > b_1 - 1$, a contradiction. Therefore Γ is locally a Petersen graph, and, since $d \geqslant 3$, J.I. Hall's classification of these graphs (Theorem 1.16.5) shows that Γ is either (D4) or (A8).
Now suppose that Γ contains some induced quadrangle. Let $(u_0, \cdots, u_d)$ be the standard sequence corresponding to the eigenvalue $\theta_1 = b_1 - 1$. By (25), $u_0 - 2u_1 + u_2 = 0$, so that by Proposition 4.4.9 (i),

$$u_i = 1 - i(\lambda+2)/k \quad \text{for } i \geqslant 0. \tag{38}$$

In particular, the sequence $t_1 = \cdots = t_d = 1$ is a t-sequence of Γ. From $a_i + b_i + c_i = k$, $b_d = 0$, and the recurrence (25) we get

$$a_i = (\lambda+2)i - 2c_i, \quad b_i = c_i + k - (\lambda+2)i, \quad c_d + k = (\lambda+2)d. \tag{39}$$

By Proposition 4.4.2, Γ has an integral affine (2,4)-representation in $\mathbb{R}^f$ where f is the multiplicity of θ_1. Therefore Γ is a strict root graph, and we may apply the results of §3.15. Suppose first that $\mathbb{L}(\Gamma) \cong E_f$ ($f \in \{6,7,8\}$). Since $\mu > 1$, Proposition 3.15.2 shows that we have one of the graphs of Petersen, Chang, Schläfli or Gosset.
Now suppose that $\mathbb{L}(\Gamma) \ncong E_6, E_7, E_8$. If $\mu > 2$, then Theorem 3.15.1 shows that (iv) or (v) holds. And if $\mu = 2$, then Theorem 3.15.4 and Proposition 1.12.1 (iv), together with the fact that the icosahedron contains no quadrangles, leave only case (iii). □

Remarks. (i) Several otherwise feasible intersection arrays, such as $\{10,6,3,1;1,3,6,10\}$ and $\{22,15,9,4;1,3,6,10\}$ have eigenvalue $b_1 - 1$ and are ruled out by this theorem.
(ii) If Γ is a distance-regular graph with a standard sequence (38) then $\theta = ku_1 = k - \lambda - 2 = b_1 - 1$ is an eigenvalue of Γ. Since the t-sequence is constant, the tridiagonal matrix T defined in §4.1B has the all-one vector $\mathbf{1}$ as eigenvector, with

associated eigenvalue θ. Thus $\mathbf{1}$ is a Perron vector of the irreducible nonnegative matrix $kI + T$, showing that θ is the largest eigenvalue of T. Therefore $\theta = b_1 - 1$ is the second largest eigenvalue of Γ.

On the other hand, the two locally Petersen graphs (D4), with intersection array $\{10,6,4;1,2,5\}$ and (A8) with intersection array $\{10,6,4,1;1,2,6,10\}$ have second largest eigenvalue $\theta = b_1 - 1 = 5$. However, the corresponding standard sequences are $(1,\frac{1}{2},0,-\frac{1}{4})$ and $(1,\frac{1}{2},0,-\frac{1}{4},-\frac{1}{2})$, so that (38) is violated for $i \geqslant 3$.

Chapter 5
Parameter Restrictions for Distance-Regular Graphs

In this chapter we collect most of the restrictions on intersection arrays of distance-regular graphs known to us. (A few very basic facts have already been mentioned in §4.1D.) Some of these restrictions are important tools in the theoretical investigation of the properties of distance-regular graphs, like the unimodality of the sequence $(k_i)_i$ discussed in §5.1. (We already used this on several occasions.) Various bounds on the diameter in terms of the valency are theoretically important. First we have Terwilliger's diameter bound for the case where the graph contains a quadrangle; next Ivanov's theory, which yields a bound on the diameter for arbitrary distance-regular graphs with fixed numerical girth, and finally the work by Bannai & Ito, who strive to remove the dependency on the girth from these bounds. Also Godsil proved diameter bounds, but this time in terms of a multiplicity.
Classifying distance-regular graphs often is a two-stage process - first one finds feasible arrays, and next all graphs with these arrays. The diameter bounds (will) make it possible to generate a complete list of feasible intersection arrays for a given valency; dozens of other restrictions speed up this generating process, or can be used to rule out proposed intersection arrays. A diameter bound in terms of valency only is not (yet) available for arbitrary k, but in the case of cubic graphs Biggs, Boshier & Shawe-Taylor succeeded in obtaining a complete classification using the 'circuit chasing' technique described in §5.10.

5.1 Unimodality of the sequence $(k_i)_i$

5.1.1. Proposition (cf. Taylor & Levingston [711]). *Let Γ be distance-regular with $d \geqslant 3$.*

(i) *The sequence $1 = k_0, k_1, \cdots, k_d$ is unimodal, i.e., there exist h, l with $1 \leqslant h \leqslant l \leqslant d$ such that $1 < k_1 < \cdots < k_h = \cdots = k_l > \cdots > k_d$.*

(ii) $k_i \leqslant k_j$ $(0 \leqslant i \leqslant j, i+j \leqslant d)$.

(iii) *If $k_i = k_j$ $(0 \leqslant i < j,\ i+j \leqslant d)$, then $b_i = c_j$, $b_{i+1} = c_{j-1}$, $\cdots$, $b_{j-1} = c_{i+1}$, and in particular $k_{i+1} = k_{j-1}$.*

(iv) *If $k_i = k_{i+1}$, then $k_i \geqslant k_j$ for all j.*

(v) *If $k_{j-1} = k$ $(3 \leqslant j \leqslant d)$, then either Γ is a polygon $(k = 2)$ or Γ is an antipodal 2-cover ($j = d$ and $k_d = 1$).*

Proof. (i)-(iv) are trivial consequences of $c_{i+1} \geqslant c_i$, $b_{i+1} \leqslant b_i$, $c_i \leqslant b_j$ $(i+j \leqslant d)$ and $k_{i+1} = b_i k_i / c_{i+1}$. (v) is a special case of the following lemma. □

Remark. A slight strengthening of (ii) is given below in Corollary 5.9.7.

5.1.2. Lemma. *Let Γ be a regular connected graph with valency k and diameter $d \geqslant e$, where $e \geqslant 3$. Assume $b_j(\Gamma) = b_j$ $(1 \leqslant j \leqslant e-2)$ and $c_j(\Gamma) = c_j$ $(2 \leqslant j \leqslant e-1)$. Then $k_{e-1}(\Gamma) = k_{e-1} \geqslant k$. Moreover, $k_{e-1} = k$ if and only if either Γ is a polygon $(k = 2)$ or an antipodal 2-cover (distance-regular with $d = e$, $k_d = 1$).*

Proof. We have $b_j \geqslant c_{e-j}$ $(1 \leqslant j \leqslant e-2)$ and

$$k_{e-1} = k \cdot \frac{b_1 b_2 \cdots b_{e-2}}{c_{e-1} c_{e-2} \cdots c_2}$$

so that $k_{e-1} \geqslant k$. If $k_{e-1} = k$, then $b_j = c_{e-j}$ $(1 \leqslant j \leqslant e-2)$. Let $d(\alpha,\beta) = e$ and let W be the union of all geodesics between α and β (except for the points α, β themselves). Let $V = \{\gamma \in \Gamma_e(\beta) \mid \gamma \neq \alpha, \Gamma(\gamma) \text{ meets } W\}$ and $X = \{\chi \in \Gamma_e(\alpha) \mid \chi \neq \beta, \Gamma(\chi) \text{ meets } W\}$. The hypothesis $b_j = c_{e-j}$ $(1 \leqslant j \leqslant e-2)$ implies that if $\delta \in W$ then $\Gamma(\delta) \subseteq V \cup W \cup X \cup \{\alpha,\beta\}$. Suppose that $V \neq \varnothing$ and choose $\gamma \in V$. Choose $\epsilon \in W \cap \Gamma_{e-1}(\gamma) \cap \Gamma(\beta)$. Since $b_1 = c_{e-1}$ we have $\alpha \sim \gamma$. All geodesics between ϵ and γ lie entirely within $W \cup \{\gamma\}$ so $\Gamma(\gamma) \cap \Gamma_{e-2}(\epsilon) \subseteq \Gamma(\alpha)$ and $c_{e-1} \leqslant \lambda$ so that $k = 1 + \lambda + b_1 \leqslant 2\lambda + 1$. Let $\delta \in W \cap \Gamma(\gamma)$ then $\Gamma(\gamma) \cap \Gamma(\delta) \subseteq \{\alpha\} \cup \Gamma(\alpha)$, but $\Gamma(\alpha)$ is regular of valency λ, so any point in $\Gamma(\alpha)$ adjacent to one of γ, δ is also adjacent to the other. Varying γ, δ we see that $\{\gamma,\delta\} \cup (\Gamma(\gamma) \cap \Gamma(\delta)) \setminus \{\alpha\}$ is a clique of size $\lambda+1$ in $\Gamma(\alpha)$. Now $k > \lambda+1$ since Γ is not a clique; thus $\Gamma(\alpha)$ has at least two components, and $k \geqslant 2(\lambda+1)$, contradiction. This contradiction shows that $V = X = \varnothing$, so that $W \cap \Gamma(\alpha)$ is a union of components of $\Gamma(\alpha)$. If $\zeta \in \Gamma(\alpha) \setminus W$ then choose $\eta \in \Gamma_2(\beta) \cap \Gamma_{e-2}(\alpha)$. Now $d(\zeta,\eta) = e-1$ and $\Gamma(\zeta) \cap \Gamma_{e-2}(\eta) = \{\alpha\}$ so $b_1 = c_{e-1} = 1$ and Γ is a polygon. Otherwise $\Gamma(\alpha) \subseteq W$ and Γ has diameter e. We have $v = 1 + k + k_2 + \cdots + k_{e-1} + 1$ so $b_{e-1}(\Gamma) = 1$ and $c_e(\Gamma) = k$, and Γ is distance-regular. □

Remark. The case $e = 3$ of this lemma is due to Taylor & Levingston [711] in the distance-regular case, and to Neumaier [559] in the amply regular case. The proofs are variations on the same theme.

The following corollary, a trivial consequence of Proposition 5.1.1, is of some use when generating parameter sets of distance-regular graphs with $v \leqslant v_0$ for some fixed v_0. (For an application, see the discussion following Corollary 5.9.7 below.)

5.1.3. Corollary. *Put $s_i = \sum_{j=0}^{i-1} k_j$. Then*

$$v \geqslant \begin{cases} 2s_i + (d+1-2i)k_i & \text{if } d \geqslant 2i, \\ s_i + s_{d-i} + k_i & \text{otherwise.} \end{cases}$$ □

Problems. (i) Does $b_i = c_j > 1$, $d \geqslant i+j$ imply $b_{i+1} = c_{j-1}$? This would imply that $k_i < k_{i+1}$ for $k > 2$, $d > 2i+1$, and thus considerably strengthen the corollary.
(ii) With h, l as in 5.1.1 (i), is it true that if $k_i = k_j$ for $i < j$, $i+j \leqslant d$, then either $h \leqslant i < j \leqslant l$ or Γ is antipodal with diameter $d = i+j$?

5.2 Diameter bounds by Terwilliger

A useful set of inequalities resulting in a tight diameter bound for distance-regular graphs containing a quadrangle was given by TERWILLIGER [717]. The inequalities improve the relations $c_i - b_i \geqslant c_{i-1} - b_{i-1}$ which are immediate from Proposition 4.1.6.

5.2.1. Theorem (TERWILLIGER [717]). *Let Γ be a distance-regular graph. If Γ contains a quadrangle, then*

$$c_i - b_i \geqslant c_{i-1} - b_{i-1} + \lambda + 2 \quad (i = 1, \cdots, d). \tag{1}$$

In particular, (1) *holds if* $\lambda(\lambda-1) < b_1(\mu-1)$ (and if $1 < c_3 < 2\mu$, cf. 5.4.1). *Equality holds for i if and only if for every quadrangle $Q = \alpha\beta\gamma\delta$ and every $\epsilon \in \Gamma$ at distance $i-1$ from Q, we have $d(\alpha,\epsilon)+d(\gamma,\epsilon) = d(\beta,\epsilon)+d(\delta,\epsilon)$.*

Proof. For $i = 1$, (1) is obvious. Let $i > 1$. Let $Q = \alpha\beta\gamma\delta$ be a quadrangle where $\alpha \not\sim \gamma$, $\beta \not\sim \delta$, and let r_{hij} be the cardinality of $\Gamma_h(\alpha) \cap \Gamma_i(\beta) \cap \Gamma_j(\gamma)$. Then

$$r_{i,i-1,i} \geqslant p^2_{i-1,i+1}(\beta,\delta),$$
$$r_{i-1,i,i-1} \geqslant p^2_{i-2,i}(\delta,\beta),$$

and

$$p^1_{i-1,i}(\alpha,\beta)+p^1_{i-1,i}(\beta,\gamma)=p^2_{i-2,i}(\alpha,\gamma)+p^2_{i-1,i+1}(\alpha,\gamma)+p^2_{i-1,i}(\alpha,\gamma)+r_{i,i-1,i}+r_{i-1,i,i-1},$$

so that

$$2p^1_{i-1,i} \geqslant 2p^2_{i-2,i}+2p^2_{i-1,i+1}+p^2_{i-1,i}.$$

By Lemma 4.1.7, we easily find

$$p^1_{i-1,i} = \frac{k_i c_i}{k_2 c_2} b_1, \quad p^2_{i-1,i} = \frac{k_i c_i}{k_2 c_2}(a_i + a_{i-1} - \lambda),$$

$$p^2_{i-2,i} = \frac{k_i c_i}{k_2 c_2} c_{i-1}, \quad p^2_{i-1,i+1} = \frac{k_i c_i}{k_2 c_2} b_i,$$

whence $2b_1 \geqslant 2c_{i-1} + 2b_i + (a_i + a_{i-1} - \lambda)$. Using $b_1 = k-1-\lambda$, $a_i = k - b_i - c_i$, $a_{i-1} = k - b_{i-1} - c_{i-1}$, this simplifies to (1). The equality statement is immediate. Finally, suppose $\lambda(\lambda-1) < b_1(\mu-1)$. Let $\gamma\delta$ be an edge. There are $b_1(\mu-1)$ edges $\alpha\beta$ such that $\alpha \sim \delta$, $\alpha \not\sim \gamma \sim \beta$, and only $\lambda(\lambda-1)$ of them may have $\beta \sim \delta$; therefore there is a quadrangle $\alpha\beta\gamma\delta$. □

Remarks. (i) For $i = 2$, the bound (1) reduces to the inequality in Corollary 1.2.4 (i).
(ii) The last argument of the above proof shows that some 2-claw is in $q \geqslant \mu - 1 - [\lambda(\lambda-1)/b_1]$ quadrangles, but by Theorem 1.2.3 (i) we have $q \leqslant \max\{0, k-2\lambda-3+\mu-b_2\} = \max\{0, \mu-2-\lambda+b_1-b_2\}$, so that

$$[\lambda(\lambda-1)/b_1] \geqslant \min(\mu-1, b_2-b_1+\lambda+1). \tag{1a}$$

For example, this implies the nonexistence of distance-regular graphs with intersection arrays $\{35,30,26;1,3,10\}$ ($v = 1296$) and $\{49,42,36;1,3,14\}$ ($v = 2500$). On the other hand, (1a) holds with equality for many Hamming graphs.
(iii) In case of equality for any i, there are no $K_{2,3}$'s, i.e., μ-graphs have no 3-cocliques.

In particular, $\mu \leqslant 2\lambda+2$, and, if equality holds for $i = 2$, then even $\mu \leqslant \lambda+2$.
(iv) If Γ contains no quadrangle, then it is a Terwilliger graph; for $\mu > 1$, very strong restrictions on such graphs are derived in §1.16. In particular, the restriction '$k \geqslant 50(\mu-1)$ unless $\iota(\Gamma) = \{5,2,1;1,2,5\}$, $\{10,6,4;1,2,5\}$ or $\{10,6,4,1;1,2,6,10\}$' for distance-regular graphs without quadrangles (Corollary 1.16.6) rules out several feasible intersection arrays which violate one of the inequalities in (1), such as $\{21,12,8;1,3,8\}$ ($v = 190$), $\{27,16,9;1,3,12\}$ ($v = 280$), $\{24,18,16;1,2,9\}$ ($v = 625$), $\{35,28,22;1,2,14\}$ ($v = 1296$), $\{38,30,24;1,2,15\}$ ($v = 1521$) of diameter 3, and $\{21,14,9,3;1,3,6,21\}$ ($v = 288$), $\{63,48,45,3;1,3,16,45\}$ ($v = 4096$) of diameter 4.
(v) The inequalities (1) are equivalent to $T\mathbf{1} \leqslant (b_1-1)\mathbf{1}$ for the tridiagonal matrix T introduced in §4.1B. Since $T+kI$ is nonnegative we conclude that, if a quadrangle exists, the largest eigenvalue θ_1 of T is $\leqslant b_1-1$; this provides an alternative proof of Proposition 4.4.9 (i).

5.2.2. Corollary. *Let Γ be a distance-regular graph of diameter d and valency k. If Γ contains a quadrangle, then $d \leqslant k$, and, more precisely,*

$$d \leqslant (k+c_d)/(\lambda+2), \tag{2}$$

with equality if and only if equality holds in (1) for all i.

Proof. Adding the inequalities in (1) gives

$$c_d = c_d - b_d \geqslant c_0 - b_0 + d(\lambda+2) = -k + d(\lambda+2),$$

hence (2) and the statement about equality. The bound for d is at most k as $\lambda \geqslant 0$ and $c_d \leqslant k$. □

The diameter bound of the corollary is tight, and there are infinite families of distance-regular graphs achieving the bound. In fact we are able to characterize the equality case.

5.2.3. Theorem (cf. Terwilliger [725]). *Let Γ be a distance-regular graph with diameter $d \geqslant (k+c_d)/(\lambda+2)$. Then one of the following holds:*

(i) *Γ is a Terwilliger graph,*

(ii) *Γ is strongly regular with smallest eigenvalue -2,*

(iii) *Γ is a Hamming graph, a Doob graph, a locally Petersen graph, a Johnson graph, a half cube, or the Gosset graph.*

Proof. Suppose Γ contains a quadrangle. By the corollary $d \geqslant (k+c_d)/(\lambda+2)$ implies that Γ has parameters (39) (of §4.4), and hence has $\theta = b_1 - 1$ as an eigenvalue with standard sequence (38) (of §4.4). Hence Theorem 4.4.11 applies. □

5.2.4. Corollary. *Let Γ be a distance-regular graph with diameter $d \geqslant 2k/(\lambda+2)$. Then one of the following holds:*

(i) *Γ is a Terwilliger graph.*

(ii) *Γ is a $K_{m\times 2}$ ($m = 1+k/2$), a d-cube, a Johnson graph $J(2d,d)$, a half cube $D_{n,n}$ ($n = 2d$), or the Gosset graph $E_7(1)$.*

Proof. If Γ contains a quadrangle then we must have $c_d = k$, and the preceding theorems leave only the stated possibilities. □

Problem. Is the icosahedron the only amply regular graph with $\mu > 1$ and diameter $d > 2k/(\lambda+2)$?

VAN BON [98] remarked that the diameter bound can be used to show nonexistence of antipodal covers: if $\tilde{\Gamma}$ is an antipodal cover of Γ containing a quadrangle, then we find $\tilde{d} \leqslant 2k/(\lambda+2)$ for $\tilde{\Gamma}$, so $d \leqslant k/(\lambda+2)$. For example, for $H(d,q)$ we have $k = d(q-1)$, $\lambda = q-2$ and distance-regular antipodal covers cannot contain a quadrangle; in particular, no distance-regular antipodal covers of diameter $\tilde{d} \geqslant 5$ exist. (And in fact the only antipodal cover of a Hamming graph is the octagon, a 2-cover of the quadrangle.)

For $\mu = 1$ (numerical girth at least 5) it is much more difficult to get a diameter bound (cf. §5.9). However, TERWILLIGER [715] proves a number of inequalities for the parameters of distance-regular graphs with larger numerical girth which, in certain cases, yield a diameter bound. We refer to his paper for details and content ourselves with the following consequence of his results.

5.2.5. Theorem. *Let Γ be a distance-regular graph with diameter d and girth $2s$ or $2s-1$ (i.e., $a_0 = \cdots = a_{s-2} = 0$, $c_1 = \cdots = c_{s-1} = 1$, and $a_{s-1} > 0$ or $c_s > 1$). If $a_{s-1} \leqslant 2c_s - 2$, then*

$$c_i - b_{i-1} \geqslant c_{i-s+1} - b_{i-s} + 2 \text{ for } s \leqslant i \leqslant d; \tag{3a}$$

if moreover $a_{s-1} \leqslant c_s$, then

$$c_i \geqslant c_{i-s+1} + 1 \text{ for } s \leqslant i \leqslant d. \tag{3b}$$

In particular

$$a_{s-1} \leqslant 2c_s - 2 \Rightarrow d \leqslant (s-1)(k-1)+1, \tag{3c}$$

and equality is possible only if $c_s = 2$.

Proof. In Theorem 2.2 of TERWILLIGER [715], put $x = i$, $y = i-1$ in (8) and $x = i$, $y = s$ in (9) to get (3a) and (3b). An induction argument yields (3c), and (11) in Theorem 2.2 of [715] shows that equality in (3c) forces $c_s = 2$. □

Note that the doubled Odd graphs have girth 6, diameter $d = 2k-1$ and $c_3 = 2$, so that the diameter bound in (3c) can be tight for $s = 3$. The inequalities (3b) generalize the inequalities $c_{i+1} > c_i$ if $\lambda \leqslant \mu \neq 1$ which follow by the argument leading to Proposition 1.9.1; they also improve earlier inequalities of TERWILLIGER [713] for bipartite distance-regular graphs.

5.3 Godsil's diameter bound. Graphs with $b_i = 1$

We now give another set of inequalities (implicitly in GODSIL [325]) which leads to a diameter bound in terms of the multiplicity of a nontrivial eigenvalue.

5.3.1. Lemma. *Let* Γ *be a distance-regular graph with* $k > 2$ *and* $b_i = 1$ *for some* i. *Then* Γ *has diameter* $d < 3i$ *and we have*

$$c_j = 1,\ c_{i+j} > a_j \ \text{ if } d \geqslant i+j, \tag{3d}$$

$$c_{2i} > 1 \ \text{ if } d \geqslant 2i, \tag{3e}$$

$$a_j = 0 \ \text{ if } d \geqslant 2i+j, \tag{3f}$$

$$a_i \geqslant (\mu - 1)a_{i+1} + \lambda. \tag{3g}$$

Proof. Let $\gamma_0 \sim \gamma_1 \sim \cdots \sim \gamma_d$ be a geodetic path. Since $b_i = 1$ implies $b_{i+j} = 1$ for $j \geqslant 0$, γ_{i+j} is the unique vertex in $\Gamma_{i+j}(\gamma_0) \cap \Gamma_j(\gamma_i)$. Hence γ_{i+j-1} and the a_j vertices in $\Gamma(\gamma_{i+j}) \cap \Gamma_j(\gamma_i)$ must have distance $i+j-1$ from γ_0 so that $a_j < c_{i+j}$. This proves the second half of (3d), and the first half follows from $c_j \leqslant b_{d-j} \leqslant b_i = 1$ if $d \geqslant i+j$. If $d \geqslant 2i$ then $c_{2i} = 1$ would imply $k = a_i + b_i + c_i < c_{2i} + b_i + c_i = 3$, contradiction, hence $c_{2i} > 1 = b_i$, giving (3e) and $d < 3i$ by 4.1.6 (ii). If $d \geqslant 2i+j$, then $a_j < c_{i+j} = 1$ by (1) so that $a_j = 0$, and (3f) holds. Finally, consider the paths $\gamma_i \sim \alpha \sim \beta \sim \gamma_{i+1}$ where $\alpha \in \Gamma_i(\gamma)$, $\beta \in \Gamma_{i+1}(\gamma)$. Since $b_i = 1$, we have $\Gamma(\gamma_{i+1}) \cap \Gamma(\gamma_i) \subseteq \Gamma_i(\gamma)$ and, given α there is at most one such β; hence there are at least $(\mu - 1)a_{i+1}$ points α in $\Gamma_i(\gamma) \cap \Gamma(\gamma_i) \setminus \Gamma(\gamma_{i+1})$. □

(See also 4.3.2 (iii) for consequences of $b_{d-1} = 1$.)

Remark (Lambeck [472]). If $i > 1$, $b_i = 1$, $d = 3i - 1$, then we have the dodecahedron.

5.3.2. Theorem (Godsil [325]). *There are only finitely many coconnected distance-regular graphs with an eigenvalue of fixed multiplicity* $f > 2$. *Any such graph has diameter* $d \leqslant 3f - 4$ *and valency* $k \leqslant \frac{1}{2}(f-1)(f+2)$.

Proof. Let $f > 2$. If $d \geqslant 3f - 3$, then the lemma implies $b_{f-1} > 1$, and Proposition 4.4.8 (iii) gives a contradiction. Hence $d \leqslant 3f - 4$, and the same proposition gives $k = b_0 \leqslant \frac{1}{2}(f-1)(f+2)$. Since diameter and valency are bounded in terms of f, the number of vertices is bounded, so that there are only finitely many distance-regular graphs with an eigenvalue of multiplicity f. □

Remarks. (i) If Γ is not coconnected, then Γ is complete multipartite (Lemma 1.1.7). All graphs $K_{(f+1)\times m}$ have an eigenvalue of multiplicity f (namely $-m$).
(ii) By combining the information in the above lemma and in Proposition 4.4.8 (iii), it is not difficult to show that the only coconnected distance-regular graphs with an eigenvalue of multiplicity $f = 3$ (i.e., those with a spanning representation in $\mathbb{R}^3$) are platonic solids. (And conversely, each of the 5 platonic solids has an eigenvalue of multiplicity $f = 3$.) The dodecahedron ($f = 3, d = 5$) shows that the diameter bound can be tight, but Godsil [325] shows that it is the only distance-regular graph with $d = 3f - 4$. The icosahedron ($f = 3, k = 5$) shows that also the valency bound can be tight.

5.4 Restrictions for $\mu > 1$

Whereas Terwilliger's inequalities give a lower bound on the increment $(c_i - b_i) - (c_{i-1} - b_{i-1})$, some inequalities of GARDINER [314] concern the increase of $c_i - c_{i-1}$ for the smallest case $i = 3$. Here, we give a more general result based on Gardiner's ideas.

5.4.1. Theorem. *Let Γ be distance-regular of diameter $d \geqslant 3$. If $\mu > 1$, then one of the following holds.*

(i) $c_3 \geqslant \frac{3}{2}\mu$.

(ii) $c_3 \geqslant \mu + b_2$, $d = 3$.

Proof. Without loss of generality, assume $\mu > 1$, $c_3 < 2\mu$. We first show that Γ contains a quadrangle. Indeed, suppose no quadrangle exists. Choose vertices γ, δ at distance 3, and put $A = \Gamma(\gamma) \cap \Gamma_2(\delta)$, $B = \Gamma_2(\gamma) \cap \Gamma(\delta)$. If A contains nonadjacent vertices ζ and ζ', then all their common neighbours are in $\Gamma(\gamma)$; therefore the sets $\Gamma(\zeta) \cap \Gamma(\delta)$ and $\Gamma(\zeta') \cap \Gamma(\delta)$ are disjoint subsets of B, giving the contradiction $c_3 = |B| \geqslant 2\mu$. Hence A is a clique, and, by symmetry, B is also a clique. If there are nonadjacent vertices $\zeta \in A$, $\beta \in B$ then they have at least 2μ common neighbours, contradiction. Therefore $A \cup B$ is a clique. Since $\mu > 1$, $\Gamma(\gamma)$ is connected of valency λ, but not a clique; hence, if $\epsilon \in A$ then $\{\gamma,\epsilon\}^{\perp}$ is not a $(\lambda+2)$-clique. Among the pairs (α,γ') of nonadjacent vertices in $\{\gamma,\epsilon\}^{\perp}$ there is at least one which contains a vertex of A, because otherwise such a pair would have at least $\mu+1$ common neighbours (namely γ and the vertices in A). Thus we can choose such a pair with $\alpha \in A$, $\gamma' \notin A$. Now $d(\gamma',\delta) = 3$; hence $A' = \Gamma(\gamma') \cap \Gamma_2(\delta)$, $B' = \Gamma_2(\gamma') \cap \Gamma(\delta)$ and $A' \cup B'$ are cliques. It follows that $B' = \{\delta,\epsilon\}^{\perp} = B$. Since $|A'| = |A|$ and $\alpha \in A \setminus A'$ (since $\alpha \not\sim \gamma'$), there is a vertex $\alpha' \in A' \setminus A$ (so that $\alpha' \not\sim \gamma$). Since α and α' have at least $\mu+1$ common neighbours (namely ϵ and the vertices in B), they must be adjacent, and then $\alpha'\alpha\gamma\gamma'$ is a quadrangle, contradiction.
Thus, Γ contains a quadrangle, say $\gamma \sim \delta \sim \gamma' \sim \delta' \sim \gamma$. Pick a vertex $\zeta \in \Gamma(\gamma) \cap \Gamma_3(\gamma')$. If (ii) does not hold, then either $d \geqslant 4$ or $b_2 > c_3 - \mu$, and in both cases it is possible to find a vertex $\zeta' \in \Gamma_3(\gamma) \cap \Gamma(\gamma') \setminus \Gamma_2(\zeta)$. Now δ and δ' both have μ neighbours in each of $\Gamma(\zeta)$ and $\Gamma(\zeta')$, and therefore have at least $2\mu - c_3$ common neighbours in each of these sets. Since these two sets are disjoint, we have $\mu \geqslant 2(2\mu - c_3)$, which is equivalent to (i). □

Examples. The intersection arrays $\{55,54,44,40; 1,11,15,55\}$ ($v = 1694$) and $\{12,11,9,8,3,1; 1,3,4,9,11,12\}$ ($v = 270$) are feasible but are ruled out by this theorem.

5.4.2. Corollary. *If $\mu > 2$, then $c_3 \geqslant \mu+2$ unless the intersection array is $\{k,k-1,1; 1,k-1,k\}$ (cf. Corollary 1.5.4).*

Proof. If $\mu > 2$ and $c_3 \leqslant \mu+1$, then the theorem implies $b_2 = 1$, $c_3 = \mu+1$, $d = 3$, and Lemma 5.3.1(3g) shows that $k - \mu - 1 = a_2 \geqslant (\mu-1)a_3 + \lambda = (\mu-1)(k-\mu-1)+\lambda$, so that $k = \mu+1$, $\lambda = a_2 = a_3 = 0$. □

Remarks. (i) GARDINER [314] also states that if $\mu = 2$, $c_3 = 3$, then $b_2 \geqslant 2$, but his proof does not exclude an intersection array $\{k,k-1,1; 1,2,3\}$. However, this array is feasible only for $k = 3$ (for $k > 4$ we have $p^3_{33} < 0$, and for $k = 4$ the multiplicities are non-integral). Thus, if $c_3 = 3$, $\mu = 2$, $b_2 = 1$, then $k = 3$ and we have the cube

(which exists), i.e., the conclusion of the corollary also holds for $\mu = 2$.
(ii) One may also derive restrictions for the situation $\mu = 1$, $c_3 = 2$. See Remark (iv) in §13.6.
(iii) If $1 < c_3 < 2\mu$, then the proof of Theorem 5.4.1 shows that Γ contains a quadrangle, and the restrictions of Theorem 5.2.1 apply.

Theorem 5.4.1 shows that $c_3 = c_2$ implies $\mu = 1$. Similarly we have

5.4.3. Proposition (GARDINER [314]). *If $b_2 = b_1$ then $\mu = 1$.*

Proof. Assume that the graph Γ satisfies $b_2 = b_1$ and $\mu > 1$. Let $d(\gamma,\delta) = 2$ and $\alpha, \beta \in \Gamma(\gamma) \cap \Gamma(\delta)$. Since $b_2 = b_1$, we have $\Gamma_2(\alpha) \cap \Gamma(\delta) = \Gamma_3(\gamma) \cap \Gamma(\delta)$, and in particular $\alpha \sim \beta$. Thus, μ-graphs are cliques. Let $E = \Gamma_2(\gamma) \cap \Gamma(\delta)$. If $\epsilon \in E$ then since $b_1 = b_2$ we have $\epsilon \sim \alpha$ and $\epsilon \sim \beta$. Consequently, of the $\lambda - 2$ common neighbours of α and β distinct from γ and δ, there are a_2 in E and $\mu - 2$ in $\Gamma(\gamma) \cap \Gamma(\delta)$. It follows that $\lambda \geqslant \mu + a_2$, i.e., $b_1 \leqslant b_2 - 1$, contradiction. □

5.4.4. Proposition (GARDINER [314]). *If $b_1 = b_2 = \cdots = b_{i+1}$, then $c_{i+1} = c_i = \cdots = c_1 = 1$.*

Proof. Induction on i. By the previous proposition we may assume $i \geqslant 2$. Now $\mu = 1$, and we have singular lines of size $\lambda + 2$. But if $d(\gamma,\delta) = i+1$, then, since $b_{i+1} = b_i$, all singular lines on δ except one hit $\Gamma_{i+2}(\gamma)$, while the last one has a single point in $\Gamma_i(\gamma)$. Thus, $c_{i+1} = 1$. □

Problem. Does $c_i = b_1$ imply $b_i = 1$ for $i < d$? (This would follow from an affirmative solution of Problem (i) following 5.1.3.)

5.5 Further restrictions from counting arguments

Here we collect several further inequalities involving the parameters of a distance-regular graph, derived by simple counting arguments.

5.5.1. Proposition. *Let Γ be a distance-regular graph of diameter d and assume $\lambda > 0$. Then, for $i \in \{1, \cdots, d-1\}$:*

(i) $2a_i \geqslant \lambda + 1$,

(ii) $a_i + a_{i+1} \geqslant \lambda + 1$,

(iii) $b_i + c_{i+1} \geqslant \lambda + 2$.

Proof. Let γ, δ be two points of Γ at distance $i+1$, and let $\gamma = \gamma_0, \gamma_1, \cdots, \gamma_i, \gamma_{i+1} = \delta$ be a geodesic from γ to δ.
(i) The set $\Gamma(\gamma_i) \cap \Gamma(\gamma_{i+1})$ of size λ is covered by the two sets $\Gamma_i(\gamma) \cap \Gamma(\gamma_i)$, $\Gamma_i(\gamma_1) \cap \Gamma(\gamma_{i+1})$ of sizes a_i each. Thus, $2a_i \geqslant \lambda$. Suppose now that (i) is false. Then $2a_i = \lambda > 0$ for some $i \in \{2, \cdots, d-1\}$. Now, the above inclusions are equalities so that every vertex in $\Gamma(\gamma_i) \cap \Gamma_i(\gamma)$ is adjacent to γ_{i+1}, and every vertex in $\Gamma(\gamma_{i+1}) \cap \Gamma_i(\gamma_1)$ is adjacent to γ_i, while $\Gamma(\gamma_{i+1}) \cap \Gamma_i(\gamma) \cap \Gamma_i(\gamma_1) = \varnothing$, so that $\Gamma(\gamma_{i+1}) \cap \Gamma_i(\gamma_1) \subseteq \Gamma_{i+1}(\gamma)$. It also follows that $c_i \geqslant a_i + 1$. Varying γ_{i+1} over $\Gamma(\gamma_i) \cap \Gamma_{i+1}(\gamma)$, we see that any point of the latter set is adjacent to any point of $\Gamma(\gamma_i) \cap \Gamma_i(\gamma)$, and varying γ_i, that any point of $\Gamma(\gamma_{i+1}) \cap \Gamma_i(\gamma)$ is adjacent to all

points of $\Gamma(\gamma_{i+1}) \cap \Gamma_i(\gamma_1)$. In particular, two points of $\Gamma(\gamma_{i+1}) \cap \Gamma_i(\gamma_1)$ have at least $c_i + 1$ common neighbours. Since $\mu \leq c_i$, it follows that $\Gamma(\gamma_{i+1}) \cap \Gamma_i(\gamma_1)$ is a clique of size a_i and $\Gamma(\gamma_i) \cap \Gamma_{i+1}(\gamma)$ is a union of (a_i+1)-cliques. Now let $\beta \in \Gamma(\gamma_i) \cap \Gamma_i(\gamma)$; then γ_i and β have $2a_i$ common neighbours, so that $\Gamma(\gamma_i) \cap \Gamma_{i+1}(\gamma)$ is an (a_i+1)-clique itself. In particular we have $b_i = a_i + 1$. But then any two points from this set have $\lambda \geq c_i + a_i - 1 \geq 2a_i$ common neighbours. Thus equality holds, and $c_i = a_i + 1$, and $(\Gamma(\gamma_i) \cap \Gamma_i(\gamma)) \cup (\Gamma(\gamma_{i+1}) \cap \Gamma_i(\gamma_1))$ is a clique of size $b_i + c_i = \lambda + 2$. Now (using the Theorems 1.2.3 and 5.4.1) $k = a_i + b_i + c_i = \frac{3}{2}\lambda + 2$, $b_1 = \frac{1}{2}\lambda + 1 = \mu$, $i = 2$, $a_2 = \lambda$, a contradiction. Hence (i).

(ii) By Lemma 4.1.7, the size of $\Gamma_{i+1}(\gamma) \cap \Gamma_2(\gamma_i)$ is $p^i_{2,i+1} = b_i(a_i + a_{i+1} - \lambda)/\mu$ so that $a_i + a_{i+1} \geq \lambda$. Suppose that $a_i + a_{i+1} = \lambda$. Then by (i) we have $i = d-1$. Since $p^{d-1}_{2,d} = 0$, each vertex of $\Gamma(\delta) \cap \Gamma_{d-1}(\gamma)$ is adjacent to each vertex of $\Gamma(\delta) \cap \Gamma_d(\gamma)$, so that $\overline{\Gamma(\delta)}$ is disconnected. But this contradicts Lemma 1.1.7.

(iii) This follows since the set $\{\gamma_i, \gamma_{i+1}\} \cup (\Gamma(\gamma_i) \cap \Gamma(\gamma_{i+1}))$ is covered by the two sets $\Gamma_i(\gamma) \cap \Gamma(\gamma_{i+1})$ and $\Gamma_{i+1}(\gamma) \cap \Gamma(\gamma_i)$ of respective sizes c_{i+1} and b_i. □

For a strengthening of part (ii), see inequality (4) following Lemma 5.5.5. The inequality (iii) is due to TAYLOR & LEVINGSTON [711]; it is much weaker than (1) but holds without the assumption of the existence of a quadrangle, and, in particular, for $\mu = 1$. For $i = 1$, the inequality in (iii) comes down to $k \geq 2\lambda + 3 - \mu$. In (7) of §1.2 we have seen that this inequality holds for any amply regular graph and that the case of equality can be completely characterized. For $i > 1$, equality in (iii) implies that $\Gamma_i(\gamma) \cap \Gamma(\delta)$ is a clique whenever $d(\gamma,\delta) = i+1$. Thus we may apply the following

5.5.2. Lemma. *Let Γ be a distance-regular graph, and let $1 < i < d$. Suppose that $\Gamma_i(\gamma) \cap \Gamma(\delta)$ is a clique whenever $d(\gamma,\delta) = i+1$. Then*

(i) *$\Gamma_j(\gamma) \cap \Gamma(\delta)$ is a clique for any two vertices γ, δ with $d(\gamma,\delta) = j+1$, $1 \leq j \leq i$.*

(ii) $\mu = c_3 = \cdots = c_{i+1} = 1$.

Proof. (i) is obvious; (ii) will be proved by induction on i. Let $\gamma = \gamma_0, \gamma_1, \cdots, \gamma_i, \gamma_{i+1} = \delta$ be a geodesic between γ and δ. Assume that $c_i < c_{i+1}$. For $\alpha, \beta \in \Gamma$, put $C(\alpha,\beta) := \Gamma_{j-1}(\alpha) \cap \Gamma(\beta)$, where $j = d(\alpha,\beta)$. Since $c_i < c_{i+1}$ there is a point $\epsilon \in C(\gamma,\delta) \setminus C(\gamma_1,\delta)$; choose $\zeta \in C(\epsilon,\gamma)$. By the hypothesis $\gamma_1 \sim \zeta$, so that $d(\gamma_1,\epsilon) = i$. If $i > 2$ then by induction we have $c_i = 1$, but ϵ has two neighbours (viz. γ_i and a point on a geodesic to ζ) at distance $i-1$ to γ_1, contradiction. So $i = 2$. Both γ_1 and ζ have μ neighbours in $C(\gamma,\delta)$, and since $\gamma_1 \not\sim \epsilon \in \Gamma(\zeta) \cap \Gamma(\delta)$ we have $\zeta \not\sim \gamma_2' \in \Gamma(\gamma_1) \cap \Gamma(\delta)$ for some vertex γ_2'. On the other hand, $\Gamma(\gamma_1) \cap \Gamma(\epsilon)$ is a clique and contains ζ and γ_2'. Contradiction again. Thus $c_{i+1} = c_i$, and in particular $c_3 = \mu$, but by Theorem 5.4.1 this implies $\mu = 1$. □

5.5.3. Corollary. *Suppose that Γ is distance-regular and that $b_i + c_{i+1} = \lambda + 2$ for some i, $2 \leq i \leq d-1$. Then $\mu = c_3 = \cdots = c_{i+1} = 1$, and either $k = 2(\lambda+1)$ or $k \geq (\lambda+1)(\lambda+3)$.*

Proof. The first part is just part (ii) of the lemma above. Since $\mu = 1$, we have lines of size $\lambda + 2$ and in particular $(\lambda+1)|k$. Write $k = r(\lambda+1)$. Since for $d(\gamma,\delta) = i$ the set $\Gamma_{i+1}(\gamma) \cap \Gamma(\delta)$ is a clique of size $b_i = \lambda + 1$, it follows that $\{\delta\} \cup (\Gamma_{i+1}(\gamma) \cap \Gamma(\delta))$ is a line. Thus, one line on δ has a point in $\Gamma_{i-1}(\gamma)$, one line on δ (say L) has all but one of its points in $\Gamma_{i+1}(\gamma)$, and the remaining $r-2$ lines on δ

are contained in $\Gamma_i(\gamma)$. Suppose $r \geqslant 3$ and let M be a line on γ contained in $\Gamma_i(\delta)$. Choose $\alpha \in M \setminus \{\gamma\}$ and $\epsilon \in L \setminus \{\delta\}$. If $d(\alpha,\epsilon) < i+1$, then ϵ has at least two neighbours in $\Gamma_i(\gamma)$, viz. δ and a point on a geodesic from ϵ to α, so that $c_{i+1} > 1$, contradiction. Thus each point of $L \setminus \{\delta\}$ has distance $i+1$ to each point of M. For $\alpha \in M$, let L_α be the unique line on δ meeting $\Gamma_{i-1}(\alpha)$, and put $\alpha' = \Gamma_{i-1}(\alpha) \cap L_\alpha$. If α, $\beta \in M$ and $\alpha' = \beta'$, then $c_i > 1$, impossible. If $L_\alpha = L_\beta$, then again $c_i > 1$ (since a geodesic from β' to α may start with α' or head towards β), contradiction. Thus, δ is on at least $|M|+1 = \lambda+3$ distinct lines. □

Remark. In the situation of the corollary, $(\lambda+1)|k$ since $\mu = 1$. If $k = 2(\lambda+1)$, then Γ is a polygon or the line graph of a regular generalized polygon (cf. Proposition 4.3.4 and Theorem 4.2.16); there are also examples with $\lambda = 0$, $b_i = 1$ such as the Biggs-Smith graph (D14). We do not know of examples with $\lambda > 0$, $k/(\lambda+1) > 2$ - it seems somewhat peculiar that a_i $(= k-\lambda-2)$ would be so much larger than b_i $(= \lambda+1)$, but it is not obvious how to derive a contradiction. We can show that the only example with $i = 2$ and $k/(\lambda+1) > 2$ is the dodecahedron.

Our next proposition generalizes Lemma 3.3 from IVANOV et al. [282]; a slightly weaker result has been given independently by NOMURA [569].

5.5.4. Proposition (BROUWER & LAMBECK [133]). *Suppose the intersection array of a distance-regular graph satisfies $a_i \neq 0$, and define $a_{d+1} = 0$.*

(i) *If $i < d$, then $b_i \leqslant a_i + \dfrac{a_{i+1}b_i}{a_i}$, and equality implies $a_{i-1} = a_i$.*

(ii) *If $i > 1$, then $c_i \leqslant a_i + \dfrac{a_{i-1}c_i}{a_i}$, and, for $i < d$, equality implies $a_{i+1} = a_i$.*

(iii) *If $1 \leqslant i \leqslant d$, then $b_i + c_i \leqslant a_i + \dfrac{a_{i+1}b_i}{a_i} + \dfrac{a_{i-1}c_i}{a_i}$, and equality holds if and only if for any four vertices α, β, γ, δ of Γ such that $\alpha \sim \beta$, $\gamma \sim \delta$, and $d(\alpha,\gamma) = i$, the three distances $d(\alpha,\delta)$, $d(\beta,\gamma)$, $d(\beta,\delta)$ are not all equal. In particular, equality implies $a_{i+1} \leqslant a_i$ and $a_{i-1} \leqslant a_i$.*

Proof. Count 4-tuples $(\alpha,\beta,\gamma,\delta)$ with $\alpha \sim \beta$, $\gamma \sim \delta$, $d(\gamma,\alpha) = d(\gamma,\beta) = i$, $d(\delta,\alpha) \neq i$. Clearly, there are $vk_ia_i(b_i+c_i)$ such 4-tuples. We find the inequality

$$vk_ia_i(b_i+c_i) \leqslant vk_ia_i^2 + vk_{i+1}a_{i+1}c_{i+1} + vk_{i-1}a_{i-1}b_{i-1}$$

by covering this collection of 4-tuples with the three sets

$$\{(\alpha,\beta,\gamma,\delta) \mid \alpha \sim \beta, \gamma \sim \delta, d(\gamma,\alpha) = d(\gamma,\beta) = i, d(\delta,\beta) = i\},$$
$$\{(\alpha,\beta,\gamma,\delta) \mid \alpha \sim \beta, \gamma \sim \delta, d(\delta,\alpha) = d(\delta,\beta) = i+1, d(\gamma,\alpha) = i\},$$
$$\{(\alpha,\beta,\gamma,\delta) \mid \alpha \sim \beta, \gamma \sim \delta, d(\delta,\alpha) = d(\delta,\beta) = i-1, d(\gamma,\alpha) = i\}.$$

This proves the inequality in (iii); those in (i), (ii) follow similarly by restricting $d(\alpha,\delta)$ to be $i+1$ and $i-1$, respectively. Equality in (iii) is equivalent with: if $\alpha \sim \beta$, $\gamma \sim \delta$, $d(\alpha,\gamma) = i$, then the three distances $d(\alpha,\delta)$, $d(\beta,\gamma)$, $d(\beta,\delta)$ are not all equal. With $d(\gamma,\beta) = d(\delta,\beta) = i \pm 1$ this implies $a_{i\pm1} \leqslant a_i$ (varying δ). Similarly, we find $a_{i-1} \leqslant a_i$ in case of equality in (i), but comparing with (iii) yields the opposite inequality, so $a_{i-1} = a_i$. Likewise $a_{i+1} = a_i$ in case of equality in (ii). □

Examples. Condition (ii) rules out an infinite family of feasible intersection arrays, see

Proposition 9.1.9. There exists a distance-transitive graph on 990 vertices with automorphism group $3{\cdot}M_{22}.2$ and intersection array $\{7,6,4,4,4,1,1,1;1,1,1,2,4,4,6,7\}$ (cf. §11.4D). This graph has equality in (i) and (iii) for $i = 3$ and in (ii) and (iii) for $i = 5$. There exists a distance-transitive graph on 506 vertices with automorphism group M_{23} and intersection array $\{15,14,12;1,1,9\}$ (cf. §11.4B); the fact that it has equality in (ii) leads to a uniqueness proof. Similarly, there exists a distance-transitive graph on 280 vertices with automorphism group $P\Gamma L(3,4)$ and intersection array $\{9,8,6,3;1,1,3,8\}$ (cf. §11.3G); the fact that it has equality in (iii) implies that it contains Petersen subgraphs, which is the starting point for a uniqueness proof (cf. §4.3C).

The inequality (ii) can be sharpened in case $i = 2$.

5.5.5. Lemma.

(i) *If* $\lambda = 0$ *and* $a_2 > 0$, *then* $a_2 = \mu = 1$, *or* $a_2 \geqslant \mu + 1$.

(ii) *If singular lines have size* $\lambda+2$, *then* $a_2 \geqslant \mu\lambda$; *in particular, if* $\lambda = 1$, *then* $a_2 \geqslant \mu$.

(iii) *If* $\lambda \geqslant 1$ *and* $a_2 > 0$, *then* $a_2 \geqslant \mu - \lambda + 1$.

(iv) *If* $d \geqslant 4$ *and* $\lambda > 0$, *then* $a_2 \geqslant \mu$.

Proof. (ii) is just a repetition of Proposition 4.3.3, and (iv) will follow from Proposition 5.5.6 (i) below (since $d \geqslant 4$ implies $b_2 \geqslant c_2$).
Let $d(\gamma,\delta) = 2$, and put $C = \Gamma(\gamma) \cap \Gamma_2(\delta)$ and $D = \Gamma(\delta) \cap \Gamma_2(\gamma)$, so that $|C| = |D| = a_2$.
(i) If $\lambda = 0$ and $\epsilon \in D$, then $\gamma^\perp \cap \epsilon^\perp \subseteq C$ so that $\mu \leqslant a_2$. If $\mu = a_2$, then each point of C is adjacent to each point of D, but since two vertices in D must be nonadjacent (since $\lambda = 0$) and also have δ as a common neighbour, we find a contradiction unless $a_2 = \mu = 1$. This proves (i).
(iii) If $\epsilon \in D$, then $\gamma^\perp \cap \epsilon^\perp \subseteq C \cup A$, where $A = \gamma^\perp \cap \delta^\perp$. It follows that $\mu \leqslant a_2 + \lambda$. Suppose equality holds. If $d = 2$ then by Theorem 1.3.1 the eigenvalues r, s are the roots of $x^2 + a_2 x - a_2 = 0$, and hence are irrational. Now we have the pentagon. If $d > 2$ and $\lambda > 0$, then we can find a $K_{1,1,2}$ on the points $\alpha, \beta, \delta, \epsilon$ with $d(\alpha,\beta) = 2$ (otherwise (ii) applies) and choose $\gamma \in \alpha^\perp \cap \Gamma_3(\beta)$. For this choice of γ, δ, ϵ we see that $\gamma^\perp \cap \epsilon^\perp$ has at most $\lambda - 1$ points in $A = \gamma^\perp \cap \delta^\perp$. This proves (iii). □

Remarks. Part (i) is sharp: the pentagon and the dodecahedron have $a_2 = \mu = 1$ and $\lambda = 0$, and the folded 5-cube has $\mu = 2$, $a_2 = 3$. There are two other feasible arrays with $a_2 = \mu = 1$ and $\lambda = 0$, namely $\{5,4,3;1,1,2\}$ ($v = 56$) and $\{5,4,3,3;1,1,1,2\}$ ($v = 176$), but no corresponding graphs are known. Also (ii) is sharp, as is witnessed by any near polygon. The triangular graph $T(5)$ has equality in (iii). In case $d = 3$, probably the extremal case is given by the Taylor graphs; for example, the Gosset graph has $\lambda = a_2 = 10$, $\mu = 16$.

Part (iii) is slightly stronger than the case $i = 2$ of the inequality

$$a_{i-1} + a_i \geqslant \lambda + \mu - \frac{\lambda\mu}{c_i} \quad \text{if } a_i \neq 0. \tag{4}$$

(Proof of (4): let $\delta\epsilon$ be an edge in $\Gamma_i(\gamma)$, then the points in $\Gamma_{i-1}(\gamma) \cap \delta^\perp$ have distance 1 or 2 to ϵ, so that $c_i \leqslant \lambda + p^i_{i-1,2} = \lambda + c_i(a_{i-1} + a_i - \lambda)/\mu$.)

This inequality is often stronger than that of Lemma 5.5.1 (ii).

BIGGS [75] gives in his feasibility condition V results that are slightly weaker than (i) and (ii) above, and also the restriction: if $\mu = 1$ and $a_2 = 2$ then $3 \mid k$.

Problem. Is it true that if $d \geqslant 3$ and $a_2 > 0$ then $\mu \leqslant a_2 + \lambda/2$ except for the Gosset graph?

Terwilliger gave inequalities that are usually much stronger. An example is:

5.5.6. Proposition (Cf. TERWILLIGER [722], formulas (14) and (26)). *Suppose* $\lambda \neq 0$.

(i) *For* $i = 1, \cdots, d-1$ *we have*

$$a_i \geqslant \min(b_i, c_i),$$

$$a_i(\lambda + 1 - a_i) \leqslant \lambda \cdot \max(b_i, c_i).$$

(ii) *If* $a_i = \min(b_i, c_i)$, *then for every vertex* α *at distance* i *to some vertex of a triangle* $\beta\gamma\delta$, *if* $b_i > c_i$, *the three distances of* α *to the vertices of the triangle are* $i, i, i-1$ *or* $i+1, i+1, i$, *and if* $b_i < c_i$ *they are* $i-1, i-1, i$ *or* $i, i, i+1$.

Proof. For $\alpha, \beta, \gamma \in \Gamma$ with $d(\alpha,\gamma) = i$, $d(\alpha,\beta) = i-1$, $d(\beta,\gamma) = 1$, define $z_i(\alpha,\beta,\gamma) := |\Gamma_{i-1}(\alpha) \cap \beta^\perp \cap \gamma^\perp|$, and let z_i be the average value of $z_i(\alpha,\beta,\gamma)$ over all such α, β, γ. (There are $vk_ic_i = vk_{i-1}b_{i-1}$ such triples.) We first observe that

$$z_i \leqslant z_{i+1} \quad (1 \leqslant i \leqslant d-1). \tag{5}$$

Indeed, $vk_ic_iz_i$ is the number of 4-sets $\{\alpha,\beta,\gamma,\delta\}$ with α, β, γ as above and $\delta \in \Gamma_{i-1}(\alpha) \cap \beta^\perp \cap \gamma^\perp$. For each such 4-set we find b_i 4-sets $\{\alpha',\beta,\gamma,\delta\}$ where $\alpha' \in \alpha^\perp \cap \Gamma_{i+1}(\gamma)$, and each such 4-set is found at most c_i times, so $vk_ib_iz_{i+1}c_i \geqslant vk_ic_iz_ib_i$.
Also,

$$\lambda(a_i - c_i) \geqslant b_iz_{i+1} - c_iz_i \geqslant \lambda(a_i - c_i) - a_i(a_i - 1). \tag{6}$$

Indeed, the first inequality is $\lambda a_i + c_iz_i \geqslant \lambda c_i + b_iz_{i+1}$, which says that on the average the number of edges in $\Gamma(\gamma)$ with at least one end point in $\Gamma_i(\alpha)$ or contained in $\Gamma_{i-1}(\alpha)$ is at least the number of edges with at least one end point in $\Gamma_{i-1}(\alpha)$ or meeting both $\Gamma_i(\alpha)$ and $\Gamma_{i+1}(\alpha)$, and this is obviously true. The difference between both sides is the average number of edges contained entirely within $\Gamma_i(\alpha)$, which is at most $a_i(a_i - 1)$.
By (6) we have $\lambda(a_i - c_i) \geqslant (b_i - c_i)z_i$. Since $0 \leqslant z_i \leqslant \lambda$, we find $a_i \geqslant \min(b_i,c_i)$. Similarly, the inequality $\lambda(a_i - c_i) \leqslant (b_i - c_i)z_{i+1} + a_i(a_i - 1)$ yields $a_i(\lambda + 1 - a_i) \leqslant \lambda.\max(b_i,c_i)$.
When $a_i = \min(b_i,c_i)$, then $\lambda(a_i - c_i) = b_iz_{i+1} - c_iz_i$ and we find that $\Gamma_i(\alpha)$ has no triangles, and either $b_i = c_i$, $z_{i+1} = z_i$, or $b_i > c_i$, $z_i = z_{i+1} = 0$ or $b_i < c_i$, $z_i = z_{i+1} = \lambda$. □

TERWILLIGER [722] gives a long list of complicated inequalities including (5) and (6); these involve many further parameters, and we do not know parameter sets that are ruled out by them.

5.5.7. Proposition (cf. GARDINER [314]). *Given a non-bipartite distance-regular graph, there are numbers* i_1, i_2 *such that* $a_l \neq 0$ *if and only if* $i_1 \leqslant l \leqslant i_2$. *Moreover,* $i_1 + i_2 \geqslant d$.

Proof. If $a_i \neq 0$, $a_{i+1} = 0$, then by 5.5.4(i) we see that

$$b_i \leqslant a_i = k - b_i - c_i \leqslant k - b_{i+1} - c_i = c_{i+1} - c_i$$

(and in particular that $i \geqslant \frac{1}{2}d$, otherwise $c_{i+1} \leqslant b_i$). Similarly, if $a_j \neq 0$, $a_{j-1} = 0$, then by 5.5.4(ii) we see that $c_j \leqslant a_j \leqslant b_{j-1} - b_j$. But if $j > i$, then $b_i < c_{i+1} \leqslant c_j < b_{j-1} \leqslant b_i$ yields a contradiction. This shows the existence of i_1, i_2. Now $\Gamma_{\geqslant i_2}(x)$ is bipartite, so if $d(\gamma,\delta) = d$, $d(\gamma,\epsilon) = i_2 + 1$, $d(\epsilon,\delta) = d - i_2 - 1$, then $\Gamma(\epsilon) \cap \Gamma_{d-i_2-1}(\delta) = \varnothing$, i.e. $d - i_2 - 1 < i_1$. □

Remarks. (i) This proposition was independently found by FISHER [289].
(ii) The proposition expresses a weak unimodal property of the a_i. The sequence $(a_i)_i$ is not necessarily unimodal. In fact the example on 990 vertices mentioned above has $(a_i) = (0,0,2,2,1,2,2,0,0)$. This example is an antipodal 3-cover; we do not know of primitive examples with non-unimodal (a_i).
(iii) The implication $a_{d-1} = 0 \Rightarrow \lambda = 0$ follows from the stronger inequality $2a_{d-1} \geqslant \lambda$. Similarly, if $d \geqslant 4$, then the implication $a_{d-2} (= a_{d-1} = a_d) = 0 \Rightarrow a_2 = 0$ can be strengthened to

$$a_2 \leqslant a_{d-2} + \min(a_{d-1}, a_d) + \frac{a_{d-1}}{\mu} \min(b_{d-2}, c_d);$$

more complicated expressions bound a_i in terms of $a_{d-i}, \cdots, a_d$.

5.6 Graphs with small k_d

By definition (cf. §4.2), a distance-regular graph Γ is antipodal when Γ_d is a disjoint union of cliques, i.e., when $p^j_{dd} = 0$ for $j \neq 0, d$. By Theorem 4.2.1 this can be expressed in the b_j and c_j as $b_j = c_{d-j}$ $(0 \leqslant j \leqslant d-1,\ j \neq [d/2])$. We shall show various propositions to the effect that if k_d is sufficiently small, then Γ must be antipodal. Trivially, if $k_d = 1$, then Γ is an antipodal 2-cover (and conversely). More generally, an antipodal r-cover has $k_d = r-1$ which is usually rather small compared with k. On the other hand, if Γ is not antipodal, then k_d cannot be too small:

5.6.1. Proposition. *If the distance-regular graph* Γ *is not antipodal, then*

$$k \leqslant k_d \cdot \max(a_d, k_d - 1 - a_d) \leqslant k_d(k_d - 1). \tag{7}$$

Moreover, at least one of the following holds:

(i) $k_2 \leqslant k_d\,(k_d - 1 - a_d)$.

(ii) $k \leqslant a_d \cdot \min(k_d, \lambda + 2 - a_d)$.

(iii) $k/(\lambda+1)$ *is an integer* $\leqslant k_d$, *and* $a_d = \lambda + 1$, $b_{d-1} = 1$.

Proof. We first note that $k_d = \sum_i p^d_{id}$ whence

$$p^d_{id} \leqslant k_d - p^d_{0d} - p^d_{1d} = k_d - 1 - a_d \quad \text{for } i > 1. \tag{8}$$

By assumption, the set I of distances $i \neq 0, d$ realized in $\Gamma_d(\gamma)$ for some $\gamma \in \Gamma$ is nonempty. For every $i \in I$ we have $k_i \leqslant k_i p^i_{dd} = k_d p^d_{id}$, and therefore

$$k \leqslant k_d a_d \qquad \text{if } 1 \in I, \tag{9}$$

$$k_i \leqslant k_d(k_d - 1 - a_d) \ \text{ if } 1 \neq i \in I. \tag{10}$$

Since $k \leqslant k_i$ for $i \neq 0,d$, this implies (7).

Suppose that (i) is violated. Then (10) implies $2 \notin I$, whence $p^d_{2d} = 0$. Lemma 4.1.7 gives

$$0 = \mu p^d_{2d} = c_d b_{d-1} + a_d(a_d - \lambda) - k = c_d(b_{d-1} - 1) + a_d(a_d - \lambda - 1). \tag{11}$$

If $b_{d-1} > 1$, then this implies $a_d \neq 0$ and $k = c_d + a_d \leqslant c_d(b_{d-1} - 1) + a_d \leqslant a_d(\lambda + 2 - a_d)$. Since $a_d \neq 0$ we have $1 \in I$, and together with (9) we find (ii). If $b_{d-1} = 1$, then (11) implies $a_d \in \{0, \lambda+1\}$. If $a_d = \lambda + 1$, then by Proposition 4.3.2, all singular lines have size $\lambda+2$, and hence $k/(\lambda+1)$ is an integer. By (9), this integer is $\leqslant k_d$, and (iii) holds. On the other hand, if $a_d = 0$, then $k_d(k_d - 1) < k_2 \leqslant k(k-1)$ since (i) is violated, whence $k_d < k$ and $k_{d-1} = k_{d-1}b_{d-1} = k_d c_d = k_d k > k_d^2 > k_d(k_d - 1 - a_d)$. Now $k_i \geqslant \min(k_2, k_{d-1}) > k_d(k_d - 1 - a_d)$ for all $i \neq 0, 1, d$; hence (9) and (10) imply that $I = \varnothing$, contradiction. Hence no further case arises. □

Example (cf. SMITH [683]). There is no bipartite distance-regular graph with intersection array $\{4,3,3,2,1; 1,1,2,3,4\}$. (Here $d = 5$ and $k_5 = 3$.)

Remark (J.T.M. van Bon). More generally, a conference graph cannot be the halved graph of a bipartite distance-regular graph Γ. For otherwise $\Gamma(\gamma)$ would be a clique in a bipartite half of Γ; applying the Hoffman bound we then find $\mu = 1$, so that Γ has intersection array $\{k, k-1, k-1, \frac{1}{2}k, 1; 1, 1, \frac{1}{2}k, k-1, k\}$, violating the above proposition. Also, a conference graph does not have distance-regular antipodal r-covers for $r > 1$, except for the pentagon, which is covered by the decagon.

5.6.2. Corollary. *If the distance-regular graph Γ is not antipodal, then*

(i) *If $\lambda^2 < 4b_1$ then $k_2 \leqslant k_d(k_d - 1)$ or $b_{d-1} = 1$.*

(ii) *If $\lambda = 0$ then $k_2 \leqslant k_d(k_d - 1)$.*

(iii) *If $\lambda = 0$ and $\mu = 1$, then $k \leqslant k_d$.*

Proof. (i) If $\lambda^2 < 4b_1$, then $4k > (\lambda+2)^2 \geqslant 4a_d(\lambda + 2 - a_d)$ so that case (ii) of the proposition cannot occur. If $\lambda = 0$, then case (iii) of the proposition implies $k_2 \leqslant k(k-1) \leqslant k_d(k_d - 1)$, giving (ii). Finally, if $\lambda = 0$ and $\mu = 1$, then $k_2 = k(k-1)$ and (iii) follows from (ii). □

Remarks. Part (iii) is due to YOSHIZAWA [798]. The regular thin near octagon (B10) and the incidence graph of the complements of projective planes satisfy the inequality in (ii) with equality. The halved $(2d+1)$-cubes have $b_{d-1} = 3$, $k_d = 2d+1$, $k_2 = \binom{2d+1}{4} > k_d(k_d - 1)$ for $d > 3$; so the inequality in (i) is not valid without the assumption $\lambda^2 < 4b_1$.

The extremal graphs in Proposition 5.6.1 are precisely the distance 2 graphs of generalized Odd graphs (cf. §4.2D):

5.6.3. Proposition. *Let Γ be distance-regular of diameter $d \geqslant 3$, but not antipodal. Then $k = k_d(k_d - 1)$ if and only if $k_d = a_d + 1$, i.e., if and only if Γ_d is a generalized Odd graph.*

Proof. Suppose that $k = k_d(k_d - 1)$. If $a_d = 0$, then $k_2 \leqslant k_d(k_d - 1)$ by the second part of Proposition 5.6.1, and $k_2 = k$ since $k_2 \geqslant k$. By Proposition 5.1.1 (v), Γ must be a polygon with an odd number of vertices as Γ is not antipodal, and $k_d = a_d + 1 = 2$. If $a_d \neq 0$, then (7) implies $a_d = k_d - 1$. Conversely, suppose that $a_d = k_d - 1 \neq 0$. (If $k_d = 1$ then Γ is antipodal.) Then $p^1_{dd} = 1$. Hence for fixed $\delta \in \Gamma$, the $k_d(k_d - 1)$ vertices contained in the union of the sets $\Gamma_d(\gamma) \cap \Gamma(\delta)$ for $\gamma \in \Gamma_d(\delta)$ are distinct; therefore $k \geqslant k_d(k_d - 1)$, and by (7) we have equality. □

The following inequality sometimes rules out intersection arrays with too small a k_d:

5.6.4. Lemma. *If $\lambda = 0$ then $k_i \geqslant 2a_i$ $(i \leqslant d)$.*

Proof. Apply Turán's theorem (TURÁN [756], cf. e.g. BOLLOBÁS [97], §8) to $\Gamma_i(\gamma)$. □

For example, there is no distance-regular graph with intersection array $\{8,7,2; 1,4,4\}$. Similarly, one has inequalities like

$$k_i \geqslant 1 + a_i + a_i(a_i - \lambda - 1)/\mu.$$

The following bound is useful in connection with Corollary 5.1.3.

5.6.5. Lemma. *If $c_i < k$, then $k_i \geqslant \max(1 + k - c_i, k/(k - c_i))$.*

Proof. If $i < d$ then $k_i \geqslant k$, so the inequality holds. If $i = d$ then $k_i + c_i \geqslant 1 + a_i + c_i = 1 + k$ gives $k_i \geqslant 1 + k - c_i$, and $a_i = k - c_i \neq 0$ implies $1 \leqslant p^1_{ii} = k_i a_i / k = k_i(k - c_i)/k$, giving $k_i \geqslant k/(k - c_i)$. □

5.7 The case $p^2_{dd} = 0$

Antipodal graphs have $b_{d-1} = 1$, $a_d = 0$, and consequently $p^2_{dd} = 0$. Let us survey the known nonantipodal graphs with $p^2_{dd} = 0$. First of all we have the graphs where $\Gamma_d(\gamma)$ is a clique, i.e., those with $k_d = a_d + 1$. These are just the distance-2 graphs of generalized Odd graphs, see §4.2D. The known examples are:

(i) The Johnson graph $J(2m+1, m)$ has $k = m(m+1)$, $k_d = m+1$, $a_d = m$, $b_{d-1} = 2$, $d = m$, $\lambda = 2m - 1$.

(ii) The halved Hamming graph $\frac{1}{2}\cdot 2^{2m+1}$ has $k = m(2m+1)$, $k_d = 2m+1$, $a_d = 2m$, $b_{d-1} = 3$, $d = m$, $\lambda = 4m - 2$.

(iii) The $2^{10}.M_{22}$-graph Γ with $v = 1024$ and $\iota(\Gamma) = \{231,160,6;1,48,210\}$ (cf. §11.3F) has $d = 3$, $k = 231$, $k_3 = 22$, $a_3 = 21$, $b_2 = 6$, $\lambda = 70$.

(iv) The $2^{11}.M_{23}$-graph Γ with $v = 2048$ and $\iota(\Gamma) = \{253,210,3;1,30,231\}$ (cf. §11.3E) has $d = 3$, $k = 253$, $k_3 = 23$, $a_3 = 22$, $b_2 = 3$, $\lambda = 42$.

All other known graphs with $p^2_{dd} = 0$ have $b_{d-1} = 1$. They are

(i) The Biggs-Smith graph (cf. §13.4) has $d = 7$, $k = 3$, $k_7 = 8$, $p^i_{77} = 0$ for $i \in \{1,2,4,6\}$.

(ii) The Livingstone graph Γ (cf. §13.5) has $d = 4$, $k = 11$, $k_4 = 12$, $p^i_{44} = 0$ for $i \in \{1,2,4\}$ (i.e., $\Gamma_4(\gamma)$ is a clique in Γ_3).

(iii) The Coxeter graph Γ (cf. §12.3) has $d = 4$, $k = 3$, $k_4 = 6$ and for each $\gamma \in \Gamma$, $\Gamma_4(\gamma)$ is isomorphic to $3K_2$.

Proposition 5.7.1 below shows that graphs with ($d \geqslant 3$ and) $b_{d-1} = 1$ and $k_d = a_d + 1$ correspond to regular thin near $(2d+1)$-gons with ($a_1 = a_2 = \cdots = a_{d-1} = 0$, $a_d > 0$ and) $c_2 = c_3 = 1$.
For $d = 3$ these are Moore graphs, and the only example is the heptagon. For $d = 4$ one knows that $c_4 \geqslant 3$ unless we have the enneagon (Roos & van Zanten [628]). Otherwise almost nothing is known.

5.7.1. Proposition. *Assume $p^2_{dd} = 0$ (i.e., $a_d(\lambda - a_d + 1) = c_d(b_{d-1} - 1)$). Then we have for any vertex $\infty \in \Gamma$:*

(i) *$\Gamma_d(\infty)$ is a union of cliques of size $a_d + 1$ so that $(a_d + 1) | k_d$ and $a_d \leqslant \lambda + 1$. Also, $\Gamma_d(\infty) \cap \Gamma(\gamma)$ is a clique of size b_{d-1} for $\gamma \in \Gamma_{d-1}(\infty)$, so that $b_{d-1} \leqslant a_d + 1$.*

(ii) *If $b_{d-1} = a_d + 1$, then $b_{d-1} = 1$, $a_d = 0$.*

(iii) *$b_{d-1} \leqslant \mu$.*

(iv) *If $b_{d-1} > 1$, then $a_d \leqslant \lambda - 1$. If $b_{d-1} = 1$ and $a_d > 0$, then Γ is the collinearity graph of a partial linear space with lines of size $a_d + 1 = \lambda + 2$ (and Proposition 4.3.3 applies).*

Proof. (i) is obvious. If $b_{d-1} = a_d + 1$ and $a_d > 0$, then any two vertices in a component of $\Gamma_d(\infty)$ have precisely the same neighbours, and $\lambda \geqslant k - 1$, i.e., Γ is a clique - contradiction. So either $b_{d-1} = 1$ or we find b_{d-1} points in some μ-graph, proving (iii). Assume $b_{d-1} > 1$, then $a_d \leqslant \lambda$. If $a_d = \lambda$ then (since $p^2_{dd} = 0$) $a_d = c_d(b_{d-1} - 1)$; if $b_{d-1} \geqslant 3$, then $\lambda = a_d \geqslant 2c_d$ so $\lambda \geqslant \frac{2}{3}k$ and $\frac{1}{3}k + 3 \leqslant \mu \leqslant c_d$ by Theorem 1.2.3, contradiction; hence $b_{d-1} = 2$, $a_d = c_d = \lambda = k/2$, and $\mu \geqslant 3$; since $\mu p^{d-1}_{d2} = b_{d-1}(a_{d-1} + a_d - \lambda)$, we find if $\mu \geqslant 4$ that $2a_{d-1} \geqslant 4(a_d + 1 - b_{d-1})$ and $a_{d-1} \geqslant k - 2$, impossible; thus $\mu = 3$ and Γ is either the icosahedron or a line graph of girth at least five; but the icosahedron has $\mu = 2$ and a line graph of girth at least five has $\mu = 1$, a contradiction in each case. Thus $a_d \leqslant \lambda - 1$. The rest is clear. □

Remark. Only very few primitive graphs are known in which $b_j = 1$ for some j (and hence $b_i = 1$ for $d > i \geqslant j$ by Proposition 4.1.6): the Coxeter graph (D9), the Livingstone graph (D12), and the Biggs-Smith graph (D14). No primitive graph is known for which $b_2 = 1$. In general we have the result of Lemma 5.3.1. No examples are known of graphs Γ with $p^2_{dd} = 0$ such that $b_{d-1} > 1$ and $\Gamma_d(\gamma)$ is a union of (several) cliques, i.e., $k_d > a_d + 1$, or such that $b_{d-1} = 1$ and $k_d = a_d + 1 > 1$.

Some immediate questions arise.
Problems. (i) Does $b_2 = 1$ imply that Γ is antipodal? (If $d \geqslant 5$ and $b_2 = 1$ then we have seen after Corollary 4.3.12 that Γ is the dodecahedron.)
(ii) Does $b_i = 1$, $\lambda > 0$ imply that Γ is antipodal?
(iii) Does $b_i > 1$ imply $c_i < b_1$?
(iv) If $p^2_{dd} = 0$, then does $b_i > 1$ imply $k_d = a_d + 1$? And does $b_i = 1$, $k_d = a_d + 1$ imply $k_d = 1$?

5.8 A lower bound for p^2_{22}

NEUMAIER [556] gave (in his Lemma 1.6) an inequality involving the size of the intersection of 'regular cliques' in an edge-regular graph. As a corollary he derived inequalities on the parameters of a partial λ-geometry. Here we generalize these ideas and results.

Let α, β be two vertices at distance two in an amply regular graph Γ with parameters (v,k,λ,μ), and let γ run over the $p = p^2_{22}(\alpha,\beta)$ vertices at distance two from both α and β. For such γ, put $n_\gamma = |\alpha^\perp \cap \beta^\perp \cap \gamma^\perp|$. By obvious counting arguments we find

$$\sum_\gamma 1 = p,$$

$$\sum_\gamma n_\gamma \geqslant \mu(k-2-2\lambda),$$

$$\sum_\gamma n_\gamma(n_\gamma - 1) \leqslant \mu(\mu-1)(\max(\lambda,\mu)-2)$$

and hence for all integral $i \geqslant 0$:

$$0 \leqslant \sum_\gamma (n_\gamma - i)(n_\gamma - i - 1) \leqslant \mu(\mu-1)(\max(\lambda,\mu)-2) - 2i\mu(k-2-2\lambda) + i(i+1)p. \quad (12)$$

and (slightly weaker) for all real $i \geqslant 0.5$:

$$0 \leqslant \sum_\gamma (n_\gamma - i)^2 \leqslant \mu(\mu-1)(\max(\lambda,\mu)-2) - (2i-1)\mu(k-2-2\lambda) + i^2 p.$$

The latter inequality proves the following proposition.

5.8.1. Proposition. *Let Γ be amply regular with parameters (v,k,λ,μ), where $k > 2(\lambda+1)$, and let $p = \min_{d(\alpha,\beta)=2} p^2_{22}(\alpha,\beta)$. Then*

$$p\left(1 + \frac{(\mu-1)(\max(\lambda,\mu)-2)}{k-2-2\lambda}\right) \geqslant \mu(k-2-2\lambda). \quad (13)$$

When equality holds, p divides $\mu(k-2-2\lambda)$. □

This bound tends to be strong when $\lambda = 0$ and μ is not too large. Many parameter sets for partial λ-geometries (bipartite distance-regular graphs of diameter 4) are ruled out. For example, the intersection arrays $\{2\mu, 2\mu-1, \mu, 1; 1, \mu, 2\mu-1, 2\mu\}$ (belonging to Hadamard graphs, see §1.8) violate (12) for $i = \frac{1}{2}(\mu-1) > 0$; thus, either $\mu = 1$ or μ is even, and we have reproved Corollary 1.8.2. A non-bipartite example is the intersection array $\{21,20,18,6; 1,3,6,12\}$ ($\lambda = 0$, $p^2_{22} = 49$) which is ruled out by (13) (but also fails the Krein conditions).

5.8.2. Corollary. *Let Γ be distance-regular of diameter $d \geqslant 3$ with $\lambda = a_2 = 0$, $k > 2$. Then $k \geqslant (\mu-1)(\frac{\mu}{\alpha} - \mu + 2) + 2$, where $\alpha = (c_3-1)/(k-2)$.*

Proof. We have $p = p^2_{22} = (\mu(k-2) + (k-\mu)(c_3-1))/\mu = (k-2)(1+\alpha(k-\mu)/\mu)$, and the inequality (13) becomes

$$\left(1+\alpha\frac{k-\mu}{\mu}\right)\left(1+\frac{(\mu-1)(\mu-2)}{k-2}\right) \geqslant \mu$$

since $k > 2$. Multiplying by $k-2$ we find a quadratic inequality for k (considering μ and α fixed) with solution $k \leqslant \mu$ or $k \geqslant (\mu-1)(\frac{\mu}{\alpha} - \mu + 2) + 2$. □

5.8.3. Corollary. *Let Γ be a distance-regular bipartite antipodal 2-cover of diameter 6. Then Γ is either the 12-gon or the 6-cube.*

Proof. Γ has eigenvalue 0 with multiplicity $2m$, where $m = \frac{(k-1)(k^2-k+\mu)}{k(\mu+1)-\mu}$ is the multiplicity of the eigenvalue $-k/\mu$ in its halved graph. Since $\mu < k$ and m is integral, we must have $k = \mu(\mu+1)$ and Γ has parameters $(c_i)_i = (1, \mu, \frac{1}{2}\mu(\mu+1), \mu^2, \mu^2+\mu-1, \mu^2+\mu)$ (belonging to antipodal 2-covers of the incidence graphs of square 2-designs with parameters 2-$(\mu^2(\mu+2), \mu(\mu+1), \mu)$; such designs are known for prime powers μ). In this case we have $p^2_{22} = (\mu+2)(k-2)/2$, hence equality holds in (13) and we find for $k > 2$ that $(\mu+2) \mid 2\mu$, so that $\mu = 2$ and we have the 6-cube. □

5.9 Ivanov-Ivanov Theory

Only three distance-regular graphs with $d \geqslant 2k$ are known: Tutte's 12-cage ((G3) for $q = 2$), the Biggs-Smith graph (D11), and the Foster graph (B13), all of valency 3. The explanation of this fact, and the construction of a tight diameter bound, is one of the main open problems about distance-regular graphs. This problem may be approached in two steps, first bounding the diameter d in terms of the valency k and the (numerical) girth g, and next bounding g (either absolutely or in terms of k).

Concerning the first step, the following has been done. Terwilliger [713] gives a bound on d in terms of k and g for bipartite distance-regular graphs (see Theorem 5.2.5 (3c)). For graphs containing a quadrangle, a tight bound on d was given by Terwilliger, cf. Corollary 5.2.2. In the general case, some progress has been made by A.A. Ivanov and his colleagues (see A.A. Ivanov [430] and A.A. Ivanov, A.V. Ivanov & I.A. Faradjev [282]). They develop a theory enabling one to obtain strong restrictions on the intersection array of a distance-regular graph of diameter d when it is known that for some j $(1 \leqslant j < d)$ we have $(c_j, a_j, b_j) = (c_{j+1}, a_{j+1}, b_{j+1})$. Since $c_j - b_j$ increases monotonically with j, from $-k$ to at most k (and since $c_j - b_j = c_{j+1} - b_{j+1}$ implies $(c_j, a_j, b_j) = (c_{j+1}, a_{j+1}, b_{j+1})$), there are at most $2k$ distinct triples (c_j, a_j, b_j), so their theory leads to information on graphs with large diameter, say with $d \geqslant 2k$. In particular, the diameter is shown to be bounded in terms of valency and numerical girth (cf. Theorem 5.9.8).
A typical result is
Theorem (A.A. Ivanov [430]). *If Γ is a distance-regular graph of numerical girth g, then $d \leqslant g \cdot 4^{k-1}$.*
In this section we shall look at this theory in some detail. Most of what follows is taken from the second paper cited above. (A.A. Ivanov informed us that 5.9.1-5.9.5 and 5.9.9 are due to A.V. Ivanov.)
Concerning the second step, Bannai & Ito [35, 38, 36, 40, 41, 37] work on the more general project of showing that the diameter is bounded in terms of the valency alone. While this section is mostly of a combinatorial nature, Bannai & Ito proceed analytically, and study the location of the eigenvalues of the tridiagonal matrix L_1. A partial result is for example:

Theorem (BANNAI & ITO [36]). *There exists a function f such that bipartite distance-regular graphs Γ of valency $k > 2$ have diameter at most $f(k)$.*

For general Γ they show that the interval of i's for which k_i is maximal is bounded in length:

Theorem (BANNAI & ITO [35]). *For a distance-regular graph Γ that is not a polygon, the number of i such that $c_i = b_i$ is at most $10k2^k$.*

For the remainder of this section, let Γ be a distance-regular graph of diameter d. For two vertices γ, δ of Γ, define $A^\gamma(\delta)$ to be the vector of length k indexed by the neighbours of γ and with entries $A^\gamma(\delta)_\alpha = d(\delta,\alpha)$ for $\alpha \sim \gamma$. Clearly, if $d(\gamma,\delta) = i$ then $A^\gamma(\delta)$ has c_i entries $i-1$, a_i entries i and b_i entries $i+1$. Also define $A(\gamma,\delta)$ to be the $k \times k$ matrix with rows indexed by neighbours of γ and columns indexed by the neighbours of δ with entries $A(\gamma,\delta)_{\zeta,\eta} = d(\zeta,\eta)$ for $\zeta \sim \gamma$, $\eta \sim \delta$. This matrix is naturally partitioned into 9 submatrices (assume $d(\gamma,\delta) = i$):

c_i	a_i	b_i		
A_{11} $i-2, i-1, i$	A_{12} $i-1$ or i	A_{13} i	c_i	Each row has entries $(i-2)^{c_{i-1}}(i-1)^{a_{i-1}}i^{b_{i-1}}$.
A_{21} $i-1$ or i	A_{22} $i-1, i, i+1$	A_{23} i or $i+1$	a_i	Each row has entries $(i-1)^{c_i}i^{a_i}(i+1)^{b_i}$.
A_{31} i	A_{32} i or $i+1$	A_{33} $i, i+1, i+2$	b_i	Each row has entries $i^{c_{i+1}}(i+1)^{a_{i+1}}(i+2)^{b_{i+1}}$.

Clearly, A_{13} and A_{31} have only entries i; A_{12} and A_{21} only entries i or $i-1$; and A_{23} and A_{32} only entries i or $i+1$.

5.9.1. Lemma. *Let $\zeta, \eta \in \Gamma_i(\gamma)$, and $\zeta \sim \eta$. Then:*

(i) *If $(c_{i-1},a_{i-1},b_{i-1}) = (c_i,a_i,b_i)$, then the vectors $A^\gamma(\zeta)$ and $A^\gamma(\eta)$ have entries $i-1$ in the same positions.*

(ii) *If $(c_{i+1},a_{i+1},b_{i+1}) = (c_i,a_i,b_i)$, then the vectors $A^\gamma(\zeta)$ and $A^\gamma(\eta)$ have entries $i+1$ in the same positions.*

(iii) *If $(c_{i-1},a_{i-1},b_{i-1}) = (c_{i+1},a_{i+1},b_{i+1})$, then $A^\gamma(\zeta) = A^\gamma(\eta)$.*

Proof. Consider $A(\gamma,\zeta)$. □

Let $d_m = d_m(i)$ be the minimal distance $d(\zeta,\eta)$ such that for some γ we have $\zeta, \eta \in \Gamma_i(\gamma)$ and $A^\gamma(\zeta) \neq A^\gamma(\eta)$. The following lemma will imply that usually $d_m \leqslant 2$.

5.9.2. Lemma. *If $i < d$ and $d_m > 2$, then $b_i = b_{i-1}$, and either $c_i = c_{i-1}$ or $c_i \geqslant c_{i-1} + b_{i-1}$.*

Proof. Choose vertices γ,α with $d(\gamma,\alpha) = i-1$ and a vertex $\delta \in \Gamma(\gamma)$ with $d(\delta,\alpha) = i$. Since $d > i$ there is a vertex $\eta \sim \alpha$ with $d(\delta,\eta) = i+1$. Since $d_m > 2$, any other vertex $\zeta \in \Gamma_i(\gamma) \cap \Gamma(\alpha)$ has $A^\gamma(\zeta) = A^\gamma(\eta)$ and in particular $d(\delta,\zeta) = i+1$ so that

$b_i = b_{i-1}$. If moreover $(c_{i-1},a_{i-1},b_{i-1}) \neq (c_i,a_i,b_i)$, then $c_i > c_{i-1}$. Now consider the matrix $A(\gamma,\alpha)$:

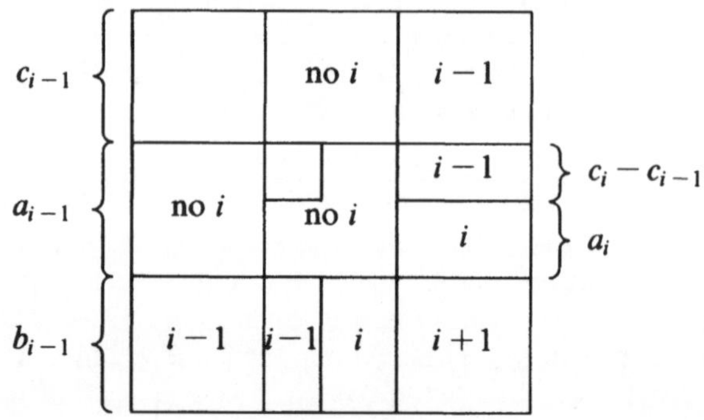

Interchanging the rôles of γ and α we find that the b_{i-1} entries i in each of the $c_i - c_{i-1}$ rows of $A(\gamma,\alpha)$ must be situated in one of the $c_i - c_{i-1}$ columns of $A(\gamma,\alpha)$, so that $b_{i-1} \leqslant c_i - c_{i-1}$. □

5.9.3. Lemma. *Assume $d_m \leqslant 2$ and $(c_i,a_i,b_i) = (c_{i+1},a_{i+1},b_{i+1})$. Then there are points γ,ζ,η such that $\zeta,\eta \in \Gamma_i(\gamma)$ and $d(\zeta,\eta) = d_m$ and for some $\delta \in \Gamma(\gamma)$ we have $A^\gamma(\zeta)_\delta = i-1$, $A^\gamma(\eta)_\delta \geqslant i$, i.e., $d(\zeta,\delta) = i-1$ and $d(\eta,\delta) \geqslant i$.*

Proof. If $d_m = 1$, then the result follows from Lemma 5.9.1 (ii). So assume $d_m = 2$. If $\zeta \sim \alpha \sim \eta$, then $\alpha \in \Gamma_{i-1}(\gamma)$ (not $\alpha \in \Gamma_i(\gamma)$, otherwise $d_m = 1$; and not $\alpha \in \Gamma_{i+1}(\gamma)$, otherwise $A^\gamma(\zeta) = A^\gamma(\alpha) - 1 = A^\gamma(\eta)$). Let $\theta \in \Gamma(\alpha) \cap \Gamma_i(\gamma)$ and consider the matrix $A(\gamma,\theta)$:

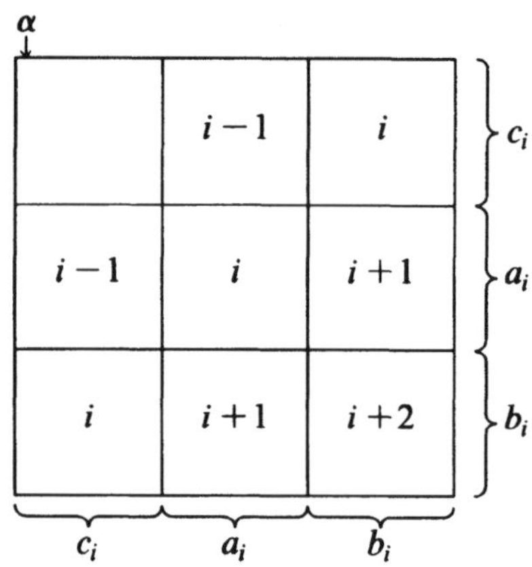

Since $(c_i,a_i,b_i) = (c_{i+1},a_{i+1},b_{i+1})$ the submatrices A_{3j} and A_{j3} ($j = 1,2,3$) are constant as shown. Since $d_m = 2$, the submatrices A_{j2} ($j = 1,2,3$) are constant as shown. We find that, if $A^\gamma(\alpha)_\delta = i$, then $A^\gamma(\theta)_\delta \neq i$ (for $\delta \sim \gamma$). Next also consider the matrix $A(\gamma,\alpha)$:

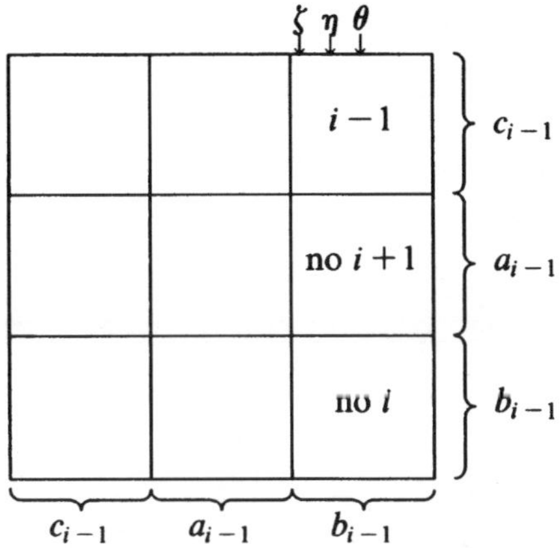

Now, if $A^\gamma(\zeta)$ and $A^\gamma(\eta)$ have the same entries $i-1$, then they coincide, contradiction. □

5.9.4. Lemma. *Assume that $(c_i,a_i,b_i) = (c_{i+j},a_{i+j},b_{i+j})$ for some $j \geqslant 1$, and that $d_m \leqslant 2$. Let $\gamma, \zeta, \eta, \delta$ be chosen as in the previous lemma, and choose $\theta \in \Gamma_j(\eta) \cap \Gamma_{i+j}(\gamma)$. Then $d(\theta,\zeta) = j+d_m$. Moreover, if $d_m = 1$, then $\Gamma(\zeta) \cap \Gamma_j(\theta) \subseteq \Gamma_i(\gamma) \cap \Gamma_i(\delta)$ (and hence $c_{j+1} \leqslant a_i$ and $c_{j+1} \leqslant b_{i-1} - b_i$).*

Proof. Clearly $d(\theta,\zeta) \leqslant j+d_m$. But if $d(\theta,\zeta) < j+d_m$ then $d_m = 2$, $d(\theta,\zeta) = j+1$, and in a shortest path $\theta_0 = \zeta, \theta_1, \cdots, \theta_{j+1} = \theta$ there is precisely one edge $\{\theta_h,\theta_{h+1}\}$ that does not increase the distance to γ. Now $A^\gamma(\theta_h) = A^\gamma(\zeta)+h$ and $A^\gamma(\theta_{h+1}) = A^\gamma(\theta)-(j-h) = A^\gamma(\eta)+h$ so $A^\gamma(\theta_h) \neq A^\gamma(\theta_{h+1})$. If $h = 0$, this contradicts $d_m = 2$, and if $0 < h < j$, this contradicts Lemma 5.9.1 (iii), and if $h = j$, this contradicts Lemma 5.9.1 (i) and the choice of ζ, η. The same argument proves the 'moreover' part of the statement. □

5.9.5. Theorem. *Assume $(c_{i-1},a_{i-1},b_{i-1}) \neq (c_i,a_i,b_i) = (c_{i+j},a_{i+j},b_{i+j})$ with $i, j \geqslant 1$ (so that $d \geqslant i+j+1$).*

(i) *If $d_m > 2$, then $b_i = b_{i-1}$ and $c_i \geqslant c_{i-1}+b_{i-1}$ (so that $d < 2i-1$).*

(ii) *If $d_m = 2$, then $c_i \geqslant c_{i-1}+c_{j+1}$ (so that $i = 1$ or $j+1 < i$), $b_{i-1} \geqslant 2$, $a_i \leqslant a_{i-1}$, and if, in addition, $c_i < c_{i-1}+c_{j+2}$, then $c_{i+j+1} > c_i$ and $c_{j+2} \geqslant 2c_{j+1}$.*

(iii) *If $d_m = 1$, then $b_j \geqslant b_i + c_{i-1}$ (so that $i = 1$ or $j < i$), $a_i \geqslant c_{j+1}$, $b_{i-1} \geqslant b_i + c_{j+1}$, and if, in addition, $c_i < c_{i-1}+c_{j+1}$, then $b_j \geqslant b_i+c_i$, and if moreover $b_{j+1} < b_i+c_i$, then $c_{i+j+1} \geqslant c_i+c_{j+1}$.*

Remark. The statement is trivial for $i = 1$.

Proof. If $d_m > 2$, then (i) holds by Lemma 5.9.2. If $d_m = 2$, then (with $\gamma, \zeta, \eta, \delta$ chosen as in Lemma 5.9.3) for a common neighbour α of ζ and η, we have $\alpha \in \Gamma_{i-1}(\gamma)$ so that $b_{i-1} \geqslant 2$. Choose $\theta_1 \in \Gamma_{j-1}(\eta) \cap \Gamma_{i+j-1}(\gamma)$. Then $\Gamma(\zeta) \cap \Gamma_j(\theta_1) \subseteq \Gamma_{i-1}(\gamma)\setminus\Gamma_{i-2}(\delta)$ so that $c_{j+1} \leqslant c_i - c_{i-1}$. Next, $a_i \leqslant a_{i-1}$ follows immediately from $d_m = 2$. Now choose $\theta \in \Gamma_j(\eta) \cap \Gamma_{i+j}(\gamma)$ and assume $c_{j+2} > c_i - c_{i-1}$. Then some shortest path from ζ to θ passes through a vertex $\pi \in \Gamma_{i+j+1}(\gamma)$ and $\delta \in \Gamma(\gamma) \cap \Gamma_{i+j}(\pi)\setminus\Gamma_{i-1}(\eta)$, so that $c_{i+j+1} - c_i \geqslant 1$; also, the c_{j+1} neighbours of ζ on a shortest path to θ passing through π are distinct from the c_{j+1} neighbours of ζ on

such a path passing through a vertex $\theta_1 \sim \theta$, $\theta_1 \in \Gamma_{i+j-1}(\gamma)$, so $c_{j+2} \geq 2c_{j+1}$.
If $d_m = 1$, then $a_i \geq c_{j+1}$ and $b_{i-1} \geq b_i + c_{j+1}$ by Lemma 5.9.4. Take $\gamma, \delta, \zeta, \eta, \theta_1, \theta$ as before. (This time $\zeta \sim \eta$.) Now $(\Gamma(\zeta) \cap \Gamma_{i-2}(\delta)) \cup (\Gamma(\zeta) \cap \Gamma_{i+1}(\gamma)) \subseteq \Gamma_{j+1}(\theta_1)$ so that $c_{i-1} + b_i \leq b_j$. The c_{j+1} vectors $A^\gamma(\rho)$ with $\rho \in \Gamma(\zeta) \cap \Gamma_j(\theta)$ are all equal (to $A^\gamma(\eta)$), and since $A^\gamma(\zeta)$ and $A^\gamma(\eta)$ have the same entries $i+1$ there is a point $\delta_1 \sim \gamma$ such that $d(\delta_1,\zeta) = i$, $d(\delta_1,\rho) = i-1$ for each such vertex ρ. Now the triple (ζ,γ,δ_1) has the same properties as we assumed for (γ,ζ,η); moreover, $A^\zeta(\gamma)$ and $A^\zeta(\delta_1)$ have at most $c_i - c_{j+1}$ common entries $i-1$. Without loss of generality we may assume that (γ,ζ,η) was chosen in such a way that $A^\gamma(\zeta)$ and $A^\gamma(\eta)$ have at most $c_i - c_{j+1}$ common entries $i-1$.
But if there is a point $\sigma \in \Gamma(\zeta) \cap \Gamma_{i-1}(\gamma) \cap \Gamma_{j+1}(\theta)$, then $A^\gamma(\zeta)$ and $A^\gamma(\eta)$ have at least c_{i-1} common entries $i-1$. Thus, if $c_i < c_{i-1} + c_{j+1}$, then there is no such point σ, and $(\Gamma(\zeta) \cap \Gamma_{i-1}(\gamma)) \cup (\Gamma(\zeta) \cap \Gamma_{i+1}(\gamma)) \subseteq \Gamma_{j+1}(\theta_1)$ so that $c_i + b_i \leq b_j$. If the left hand side of the above inclusion also is contained in $\Gamma_{j+2}(\theta)$, then $c_i + b_i \leq b_{j+1}$. Otherwise, there is a geodesic from θ to some point of $\Gamma(\zeta) \cap \Gamma_{i+1}(\gamma)$ over some point $\tau \in \Gamma_{i+j+1}(\gamma)$. But $A^\gamma(\tau)$ has entries $i+j$ in all the at least $2c_i - (c_i - c_{j+1}) = c_i + c_{j+1}$ positions where at least one of $A^\gamma(\zeta)$ or $A^\gamma(\eta)$ has an entry $i-1$. Thus $c_{i+j+1} \geq c_i + c_{j+1}$. □

5.9.6. Corollary. *If $i > 1$ and $(c_{i-1},a_{i-1},b_{i-1}) \neq (c_i,a_i,b_i) = (c_{i+j},a_{i+j},b_{i+j})$, then*

$$j+1 \leq i, \tag{14}$$

$$c_{j+1} \leq \max(c_i - c_{i-1}, b_{i-1} - b_i), \tag{15}$$

$$b_j \geq c_{i-1} + c_{j+1}. \tag{16}$$

In particular, if, in addition, $i \geq j+2$, then

$$k_{j+1} \geq 2k_j. \tag{17}$$

Proof. We apply Proposition 4.1.6 to the various cases of the theorem. In case (i) we have (14) since $c_i > b_{i-1}$ gives $2i-1 > d \geq i+j+1$, and (15) since $c_{j+1} \leq b_{i-1} \leq c_i - c_{i-1}$. In case (ii) we have (14) since $c_i > c_{j+1}$, and (15) since $c_{j+1} \leq c_i - c_{i-1}$. Since $b_j \geq c_i \geq c_{i-1} + c_{j+1}$, we also have (16) in these cases. In case (iii) we have (14) since $b_j > b_i$, (15) since $c_{j+1} \leq b_{i-1} - b_i$, and (16) since $b_j \geq c_{i-1} + b_i \geq c_{i-1} + c_{j+1}$. Finally, if $i \geq j+2$, then (16) implies $b_j \geq 2c_{j+1}$, giving (17). □

5.9.7. Corollary. *If $k > 2$, then $k_{i+1} > k_i$ for all $i < d/3$.*

Proof. Suppose that $k > 2$, but $k_{l+1} \leq k_l$ for some $l < d/3$. We may assume that l is the smallest such index. Proposition 5.1.1 implies $k_l = k_{l+1} = \cdots = k_{d-l}$ and $b_l = b_{l+1} = \cdots = b_{d-l-1} = c_{l+1} = c_{l+2} = \cdots = c_{d-l}$.
CASE 1. $c_l < c_{l+1}$. Then the theorem applies with $i = l+1$, $j = d-2l-2$. But (i) is impossible since $c_i = b_{i-1}$, (ii) is impossible since $c_{i+j+1} = c_i$ and $c_i \leq c_{j+2}$ (because $j+2-i = d-3l-1 \geq 0$), and (iii) is impossible since $b_i = b_{i-1}$; contradiction.
CASE 2. $c_l = c_{l+1}$. Then $l > 1$, since otherwise $b_1 = c_1 = 1$ and Γ would be a polygon. Hence the theorem applies with $i = l$, $j = d-2l-1$. Now (i) is impossible, since $b_i = b_{i-1}$ would imply $k_l = k_{l-1}$, contradicting the minimality of l. And (ii) is impossible, since $c_{i+j+1} = c_i$ and $c_i \leq c_{j+2}$ (because $j+2-i > 0$). Finally, (iii) is impossible, since $b_j \leq b_i$ (because $j-i \geq 0$). Contradiction. □

Remarks. (i) This corollary was found independently by ALDRED [6], Theorem 2.11.
(ii) The dodecahedron and the Desargues graph (the bipartite double of the Petersen graph) satisfy $k_{i+1} \leqslant k_i$ for $i \geqslant d/3$, but it seems likely that the right bound in 5.9.7 is of the order of $d/2$.

Problem. Show that $k_{i+1} > k_i$ for $k > 2$, $2i+1 < d$. (Cf. Problem 5.1.3 (i).)

This result, combined with Corollary 5.1.3, can serve to derive an upper bound for d when an upper bound for v is given. For example, all distance-regular graphs with valency $k \geqslant 3$ and $v \leqslant 4096$ vertices have diameter $d < 64$. (In fact the largest diameter occurring is 13, for the doubled Odd graph with $v = 3432$ and $k = 7$.)

[Indeed, we may obtain $v > d^2$ for $d \geqslant 20$ and $k > 2$; let us sketch the arguments. For $1 \leqslant i < \frac{1}{3}d$ we have $k_{i+1} = b_i k_i / c_{i+1} > k_i$, so that

$$k_{i+1} \geqslant \lceil k_i(1+\frac{1}{c_{i+1}})\rceil \geqslant \max(k_i+2, k_i(1+\frac{1}{k-1}))$$

for these i. If $k_2 < 2k$, then by Corollary 5.9.6 either all triples (c_j,a_j,b_j) ($j \geqslant 3$) are distinct, or $(1,\lambda,b_1) = (c_2,a_2,b_2)$. But in the latter case $c_2 = 1$, $k_2 = b_1 k$, $b_1 = 1$, $k = 2$ (by Lemma 5.3.1), contradiction. Thus we are in the former case, and from $2c_2 > b_1 \geqslant b_2$ we find that

$$0 \leqslant 2(2c_2 - b_2 - 1) + (k - b_2 - c_2) = k - 2 + 3(c_2 - b_2),$$

so that $c_j - b_j$ $(3 \leqslant j \leqslant d)$ can take at most $(4k-2)/3$ distinct values, and $d \leqslant \frac{4}{3}(k+1)$, i.e., $k \geqslant \frac{3}{4}d - 1$. Now with $t = [d/3]$ we find

$$v \geqslant 2 + (d-1)k + 2t(d-t-2) > \tfrac{43}{36}d(d-3) > d^2$$

for $d \geqslant 19$. Now let $k_2 \geqslant 2k$. Then $k_{i+1} \geqslant k_i + 3$ for $2 \leqslant i < \frac{1}{3}d$. If $k_3 < 2k_2$, then by Corollary 5.9.6 at most two consecutive triples (c_j,a_j,b_j) ($j \geqslant 4$) coincide, unless $(1,\lambda,b_1) = (c_3,a_3,b_3)$, but the latter would imply that $k_2/k = k_3/k_2$, contradiction. Thus, $d \leqslant 2(4k-2)/3+3$, i.e., $k \geqslant (3d-5)/8$. As before we find

$$v \geqslant 2 + 2k + (d-3)k_2 + 3(t-1)(d-t-3) > \tfrac{17}{12}d(d-6) + 14 > d^2$$

for $d \geqslant 19$. Now let $k_3 \geqslant 2k_2$. Then $k_{i+1} \geqslant k_i + 5$ for $3 \leqslant i < \frac{1}{3}d$. It follows that

$$v \geqslant 2 + 2k + 2k_2 + (d-5)k_3 + 5(t-2)(d-t-4) > d^2$$

for $d \geqslant 20$.

In view of the existence of distance-regular graphs with $d = 5$, $v = 20$ or with $d = 8$, $v = 90$, a bound like $v > d^2$ is perhaps not totally ridiculous, but of course one wants a bound that grows exponentially with d. For fixed k the inequality $k_{i+1} \geqslant k(1+\frac{1}{k-1})^i$ for $1 \leqslant i < d/3$ does the trick.]

Problem. Let $k > 2$. Prove or disprove: (i) $v \geqslant 2^d$ if d is even, with equality only for d-cubes; (ii) $v \geqslant 2\binom{2e+1}{e}$ if $d > 2e$, with equality only for the doubled Odd graphs and the dodecahedron; (iii) $d \leqslant 2k-1$ if $k > 3$, with equality only for the doubled Odd graph.

5.9.8. Theorem (A.A. IVANOV [430]). *The diameter d of a distance-regular graph Γ of valency $k > 2$ and numerical girth g satisfies $d < g.\,2^{2k-3}$.*

Proof. The number of distinct triples (c_i,a_i,b_i) $(1 \leqslant i \leqslant d)$ is at most $2k-1$ since $c_i - b_i$ increases from $1 - b_1 \geqslant 2 - k$ to $c_d \leqslant k$. If $(1,a_1,b_1) = (c_{r-1},a_{r-1},b_{r-1}) \neq (c_r,a_r,b_r)$ (so that $g = 2r$ or $g = 2r+1$) then, by induction and using (14) above, we find that the first occurrence of the j-th such triple is for $i \leqslant 2^{j-2}r$, and finally that $d \leqslant 2^{2k-2}r$. □

Remark. If Γ is bipartite, then 5.9.5 (ii) gives the much more satisfactory bound $d \leqslant k(r-1)$. (Indeed, by induction one proves that $c_i \geqslant i/(r-1)$ for all i.) About the same bound was given by Terwilliger, see 5.2.5 (3c).

5.9.9. Theorem. *Assume*

$$(c_{i-1},a_{i-1},b_{i-1}) \neq (c_i,a_i,b_i) = (c_{i+j},a_{i+j},b_{i+j}) \neq (c_{i+j+1},a_{i+j+1},b_{i+j+1})$$

with $i > 1$, $j \geqslant 1$. *Assume moreover that* $(1,\lambda,b_1) = (c_{r-1},a_{r-1},b_{r-1}) \neq (c_r,a_r,b_r)$, *i.e., that* Γ *has numerical girth either* $2r$ *or* $2r+1$. *Then*

(i) *If* $c_i = 1$, *then* $j \leqslant r-1$ *and if equality holds, then* $c_{i+j+1} > 1$;

(ii) *If* $b_i = 1$ *and* $d > i+j+1$, *then* $j \leqslant r-1$ *and if equality holds, then* $b_{i-1} > 1$.

Proof. (i) If $c_i = 1$, then $d_m = 1$. Assume $j > r-1$. Take $\gamma, \delta, \zeta, \eta$ as before and choose $\theta \in \Gamma_{r-1}(\zeta) \cap \Gamma_{i+r-1}(\gamma)$. Since $i > j \geqslant r-1$ and $c_i = 1$ we have $c_r = c_{r-1} = 1$, so $a_r > a_{r-1}$ and we can find a neighbour θ_1 of θ in $\Gamma_r(\eta) \setminus \Gamma_{r-1}(\zeta)$. Now $\theta_1 \notin \Gamma_{i+r}(\gamma)$, since otherwise $A^\gamma(\zeta) = A^\gamma(\theta) = A^\gamma(\theta_1) = A^\gamma(\eta)$, contradiction. Also $\theta_1 \notin \Gamma_{i+r-2}(\gamma)$, since otherwise the geodesic from θ_1 to η would have to pass through $\Gamma_{i-1}(\gamma)$, and we find a contradiction with the choice of δ, ζ, η and $c_i = 1$. So $\theta_1 \in \Gamma_{i+r-1}(\gamma)$. Since $c_r = 1$, there is a unique point $\zeta_1 \in \Gamma(\eta) \cap \Gamma_{r-1}(\theta_1)$ and by Lemma 5.9.4 we have $\zeta_1 \in \Gamma_i(\gamma)$. Now $A^\gamma(\zeta) = A^\gamma(\theta) = A^\gamma(\theta_1) = A^\gamma(\zeta_1)$ so $\Gamma(\eta) \cap \Gamma_{i-1}(\delta) \supseteq \{\zeta,\zeta_1\}$, i.e., $c_i \geqslant 2$, contradiction. Hence $j \leqslant r-1$. If $j = r-1$, then we find the same contradiction, unless $\theta_1 \in \Gamma_{i+r}(\gamma)$ and $c_{i+r} > 1$.
(ii) If $b_i = 1$, then an argument dual to the preceding one works. There is a vertex γ and adjacent vertices ζ, η in $\Gamma_{i+j}(\gamma)$ with $d(\zeta,\delta) = i+j+1$, $d(\eta,\delta) = i+j$ for some $\delta \sim \gamma$, for otherwise the matrix $A(\gamma,\zeta)$ would look like

	c_i	a_i	$b_i = 1$
c_i	$i+j-2$	$i+j-1$	$i+j$
a_i	$i+j-1$	$i+j$	$i+j+1$
$b_i = 1$	$i+j$	$i+j+1$	e

and with $e = i+j+2$ we get $(c_{i+j},a_{i+j},b_{i+j}) = (c_{i+j+1},a_{i+j+1},b_{i+j+1})$, and otherwise we have $b_{i+j+1} = 0$, i.e., $d = i+j+1$.
Now assume that $j > r-1$ and choose $\theta \in \Gamma_{i+j-r+1}(\gamma) \cap \Gamma_{r-1}(\zeta)$, $\theta_1 \in \Gamma(\theta) \cap \Gamma_r(\eta) \setminus \Gamma_{r-1}(u)$, then $\theta_1 \in \Gamma_{i+j-r+1}(\gamma)$. (Note that we can choose θ_1 since in this case there is a unique geodesic from θ to η so that $c_r = 1$ and $a_r > a_{r-1}$.) Find $\zeta_1 \in \Gamma(\eta) \cap \Gamma_{r-1}(\theta_1)$, then $\zeta_1 \in \Gamma_{i+j}(\gamma) \cap \Gamma_{i+j+1}(\delta)$ so that $b_{i+j} > 1$, contradiction. If $j = r-1$, then we find the same contradiction, unless $\theta_1 \in \Gamma_{i+j-r}(\gamma) = \Gamma_{i-1}(\gamma)$ and $b_{i-1} > 1$. □

Remark. A.V. Ivanov [pers. comm.] further proved that in the situation of the above theorem $j \geqslant r$ implies $c_r = 1$, $a_i \geqslant 1$, $b_i \leqslant \frac{1}{2}(b_1 - c_{i-1})$ and $c_i \leqslant \frac{1}{2}(b_1 - b_{i+j+1})$. He conjectures that in fact $j \geqslant r$ is impossible (which would yield a diameter bound $d \leqslant kg$).

5.10 Circuit chasing

In BIGGS, BOSHIER & SHAWE-TAYLOR [80] an ingenious technique for the study of distance-regular graphs with large girth is developed. The idea is to write down the distance distribution ('profile') of a circuit with respect to one of its edges, and to derive from that the profile around the next edge. Continuing in this vein one may walk around the circuit. If, upon returning to the first edge, one finds a profile different from the original one, then this contradiction shows that the graph does not exist. As an example we show part of the derivation of the following result.

5.10.1. Theorem (NOMURA & BOSHIER [112]). *Let Γ be a distance-regular graph with $(c_i,a_i,b_i) = (1,0,k-1)$ for $1 \leqslant i \leqslant r$ and $(c_i,a_i,b_i) = (1,1,k-2)$ for $r+1 \leqslant i \leqslant r+s$. If $s \geqslant 3$, then r is divisible by 3, and if $r > 0$ then $s \leqslant 4$.*

Proof. We may assume $r > 0$, $s \geqslant 3$. Fix an edge $\gamma\delta$ of Γ and draw the distance distribution diagram of Γ with respect to (the vertices of) this edge. We find

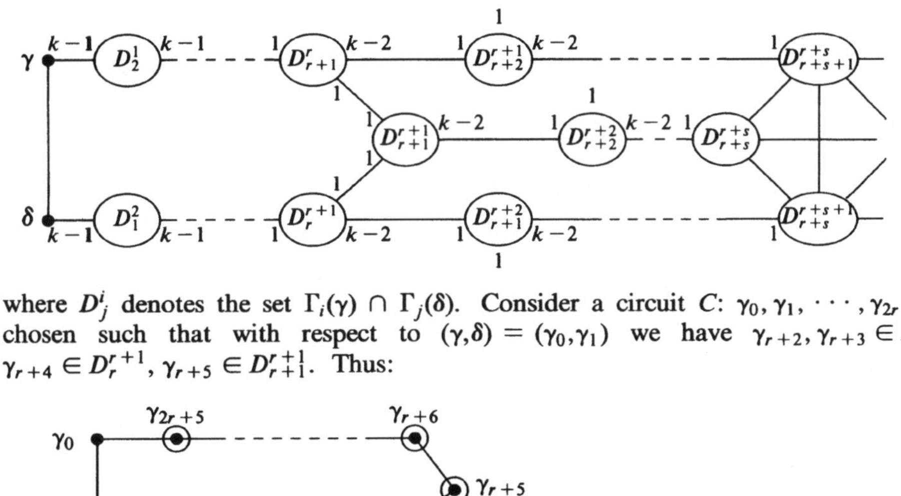

where D^i_j denotes the set $\Gamma_i(\gamma) \cap \Gamma_j(\delta)$. Consider a circuit C: $\gamma_0, \gamma_1, \cdots, \gamma_{2r+5}, \gamma_0$ chosen such that with respect to $(\gamma,\delta) = (\gamma_0,\gamma_1)$ we have $\gamma_{r+2}, \gamma_{r+3} \in D^{r+2}_{r+1}$, $\gamma_{r+4} \in D^{r+1}_r$, $\gamma_{r+5} \in D^{r+1}_{r+1}$. Thus:

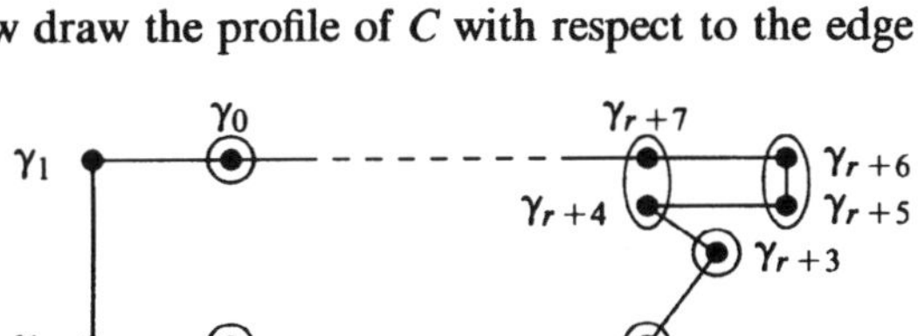

Now draw the profile of C with respect to the edge $\gamma_1\gamma_2$. We find:

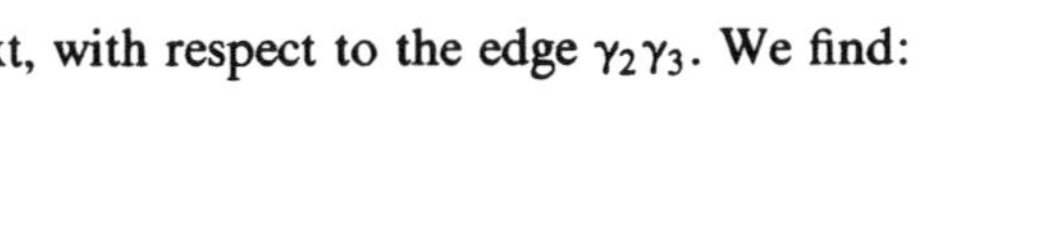

Next, with respect to the edge $\gamma_2\gamma_3$. We find:

In the next step we find with respect to $\gamma_3\gamma_4$ the same profile as with respect to $\gamma_0\gamma_1$, but with all indices shifted by three. It follows immediately that $2r+6\equiv 0 \pmod 3$, i.e., $3\,|\,r$.

To get another condition on r when $s > 4$, Nomura & Boshier took the $(2r+13)$-cycle

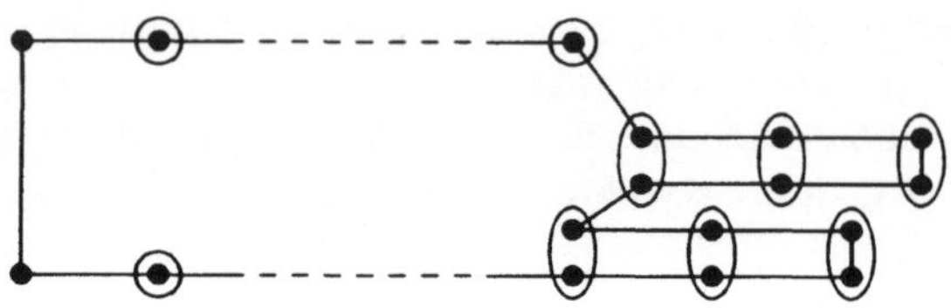

and found $2r+13\equiv 0 \pmod{15}$, contradicting $r\equiv 0 \pmod 3$. [Here the argument is more complicated than in the previous case - not at all steps is the profile with respect to some edge determined uniquely by the profile with respect to the previous edge, and one has to distinguish cases.] □

(Note that the dodecahedron and the Biggs-Smith graph provide examples with $(r,s) = (1,2)$ and $(3,3)$, respectively. E.W. Lambeck [pers. comm.] showed that if $s = 4$ and $r > 0$ then $r\equiv 36 \pmod{90}$.)

Using this technique, Biggs, Boshier & Shawe-Taylor [80] succeeded in completing the classification of cubic distance-regular graphs (cf. Theorem 7.5.1). Using the distance distribution around an edge (but not this 'circuit chasing' technique), Nomura [571] determined all distance-regular graphs with valency 4 and girth 3.

Chapter 6
Classification of the Known Distance-Regular Graphs

In this chapter we classify distance-regular graphs with diameter $d \geqslant 3$ into the following four (non-exclusive) classes:

(i) graphs with classical parameters,

(ii) partition graphs,

(iii) regular near polygons,

(iv) the remaining distance-regular graphs.

Each class is subdivided into primitive and imprimitive graphs. For each class, we give some general properties of the graphs and list the known examples. It is remarkable that each known distance-regular graph of diameter $d > 8$ belongs to at least one of the first three classes.

Details on the known distance-regular graphs of diameter $d \geqslant 3$ and proofs of their distance-regularity are given in Chapters 9-13. We do *not* discuss here distance-regular graphs of diameter $d \leqslant 2$; they are precisely the connected strongly regular graphs, and for information on the known strongly regular graphs we refer to the surveys of Hubaut [413] and Brouwer & van Lint [134], and to the papers quoted after Theorem 7.7.1 below.

6.1 Graphs with classical parameters

Most distance-regular graphs related to classical groups and groups of Lie type have an intersection array whose parameters can be expressed in terms of the diameter d and three other parameters b, α, β as follows:

$$b_i = \left(\left[{d \atop 1}\right] - \left[{i \atop 1}\right]\right)\left(\beta - \alpha\left[{i \atop 1}\right]\right), \tag{1a}$$

$$c_i = \left[{i \atop 1}\right]\left(1 + \alpha\left[{i-1 \atop 1}\right]\right) \tag{1b}$$

($i = 0, \cdots, d$), where

$$\left[{i \atop l}\right] = \begin{cases} \prod_{j=0}^{l-1} \dfrac{i-j}{l-j} = \binom{i}{l} & \text{if } b = 1, \\ \prod_{j=0}^{l-1} \dfrac{b^i - b^j}{b^l - b^j} & \text{if } b \neq 1 \end{cases} \tag{2}$$

are the Gaussian binomial coefficients with basis b. (In cases of ambiguity we shall indicate the basis by an index and write $\left[{i \atop h}\right]_b$ for $\left[{i \atop h}\right]$.) From (1a) and (1b) we easily obtain formulae for k, μ and a_i:

$$k = {d \brack 1}\beta, \quad \mu = (1+b)(1+\alpha), \tag{1c}$$

$$a_i = {i \brack 1}\left(\beta - 1 + \alpha\left({d \brack 1} - {i \brack 1} - {i-1 \brack 1}\right)\right). \tag{1d}$$

We shall say that a distance-regular graph has *classical parameters* (d,b,α,β) if its diameter is d and its intersection array is given by (1). The parameters of the known distance-transitive graphs with $d \geqslant 3$ and classical parameters are given in Table 6.1. A few families of distance-regular graphs with classical parameters are known which are not distance-transitive (see Table 6.2). By the theorem below, all distance-regular graphs with classical parameters (d,b,α,β) with $b = 1$ are known and occur in Table 6.1 or 6.2. For graphs with $\alpha = 0$, cf. Theorem 9.4.3. All graphs with classical parameters are Q-polynomial (Corollary 8.4.2). This also leads to closed formulae for the eigenvalues and multiplicities (Corollary 8.4.2 and Theorem 8.4.3). The graphs with $\alpha + 1 = b$ are formally self-dual and have $v = (\beta+1)^d$ vertices, cf. Corollary 8.4.4.

Table 6.1. *Distance-transitive graphs of diameter* $\geqslant 3$ *with classical parameters (d, n, m positive integers, q, r prime powers; in (C4), (C9), (C11) either $m = n = 2d+1$ or $m+1 = n = 2d$)*

	name	§	d	b	$\alpha+1$	$\beta+1$
(C1)	Johnson graph $J(n,d)$, $n \geqslant 2d$	9.1	d	1	2	$n-d+1$
(C2)	Grassmann graph, $n \geqslant 2d$	9.3	d	q	$q+1$	${n-d+1 \brack 1}$
(C3)	Dual polar graph $e \in \{0,\frac{1}{2},1,\frac{3}{2},2\}$	9.4	d	q	1	q^e+1
	$U(2d,r)$ (see §6.2)	9.4	d	$-r$	$\frac{1+r^2}{1-r}$	$\frac{(1-(-r)^{d+1})}{1-r}$
(C4)	Half dual polar graph $D_{n,n}(q)$	9.4C	d	q^2	${3 \brack 1}_q$	${m+1 \brack 1}_q$
(C5)	Exceptional Lie graph $E_{7,7}(q)$	10.7	3	q^4	${5 \brack 1}_q$	${10 \brack 1}_q$
	Gosset graph $E_7(1)$	3.11	3	1	5	10
(C6)	Triality graph ${}^3D_{4,2}(q)$	10.7	3	$-q$	$\frac{1}{1-q}$	${3 \brack 1}_q$
(C7)	Witt graph M_{24}	11.4A	3	-2	-3	11
(C7a)	Witt graph M_{23}	11.4B	3	-2	-1	6
(C8)	Hamming graph $H(d,n)$	9.2	d	1	1	n
(C9)	Halved cube $\frac{1}{2}H(n,2)$	9.2D	d	1	3	$m+1$
(C10)	Bilinear forms graph, $n \geqslant d$	9.5A	d	q	q	q^n
(C11)	Alternating forms graph	9.5B	d	q^2	q^2	q^m
(C12)	Hermitean forms graph, $q = r^2$	9.5C	d	$-r$	$-r$	$-(-r)^d$
(C13)	Affine $E_6(q)$ graph	10.8	3	q^4	q^4	q^9
(C14)	Extended ternary Golay code graph	11.3A	3	-2	-2	9

(The graphs below the bar have a regular abelian group of automorphisms; all except the halved cubes are self-dual; the duals of the halved cubes are the folded cubes.)

Table 6.2. *Other distance-regular graphs with classical parameters (In (C11a), unlike (C11), $m = n = 2d-1$ or $m-1 = n = 2d$)*

	name	§	d	b	$\alpha+1$	$\beta+1$
(C3a)	Pseudo $D_m(q)$ graphs	9.4.5	d	q	1	2
(C4a)	Dist. 1-or-2 in symplectic dual polar graph	9.4C	d	q^2	$\left[{3 \atop 1}\right]_q$	$\left[{m+1 \atop 1}\right]_q$
(C8a)	Doob graph	9.2B	d	1	1	4
(C11a)	Quadratic forms graph	9.6	d	q^2	q^2	q^m

6.1.1. Theorem. *All distance-regular graphs with classical parameters (d,b,α,β) and $b = 1$ are known (and occur in Table 6.1 or 6.2 when $d > 2$). In particular, the Hamming graphs $H(d,n)$ ($n \neq 4$), the Johnson graphs $J(n,d)$ ($(n,d) \neq (8,2)$) and the half cubes $D_{n,n}(1)$ are characterized by their intersection array.*

Proof. Let Γ have classical parameters with $b = 1$. Then $\theta = b_1 - 1$ is an eigenvalue corresponding to the standard sequence $u_i = 1-i(\lambda+2)/k$ $(i = 0, \cdots, d)$, and Theorem 4.4.11 applies. Case (ii) in the conclusion of that theorem is impossible since then we would have $\mu = 1$, $d \geqslant 3$, and therefore $c_3 = 0$. □

6.1.2. Proposition. *The intersection numbers of every distance-regular graph of diameter $d \geqslant 3$ with classical parameters satisfy*

$$c_{i+1} > c_i \quad \text{for } i = 0, \cdots, d-1;$$

in particular the geometric girth of such a graph cannot exceed 4.

Proof. We first note that $c_2 = (1+\alpha)(b+1)$. Now if $c_{i+1} = c_i$ then $\left[{i+1 \atop 1}\right](1+\alpha\left[{i \atop 1}\right]) = \left[{i \atop 1}\right](1+\alpha\left[{i-1 \atop 1}\right])$ which simplifies to $b = -\alpha(b+1)\left[{i \atop 1}\right] = (b+1-c_2)\left[{i \atop 1}\right]$. But c_2 is an integer, and b and $\left[{i \atop 1}\right]$ are coprime integers; therefore $\left[{i \atop 1}\right] = \pm 1$ which forces $b = 1$, $i = 1$. But then $\alpha = -\frac{1}{2}$, $c_3 = 0$, contradiction. Since the c_i are nondecreasing, it follows that $c_{i+1} > c_i$ for all i. In particular, $c_2 > 1$, which implies that the geometric girth is at most 4. □

6.2 Computation of classical parameters

6.2.1. Proposition. *Let Γ be a distance-regular graph of diameter $d \geqslant 3$ with classical parameters (d,b,α,β). Then the basis b is an integer $\neq 0, -1$ and we have one of the following cases:*

(i) $a_i \neq \lambda c_i$ *for* $i = 2, \cdots, d$, *and*

$$b = \frac{a_2 c_3 - \mu a_3}{\lambda c_3 - a_3}. \tag{3}$$

(ii) $a_i = \lambda c_i$ *for* $i = 0, \cdots, d$ *and*

$$b \in \{\mu-1, -\lambda-1\}. \tag{4}$$

In both cases, α and β are given by

$$\alpha = \mu/(1+b)-1 \quad \textit{and} \quad \beta = k/\left[{d \atop 1}\right] = \lambda+1-\alpha(\left[{d \atop 1}\right]-1). \tag{5}$$

Proof. First we note that $b_2 \neq 0, c_2 \neq 0$, $c_i \neq 0$ imply that $b \neq 0, -1$ and

$$\left[{i \atop 1}\right] \neq 0 \quad \text{for } i = 1, \cdots, d. \tag{6}$$

With the abbreviation

$$\epsilon := (b+1+\lambda)(b+1-\mu) \tag{7}$$

we find from (1b) and (1d) that

$$\lambda = \beta-1+\alpha(\left[{d \atop 1}\right]-1), \quad \mu = (b+1)(1+\alpha), \quad \epsilon = -\alpha(b+1+\lambda)(b+1). \tag{8}$$

Now (8), (1b) and (1d) imply

$$\begin{aligned} a_i-\lambda c_i &= \left[{i \atop 1}\right](\lambda-\alpha(\left[{i \atop 1}\right]+\left[{i-1 \atop 1}\right]-1))-\lambda\left[{i \atop 1}\right](1+\alpha\left[{i-1 \atop 1}\right]) \\ &= -\alpha\left[{i \atop 1}\right](\left[{i \atop 1}\right]+\left[{i-1 \atop 1}\right]-1+\lambda\left[{i-1 \atop 1}\right]) = -\alpha\left[{i \atop 1}\right]\left[{i-1 \atop 1}\right](b+1+\lambda), \end{aligned}$$

whence by (8),

$$a_i-\lambda c_i = \epsilon\left[{i \atop 1}\right]\left[{i-1 \atop 1}\right]/(b+1) = \epsilon\left[{i \atop 2}\right]. \tag{9}$$

In particular, for $i = 2$,

$$\epsilon = a_2-\lambda c_2 = a_2-\lambda\mu \tag{10}$$

is an integer, and by (7), the basis b is an algebraic integer. We now consider the cases $\epsilon \neq 0$ and $\epsilon = 0$ separately. If $\epsilon \neq 0$ then by (9) and (6),

$$a_i \neq \lambda c_i \text{ for } i = 2, \cdots, d. \tag{11}$$

Moreover, (1b) and (8) imply

$$c_3 = \left[{3 \atop 1}\right](\mu-b). \tag{12}$$

If we combine (12), (9), and (10) we obtain $c_3(a_2-\lambda\mu) = \left[{3 \atop 1}\right](\mu-b)\epsilon = (\mu-b)(a_3-\lambda c_3)$. Therefore $b(\lambda c_3-a_3) = a_2c_3-\mu a_3$ which implies (3). In particular, b is rational, hence an integer. But if $\epsilon = 0$ then (9) implies $a_i = \lambda c_i$ for all i, and (7) implies $b \in \{\mu-1, -\lambda-1\}$ which gives (4). In both cases, (5) follows from (8) and (1c) by solving for α and β. □

6.2.2. Corollary. *The parameters of a distance-regular graph of diameter $d \geqslant 3$ with classical parameters are uniquely determined by the intersection array, with the exception of the parameter combinations*

$$(d,b,\alpha,\beta) = (d,q^2,0,q),$$

and

$$(d,b,\alpha,\beta) = (d,-q,\frac{q}{1-q}(1+q),\frac{q}{1-q}(1+(-q)^d))$$

(for integers d, $q > 1$) which belong to the same intersection array.

Proof. If two different parameter sets (d,b,α,β) and (d,b',α',β') define the same intersection array then we are in case (ii) of the preceding proposition, so that

$$b = \mu - 1,\ b' = -q, \text{ where } q = \lambda + 1.$$

Since $\lambda \geqslant 0$ and $b' \neq 0, -1$, the number q is an integer > 1. Now (12) implies

$$c_3 = \left[{3 \atop 1}\right]_b(\mu - b) = (1+b+b^2)(\mu - b) = \mu^2 - \mu + 1,$$

but also

$$c_3 = \left[{3 \atop 1}\right]_{b'}(\mu - b') = (1+b'+b'^2)(\mu - b') = (q^2 - q + 1)(\mu + q).$$

Equating the two expressions for c_3 gives a quadratic equation for μ with solutions $1-q$ and q^2+1. Since $\mu > 0$ we must have $\mu = q^2+1$, $b = q^2$. Now (5) with $\lambda = q-1$, $\mu = q^2+1$ and $b \in \{q^2, -q\}$ gives the stated parameter combinations. It is easily checked that they both belong to the same intersection array. □

Remark. For q a prime power, the intersection arrays of the above corollary are realized by the unitary dual polar graphs $U(2d, q^{1/2})$ (series (C3) above). It follows that these graphs have two Q-polynomial structures.

6.3 Imprimitive graphs with classical parameters; partition graphs

6.3.1. Proposition. *Let* Γ *be a distance-regular graph with classical parameters* (d, b, α, β) *and diameter* $d \geqslant 3$.

(i) Γ *is bipartite if and only if* $\alpha = 0$ *and* $\beta = 1$.

(ii) Γ *is antipodal if and only if* $b = 1$ *and* $\beta = 1 + \alpha(d-1)$. *In this case* Γ *is an antipodal double cover of its folded graph.*

Proof. By Proposition 6.2.1, the conditions are sufficient. Necessity follows from the same proposition in case (i) from $a_1 = a_2 = 0$ and (1d) and in case (ii) from $b_0 = c_d$, $b_{d-1} = c_1 = 1$, (1a) and (1b) by using the fact that b is an integer $\neq -1$ (Proposition 6.2.1). □

It follows that the only bipartite graphs among the graphs listed in Tables 6.1 and 6.2 are (C3) for $e = 0$, (C8) for $n = 2$, and (C3a). Now the halved graphs of (C3) with $e = 0$ are isomorphic to (C4), those of (C8) with $n = 2$ are isomorphic to (C9), and those of (C3a) are complete graphs; hence no new graphs arise by halving.

The situation is different for antipodal graphs. By the above proposition the only antipodal graphs in Tables 6.1 and 6.2 are (C1) for $n = 2d$, (C5) for $q = 1$, (C8) for $n = 2$, and (C9) for $m = 2d$. It happens that their folded graphs can be described in terms of partitions, see Table 6.5. (The folded graph of (C5) for $q = 1$ is a complete graph, hence uninteresting.) If we write m for the diameter of the antipodal graph and d for the diameter of the folded graph then Proposition 6.2.1 (ii) implies that the folded graph of an antipodal distance-regular graph with classical parameters $(m, 1, \alpha, 1+\alpha(m-1))$ has an intersection array given by

$$b_i = (m-i)(1+\alpha(m-1-i)),\quad c_i = i(1+\alpha(i-1)) \text{ for } i < d, \tag{13a}$$

$$b_d = 0, \quad c_d = \gamma d(1 + \alpha(d-1)), \tag{13b}$$

$$\gamma = 2 \text{ if } m = 2d, \quad \gamma = 1 \text{ if } m = 2d+1. \tag{13c}$$

We call a distance-regular graph of diameter d whose intersection array satisfies (13) a *pseudo partition graph.* But apart from the foldings of classical graphs listed in the table below no pseudo partition graphs are known.

Table 6.3. *Partition graphs with parameters* (m, α)

	partitions of	group	§	α	folding of
(P1)	m-set into two sets	$2^{m-1}.\mathrm{Sym}_m$	9.2D	0	cube (C8), $n = 2$
(P2)	$2m$-set into two m-sets	Sym_{2m}	9.1D	1	Johnson graph (C1), $n = 2m$
(P3)	$2m$-set into two even sets	$2^{2m-2}.\mathrm{Sym}_{2m}$	9.2D	2	half cube (C9), $m = 2d$

6.3.2. Theorem (BUSSEMAKER & NEUMAIER [155]). *Let* Γ *be a pseudo partition graph with parameters* (m,α).

(i) *If* $m \geqslant 8$, *then* $\alpha \in \{0,1,2\}$.

(ii) *If* $m \geqslant 307$, *then* Γ *is one of the graphs listed in Table 6.3.*

About the proof. See the reference given. If Γ has the parameters of the folded half cube, Theorem 4.4.3 (i) yields the lower bound $-2-2/(m-3)$ for the smallest eigenvalue of $\Gamma(\gamma)$, and by Theorem 3.12.5 (ii) this smallest eigenvalue must be $\geqslant -2$ for $m \geqslant 307$. Now using Theorem 3.12.2 one can show that Γ must be locally triangular, and then (ii) follows. □

Problem. Improve the bound on m in part (ii); it seems much too large.

A pseudo partition graph is never antipodal. It is bipartite if and only if $\alpha = 0$; in this case the halved graphs are pseudo partition graphs with $\alpha = 2$. This applies for the folded cubes, but halved folded cubes and folded half cubes are isomorphic; hence no further graphs arise. In theorem 8.4.5 we shall see that all pseudo partition graphs are Q-polynomial.

6.4 Regular near polygons

A connected graph Γ of diameter $d \geqslant 2$ is called a *near polygon* if the following two axioms hold.

NP1. There are no induced subgraphs of shape $K_{1,1,2}$:

NP2. If $\gamma \in \Gamma$ and L is a singular line with $d(\gamma,L) < d$, then there is a unique point on L nearest to γ.

Such a graph Γ is also called a *near n-gon,* where $n = 2d+1$ if there is a point at distance d from some singular line, and $n = 2d$ otherwise. (Near 6-gons are commonly called *near hexagons*, and similarly one uses the term *near octagon* instead of near 8-gon, etc.) If in addition Γ is distance-regular, then we call Γ a *regular near polygon* (or *regular near n-gon).* We also call a partial linear space a near polygon if its point graph is a near polygon; in this case, lines and singular lines are the same. A near polygon is called *thin* when its lines have size two, i.e., when $\lambda = 0$.

Axiom NP1 ensures that singular lines and maximal cliques coincide (cf. §1.1.2), and axiom NP2 is modelled after a property of ordinary polygons. Note that for a near $(2d+1)$-gon, if $\gamma \in \Gamma$ and L is a singular line with $d(\gamma,L) = d$, then *all* points on L are nearest to γ. For even $n = 2d$, the concept of a near n-gon was introduced by Shult & Yanushka [673]. Regular near pentagons (the case $n = 5$) were considered in some detail by Cameron [164] under the name *proper partial quadrangles;* since they have diameter two we do not discuss them further. All bipartite graphs satisfy NP1 and NP2 trivially and hence are near $2d$-gons with singular lines of size two. Conversely, if all singular lines of a near n-gon Γ with n even have two points only, Γ is bipartite.

It is interesting that, given axiom NP1, axiom NP2 is equivalent to a purely arithmetical restriction on the intersection array.

6.4.1. Theorem. *A distance-regular graph* Γ *with intersection array* $\iota(\Gamma) = \{b_0, \cdots, b_{d-1}; 1, c_2, \cdots, c_d\}$ *is a regular near polygon if and only if axiom NP1 holds and there is an integer* λ *such that*

$$b_i = k-(\lambda+1)c_i \quad \textit{for } i = 0, \cdots, d-1. \tag{14}$$

In this case,

$$k \geqslant (\lambda+1)c_d, \tag{15}$$

and Γ *is a near* $2d$*-gon if* (15) *holds with equality and a near* $(2d+1)$*-gon otherwise.*

Proof. Assume first that NP1 and (14) hold. For $i = 1$, (14) implies $\lambda = a_1$; then NP1 shows that all singular lines have size $\lambda+2$. By definition of a singular line, NP2 holds for those pairs (γ,L) with $d(\gamma,L) \leqslant 1$. Now assume that for some $j < d$, NP2 holds for all pairs (γ,L) with $d(\gamma,L) < j$. Let (γ,L) be a fixed pair with $d(\gamma,L) = j$. By definition of the distance, there is $\delta \in L$ with $d(\gamma,\delta) = j$, but no $\delta' \in L$ with $d(\gamma,\delta') = j-1$. The c_j points in $\Gamma_{j-1}(\gamma)$ adjacent with δ define c_j lines L' through δ with $d(\gamma,L') = j-1$. By our inductive assumption, each such line contains just one point of $\Gamma_{j-1}(\gamma)$ and $(\lambda+2)-2 = \lambda$ points of $\Gamma_j(\gamma)\setminus\{\delta\}$; in particular these lines are distinct and carry λc_j distinct points of $\Gamma_j(\gamma) \cap \Gamma(\delta)$. Since (14) implies $a_j = \lambda c_j$, this accounts for all points of $\Gamma_j(\gamma) \cap \Gamma(\delta)$. Hence the other lines through δ, in particular the line L, intersect $\Gamma_j(\gamma)$ in $\{\delta\}$ only. Therefore δ is the unique point of L nearest to γ. Now NP2 follows by induction, whence Γ is a near polygon.
Conversely, assume that Γ is a near polygon. Let $\gamma \in \Gamma$, $\delta \in \Gamma_j(\gamma)$. As before, the c_j points in $\Gamma_{j-1}(\gamma)$ adjacent with δ define c_j lines through δ carrying λc_j points of $\Gamma_j(\gamma) \cap \Gamma(\delta)$. The remaining lines through δ have distance j from γ. If $j < d$, these lines intersect $\Gamma_j(\gamma)$ in $\{\delta\}$, by NP2. Therefore, $a_j = \lambda c_j$ for all $j < d$, and this implies (14).
But if $j = d$, then these lines must be completely inside $\Gamma_j(\gamma)$, and if there are u such lines, then $a_j = \lambda c_j + u(\lambda+1)$. Hence $k = a_d+b_d+c_d = (\lambda+1)(c_d+u)$. This implies (15). Equality holds in (15) if and only if $u = 0$, i.e., if and only if Γ is a near $2d$-gon. □

Remarks. (i) The proof shows that the parameters of a regular near polygon satisfy the divisibility condition

$$(\lambda+1) \mid k. \tag{16}$$

Indeed, $s+1 := \lambda+2$ is the size of a line, and $t+1 := k/(\lambda+1)$ is the number of lines on each point. We have $c_d \leqslant t+1$ with equality in case of a regular near $2d$-gon. In view of (14) above, the intersection array of a regular near polygon is determined by s, t, and the c_i $(2 \leqslant i \leqslant d)$, and usually one gives as parameters $(s, t_2, \cdots, t)$, where $t_i := c_i - 1$.
(ii) All distance-regular graphs whose intersection arrays satisfy (14) with $\lambda \leqslant 1$ or $\mu = 1$ are regular near polygons (since NP1 holds trivially).
(iii) This theorem is applied in Theorem 9.4.3 to show that the graphs with classical parameters (d, b, α, β) with $\alpha = 0$ and singular lines of size $\lambda+2 > 2$ are known.

6.5 Generalized polygons

An important class of regular near polygons is defined by the parameters

$$k = s(t+1),\ \lambda = s-1,\ c_i = 1\ (i = 1, \cdots, d-1),\ c_d = t+1; \tag{17}$$

regular near polygons with these parameters are called *generalized 2d-gons of order* (s,t). More generally, a *generalized n-gon* is a partial linear space $(P, \mathfrak{L})$ with the property that the incidence graph has diameter n and any two vertices γ, δ at distance $i < n$ have $c_i(\gamma, \delta) = 1$. It is said to be *of order* (s,t) if all of its lines have size $s+1$ and all of its points are incident to exactly $t+1$ lines. It is said to be *thick* when each line contains at least three points, and each point is on at least three lines. In this case there are s, t such that $(P, \mathfrak{L})$ is of order (s,t).
For even n, the point graph of a generalized n-gon of order (s,t) is distance-regular with the parameters of (17); since in this case the partial linear space can be recovered from the point graph, the two definitions are equivalent. If n is odd, then the point graph of $(P, \mathfrak{L})$ is a regular near polygon with parameters as in (17), except that $c_d = 1$. In this case, it is easily seen that we must have $s = t$. For $n = 2$, $(P, \mathfrak{L})$ is linear space in which each point is incident to each line; for $n = 3$, the partial linear space is a projective plane of order $s = t$, and for $n = 4$, it is a generalized quadrangle of order (s,t), cf. §1.15. The latter are treated in detail in the book by PAYNE & THAS [588].

The (singular) lines of a generalized n-gon of order (s,t) form the points of a generalized n-gon of order (t,s). If $s = t$, the incidence graph of Γ is a generalized $2n$-gon of order $(1,t)$. Conversely, it is easy to see that the generalized $2n$-gons of order $(1,t)$ and $(t,1)$ are precisely the incidence graphs of generalized n-gons with $s = t$ and their line graphs. Finally, a generalized n-gon of order $(1,1)$ is just an n-gon.

Generalized polygons were introduced by TITS in [743]. They have received considerable attention since then. Although, formally, d is unbounded, a famous theorem of FEIT & G. HIGMAN [283] asserts that, apart from the ordinary polygons (R1), finite examples can exist only for $n = 3, 4, 6, 8$ or 12.

6.5.1. Theorem. *A finite regular generalized n-gon of order* (s,t) *has* $n \in \{3,4,6,8,12\}$ *unless* $s = t = 1$. *If* $s > 1$ *and* $t > 1$, *then*

(i) $n \neq 12$;

(ii) *if* $n = 4$, *then* $s \leqslant t^2$, $t \leqslant s^2$;

(iii) *if* $n = 6$, *then* st *is a square and* $s \leqslant t^3$, $t \leqslant s^3$.

(iv) *if* $n = 8$, *then* $2st$ *is a square and* $s \leqslant t^2$, $t \leqslant s^2$;

If $s = 1$, *then* n *is even and the point graph is the incidence graph of a generalized* $\frac{n}{2}$*-gon of order* (t,t).

Apart from the inequalities, this is the original Feit-Higman Theorem. An elegant proof was given by KILMOYER & SOLOMON [465]. Another one, along the same lines, but involving orthogonality relations among Chebychev polynomials rather than representation theory of algebras, can be found in STANTON [699]. The inequalities in (ii) and (iv) are due to D.G. HIGMAN [381, 382], and (iii) to Haemers, cf. HAEMERS & ROOS [342]. The following full proof of the theorem is taken from ROOS [624].

Proof. Denote by Γ the point graph of the generalized n-gon; it has diameter $d = [n/2]$. Let θ be an eigenvalue of the adjacency matrix of Γ, and let $(u_l(\theta))_{0 \leqslant l \leqslant d}$ be the standard sequence corresponding to θ, cf. §4.1B. Choosing $\eta \in \mathbf{C}$ such that

$$\theta = \sqrt{st}(\eta + \eta^{-1}) + s - 1 \tag{18}$$

and using the recursion (13) of §4.1B, the $u_l(\theta)$ can be shown to satisfy

$$u_l(\theta) = \frac{t(\eta^{l+1} - \eta^{-l-1}) + (s-1)\sqrt{s^{-1}t}(\eta^l - \eta^{-l}) - (\eta^{l-1} - \eta^{-l+1})}{(1+t)(\eta - \eta^{-1})(\sqrt{st})^l} . \tag{19}$$

for $0 \leqslant l \leqslant d$. The last step in the recursion reads:

$$c_d u_{d-1}(\theta) = (\theta - k + c_d)u_d(\theta) \tag{19a}$$

The eigenvalue $\theta = k$ corresponds to $\eta = \sqrt{st}$. Substitution of $u_l(\theta)$ in (19a) by use of (19) for $l = d-1, d$, and dividing out $\theta - k$ yields

$$t(\eta^{d+1} - \eta^{-d-1}) + (c_d + s - 1)\sqrt{s^{-1}t}(\eta^d - \eta^{-d}) + (c_d - 1)(\eta^{d-1} - \eta^{-d+1}) = 0 .$$

If $n = 2d$ is even, then $c_d = 1+t$, and the equation can be further simplified to

$$(\eta^d - \eta^{-d})(s + t + (\eta + \eta^{-1})\sqrt{st}) = 0$$

Thus, $\theta = -1-t$ is an eigenvalue corresponding to the root $\eta = -\sqrt{s^{-1}t}$ of the above equation. The remaining roots are n^{th} roots of unity (distinct from ± 1), leading to the eigenvalues $\theta = s - 1 + 2\sqrt{st}\cos(2\pi l/n)$ $(l = 1, \cdots, d-1)$.

If $n = 2d+1$ is odd, then $c_d = 1$, $s = t$, and the eigenvalues $\theta \neq k$ are similarly found to be $\theta = s - 1 + 2s\cos(2\pi l/n)$, corresponding to $\eta = \exp(2\pi i l/n)$ $(l = 1, \cdots, d)$.

By (1c) of §4.1A, the valencies are

$$k_i = s^i t^{i-1}(t+1)/c_i \quad (1 \leqslant i \leqslant d). \tag{20}$$

Let n be even. From (19) and (20) it is tedious but straightforward to compute for $\theta = s-1+2\sqrt{st}\cos(2\pi l/n)$:

$$\sum_{i=0}^{d} k_i u_i(\theta)^2 = \frac{n}{t+1}[1+\frac{s(t-1)^2+t(s-1)^2+2(s-1)(t-1)\sqrt{st}\cos(2\pi l/n)}{4st\sin^2(2\pi l/n)}].$$

This must be a rational number in view of Theorem 4.1.4. Adding the expressions for l and $n/2-l$ we see that this implies that $(\eta-\eta^{-1})^2 = 4\sin^2(2\pi l/n)$ is a rational algebraic integer and hence integral, whence $d \in \{1,2,3,4,6\}$ and $n \in \{2,4,6,8,12\}$. In the case where $s > 1$ and $t > 1$, we must also have that $\sqrt{st}\cos(2\pi l/n)$ is rational for $1 \leqslant l \leqslant d-1$, and this leads to the conclusion that $n \neq 12$, that st is a square if $n = 6$, and that $2st$ is a square if $n = 8$.

Suppose now that n is odd. Then $s = t$ and the same expression can be computed (even more tediously):

$$\sum_{i=0}^{d} k_i u_i(\theta)^2 = \frac{n}{t+1}[1+\frac{(t-1)^2}{2t(1-\cos(2\pi l/n))}].$$

Rationality of this number implies that $\cos(2\pi l/n)$ is rational if $t > 1$. Thus, $\eta+\eta^{-1} = 2\cos(2\pi l/n)$ must be integral, which can only happen if $n = 3$.

The inequalities in the theorem will be derived from the Krein conditions, cf. Theorem 2.3.2 and the following Remark (ii). Although we do not need all of this information, let us write down all eigenvectors u_j, in the form of the matrix $Q\Delta_\mu^{-1}$, that has these eigenvectors as columns (cf. Proposition 2.2.2).

For $n = 4$, we have eigenvalues $\theta_0 = s(t+1)$, $\theta_1 = s-1$, and $\theta_2 = -1-t$. Next,

$$Q\Delta_\mu^{-1} = \begin{bmatrix} 1 & 1 & 1 \\ 1 & (s-1)/s(t+1) & -1/s \\ 1 & -1/st & 1/s^2 \end{bmatrix}.$$

The Krein condition $q_{22}^2 \geqslant 0$ yields

$$0 \leqslant \sum_{i=0}^{2} k_i (u_2)_i^3 = 1-(t+1)/s^2+t/s^4 = (s^2-1)(s^2-t)/s^4,$$

whence $t \leqslant s^2$ if $s > 1$. If $t > 1$, then $s \leqslant t^2$ follows by applying the above to the dual generalized 4-gon.

Similarly, if $n = 6$, the eigenvalues are $\theta_0 = s(t+1)$, $\theta_1 = s-1+\sqrt{st}$, $\theta_2 = s-1-\sqrt{st}$, $\theta_3 = -1-t$. The matrix $Q\Delta_\mu^{-1}$ is

$$\begin{bmatrix} 1 & 1 & 1 & 1 \\ 1 & (s-1+\sqrt{st})/s(t+1) & (s-1-\sqrt{st})/s(t+1) & -1/s \\ 1 & (-s+(s-1)\sqrt{st})/s^2t(t+1) & (-s-(s-1)\sqrt{st})/s^2t(t+1) & 1/s^2 \\ 1 & -1/st\sqrt{st} & 1/st\sqrt{st} & -1/s^3 \end{bmatrix}.$$

The Krein condition $q_{33}^2 \geqslant 0$ gives

$$0 \leqslant \sum_{i=0}^{3} k_i (u_2)_i (u_3)_i^2 = 1 + s^{-2}(s-1-\sqrt{st}) - s^{-4}(s+(s-1)\sqrt{st}) + s^{-6} s\sqrt{st}$$
$$= s^{-5}(s+1)(s^2-1)(s^2-\sqrt{st}),$$

whence the inequality $\sqrt{st} \leqslant s^2$ for $s > 1$, implying $t \leqslant s^3$, and dually $t \leqslant s^3$ if $t > 1$.

Finally, if $n = 8$, the eigenvalues become $\theta_0 = s(t+1)$, $\theta_1 = s-1+\sqrt{2st}$, $\theta_2 = s-1$, $\theta_3 = s-1-\sqrt{2st}$, and $\theta_4 = -1-t$. The matrix $Q\Delta_\mu^{-1}$ is

$$\begin{bmatrix} 1 & 1 & 1 & 1 & 1 \\ 1 & (s-1+\sqrt{2st})/s(t+1) & (s-1)/s(t+1) & (s-1-\sqrt{2st})/s(t+1) & -1/s \\ 1 & (s(t-1)+(s-1)\sqrt{2st})/s^2t(t+1) & -1/st & (s(t-1)-(s-1)\sqrt{2st})/s^2t(t+1) & 1/s^2 \\ 1 & ((s-1)st - s\sqrt{2st})/s^3t^2(t+1) & -(s-1)/s^2t(t+1) & ((s-1)st + s\sqrt{2st})/s^3t^2(t+1) & -1/s^3 \\ 1 & -1/s^2t^2 & 1/s^2t^2 & -1/s^2t^2 & 1/s^4 \end{bmatrix},$$

and the Krein condition $q_{44}^4 \geqslant 0$ yields

$$0 \leqslant \sum_{i=0}^{4} k_i (u_4)_i^3 = 1 - s^{-2}(t+1) + s^{-4}t(t+1) - s^{-6}t^2(t+1) + s^{-8}t^3$$
$$= s^{-8}(s^2-t)(s^2-1)(s^4+t^2).$$

Hence, for $s > 1$, we get $t \leqslant s^2$, and dually $s \leqslant t^2$ if $t > 1$. □

Let us record some information obtained in (or directly following from) the above proof.

Table 6.4. *Spectra of generalized n-gons of order* (s,t)

n	eigenvalues	multiplicities
4	$s(t+1)$	1
	$s-1$	$st(s+1)(t+1)/(s+t)$
	$-t-1$	$s^2(st+1)/(s+t)$
		$v = (1+s)(1+st)$
6	$s(t+1)$	1
	$s-1 \pm \sqrt{st}$	$\frac{1}{2}(1+t)stv / (s(t-1)^2 + t(s-1)^2 + 3st \pm (s-1)(t-1)\sqrt{st})$
	$-t-1$	$s^3 v / (s+1)(s^2+st+t^2)$
		$v = (1+s)(1+st+s^2t^2)$
8	$s(t+1)$	1
	$s-1 \pm \sqrt{2st}$	$\frac{1}{4}(t+1)stv / (s(t-1)^2 + t(s-1)^2 + 2st \pm (s-1)(t-1)\sqrt{2st})$
	$s-1$	$\frac{1}{2}(t+1)stv / (s(t-1)^2 + t(s-1)^2 + 4st)$
	$-t-1$	$s^4 v / (s+1)(s+t)(s^2+t^2)$
		$v = (1+s)(1+st)(1+s^2t^2)$

Remarks. (i) W. Halve (personal communication referred to in Roos [624]) has shown that all multiplicities here are integral provided that the eigenvalues and the multiplicity of $-t-1$ are integral.
(ii) The ordinary n-gon (generalized n-gon of order $(1,1)$) has eigenvalues $2\cos(2\pi j/n)$ ($j = 0, \cdots, n-1$) with multiplicities 1 if $j = 0$ or $j = \frac{1}{2}n$, and 2 otherwise.

The known finite thick generalized polygons with diameter > 2 are graphs of Lie type associated with the groups $G_2(q)$, ${}^3D_4(q)$ and ${}^2F_4(q)$ (cf. Theorem 10.7.2). KANTOR ([459], p. 102) conjectures that there are no other finite thick generalized hexagons and generalized octagons than these. The orders of the known examples with diameter and valency larger than two are listed in Table 6.5. The generalized hexagons of order (q,q^3) are classical, with parameters (C6).

Table 6.5. *Parameters of the known generalized polygons of diameter and valency > 2 (q a prime power, r an odd power of 2)*

		§	n	d	order (s,t)
(G1)	Generalized hexagons	10.7	6	3	$(1,q), (q,1), (q,q), (q,q^3), (q^3,q)$
(G2)	Generalized octagons	10.7	8	4	$(1,q), (q,1), (r,r^2), (r^2,r)$
(G3)	Generalized dodecagons	-	12	6	$(1,q), (q,1)$

We shall sometimes use the notations $GQ(s,t)$, $GH(s,t)$, $GO(s,t)$ to denote an arbitrary generalized quadrangle, resp. hexagon, resp. octagon of order (s,t).

Geometric characterizations of the classical generalized hexagons were given by RONAN [616, 617, 618] (see also THAS [733]); detailed parameter information can be found in HAEMERS [340].

COHEN [202] constructs three generalized hexagons $GH(2,1)$, $GH(2,2)$ and $GH(2,8)$ inside the projective planes $PG(2,\mathbf{C})$, $PG(2,\mathbf{H})$, $PG(2,\mathbf{O})$, respectively.

Concerning uniqueness, the following is known. Projective planes $PG(2,q)$ (and hence generalized hexagons of order $(1,q)$ and $(q,1)$) are unique for $q \leqslant 8$ (by MACINNES [512], HALL [352] and HALL, SWIFT & WALKER [356]). Generalized quadrangles of order (q,q) (and hence generalized octagons of order $(1,q)$ and $(q,1)$) are unique for $q \leqslant 4$ (in case of $GQ(3,3)$: unique up to duality), see Theorem 1.15.2 for $q = 2$ and PAYNE [586, 587] and PAYNE & THAS [588], pp. 129-147, for $q = 3, 4$. The generalized hexagons of orders (2,2), (2,8) and (8,2) (and hence generalized dodecagons of orders (1,2) and (2,1)) are unique (in case of $GH(2,2)$: unique up to duality) by COHEN & TITS [207].

Concerning distance transitivity we can say the following. The known finite thick generalized polygons with diameter > 2 are distance-transitive. By Theorem 4.2.16, a generalized polygon of order $(s,1)$ is distance-transitive if and only if its dual (of order $(1,s)$) is. But a generalized $2n$-gon Γ of order $(1,s)$ is the incidence graph of a generalized n-gon Δ of order (s,s), and Γ is distance-transitive if and only if Δ is distance-transitive and is self-dual (admits a correlation). Now the known generalized quadrangles $GQ(q,q)$ are the orthogonal generalized quadrangles $B_2(q)$ and the symplectic generalized quadrangles $C_2(q)$, and the former are duals of the latter. They are self-dual if and only if q is a power of 2 (BENSON [50]), but they have a polarity only when q is an odd power of 2.

[For arbitrary q we have the result (PAYNE [585]) that if some $GQ(q,q)$ has a polarity, and $q \neq 1$, then $2q$ is a square, and the set of absolute points is an ovoid.]

The known generalized hexagons $GH(q,q)$ are those associated with the group $G_2(q)$, and these are self-dual if and only if q is a power of 3. (But they have a polarity only when q is an odd power of 3.)

[For arbitrary q we have the result (OTT [581]) that if some $GH(q,q)$ has a polarity, and $q \neq 1$, then $3q$ is a square.]

It is very likely that every finite thick distance-transitive generalized $2d$-gon with $d \geqslant 3$ is known. A proof of this fact is expected to emerge from the prospective classification of all primitive distance-transitive graphs. See §7.7 for a further discussion. The case of $GH(p,p)$ with prime p has been settled already:

6.5.2. Theorem (YANUSHKA [796]). *Let p be a prime. Suppose Γ is a primitive distance-transitive graph with intersection array $\{p(p+1), p^2, p^2; 1, 1, p+1\}$. Then Γ is isomorphic to the collinearity graph of the classical generalized hexagon associated with $G_2(p)$ or its dual* - (G1). □

In the infinite case generalized m-gons exist for all m (for example, there are 'free' constructions), but when one imposes the so-called Moufang property (satisfied in the classical examples, cf. TITS [748]) then necessarily $m \in \{2,3,4,6,8\}$ (TITS [749, 751], WEISS [776]).

6.6 Other regular near polygons

Six families of regular near polygons are known which contain graphs of arbitrarily large diameter. Their parameters are listed in Table 6.6. Among them are two families of graphs with classical parameters, (R5) = (C8) and (R6) = (C3), and a family of partition graphs, (R4) = (P1).

Table 6.6. *Parameters of families of regular near n-gons with unbounded diameter d (s a positive integer, q a prime power, $\gamma = 1$ if $n = 2d+1$, $\gamma = 2$ if $n = 2d$; in (R3): $d = 2e+1$; in (R6): $2e \in \{0,1,2,3,4\}$)*

	name	§	n	k	λ	c_i
(R1)	Polygons	1.0	n	2	0	$c_i = 1 (i < d), c_d = \gamma$
(R2)	Odd graphs	9.1D	$2d+1$	$d+1$	0	$c_{2i} = c_{2i-1} = i$
(R3)	Doubled Grassmann graphs	9.3C	$2d$	$\left[{e+1 \atop 1}\right]_q$	0	$c_{2i} = c_{2i-1} = \left[{i \atop 1}\right]_q$
	Doubled Odd graphs	9.1D	$2d$	$e+1$	0	$c_{2i} = c_{2i-1} = i$
(R4)	Folded cubes (P1)	9.2D	n	d	n	0
(R5)	Hamming graphs (C8)	9.2	$2d$	d	sd	$s-1$
(R6)	Dual polar graphs (C3)	9.4	$2d$	d	$q^e\left[{d \atop 1}\right]_q$	q^e-1

Apart from the generalized polygons and the near polygons of Table 6.6 we know a few primitive sporadic regular near polygons, and a large number of imprimitive ones (see §6.8). The parameters of the known primitive sporadic regular near polygons of diameter at least 3 are listed in Table 6.7; among them are two graphs (N1) = (C7), (N2) = (C14) with classical parameters.

Table 6.7. *Parameters of sporadic primitive regular near polygons*
$b_i = k - (\lambda + 1)c_i$

	group	§	n	d	v	k	λ	$(c_1, \cdots, c_d)$	Q-polynomial
(N1) = (C7)	M_{24}	11.4A	6	3	759	30	1	(1,3,15)	yes
(N2) = (C14)	$3^6.M_{12}$	11.3A	6	3	729	24	1	(1,2,12)	yes
(N3)	$2^{10}.M_{22}.2$	11.3F	7	3	1024	22	0	(1,2,6)	no
(N4)	$2^{11}.M_{23}$	11.3E	7	3	2048	23	0	(1,2,3)	no
(N5)	Aut J_2	13.6	8	4	315	10	1	(1,1,4,5)	no

6.6.1. Theorem (BROUWER & WILBRINK [138]). *The parameters of any regular near 2d-gon satisfy at least one of the four restrictions* $d \leqslant 3$, $\lambda = 0$, $\mu = 1$, $c_3 = \mu^2 - \mu + 1$. (But see below.) □

Remark. In fact, [138] contains a structure theory for (not necessarily regular) near $2d$-gons with lines of size at least 3 and such that any two points at distance 2 have at least two common neighbours. It is shown that for any two points x, y at distance i, say, there is a unique geodetically closed sub near $2i$-gon containing them. Thus, in order to prove Theorem 6.6.1, it suffices to consider regular near octagons. In [138] a contradiction is derived from the assumption $d = 4$, $\lambda > 0$, $\mu > 1$, $c_3 \neq \mu^2 - \mu + 1$; however, E.W. Lambeck discovered an arithmetical error in the proof [in line 6 on p. 172], which means that the situation $q := \mu - 1 \geqslant 3$, $s = q^2$, $q^4 + q^3 + 2q \leqslant c_3 - 1 \leqslant q^4 + q^3 + q^2 + 2q + 18$ has not yet been ruled out. (No surviving feasible parameter sets are known. In the notation of [138], p. 172, we must have $a_{\max} \in \{q^6 + q^3 + 1, q^6 + 2q^3 + 1\}$.)

Problem. Fill this gap again.

6.7 Moore graphs

For a regular graph of valency k and diameter d one has the inequality (due to MOORE [546])

$$v \leqslant 1 + k + k(k-1) + \cdots + k(k-1)^{d-1}, \tag{21a}$$

and graphs for which equality holds are called *Moore graphs*. The girth g of a Moore graph is odd and satisfies $g = 2d + 1$. One may also take this as a defining property: SINGLETON [677] shows that a connected graph with diameter d and girth $g = 2d + 1$ necessarily is regular and hence is a Moore graph.

(Moore graphs can also be defined by equality in a lower bound for v: by TUTTE [759], the number v of vertices of a graph of valency k and girth g satisfies

$$v \geqslant \begin{cases} 1 + k + k(k-1) + \cdots + k(k-1)^{(g-3)/2} & \text{if } g \text{ is odd,} \\ 2(1 + (k-1) + \cdots + (k-1)^{(g-2)/2}) & \text{if } g \text{ is even,} \end{cases} \tag{21b}$$

and graphs of odd girth for which equality holds are called Moore graphs. Note that the equality case for even girth corresponds to the regular generalized polygons with $s = 1$.)

6.7.1. Theorem. *A Moore graph with valency $k = 2$ is a polygon, and each $(2d+1)$-gon is a Moore graph. A Moore graph with valency $k \geqslant 3$ has diameter 2 and $k \in \{3,7,57\}$.*

Proof. See DAMERELL [240] and BANNAI & ITO [29]. Damerell's proof consists of showing that there must be an irrational eigenvalue, and that this eigenvalue cannot have integral multiplicity. Bannai & Ito obtained the same result by similar but more complicated methods, partly relying on some computer work by FRIEDMAN [302]. □

Remarks. (i) There are unique examples for $k = 3$ (the Petersen graph) and $k = 7$ (the Hoffman-Singleton graph); these have a distance-transitive automorphism group (cf. HOFFMAN & SINGLETON [401] and Section 13.1). No example with $k = 57$ is known, and no such example can be distance-transitive (ASCHBACHER [10]).
(ii) For arbitrary graphs, one may ask for the maximum value of v, given k and d, or for the minimum value of v, given k and g. The latter problem is that of the construction of so-called 'cages', and is discussed in §6.9. The former problem has been studied in the context of computer networks where each processor is connected to only a few others and the diameter of the network should be small. If we call the difference between both sides of (21a) the *defect* of the graph, then ERDŐS, FAJTLOWICZ & HOFFMAN [277] (for $d = 2$), and BANNAI & ITO [32] (for arbitrary d) show that the only regular graph of defect one is the 4-gon. For further results, see DELORME & FARHI [248], BERMOND, DELORME & FARHI [57], DELORME [247], GÓMEZ & FIOL [334], ALEGRE, FIOL & YEBRA [7], BERMOND, DELORME & QUISQUATER [58].
(iii) BRIDGES & TOUEG [117] introduced a concept of directed Moore graph (regular with indegree and outdegree k, with $v = 1+k+k^2+\cdots+k^d$) and showed that such graphs exist only for $d \leqslant 1$ or $k \leqslant 1$.

6.8 Moore geometries

The concept of a Moore graph was generalized by BOSE & DOWLING [106]; they defined a *Moore geometry of diameter d* as a point-line incidence system satisfying

(i) every point is in exactly $t+1$ lines;

(ii) every line contains exactly $s+1$ points;

(iii) any two distinct points are joined by a unique shortest path in the incidence graph, and any other path joining them has length more than d.

6.8.1. Theorem. *A Moore geometry is either a $(2d+1)$-gon ($s = t = 1$), or has diameter at most two.*

Proof. FUGLISTER [305, 306] eliminates the cases $d = 3$ and $d = 4$, DAMERELL & GEORGIACODIS [243] settle $d > 12$ and DAMERELL [242] handles $4 < d \leqslant 12$. □

Moore geometries with $s = 1$ are Moore graphs. Moore geometries of diameter one are just Steiner systems $S(2,s+1,v)$, i.e., balanced incomplete block designs with $(r,k,\lambda) = (t+1,s+1,1)$. Moore geometries of diameter two have been studied by BOSE & DOWLING [106]. Almost nothing is known (apart from the usual conditions on strongly regular graphs); the pentagon is the only known example. KANTOR [456] has a concept of 'Moore geometry' that is slightly more general than our 'Moore geometry of diameter two', and shows that if a rank 3 group acts, then both concepts coincide with 'Moore graph of diameter two' (see §1.17).

Both Moore graphs and collinearity graphs of Moore geometries are distance-regular. More generally one may consider distance-regular graphs of diameter d with intersection array

$$\iota_d(s,t,c) = \{s(t+1), st, st, \cdots, st; 1, 1, \cdots, 1, c\}.$$

The Moore graphs and Moore geometries correspond to $\iota_d(1,t,1)$ and $\iota_d(s,t,1)$, respectively. Incidence graphs of regular generalized polygons with $s = t$ correspond to $\iota_d(1,t,t+1)$; collinearity graphs of (regular) generalized $2d$-gons correspond to $\iota_d(s,t,t+1)$, and those of (regular) generalized $(2d+1)$-gons to $\iota_d(s,s,1)$.

Remark. The definition of a generalized polygon is symmetric in points and lines, and in the case of a regular generalized polygon with odd girth one immediately sees $s = t$. The definition of a Moore geometry requires that points be joined by a unique shortest path but has no requirements on lines; here one immediately gets $s \leqslant t$. The array $\iota_d(1,t,t+1)$ was treated in SINGLETON [676].

The case of $\iota_d(1,t,c)$ was studied by Bannai & Ito and by Damerell & Georgiacodis (who called such graphs Moore polygons). Their results follow.

6.8.2. Theorem. *Let Γ be a distance-regular graph with intersection array $\iota_d(1,t,c)$, with $t > 1$. Then:*

(i) *If we order the eigenvalues θ_j in increasing order, then the corresponding multiplicities f_j form a unimodal sequence, i.e., satisfy $f_0 < f_1 < \cdots < f_i$ and $f_{i+1} > f_{i+2} > \cdots > f_d = 1$ for some i.*

(ii) *More precisely, we have*

$$f_0 < f_{d-1} \leqslant f_1 < f_{d-2} \leqslant f_2 < \cdots$$

(i.e., $f_j < f_{d-j-1} \leqslant f_{j+1}$ for $0 \leqslant j \leqslant \frac{1}{2}d - 1$), and the equality $f_j = f_{d-j}$ $(j < d-j)$ holds if and only if either $c = t+1$ or $(d,t,c,j) = (3,3,2,1)$; in particular, if $c \neq t+1$, then all eigenvalues of Γ are rational integers.

(iii) *If $c \neq t+1$, then $d \leqslant 5$.*

(iv) *If c and t are both odd, then $d \leqslant 2$.*

Proof. (i) See BANNAI & ITO [30]. (ii) See BANNAI & ITO [31]. (iii) See DAMERELL & GEORGIACODIS [244]; (iii) rests on (ii), which has a long and tedious proof. (iv) See GEORGIACODIS [320]. □

Remarks. If $(d,t,c,j) = (3,3,2,1)$, then Γ is the Odd graph O_4 with $v = 35$ and spectrum $4^1\, 2^{14}\, (-1)^{14}\, (-3)^6$.

If $c = t+1$, then we have $d \leqslant 6$ (since regular generalized octagons cannot have $s = t > 1$).

In the special case $c = 2$, ROOS & VAN ZANTEN [628] show that $d \leqslant 3$ and if $d = 3$, then $t = \frac{1}{2}a(a+1)$ with $a \in \{2, 3, 5, 10, 17, 27, 38, 82, 115, 577\}$. The only known example with $c = 2$ arises from $a = 2$ and is the Odd graph O_4. The same authors also investigated other cases, like $\iota_d(s,t,s+1)$ (cf. ROOS & VAN ZANTEN [629, 630, 631] and ROOS, VAN ZANTEN & COSTER [632], where they show that the eigenvalues are at most quadratic, and that this array is not feasible for $d = 4$ or $d > 5$ when $st > 1$. VAN ZANTEN [799] shows that all eigenvalues of graphs with intersection array $\iota_d(s,t,c)$, $st > 1$, have degree at most three over $\mathbb{Q}$, and FUGLISTER [307] shows that these graphs have diameter $d \leqslant 13$. A short proof of the impossibility of $\iota_d(1,t,c)$ for $d = 24, 25$ using Newton polygons was given by GEORGIACODIS [321].

6.9 Cages

Now that we have discussed generalized polygons and Moore graphs, let us digress for a moment from our main topic and look at the related subject of 'cages'. Let $k \geqslant 2$, $g \geqslant 3$. A *(k,g)-cage* is a regular graph of valency k, girth g and minimum number of vertices $v = v(k,g)$. The problem of determining $v(k,g)$ was posed in 1959 by F. Kárteszi who noticed that $v(3,5) = 10$ was realized by the Petersen graph. SACHS [635] first showed that $v(k,g)$ is finite, and ERDŐS & SACHS [279] gave an upper bound. The best known (very crude) upper bounds for general k and g can be found in SAUER [636]. A lower bound for $v(k,g)$ was given in (21b) above, and we saw that the case of equality corresponds to Moore graphs (for odd girth) or incidence graphs of regular generalized polygons with $s = t$ (for even girth).
Thus:

(i) the $(2,g)$-cage is the g-circuit C_g, and $v(2,g) = g$;

(ii) the $(k, 3)$-cage is the complete graph K_{k+1}, and $v(k, 3) = k+1$;

(iii) the $(k, 4)$-cage is the complete bipartite graph $K_{k,k}$, and $v(k, 4) = 2k$.

If $k = q+1$ for a prime power q, then

(iv) a $(k, 6)$-cage is the incidence graph of a projective plane $PG(2,q)$, and $v(k, 6) = 2(q^2+q+1)$;

(v) a $(k, 8)$-cage is the incidence graph of a generalized quadrangle $GQ(q,q)$, and $v(k, 8) = 2(q+1)(q^2+1)$;

(vi) a $(k, 12)$-cage is the incidence graph of a generalized hexagon $GH(q,q)$, and $v(k, 12) = 2(q+1)(q^4+q^2+1)$.

The two known Moore graphs yield

(vii) the $(3,5)$-cage is the Petersen graph (PETERSEN [592]), and $v(3,5) = 10$;

(viii) the $(7,5)$-cage is the Hoffman-Singleton graph (HOFFMAN & SINGLETON [401]), and $v(7,5) = 50$.

(The $(3,6)$-cage is also known as the Heawood graph ($v = 14$), cf. HEAWOOD [366], and the $(3,8)$-cage as the Tutte-Coxeter graph ($v = 30$), cf. TUTTE [757]. For the $(k, 6)$-cages, cf. KÁRTESZI [461]; for the $(k, 8)$- and $(k, 12)$-cages, cf. BENSON [49].)

In a few other cases the extremal graph is known.

- One has $v(3,7) = 24$, and the unique $(3,7)$-cage Γ is known as the McGee graph. (It was found by Sachs (unpublished, cf. KÁRTESZI [461]) and by MCGEE [528], and shown to be unique in TUTTE [759].) A description: Take as vertex set $X = \mathbf{Z}_8 \times \{-1,0,1\}$; join $(a, 0)$ to $(a, \pm 1)$ and $(a+4,0)$; furthermore, join $(a, \pm 1)$ to $(a \pm 1, \mp 1)$, $(a \pm 3, \mp 1)$. AutΓ has order 48, and has two orbits on X: for $\gamma = (a, 0)$ we find that $\Gamma_4(\gamma)$ is an edge, while $\Gamma_4(\gamma)$ is a non-edge for $\gamma = (a, \pm 1)$. The graphs induced on the two orbits are $4K_2$ and $2C_8$.

- One has $v(4,5) = 19$, and the unique $(4,5)$-cage is known as the Robertson graph (ROBERTSON [614]). A description: Take as vertex set $X = \{a_i\ (i \in \mathbf{Z}_{12}),\ b_i\ (i \in \mathbf{Z}_4),$

c_i ($i \in \mathbf{Z}_3$)}; and join a_i to $a_{i\pm1}$, b_i, c_i; furthermore, join b_i to b_{i+2}. Aut Γ has order 24, and has three orbits on X: the two vertices in $\Gamma_3(\gamma)$ have distance 1, 2, 3 for $\gamma = b_i, a_i, c_i$, respectively. (For some remarks on the spectrum, see BROWN [139].)

- One has $v(5,5) = 30$ (KÁRTESZI [461]). One of the three known (5,5)-cages Γ is known as the Robertson-Wegner graph (WEGNER [772]). A description: Take as vertex set $X = \{a_i, b_i, c_i \ (i \in \mathbf{Z}_{10})\}$; join a_i to $a_{i\pm2}$, b_i, c_{i+2}, c_{i+3}; furthermore, join b_i to $b_{i\pm1}$, c_{i-4}, c_{i-1}; and c_i to c_{i+5}. Aut Γ has order 20 and has three orbits on X (an additional automorphism is given by $a_i \mapsto a_{-i}$, $b_i \mapsto b_{-i}$, $c_i \mapsto c_{5-i}$). A nonisomorphic (5,5)-cage Γ' may be obtained as an induced subgraph of the Hoffman-Singleton graph (on 3 pentagons and 3 pentagrams in the 5 pentagons + 5 pentagrams description of this graph, cf. construction A in §13.1; here Aut Γ' has order 10 and has four orbits on the vertex set).

- One has $v(6,5) = 40$ and the unique (6,5)-cage is a subgraph of the Hoffman-Singleton graph (the complement of a Petersen graph), see O'KEEFE & WONG [576] and WONG [792].

- One has $v(7,6) = 90$ and the unique (7,6)-cage Γ is the unique 3-fold cover with girth 6 of the incidence graph of the points and planes of $PG(3,2)$; it is the incidence graph of the elliptic semiplane found by BAKER [21], cf. §1.10. It was rediscovered by O'KEEFE & WONG [578] and ITO [427] found that Aut $\Gamma \cong 3\cdot\mathrm{Alt}(7)$. Apparently the elliptic semiplane is not self-dual, and Aut Γ has two orbits on Γ. The point stabilizer is $L_3(2)$; note that $3\cdot\mathrm{Alt}(7)$ has two conjugacy classes of subgroups $L_3(2) \cong L_2(7)$.

- Finally, O'KEEFE & WONG [577] found $v(3,10) = 70$; see also BALABAN [23, 24]. There are three distinct graphs achieving $v(3,10) = 70$ (WONG [794]); these graphs are bipartite.

For a survey on the topic of cages, see WONG [793].

If we call the difference between both sides of (21b) the *excess* of the graph, then BANNAI & ITO [32] showed that no regular graph with odd girth $\geqslant 5$ has excess one. For further results on graphs with small excess, see BIGGS [76, 78], BIGGS & HOARE [82], BIGGS & ITO [83] and KOVÁCS [468]. For an explicit construction of graphs with large girth, cf. MARGULIS [517], IMRICH [419], BIGGS [79] and LUBOTZKY, PHILLIPS & SARNAK [511].

6.10 The remaining primitive graphs

A number of primitive graphs are known which do not fit into one of the preceding families. Their parameters are given in Table 6.8. (Note that the case $q = 3$ of (D1) is a generalized hexagon of order (2,2), and that the case $j = r = 2$ of (D0) is the folded 7-cube.) All are distance-transitive except perhaps some in (D0), but those in (D1) only for $q \leqslant 4$; none is Q-polynomial, except for the folded 7-cube.

Table 6.8. *Other primitive distance-regular graphs ($q \geqslant 3$ a prime power, r a power of 2, $j \geqslant 1$)*

	group	§	d	v	intersection array
(D0)		11.2	3	r^{4j+2}	$\{r^{2j+1}-1, r^{2j+1}-r, r^{2j}(r-1)+1; 1, r, r^{2j}-1\}$
(D1)	$P\Gamma U_3(q^2)$	12.4	3	$q^2(q^2-q+1)$	$\{q^2-q, q^2-q-2, q+1; 1, 1, q^2-2q\}$
(D2)	Aut S_6	13.1A	3	36	{5,4,2; 1,1,4}
(D3)	$L_2(19)$	13.3	3	57	{6,5,2; 1,1,3}
(D4)	$L_2(25).2$	12.2	3	65	{10,6,4; 1,2,5}
(D5)	$P\Sigma U_3(5^2)$	13.1	3	175	{12,6,5; 1,1,4}
(D6)	$L_2(16).2$	12.1	3	68	{12,10,3; 1,3,8}
(D7)	$2^{10}.M_{22}$	11.3F	3	1024	{231,160,6; 1,48,210}
(D8)	$2^{11}.M_{23}$	11.3E	3	2048	{253,210,3; 1,30,231}
(D9)	$L_2(7).2$	12.3	4	28	{3,2,2,1; 1,1,1,2}
(D10)	$M_{22}.2$	11.4C	4	330	{7,6,4,4; 1,1,1,6}
(D11)	$L_3(4)$	11.4F	4	280	{9,8,6,3; 1,1,3,8}
(D12)	J_1	13.5	4	266	{11,10,6,1; 1,1,5,11}
(D13)	Suz	13.7	4	22880	{280,243,144,10; 1,8,90,280}
(D14)	$L_2(17)$	13.4	7	102	{3,2,2,2,1,1,1; 1,1,1,1,1,1,3}

6.11 Bipartite distance-regular graphs; imprimitive regular near polygons

We remarked already that every bipartite graph is a near polygon; hence all bipartite distance-regular graphs are imprimitive regular near polygons. Here we show the converse.

6.11.1. Proposition. *A regular near polygon is imprimitive if and only if it is bipartite.*

Proof. By Theorem 4.2.1, an imprimitive regular near polygon Γ is bipartite or antipodal. If Γ is antipodal, then $c_d = b_0 = k$, whence by (15), $\lambda = 0$, $a_d = 0$. But, then (14) implies $b_i = k - c_i$, $a_i = 0$ for all $i \leqslant d-1$. Hence Γ is bipartite. □

Among the generalized polygons in Table 6.5, precisely those with $s = 1$ are bipartite. Their halved graphs are complete graphs if $2d = 6$, generalized quadrangles if $2d = 8$, and generalized hexagons if $2d = 12$.

The bipartite graphs in Table 6.6 are easily recognized by the condition

$$n = 2d, \lambda = 0. \tag{22}$$

They are the ordinary polygons with $2d$ vertices, the doubled Odd and Grassmann graphs, the folded cubes of even dimension, the binary Hamming graphs (d-cubes) and the dual polar graphs with $e = 0$. Their halved graphs are polygons, Johnson and Grassmann graphs, partition graphs (P3), and the graphs (C4), respectively.

The other known bipartite distance-regular graphs are listed in Table 6.9. Two general families occur: The graphs in (B0) are the incidence graphs of *square* 2-

$(1+k(k-1)/\mu,k,\mu)$ *designs,* cf. Theorem 1.6.1. The graphs (B1) come from Kasami codes, cf. §11.2. The graphs (B2) - (B7) are incidence graphs of partial μ-geometries, cf. §1.7. The graphs in (B2) are the incidence graphs of symmetric (m,μ)-nets; in particular, Hadamard graphs belong to (B2).
The graphs (B8) - (B13) are bipartite doubles; (B8) is the bipartite double of the Hoffman-Singleton graph, (B9) of the Gewirtz graph, (B10) of the complement of the block graph of the Steiner system $S(3,6,22)$, (B11) of the Higman-Sims graph, and (B12) and (B13) of (N3) and (N4), respectively.
The halved graphs of these graphs are of diameter at most 2 for (B0)-(B11); the halved graphs of (B12) and (B13) are (D7) and (D8), and the halved graph of (B14) is the antipodal graph (A7) (see Table 6.10).

Table 6.9. *Parameters of bipartite distance-regular graphs (thin regular near polygons)*

	group	§	d	v	$(c_1,\cdots,c_d)$
(B0)		1.6	3	$2(1+k(k-1)/\mu)$	$(1,\mu,k)$
					square 2-$(v/2,k,\mu)$ design
(B1)		11.2	4	$2q^{4j+2}$	$(1,q,q^{2j}-1,q^{2j+1})$
					q power of 2, $j \geqslant 1$
(B2)		1.7	4	$2m^2\mu$	$(1,\mu,m\mu-1,m\mu)$
					symmetric (m,μ)-net
(B3)		12.6	4	$2(q^2+1)(q^2-q+1)$	$(1,1,(q-1)^2,q^2-q+1)$
					q power of 2
(B4)	$3^4.S_6$	11.5	4	162	(1,1,2,6)
(B5)	$P\Sigma U_3(5^2)$	13.1	4	100	(1,5,12,15)
(B6)	$M_{12}.2$	11.4F	4	288	(1,2,5,12)
(B7)	$2^{12}.M_{24}$	11.3D	4	4096	(1,2,3,24)
(B8)	$2\times P\Sigma U_3(5^2)$	13.1	5	100	(1,1,6,6,7)
(B9)	$2\times L_3(4).4$	11.4G	5	112	(1,2,8,9,10)
(B10)	$2\times M_{22}.2$		5	154	(1,4,12,15,16)
(B11)	$2\times$HiS.2		5	200	(1,6,16,21,22)
(B12)	$2\times 2^{10}.M_{22}.2$	11.3F	7	2048	(1,2,6,16,20,21,22)
(B13)	$2\times 2^{11}.M_{23}$	11.3E	7	4096	(1,2,3,20,21,22,23)
(B14)	$3\cdot\mathrm{Aut}\,S_6$	13.2A	8	90	(1,1,1,1,2,2,2,3)

6.12 Antipodal distance-regular graphs

Proposition 4.2.2 shows that any primitive distance-regular graph has only a limited (finite) number of antipodal distance-regular covers. In Table 6.10 below we list the nonbipartite antipodal distance-regular graphs known to us. We have already met the dodecahedron (A13) and the Taylor graphs (A2) in Chapter 1.

In §9.1D we discuss the covers of Johnson schemes. Some antipodal covers of coset graphs are discussed in §11.3H. Some antipodal covers derived from finite geometries are constructed in §12.5. A generalization of the construction of Taylor graphs is discussed in §12.7, and produces graphs with parameters as given under (A3), (A4) and (A5). In VAN BON & BROUWER [100] the antipodal covers of the classical families of

Table 6.10. *Some nonbipartite antipodal r-covers of distance-regular graphs*

	group	§	d	$v = r\times\overline{v}$	intersection array
(A0)		12.5	3	$(s+1)\times(st+1)$	$\{st, s(t-1), 1;\ 1, t-1, st\}$
	$GQ(s,t)$ minus spread				
(A1)		12.5	3	$r\times r\mu$	$\{r\mu-1, (r-1)\mu, 1;\ 1, \mu, r\mu-1\}$
	cover of a symplectic line; $r = p^i$, $\mu = p^{i+2j}$, $i,j \geqslant 0$; for $r = \mu$, see also §11.2				
(A2)		1.5	3	$2\times(k+1)$	$\{k, \mu, 1;\ 1, \mu, k\}$
	(complement of) regular two-graph on $k+1$ vertices and μ coherent triples/pair				
(A3)	$L_2(q)$	12.7A, 12.5	3	$r\times(q+1)$	$\{r\mu+1, (r-1)\mu, 1;\ 1, \mu, r\mu+1\}$
	cover of an oval; $q = r\mu+1$ prime power, $\mu(r-1)$ even				
(A4)		12.7B	3	3×29	$\{28,18,1;\ 1,9,28\}$
(A5)		12.7C	3	3×497	$\{496,330,1;\ 1,165,496\}$
(A6)	Sym(7)	13.1	3	6×7	$\{6,5,1;\ 1,1,6\}$
(A7)	3·Sym(6)	13.2A	4	3×15	$\{6,4,2,1;\ 1,1,4,6\}$
(A8)	3·Sym(7)	13.2B	4	3×21	$\{10,6,4,1;\ 1,2,6,10\}$
(A9)	$3\cdot P\Omega_6^-(3).2$	13.2C	4	3×126	$\{45,32,12,1;\ 1,6,32,45\}$
(A10)	$3\cdot P\Omega_7^-(3).2$	13.2D	4	3×378	$\{117,80,24,1;\ 1,12,80,117\}$
(A11)	$3\cdot Fi_{24}$	13.2	4	3×306936	$\{31671,28160,2160,1;\ 1,1080,28160,31671\}$
(A12)	2_+^{1+4}.Alt(5)	9.2E	4	2×16	$\{5,4,1,1;\ 1,1,4,5\}$
(A13)	$2\times$Alt(5)	1.0, 10.4	5	2×10	$\{3,2,1,1,1;\ 1,1,1,2,3\}$
(A14)		11.3H	6	2×1024	$\{22,21,20,3,2,1;\ 1,2,3,20,21,22\}$
(A15)		11.3H	6	2×512	$\{21,20,16,6,2,1;\ 1,2,6,16,20,21\}$
(A16)		11.3H	6	4×512	$\{21,20,16,9,2,1;\ 1,2,3,16,20,21\}$
(A17)		11.3H	4	3×81	$\{20,18,4,1;\ 1,2,18,20\}$
(A18)		11.3H	5	3×243	$\{22,20,18,2,1;\ 1,2,9,20,22\}$
(A19)	$3\cdot M_{22}.2$	11.4D	8	3×330	$\{7,6,4,4,4,1,1,1;\ 1,1,1,2,4,4,6,7\}$

distance-regular graphs are determined. In §11.4G we find the antipodal covers of the Gewirtz graph.

Problem. Surely, many more antipodal distance-regular graphs exist. Find them.

Chapter 7
Distance-Transitive Graphs

There are only finitely many distance-transitive graphs with given valency > 2. This result was first shown in CAMERON, PRAEGER, SAXL & SEITZ [183] by use of the classification of finite simple groups. Below we give a proof due to WEISS [779] which is independent of this classification. A basic ingredient to the proof of Weiss' theorem is the celebrated Thompson-Wielandt Theorem. The proof of the latter theorem requires group-theoretic preparation which can be found in Section 7.1. The Thompson-Wielandt Theorem is the content of Section 7.2 and Weiss' theorem is in Section 7.3. Subsequently we discuss results in the cases of large girth (Section 7.4), small valency (Section 7.5), and imprimitive graphs (Section 7.6). The state of the art in overall classification and a few related results are given in the final sections (7.7-8).

7.1 Some elementary group theory

We need some group theoretic notions, which we shall briefly review. See GORENSTEIN [338] and SUZUKI [704] for details.

If H is a subgroup of G, we write $H \leqslant G$. If H is a normal subgroup of G, we write $H \trianglelefteq G$. If H is a subnormal subgroup (i.e. is the first member of a chain of normal subgroup embeddings leading to G), we write $H \trianglelefteq\trianglelefteq G$. The center of G is denoted by $Z(G)$. If H, K are subgroups of G, then $[H,K]$ stands for the subgroup generated by all commutators of the form $[h,k] = hkh^{-1}k^{-1}$ for $h \in H$, $k \in K$. Thus, $[G,G]$ is the commutator subgroup of G. A group K is called *perfect* if $[K,K] = K$ and *quasisimple* if it is perfect and $K/Z(K)$ is simple. By $E(G)$ we denote the *layer* of G: the (normal) subgroup generated by all *components*, i.e., all quasisimple subnormal subgroups K of G.

7.1.1. Lemma. *Let G be a group with subgroups K, L, and M. Then*

(i) *Both K and L normalize $[K,L]$.*

(ii) *If $[[K,L],M] = [[L,M],K] = 1$, then $[[M,K],L] = 1$.*

(iii) *If M is a proper normal subgroup of G and G is quasisimple, then $M \leqslant Z(G)$.*

(iv) *If K is a component of G and $[K,[K,L]] = 1$, then $[L,K] = 1$.*

Proof. (This is well known, see e.g. GORENSTEIN [338], Theorems 2.1, 2.3 for (i) and (ii). Also ASCHBACHER [14], §8, and SUZUKI [704], Vol. 2, Ch. 4, §1.)
(i) Clearly, $[K,L] = [L,K]$. For $g \in L$ we have $g[k,l]g^{-1} = [k,g]^{-1}[k,gl] \in [K,L]$ for $k \in K$ and $l \in L$.
(ii) We have to show that if $k \in K$, $l \in L$, $m \in M$, then $l[m,k] = [m,k]l$. But

$$l[m,k] = ml[l^{-1},m^{-1}]km^{-1}k^{-1} = mlk[l^{-1},m^{-1}]m^{-1}k^{-1} =$$
$$= mk[k^{-1},l]m^{-1}lk^{-1} = mkm^{-1}[k^{-1},l]lk^{-1} = [m,k]l.$$

(iii) Observe that $MZ(G)/Z(G)$ is a normal subgroup of $G/Z(G)$ which must be proper since $MZ(G) = G$ would imply that $G/M = MZ(G)/M \cong Z(G)/(M \cap Z(G))$ is abelian, so that $M \geqslant [G,G] = G$. Finally, (iv) is a direct consequence of (ii). □

7.1.2. Theorem. *Suppose G is a finite group. Then its layer $E(G)$ is a central product of its components, i.e., $E(G)/Z(G)$ is the direct product of the groups $K/Z(K)$, for K ranging over all components of G.*

Proof. The proof is by induction on $|G|$. Let K and L be components of G. Then there are subnormal chains:

$$K = K_0 \trianglelefteq K_1 \trianglelefteq \cdots \trianglelefteq K_m = G$$
$$L = L_0 \trianglelefteq L_1 \trianglelefteq \cdots \trianglelefteq L_n = G.$$

If $G = K$, then $K = L$ due to (iii) of the above lemma, for otherwise, we would have $L \leqslant Z(K)$, so that $L = [L,L] = 1$, which is absurd. Therefore, we may assume $m \geqslant 1$, and, similarly, $n \geqslant 1$. Now $[K,L]$ is a subgroup of $[E(K_{m-1}),L]$. But $E(K_{m-1})$ is normal in G, so $[E(K_{m-1}),L] = E(K_{m-1})$, and $[K,L]$ is a subgroup of $E(K_{m-1})$. Suppose now that $1 \neq [K,L]$. Then, by (iv) of the above lemma, also $1 \neq [[K,L],K]$, and $1 \neq [[[K,L],K],K]$. But K is a normal subgroup of $E(K_{m-1})$ by the induction hypothesis, so $[[K,L],K]$ is a nontrivial normal subgroup of K, not contained in the center of K. By (iii) of the above lemma it follows that $[[K,L],K] = K$. On the other hand, by (i) of the lemma, $[K,L]$ is normalized by K, whence $[[K,L],K] \leqslant [K,L]$. Thus, $K \leqslant [K,L]$, and, by symmetry of K and L, also $L \leqslant [L,K]$. But then K and L are two noncommuting components of K_{m-1}, whence $K = L$ by the induction hypothesis. □

Let G be a finite group. For any prime p, there exist a maximal normal subgroup of G of order a power of p, denoted by $O_p(G)$, and a minimal normal subgroup of G whose index in G is a power of p, denoted by $O^p(G)$. Both are characteristic subgroups of G. The product over all p of $O_p(G)$ is called the *Fitting subgroup*, notation $F(G)$ of G. The product of $E(G)$ and $F(G)$ is a characteristic subgroup of G with the property that it is trivial if and only if G is trivial. (It is called the *generalized Fitting subgroup*, cf. HUPPERT & BLACKBURN [417], Ch. X, §13.)

7.1.3. Proposition. *Let $H \trianglelefteq\trianglelefteq G$ and let p be a prime. Then:*

(i) *If $|G/H|$ is a power of p, then $O^p(G) = O^p(H)$.*

(ii) $O_p(G) \leqslant N_G(O^p(H))$.

(iii) $O_p(H) \leqslant O_p(G)$.

(iv) $E(G) \leqslant N_G(H)$.

Proof. (i) Consider a chain $H \trianglelefteq H_1 \trianglelefteq \cdots \trianglelefteq H_n = G$. Since $O^p(H)$ is a characteristic subgroup of H, we have $O^p(H) \trianglelefteq H_1$. Furthermore, $|H_1/O^p(H)| = |H_1/H|\cdot|H/O^p(H)|$ is a power of p, so, by definition of $O^p(.)$, we have $O^p(H_1) \leqslant O^p(H)$. As $O^p(H_1) \trianglelefteq H_1$, we also have $O^p(H_1) \trianglelefteq H$, and $|H/O^p(H_1)|$, being a divisor of $|H_1/O^p(H_1)|$, is a power of p. Therefore,

$O^p(H) \leqslant O^p(H_1)$, whence $O^p(H) = O^p(H_1)$. We can finish by induction on n.
(ii) Consider $X = O_p(G)O^p(H)$. Since $O_p(G) \trianglelefteq G$, it is a normal subgroup of X. Moreover, $O^p(H) \trianglelefteq H \trianglelefteq\trianglelefteq G$, so $O^p(H) \trianglelefteq\trianglelefteq X$. As $|X/O^p(H)|$ is a divisor of $|O_p(G)|$, it follows from (i) that $O^p(X) = O^p(O^p(H)) = O^p(H)$. Thus $O_p(G) \leqslant X \leqslant N_G(O^p(H))$, proving (ii).
(iii) Let $H \trianglelefteq H_1 \trianglelefteq \cdots \trianglelefteq H_n = G$ be a normal chain. We proceed by induction on n. As $O_p(H)$ is characteristic in H, we have $O_p(H) \trianglelefteq H_1$, whence $O_p(H) \leqslant O_p(H_1)$. The induction hypothesis yields $O_p(H_1) \leqslant O_p(G)$, and concatenation of the last two inclusions ends the proof of (iii).
(iv) Let K be a component of G and let $H \trianglelefteq H_1 \trianglelefteq \cdots \trianglelefteq H_n = G$ be a subnormal chain. In view of induction, it suffices to show that we have either $[H_{n-1},K] = 1$ or $K \leqslant H_{n-1}$. Since $[K,H_{n-1}] \trianglelefteq H_{n-1} \trianglelefteq G$ and $[K,H_{n-1}] \leqslant [E(G),H_{n-1}] \leqslant E(G)$, the subgroup $[K,H_{n-1}]$ is subnormal in $E(G)$, whence normal in $E(G)$. If K is not contained in $[K,H_{n-1}]$, then, by the theorem, since the latter group is normalized by K, we must have $[[K,H_{n-1}],K] = 1$. By (iv) of the lemma, this implies $[K,H_{n-1}] = 1$. Hence (iv). □

7.2 The Thompson-Wielandt Theorem

Let t be a natural number. We say that a group G acts *locally t-arc transitively* on a graph Γ if G acts on Γ as a group of automorphisms in such a way that for each vertex γ of Γ, the stabilizer G_γ acts transitively on the set of *reduced paths* of Γ starting at γ and having length t (i.e., paths $\gamma = \gamma_0, \gamma_1, \cdots, \gamma_t$ where $\gamma_{i+1} \neq \gamma_{i-1}$ for $1 \leqslant i \leqslant t-1$). Note that, in particular if Γ is bipartite, local t-arc transitivity need not imply transitivity of G on the vertex set of Γ. If S is a set of vertices or a path of Γ and m is a natural number, then $G_S^{(m)}$ denotes the subgroup of G consisting of all elements fixing each vertex at distance at most m to some member of S. We shall use notation like $G_{\beta,\gamma}^{(m)}$ for $G_{\{\beta,\gamma\}}^{(m)}$, omitting the braces whenever S is explicitly given as a list of vertices.

7.2.1. Proposition. *Let Δ be a connected graph all of whose vertices have valency at least two. Suppose $G \leqslant \mathrm{Aut}\,\Delta$ is locally t-arc transitive for some $t \geqslant 1$. Then, for all $\beta, \gamma, \delta, \epsilon \in \Delta$ with $\beta \sim \gamma \sim \delta$, $\beta \neq \delta$ and all $m, n \in \mathbf{N}$, the following hold.*

(i) *If $m \leqslant t$, then G is locally m-arc transitive.*

(ii) *Each vertex of Δ is in the same $\langle G_\beta, G_\gamma\rangle$-orbit as β or γ. If Δ is not bipartite, G is transitive (and edge-transitive) on Δ.*

(iii) *If $t \geqslant 2$, the stabilizer G_γ is doubly transitive on $\Delta(\gamma)$. In particular, if γ has valency > 2, then $G_\gamma = \langle G_{\gamma,\beta}, G_{\gamma,\delta}\rangle$.*

(iv) *If $H \leqslant G_\gamma$, then H normalizes $G_\gamma^{(n)}$. In particular, if $d(\gamma,\epsilon) \leqslant m$, then $G_\epsilon^{(m)}$ normalizes $G_\gamma^{(n)}$.*

(v) *If $H \leqslant G_{\beta,\gamma}$ is normalized by both G_β and G_γ, then $H = 1$.*

(vi) *Suppose $t \geqslant 2$. If $H \leqslant G_{\gamma,\delta}$ is normalized by both G_β and G_δ, then $H = 1$.*

(vii) *Suppose $t \geqslant 2$. If γ has valency > 2 and $G_{\gamma,\delta}$ normalizes $G_{\beta,\gamma}^{(1)}$, then $G_{\beta,\gamma}^{(1)} = G_\gamma^{(2)}$.*

(viii) *Suppose $t \geqslant 2$ and $m \geqslant 1$. If $G_\beta^{(m)} = G_\delta^{(m)}$, then $G_\beta^{(m)} = 1$.*

Proof. (i) Any reduced path of length m can be extended to a reduced path of length

t, since Δ has valency $\geqslant 2$.
(ii) Observe that G_α is transitive on $\Delta(\alpha)$ for each $\alpha \in \Delta$ according to (i). Applying this observation to $hG_\gamma h^{-1} = G_{h(\gamma)}$ for $h \in G_\beta$, we get that β and all vertices of $\Delta_2(\beta)$ belong to the same $\langle G_\beta, G_\gamma\rangle$-orbit, and similarly for γ. Statement (ii) now follows from connectedness of Δ.
(iii) Since $t \geqslant 2$, the subgroup $G_{\beta,\gamma}$ is transitive on $\Delta(\gamma)\setminus\beta$. As G_γ is transitive on $\Delta(\gamma)$, this yields (iii).
(iv) If $\alpha \in \Delta_{\leqslant n}(\gamma)$, $g \in G_\gamma^{(n)}$, and $h \in H$, then $h(\gamma) = \gamma$, so $h(\alpha) \in \Delta_{\leqslant n}(\gamma)$, whence $gh(\alpha) = h(\alpha)$, and $h^{-1}gh(\alpha) = \alpha$. Thus $h^{-1}gh \in G_\gamma^{(n)}$, proving (iv).
(v) We have $H \trianglelefteq \langle G_\beta, G_\gamma\rangle$. Since β, γ are fixed vertices of H and $\langle G_\beta, G_\gamma\rangle$ stabilizes the set of all fixed vertices of H, it follows from (ii) that H fixes every vertex of Δ.
(vi) We show that H fixes β. If γ has valence 2, this is obvious; otherwise this follows since (by (iii)) $G_\gamma = \langle G_{\gamma,\delta}, G_{\gamma,\beta}\rangle$ normalizes H, and H fixes δ. Now, using $H \trianglelefteq \langle G_\beta, G_\delta\rangle$, we can finish as in (v), using $t \geqslant 2$ to derive that each vertex of Δ is conjugate by an element of $\langle G_\beta, G_\delta\rangle$ to β, γ, or δ.
(vii) In view of (iv) and (iii), the group $G_{\beta,\gamma}^{(1)}$ is normal in $\langle G_{\beta,\gamma}, G_{\gamma,\delta}\rangle = G_\gamma$. Now $G_{\beta,\gamma}^{(1)}$ has a fixed vertex in $\Delta_2(\gamma) \cap \Delta_1(\beta)$, and G_γ is transitive on $\Delta_2(\gamma)$. Therefore, all vertices of $\Delta_2(\gamma)$ are fixed by $G_{\beta,\gamma}^{(1)}$, whence $G_{\beta,\gamma}^{(1)} \leqslant G_\gamma^{(2)}$. The other inclusion is obvious.
(viii) By (iv), we have $G_\beta^{(m)} \trianglelefteq \langle G_{\beta,\gamma}, G_{\delta,\gamma}\rangle$. Thus, by (iii), $G_\beta^{(m)} \leqslant G_{\beta,\gamma}$ is normalized by both G_β and G_γ (observe that, if the valency of γ is 2, then this conclusion is still valid, despite the fact that G_γ is not generated by $G_{\beta,\gamma}$ and $G_{\delta,\gamma}$), and we can finish by (v). □

7.2.2. Theorem (THOMPSON [735]-WIELANDT [784]; cf. KNAPP [466], WEISS [775], FAN [281]). *Suppose Δ is a connected graph all of whose vertices have valency at least two. If $G \leqslant \operatorname{Aut}\Delta$ is locally 2-arc transitive on Δ with $G_\gamma^{(2)} \neq 1$ for all $\gamma \in \Delta$, then each edge contains a vertex β such that $G_\beta^{(2)}$ is a nontrivial p-group for some prime p. Also, for $\gamma \in \Delta(\beta)$, we have $E(G_\gamma) = E(G_\beta) = 1$, and either $G_\gamma^{(2)}$ is a p-group, or $O_p(G_\gamma^{(2)}) \neq 1$ and $G_\beta^{(2)} = G_\beta^{(3)}$.*

Proof. Let $\beta\gamma$ be an edge of Δ and p a prime number. We proceed in four steps.

STEP 1. *If $O^p(G_\beta^{(2)}) \neq 1$, then $O_p(G_\gamma) = O_p(G_{\beta,\gamma}) = O_p(G_\gamma^{(1)})$.*

Let $\delta \in \Delta(\gamma)$. Suppose $O_p(G_\gamma) \cup O_p(G_{\beta,\gamma})$ is not contained in G_δ. Then there is $g \in O_p(G_\gamma) \cup O_p(G_{\beta,\gamma})$ with $g(\delta) \neq \delta$. Since $G_\delta^{(2)} \trianglelefteq G_\gamma^{(1)} \trianglelefteq G_\gamma$ by Proposition 7.2.1 (iv), we get $G_\delta^{(2)} \trianglelefteq\trianglelefteq G_\gamma$ and $G_\delta^{(2)} \trianglelefteq\trianglelefteq G_{\beta,\gamma}$. It follows from Proposition 7.1.3 (ii) that $O_p(G_\gamma)$ and $O_p(G_{\beta,\gamma})$ both normalize $O^p(G_\delta^{(2)})$, whence $O^p(G_\delta^{(2)}) = gO^p(G_\delta^{(2)})g^{-1} = O^p(G_{g(\delta)}^{(2)})$. Thus, $O^p(G_\delta^{(2)}) = 1$ by (vi) of the proposition, contradicting the hypothesis as β and δ are in the same G-orbit. Therefore, $O_p(G_\gamma)\cdot O_p(G_{\beta,\gamma}) \leqslant G_\gamma^{(1)}$. Now $O_p(G_\gamma) \trianglelefteq G_{\beta,\gamma}$ and $O_p(G_{\beta,\gamma}) \trianglelefteq G_\gamma^{(1)}$, and, clearly, $G_\gamma^{(1)} \trianglelefteq G_\gamma$, so by Proposition 7.1.3 (iii) we have $O_p(G_\gamma) \leqslant O_p(G_{\beta,\gamma}) \leqslant O_p(G_\gamma^{(1)}) \leqslant O_p(G_\gamma)$, establishing Step 1.

STEP 2. $E(G_\beta) = E(G_\gamma) = E(G_{\beta,\gamma}) = 1$.

By the hypothesis, $G_\beta^{(2)} \neq 1$, so there is a prime p for which the condition in Step 1 is satisfied. By the same argument as in the proof of Step 1 (replacing O_p by E and using Proposition 7.1.3 (iv)), it follows that $E(G_\gamma) = E(G_{\beta,\gamma}) = E(G_\gamma^{(1)})$. Since the conclusion is independent of p, the hypothesis $G_\gamma^{(2)} \neq 1$ leads to $E(G_\beta) = E(G_{\beta,\gamma})$

$= E(G_\beta^{(1)})$ in the same way as before, and Step 2 follows by application of Proposition 7.2.1 (v) to $E(G_\beta) = E(G_{\beta,\gamma}) = E(G_\gamma)$.

STEP 3. *There is a prime p and a vertex $\alpha \in \{\beta,\gamma\}$ such that $G_\alpha^{(2)}$ is a nontrivial p-group.*

Suppose the contrary. Then, by Step 1, for each prime number p, we have $O_p(G_\beta) = O_p(G_{\beta,\gamma}) = O_p(G_\gamma)$, whence $O_p(G_\beta) = 1$. (Use Proposition 7.2.1 (v) again.) Due to Step 2, we obtain $E(G_\beta)F(G_\beta) = 1$, whence $G_\beta = 1$, which is clearly absurd. Hence Step 3.

STEP 4. *Suppose that $G_\beta^{(2)}$ is a nontrivial p-group, and that $G_\gamma^{(2)}$ is not a p-group. Then $G_\beta^{(2)} = G_\beta^{(3)}$ and $O_p(G_\gamma^{(2)}) \neq 1$.*

Let $\delta \in \Delta(\gamma) \setminus \{\beta\}$. Then $G_\beta^{(2)} \trianglelefteq G_\gamma^{(1)} \trianglelefteq G_{\gamma,\delta}$ (use Proposition 7.2.1 (iv)), so $G_\beta^{(2)} \leqslant O_p(G_{\gamma,\delta})$. Since $G_\gamma^{(2)}$ is not a p-group, $O^p(G_\gamma^{(2)}) \neq 1$, whence, by Step 1, $O_p(G_{\gamma,\delta}) = O_p(G_\delta^{(1)}) \leqslant G_\delta^{(1)}$. Consequently, $G_\beta^{(2)} \leqslant G_\delta^{(1)}$. By transitivity, varying γ and δ, we obtain $G_\beta^{(2)} = G_\beta^{(3)}$. As this is a p-group (cf. Step 3) which is normal in $G_\gamma^{(2)}$, we are done. □

Remark. The technique of establishing a prime p as the 'characteristic' of a locally t-arc transitive graph goes back to GARDINER [309].

7.3 A diameter bound for distance-transitive graphs

The following theorem shows that, for locally t-arc transitive graphs with large t, the order of a vertex stabilizer is bounded by a function of t and the girth. Its main application (stated in the subsequent corollary) is the existence of a diameter bound for distance-transitive graphs in terms of the valency. The idea of Weiss' proof is that, for distance-transitive graphs Γ, either the girth is small, in which case Ivanov's bounds for the diameter in terms of valency and girth prevail (cf. §5.9), or the girth is large, G is locally t-arc transitive (on the incidence graph of Γ) for a large value of t, and one can show that $G_S^{(5)} = 1$ for reduced paths S of length 14, so that the order of G_γ can be bounded by a function of the valency, in which case it is immediate from distance-transitivity that the diameter is bounded by a function of the valency.

7.3.1. Theorem (WEISS [779]). *Let $t \geqslant 8$ and $g \leqslant 2t+11$ be integers. Suppose that Δ is a finite connected graph all of whose vertices have valency at least two, with girth $g \geqslant 3$, and that G is a group of automorphisms of Δ which is locally t-arc transitive. Then $G_S^{(5)} = 1$ for every reduced path S of length 14.*

Proof. Let $l \leqslant t$ be maximal with the property that there is a reduced path $S = \gamma_0, \cdots, \gamma_l$ such that $G_{S\setminus\{\gamma_l\}}^{(5)}$ is not contained in $G_{\gamma_l}^{(5)}$.

Consider a reduced path $\gamma_0,\gamma_1,\gamma_2$ in Δ. If $G_{\gamma_0}^{(5)} \leqslant G_{\gamma_1}^{(5)} \leqslant G_{\gamma_2}^{(5)}$, then by equality of orders of the two groups at the extremes of the chain, we have $G_{\gamma_0}^{(5)} = G_{\gamma_1}^{(5)} = G_{\gamma_2}^{(5)}$, whence, by Proposition 7.2.1 (viii), $G_{\gamma_0}^{(5)} = 1$, and we are done. Therefore, we may restrict attention to the case where $l \geqslant 1$. We proceed in four steps.

STEP 1. *If $l \leqslant t-1$, then $G_S^{(5)} = 1$ for every reduced path S of length l.*

For, by definition of l, each reduced path $S = \gamma_0, \cdots, \gamma_{l+1}$ satisfies $G^{(5)}_{S \setminus \{\gamma_{l+1}\}} \leqslant G^{(5)}_{\gamma_{l+1}}$, whence, by connectedness of Δ, we have $G^{(5)}_{S \setminus \{\gamma_{l+1}\}} \leqslant G^{(5)}_{\delta}$ for all $\delta \in \Delta$.

STEP 2. *Each edge of Δ has a vertex γ with $G^{(e)}_{\gamma} \neq 1$, where $e = [(l+7)/2]$.*

By definition of l, there is a reduced path $R = \gamma_0, \cdots, \gamma_l$ such that

$$|G^{(5)}_R| < |G^{(5)}_{R \setminus \gamma_l}|. \tag{1}$$

Extend R to a reduced path $\gamma_0, \cdots, \gamma_{l+3}$. According to the Thompson-Wielandt Theorem 7.2.2 (which we may apply as we are clearly done if $G^{(2)}_{\alpha} = 1$ for some $\alpha \in \Delta$), there is a unique prime p such that $O_p(G^{(2)}_{\delta}) \neq 1$ for each $\delta \in \Delta$. Write $P = O_p(G^{(1)}_{\gamma_0})$. By Propositions 7.2.1 (iv) and 7.1.3 (iii), we have $O_p(G^{(2)}_{\gamma_0}) \leqslant O_p(G^{(1)}_{\gamma_0}) = P$, so P is nontrivial. In particular, the center $Z(P)$ of P contains a nontrivial element, say a. Let $m \leqslant l+3$ be maximal such that a fixes $\gamma_0, \cdots, \gamma_m$. We claim that $m \geqslant e$.

Suppose that $m \leqslant 3$. Observe that $G^{(5)}_{\gamma_i} \leqslant P$ for $i = 0, \cdots, 4$ as $G^{(5)}_{\gamma_i} \trianglelefteq G^{(4)}_{\gamma_{i-1}} \trianglelefteq \cdots \trianglelefteq G^{(5-i)}_{\gamma_0}$ and $G^{(5)}_{\gamma_i}$ is a p-group. As a commutes with $G^{(5)}_{\gamma_i}$ for each $i \in \{0,1,2,3,4\}$, we have $G^{(5)}_{\gamma_{m+1}} = aG^{(5)}_{\gamma_{m+1}}a^{-1} = G^{(5)}_{a(\gamma_{m+1})}$. Due to Proposition 7.2.1 (vi) and the choice of m, this yields $G^{(5)}_{\gamma_{m+1}} = 1$, which is absurd. Therefore, $m \geqslant 4$.

Suppose now that $m < e$. Then $m < l+3$, and $\gamma_{l+3}, \cdots, \gamma_{m+1}, \gamma_m, a(\gamma_{m+1}), \cdots, a(\gamma_{l+3})$ is a reduced path of length $2(l+3-m)$. Since $G^{(5)}_{\gamma_4, \cdots, \gamma_{l+3}} \leqslant G^{(5)}_{\gamma_4} \leqslant P$, we have $G^{(5)}_{\gamma_4, \cdots, \gamma_{l+3}} = G^{(5)}_{a(\gamma_4), \cdots, a(\gamma_{l+3})}$ and because $m \geqslant 4$ it follows that $G^{(5)}_{\gamma_4, \cdots, \gamma_{l+3}} \leqslant G^{(5)}_{\gamma_{l+3}, \cdots, \gamma_{m+1}, \gamma_m, a(\gamma_{m+1}), \cdots, a(\gamma_{l+3})}$. But $\gamma_4, \cdots, \gamma_{l+3}$ and $\gamma_0, \cdots, \gamma_{l-1} = R \setminus \gamma_l$ are paths in the same G-orbit, so by (1) we find

$$|G^{(5)}_R| < |G^{(5)}_{\gamma_{l+3}, \cdots, \gamma_{m+1}, \gamma_m, a(\gamma_{m+1}), \cdots, a(\gamma_{l+3})}|. \tag{2}$$

If l is odd, then R is in the same orbit as the initial part of $\gamma_{l+3}, \cdots, \gamma_{m+1}, \gamma_m, a(\gamma_{m+1}), \cdots, a(\gamma_{l+3})$ of the same length. Therefore, comparing lengths of the paths involved, we get $2(l+3-m) < l$. This implies $m \geqslant (l+7)/2 = e$, contradicting the assumption $m < e$.

If l is even, then by the same argument, using the inequality $|G^{(5)}_R| < |G^{(5)}_{\gamma_{l+2}, \cdots, \gamma_{m+1}, \gamma_m, a(\gamma_{m+1}), \cdots a(\gamma_{l+3})}|$ (which is a direct consequence of (2)), we get $2(l+3-m)-1 < l$ so that $m \geqslant (l+6)/2 = e$, again a contradiction.

Thus $m \geqslant e$, indeed, whence $a \in G_{\gamma_0, \cdots, \gamma_e}$. This establishes $Z(P) \leqslant G_{\gamma_0, \cdots, \gamma_e}$. Since G_{γ_0} is transitive on all reduced paths of length e starting at γ_0 and normalizes $Z(P)$, we obtain $Z(P) \leqslant G^{(e)}_{\gamma_0}$. Our claim now follows from $Z(P) \neq 1$.

STEP 3. $l \leqslant t-1$.

Since $t-1 \geqslant 7$, we may assume $l \geqslant 8$. Then, clearly, $g \geqslant t \geqslant l > e$ (unless Δ is a polygon, in which case there is nothing to show). In view of Step 2 and Proposition 7.2.1 (v), we have $G^{(e)}_{\gamma_0} \neq G^{(e)}_{\gamma_1}$ for every edge $\gamma_0\gamma_1$ of Δ. Let $\delta_0\delta_1$ be an edge with $G^{(e)}_{\delta_0}$ not contained in $G^{(e)}_{\delta_1}$. Extend the edge to a minimal circuit $\delta_0, \delta_1, \cdots, \delta_g = \delta_0$ in Δ. Since G_{δ_0} is transitive on the set of reduced paths of length e starting at δ_0, there must be $b \in G^{(e)}_{\delta_0}$ with $b(\delta_{e+1}) \neq \delta_{e+1}$. Now $b(\delta_i) = \delta_i$ if $i \leqslant e$ or $i \geqslant g-e$, so $g \geqslant 2e+1$ and the path $\delta_e, \delta_{e+1}, \cdots, \delta_{g-e}, b(\delta_{g-e-1}), \cdots, b(\delta_{e+1})$ is a circuit of

length $2(g-2e)$. Consequently, $2(g-2e) \geqslant g$. Using $g \leqslant 2t+11$ from the hypothesis, we obtain $t-1 \geqslant (g-13)/2 \geqslant 2[(l+7)/2]-13/2$, whence $l \leqslant t-1$.
In view of Steps 1 and 3, it suffices for the proof of the theorem to establish

STEP 4. $l \leqslant 14$.

Let $\delta_0, \cdots, \delta_g$ be as in Step 3. Take $j \in \{1,2\}$ such that $e+j$ is even, and consider the group $H = [G_{\delta_0}^{(e)}, G_{\delta_{e+j}}^{(e)}]$. Since $G_{\delta_0}^{(e)}$ normalizes $G_{\delta_e}^{(e-j)}$ and is contained in $G_{\delta_j}^{(e-j)}$, while $G_{\delta_{e+j}}^{(e)}$ normalizes $G_{\delta_j}^{(e-j)}$ and is contained in $G_{\delta_e}^{(e-j)}$, we have $H \leqslant G_{\delta_e,\delta_j}^{(e-j)}$.
Now suppose that $l \geqslant 15$. Then $e-j = [(l+7)/2]-j \geqslant 10$, so $G_{\delta_e,\delta_j}^{(e-j)} \leqslant G_{\alpha_{e-j-5},\cdots,\alpha_1,\delta_j,\cdots,\delta_e,\beta_1,\cdots,\beta_{e-j-5}}^{(5)}$ if $\alpha_{e-j-5}, \cdots, \alpha_1, \delta_j, \cdots, \delta_e, \beta_1, \cdots, \beta_{e-j-5}$ is a reduced path in Δ. As the length of this path is $3(e-j)-10 \geqslant l$, it follows from Steps 1 and 3 and the above inclusion that $G_{\delta_e,\delta_j}^{(e-j)} = 1$, whence $H = 1$. This means that $G_{\delta_0}^{(e)}$ and $G_{\delta_{e+j}}^{(e)}$ commute. Taking $b \in G_{\delta_0}^{(e)}$ with $b(\delta_{e+1}) \neq \delta_{e+1}$ (as in Step 3) we get $G_{\delta_{e+j}}^{(e)} = G_{b(\delta_{e+j})}^{(e)}$. Now $e+j$ is even, so the (unique) reduced path from δ_{e+j} to $b(\delta_{e+j})$ of length $2j$ is in the same G-orbit as $\delta_0, \cdots, \delta_{2j}$. Therefore, $G_{\delta_0}^{(e)} = G_{\delta_{2j}}^{(e)} \leqslant G_{\delta_{e+1}}$, in contradiction with the existence of b. Thus, $l \leqslant 14$ and the theorem is proved. □

7.3.2. Corollary (CAMERON, PRAEGER, SAXL & SEITZ [183], WEISS [779]). *The diameter of a distance-transitive graph of valency $k \geqslant 3$ is at most $(k^6)!2^{2k}$. In particular, there are only finitely many distance-transitive graphs of a given valency $\geqslant 3$.*

Proof. In view of Ivanov's Theorem 5.9.8, it suffices to show that for all distance transitive graphs with valency $k \geqslant 3$ and $i \in \mathbb{N}$ satisfying $c_i - b_i > c_{i-1} - b_{i-1} = \cdots = c_1 - b_1$, we have $i \leqslant k^6!$. To this end consider the incidence graph Δ of the design consisting of the points and the singular lines of Γ. Then a vertex $\delta \in \Delta$ has valency $k/(\lambda+1)$ or $\lambda+2$, so that always $2 \leqslant k(\delta) \leqslant k$. Moreover, Δ has girth $4i$ or $4i+2$ according as $c_i > c_{i-1}$ or not. Now $G = \operatorname{Aut}\Gamma$ is distance-transitive on Γ, and has a natural action on Δ. For this action, if δ is a vertex of Δ, the stabilizer G_δ is transitive on the set of reduced paths in Δ of length $2i-3$ if $c_i > c_{i-1}$ and of length $2i-1$ otherwise. Thus, (if $i \geqslant 6$) Δ and G satisfy the hypotheses of the above theorem. Hence $G_S^{(5)} = 1$ for every reduced path S of length 14 in Δ. But if $\delta \in \Delta$, then $G_\delta^{(12)} \leqslant G_S^{(5)}$ for a suitable path S of length 14 with δ in the middle. Thus, choosing $\delta \in \Gamma$, we obtain that the pointwise stabilizer of $\Gamma_{\leqslant 6}(\delta)$ is trivial. Consequently, G_δ has a faithful permutation representation on $\Gamma_{\leqslant 6}(\delta)$, so that $|G_\delta| \leqslant |\Delta_6(\delta)|! \leqslant k^6!$. On the other hand, the stabilizer G_δ of $\delta \in \Gamma$ is transitive on $\Gamma_i(\delta)$, a set of size $k_i = kb_1^{i-1} \geqslant k2^{i-1}$, whence $i \leqslant 2^i \leqslant k_i \leqslant |G_\delta| \leqslant k^6!$, as required for the application of Ivanov's Theorem. □

Remarks. (i) CAMERON et al. [183] use the classification of finite simple groups to prove the Sims conjecture, i.e., the existence of an integral function f having the property that, whenever G is a primitive permutation group on a finite set Ω and d is the length of any G_α-orbit in $\Omega \setminus \{\alpha\}$ for an $\alpha \in \Omega$, then $|G_\alpha| \leqslant f(d)$. The second statement of the above corollary is derived from this result. CAMERON [173] sketched how an elementary proof of the above corollary can be obtained by use of the classification of infinite distance-transitive graphs of finite valency (MACPHERSON [513]) once a result such as Weiss' Theorem (not available at that time) is available.
(ii) The bound of the corollary is unrealistically high; for $k = 3$ we have $d \leqslant 8$, and in all known distance-transitive graphs of valency $k > 3$ we have $d \leqslant 2k-1$, with

equality for the doubled Odd graphs. In the special cases $k = 3, 4$, a much better bound on the diameter was already known to SMITH [683], who used the result as a first step in the classification of all distance-transitive graphs with these valencies (cf. §7.6).

7.4 The case of large girth

The following beautiful theorem of Weiss depends on the classification of doubly transitive permutation groups.

7.4.1. Theorem (WEISS [778]). *Suppose Γ is a distance-transitive graph of valency $k \geqslant 3$ and girth $g \geqslant 9$. Then Γ is isomorphic to one of the following graphs:*
(D14) - *The Biggs-Smith graph on* 102 *vertices* ($g = 9$)*;*
(B14) - *The Foster graph on* 90 *vertices* ($g = 10$)*;*
(G3) - *The incidence graph of the generalized hexagon associated with $G_2(q)$, q a power of* 3 ($g = 12$). □

Problem. Extend Weiss' theorem to the case of geometric girth $g \geqslant 9$. No further examples are expected.

7.5 Graphs with small valency

For valency $\leqslant 13$, all distance-transitive graphs are known. For cubic graphs we even know all distance-regular graphs.

7.5.1. Theorem (BIGGS, BOSHIER & SHAWE-TAYLOR [80]). *Up to isomorphism the distance-regular graphs with valency* 3 *are: K_4 ($v = 4$), $K_{3,3}$ ($v = 6$), O_3 (the Petersen graph, $v = 10$), each of diameter $\leqslant 2$, and the following:*

label	name	intersection array	v
(C8)	The cube	{3,2,1; 1,2,3}	8
(G1)	The Heawood graph	{3,2,2; 1,1,3}	14
(B2)	The Pappus graph	{3,2,2,1; 1,1,2,3}	18
(D9)	The Coxeter graph	{3,2,2,1; 1,1,1,2}	28
(G2)	Tutte's 8-cage	{3,2,2,2; 1,1,1,3}	30
(A11)	The dodecahedron	{3,2,1,1,1; 1,1,1,2,3}	20
(R3)	The Desargues graph	{3,2,2,1,1; 1,1,2,2,3}	20
(G3)	Tutte's 12-cage	{3,2,2,2,2,2; 1,1,1,1,1,3}	126
(D14)	The Biggs-Smith graph	{3,2,2,2,1,1,1; 1,1,1,1,1,1,3}	102
(B14)	The Foster graph	{3,2,2,2,2,1,1,1; 1,1,1,1,2,2,2,3}	90

Each of these graphs is uniquely determined by the intersection array. All graphs except for Tutte's 12*-cage are distance-transitive.* □

Remark. The proof of the above result rests on some estimates obtained by BANNAI & ITO [38]. The distance-transitive case was handled in BIGGS & SMITH [84]; see also GARDINER [312]. The bipartite distance-regular case was settled by ITO [428]. For early references, cf. HEAWOOD [366], COXETER [221], TUTTE [757, 758], SMITH [679].

All primitive permutation groups with a suborbit of length 3 are determined in WONG [795]. GOLDSCHMIDT [333] determines the possible point stabilizers in finite connected edge-transitive cubic graphs.

7.5.2. Theorem. *For every integral value of k, the following graphs of valency k are distance-transitive.*
(i) *The complete graph K_{k+1}, with diameter $d = 1$, $v = k+1$ vertices and intersection array $\{k; 1\}$.*
(ii) *The complete bipartite graph $K_{k,k}$, with diameter $d = 2$, $v = 2k$ vertices and intersection array $\{k,k-1; 1,k\}$.*
(iii) *The complement of the $2\times(k+1)$ grid, with diameter $d = 3$, $v = 2(k+1)$ vertices and intersection array $\{k,k-1,1; 1,k-1,k\}$.*
(iv) *The folded k-cubes $\square_k$, with diameter $d = [k/2]$, $v = 2^{k-1}$ vertices and intersection array $\{k,k-1,\cdots,k+1-d;1,2,\cdots,d-1,c_d\}$ with $c_d = d$ if k is odd, and $c_d = k$ if k is even.* (P1)
(v) *The Odd graph O_k, with diameter $d = k-1$, $v = \binom{2k-1}{k}$ vertices and intersection array $\{k,k-1,k-1,k-2,k-2,k-3,\cdots; 1,1,2,2,3,3,\cdots\}$.* (R2)
(vi) *The k-cubes $Q_k = H(k,2)$, with diameter $d = k$, $v = 2^k$ vertices and intersection array $\{k,k-1,k-2,\cdots,1; 1,\cdots,k-1,k\}$.* (C8)
(vii) *The bipartite double of the Odd graph O_k, with diameter $d = 2k-1$, $v = 2\binom{2k-1}{k}$ vertices and intersection array*

$$\{k,k-1,k-1,k-2,k-2,\cdots,1,1; 1,1,2,2,3,\cdots k-1,k\}. \text{ (R3)}$$

Each of these graphs is uniquely determined by its intersection array, apart from the folded 6-cube which is unique as a distance-transitive graph.

Proof. Distance transitivity is straightforward. Uniqueness is trivial for (i), (ii) and (iii) - in the last case use that the graph is antipodal. The cases of the Odd graph and its bipartite double are handled in §9.1D and the cases of the k-cube and its quotient are handled in §9.2D. □

Remark. There are 3 nonisomorphic 2-(16,6,2) designs, each of which is self-dual, so we find three distance-regular graphs with intersection array {6,5,4; 1,2,6}. Only one of these is distance-transitive.

7.5.3. Theorem. *Up to isomorphism, the only distance-transitive graphs with valency k, $4 \leqslant k \leqslant 13$ not mentioned in the previous theorem are the following.*

(i) *For $k = 4$* (D.H. SMITH [679, 681, 682, 683]; *see also* GARDINER [317]): *$K_{3\times 2}$ (the octahedron, $v = 6$) and $3\times3 = \overline{3\times3} = L(K_{3,3}) = H(2,3) = QR(9)$ ($v = 9$), both of diameter 2, and*

label	name	intersection array	v
(A0)	line graph of Petersen graph	{4,2,1; 1,1,4}	15
(G1)	line graph of Heawood graph ($GH(2,1)$)	{4,2,2; 1,1,2}	21
(B0)	incidence graph of 2-(7,4,2) design (distance 3 graph of Heawood graph)	{4,3,2; 1,2,4}	14
(G2)	gen. hexagon of order (1,3)	{4,3,3; 1,1,4}	26
(G2)	line graph of Tutte's 8-cage ($GO(2,1)$)	{4,2,2,2; 1,1,1,2}	45
(B2)	$AG(2,4)$ minus a parallel class	{4,3,3,1; 1,1,3,4}	32
(G3)	gen. dodecagon of order (1,3)	{4,3,3,3,3,3; 1,1,1,1,1,4}	728

Each of these graphs is uniquely determined by the intersection array, except perhaps the gen. dodecagon of order (1,3) *(which is unique as distance-transitive graph).*

(ii) *For* $k = 5$ (IVANOV, IVANOV & FARADJEV [282]; see also GARDINER [316] and GARDINER & PRAEGER [319]): *(no further graphs of diameter* 2 *and)*

label	name	intersection array	v
(A1)	icosahedron	{5,2,1; 1,2,5}	12
(D2)	Sylvester graph (double six)	{5,4,2; 1,1,4}	36
(B0)	incidence graph of 2-(11,5,2) design	{5,4,3; 1,2,5}	22
(G1)	gen. hexagon of order (1,4)	{5,4,4; 1,1,5}	42
(A12)	Wells graph	{5,4,1,1; 1,1,4,5}	32
(B2)	$AG(2,5)$ minus a parallel class	{5,4,4,1; 1,1,4,5}	50
(G2)	gen. octagon of order (1,4)	{5,4,4,4; 1,1,1,5}	170

(iii) For $k = 6$ (IVANOV, IVANOV & FARADJEV [282]; see also GARDINER & PRAEGER [318]): $K_{4\times 2}$ ($v = 8$), $K_{3\times 3}$ ($v = 9$), $L(K_5) = T(5)$ ($v = 10$), $QR(13)$ *(the Paley graph on* $v = 13$ *vertices),* $\overline{T(6)} = GQ(2,2)$ ($v = 15$), $4\times 4 = L(K_{4,4}) = H(2,4)$ ($v = 16$), *each of diameter* 2, *and*

label	name	intersection array	v
(G1)	gen. hexagon of order (3,1)	{6,3,3; 1,1,2}	52
(C8)	$H(3,3)$	{6,4,2; 1,2,3}	27
(G1)	gen. hexagon of order (2,2) (two graphs)	{6,4,4; 1,1,3}	63
(A6)	$\Gamma_2(\gamma)$ in Hoffman-Singleton graph	{6,5,1; 1,1,6}	42
(D3)	Perkel graph	{6,5,2; 1,1,3}	57
(B0)	incidence graph of 2-(11,6,3) design	{6,5,3; 1,3,6}	22
(G1)	gen. hexagon of order (1,5)	{6,5,5; 1,1,6}	62
(A7)	3-cover of GQ(2,2) = halved Foster graph	{6,4,2,1; 1,1,4,6}	45
(B2)	$3{\cdot}K_{6,6}$ from $RT[6,2;3]$, cf. §13.2A	{6,5,4,1; 1,2,5,6}	36
(G3)	gen. dodecagon of order (3,1)	{6,3,3,3,3,3; 1,1,1,1,1,2}	1456

(iv) For $k = 7$ (IVANOV, IVANOV & FARADJEV [282]): *the Hoffman-Singleton graph* ($v = 50$), *and*

label	name	intersection array	v
(B0)	points and planes in $PG(3,2)$	{7,6,4; 1,3,7}	30
(D8)	2-residual of S(5,8,24)	{7,6,4,4; 1,1,1,6}	330
(B2)	$AG(2,7)$ minus a parallel class	{7,6,6,1; 1,1,6,7}	98
(D7)	doubled Hoffman-Singleton graph	{7,6,6,1,1; 1,1,6,6,7}	100
(R3)	doubled Grassmann graph	{7,6,6,4,4; 1,1,3,3,7}	310
(A19)	Ivanov-Ivanov-Faradjev graph	{7,6,4,4,4,1,1,1; 1,1,1,2,4,4,6,7}	990

(v) *For* $k = 8$ (IVANOV & IVANOV [434]): $K_{5\times 2}$ ($v = 10$), $K_{3\times 4}$ ($v = 12$), $L(K_6) = T(6)$ ($v = 15$), $QR(17)$ ($v = 17$), $5\times 5 = L(K_{5,5}) = H(2,5)$ ($v = 25$), *each of diameter* 2, *and*

label	name	intersection array	v
(G1)	gen. hexagon of order (4,1)	{8,4,4; 1,1,2}	105

label	name	intersection array	v
(A0)	$GQ(2,4)$ minus spread	$\{8,6,1;1,3,8\}$	27
(B0)	incidence graph of 2-(15,8,4) design	$\{8,7,4;1,4,8\}$	30
(G1)	gen. hexagon of order (1,7)	$\{8,7,7;1,1,8\}$	114
(G2)	gen. octagon of order (4,1)	$\{8,4,4,4;1,1,1,2\}$	425
(C8)	$H(4,3)$	$\{8,6,4,2;1,2,3,4\}$	81
(B2)	$2{\cdot}K_{8,8}$, Hadamard graph	$\{8,7,4,1;1,4,7,8\}$	32
(B2)	$4{\cdot}K_{8,8}$ from $RT[8,2;4]$	$\{8,7,6,1;1,2,7,8\}$	64
(B2)	$AG(2,8)$ minus a parallel class	$\{8,7,7,1;1,1,7,8\}$	128

(vi) *For* $k = 9$ (Ivanov & Ivanov [434]): $K_{4\times 3}$ ($v = 12$) *and* $\overline{4\times 4}$ ($v = 16$) *of diameter 2, and*

label	name	intersection array	v
(C1)	Johnson graph $J(6,3)$	$\{9,4,1;1,4,9\}$	20
(C8)	$H(3,4)$	$\{9,6,3;1,2,3\}$	64
(B0)	incidence graph of 2-(13,9,6) design	$\{9,8,3;1,6,9\}$	26
(G1)	gen. hexagon of order (1,8)	$\{9,8,8;1,1,9\}$	146
(B2)	$3{\cdot}K_{9,9}$ from $RT[9,3;3]$	$\{9,8,6,1;1,3,8,9\}$	54
(D11)	unitals in $PG(2,4)$	$\{9,8,6,3;1,1,3,8\}$	280
(B2)	$AG(2,9)$ minus a parallel class	$\{9,8,8,1;1,1,8,9\}$	162
(G2)	gen. octagon of order (1,8)	$\{9,8,8,8;1,1,1,9\}$	1170

(vii) *For* $k = 10$ (Ivanov & Ivanov [434]): $K_{6\times 2}$ ($v = 12$), $K_{3\times 5}$ ($v = 15$), *the Clebsch graph* ($v = 16$), $T(7)$ *and* $\overline{T(7)}$ ($v = 21$), *the complement of the Schläfli graph* ($v = 27$), $6\times 6 = L(K_{6,6}) = H(2,6)$ ($v = 36$), *the Gewirtz graph* ($v = 56$), *each of diameter 2, and*

label	name	intersection array	v
(G1)	gen. hexagon of order (5,1)	$\{10,5,5;1,1,2\}$	186
(D4)	$L_2(25).2$ (locally Petersen)	$\{10,6,4;1,2,5\}$	65
(B0)	incidence graph of 2-(16,10,6) design	$\{10,9,4;1,6,10\}$	32
(G1)	gen. hexagon of order (1,9)	$\{10,9,9;1,1,10\}$	182
(A8)	$3{\cdot}\mathrm{Sym}(7)$ (locally Petersen)	$\{10,6,4,1;1,2,6,10\}$	63
(N5)	Hall-Janko near octagon	$\{10,8,8,2;1,1,4,5\}$	315
(G2)	gen. octagon of order (2,4)	$\{10,8,8,8;1,1,1,5\}$	1755
(C8)	$H(5,3)$	$\{10,8,6,4,2;1,2,3,4,5\}$	243
(B9)	doubled Gewirtz graph	$\{10,9,8,2,1;1,2,8,9,10\}$	112
(G3)	gen. dodecagon of order (1,9)	$\{10,9,9,9,9,9;1,1,1,1,1,10\}$	132860

(viii) For $k = 11$ (Ivanov & Ivanov [434]): *(no further graphs of diameter* 2 *and)*

label	name	intersection array	v
(D12)	Livingstone graph for J_1	$\{11,10,6,1;1,1,5,11\}$	266
(B2)	$AG(2,11)$ minus a parallel class	$\{11,10,10,1;1,1,10,11\}$	242

(ix) *For* $k = 12$ (Ivanov & Ivanov [434]): $K_{7\times 2}$ ($v = 14$), $K_{5\times 3}$ ($v = 15$), $K_{4\times 4}$ ($v = 16$), $K_{3\times 6}$ ($v = 18$), $QR(25)$ ($v = 25$), $L(K_8) = T(8)$ ($v = 28$), $GQ(3,3)$

(v = 40) *(two graphs)*, $GQ(4,2)$ (v = 45), $7\times7 = L(K_{7,7}) = H(2,7)$ (v = 49), *each of diameter* 2, *and*

label	name	intersection array	v
(C1)	Johnson graph $J(7,3)$	{12,6,2; 1,4,9}	35
(D5)	line graph of Hoffman-Singleton graph	{12,6,5; 1,1,4}	175
(C8)	$H(3,5)$	{12,8,4; 1,2,3}	125
(G1)	gen. hexagon of order (3,3)	{12,9,9; 1,1,4}	364
(D6)	$P\Sigma L(2,16)$	{12,10,3; 1,3,8}	68
(D1)	$P\Gamma U(3,4)$ on nonisotropic points	{12,10,5; 1,1,8}	208
(G1)	gen. hexagon of order (1,11)	{12,11,11; 1,1,12}	266
(G2)	gen. octagon of order (4,2)	{12,8,8,8; 1,1,1,3}	2925
(C8)	$H(4,4)$	{12,9,6,3; 1,2,3,4}	256
(B2)	$2{\cdot}K_{12,12}$, Hadamard graph	{12,11,6,1; 1,6,11,12}	48
(C8)	$H(6,3)$	{12,10,8,6,4,2; 1,2,3,4,5,6}	729

(x) For k = 13 (Ivanov & Ivanov [434]): *(no further graphs of diameter* 2 *and)*

label	name	intersection array	v
(A3)	Taylor graph	{13,6,1; 1,6,13}	28
(B0)	points and planes in $PG(3,3)$	{13,12,9; 1,4,13}	80
(B2)	$AG(2,13)$ minus a parallel class	{13,12,12,1; 1,1,12,13}	338
(R3)	doubled Grassmann graph	{13,12,12,9,9; 1,1,4,4,13}	2420

Remark. Nomura [571] showed that there are precisely 7 distance-regular graphs with valency four and girth three; all occur in the above list under (i). Two more graphs of valency four are known that are distance-regular but not distance-transitive, namely the (unique) generalized octagon of order (1,3) on 80 vertices, and the (unique) generalized dodecagon of order (2,1) on 189 vertices. Bannai & Ito [37] announce that there are only finitely many distance-regular graphs with k = 4.

7.6 Imprimitive distance-transitive graphs

We have seen in Chapter 1 that all bipartite and certain antipodal distance-regular graphs of diameter at most 4 correspond to certain designs. As it turns out, distance-transitivity for the graphs implies some kind of 2-transitivity for the design. E.g., if Γ is an antipodal distance-transitive graph of diameter 3, then $\mathrm{Aut}\,\Gamma$ acts 2-transitively on the sets of antipodal vertices. And if Γ is a distance-transitive symmetric net, then $\mathrm{Aut}\,\Gamma$ acts 2-transitively on the classes of the (complete multipartite) halved graphs of Γ; in particular this applies to Hadamard graphs. Now that all 2-transitive permutation groups are known, there are good prospects for a complete classification, and indeed, in the case of bipartite distance-transitive graphs of diameter 3 this has been achieved already (Theorem 7.6.2). Useful material related to this problem may be Shult [671] and Hale & Shult [344] for antipodal double covers of complete graphs, and Jungnickel [448] for symmetric nets. (For Hadamard graphs, see below.)

A. 2-transitive square designs

7.6.1. Proposition (VIJAYAN [766]). *The incidence graph Γ of a square 2-design D (cf. §1.6) is distance-transitive if and only if D is self-dual and* Aut D *is doubly transitive on the point set of D.*

Proof. Suppose G is distance-transitive on Γ. Then, for each $\gamma \in \Gamma$, the stabilizer G_γ is transitive on $\Gamma_2(\gamma)$. But $\{\gamma\} \cup \Gamma_2(\gamma)$ is the point (or block) set of D. Since clearly we can find an element of G moving γ to some element of $\Gamma_2(\gamma)$, this shows that G is doubly transitive on the point (or block) set of D. Also, an element of G moving γ to an element of $\Gamma(\gamma)$ induces a duality on D, showing that D is self-dual.
Conversely, if $G = \mathrm{Aut}\, D$ is doubly transitive on the point set of D, then the stabilizer in G of a point of D has two orbits on the point set of D and hence also two orbits on the set of blocks, i.e., is transitive on the set of all blocks containing this point as well as on the set of all blocks not containing it. Due to the existence in G of a correlation of D, it follows from the double transitivity that G_γ is transitive on $\Gamma_2(\gamma)$ for each $\gamma \in \Gamma$ and from the transitivity of a point stabilizer on the two above mentioned block sets that G_γ is transitive on $\Gamma(\gamma)$ and $\Gamma_3(\gamma)$ for each $\gamma \in \Gamma$, whence G is distance-transitive on Γ. □

KANTOR [457] has determined the square designs with 2-transitive automorphism groups. Using the classification of 2-transitive groups, he finds:

7.6.2. Theorem. *Let D be a square 2-(v,k,λ) design with $v \geqslant 2k > 2$ such that* Aut D *is 2-transitive on points. Then D is one of the following:*

(i) *the design of the points and hyperplanes in a projective space* ($v = \dfrac{q^d-1}{q-1}$, $k = \dfrac{q^{d-1}-1}{q-1}$, $\lambda = \dfrac{q^{d-2}-1}{q-1}$);

(ii) *the unique biplane with $v = 11$, $k = 5$ and $\lambda = 2$, the Paley design for $q = 11$* (cf. §7.6B);

(iii) *Higman's design with $v = 176$, $k = 50$ and $\lambda = 14$* (cf. §11.4E);

(iv) *a design with $v = 2^{2m}$, $k = 2^{m-1}(2^m - 1)$ and $\lambda = 2^{m-1}(2^{m-1} - 1)$, of which there is exactly one for each $m \geqslant 2$* (cf. KANTOR [455]). □

In these four cases Aut D is $P\Gamma L(d,q)$, $PSL(2,11)$, HS and $2^{2m}.Sp(2m,2)$, respectively.

The complements of the designs under (iv) have parameters $v = 2^{2m}$, $k = 2^{m-1}(2^m + 1)$ and $\lambda = 2^{m-1}(2^{m-1} + 1)$; an easy construction of these is obtained by taking the tensor product of m copies of the Hadamard matrix $J - 2I$ of order 4; this yields the $(-1,1)$ incidence matrix of these designs.
More generally, let V be the $2m$-dimensional vector space over $\mathbb{F}_2$, and let Q be a non-degenerate quadratic form on V. Then V together with the set of blocks $\{B_u \mid u \in V\}$, where $B_u := \{v \in V \mid Q(v-u) = 0\}$, is the square 2-design of (iv) in case Q is elliptic, and its complement in case Q is hyperbolic. (The previous construction is the special case where V is the direct sum of m hyperbolic lines.)

7.6.3. Corollary. *A bipartite graph of diameter 3 is distance-transitive if and only if it is either the incidence graph of one of the above designs or of the complement of one of the above designs or a complete bipartite graph minus a 1-factor (i.e., $\overline{n \times 2}$).*

Proof. This follows from §1.6 and Kantor's theorem, since each of the above designs is self-dual. □

B. 2-transitive Hadamard matrices

For two Hadamard matrices $H = (h_{ij})_{ij}$ and $H' = (h'_{ij})_{ij}$, we write $H \approx H'$ when H' can be obtained from H by suitably multiplying rows and columns by ± 1, i.e., when $h_{ij}h_{1j}h_{i1}h_{11} = h'_{ij}h'_{1j}h'_{i1}h'_{11}$ for all i, j. Let $\operatorname{Aut} H$ be the group of all pairs (σ,τ) of permutations of rows and columns such that $H^{(\sigma,\tau)} := (h_{\sigma(i)\tau(j)})_{ij} \approx H$. We say that H and H' are H-equivalent, written $H \underset{H}{\sim} H'$, if $H^{(\sigma,\tau)} \approx H'$ for some pair (σ,τ) of permutations of rows and columns.

7.6.4. Proposition.

(i) *A Hadamard graph Γ determines the associated Hadamard matrix H up to transposition, permutation of rows and columns, and multiplication of rows and columns by* ± 1.

(ii) *If a Hadamard graph Γ is distance-transitive then each associated Hadamard matrix H satisfies $H \underset{H}{\sim} H^{\mathrm{T}}$ and* $\operatorname{Aut} H$ *acts 2-transitively on the rows and on the columns of H.* □

Let us describe the known 2-transitive Hadamard matrices H of order > 2. If H is a Hadamard matrix of order $4n$, then we may assume that H has been normalized such that its first row and column consist of ones only; upon deleting this row and column, we find a square matrix N of order $4n-1$, the $(-1,1)$ point-block incidence matrix of a *Hadamard 2-design*, that is, a square 2-$(4n-1,2n-1,n-1)$ design. It suffices to describe these Hadamard 2-designs. Let I denote the incidence relation.

(i) $PG(d-1,2)$: the points and hyperplanes of the projective geometry of dimension $d-1$ over $\mathbb{F}_2$; here $4n-1 = 2^d-1$, and $\operatorname{Aut} H \cong 2^{2d}.PSL(d,2)$. This is the only case where $\operatorname{Aut} H$ is 2-transitive on rows but not faithful on columns: the group induced on the columns is $2^d.PSL(d,2)$ (Kantor [453]).

(ii) (Paley [582]) The Paley design: $4n-1 = q$, where q is a prime power. The points and the blocks are the elements of $\mathbb{F}_q$, and $x\,I\,y$ if and only if $x+y$ is a nonzero square. For $q = 3,7$ we find $PG(1,2)$ and $PG(2,2)$ again. We have $\operatorname{Aut} H \cong M_{12}$ for $q = 11$ (Hall [354]), and $\operatorname{Aut} H \cong P\Sigma L(2,q)$ for $q > 11$ (Kantor [453]). For $q > 11$, these designs are the only Hadamard 2-designs with a group transitive on the point-block flags, but not 2-transitive on the points (Kantor [452]).

(iii) (Ito & Leon [426]) $4n = 36$; the points and the blocks are the lines of $PG(3,2)$, and $l\,I\,m$ if and only if either $l = m$ or $l \cap m = \varnothing$. We have $\operatorname{Aut} H \cong Sp(6,2)$ ($\cong \Omega(7,2)$).

This last example is related to the unique completely regular two-graph on 36 points. (See Neumaier [557] for the definition and properties of completely regular two-graphs.) Indeed, suppose that $(X,\mathfrak{D})$ is a regular two-graph on v points such that $\lambda(\mathfrak{D}) = \frac{1}{2}v-2$. Then we may define a Hadamard matrix H_a for each $a \in X$ by putting $H_a(x,y) = 1$ if the set $\{a,x,y\}$ is coherent (and in particular when $a = x$ or $a = y$ or $x = y$), and $H_a(x,y) = -1$ otherwise. The matrix H_a is symmetric with constant diagonal. Since for $a, b \in X$ we have $H_a \approx H_b$, the Hadamard graph $\Gamma = \Gamma(H_a)$ (cf. §1.8) does not depend on the choice of $a \in X$.

Now consider $PG(6,2)$ provided with a nondegenerate quadric Q. Let $(X,\mathfrak{D})$ be the regular two-graph whose points are the 36 hyperbolic hyperplanes, and where

$\{x,y,z\} \in \mathfrak{D}$ if and only if x, y, z are distinct and $Q|_{x \cap y \cap z}$ is nondegenerate. Then for $a \in X$ the Hadamard matrix H_a is isomorphic to the matrix constructed under (iii). (Note that $\Omega(7,2)$ acts 2-transitively on X; for any two distinct hyperbolic hyperplanes x, y the intersection $x \cap y$ has a 1-point radical.)

Another description may be given by noting that $Sp(6,2)\times 2$ is the Weyl group of type E_7. The root system of type E_7 contains 36 sub root systems of type A_7, and any two such A_7 systems meet in $A_3 \oplus A_3$. Any three A_7 meet in either $A_2 \oplus A_2$ or $4A_1$. Calling a triple of A_7 systems coherent when the former happens, we find our regular two-graph again.

7.6.5. Theorem. *The distance-transitive Hadamard graphs are precisely those obtained from the examples* (i) *and from example* (ii) *for* $q = 11$.

Proof. Ito [423] and Ito & Kimura [425] study the Hadamard matrices H with a group that is 2-transitive on the rows. They find that either we have one of the examples (i)-(iii) above, or $\mathrm{Aut}\,H$, viewed as permutation group on the rows and columns of the matrix $\begin{bmatrix} H & -H \\ -H & H \end{bmatrix}$, acts rank 4 on the rows of this matrix. But for the Hadamard graph $\Gamma(H)$ this means that the stabilizer of a vertex γ has two orbits on $\Gamma_2(\gamma)$.
Concerning examples (i)-(iii), it is easy to see that (i) yields a distance-transitive graph for all d; for (ii)-(iii), in order to get an edge-transitive graph, the stabilizer of a row must still be transitive on the columns, but that is only the case in example (ii) for $q = 11$. □

C. 2-transitive regular two-graphs

Recall that a Taylor graph is a distance-regular antipodal 2-cover of a complete graph, with diameter 3, i.e., a distance-regular graph with intersection array $\{k,\mu,1;1,\mu,k\}$ on $2(k+1)$ vertices. These graphs correspond to non-complete regular two-graphs (cf. §1.5); the graph belonging to the complementary two-graph (when not complete) has intersection array $\{k,\lambda,1;1,\lambda,k\}$, where $\lambda = k-\mu-1 > 0$. A Taylor graph Γ is completely determined by any local graph $\Gamma(\gamma)$.
From Theorem 1.5.3 we see that distance-transitive Taylor graphs are equivalent to non-complete regular 2-graphs with 2-transitive automorphism group. Let us describe the known examples. (See also Taylor [710] and Buekenhout & Hubaut [148].)

(i) $\Gamma \cong \overline{2\times n}$ $(n \geqslant 3)$, $\mathrm{Aut}\,\Gamma \cong 2\times\mathrm{Sym}(n)$, $k = n-1$, $\mu = n-2$.

(ii) Γ is locally the Paley graph of order q, where $q = 4t+1$ is a prime power, $\mathrm{Aut}\,\Gamma \cong 2\times P\Sigma L(2,q)$, $k = q$, $\mu = (q-1)/2$ (cf. Proposition 12.5.3). [For $q = 5$ we find the icosahedron; for $q = 9$ the Johnson graph $J(6,3)$.]

(iii) Γ is locally the graph on the points of a nondegenerate quadric in $PG(2m-1,2)$, two points being adjacent when they are orthogonal (or not). Now $\mathrm{Aut}\,\Gamma \cong 2\times Sp(2m,2)$, $k = 2^{2m-1}\pm 2^{m-1}-1$, and $\mu = 2^{2m-2}$ (or $\mu = 2^{2m-2}\pm 2^{m-1}-2$), where the upper (lower) sign is for a hyperbolic (elliptic) quadric.

(iv) Γ is locally the graph on the points of a nondegenerate quadric in $PG(2m,2)$, two points being adjacent when they are orthogonal (or not). $\mathrm{Aut}\,\Gamma \cong 2\times 2^{2m}.Sp(2m,2)$, $k = 2^{2m}-1$, $\mu = 2^{2m-1}$ (or $\mu = 2^{2m-1}-2$).

(v) For an odd prime power q, let H be a Hermitean form on the projective plane $PG(2,q^2)$. Define a regular two-graph on the set $X = \{x \mid H(x,x) = 0\}$ by taking the triples $\{x,y,z\}$ for which $H(x,y)H(y,z)H(z,x)$ is a nonsquare (when $q \equiv 1 \pmod 4$) or a square (when $q \equiv 3 \pmod 4$) in the field $GF(q^2)$. The corresponding graph Γ admits $2 \times P\Gamma U(3,q^2)$ as a group of automorphisms, and has $k = q^3$, $\lambda = (q-1)(q^2+1)/2$, $\mu = (q+1)(q^2-1)/2$. [For $q = 3$ we have $\mathrm{Aut}\,\Gamma \cong 2 \times Sp(6,2)$ and we find again the example given under (iii) (for $m = 3$, elliptic case).]

(vi) For $q = 3^{2m+1}$ the group ${}^2G_2(q)$ also has a 2-transitive representation on q^3+1 points, leading to a regular two-graph and a Taylor graph with the same parameters as those given under (v).

(vii) Γ is the graph described in §11.4E; $\mathrm{Aut}\,\Gamma \cong 2 \times HS$, $k = 175$, $\mu = 72$ or 102.

(viii) Γ is the graph described in §11.4H; $\mathrm{Aut}\,\Gamma \cong 2 \times Co_3$, $k = 275$, $\mu = 112$ or 162.

Using the classification of 2-transitive permutation groups (of even degree), it should be relatively straightforward to show that this list of examples is complete.

7.7 Towards the classification of all distance-transitive graphs

Concerning the classification of all distance-transitive graphs, there are good hopes that this project might be completed in the not too distant future. In a first approximation, one can restrict the search to primitive distance-transitive graphs, since the existence of an imprimitive distance-transitive graph would, by the results of §4.2, imply the existence of a primitive one with a related group from which one might reconstruct the imprimitive doubles and/or covers, doubles of covers and covers of doubles.

In the case of diameter ≤ 2, everything is known. Indeed, if Γ is a clique, then G acts distance-transitively on Γ if and only if it acts 2-transitively on the vertex set of Γ. As a consequence of the classification of finite simple groups, all 2-transitive permutation groups are known (cf. Cameron [169], Cohen & Zantema [208] Hering [378, 379], Kantor [458] and Liebeck [491].) If Γ has diameter 2 (i.e., is strongly regular), then G acts distance-transitively on Γ if and only if it acts transitively with permutation rank 3 on the vertex set of Γ. This case has been settled completely by Foulser [294], Foulser & Kallaher [295], Kantor & Liebler [460], Liebeck [490, 491], and Liebeck & Saxl [494].

Problem. Using the same techniques used for the rank 3 case, determine all primitive permutation groups of rank ≤ 5.

The following theorem, proved by use of the classification of finite simple groups, shows for which kind of groups primitive distance-transitive graphs still unknown to us may arise.

7.7.1. Theorem (Praeger, Saxl & Yokoyama [597]). *Let Γ be a primitive distance-regular graph with a distance-transitive group G of automorphisms of Γ. Then one of the following holds.*

(i) *Γ is a Hamming graph or, in case $d = 2$, the complement of a Hamming graph, and G is a wreath product.*

(ii) *G is almost simple (i.e., there is a nonabelian simple group X such that* $X \leqslant G \leqslant \mathrm{Aut}\, X$*).*

(iii) *G has an elementary abelian normal subgroup which is regular on* Γ*. Thus, there are numbers m and p (p prime) such that the vertex set of* Γ *can be identified with the set underlying* $\mathbf{F}_p^m$ *and G is a subgroup of* $AGL(m,p)$ *containing all translations.*

In case (i), Γ is completely determined (but G is not).
Case (ii) has been handled for the alternating and symmetric groups by SAXL [638] (for $\mathrm{Sym}(n)$, $n > 18$), IVANOV [431] (for $\mathrm{Sym}(n)$; he even treats the nonprimitive case, which is much more difficult), and LIEBECK, PRAEGER & SAXL [492]. In case $d > 2$ and $X = \mathrm{Alt}(n)$, $n \geqslant 5$ one finds the Johnson graphs, the Odd graphs and the Double Johnson graphs, each with full automorphism group $\mathrm{Sym}(n)$, and for $n = 6$ also the two graphs with intersection arrays $\{5,4,2;1,1,4\}$ ($v = 36$) (cf. §13.1A) and $\{4,2,2,2;1,1,1,2\}$ ($v = 45$) (the line graph of Tutte's 8-cage, the thin generalized 8-gon of order (2,1)), both with full automorphism group $\mathrm{Aut}(\mathrm{Alt}(6))$.
For case (ii) with $X = PSL(n,q)$, $n \geqslant 8$, see INGLIS, LIEBECK & SAXL [421], who determine the multiplicity-free permutation representations. The only distance-transitive graphs here are the Grassmann graphs. VAN BON & COHEN [101] handled $PSL(n,q)$ for $2 \leqslant n \leqslant 7$ (besides Grassmann graphs, one finds here the generalized hexagons of order $(q, 1)$ for $n = 3$, and various sporadic examples with $n \leqslant 4$). Case (ii) has been handled for the classical groups of dimension at least 13 in INGLIS [420]. Some information concerning Chevalley groups is contained in Theorem 10.9.4. Concerning the sporadic groups, VAN BON, COHEN & CUYPERS [102] treat the Held group (and find no distance-transitive graphs for it).

Problem. Give an elementary proof of Theorem 7.7.1. The following particular case seems to be hard. Let $G = (T \times T){:}\langle\tau\rangle$, where T is a nonabelian simple group and $\tau \in G$ is an involution normalizing $T \times T$ such that $(t_1,t_2)^\tau = (t_2,t_1)$ for $t_i \in T$. Let Γ be a graph with vertex set T on which G acts via $t^{(t_1,t_2)} = t_1^{-1} t t_2$ and $t^\tau = t^{-1}$, and in which γ and δ are adjacent whenever (γ,δ) belongs to a distinguished self-paired orbital - equivalently, whenever $\delta^{-1}\gamma \in C \cup C^{-1}$ for some fixed conjugacy class C of T. Prove that G does not act distance-transitively on Γ.

LIEBECK, PRAEGER & SAXL [493] show that if a group G acts distance-transitively and primitively on a graph Γ of diameter $d > 2$, and Γ is not a Hamming graph, then $\mathrm{soc}\, G \trianglelefteq \mathrm{Aut}\,\Gamma$ - in other words, $\mathrm{Aut}\,\Gamma$ cannot be much larger than G.

The following lemma shows that, for a given group G, the stabilizer in G of a vertex in a graph on which G acts distance-transitively has fairly small index in G.

7.7.2. Lemma (cf. SAXL [638]). *If G is a distance-transitive group of automorphisms of a graph* Γ *of diameter d on v vertices, then* $v \leqslant \min(\sqrt{(d+1)|G|},\ 1+t+\sqrt{s|G|})$ *and* $d \leqslant r-1$*, where r (resp. s) stand for the number of irreducible real (resp. symplectic) characters of G, and t is the number of involutions in G.*

Proof. Let $\pi = \sum_{i=0}^{d} \chi_i$ be the decomposition of the permutation character π of G on Γ in irreducibles χ_i ($0 \leqslant i \leqslant d$). Then all χ_i are distinct (see Proposition 4.1.11 (ii)) so

$$v^2 = \pi(1)^2 = (\sum_{i=0}^{d} \chi_i(1))^2 \leqslant (d+1)\sum_{i=0}^{d} \chi_i(1)^2 \leqslant (d+1)|G|$$

and $d+1 \leqslant r$ as all χ_i are real (again, see Proposition 4.1.11 (ii)). Furthermore, the following relation between irreducible real and symplectic characters is well known (cf. ISAACS [422], p. 51):

$$\sum_{\chi \, real} \chi(1) = \sum_{\chi \, symplectic} \chi(1) + t + 1.$$

Hence

$$v = \sum_{i=0}^{d} \chi_i(1) \leqslant \sum_{\chi \, symplectic} \chi(1) + t + 1 \leqslant \sqrt{s} \sqrt{\sum_{\chi \, symplectic} \chi(1)^2} + t + 1$$

$$\leqslant \sqrt{s}\sqrt{|G|} + t + 1.$$

□

7.8 Further transitivity in graphs

A connected graph Γ is called *n-transitive* if, for any two ordered n-tuples $(\gamma_1, \cdots, \gamma_n)$ and $(\delta_1, \cdots, \delta_n)$ of vertices satisfying $d(\gamma_i,\gamma_j) = d(\delta_i,\delta_j)$ for all i, j, there is an automorphism of Γ sending γ_i to δ_i for all i. Thus, the distance-transitive graphs are precisely the 2-transitive graphs. MEREDITH [533] studied 3-transitive graphs; among other things, he showed that these have girth 3 or 4 (when not a polygon) and that Hamming cubes and folded cubes are 3-transitive but not 4-transitive, and gave further parameter conditions. Using this work, CAMERON [168] determined all 6-transitive graphs; these are: (i) complete multipartite graphs, (ii) graphs $\overline{2 \times m}$ ($K_{m,m}$ minus a 1-factor), (iii) polygons, (iv) 3×3 ($= L(K_{3,3})$), (v) the icosahedron, (vi) the Johnson graph $J(6,3)$. All of these graphs are n-transitive for all n. (In the above result, the assumption on $\mathrm{Aut}\,\Gamma$ may be replaced by a combinatorial condition. Call a connected graph Γ *n-regular* if, for any set S of at most n vertices, the number of vertices adjacent to every vertex of S depends only on the isomorphism type of the induced subgraph on S. Clearly, any n-transitive graph is locally $(n-1)$-regular, and, as Cameron remarks in a 'Note added in proof', all 5-regular graphs are known; see also BUCZAK [142].)

Terwilliger calls a graph Γ triply regular when for any 3-set $\{\gamma,\delta,\epsilon\}$ of vertices of Γ the number $a_{ijk} = |\Gamma_i(\gamma) \cap \Gamma_j(\delta) \cap \Gamma_k(\epsilon)|$ only depends on $d(\gamma,\delta)$, $d(\gamma,\epsilon)$ and $d(\delta,\epsilon)$. Clearly, a 3-transitive graph is triply regular. He announces the complete determination of triply regular graphs of diameter at least 6 (these are the polygons, the n-cubes, the folded $(2m+1)$-cubes, and the halved cubes) and gives parameter restrictions for $3 \leqslant d \leqslant 5$; unknown cases correspond to imprimitive graphs.

Cubic graphs Γ with as little transitivity as is possible in a vertex-transitive graph (i.e., with $|\mathrm{Aut}\,\Gamma| = |\Gamma|$) are studied in COXETER, FRUCHT & POWERS [226].

We only list a few more references for related notions of transitivity on graphs. CAMERON [163, 166, 174], PRAEGER [596], PRAEGER & XU [598], WEISS [777], YAP [797].

Distance-transitive digraphs

Some work has been done in the directed case; in marked contrast to what we find for (undirected) distance-transitive graphs, it seems that there are only very few examples of distance-transitive digraphs (digraph = directed graph). Let us briefly sketch these developments. LAM [471] introduced a concept of *distance-transitive digraph* (by requiring that there be an automorphism mapping a pair of vertices (α,β) to a pair (γ,δ) whenever $d(\alpha,\beta) = d(\gamma,\delta)$) and gave some elementary properties and examples (directed circuits, Paley tournaments on $q = 4t+3$ vertices, and coclique extensions of these). DAMERELL [241] generalized this concept to that of (strongly) distance-regular digraph and noted that these digraphs give rise to commutative coherent configurations. He proved for the girth g that $g \in \{2,d,d+1\}$, and that the examples with $d = g$ arise from those with $d = g-1$ by coclique extension. [He suggests that the scarcity of distance-transitive digraphs is caused by an unnaturally strong definition, and that the concept of (weakly) distance-transitive digraph, where an automorphism mapping (α,β) to (γ,δ) is required only when both $d(\alpha,\beta) = d(\gamma,\delta)$ and $d(\beta,\alpha) = d(\delta,\gamma)$, would be more appropriate.] BANNAI, CAMERON & KAHN [28] show in the distance-transitive case that there are no unknown examples of odd girth. LIEBLER & MENA [496] construct an infinite family of edge-transitive distance-regular digraphs with $d = 3$, $g = 4$ (see also ENOMOTO & MENA [276]). LEONARD [488, 489] shows that in the metric & cometric case the schemes are self-dual. DUVAL [272] considers the case $d = g = 2$.

Infinite distance-transitive graphs

According to MACPHERSON [513] any infinite distance-transitive graph of finite valency k is isomorphic to the halved graph of an infinite distance-biregular tree with valencies $s+1$ and $t+1$, for some s, t satisfying $k = s(t+1)$. In contrast to this result, allowing k to be infinite gives rise to several non-trivial distance-transitive graphs, as the following construction by Cameron [pers. comm., 1988] shows. First, let us introduce some (temporary) terminology.

A metric space with integral valued distance will be called a *metric scheme*. We shall use the same symbol for a metric scheme and its underlying set of points. A *subscheme* of a metric scheme M is a subset of M with the induced distance. A metric scheme M is said to be *highly transitive* if any isomorphism between finite subschemes of M can be extended to an automorphism of M.
Let $\mathcal{C}$ be a class of metric schemes. It is said to have the *amalgamation property* if, for any $A, B, C \in \mathcal{C}$ with embeddings $f_1 : A \to B$, $f_2 : A \to C$, there exist embeddings $g_1 : B \to D$ and $g_2 : C \to D$ such that $g_1 f_1 = g_2 f_2$.

7.8.1. Theorem (FRAÏSSÉ [296], a special case). *Let $\mathcal{C}$ be a class of finite metric schemes. Then $\mathcal{C}$ is the class of all finite metric schemes isomorphic to finite subschemes of some countable highly transitive metric scheme M if and only if $\mathcal{C}$ is closed under isomorphism, closed under taking subschemes, has at most countably many non-isomorphic members, and has the amalgamation property. Moreover, if M exists, it is unique; that is, two countable highly transitive metric schemes with up to isomorphism the same finite metric subschemes are isomorphic.* □

Cameron observed that the theorem can be put into use for constructing distance-transitive graphs by pointing out classes $\mathcal{C}$ of metric schemes that satisfy the conditions of the theorem. Usually, the amalgamation property is the only non-trivial condition that needs to be verified.

7.8.2. Proposition. *Let $d \in \mathbb{N} \cup \{\infty\}$. The class $\mathcal{C}_d$ of finite metric spaces of diameter at most d has the amalgamation property.*

Proof. By induction from the case where $|B| = |C| = |A|+1$ and $A = B \cap C$. Write $\{b\} = B \setminus A$, and $\{c\} = C \setminus A$. We have to show that we can define the distance $t := d(b,c)$ between b and c in such a way that $D = B \cup C$ becomes a metric scheme. For each $a \in A$, the value t must satisfy

$$|d(a,b) - d(a,c)| \leqslant t \leqslant d(a,b) + d(a,c).$$

Besides, the definition of $\mathcal{C}_d$ requires $t \leqslant d$. As these are all constraints to be imposed on t, it suffices to verify that such a t exists. If not, then either there exist $a_1, a_2 \in A$ with

$$|d(a_1,b) - d(a_1,c)| > d(a_2,b) + d(a_2,c),$$

or

$$|d(a_1,b) - d(a_1,c)| > d.$$

The latter is clearly impossible in view of the diameters of B and C; if the former holds, then, assuming (without loss of generality) that $d(a_1,b) > d(a_1,c)$, and using the triangle inequality in B and C respectively, we get

$$d(a_1,a_2) \leqslant d(a_1,c) + d(a_2,c) < d(a_1,b) - d(a_2,b) \leqslant d(a_1,a_2),$$

a contradiction. Hence, we can choose t as required, proving the amalgamation property for the particular choice of A, B, and C. □

If Γ is a graph, it determines a metric scheme $M(\Gamma)$ by letting distance be the graph-theoretical distance. Conversely, if M is a metric scheme, we can distill a graph $\Gamma(M)$ with vertex set M by letting two vertices be adjacent if and only if their distance is 1. Clearly there are metric schemes M, for which $M(\Gamma(M))$ is not isomorphic to M. (For instance if distance 2 is realized and distance 1 is not realized within M.) However, if Γ is a graph, we have $\Gamma(M(\Gamma)) \cong \Gamma$.

7.8.3. Theorem (Cameron) *For any $d \in \mathbb{N} \cup \{\infty\}$, there exists a countable primitive distance-transitive graph of diameter d.*

Proof. Observe that the class $\mathcal{C}_d$ of the above proposition satisfies all conditions of Fraïssé's theorem. Hence there exists a countable highly transitive metric scheme M such that $\mathcal{C}_d$ is the class of all finite metric schemes isomorphic to finite metric subschemes of M. Take $\Gamma = \Gamma(M)$. We assert that the graph-theoretic distance in Γ coincides with the distance of the metric scheme M. For, let $x, y \in M$. Then by the triangle inequality, $i := d_M(x,y) \leqslant d_\Gamma(x,y)$. On the other hand, M contains a $(i+1)$-point subscheme M_0 such that $\Gamma(M_0)$ is a path of length i, say with end points x' and y', and such that $M(\Gamma(M_0)) = M_0$. Now $\{x,y\}$ is isometric with $\{x',y'\}$, and by high transitivity of M there is an isometry of M mapping $\{x',y'\}$ to $\{x,y\}$. Since any isometry of M is an automorphism of Γ, there is a Γ-geodesic from x to y of length i, so that $d_\Gamma(x,y) = d_M(x,y)$.
The distance transitivity of Γ is immediate from the high transitivity of M. To show

that Γ is primitive, we suppose that distance i pertains to an invariant relation. For any $j \leqslant \min(2i,d)$, there is a triangle with sides i,i,j in $\mathcal{C}_d$, hence embedded in M and so in Γ; thus distance j also belongs to the invariant relation. In particular, taking $j = 1$ and applying transitivity, we get that every pair of vertices belongs to the invariant relation; hence Γ is primitive. □

Remarks. (i) For $d = 2$, the random countable graph emerges.
(ii) Taking other natural classes of metric schemes, other distance-transitive graphs can be obtained (e.g. taking $\mathcal{C}_d$ as above but forbidding $n-1$-simplices leads to graphs without K_n). Also non-primitive countable distance-transitive graphs can be constructed.

Chapter 8
Q-polynomial Distance-Regular Graphs

In this chapter we discuss in detail the relations between the parameters of Q-polynomial distance-regular graphs and their Q-sequences. We start with a reformulation of the Q-polynomial property defined in §4.1E, which is the source of most of our results. This leads to a three-term recurrence relation for Q-sequences (Theorem 8.1.2) discovered by LEONARD [485], and a representation of the intersection array of Q-polynomial graphs in terms of 5 parameters only (Proposition 8.1.5). We mention Leonard's explicit parameter formulae and results of BANNAI & ITO [33] on the integrality of eigenvalues for large diameters based on these formulae. We also determine which of the known distance-regular graphs are Q-polynomial.

8.1 Leonard's characterization of Q-polynomial graphs

As a tool for the characterization of the Q-sequences of distance-regular graphs we prove here a reformulation of the Q-polynomial property which is basic to the whole chapter.

Recall (cf. §2.7 and §4.1E) that an association scheme (with eigenmatrices P, Q) is called *Q-polynomial* with *Q-sequence* $(\sigma_0, \cdots, \sigma_d)$ when there are polynomials q_j of degree j such that $Q_{ij} = q_j(\sigma_i)$ for $0 \leqslant i, j \leqslant d$.

8.1.1. Theorem. *Let Γ be a distance-regular graph of diameter d with intersection numbers a_i, b_i, c_i. Then the following statements concerning a sequence $(\sigma_0, \cdots, \sigma_d)$ are equivalent (we use the convention $0^0 = 1$, and the undefined quantities σ_{-1} and σ_{d+1} are treated as indeterminates):*

(i) *$(\sigma_0, \cdots, \sigma_d)$ is a Q-sequence of Γ.*

(ii) *$\sigma_0, \cdots, \sigma_d$ are $d+1$ distinct numbers, and there are polynomials $r_l(x)$ of degree at most l ($l = 0, \cdots, d-1$) such that*

$$\sigma_{i-1}^l c_i + \sigma_i^l a_i + \sigma_{i+1}^l b_i = r_l(\sigma_i) \qquad (1_{il})$$

for $i = 0, \cdots, d$ and $l = 0, \cdots, d-1$.

(iii) *$\sigma_0, \cdots, \sigma_d$ are $d+1$ distinct numbers, and there are polynomials $\phi_l(x)$ of degree at most l ($l = 1, \cdots, d-1$) such that*

$$b_i = \frac{\phi_2(\sigma_i) - \sigma_{i-1}\phi_1(\sigma_i)}{(\sigma_{i+1} - \sigma_{i-1})(\sigma_{i+1} - \sigma_i)}, \qquad (2)$$

$$c_i = \frac{\phi_2(\sigma_i) - \sigma_{i+1}\phi_1(\sigma_i)}{(\sigma_{i-1} - \sigma_{i+1})(\sigma_{i-1} - \sigma_i)}, \tag{3}$$

$$a_i = \frac{\phi_2(\sigma_i) + (\sigma_i - \sigma_{i-1} - \sigma_{i+1})\phi_1(\sigma_i) + (\sigma_i - \sigma_{i-1})(\sigma_i - \sigma_{i+1})k}{(\sigma_i - \sigma_{i-1})(\sigma_i - \sigma_{i+1})}, \tag{4}$$

and

$$\phi_{l+1}(\sigma_i) = (\sigma_{i-1} + \sigma_{i+1})\phi_l(\sigma_i) - \sigma_{i-1}\sigma_{i+1}\phi_{l-1}(\sigma_i), \tag{5}$$

for $i = 0, \cdots, d$ *and* $l = 2, \cdots, d-2$.

Moreover, if (i)-(iii) *hold then* Γ *is Q-polynomial, and the coefficients* θ_l *of* x^l *in* $r_l(x)$ *are the eigenvalues of* Γ *in a Q-polynomial ordering* $k = \theta_0, \theta_1, \cdots, \theta_{d-1}, \theta_d$*; in particular they are distinct. The degree of* ϕ_l *is exactly* l*, and the highest coefficient of* ϕ_l *is* $\theta_l - \theta_{l-1}$.

Proof. Suppose first that $(\sigma_0, \cdots, \sigma_d)$ is a Q-sequence. Then by Proposition 4.1.8 the σ_i are distinct, and by definition there are polynomials $q_l(x)$ of degree l ($l = 0, \cdots, d$) such that the eigenmatrices can be written as

$$E_l = \sum_{i=0}^{d} q_l(\sigma_i)A_i \qquad (l = 0, \cdots, d), \tag{6}$$

where $\theta_0, \cdots, \theta_d$ is a Q-polynomial ordering of the eigenvalues. Now

$$\begin{aligned}\theta_l E_l = AE_l = \sum_{i=0}^{d} q_l(\sigma_i)AA_i &= \sum_{i=0}^{d} q_l(\sigma_i)(c_{i+1}A_{i+1} + a_iA_i + b_{i-1}A_{i-1}) \\ &= \sum_{i=0}^{d} (q_l(\sigma_{i-1})c_i + q_l(\sigma_i)a_i + q_l(\sigma_{i+1})b_i)A_i\end{aligned}$$

since $A_{-1} = A_{d+1} = 0$, $b_d = c_0 = 0$. A comparison with (6), multiplied by θ_l, shows that

$$q_l(\sigma_{i-1})c_i + q_l(\sigma_i)a_i + q_l(\sigma_{i+1})b_i = \theta_l q_l(\sigma_i) \tag{7_{il}}$$

holds for $i = 0, \cdots, d$ and $l = 0, \cdots, d-1$ (and $l = d$). Now x^l is a linear combination of $q_0(x), \cdots, q_l(x)$, and combining $(7_{i0}), \cdots, (7_{il})$ accordingly gives (1_{il}) with suitable polynomials $r_l(x) = \theta_l x^l + \cdots$ of degree at most l. The highest coefficients of the $r_l(x)$ are just the (distinct) eigenvalues of Γ.

Conversely, suppose that (ii) holds. By interpolation we can find a polynomial $r_d(x)$ such that (1_{id}) holds for $i = 0, \cdots, d$. We introduce the matrices

$$C^{(l)} = \sum_{i=0}^{d} \sigma_i^l A_i \tag{8}$$

and denote by V_l the vector space spanned by $C^{(0)}, \cdots, C^{(l)}$. Since the σ_i are distinct, $C^{(0)}, \cdots, C^{(l)}$ are linearly independent (Vandermonde determinant!). Therefore, V_l has dimension $l+1$ for $l = 0, \cdots, d$. Now, by (1_{il}),

$$\begin{aligned}AC^{(l)} = \sum_{i=0}^{d} \sigma_i^l AA_i &= \sum_{i=0}^{d} \sigma_i^l(c_{i+1}A_{i+1} + a_iA_i + b_{i-1}A_{i-1}) \\ &= \sum_{i=0}^{d} (\sigma_{i-1}^l c_i + \sigma_i^l a_i + \sigma_{i+1}^l b_i)A_i = \sum_{i=0}^{d} r_l(\sigma_i)A_i\end{aligned}$$

is a linear combination of $C^{(0)}, \cdots, C^{(l)}$ whence $AC^{(l)} \in V_l$. By induction,

$A^iC^{(l)} \in V_l$ for all $i \geqslant 0$. Since the A_i are polynomials in A, the same holds by (8) for the $C^{(i)}$; therefore $C^{(i)}C^{(l)} \in V_l$ for all $i, l \geqslant 0$. Since polynomials in A commute, the chain $\{0\} = V_{-1} \subset V_0 \subset V_1 \subset \cdots \subset V_d$ consists of commutative algebras V_i of dimension $i+1$; in particular, V_d is the space of all polynomials in A. Now for $l = 0, \cdots, d$, the space V_l is generated by V_{l-1} and a nonzero idempotent E_l. By induction on l, $AE_l = \theta_l E_l$ for some eigenvalue θ_l of A; moreover $E_0 = J$ and $\theta_0 = k$. Now $E_0, \cdots, E_d$ are linearly independent since V_d has dimension $d+1$. Therefore $E_0, \cdots, E_d$ is a complete set of primitive idempotents of the algebra of polynomials in A, and $\theta_0, \cdots, \theta_d$ is an ordering of the eigenvalues of A. Since $E_l \in V_l \setminus V_{l-1}$ it can be written in the form (6) with a polynomial q_l of degree l. Hence $\theta_0, \cdots, \theta_d$ is a Q-polynomial ordering of the eigenvalues, Γ is Q-polynomial, and $(\sigma_0, \cdots, \sigma_d)$ is a Q-sequence of Γ. This shows that (i) and (ii) are equivalent.

Now suppose that (ii) holds, and define $\phi_l(x) = r_l(x) - xr_{l-1}(x)$, $l = 1, \cdots, d-1$. From (1) we find

$$\phi_l(\sigma_i) = \sigma_{i-1}^{l-1}(\sigma_{i-1} - \sigma_i)c_i + \sigma_{i+1}^{l-1}(\sigma_{i+1} - \sigma_i)b_i \tag{9_{il}}$$

for $i = 0, \cdots, d$ and $l = 1, \cdots, d-1$. Solving (9_{i1}) and (9_{i2}) for b_i, c_i gives (2) and (3), and (4) follows since $a_i = k - b_i - c_i$, and (5) follows by combining (9_{il+1}), (9_{il}) and (9_{il-1}). Conversely, if (iii) holds one easily establishes (9), and (1) follows by defining $r_0(x) = k$ and recursively $r_l(x) = \phi_l(x) + xr_{l-1}(x)$ for $l = 1, \cdots, d-1$. Hence (ii) and (iii) are equivalent. Moreover, if (ii) holds then $\phi_l(x)$ has highest coefficient $\theta_l - \theta_{l-1} \neq 0$ so that its degree is precisely l. □

Remark. Note that σ_{-1} and σ_{d+1} do not really occur in (ii) since $c_0 = b_d = 0$. In (iii) we may replace the conditions in the cases $i = 0$ and $i = d$ by

$$\phi_l(\sigma_0) = k(\sigma_1 - \sigma_0)\sigma_1^{l-1} \tag{5a}$$

and

$$\phi_l(\sigma_d) = c_d(\sigma_{d-1} - \sigma_d)\sigma_{d-1}^{l-1} \tag{5b}$$

for $l = 1, \cdots, d-1$.

Recurrence relations for Q-sequences

8.1.2. Theorem (Leonard [485]). *Let Γ be a Q-polynomial distance-regular graph. Then every Q-sequence $(\sigma_0, \cdots, \sigma_d)$ of Γ satisfies the recurrence*

$$\sigma_{i+1} + \sigma_{i-1} = p\sigma_i + r \quad (i = 1, \cdots, d-1), \tag{10}$$

for suitable numbers p and r.

Leonard's proof makes use of orthogonal polynomials; here we give two other proofs. Our first proof uses graph representations and Terwilliger's characterization of Q-polynomial distance-regular graphs in terms of the Krein module (Theorem 2.11.6); the second proof shows that the recurrence relation (10) is already an algebraic consequence of the conditions in Theorem 8.1.1.

First proof. Let the notation be as in §2.11. By Theorem 2.11.6, $\delta(A_i\hat{A}_j) = \delta\hat{T}$ for some matrix T of the form $T = \sum_h t_{ij}^h A_h$. Using equation (23) of §2.11 and the notation $\overline{S}$ for the sum of the images $\overline{\alpha}$ of vertices $\alpha \in S$, we can express this relation entrywise in terms of the sets

$$\Gamma_{i,j}(\gamma,\delta) := \Gamma_i(\gamma) \cap \Gamma_j(\delta)$$

as

$$\overline{\Gamma_{i,j}(\gamma,\delta)} - \overline{\Gamma_{j,i}(\gamma,\delta)} = t_{ij}^h(\overline{\gamma} - \overline{\delta}) \text{ if } d(\gamma,\delta) = h. \tag{11}$$

Now consider two vertices γ, δ at distance $h = 3$, and choose $\beta \in \Gamma_{l-2}(\gamma) \cap \Gamma_{l+1}(\delta)$. Taking inner products of $\overline{\beta}$ with both sides of (11) for $i = 1$, $j = 2$, we find

$$c_3(u_{l-1} - u_l) = t_{12}^3(u_{l-2} - u_{l+1}) \quad (l = 2, \cdots, d-1).$$

Since the u_l form a Q-sequence, they are distinct. Hence $t_{12}^3 \neq 0$, and, with $p = -1 + c_3 / t_{12}^3$, we see that $u_{l+1} - pu_l + u_{l-1} = u_l - pu_{l-1} + u_{l-2}$ for $l = 2, \cdots, d-1$. Thus this expression is a constant r, and (10) holds. □

Second proof. For our second proof we note that the σ_i are distinct whence the polynomial sequences $r_l(x)$ and $\phi_l(x)$ can be extended by interpolation for all $l \geqslant d$ such that (1_{il}) and hence (9_{il}) holds for $i = 0, \cdots, d$ and all $l \geqslant 1$. From (9) we obtain the relations

$$\begin{aligned} &\phi_{j+l}(\sigma_i)\phi_k(\sigma_i) - \phi_j(\sigma_i)\phi_{k+l}(\sigma_i) = \\ &= (\sigma_{i-1}^{j-1}\sigma_{i+1}^{k-1} - \sigma_{i+1}^{j-1}\sigma_{i-1}^{k-1})(\sigma_{i-1}^l - \sigma_{i+1}^l)(\sigma_{i-1} - \sigma_i)(\sigma_{i+1} - \sigma_i)c_ib_i \end{aligned} \tag{12}$$

for $i = 0, \cdots, d$ and all $j, k, l \geqslant 1$. Since $c_0 = b_d = 0$, all polynomials $\phi_{j+l}(x)\phi_k(x) - \phi_j(x)\phi_{k+l}(x)$ are divisible by $\pi(x) = (x - \sigma_0)(x - \sigma_d)$. Therefore

$$\begin{aligned} \psi_2 &:= (\phi_3\phi_1 - \phi_2^2) / \pi, \\ \psi_3 &:= (\phi_4\phi_1 - \phi_3\phi_2) / \pi, \\ \psi_4 &:= (\phi_5\phi_1 - \phi_3^2) / \pi, \\ \psi_5 &:= (\phi_6\phi_1 + \phi_5\phi_2 - 2\phi_4\phi_3) / \pi \end{aligned}$$

are polynomials, and ψ_l has degree at most l. Evaluation of (12) at the appropriate indices now shows that

$$\psi_l(\sigma_i) = (\sigma_{i-1} + \sigma_{i+1})^{l-2}(\sigma_{i-1} - \sigma_{i+1})^2(\sigma_{i-1} - \sigma_i)(\sigma_{i+1} - \sigma_i)c_ib_i / \pi(\sigma_i) \tag{13}$$

for $i = 1, \cdots, d-1$ and $l = 2, 3, 4, 5$. This implies

$$\psi_l(\sigma_i) = (\sigma_{i-1} + \sigma_{i+1})^{l-2}\psi_2(\sigma_i) \quad \text{for } l = 3, 4, 5, \tag{14}$$

and since $\sigma_0, \cdots, \sigma_d$ are distinct and $b_ic_i \neq 0$ for $i \neq 0$, d, we also get

$$\psi_2(\sigma_i) \neq 0. \tag{15}$$

Now suppose first that $d \geqslant 8$. Then (14) implies

$$\psi_2(\sigma_i)\psi_4(\sigma_i) = \psi_3(\sigma_i)^2 \quad (i = 1, \cdots, 7), \tag{16}$$

$$\psi_2(\sigma_i)\psi_5(\sigma_i) = \psi_3(\sigma_i)\psi_4(\sigma_i) \quad (i = 1, \cdots, 7). \tag{17}$$

Since $\psi_2\psi_4$ and ψ_3^2 are both polynomials of degree at most 6, (16) implies the polynomial identity

$$\psi_2\psi_4 = \psi_3^2. \tag{18}$$

By (15), ψ_2 is not identically zero. If ψ_2 has degree at most 1 then by (18), ψ_3 has degree at most 2. Therefore (17) implies the polynomial identity $\psi_2\psi_5 = \psi_3\psi_4$; combined with (18) we get $\psi_2^2\psi_5 = \psi_3^3$. But this is only possible if ψ_3 / ψ_2 is linear in x,

$\psi_3(x)/\psi_2(x) = px+r$, and (10) follows. And if ψ_2 has degree 2 then either ψ_3/ψ_2 is a linear polynomial, and (10) follows again, or ψ_2 does not divide ψ_3, and there is a linear polynomial π_1 and an at most quadratic polynomial π_2 such that

$$\psi_2 = \pi_1^2\,,\quad \psi_3 = \pi_1\pi_2\,,\quad \psi_4 = \pi_2^2\,. \tag{19}$$

Since by (15), $\pi_1(\sigma_i) \neq 0$ for $i = 1, \cdots, 7$ we find from (17) and (19) that

$$\pi_1(\sigma_i)\psi_5(\sigma_i) = \pi_2(\sigma_i)^3 \quad (i = 1, \cdots, 7). \tag{20}$$

Again (20) implies the polynomial identity $\pi_1\psi_5 = \pi_2^3$; therefore π_1 divides π_2 and ψ_2 divides ψ_3, contradiction. Hence (10) holds in general.

The case $d \leqslant 3$ is trivial. The cases $d \in \{4,5,6,7\}$ can be handled by long arithmetical calculations which we do not give here; in each case certain exceptional possibilities for the $\phi_l(x)$ satisfying (5) lead to multiple eigenvalues θ_l and are thus impossible. □

Remarks. (i) The constants t_{ij}^h in (11) can be determined by taking inner products with $\overline{\gamma}$. One finds the explicit values $t_{ij}^h = p_{ij}^h(u_i - u_j)/(u_0 - u_k)$.
(ii) For $\beta \in \Gamma_r(\gamma) \cap \Gamma_s(\delta)$, taking inner products of $\overline{\beta}$ with both sides of (11) gives a restriction on the distance distribution of vertices of $\Gamma_{i,j}(\gamma,\delta)$ and $\Gamma_{j,i}(\gamma,\delta)$ from β. Thus, (11) will probably be an important step in the characterization problem for Q-polynomial distance-regular graphs. For an application, see Theorem 8.5.1 below; cf. also TERWILLIGER [716, 726, 728]. (Compare also the rôle played by Proposition 4.4.9 (i) - and the related Proposition 3.14.4 (ii) - in Theorem 4.4.11.)
(iii) In the particular case where $h = |r \pm s|$, the distance distribution of vertices of $\Gamma_{i,j}$ and $\Gamma_{j,i}$ from $\beta \in \Gamma_r(\gamma) \cap \Gamma_s(\delta)$ is already determined, and one gets recurrence relations as in the above proof; however, these are already implied by (10).

For sequences satisfying (10) it is easy to decide whether they are Q-sequences of a given distance-regular graph Γ.

8.1.3. Proposition. *Let $\sigma_0, \cdots, \sigma_d$ be numbers satisfying*

$$\sigma_{i+1} + \sigma_{i-1} = p\sigma_i + r \quad (i = 1, \cdots, d-1). \tag{21}$$

Then there is a number s such that

$$\sigma_{i+1}\sigma_{i-1} = \sigma_i^2 - r\sigma_i - s \quad (i = 1, \cdots, d-1). \tag{22}$$

In particular, a distance-regular graph Γ has $(\sigma_0, \cdots, \sigma_d)$ as a Q-sequence if and only if the σ_i are distinct and the intersection array of Γ can be expressed in the form (2), (3) with a linear polynomial $\phi_1(x)$ and a quadratic polynomial $\phi_2(x)$; equivalently, if (9_{il}) holds for $l = 1$, 2 and $i = 0, \cdots, d$.

Proof. Put $s_i := \sigma_i^2 - r\sigma_i - \sigma_{i+1}\sigma_{i-1}$. For $i = 1, \cdots, d-2$, (21) implies

$$\begin{aligned} s_{i+1} &= \sigma_{i+1}^2 - r\sigma_{i+1} - \sigma_{i+2}\sigma_i = \sigma_{i+1}^2 - r\sigma_{i+1} - (p\sigma_{i+1} + r - \sigma_i)\sigma_i \\ &= \sigma_i^2 - r\sigma_i - \sigma_{i+1}(p\sigma_i + r - \sigma_{i+1}) = \sigma_i^2 - r\sigma_i - \sigma_{i+1}\sigma_{i-1} \\ &= s_i \end{aligned}$$

so that s_i does not depend on i. This proves (22). The remainder follows by Theorem 8.1.1 since the ϕ_l defined by

$$\phi_{l+1}(x) := (px + r)\phi_l(x) - (x^2 - rx - s)\phi_{l-1}(x) \tag{23}$$

for $l \geqslant 2$ satisfy (5); note that (4) follows from (2), (3), and $a_i + b_i + c_i = k$. □

8.1.4. Corollary. *If $k = \theta_0, \theta_1, \cdots, \theta_d$ is the Q-polynomial ordering of the eigenvalues of a distance-regular graph Γ corresponding to a Q-sequence satisfying* (10)*, then there are constants r^*, s^* such that*

$$\left.\begin{aligned} \theta_{l+1} + \theta_{l-1} &= p\theta_l + r^* \\ \theta_{l+1}\theta_{l-1} &= \theta_l^2 - r^*\theta_l - s^* \end{aligned}\right\} (l = 1, \cdots, d-1). \tag{24}$$

Proof. If we compare the coefficients of x^{l+1} in (23), we find from Theorem 8.1.1 that $\theta_{l+1} - \theta_l = p(\theta_l - \theta_{l-1}) - (\theta_{l-1} - \theta_{l-2})$; so $\theta_{l+1} - p\theta_l + \theta_{l-1} = \theta_l - p\theta_{l-1} + \theta_{l-2}$ is a constant r^*, and the same reasoning as in the proposition shows that $\theta_l^2 - r^*\theta_l - \theta_{l+1}\theta_{l-1}$ is also a constant s^*. □

Remark. The dual form of the equations in 8.1.3 and 8.1.4 is essentially due to the fact that Q-sequences and Q-polynomial eigenvalue orderings are dual objects in the sense of Chapter 2. In fact, for $d > 8$, LEONARD [487] derives the present results in a completely self-dual way in terms of pairs of orthogonal polynomials (cf. Remark (ii) after Proposition 8.3.2 below). The restriction on d was later removed by BANNAI & ITO [33].

Reduction of parameters

LEONARD [486, 487] showed that all parameters of a P- and Q-polynomial association scheme can be expressed in only a few of them. In particular, LEONARD [486] shows that it suffices to know the five parameters k, b_1, b_2, μ and θ_1. From the above we find a similar result.

8.1.5. Proposition. *The intersection numbers of a Q-polynomial distance-regular graph Γ are determined by the five numbers*

$$d,\quad k,\quad c_d,\quad b := \frac{b_1}{\theta+1},\quad b' := \frac{b_2}{\theta - 1 - \lambda + \mu - b},$$

where $\theta_0 = k$, $\theta_1 = \theta$, $\theta_2, \cdots, \theta_d$ is a Q-polynomial ordering of the eigenvalues of Γ. (The resulting formulae for a_i, b_i, c_i are not pleasant, however.)

Proof. Let $(u_0, \cdots, u_d)$ be a standard Q-sequence of Γ. Then the u_i are distinct, and the $t_i = (u_{i-1} - u_i)/(u_0 - u_1)$ are nonzero, $t_1 = 1$. From the recurrence for t-sequences ((17) in §4.1B) we find that $1 + \theta = b_1 t_2 \neq 0$ whence $t_2 = b^{-1}$, and $(\theta - 1 - \lambda + \mu - b)\cdot b^{-1} = b_2 t_3 \neq 0$ whence $t_3 = (bb')^{-1}$. Now the $\sigma_i = (u_0 - u_i)/(u_0 - u_1) = \sum_{l=1}^{i} t_l$ form a Q-sequence of Γ, and by Theorem 8.1.2 we have $\sigma_{i+1} + \sigma_{i-1} = p\sigma_i + r$. For $i = 1, 2$, this implies $p = b + b'^{-1}$, $r = 1 - b + b^{-1} - b'^{-1}$, so that the whole Q-sequence $(\sigma_0, \cdots, \sigma_d)$ is determined by b and b'. Now (2) for $i = 0, d$ and (3) for $i = 0, 1, d$ give 5 linear equations for the coefficients w_i of $\phi_2(x) = w_1 x^2 + w_2 x + w_3$ and $\phi_1(x) = w_4 x + w_5$ with nonzero determinant (take for the indeterminates σ_{-1} and σ_{d+1} values distinct from the other σ_i); therefore $\phi_1(x)$ and $\phi_2(x)$, and hence, by Theorem 8.1.1, all intersection numbers, are determined by b, b', d, $b_0 = k$, and c_d. □

It is possible to give explicit formulae for the σ_i satisfying (10); however, this may involve irrational or even complex numbers although the σ_i are, usually, rational.

8.1.6. Lemma. *The general solution of* (10) *is*

$$\sigma_i = \alpha_1 q^i + \alpha_2 q^{-i} + \alpha_3 \quad \text{if } p \notin \{2, -2\},$$

where q and q^{-1} are the zeros of the characteristic equation $q^2 - pq + 1 = 0$, and

$$\sigma_i = \alpha_1 i^2 + \alpha_2 i + \alpha_3 \quad \text{if } p = 2,$$

$$\sigma_{2i} = \alpha_1 + \alpha_3 i, \quad \sigma_{2i+1} = \alpha_2 - \alpha_3 i \quad \text{if } p = -2.$$

Proof. Direct verification. □

Remarks. (i) Lengthy explicit formulae for the intersection arrays, eigenvalues and multiplicities in terms of this number q are derived in BANNAI & ITO [33] from the formulae given by LEONARD [485]. We shall derive equivalent but simpler formulae for eigenvalues and multiplicities in terms of the intersection array and Q-sequence in §8.3.
(ii) TERWILLIGER [723] shows that for $d \geqslant 3$ the case $p = -2$ occurs if and only if Γ is either the d-cube, the Odd graph O_{d+1}, or the folded $(2d+1)$-cube.

8.2 Imprimitive Q-polynomial distance-regular graphs

In Chapter 2 we have seen that imprimitivity and dual imprimitivity are the same concepts. Indeed, a bipartite Q-polynomial distance-regular graph Γ is 'Q-antipodal', and the 'dual spectrum', i.e., the Q-sequence, of the halved graphs of Γ is a subset of the dual spectrum of Γ. Similarly, an antipodal Q-polynomial distance-regular graph Γ is 'Q-bipartite', and the dual eigenvalues of the folded graph are (up to a linear transform) the squares of the dual eigenvalues of Γ.

[A graph is called *Q-bipartite* when for some Q-polynomial ordering $012\cdots$ we have $q_{ij}^k = 0$ whenever $i+j+k$ is odd. If this holds for one Q-polynomial ordering, then for all. A graph is called *Q-multipartite* when for some Q-polynomial ordering $012\cdots$ we have $q_{ij}^k = 0$ whenever precisely one of i, j, k is odd.]

Let us first look at the bipartite case.

8.2.1. Theorem. *Let Γ be a bipartite distance-regular graph of diameter d. Then:*

(i) *Γ is Q-polynomial if and only if some standard sequence $(\sigma_0, \cdots, \sigma_d)$ of Γ consists of distinct elements satisfying the recurrence* (10) *for suitable numbers p and r.*

If this is the case, then:

(ii) *The statement of Corollary 8.1.4 holds with $r^* = 0$.*

(iii) *The halved graphs of Γ are Q-polynomial with Q-sequence $(\sigma_0, \sigma_2, \cdots, \sigma_{2m})$, where $m = [d/2]$.*

Proof. (i) Necessity follows from Theorem 8.1.2 and Proposition 4.1.8. Conversely, suppose that $(\sigma_0, \cdots, \sigma_d)$ is a standard sequence of Γ consisting of distinct numbers satisfying (10). Since $a_i = 0$ we have $b_i + c_i = k$, and the recurrence for a standard sequence becomes $b_i\sigma_{i+1} + c_i\sigma_{i-1} = \theta\sigma_i$; therefore

$$b_i = \frac{\theta\sigma_i - k\sigma_{i-1}}{\sigma_{i+1} - \sigma_{i-1}}, \quad c_i = \frac{\theta\sigma_i - k\sigma_{i+1}}{\sigma_{i-1} - \sigma_{i+1}}.$$

Using (22) and

$$\phi_1(x) = (\theta-k)x, \quad \phi_2(x) = \theta((p-1)x^2+rx)-k(x^2-rx-s)$$

it is easy to verify (2) and (3). By Proposition 8.1.3, the σ_i form a Q-sequence, whence Γ is Q-polynomial.
(ii) With $\pi(x) = (x-\sigma_0)(x-\sigma_d)$ and $\psi_{2l} = (r_{l+2}r_l - r_{l+1}^2)/\pi$ we find $\psi_{2l} = \sigma_{i-1}\sigma_{i+1}\psi_{2l-2}$; comparing highest coefficients shows that for Q-polynomial bipartite distance-regular graphs one has

$$\theta_l^2 - \theta_{l-1}\theta_{l+1} = s^* \qquad (l = 1, \cdots, d-1)$$

for some constant s^*. Thus (24) holds with $r^* = 0$.
(iii) Applying $b_j\sigma_{j+1} + c_j\sigma_{j-1} = \theta\sigma_j$ with $j = 2i-1, 2i+1, 2i$ we find

$$b_{2i}b_{2i+1}\sigma_{2i+2} + (c_{2i}b_{2i-1} + b_{2i}c_{2i+1})\sigma_{2i} + c_{2i}c_{2i-1}\sigma_{2i-2} = \theta^2\sigma_{2i}$$

so that $b_i'\sigma_{i+1}' + a_i'\sigma_i' + c_i'\sigma_{i-1}' = \theta'\sigma_i'$, where $b_i' = b_{2i}b_{2i+1}/\mu$, $c_i' = c_{2i}c_{2i-1}/\mu$, $a_i' = (c_{2i}b_{2i-1} + b_{2i}c_{2i+1} - k)/\mu$ and $\theta' = (\theta^2-k)/\mu$ are the parameters of the halved graphs of Γ, and $\sigma_i' = \sigma_{2i}$. This proves (iii). □

A very simple consequence is the following.

8.2.2. Corollary. *Every bipartite distance-regular graph Γ of diameter $d = 3$ is Q-polynomial and has two distinct standard Q-sequences, unless $\Gamma \cong \overline{2\times(k+1)}$ (the complete bipartite graph $K_{k+1,k+1}$ minus a 1-factor), in which case there is only one standard Q-sequence.*

Proof. We find $u = (u_0,u_1,u_2,u_3) = (1, \frac{\theta}{k}, \frac{\theta^2-k}{k(k-1)}, \frac{\theta^2-k}{\theta(k-1)})$. All entries are distinct if and only if $\theta \neq \pm k, -1$ so there are two choices for θ unless Γ has eigenvalue -1. □

Next, we look at the antipodal case.

8.2.3. Proposition. *An antipodal distance-regular graph Γ of diameter $d = 3$ is Q-polynomial if and only if it has an intersection array of the form*

$$\iota(\Gamma) = \{k,\mu,1;\ 1,\mu,k\}\ .$$

In this case there are precisely two standard Q-sequences, namely

$$\sigma_0 = 1,\ \sigma_1 = z,\ \sigma_2 = -z,\ \sigma_3 = -1$$

for the two solutions z of the equation

$$kz^2 - (k-1-2\mu)z - 1 = 0\ . \tag{25}$$

Proof. By Proposition 4.2.2, the intersection array for a $(t+1)$-cover of K_{k+1} is $\iota(\Gamma) = \{k,t\mu,1;\ 1,\mu,k\}$; the minimal polynomial of its adjacency matrix A is

$$\mu k(v_0(x)+v_1(x)+v_2(x)+v_3(x)) = (x-k)(x+1)(x^2-qx-k),$$

where $q = k-1-(t+1)\mu$. The standard sequence for $\theta = -1$ has $u_0 = u_3 = 1$, $u_1 = u_2 = -1/k$, and cannot be a Q-sequence since it contains repeated elements. The standard sequence for a zero θ of the quadratic factor x^2-qx-k is

$$u_0 = 1\,,\ u_1 = \frac{\theta}{k}\,,\ u_2 = -\frac{\theta}{kt}\,,\ u_3 = -\frac{1}{t}\,,$$

and it is easy to check that condition (ii) of Theorem 8.1.1 holds if and only if $t = 1$. In this case, the two zeros θ of $x^2 - qx - k$ correspond to the two solutions $z = \theta / k$ of (25). □

8.2.4. Theorem. *Let Γ be an antipodal distance-regular graph of diameter $d \geqslant 3$. If Γ is Q-polynomial then the intersection numbers satisfy*

$$b_i = c_{d-i} \quad \text{for } i = 0, \cdots, d, \tag{26}$$

and every standard Q-sequence $(\sigma_0, \cdots, \sigma_d)$ satisfies

$$\sigma_i + \sigma_{d-i} = 0 \quad \text{for } i = 0, \cdots, d. \tag{27}$$

Moreover, Γ is then a double cover of its folded graph $\overline{\Gamma}$, and $\overline{\Gamma}$ is also Q-polynomial, with Q-sequence $(\sigma_0^2, \cdots, \sigma_m^2)$, where $m = [d/2]$.

Proof. For $d = 3$ this follows from the preceding proposition; hence we assume that $d \geqslant 4$, and suppose that Γ is a $(t+1)$-cover of its folded graph $\overline{\Gamma}$. By Proposition 4.2.2, (26) holds with the possible exception of the case $i = m := [d/2]$, where

$$b_m = tc_{d-m}\,. \tag{28}$$

Let $(\sigma_0, \cdots, \sigma_d)$ be a standard Q-sequence of Γ. Then the σ_i are distinct, $\sigma_0 = 1$, $\sigma_1 = \theta / k$, and from the recurrence of a standard sequence we obtain with $a_i + b_i + c_i = k$ the relations

$$(\sigma_{i-1} - \sigma_i)c_i + (\sigma_{i+1} - \sigma_i)b_i = (\theta - k)\sigma_i \tag{29_i}$$

for $i = 0, \cdots, d$. Now (29_d) and (26) for $i = 0$ imply $k\sigma_{d-1} = \theta\sigma_d$, hence $\sigma_{d-1} = \sigma_d\sigma_1$. Using induction with (29_i) for $i = j-1,\ d+1-j$ and (26) for $i = j-1\ (\neq m)$ we get

$$\sigma_{d-i} = \sigma_d\sigma_i \quad \text{for } i = 0, \cdots, m. \tag{30}$$

Now (28) and (30) imply

$$b_m = tc_m,\ \sigma_m = 0 \quad \text{if } d = 2m \tag{31a}$$

since $\sigma_d \neq \sigma_0 = 1$; and from (26) and (28) we have

$$c_m = b_{m+1},\ b_m = tc_{m+1} \quad \text{if } d = 2m+1. \tag{31b}$$

Comparing (30) with (29_m) and (29_{d-m}) yields in both cases that $\sigma_d = -1/t$, so that (30) can be rewritten as

$$\sigma_i + t\sigma_{d-i} = 0 \quad \text{for } i = 0, \cdots, m. \tag{32}$$

Now suppose that $t > 1$. Since the σ_i are distinct, (32) implies

$$\sigma_d \neq 0, 1, -1, \quad \sigma_1 \neq 1\,. \tag{33}$$

By Theorem 8.1.1, there is a quadratic polynomial $r_2(x)$ such that

$$\sigma_{i-1}^2 c_i + \sigma_i^2 a_i + \sigma_{i+1}^2 b_i = r_2(\sigma_i)\,. \tag{34_i}$$

If we insert (26) and (32) into (34_{d-i}) and compare with (34_i) we find $r_2(\sigma_i) = t^2 r_2(-\sigma_i / t)$ for $i < m$, in particular for $i = 0, 1$, so that by (33), $r_2(x)$ has the form rx^2 for some number r. With $a_i + b_i + c_i = k$, (34) therefore gives

$$(\sigma_{i-1}^2 - \sigma_i^2)c_i + (\sigma_{i+1}^2 - \sigma_i^2)b_i = (r-\theta)\sigma_i^2 . \qquad (35_i)$$

If $d = 2m$ then (31a) and (35) give the contradiction $0 = (\sigma_{m-1}^2 + t\sigma_{m+1}^2)b_m$. And if $d = 2m+1$ then (31b) and (32) give readily a contradiction with (35_m) and (35_{m+1}). Therefore $t = 1$, Γ is a double cover of $\overline{\Gamma}$, and (28), (32) imply (26) and (27). To show that $\overline{\Gamma}$ is Q-polynomial we shall use the recurrence

$$q_l(\sigma_{i-1})c_i + q_l(\sigma_i)a_i + q_l(\sigma_{i+1})b_i = \theta_l q_l(\sigma_i) \qquad (36_i)$$

derived in the proof of Theorem 8.1.1. Since (26) implies $a_i = a_{d-i}$ we deduce from (36_i) and (36_{d-i}) that the polynomials $\overline{q}_l(x)$ of degree l ($l = 0, \cdots, m$) defined by

$$\overline{q}_l(x^2) := \tfrac{1}{2}(q_{2l}(x) + q_{2l}(-x)) \qquad (37)$$

satisfy the recurrence

$$\overline{q}_l(\sigma_{i-1}^2)c_i + \overline{q}_l(\sigma_i^2)a_i + \overline{q}_l(\sigma_{i+1}^2)b_i = \theta_{2l}q_l(\sigma_i^2) \qquad (38)$$

for $i = 0, \cdots, m$. If now $\overline{A_i}$ denotes the adjacency matrix of the graph $\overline{\Gamma}_i$ then, using Proposition 4.2.2 again, it is easy to see that the matrices

$$\overline{E}_{2l} = \sum_{i=0}^{m} \overline{q}_l(\sigma_i^2)\overline{A_i}$$

are the eigenmatrices of $\overline{\Gamma}$. This implies that $\overline{\Gamma}$ is Q-polynomial with (not necessarily standard) Q-sequence $(\sigma_0^2, \cdots, \sigma_m^2)$. □

8.3 Further results on Q-polynomial graphs

This section contains a number of miscellaneous results on Q-polynomial graphs (see also Terwilliger [723, 727] for further recent results). In particular, we give explicit formulae for the eigenvalues and multiplicities of a Q-polynomial graph, and bounds for the girth and diameter of such a graph.

Q-polynomial distance-regular graphs as extremal graphs

8.3.1. Theorem (Terwilliger [728]). *Let $u_0, \cdots, u_d$ be the standard sequence of a distance-regular graph corresponding to the eigenvalue θ. Then:*

(i) *If $u_{i-1}, u_i \neq u_0$ then*

$$c_i\left[u_2 - u_i - \frac{(u_1 - u_{i-1})^2}{u_0 - u_i}\right] + b_{i-1}\left[u_2 - u_{i-1} - \frac{(u_1 - u_i)^2}{u_0 - u_{i-1}}\right] \qquad (39)$$
$$\geqslant (k-\theta)(u_1 + u_2 - u_{i-1} - u_i) - (\theta+1)(u_0 - u_2);$$

(ii) *$u_0, \cdots, u_d$ is a Q-sequence if and only if $u_i \neq u_0$ for $i \neq 0$, and (39) holds with equality for $i = 2, \cdots, d$.* □

Remark. It is not known whether these inequalities are consequences of the Krein conditions; this is the case, however, for $d = 3$.

Explicit formulae for eigenmatrices, eigenvalues, and multiplicities

For a Q-polynomial distance-regular graph Γ with Q-sequence $(\sigma_0, \cdots, \sigma_d)$, closed formulae for the eigenvalues of Γ can be given in terms of the symbol $\left[{i \atop l}\right]$ defined by $\left[{i \atop 0}\right] := 1$ and

$$\left[{i \atop l}\right] := \frac{(\sigma_{d-i} - \sigma_d)(\sigma_{d-i} - \sigma_{d-1}) \cdots (\sigma_{d-i} - \sigma_{d-l+1})}{(\sigma_{d-l} - \sigma_d)(\sigma_{d-l} - \sigma_{d-1}) \cdots (\sigma_{d-l} - \sigma_{d-l+1})} \tag{40}$$

if $l > 0$. The symbol is invariant under linear transformations of the sequence σ_i. We note that $\left[{i \atop l}\right] = p_l(\sigma_{d-i})$, where p_l is the unique polynomial of degree l with $p_l(\sigma_{d-l}) = 1$ whose zeros are $\sigma_d, \sigma_{d-1}, \cdots, \sigma_{d-l+1}$. In particular, for $0 \leqslant i, l \leqslant d$ we have

$$\left[{i \atop l}\right] = 0 \text{ if (and only if) } i < l, \quad \left[{l \atop l}\right] = 1 .$$

We also note that if the σ_i are normalized such that $\sigma_d = 0$, $\sigma_{d-1} = 1$ then

$$\sigma_i = \left[{d-i \atop 1}\right] \quad \text{for } i = 0, \cdots, d.$$

8.3.2. Proposition (cf. Leonard [485]). *There are constants $\alpha_0, \cdots, \alpha_d$ such that*

$$\frac{P_{ji}}{k_i} = \frac{Q_{ij}}{f_j} = \alpha_j \sum_{l=0}^{j} \Big(\prod_{m=0}^{l-1} \frac{\theta_j - \theta_m}{c_{d-m}} \Big) \left[{d-i \atop l}\right]$$

for $i, j = 0, \cdots, d$.

Proof. The equality $f_j P_{ji} = k_i Q_{ij}$ was given in Lemma 2.2.1(iv). Let us define the matrices $N_{-1} = 0$ and

$$N_l = \sum_{i=0}^{d-l} \left[{d-i \atop l}\right] A_i \quad (l = 0, \cdots, d).$$

We shall prove by induction on l that

$$AN_l = \theta_l N_l + c_{d+1-l} N_{l-1} \tag{41}$$

for $l = 0, \cdots, d$. Clearly, this holds for $l = 0$ since $N_0 = J$ and $\theta_0 = k$. Suppose that (41) holds for $l < j$. Since $\left[{d-i \atop l}\right]$ is a polynomial of degree l in σ_i we can write $q_j(\sigma_i)$ as a linear combination

$$Q_{ij} = q_j(\sigma_i) = v \sum_{l=0}^{j} \alpha_{jl} \left[{d-i \atop l}\right] \tag{42}$$

with suitable coefficients α_{jl}, and $\alpha_{jj} \neq 0$, giving

$$E_j = \sum_{l=0}^{j} \alpha_{jl} N_l.$$

Now

$$0 = AE_j - \theta_j E_j = \sum_{l=0}^{j} \alpha_{jl}(AN_l - \theta_j N_l) =$$

$$= \alpha_{jj}(AN_j - \theta_j N_j) + \sum_{l=0}^{j-1} \alpha_{jl}(c_{d+1-l} N_{l-1} - (\theta_j - \theta_l) N_l)$$

implies

$$AN_j = \theta_j N_j + c_{d+1-j} N_{j-1} - \frac{1}{\alpha_{jj}} \sum_{l=0}^{j-1} (\alpha_{j,l+1} c_{d-l} - \alpha_{jl}(\theta_j - \theta_l)) N_l. \tag{43}$$

On the other hand,

$$AN_j = \sum_{i=0}^{d-j} \left[{d-i \atop j}\right] AA_i = \sum_{i=0}^{d-j} \left[{d-i \atop j}\right] (c_{i+1}A_{i+1} + a_i A_i + b_{i-1}A_{i-1}) \tag{44}$$

$$= \sum_{i=0}^{d-j+1} \left(\left[{d+1-i \atop j}\right] c_i + \left[{d-i \atop j}\right] a_i + \left[{d-1-i \atop j}\right] b_i \right) A_i ,$$

and a comparison of the (γ,δ)-entries with $d(\gamma,\delta) = 0, 1, \cdots, j-1$ in (43) and (44) yields

$$\alpha_{j,l+1} c_{d-l} - \alpha_{jl}(\theta_j - \theta_l) = 0 \quad (l = 0, \cdots, j-1). \tag{45}$$

Therefore (43) reduces to the case $l = j$ of (41), showing that (41) holds generally. Writing $\alpha_j = \alpha_{j0} v / f_j$, (45) yields

$$\alpha_{jl} = f_j \alpha_j \prod_{m=0}^{l-1} \frac{\theta_j - \theta_m}{c_{d-m}},$$

and insertion into (42) yields the assertion. □

Remarks. (i) Since $P_{j0} = k_0 = 1$, the value of α_j is obtained by putting $i = 0$.
(ii) Considered as polynomials in σ_i, the expressions in this proposition are members of the family of orthogonal polynomials discovered by Askey & Wilson [16], and in fact the above results together with those of §8.1 can be viewed more abstractly as a characterization of the Askey-Wilson polynomials as those families of orthogonal polynomials having duals. See Leonard [485].
(iii) In some important special cases, the matrices N_l have a natural combinatorial interpretation; see Neumaier [552].

8.3.3. Corollary. *The eigenvalue θ_l $(l > 0)$ of a Q-polynomial distance-regular graph satisfies the relations*

$$\theta_l \left[{i \atop l}\right] = a_{d-i} \left[{i \atop l}\right] + b_{d-i} \left[{i-1 \atop l}\right] + c_{d-i} \left[{i+1 \atop l}\right] - c_{d+1-l} \left[{i \atop l-1}\right]$$

$(i = l-1, \cdots, d)$; in particular

$$\theta_l = a_{d-l} + c_{d-l} \left[{l+1 \atop l}\right] - c_{d+1-l} \left[{l \atop l-1}\right],$$

$$\theta_l \left[{d \atop l}\right] = k \left[{d-1 \atop l}\right] - c_{d+1-l} \left[{d \atop l-1}\right].$$

Proof. Compare the entries in (41) and (44). □

8.3.4. Proposition (cf. Leonard [485]). *The multiplicities of a Q-polynomial distance-regular graph are given by*

$$f_i = \frac{\alpha_0 \cdots \alpha_{i-1}(\theta_{i-1} - \theta_{i+1})}{\beta_1 \cdots \beta_i (\theta_{-1} - \theta_1)},$$

where

$$\alpha_i = (1+b)k + \theta_i(\lambda - \theta_i + b\theta_{i-1}),$$

$$\beta_i = (1+b)k + \theta_i(\lambda - \theta_i + b\theta_{i+1}),$$

with arbitrary $\theta_{-1} \neq \theta_1$, $\theta_{d+1} \neq \theta_{d-1}$, *and*

$$b := b_1 / (\theta_1 + 1).$$

Proof. By Theorem 2.7.1, the association scheme defined by the distances is cometric; in particular, $q_{1j}^h = 0$ if $|j-h| > 1$. Put $a_l^* = q_{1l}^l$, $b_l^* = q_{1l+1}^l$, $c_l^* = q_{1l-1}^l$. Then Lemma 2.3.1 yields

$$c_1^* = 1, \quad c_{j+1}^* f_{j+1} = b_j^* f_j \tag{46}$$

and ($i = 1, k = l$ in 2.3.1 (viii))

$$P_{lm} Q_{m1} = c_l^* P_{l-1,m} + a_l^* P_{lm} + b_l^* P_{l+1,m}. \tag{47_m}$$

Since the P_{li} are the eigenvalues of A_i corresponding to the idempotent E_l, we have

$$P_{l0} = 1, \quad P_{l1} = \theta_l, \quad P_{l2} = v_2(\theta_l) = (\theta_l^2 - \lambda\theta_l - k)/\mu,$$

and in terms of a standard sequence $(u_0, \cdots, u_d)$ we have

$$Q_{m1} = u_m f_1$$

and $b = (u_0 - u_1)/(u_1 - u_2)$. Subtracting P_{lm} times (47_0) from (47_m) for $m = 1, 2$ gives

$$\theta_l(u_0 - u_1) f_1 = c_l^*(\theta_l - \theta_{l-1}) - b_l^*(\theta_{l+1} - \theta_l)$$

$$(\theta_l^2 - \lambda\theta_l - k)(u_0 - u_2) f_1 = c_l^*(\theta_l - \theta_{l-1})(\theta_l + \theta_{l-1} - \lambda) - b_l^*(\theta_{l+1} - \theta_l)(\theta_{l+1} + \theta_l - \lambda).$$

Solving this for c_l^*, b_l^* gives for $l \geqslant 1$:

$$\begin{aligned}(\theta_{l+1} - \theta_l)(\theta_{l+1} - \theta_{l-1}) b_l^* / f_1 &= (k + \theta_l\theta_{l-1})(u_0 - u_1) + (\theta_l^2 - \lambda\theta_l - k)(u_2 - u_1) \\ &= (u_1 - u_2)\alpha_l, \\ (\theta_l - \theta_{l-1})(\theta_{l+1} - \theta_{l-1}) c_l^* / f_1 &= (k + \theta_l\theta_{l+1})(u_0 - u_1) + (\theta_l^2 - \lambda\theta_l - k)(u_2 - u_1) \\ &= (u_1 - u_2)\beta_l.\end{aligned}$$

Insertion into (46) and slight simplification proves the assertion by induction on i. □

Integrality of eigenvalues

Bannai & Ito [33] (Theorems 3.7.10 and 3.7.11) used Leonard's representation of the parameters of a Q-polynomial distance-regular graph in terms of the complex number q mentioned at the end of §8.1 to show the following result.

8.3.5. Theorem. *Let* Γ *be a* Q*-polynomial distance-regular graph of valency* $k > 2$ *(i.e., not a polygon) and diameter* $d \geqslant 34$. *Then* Γ *has at most two distinct standard* Q*-sequences; moreover, all eigenvalues are integral.* □

The bound $d \geqslant 34$ seems much larger than necessary, and we conjecture that it can be replaced by $d \neq 4$. The case $d = 4$ is possibly exceptional; indeed, there is an infinite sequence of feasible intersection arrays which, if realized, would belong to (formally self-dual and hence) Q-polynomial distance-regular graphs with nonintegral eigenvalues: As one easily checks, the graphs on $v = 8\mu^2(\mu+1)$ vertices with intersection arrays

$$\{2\mu^2+\mu,\ 2\mu^2-\mu-1,\ \mu^2,\ \mu;\ 1,\ \mu,\ \mu^2-\mu,\ 2\mu^2+\mu\} \tag{48}$$

have spectrum

$$(2\mu^2+\mu)^1\ ,\quad \mu(1\pm\sqrt{2\mu+2})^{2\mu^2+\mu}\ ,\quad \mu^{(2\mu+1)^2(\mu-1)}\ ,\quad -\mu^{(2\mu+1)^2\mu}\ ;$$

and the standard sequences

$$u_0 = 1\ ,\ u_1 = \frac{1\pm\sqrt{2\mu+2}}{2\mu+1}\ ,\ u_2 = \frac{1}{2\mu+1}\ ,\ u_3 = \frac{-1}{2\mu+1}\ ,\ u_4 = \frac{1\mp\sqrt{2\mu+2}}{2\mu+1}$$

belonging to the eigenvalues $\mu(1\pm\sqrt{2\mu+2})$ are Q-sequences. These graphs have two P-polynomial structures: also Γ_4 is a distance-regular graph (with the same parameters); this would be an example of case (ii) in Proposition 4.2.15. For $\mu \leqslant 3$, geometric arguments show that no corresponding graph can exist; nothing is known for $\mu > 3$.

Problem. Show that if a distance-regular graph Γ of valency $k > 2$ has a standard Q-sequence, then the corresponding eigenvalue is θ_1 or θ_d (or θ_{d-1} if Γ is bipartite). (This would immediately imply that there are at most two distinct standard Q-sequences.)

Bounds for girth and diameter

In view of the known examples of Q-polynomial distance-regular graphs (see §8.5) it is natural to conjecture that such graphs Γ have girth $g \leqslant 6$, with equality only for the Odd graphs, and diameter $d \leqslant k$, with equality if and only if Γ is a d-cube. We are only able to prove the following weaker results.

8.3.6. Theorem. *The numerical girth of a Q-polynomial distance-regular graph of valency $k > 2$ cannot exceed 7.*

Proof. A standard Q-sequence $(\sigma_0, \cdots, \sigma_d)$ satisfies the recurrence relations $b_i\sigma_{i+1} + a_i\sigma_i + c_i\sigma_{i-1} = \theta\sigma_i$ and, by Theorem 8.1.2,

$$\sigma_{i+1} + \sigma_{i-1} = p\sigma_i + r\ ; \tag{49}$$

therefore

$$(a_i + pb_i - \theta)\sigma_i + (c_i - b_i)\sigma_{i-1} + b_i r = 0 \quad \text{for } i = 1, \cdots, d-1.$$

Now suppose that the numerical girth is at least 8, i.e., that $b_1 = b_2 = b_3$ and $c_1 = c_2 = c_3 = 1$. If $b_1 > 1$, then we get for $i = 1, 2, 3$ the relation

$$\sigma_{i-1} = b\sigma_i + c\ , \tag{50}$$

where $b = (a_1 + pb_1 - \theta)/(b_1 - 1)$, $c = b_1 r/(b_1 - 1)$. Then (49) implies $p = b + b^{-1}$, $r = c(1 - b^{-1})$ and, inductively, (50) for $i = 1, \cdots, d$. By Theorem 8.4.1 below, Γ has classical parameters, contradicting Proposition 6.1.2. On the other hand, if $b_1 = 1$, then $k = 2$ by Corollary 1.1.6. □

8.3.7. Proposition. *Let Γ be a Q-polynomial distance-regular graph, and suppose that $a_i \neq a_1$ for some index $i \neq 0$, d. Then $d \leqslant 2k - 2$.*

Proof. Theorem 8.1.2 implies the recurrence (49) for $i \neq 0$, d; by a suitable choice of σ_{-1} and σ_{d+1} this also holds for $i = 0$, d. By Theorem 8.1.1,

$$(\sigma_i-\sigma_{i-1})(\sigma_i-\sigma_{i+1})(a_i-k) = \phi_2(\sigma_i)+(\sigma_i-\sigma_{i-1}-\sigma_{i+1})\phi_1(\sigma_i) \tag{51}$$

holds for $i = 0, \cdots, d$, and by Proposition 8.1.3 we may rewrite this as

$$((p-2)\sigma_i^2+2r\sigma_i+s)(k-a_i) = \phi_2(\sigma_i)-((p-1)\sigma_i+r)\phi_1(\sigma_i)\,. \tag{52}$$

Now suppose that $d \geqslant 2k-1$. Since $0 \leqslant a_i = k-b_i-c_i \leqslant k-2$ for $i = 0, \cdots, d-1$, there is an integer a such that $a_i = a$ for three different indices $i = i_1, i_2, i_3$. Insertion into (52) gives

$$((p-2)\sigma_i^2+2r\sigma_i+s)(k-a) = \phi_2(\sigma_i)-((p-1)\sigma_i+r)\phi_1(\sigma_i) \tag{53}$$

for $i = i_1, i_2, i_3$. But both sides of (53) are quadratic polynomials in σ_i; hence this is an identity which holds for all i. A comparison of (52) and (53) now shows that

$$(\sigma_i-\sigma_{i-1})(\sigma_i-\sigma_{i+1})(a_i-a) = ((p-2)\sigma_i^2+2r\sigma_i+s)(a-a_i) = 0$$

for $i = 0, \cdots, d$. Since $\sigma_0, \cdots, \sigma_d$ are distinct we conclude that $a_i = a$ for $i = 1, \cdots, d-1$ against assumption. Therefore $d \leqslant 2k-2$. □

8.4 Graphs with classical parameters

In this section we show that all graphs with classical parameters are Q-polynomial, and give explicit formulae for their eigenvalues and multiplicities.

A characterization of graphs with classical parameters

8.4.1. Theorem. *Let Γ be a distance-regular graph of diameter d, and suppose that the sequence $(\sigma_0, \cdots, \sigma_d)$ consists of distinct elements satisfying the recurrence*

$$\sigma_{i-1} = b\sigma_i+c \quad (i = d, \cdots, 1).$$

Then σ is a Q-sequence of Γ if and only if Γ has classical parameters (d,b,α,β), for suitable numbers α and β.

Proof. By replacing σ_i by the linear transform $(\sigma_i-\sigma_d)/(\sigma_{d-1}-\sigma_d)$ if necessary, we may assume w.l.o.g. that

$$\sigma_d = 0\,, \quad \sigma_{i-1} = b\sigma_i+1 \;\; (i = d, \cdots, 1), \tag{54}$$

or, in terms of the b-ary Gaussian binomial coefficients,

$$\sigma_i = \begin{bmatrix} d-i \\ 1 \end{bmatrix}_b . \tag{55}$$

Now $\sigma_0 \neq \sigma_2$ implies $b \neq 0, -1$, and from (54) we find $\sigma_{i+1}+\sigma_{i-1} = (b+b^{-1})\sigma_i+1+b^{-1}$; therefore by Proposition 8.1.3, $(\sigma_0, \cdots, \sigma_d)$ is a Q-sequence if and only if the intersection array can be expressed as in Theorem 8.1.1 with a linear polynomial $\phi_1(x)$ and a quadratic polynomial $\phi_2(x)$.

Suppose first that the latter holds, i.e.,

$$b_i = \frac{\phi_2(\sigma_i)-\sigma_{i-1}\phi_1(\sigma_i)}{(\sigma_{i+1}-\sigma_{i-1})(\sigma_{i+1}-\sigma_i)}\,, \tag{56}$$

$$c_i = \frac{\phi_2(\sigma_i)-\sigma_{i+1}\phi_1(\sigma_i)}{(\sigma_{i-1}-\sigma_{i+1})(\sigma_{i-1}-\sigma_i)}\,. \tag{57}$$

Since (54), (56) for $i = d$ imply $\phi_2(0) = \phi_1(0)$, and (57) for $i = 0$ implies

$\phi_2(\sigma_0) = \sigma_1\phi_1(\sigma_0)$, there are constants A, B, C such that

$$\phi_1(x) = A(\sigma_0 - x) - Bx \ , \tag{58a}$$

$$\phi_2(x) = A(\sigma_0 - x) - B\sigma_1 x + Cx(\sigma_0 - x) \, . \tag{58b}$$

We now replace the indeterminates σ_{-1} and σ_{d+1} by the values defined by (55), and observe that (55) implies

$$\sigma_{i-1} - \sigma_{i+1} = (b+1)b^{d-i-1} \ , \ \sigma_{i-1} - \sigma_i = b^{d-i} \ ;$$

hence (57) becomes

$$c_i = \frac{A(1-\sigma_{i+1})(\sigma_0 - \sigma_i) + B(\sigma_{i+1} - \sigma_1)\sigma_i + C\sigma_i(\sigma_0 - \sigma_i)}{(b+1)b^{2d-2i-1}} \, .$$

With the abbreviation

$$D := (bC - A - B)/(b+1) \, , \tag{59}$$

we find after elimination of C that $c_i = [{i \atop 1}]b^{i-d}(A + D\sigma_i)$. Now $c_1 = 1$ implies

$$A = b^{d-1} - D\sigma_1 \ ; \tag{60}$$

insertion gives

$$c_i = [{i \atop 1}](1 + (b-1-D)[{i-1 \atop 1}]) \, .$$

Similarly, (56) becomes

$$b_i = \frac{A(1-\sigma_{i-1})(\sigma_0 - \sigma_i) + B(\sigma_{i-1} - \sigma_1)\sigma_i + C\sigma_i(\sigma_0 - \sigma_i)}{(b+1)b^{2d-2i-2}} \, ,$$

and after elimination of C by (59) and of A by (60) we find

$$b_i = ([{d \atop 1}] - [{i \atop 1}])(Bb^{1-d} - (b-1-D)[{i \atop 1}]) \, .$$

Therefore, Γ has classical parameters (d,b,α,β), where $\alpha = b - 1 - D$, $\beta = Bb^{1-d}$. Conversely, if Γ has classical parameters (d,b,α,β) then, defining $\phi_1(x)$ and $\phi_2(x)$ by (58) with

$$D = b - 1 - \alpha \ , \ B = b^{d-1}\beta \ , \ A = b^{d-1} - D\sigma_1 \ , \ C = b^{-1}(A + B + (b+1)D) \, ,$$

one finds in the same way that (56) and (57) hold. □

8.4.2. Corollary. *Every distance-regular graph Γ with classical parameters (d,b,α,β) is Q-polynomial, and the sequence $(\sigma_0, \cdots, \sigma_d)$ defined by*

$$\sigma_i := [{d-i \atop 1}] \quad (i = 0, \cdots, d)$$

is a Q-sequence of Γ. The eigenvalues of Γ are

$$\theta_l = [{d-l \atop 1}](\beta - \alpha[{l \atop 1}]) - [{l \atop 1}] = b^{-l}b_l - [{l \atop 1}] \quad (l = 0, \cdots, d).$$

Proof. The first part is obvious from the above theorem and (55). To find the eigenvalues, we note that the symbol $[{i \atop l}]$ introduced in (40) becomes the ordinary or Gaussian binomial coefficient $[{i \atop l}]_b$. In particular, $[{l+1 \atop l}] = [{l+1 \atop 1}]$, whence by Corollary 8.3.3,

$$\theta_l = a_{d-l} + c_{d-l}[{l+1 \atop 1}] - c_{d+1-l}[{l \atop 1}] \, . \tag{61}$$

Now we have

$$ {d \brack 1}-{i \brack 1} = b^i{d-i \brack 1}, \quad {i+1 \brack 1}-1 = b{i \brack 1}, \tag{62} $$

whence

$$ \begin{aligned} a_i + c_i &= {i \brack 1}\Big(\beta+\alpha\Big({d \brack 1}-{i \brack 1}\Big)\Big) = {i \brack 1}\Big(\beta+\alpha b^i{d-i \brack 1}\Big), \\ bc_i - c_{i+1} &= {i+1 \brack 1}-1+\alpha{i \brack 1}\Big({i \brack 1}-1\Big)-{i+1 \brack 1}\Big(1+\alpha{i \brack 1}\Big) \\ &= -1+\alpha{i \brack 1}\Big({i \brack 1}-1-{i+1 \brack 1}\Big) = -1-\alpha{i \brack 1}(b^i+1). \end{aligned} $$

Therefore (61) implies

$$ \begin{aligned} \theta_l &= (a_{d-l}+c_{d-l})+(bc_{d-l}-c_{d+1-l}){l \brack 1} \\ &= {d-l \brack 1}\Big(\beta+\alpha b^{d-l}{l \brack 1}\Big)-\Big(1+\alpha{d-l \brack 1}(b^{d-l}+1)\Big){l \brack 1} \\ &= {d-l \brack 1}\Big(\beta-\alpha{l \brack 1}\Big)-{l \brack 1} = b^{-l}b_l-{l \brack 1} \end{aligned} $$

by (62) and the definition of classical parameters. □

8.4.3. Theorem. *The multiplicities $f_0 = 1, f_1, \cdots, f_d$ of a distance-regular graph with classical parameters (d,b,α,β) are given by*

$$ f_i = \frac{\alpha_0 \cdots \alpha_{i-1}\Big(1+{d-2i \brack 1}\alpha+b^{d-2i}\beta\Big)}{\beta_1 \cdots \beta_i\Big(1+{d \brack 1}\alpha+b^d\beta\Big)}, $$

where

$$ \begin{aligned} \alpha_i &= b{d-i \brack 1}\Big(\beta-{i \brack 1}\alpha\Big)\Big(1+{d-i \brack 1}\alpha+b^{d-i}\beta\Big), \\ \beta_i &= {i \brack 1}\Big(\beta-{i \brack 1}\alpha+b^i\Big)\Big(1+{d-i \brack 1}\alpha\Big). \end{aligned} $$

Proof. This follows directly from Proposition 8.3.4 and Corollary 8.4.2 after some simplification. The present α_i, β_i correspond to $b\alpha_i/(1+b)$ and $\beta_i/(1+b)$ in Proposition 8.3.4. □

8.4.4. Corollary. *A distance-regular graph Γ with classical parameters (d,b,α,β) where $\alpha = b-1$ is formally self-dual, has $v = (\beta+1)^d$ vertices, valencies and multiplicities*

$$ k_j = f_j = {d \brack j}\prod_{i=0}^{j-1}(\beta+1-b^i) $$

and eigenvalues

$$ \theta_j = k-b^{d-j}{j \brack 1}(\beta+1). $$

□

Remark. Delsarte [253] gives the eigenvalues of the Hamming scheme, the schemes of bilinear and alternating forms, the Johnson scheme and the Grassmann scheme (in terms of hypergeometric functions).

8.4.5. Theorem. *A partition graph with parameters (m,α) is Q-polynomial with Q-sequence $\sigma_i = i(m-i)$ $(i = 0, \cdots, \alpha = [m/2])$.*

Proof. Such a graph is obtained by folding a graph Γ with classical parameters $(b,\alpha,\beta,d) = (1,\alpha,1+\alpha(m-1),m)$ and intersection numbers

$$b_i = (m-i)(1+\alpha(m-1-i)), \quad c_i = i(1+\alpha(i-1)), \quad a_i = 2\alpha i(m-i).$$

One easily verifies that $\sigma_i' = 1-2i/m$ is a standard sequence of Γ; by Theorem 8.4.1 it is a Q-sequence. Theorem 8.2.4 now shows that the folding $\overline{\Gamma}$ of Γ is Q-polynomial with Q-sequence $\sigma_i'' = (1-2i/m)^2$ and hence with Q-sequence $\sigma_i := \frac{1}{4}m^2(1-\sigma_i'')$ $= i(m-i)$. □

Remark. This theorem is also true for all pseudo partition graphs; this can be deduced from the parameters and Proposition 8.1.3.

8.5 The known Q-polynomial distance-regular graphs

We already saw that graphs with classical parameters, partition graphs and Taylor graphs are Q-polynomial. Not many other known distance-regular graphs have this property. Of the categories distinguished in Chapter 6, only the near polygons and the sporadic graphs remain to be discussed.

8.5.1. Theorem. *A regular near polygon of diameter $d \geqslant 3$ with thick lines is Q-polynomial if and only if it has classical parameters.*

Proof. Suppose that Γ is a regular near polygon with thick lines. If Γ has classical parameters then Γ is Q-polynomial by Corollary 8.4.2. Conversely, suppose that Γ is Q-polynomial. Then, with notation as in the first proof of Theorem 8.1.2, equation (11) holds. We apply this with $i = h+1$, $j = 1$. Since $a_h \neq 0$, there is a vertex $\alpha \in \Gamma_h(\gamma) \cap \Gamma(\delta)$. Since Γ is a near polygon, the singular line through α and δ is disjoint from $\Gamma_{h+1}(\gamma)$ and contains a unique $\epsilon \in \Gamma_{h-1}(\gamma)$. Thus, every $\beta \in \Gamma_{h+1,1}(\gamma,\delta)$ has distance 2 from α, and every $\beta' \in \Gamma_{1,h+1}(\gamma,\delta)$ has distance $h+1$ from δ, distance h from ϵ, and hence distance $h+1$ from α. Therefore, by taking inner products of $\overline{\alpha}$ with (11), we get

$$b_h(u_{h+1}-u_2) = t^h_{h+1,1}(u_h-u_1).$$

Similarly, choosing $\alpha \in \Gamma_{h-1}(\gamma) \cap \Gamma(\delta)$, we find

$$b_h(u_h-u_2) = t^h_{h+1,1}(u_{h-1}-u_1).$$

Since Γ is Q-polynomial, the u_i are distinct, and elimination of $t^h_{h+1,1}$ yields

$$\frac{u_h-u_1}{u_{h+1}-u_2} = \frac{u_{h-1}-u_1}{u_h-u_2}.$$

Since h was arbitrary, this expression is a constant b independent of h, and we have

$$u_{h-1}-u_1 = b(u_h-u_2) \quad \text{for } k = 1, \cdots, d.$$

Now Theorem 8.4.1 shows that Γ has classical parameters. □

Remark. A regular near polygon with thick lines and classical parameters (d,b,α,β), $d>2$ is a near $2d$-gon, and satisfies $b \in \{\mu-1, -\lambda-1\}$; this follows from $a_i = \lambda c_i$

for $i < d$ as in the proof of Proposition 6.2.1. If $b = \mu - 1$, then $\alpha = 0$, $\beta = \lambda + 1 > 0$, and by Theorem 9.4.4 we have either a Hamming graph or a dual polar space. On the other hand, if $b = -\lambda - 1$, then, writing $s = \lambda + 1$, we find $\beta = s + \alpha(s + (-s)^d)/(s+1)$, and we have as examples the unitary dual polar graphs, the triality graphs ${}^3D_4(q)$ (generalized hexagons of order (q,q^3)), and two Witt graphs (C7) = (N1), (C7a) = (N2). It is very likely that there are no other examples with $b = -\lambda - 1$.

8.5.2. Theorem. *A thin regular near polygon of diameter $d \geqslant 3$ is Q-polynomial if and only if there is a standard sequence $(\sigma_0, \cdots, \sigma_d)$ of Γ consisting of distinct elements such that the recurrence (10) holds for suitable numbers p and r, and*

$$\sigma_d + \sigma_{d-1} = p\sigma_d + r \quad \text{if } a_d \neq 0.$$

Proof. Since $a_i = 0$ for $i < d$, the proof of Theorem 8.2.1 applies with minor changes only. □

8.5.3. Corollary. *The following regular near polygons are Q-polynomial:*

(i) *Ordinary polygons* (R1) *with Q-sequences*

$$\sigma_i = \frac{1}{2}(\rho^i + \rho^{-i}) = \mathrm{Re}(\rho^i)$$

for any primitive n-th root of unity ρ.

(ii) *The Odd graphs* (R2) *with Q-sequence*

$$\sigma_{2i} = i\,, \quad \sigma_{2i+1} = k - 1 - i\,.$$

(iii) *The folded cubes* (R3).

(iv) *The near polygons with classical parameters* (R5), (R6), (N1), (N2) *and the generalized hexagons of order (q,q^3).*

(v) *The incidence graphs of square 2-designs* (B0)*; in particular the generalized hexagons of order $(1,q)$.*

(vi) *Hadamard graphs* (B2) *($m = 2$), with Q-sequences*

$$\sigma_0 = \pm\sqrt{2\mu}\,, \quad \sigma_1 = 1\,, \quad \sigma_2 = 0\,, \quad \sigma_3 = -1\,, \quad \sigma_4 = \mp 2\sqrt{2\mu}\,.$$

(vii) *The regular thin near octagon* (B4) *with Q-sequence*

$$\sigma_0 = 11\,, \quad \sigma_1 = 4\,, \quad \sigma_2 = 1\,, \quad \sigma_3 = -1\,, \quad \sigma_4 = -4\,.$$

(viii) *The regular thin near decagon* (B10) *with Q-sequence*

$$\sigma_0 = 11\,, \quad \sigma_1 = 4\,, \quad \sigma_2 = 1\,, \quad \sigma_3 = -1\,, \quad \sigma_4 = -4\,, \quad \sigma_5 = -11\,.$$

Proof. (iii), (iv) and (v) follow from Theorem 8.4.5, Corollary 8.4.2, and Corollary 8.2.2; the other cases can be verified using Theorem 8.5.1. □

Summarizing, we have established Q-polynomiality of the following distance-regular graphs:

- the graphs with classical parameters (Corollary 8.4.2),
- the partition graphs (Theorem 8.4.5),

- antipodal double covers of complete graphs (Proposition 8.2.3),
- the families of near polygons listed in Corollary 8.5.3.

No other distance-regular graph known to us is Q-polynomial; in particular, the doubled Odd graphs and doubled Grassmann graphs (R3) and the generalized hexagons of order (q,q) are not Q-polynomial, and the only Q-polynomial graphs with parameters (B2) are the Hadamard graphs.

Chapter 9
The Families of Graphs with Classical Parameters

In this chapter we discuss the known infinite families of graphs with classical parameters, except for some graphs of Lie type, treated in the next chapter. A few sporadic graphs with classical parameters can be found in Chapters 3 and 11, cf. Table 6.1.

9.1 Johnson graphs

Johnson graphs form one of the simplest families of distance-regular graphs of unbounded diameter. They can be considered as the thin case of the Grassmann graphs discussed in §9.3 and as the Coxeter graphs of type $A_{n,i}$ (cf. §10.3). We have already met Johnson graphs as an example of code graphs in §3.14.

Let X be a finite set. The *Johnson graph of the e-sets in X* has vertex set $\binom{X}{e}$, the collection of e-subsets of X. Two vertices γ, δ are adjacent whenever $\gamma \cap \delta$ has cardinality $e-1$. By abuse of notation, we shall denote this graph by $\binom{X}{e}$; when X is some unspecified n-set we write $\binom{n}{e}$ or $J(n,e)$. The graphs $J(n,2)$ are known as the *triangular graphs* and are also denoted by $T(n)$. Sometimes the graphs $J(n,3)$ are called the *tetrahedral graphs*.

9.1.1. Lemma. *Let $|X| = e+f$. Then the graphs $\binom{X}{e}$ and $\binom{X}{f}$ are isomorphic.*

Proof. The map $\gamma \mapsto X \setminus \gamma$ establishes an isomorphism. □

9.1.2. Theorem. *Let Γ be the Johnson graph of the e-sets in a finite set X of cardinality n. Then Γ has diameter $d = \min(e, n-e)$, is distance-transitive, has intersection array given by*

$$\left.\begin{aligned} b_j &= (e-j)(n-e-j) \\ c_j &= j^2 \end{aligned}\right\} \quad \text{for } 0 \leqslant j \leqslant d\ . \tag{1}$$

All of its singular lines have size $s+1=2$ when $2 \leqslant e \leqslant n-2$. (Otherwise Γ is a clique and $s+1=\binom{n}{e}$.) Moreover, Γ has classical parameters as given in Table 6.1, (C1), and

$$\operatorname{Aut}\Gamma \cong \begin{cases} 1 & \text{if } d = 0, \\ \operatorname{Sym}(X) \times \mathbf{Z}_2 & \text{if } n = 2e \geqslant 4, \\ \operatorname{Sym}(X) & \text{otherwise.} \end{cases}$$

Γ has eigenvalues and multiplicities given by

$$\theta_j = (e-j)(n-e-j)-j, \quad f_j = \binom{n}{j} - \binom{n}{j-1}.$$

Proof. Observing that $\delta \in \Gamma_j(\gamma)$ if and only if $|\gamma \cap \delta| = e-j$ and that Sym(X), the symmetric group on X, acts as a group of automorphisms of Γ, we see that Γ is distance-transitive. Also, the parameters d, b_j, c_j follow immediately. If $\gamma \sim \delta$ then $\{\gamma,\delta\}^\perp \setminus \{\gamma,\delta\}$ is the disjoint union of two cliques with respective sizes $e-1$ and $n-e-1$ so that $\{\gamma,\delta\}^{\perp\perp} = \{\gamma,\delta\}$ whenever $2 \leqslant e \leqslant n-2$. Finally, let us determine Aut Γ. For $d \leqslant 1$ the graph Γ is a clique and if $d = 1$, Aut $\Gamma \cong$ Sym(X). Suppose $d > 1$ and use induction on d. Form a new graph Δ with vertices the $(n-d+1)$-cliques of Γ, where two such cliques are adjacent when they have precisely one vertex of Γ in common. If an automorphism of Γ stabilizes all $(n-d+1)$-cliques of Γ, it fixes all vertices of Γ, for the singleton on each vertex occurs as the intersection of a pair of $(n-d+1)$-cliques. It follows that the natural morphism Aut $\Gamma \to$ Aut Δ is injective. Now if $n > 2d$, then Δ is isomorphic to the graph $\binom{X}{d-1}$ and by induction Aut $\Delta \cong$ Sym(X), so that also Aut $\Gamma \cong$ Sym(X). On the other hand, if $n = 2d$, then Δ has two components, each isomorphic to the graph $\binom{X}{d-1}$, so that Aut $\Delta \cong$ Sym$(X) \times \mathbf{Z}_2$. These two components are interchanged by the automorphism of Γ sending each e-set to its complement, so also Aut $\Gamma \cong$ Sym$(X) \times \mathbf{Z}_2$. For the spectrum, cf. 8.4.2-3. □

A. Characterizations by structure

Let $\mathfrak{X}$ be a graph or a family of graphs. A graph Γ is said to be *locally* $\mathfrak{X}$ if, for each of its vertices γ, the subgraph induced on $\Gamma(\gamma)$ is isomorphic to (a member of) $\mathfrak{X}$. Let p,q be integers. The grid on $P \times Q$ (often denoted by $p \times q$) is the graph with vertex set $P \times Q$, where $|P| = p, |Q| = q$ such that $(x,y) \sim (x',y')$ if and only if either $x = x'$ or $y = y'$ (but not both). The Johnson graphs $J(n,e)$ are locally $e \times (n-e)$, but this property alone does not suffice to characterize the Johnson graphs (see below). Several authors gave nearly equivalent characterizations of the Johnson graphs - essentially by saying that these are locally grid graphs with $\mu = 4$. The following theorem is a (trivial) common generalization of these results.

9.1.3. Theorem. *Let Γ be a connected graph such that*

(i) *for each vertex γ of Γ, the set $\Gamma(\gamma)$ is partitioned by a family of cliques $\{M_i(\gamma) \mid 1 \leqslant i \leqslant n(\gamma)\}$ such that each point in $M_i(\gamma)$ is adjacent to a unique point of $M_j(\gamma)$ for each $j \neq i$,*

and

(ii) *if $d(\gamma,\delta) = 2$, then each component of $\Gamma(\gamma) \cap \Gamma(\delta)$ has at most 4 points.*

Then Γ is a Johnson graph or is doubly covered by a Johnson graph. More precisely, in the latter case Γ is the quotient of the Johnson graph $\binom{X}{e}$ where $|X| = 2e$ by an automorphism of the form $\tau\omega$, where τ is the automorphism sending each e-set to its complement and ω is an element of order at most 2 in Sym (X) with at least 8 fixed points.

Proof. We commence by showing that $n(\gamma)$ is independent of γ and that Γ is locally grid. Observe that, given $\gamma \in \Gamma$, all cliques $M_j(\gamma)$ have the same size, say $m(\gamma)$. If $\Gamma(\gamma)$ is empty, then (by connectedness of Γ) the graph Γ has a single vertex only, and $\Gamma \cong \binom{X}{0}$ for any X. If $\Gamma(\gamma)$ is a clique (i.e., if $m(\gamma) = 1$ or $n(\gamma) = 1$), then for each

$\delta \sim \gamma$ we see that the neighbours of γ in $\Gamma(\delta)$ form a clique, so that $\Gamma(\delta)$ itself is a clique and thus Γ is locally a clique. But then Γ is a clique and $\Gamma \cong \binom{X}{1}$ for suitable X. Using (i) we see that if $d(\gamma,\delta) = 2$, then each vertex ϵ in the subgraph $\Gamma(\gamma) \cap \Gamma(\delta)$ has valency at least 2 in this subgraph. Next, using (ii), we find that if $\gamma \sim \delta$, then the subgraph $\Gamma(\gamma) \cap \Gamma(\delta)$ is the union of two cliques. [Indeed, suppose that $\delta \in M_1(\gamma)$ and $\delta \sim \epsilon_j \in M_j(\gamma)$ for $j > 1$. Now if $j > k > 1$ and $\epsilon_j \not\sim \epsilon_k$, then $\Gamma(\epsilon_j) \cap \Gamma(\epsilon_k)$ contains an induced $K_{1,3}$ on $\{\gamma,\delta,\zeta_j,\zeta_k\}$ where $\epsilon_j \sim \zeta_k \in M_k(\gamma)$ and $\epsilon_k \sim \zeta_j \in M_j(\gamma)$. By (ii) this is a full component of $\Gamma(\epsilon_j) \cap \Gamma(\epsilon_k)$ contradicting the fact that δ has valency one in this subgraph.] Thus $\Gamma(\gamma)$ is an $m(\gamma)\times n(\gamma)$ grid, and, by connectedness of Γ, we see that $m(\gamma)$ and $n(\gamma)$ are independent of γ. Also, when $d(\gamma,\delta) = 2$, then each component of $\Gamma(\gamma) \cap \Gamma(\delta)$ is a 4-gon.

The remainder of the proof is standard (for details see BLOKHUIS & BROUWER [90]; compare also DOWLING [270] and MOON [543], Theorem 1): Proceed by induction on $\min(m,n)$; just as in the proof of the previous theorem construct a graph Δ with as vertices the cliques of Γ of maximal cardinality, two such cliques being adjacent when they have precisely one vertex of Γ in common. If $m \leq n$, then Δ is locally $(m-1)\times(n+1)$ and two vertices at distance two in Δ have at most four common neighbours. Now by induction each component Δ_0 of Δ is a Johnson graph, and using the labelling of Δ_0 with $J(n+m,m-1)$, we find a labelling of Γ such that each vertex of Γ is labelled with one or two elements of $J(n+m,m)$. If each vertex of Γ has one label, then $\Gamma \cong J(n+m,m)$, otherwise we have a quotient of $J(n+m,m)$ by the automorphism interchanging the two members of each pair of labels. □

The occurrence of quotients in the conclusion of the above theorem is due to the fact that the graph $J(2e,e)$ contains two indistinguishable families of maximal cliques. If one family of maximal cliques is distinguished as the set of lines, then quotients disappear:

9.1.4. Corollary (SPRAGUE [696], TŮMA [755]). *Let the connected graph Γ be the point graph of a partial linear space such that*

(i) *for all lines l, every point $\gamma \notin l$ is collinear with precisely 0 or 2 points of l; and*

(ii) *if l_1, l_2 and m_1, m_2 are two pairs of disjoint lines such that l_i and m_j intersect for all i, $j \in \{1,2\}$, then the four intersection points are pairwise collinear.*

Then Γ is a Johnson graph. Moreover, the same holds if (ii) *is replaced by the pair of dual conditions*

(iia) *If l and m are distinct lines through a point γ, then any two lines not through γ meeting both l and m intersect.*

(iib) *If γ and δ are distinct points on a line l, then any two points not on l that are collinear with both γ and δ are collinear.*

Proof. Note that (ii) immediately implies (iib), and that (i) and (iib) imply that Γ is locally grid. Now (iia) says that for $d(\gamma,\delta) = 2$ each component of $\Gamma(\gamma) \cap \Gamma(\delta)$ is a 4-gon, while (ii) implies that for $d(\gamma,\delta) = 2$ the diameter of $\Gamma(\gamma) \cap \Gamma(\delta)$ is at most two. □

The conditions (iia) and (iib) are known as the *Pasch axiom* and *dual Pasch axiom*, respectively. They also arise in the characterization of Grassmann graphs (Theorem 9.3.5).

Just requiring that the graph be locally grid does not suffice to characterize the Johnson graphs; e.g., there are precisely two locally 3×3 grid graphs, namely the Johnson graph $J(6,3)$ and the complement of the 4×4 grid. HALL [347] characterized all locally $3\times m$ grid graphs by observing that these correspond to certain Fischer spaces (cf. §13.2). BLOKHUIS & BROUWER [90] found the (four) connected locally 4×4 grid graphs.

If Γ is the Johnson graph $J(n,d)$, then Γ_d is known as the *Kneser graph* $K(n,d)$. (We shall meet the Kneser graphs $K(2m-1,m-1)$ below under the name Odd graphs O_m; clearly $K(n,2)=\overline{T(n)}$.) HALL [349] showed that if $n \geqslant 3d+1$ any connected graph that is locally $K(n,d)$ must be isomorphic to $K(n+d,d)$. On the other hand, there are just three connected graphs that are locally $\overline{T(6)}$ (BUEKENHOUT & HUBAUT [148]: let $Sp(d,q)$ denote the collinearity graph of points and totally isotropic lines in $PG(d-1,q)$ provided with a nondegenerate symplectic form; inside $Sp(6,2)$ the complements of a hyperplane, of an elliptic quadric, and of a hyperbolic quadric are all locally $Sp(4,2)=\overline{T(6)}$; note that the third of these graphs is $\overline{T(8)}$), so that the bound $n \geqslant 3d+1$ is sharp for $d=2$.

B. Characterization by parameters

For $d \leqslant 1$ the Johnson graph is a clique, characterized by its valency $k=v-1$. For $d=2$ the Johnson graph is strongly regular and characterized by $v=\binom{n}{2}$, $k=2(n-2)$, $\lambda=n-2$, $\mu=4$ for $n\neq 8$; see Theorem 3.12.4. Uniqueness was shown for $n>8$ in CONNOR [211], for $n\leqslant 6$ in SHRIKHANDE [667, 668], and for $n=7$ in HOFFMAN [395] and CHANG [190]. For $n=8$ there are precisely three other graphs with the same parameters (CHANG [190, 191]) namely the Chang graphs described in §3.11.

For $d\geqslant 3$ the Johnson graph is characterized by its parameters (1) as was shown independently by NEUMAIER [561] and TERWILLIGER [719], cf. Corollary 6.1.1. Many special cases had been settled earlier. In case $d=3$, uniqueness was shown for $n>16$ by BOSE & LASKAR [107], by AIGNER [3, 4] for $n\leqslant 8$, and by ROLLAND [615] for $n\geqslant 9$. Independently, LIEBLER [495] settled $11\leqslant n\leqslant 16$, and MOON [542] did $n=9,10$. For $d>3$ the cases $n=2d+1$, $n=2d+2$, $n=3d$ or $n=3d+1$, $n>4d$ and $n\geqslant 20$ were handled by MOON [543, 540, 542, 544, 545], respectively.

When n is sufficiently large, it is not necessary to require distance regularity: any graph Γ with $v(\Gamma)=\binom{n}{e}$, $k(\Gamma)=e(n-e)$, $\lambda(\Gamma)=n-2$ such that any two nonadjacent points have at most 4 common neighbours is a Johnson graph whenever $n>2e(e-1)+4$ (DOWLING [270]) or $n\geqslant \max(6e-1, e^2+2e-1)$ (BROUWER [122]).

Among distance-regular graphs containing a quadrangle, $J(n,d)$ is characterized for $(n,d)\neq(8,2)$ by d, k, λ, μ and c_d (TERWILLIGER [725], see Theorem 5.2.3).

The tetrahedral graph $J(n,3)$ is probably characterized by its spectrum; at any rate it is characterized by being regular and connected, with $v=\binom{n}{3}$ vertices and with eigenvalues $3n-9$, $2n-9$, $n-7$, -3 if one also assumes that $k_2=3(n-3)(n-4)/2$ and $n>16$ (BOSE & LASKAR [108]). [CVETKOVIĆ, DOOB & SACHS [235], p. 183, state that Goethals and Seidel removed the assumption on k_2, but that seems not to be the case.]

C. Folded Johnson graphs

The only imprimitive Johnson graphs are the graphs $J(2m,m)$ which are antipodal. They are double covers of the *folded Johnson graphs* $\overline{J}(2m,m)$, which can be described as the set $\frac{1}{2}\binom{X}{m} = \binom{X}{m,m}$ of partitions of a $2m$-set X into two m-sets, two partitions being adjacent whenever their common refinement is a partition of X into four sets of sizes $1, m-1, 1, m-1$. More generally, two partitions have distance i when their common refinement is a partition of X into sets of sizes i, $m-i$, i, $m-i$ where $i \leqslant m-i$.

9.1.5. Proposition. *For $m \leqslant 3$, the folded Johnson graph is a clique. For $m \geqslant 4$, the folded Johnson graph on a $2m$-set X is distance-transitive with diameter $d = [m/2]$, intersection array given by*

$$b_i = (m-i)^2,\ c_i = i^2\ (0 \leqslant i \leqslant \tfrac{1}{2}m)\ \textit{but}\ c_d = 2d^2\ \textit{if}\ m\ \textit{is even},$$

singular line size $s+1 = 2$, and automorphism group Sym(X). *It is a partition graph (Table 6.3, (P2)). Its eigenvalues and multiplicities are*

$$\theta_i = (m-2i)^2 - 2i,\quad f_i = \binom{2m}{2i} - \binom{2m}{2i-1}.$$

□

The folded Johnson graph $\overline{J}(8,4)$ is the complement of the Grassmann graph of the lines in $PG(3,2)$.

There are at least 1853 nonisomorphic distance-regular graphs (of diameter two) with the same intersection array as $\overline{J}(8,4)$, see BUSSEMAKER et al. [154]. On the other hand, it can be shown on the basis of Proposition 3.12.5 (i) that the folded Johnson graph $\overline{J}(2m,m)$ is characterized by its intersection array if m is sufficiently large (BUSSEMAKER & NEUMAIER [155], cf. Theorem 6.3.2).

D. Odd graphs and doubled Odd graphs

Let X be a set of cardinality $2m+1$. The *doubled Odd graph* on X is the graph Γ whose vertices are the m-subsets and $(m+1)$-subsets of X, where distinct vertices γ, $\delta \in \Gamma$ are adjacent if and only if $\gamma \subset \delta$ or $\delta \subset \gamma$. It is easy to see that we have $d(\gamma,\delta) = |\gamma \setminus \delta| + |\delta \setminus \gamma|$ for all γ, $\delta \in \Gamma$. The graph Γ can be viewed as the subgraph of the binary Hamming graph $H(2m+1,2)$ consisting of all words of weight m and $m+1$.

The *Odd graph* on a $(2m+1)$-set X (denoted by O_{m+1} - the subscript gives the valency) is the graph whose vertex set is the set $\binom{X}{m}$ of m-subsets of X, where two m-sets are adjacent if and only if they are disjoint. (BIGGS & GARDINER [81] explain this name by: 'since each edge can be assigned the unique element of X which is not a member of either vertex incident with that edge - the "odd man out"'.)

9.1.6. Lemma. *Let Γ be the Odd graph on X, where $|X| = 2m+1$, and let $\Delta := \Gamma_2$. Then*

(i) *Δ is the Johnson graph $\binom{X}{m}$, and Γ can be reconstructed from Δ as $\Gamma = \Delta_m$.*

(ii) *$\Gamma \cup \Delta$ is isomorphic to the folded Johnson graph $\overline{J}(2m+2,m+1)$.*

Proof. Trivial. □

9.1.7. Proposition. *The Odd graph O_{m+1} is distance-transitive with diameter $d = m$, intersection array given by*

$$b_j = m+1-[\tfrac{1}{2}(j+1)], \quad c_j = [\tfrac{1}{2}(j+1)] \quad (0 \leqslant j \leqslant d). \qquad (2)$$

except that $b_m = 0$, and we have $\mathrm{Aut}\, O_{m+1} \cong \mathrm{Sym}(X)$. *The eigenvalues of O_{m+1} are $(-1)^i(m+1-i)$ with multiplicity $\binom{2m+1}{i}-\binom{2m+1}{i-1}$ $(i = 0, \cdots, d)$. This graph has girth 5 if $m = 2$, girth 6 if $m > 2$, and hence singular line size $s+1 = 2$. Odd graphs are characterized by their intersection array.*

Proof. The computation of parameters and automorphism group is straightforward. Uniqueness of the Odd graphs is due to MOON [543]; given the uniqueness of the Johnson graphs, it follows immediately from the lemma. □

9.1.8. Proposition. *The doubled Odd graph Γ on a $(2m+1)$-set X is distance-transitive with diameter $d = 2m+1$ and intersection array given by* (2). *Γ has singular line size $s+1 = 2$, and* $\mathrm{Aut}\,\Gamma \cong \mathrm{Sym}(X) \times \mathbf{Z}_2$. *$\Gamma$ is bipartite and antipodal, it is the bipartite double of the Odd graph; its halved graphs are the Johnson graphs $\binom{X}{m}$ and $\binom{X}{m+1}$, and its folded graph is the Odd graph O_{m+1} on X. The graph Γ is not Q-polynomial. Doubled Odd graphs are characterized by their intersection array.*

Proof. The computation of parameters and automorphism group is straightforward. The fact that Γ is not Q-polynomial was observed by BANNAI & ITO [33], p. 315. Uniqueness of the doubled Odd graph Γ (shown by HEMMETER [370]), follows from the fact that its folded graph is the Odd graph Γ' since Γ can be reconstructed uniquely from Γ' by the covering construction given in §1.11 (i.e., Γ is the bipartite double of Γ'), cf. Proposition 4.2.11. □

Doubled Odd graphs and Odd graphs are regular near polygons with parameters as given in Table 6.6, (R3) and (R4), respectively. O_3 is commonly known as the Petersen graph. We shall meet O_4 again as a subgraph of the Hoffman-Singleton graph.

9.1.9. Proposition (cf. IVANOV [431]). *The only distance-regular antipodal covers of Odd graphs O_m $(m \geqslant 4)$ are their bipartite doubles. The Petersen graph O_3 has two distance-regular antipodal covers, namely the Desargues graph (its bipartite double) and the dodecahedron.*

Proof. The standard conditions on t and r for covers $r.O_m$ or $(r.O_m)_t$ (cf. the remarks after 4.2.2) immediately yield that we have one of the possibilities mentioned, or $(3.O_3)_1$ or $2.O_4$ or $(2.O_{2t+1})_t$ with $t \geqslant 2$. The last possibility is ruled out by Proposition 5.5.4(ii) (and also by Proposition 4.2.8), $(3.O_3)_1$ has nonintegral multiplicities, and $2.O_4$ is ruled out by detailed explicit investigation along the lines of SMITH [682], who did the distance-transitive case. □

Remarks. (i) HEMMETER [370] proves that if a non-complete Johnson graph $J(n,m)$ is the halved graph of a bipartite distance-regular graph Γ, then $n = 2m+1$ and Γ is the doubled Odd graph. HEMMETER [371] shows that non-complete folded Johnson graphs or Odd graphs never are halved graphs of a distance-regular graph. Theorem 9.1.3 immediately implies that non-complete Johnson graphs do not have antipodal covers, and the only antipodal covers of non-complete folded Johnson graphs are the Johnson graphs. The only distance-regular antipodal covers of $\overline{T(n)}$ $(n \geqslant 6)$ are $3.\overline{T(6)}$ and $3.\overline{T(7)}$, see §4.2.

(ii) The graphs defined in Corollary 1.5.4 are also related to the Johnson graphs. They can be interpreted as the graphs whose vertices are the 1-sets and $(n-1)$-sets of an n-set, adjacency being symmetrized containment. These graphs are distance-transitive of diameter 3 with automorphism group $2 \times \mathrm{Sym}(n)$.
(iii) For the Erdős-Ko-Rado Theorem in Johnson schemes, see ERDŐS, KO & RADO [278], FRANKL [298] and WILSON [789].
(iv) According to IVANOV, MUZICHUK & USTIMENKO [436] (Theorem 10), if $n > \max(3m+4, 13)$, then merging classes in $J(n,m)$ does not lead to an association scheme. Cases where merging classes does lead to an association scheme with fewer classes include (Mathon, Faradjev [pers. comm.]): the distance 1-or-3 graphs of $J(7,3)$, $J(10,3)$ and $J(12,4)$, the distance 1-or-4 graphs of $J(9,4)$ and $J(11,4)$, and the distance 1-or-2-or-4 graphs of $J(13,6)$, which are all strongly regular (with parameters $(v,k,\lambda,\mu) = (35,16,6,8)$, $(120,56,28,24)$, $(495,256,136,128)$, $(126,25,8,4)$, $(330,63,24,9)$, and $(1716,882,456,450)$, respectively).

9.2 Hamming graphs

Hamming graphs form another basic family of distance-regular graphs. They can be considered as the thin case of both the dual polar graphs discussed in §9.4, and the bilinear and alternating forms graphs discussed in §9.5. They also arise as code graphs in §3.14 and (for $q = 2$) as the Coxeter graphs of type $B_{n,1}$ (cf. §10.3). In older literature one encounters the name 'lattice graph' for them.

Take X to be a finite set of cardinality $q \geqslant 2$. The *Hamming graph* Γ *(with diameter d on X)* has vertex set $X^d = \times_{i=1}^{d} X$, the cartesian product of d copies of X; two points of Γ are adjacent whenever they differ in precisely one coordinate. It follows that $\gamma,\delta \in \Gamma$ are at distance i if and only if they differ in precisely i coordinates. By abuse of notation we denote Γ by X^d; when X is unspecified, we write q^d or $H(d,q)$. In the literature one also finds the notation $L_d(q)$ for this graph. For $q = 2$, $H(d,2)$ is the d-cube, and for $d = 2$, $H(2,q)$ is the $q \times q$-grid.

9.2.1. Theorem. *Let Γ be the Hamming graph with diameter d on a finite set X of cardinality $q \geqslant 2$. Then Γ is distance-transitive, has intersection array given by*

$$b_j = (d-j)(q-1), \quad c_j = j \quad \textit{for } 0 \leqslant j \leqslant d, \tag{3}$$

and has singular line size $s+1 = q$. Moreover, Γ has classical parameters as in Table 6.1, (C8), and $\mathrm{Aut}\,\Gamma \cong \mathrm{Sym}(q)\ \mathrm{wr}\ \mathrm{Sym}(d)$, *the wreath product of* $\mathrm{Sym}(q)$ *and* $\mathrm{Sym}(d)$. *Γ has eigenvalues and multiplicities given by*

$$\theta_j = q(d-j)-d \text{ and } f_j = \binom{d}{j}(q-1)^j.$$

Proof. Straightforward. □

A. Geometric characterization

Hamming graphs do not contain induced pentagons and their μ-graphs are cocliques. This, together with some parameter information, suffices to characterize them.

9.2.2. Theorem (ENOMOTO [275]). *Let the graph Γ be amply regular with parameters (v,k,λ,μ) where $\mu > 1$, and assume that Γ has no induced $K_{1,1,2}$ or induced pentagon. Then $r := k/(\lambda+1)$ is an integer, and $|\Gamma_i(\gamma)| \leqslant \binom{r}{i}(\lambda+1)^i$ for all $\gamma \in \Gamma$ and $i \geqslant 0$. In particular, Γ has diameter $d \leqslant r$ and $v \leqslant (\lambda+2)^r$. If $v = (\lambda+2)^r$, then $\mu = 2$ and $\Gamma \cong H(r,\lambda+2)$.* □

9.2.3. Corollary. *A near $2d$-gon Γ with q^d vertices, line size $s+1 = q$, having $t+1$ lines on each point, and $c_2(\Gamma) = 2$ is a Hamming graph $H(d,q)$ (and $t+1 = d$).* □

Remark. Theorem 9.2.2 provides a characterization of the d-cube $H(d,2)$ by parameters: $\lambda = a_2 = 0$ implies the absence of $K_{1,1,2}$ and C_5. The d-cube can be characterized under much weaker assumptions; see e.g. §1.13 and §4.3B.

B. Characterization by parameters - Pseudo Hamming graphs

For $d \leqslant 1$ the Hamming graph $H(d,q)$ is a clique. For $d = 2$ we have a strongly regular graph (the $q \times q$-grid) uniquely determined by its parameters except when $q = 4$, in which case there is precisely one other graph (the Shrikhande graph, cf. §3.11) with the same parameters (SHRIKHANDE [668], MESNER [534]). In his investigation of when the direct product of distance-regular graphs is again distance-regular, DOOB [266] noticed that the direct product of a Hamming graph $H(n,4)$ with $m \geqslant 1$ copies of a Shrikhande graph is distance-regular with the same parameters as $H(n+2m,4)$; cf. Proposition 1.12.1 (iii). We call these graphs *Doob graphs*; it is easily seen that they are not distance-transitive.

For $d = 3$ it follows by LASKAR [476], DOWLING [269], AIGNER [5] and COOK [217], that any connected amply regular graph with $(v,k,\lambda,\mu) = (q^3,3(q-1),q-2,2)$ and either $q > 7$ or $a_2 = 2(q-2)$ is a Hamming graph or (in case $q = 4$) the direct product of K_4 and the Shrikhande graph.

LASKAR [477] and CVETKOVIĆ [231] showed that these same graphs also are characterized by the conditions: connected, regular, $v = q^3$, $k_2(\Gamma) = 3(q-1)^2$, and the eigenvalues are $3q-3$, $2q-3$, $q-3$, -3. The condition on k_2 may be replaced by the condition that any two vertices at distance two have at least two common neighbours; maybe this condition is entirely superfluous. [CVETKOVIĆ, DOOB & SACHS [235], p. 183, refer to a personal communication by Goethals & Seidel but it seems that the work of these two authors - mostly on tetrahedral graphs - used somewhat stronger assumptions.]

9.2.4. Theorem. *An amply regular graph Γ with parameters $(v,k,\lambda,\mu) = (q^d,d(q-1),q-2,2)$ and $a_2(\Gamma) = 2(q-2)$, $c_3(\Gamma) = 3$ is a Hamming graph or a Doob graph.*

Proof. This is implicit in EGAWA [273]. □

9.2.5. Corollary (EGAWA [273]). *A distance-regular graph with intersection array* (3) *is a Hamming graph or a Doob graph.* □

This result also follows from the representation theoretic approach, see Corollary 6.1.1.

C. Characterization by spectrum

Concerning characterization by spectrum alone, HOFFMAN [397] showed that the Hamming d-cube is characterized by its spectrum for $d \leqslant 3$, but not for $d = 4$ (and hence not for $d \geqslant 4$): for $d = 4$ there is precisely one other graph with the same spectrum as the 4-cube (see below), which, however, is not distance-regular.

9.2.6. Example. Let us describe Hoffman's graph. From the spectrum $(4^1\, 2^4\, 0^6\, (-2)^4\, (-4)^1)$ we can see that this graph is regular, connected and bipartite, i.e., it is the incidence graph of a square 1-(8,4,4) design.
Given the Fano plane $PG(2,2)$, throw away one point and the three lines on it to obtain a transversal design $TD[3,1;2]$ (cf. §A.6). Find two new symbols a, a' and take for each block B of this transversal design two new blocks $B \cup \{a\}$ and $B \cup \{a'\}$. This yields a transversal design $TD[4,2;2]$ (with $\{a,a'\}$ as 4th part), the design we were looking for. Any two blocks meet in 1, 2, or 3 points. Let N be the point-block incidence matrix of this design, then $NN^{\mathrm{T}} = 2J+2I-2D$, where D is the 0-1 matrix with $D_{xy} = 1$ if and only if $\{x,y\}$ is one of the parts. Clearly, the equation $X^2-4X = 24J$ holds for $X = NN^{\mathrm{T}}$, but by some coincidence it also holds for $X = N^{\mathrm{T}}N$. This shows that the matrix $A = \begin{bmatrix} 0 & N \\ N^{\mathrm{T}} & 0 \end{bmatrix}$ satisfies $A(A-2I)(A+2I)(A+4I) = 24J$. Since rk $N = 5$ it follows that A has spectrum $(4^1\, 2^4\, 0^6\, (-2)^4\, (-4)^1)$, but the graph Γ with adjacency matrix A is not distance-regular. (The distance distribution around a point of the design is as in the 4-cube, but around a block B we see $k_4(B) = 0$, and $c_2(B,\cdot)$ takes three values.) The matrix N also occurs in HOFFMAN [396] and earlier in CONNOR [210] as an example of a square divisible design whose dual is not a divisible design.

This example shows that one cannot see from the spectrum of a graph whether it is distance-regular. (But one can see whether it is regular and connected, and whether it is strongly regular, and whether it is bipartite distance-regular of diameter 3.)

Problem. Can one recognize 'distance regularity of diameter 3' from the spectrum? As H. Cuypers observed, it is not enough for a graph to be regular and connected with just 4 distinct eigenvalues: the line graphs (block graphs) of some near polygons (such as the near hexagon on 759 vertices) provide counterexamples.

CVETKOVIĆ [231] shows that the cubic lattice graphs $H(3,q)$ (and for $q = 4$ the unique Doob graph with the same parameters) are characterized by their spectrum and the property that any two vertices at distance two have at least two common neighbours.

D. Halved and folded cubes

The only imprimitive Hamming graphs are the n-cubes; they are both bipartite and antipodal.

The *halved n-cube* is the graph $D_{n,n}(1)$ defined in §3.14; $D_{n,n}(1)$ is distance-transitive with classical parameters (Table 6.1, (C9)). For $n \geq 3$, its intersection array is given by $d = [n/2]$,

$$b_j = \tfrac{1}{2}(n-2j)(n-2j-1), \quad c_j = j(2j-1).$$

The eigenvalues and multiplicities are

$$\theta_j = \tfrac{1}{2}(n-2j)^2 - \tfrac{1}{2}n, \quad f_j = \binom{n}{j} \quad (2j < n)$$

and if n is even, then $f_d = \frac{1}{2}\binom{n}{d}$. By NEUMAIER [561] and TERWILLIGER [725], the halved n-cubes are characterized by their intersection array; see Theorem 4.4.11.

The *folded n-cube,* often denoted by $\square_n$, is a partition graph (Table 6.3, (P1)); it can be described as the graph whose vertices are the partitions of an n-set into two subsets, two partitions being adjacent when their common refinement contains a set of size one. For $n \geq 3$, its intersection array is given by $d = [n/2]$,

$$b_j = n - j, \quad c_j = j \quad (2j < n)$$

and if n is even, then $c_d = n$. The eigenvalues and multiplicities are

$$\theta_j = n - 4j, \quad f_j = \binom{n}{2j}.$$

9.2.7. Theorem (cf. BROUWER [123], TERWILLIGER [727]). *For $n \neq 6$, the folded n-cube is uniquely characterized by its intersection array. There are precisely three nonisomorphic graphs with the intersection array $\{6,5,4;1,2,6\}$ of the folded 6-cube.*

Proof. For odd n, say $n = 2m+1$, the folded n-cube has diameter $d = m$, and satisfies $a_i = 0$ $(i < m)$ and $a_m = m+1$; its bipartite double is again the n-cube. Given any distance-regular graph with the intersection array of the folded $(2m+1)$-cube, its bipartite double is distance-regular with the intersection array of the $(2m+1)$-cube (cf. Theorem 1.11.1) and hence is the $(2m+1)$-cube. It follows that the folded $(2m+1)$-cubes are characterized by their intersection array.
For $n \geq 7$, the folded n-cube has parameters $\lambda = a_2 = 0$, $\mu = 2$, $c_3 = 3$, and the covering construction of §4.3B shows that any graph with its intersection array must be the antipodal quotient of the n-cube. Thus, also in this case the folded cubes are characterized by their intersection array.
The folded 2-cube and the folded 4-cube are K_2 and $K_{4,4}$, respectively, both uniquely determined by their parameters.
Finally, any distance-regular graph with intersection array $\{6,5,4;1,2,6\}$ is the incidence graph of a square 2-(16,6,2) design. But there are precisely three such designs, all of them self-dual (cf. HUSSAIN [418]). □

For even n, halved n-cubes and folded n-cubes are still imprimitive. A folded half cube (or halved folded cube) is a partition graph (Table 6.3, (P3)); it can be described as the graph whose vertices are the partitions of an n-set into two subsets of even size, two partitions $\{A,A'\}$ and $\{B,B'\}$ being adjacent when $\min(|A\Delta B|, |A\Delta B'|) = 2$ (where Δ denotes symmetric difference). A halved folded $2m$-cube with $m \geqslant 3$ has intersection array given by $d = [m/2]$,

$$b_j = (m-j)(2m-2j-1), \ c_j = j(2j-1) \ (2j < m)$$

and if m is even, then $c_d = m(m-1)$. The eigenvalues and multiplicities are

$$\theta_j = 2(m-2j)^2 - m, \ f_j = \binom{2m}{2j}.$$

For $n \in \{8, 10\}$, folded halved n-cubes are strongly regular (with parameters $(v,k,\lambda,\mu) = (64,28,12,12)$ and $(256,45,16,6)$, respectively), and these are not determined by their parameters; for example, the block graph of any transversal design $TD[4;8]$ or $TD[3;16]$ has the same parameters but is not isomorphic to the folded halved n-cube ($n = 8, 10$, respectively). However, it can be shown on the basis of Theorem 3.12.5 (ii) that folded halved n-cubes are characterized by their intersection array if n is sufficiently large (BUSSEMAKER & NEUMAIER [155], cf. Theorem 6.3.2).

The automorphism group of the halved n-cube is $2^{n-1}.\mathrm{Sym}(n)$ for $n > 4$, $2^4.\mathrm{Sym}(4)$ for $n = 4$, and $\mathrm{Sym}(2^{n-1})$ for $1 \leqslant n \leqslant 3$. The automorphism group of the folded n-cube is $2^{n-1}.\mathrm{Sym}(n)$ for $n > 4$, $(\mathrm{Sym}(4)\times\mathrm{Sym}(4)).2$ for $n = 4$, and $\mathrm{Sym}(2^{n-1})$ for $1 \leqslant n \leqslant 3$. The automorphism group of the halved folded $2m$-cube is $2^{2m-2}.\mathrm{Sym}(2m)$ for $m > 3$, and $\mathrm{Sym}(2^{2m-2})$ for $1 \leqslant m \leqslant 3$.

Remarks. (i) Let Γ be a Hamming graph of diameter d. The graph Γ_{d-1} is isomorphic to Γ whenever $q = 2$ and d is even. For, fix a point 0 of Γ and define the weight of a vertex of Γ to be its distance to 0. Replacing the points of odd weight by their (unique) opposite point, leads to an isomorphism between Γ and Γ_{d-1}; in other words, $H(2n, 2)$ has two (isomorphic) P-polynomial structures.
(ii) HEMMETER [370] shows that if a Hamming graph $H(d,q)$ with $d > 1$ is the halved graph of a bipartite distance-regular graph Γ, then $d = q = 2$ and Γ is the 8-gon. HEMMETER [371] shows that if the halved d-cube is the halved graph of a bipartite distance-regular graph Γ, then either Γ is isomorphic to the d-cube, or $d \leqslant 4$. (For $d \leqslant 4$ the only exception occurs for $d = 3$: both the 3-cube and $K_{4,4}$ have halved graph K_4.) Also, that non-complete folded cubes are never halved graphs. Finally, that if a non-complete folded halved $2m$-cube is the halved graph of a bipartite distance-regular graph Γ, then Γ is isomorphic to the folded $2m$-cube.
(iii) The halved 5-cube is the complement of the folded 5-cube. The folded 5-cube is isomorphic to the Hermitean forms graph on $\mathbb{F}_2^4$ (see §9.5C). The folded halved 8-cube is the complement of the alternating forms graph on $\mathbb{F}_2^4$ (see §9.5B).
(iv) Let Γ be the halved $4m$-cube. ITO [424] studies the graph Γ_m under the name 'Hadamard graph'. (Do not confuse this with our concept of Hadamard graph. The reason for his terminology is that a Hadamard matrix of order $4m$ corresponds to a $4m$-clique in Γ_m; in particular, the number of inequivalent Hadamard matrices of order $4m$ equals the number of orbits of $\mathrm{Aut}(\Gamma_m) \cong 2^{2^{4m-2}}.2^{4m-2}.\mathrm{Sym}(4m)$ (for $m > 1$) on $4m$-cliques.)
(v) For the Erdős-Ko-Rado theorem for Hamming graphs, see LIVINGSTON [506] and MOON [541].

E. Covers of cubes and folded cubes - the Wells graph

9.2.8. Proposition (BIGGS & GARDINER [81], VAN BON [98]).

(i) *$H(d,q)$ with $d, q \geqslant 2$ has no distance-regular antipodal covers except for the octagon, a double cover of the quadrangle $H(2,2)$.*

(ii) *The only distance-regular antipodal covers with diameter at least 3 of folded n-cubes are: n-cubes, a unique graph with intersection array $\{5,4,1,1;1,1,4,5\}$, and a unique graph with intersection array $\{4,3,3,1;1,1,3,4\}$ ($4K_{4,4}$, a symmetric net: $AG(2,4)$ minus a parallel class). The two sporadic graphs with the parameters of the folded 6-cube do not have distance-regular antipodal covers.*

Proof. (i) For covers of even diameter, this was already remarked after Proposition 4.2.7, and for diameter $\geqslant 5$ after Corollary 5.2.4.
(ii) In the same way one finds that any distance-regular antipodal r-cover of even diameter of $\square_n$ must have $r = 2$ if $n > 4$, and n even if $n > 5$. Also, that the two sporadic graphs with the parameters of $\square_6$ do not have antipodal distance-regular covers.
On the other hand, the inequalities $t \geqslant c_d$ and $t(r-1) \leqslant b_{d-1}$ for the parameters of a graph $(r.\Gamma)_t$ (cf. §4.2) immediately imply for $\Gamma = \square_n$ that n is odd (if $n > 2$) and $r = 2$ (if $n > 5$). Also, $n = 5$, $r = 3$ is ruled out by Theorem 5.4.1. Thus (by 1.13.1), for $n \geqslant 6$ the only possible covers are n-cubes. The unique antipodal double cover of diameter 4 of $\square_5$ is the subject of the following theorem. The unique antipodal 4-cover of diameter 4 of $\square_4 = K_{4,4}$ is derived from the incidence graph of $PG(2,4)$, cf. Remark 9.3.12(iii). Finally, there is no graph $(3.K_4)_1$, since it would have a nonintegral number of pentagons. □

9.2.9. Theorem. *There is a unique distance-regular graph Γ with intersection array $\{5,4,1,1;1,1,4,5\}$ and distance distribution diagram:*

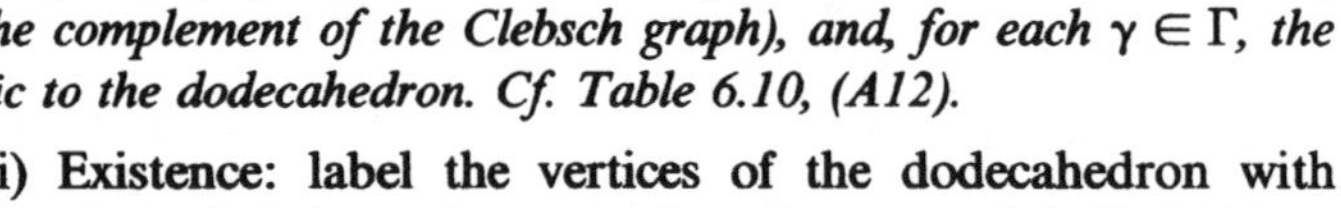

$v = 32.$

It is distance-transitive with $\mathrm{Aut}\,\Gamma \cong 2^{1+4}_{-}{:}\mathrm{Alt}(5)$ *where* $2^{1+4}_{-} = D_8 \circ Q_8$ *is an extraspecial group of order* 32 *and* $\mathrm{Alt}(5)$ *is the point stabilizer. Its spectrum is* $5^1\ \sqrt{5}^8\ 1^{10}\ (-\sqrt{5})^8\ (-3)^5$. *$\Gamma$ is the unique double cover without 4-cycles of the folded 5-cube (the complement of the Clebsch graph), and, for each $\gamma \in \Gamma$, the subgraph $\Gamma_2(\gamma)$ is isomorphic to the dodecahedron. Cf. Table 6.10, (A12).*

Proof. (i) Existence: label the vertices of the dodecahedron with ordered pairs ij ($1 \leqslant i, j \leqslant 5$, $i \neq j$) in such a way that the vertex labelled ij has distance 3 to the vertices labelled ik or kj ($1 \leqslant k \leqslant 5$, $k \neq i, j$). Now invent 12 new vertices $\infty^{\pm}$, $i^{\pm}$ ($1 \leqslant i \leqslant 5$), join $\infty^{\pm}$ to all $i^{\pm}$ and i^{+} to ij and i^{-} to ji ($j \neq i$). This yields the desired graph Γ.
(ii) Uniqueness: for $\gamma \in \Gamma$ the subgraph $\Gamma_2(\gamma)$ is a double cover of the Petersen graph, and it is easy to see that this must be the dodecahedron. Now the construction described under (i) is forced.
(iii) Group: this graph has been found and its group has been determined by ARMANIOS [8] (who refers to A.J. Well, jr. (sic); in fact the graph already occurred in the unpublished paper BIGGS & GARDINER [81]). Distance transitivity is already clear from the uniqueness proof. □

We call the graph of the theorem the Wells graph.

Using similar arguments as above, it is easy to show that the Doob graphs and the halved cubes of diameter $d \geqslant 2$ do not have distance-regular antipodal covers, and that the only distance-regular antipodal covers of the folded halved cubes are the halved cubes. (Concerning the folded halved n-cubes $\frac{1}{2}\bar{\square}_n$ ($n \geqslant 8$), these are locally $T(n)$. By NEUMAIER [561] (Propositions 2(ii) and 3) any connected locally triangular graph is the halved graph of a bipartite graph with $\mu = 2$ (cf. the remarks in the beginning of §1.13). It follows that any cover of a folded halved n-cube gives rise to a cover of the folded n-cube, but since $n \geqslant 8$, such a cover must be the n-cube.)
On the other hand, there is an important family of covers of the n-cubes:

9.2.10. Proposition (COHEN & TITS [207]). *Let Γ be the n-cube $H(n, 2)$, and $n \geqslant 2$.*

(i) *Γ has a unique 2-cover $\tilde{\Gamma}$ without quadrangles, and $\tilde{\Gamma}$ is connected and bipartite.*

(ii) *The involution τ which exchanges the pairs of points of $\tilde{\Gamma}$ having the same images in Γ, is an automorphism of $\tilde{\Gamma}$.*

(iii) *Suppose $n \neq 4$. If $\gamma \in \tilde{\Gamma}$, the points γ and $\tau\gamma$ are the only points at distance two from all the points which are at distance two from γ; thus, τ is central in* Aut $\tilde{\Gamma}$.

(iv) *If $\gamma, \delta \in \tilde{\Gamma}$ are such that their images in Γ are opposite (i.e., have distance n), then the group of all automorphisms of $\tilde{\Gamma}$ fixing γ and $\tau\gamma$ is a symmetric group on n letters whose even (resp. odd) elements fix (resp. exchange) δ and $\tau\delta$.* □

The distance distribution diagram of $\tilde{H}(n, 2)$ is

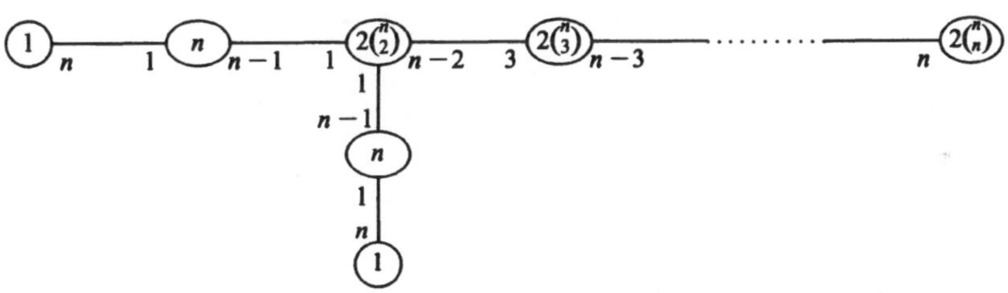

Remarks. (i) COHEN & TITS [207] did not exclude $n = 4$ for part (iii), but that is necessary: $\tilde{H}(4,2)$ is distance-regular with intersection array $\{4,3,3,1;1,1,3,4\}$, an antipodal 4-cover of $K_{4,4}$, and Aut $\tilde{H}(4,2) \cong 2^4.3.\mathrm{Sym}(4).2$ acts distance-transitively, so all three points in $\Gamma_4(\gamma)$ have the property claimed in (iii) for $\tau\gamma$; also, $Z(\mathrm{Aut}\,\tilde{H}(4,2)) = 1$.
(ii) $\tilde{H}(2,2)$ is the 8-gon; $\tilde{H}(3,2)$ occurs as component of $\Gamma_3(\infty) \cong 2\tilde{H}(3,2)$ in one of the two generalized hexagons Γ of order (2,2), cf. §12.4, Remark (iii); $\tilde{H}(4,2)$ and $\tilde{H}(5,2)$ occur in the Hall-Janko near octagon Γ, cf. §13.6: $\Gamma_3(\infty) \cong 5\tilde{H}(4,2)$ and $\Gamma_4(\infty) \cong \tilde{H}(5,2)$ for $\infty \in \Gamma$.
(iii) The folded n-cube $\bar{\square}_n$ has a 2-cover $\tilde{\bar{\square}}_n$ of girth $g > 4$ if and only if $n \equiv 0$ or 1 (mod 4), $n > 4$, and such a 2-cover is unique. $\tilde{H}(n, 2)$ is a 2-cover of $\tilde{\bar{\square}}_n$. $\tilde{\bar{\square}}_5$ is the Wells graph.

9.3 Grassmann graphs

Let F be a field, and let V be an n-dimensional vector space over F. The *Grassmann graph of the e-subspaces of V* has vertex set $\left[{V \atop e}\right]$, the collection of linear subspaces of V of dimension e. Two points γ, δ are adjacent whenever $\dim \gamma \cap \delta = e-1$.

A linear subspace of V of dimension e is called an *e-space* (of V). The 0-dimensional subspace is written as 0, and the subspace spanned by two subspaces X, Y of V is written as $X+Y$.

Remarks. (i) Let (,) be any nondegenerate symmetric bilinear form on V and define, for $X \subseteq V$,

$$X^{\perp} := \{y \in V \mid (x,y) = 0 \text{ for all } x \in X\}.$$

If X is a linear subspace of V, then $\dim X^{\perp} = n - \dim X$; if X, Y are two linear subspaces of V, then $\dim X^{\perp} - \dim X^{\perp} \cap Y^{\perp} = \dim Y - \dim X \cap Y$. This shows that the Grassmann graphs $\left[{V \atop e}\right]$ and $\left[{V \atop n-e}\right]$ are isomorphic.

(ii) It is easy to verify that the maximal cliques in the Grassmann graph $\left[{V \atop e}\right]$ are the collections of all e-spaces containing a fixed $(e-1)$-space and the collections of all e-spaces contained in a fixed $(e+1)$-space.

9.3.1. Theorem (Chow [196]). *Let Γ be the Grassmann graph $\left[{V \atop e}\right]$, and suppose that Γ is not complete, i.e., that $1 < e < n-1$. Then*

$$\operatorname{Aut}\Gamma \cong \begin{cases} P\Gamma L(V) & \text{if } n \neq 2e, \\ \operatorname{Aut} P\Gamma L(V) = P\Gamma L(V).2 & \text{if } n = 2e. \end{cases}$$

Proof. The graph Γ has maximal cliques corresponding to the e-spaces containing a given $(e-1)$-space and the e-spaces contained in a given $(e+1)$-space, respectively. Let Δ be the set of all these cliques and turn Δ into a graph by defining $\alpha \sim \beta$ for $\alpha, \beta \in \Delta$ if and only if $|\alpha \cap \beta| = 1$. Then Δ has two connected components Δ^1, Δ^2 associated with the above partitioning of cliques in two classes, and the graphs Δ^1 (respectively Δ^2) are isomorphic to the Grassmann graphs $\left[{V \atop e-1}\right]$ (respectively $\left[{V \atop e+1}\right]$). $\operatorname{Aut}\Gamma$ contains an automorphism of Γ interchanging Δ^1 and Δ^2 if and only if $n = 2e$ (cf. Remark (i) above); let G_0 be the subgroup of $\operatorname{Aut}\Gamma$ stabilizing Δ^1 and Δ^2 (setwise). By induction on $e \geqslant 2$ we find that G_0 is contained in the automorphism group of the Grassmann graph $\left[{V \atop 2}\right]$, and that G_0 permutes the points of V. By the Fundamental Theorem of Projective Geometry we see that $G_0 \cong P\Gamma L(V)$ so that $\operatorname{Aut}\Gamma$ is as claimed. □

Now suppose for the remainder of this section that $F = \mathbb{F}_q$, the finite field with q elements, q a prime power. Let $\left[{n \atop m}\right]$ be the q-ary Gaussian binomial coefficient:

$$\left[{n \atop m}\right] = \left[{n \atop m}\right]_q = \frac{(q^n-1)\cdots(q^{n-m+1}-1)}{(q^m-1)\cdots(q-1)}.$$

9.3.2. Lemma. *Let V be as above. Suppose* $0 \leqslant i, j \leqslant n$. *Then:*

(i) *The number of m-spaces of V is* $\left[{n \atop m}\right]$.

(ii) *If X is a j-space of V, then there are precisely* $q^{ij} \left[{n-j \atop i}\right]$ *i-spaces Y in V such that* $X \cap Y = 0$.

(iii) *If X is a j-space of V, then there are precisely* $q^{(i-m)(j-m)} \left[{n-j \atop i-m}\right] \left[{j \atop m}\right]$ *i-spaces Y in V such that* $X \cap Y$ *is an m-space.*

Proof. (i) There are $\prod_{i=0}^{m-1} (q^n - q^i)$ ordered m-tuples of linearly independent vectors in V, and $\prod_{i=0}^{m-1} (q^m - q^i)$ ordered bases of any m-space.
(ii) Given an ordered i-tuple of linearly independent vectors in V/X there are q^{ij} ways to lift this i-tuple to an i-tuple of (necessarily linearly independent) vectors in V.
(iii) By (i) there are $\left[{j \atop m}\right]$ m-spaces Z in X, and for each of these we find by (ii) $q^{(i-m)(j-m)} \left[{n-j \atop i-m}\right]$ i-spaces Y in V such that $X \cap Y = Z$ (i.e., $(i-m)$-spaces Y/Z in V/Z such that $(X/Z) \cap (Y/Z) = 0$). □

9.3.3. Theorem. *Let* Γ *be the Grassmann graph of the e-subspaces of V, where V is an n-dimensional vector space over* $\mathbb{F}_q$. *Then* Γ *has diameter* $d = \min(e, n-e)$ *and is distance-transitive with intersection array given by*

$$b_j = q^{2j+1}\left[{e-j \atop 1}\right]\left[{n-e-j \atop 1}\right], \quad c_j = \left[{j \atop 1}\right]^2 \quad (0 \leqslant j \leqslant d). \tag{4}$$

Γ *has classical parameters as given in Table 6.1, (C2).* Γ *has eigenvalues and multiplicities given by*

$$\theta_j = q^{j+1}\left[{e-j \atop 1}\right]\left[{n-e-j \atop 1}\right] - \left[{j \atop 1}\right]$$

and

$$f_j = \left[{n \atop j}\right] - \left[{n \atop j-1}\right].$$

If $d \leqslant 1$, *then* Γ *is a clique; otherwise,* Γ *has singular line size* $s+1 = q+1$.

Proof. Clearly, $\gamma, \delta \in \Gamma$ have distance j if and only if $\dim(\gamma \cap \delta) = e-j$. Note that $P\Gamma L(n,q)$ is a subgroup of $\operatorname{Aut}\Gamma$. As $\Gamma L(n,q)$ is transitive on ordered bases, it follows readily that Γ is distance-transitive. In order to obtain b_j and c_j, fix $\gamma, \delta \in \Gamma$ with $d(\gamma,\delta) = j$. Note that $\zeta \in \Gamma$ is in $\Gamma_{j-1}(\gamma) \cap \Gamma(\delta)$ if and only if $\gamma \cap \delta \subseteq \zeta \subseteq \gamma + \delta$, $\dim \gamma \cap \zeta = e-j+1$, and $\dim \delta \cap \zeta = e-1$. Calculating in $(\gamma+\delta)/(\gamma \cap \delta)$, we see that the number $c(\gamma,\delta)$ of such ζ equals the product of the number of 1-spaces in $\gamma/(\gamma \cap \delta)$ and the number of $(j-1)$-spaces of $\delta/(\gamma \cap \delta)$. Thus $c_j = \left[{j \atop 1}\right]\left[{j \atop j-1}\right]$, as wanted. By Lemma 9.3.2(iii), $k_j = q^{j^2} \left[{n-e \atop j}\right] \left[{e \atop j}\right]$ and b_j is determined by $k_j b_j = k_{j+1} c_{j+1}$. Let us now determine the cardinality $s+1$ of singular lines. Suppose $\gamma, \delta \in \Gamma$, $\gamma \sim \delta$. Then

$$\{\gamma,\delta\}^{\perp} = \{\zeta \in \Gamma \mid \zeta \subseteq \gamma+\delta\} \cup \{\zeta \in \Gamma \mid \zeta \cap (\gamma+\delta) = \gamma \cap \delta\}.$$

It follows that

$$\{\gamma,\delta\}^{\perp\perp} = \{\eta \in \Gamma \mid \gamma \cap \delta \subseteq \eta \subseteq \gamma+\delta\}$$

is of cardinality $q+1$, proving $s = q$. □

A. Characterization by structure

We saw above that, just as in the Johnson graphs, there are two families of maximal cliques in a Grassmann graph. Characterizations more or less explicitly assuming both families as given can be found in TALLINI [707], LO RE & OLANDA [509], BICHARA & TALLINI [63], MELONE & OLANDA [532] and BIONDI [85, 86]. For example:

9.3.4. Theorem (BICHARA & TALLINI [63]). *Let $(\mathcal{P},\mathcal{L})$ be a connected partial linear space which is not a clique. Assume*

A1. *If three points in $\mathcal{P}$ are pairwise collinear, then a singular subspace of $(\mathcal{P},\mathcal{L})$ containing them exists.*

A2. *No line in $\mathcal{L}$ is a maximal subspace. There exist two collections of maximal subspaces in $(\mathcal{P},\mathcal{L})$, say Σ and T such that any maximal subspace belongs to either Σ or T. Furthermore:*

(i) $\sigma \in \Sigma$, $\tau \in \mathrm{T}$ *implies* $\sigma \cap \tau = \emptyset$ *or* $\sigma \cap \tau \in \mathcal{L}$.

(ii) *For all* $l \in \mathcal{L}$ *there are unique* $\sigma \in \Sigma$, $\tau \in \mathrm{T}$ *with* $l \subseteq \sigma$, $l \subseteq \tau$.

(iii) *If* $\sigma, \sigma', \sigma'' \in \Sigma$ *meet pairwise at distinct points, then any element in* $\Sigma \setminus \{\sigma''\}$ *meeting* σ, σ' *at distinct points meets* σ''.

A3. *There exists a number h, a line $l \in \mathcal{L}$ and a subspace $\tau \in \mathrm{T}$ with $l \subseteq \tau$ and a complete, finite chain of subspaces of $(\mathcal{P},\mathcal{L})$ such that its length is $h+1$, its minimum is l and its maximum is τ.*

Then there is a projective space $\mathbf{P}$ such that $\mathcal{P}$ is the collection of h-subspaces of $\mathbf{P}$ and $\mathcal{L}$ is the collection of pencils of h-subspaces of $\mathbf{P}$ (i.e., the collection of sets $\{K \in \mathcal{P} \mid M \subset K \subset N\}$ where $M \subset N$, M is an $(h-1)$-subspace and N is an $(h+1)$-subspace of $\mathbf{P}$). □

Given one family of maximal cliques (called lines) we have characterizations similar to those for Johnson graphs (§9.1A):

9.3.5. Theorem (SPRAGUE [696]). *Let the connected graph Γ be the point graph of a partial linear space satisfying the Pasch axiom and the dual Pasch axiom. Suppose that for all lines l, no point $\gamma \notin l$ is collinear with precisely one point of l. Then there is a generalized projective geometry and an integer e such that Γ is the graph of the e-dimensional subspaces of the geometry, two subspaces being adjacent whenever their intersection has codimension one in each of them.* □

9.3.6. Corollary. *Let the connected graph Γ be the point graph of a partial linear space satisfying the Pasch axiom and the dual Pasch axiom. Suppose that for all lines l, no point $\gamma \notin l$ is collinear with precisely one or two points of l. Then Γ is a Grassmann graph.* □

The following result shows that it suffices to have some 'local' information. Note that all μ-graphs of Grassmann graphs over $\mathbb{F}_q$ are $(q+1)\times(q+1)$-grids.

9.3.7. Theorem (COOPERSTEIN [219], COHEN [203]). *Let Γ be a connected graph with diameter at least 3 in which singular lines have at least three points. Suppose that if $\gamma, \delta \in \Gamma$ with $d(\gamma,\delta) = 2$, then $\{\gamma,\delta\}^\perp$ together with the singular lines contained in it, is a nondegenerate generalized quadrangle, and that $|\gamma^\perp \cap L^\perp| \neq 1$ for any $\gamma \in \Gamma$ and singular line L. Then Γ is a Grassmann graph.*

Proof. Apply the Main Theorem of COHEN [203] to the partial linear space $(\Gamma,\mathfrak{L})$ where $\mathfrak{L}$ is the collection of singular lines of Γ. It follows from the remark after Proposition 3.2 and the proof of Remark 3.4 in [203] that axiom (P4) stated there may be weakened to

(P4)″ If $\gamma \in \Gamma$, $L \in \mathfrak{L}$ and $\gamma^\perp \cap L = \emptyset$, then $|\gamma^\perp \cap L^\perp| \neq 1$. □

Extending the argument of NUMATA [573] (who only treats the case of diameter 2) and using the preceding theorem one finds the following.

9.3.8. Theorem. *Let Γ be a finite connected graph such that*

(i) *if $\gamma,\delta \in \Gamma$ have distance 2, then $\{\gamma,\delta\}^\perp$ is a nondegenerate grid*

and

(ii) *if $\{\gamma,\delta,\epsilon\}$ is a coclique in Γ, then $\{\gamma,\delta,\epsilon\}^\perp$ is a coclique.*

Then Γ is either a clique, or a Johnson graph $J(n,k)$, or the quotient of the Johnson graph $J(2k,k)$ obtained by identifying a k-set with the image of its complement under the identity or an involution in Sym$(2k)$ *with at least* 10 *fixed points, or a Grassmann graph over a finite field.* □

B. Characterization by parameters

9.3.9. Theorem (RAY-CHAUDHURI & SPRAGUE [608]). *Let $(\mathcal{P},\mathfrak{L})$ be a partial linear space such that for some integer $q \geq 2$:*

(i) *each line has at least q^2+q+1 points;*

(ii) *each point is on more than $q+1$ lines;*

(iii) *if $p \in \mathcal{P}$ and $L \in \mathfrak{L}$ such that $d(p,L) = 1$, then there are precisely $q+1$ lines on p meeting L;*

(iv) *if $p, p' \in \mathcal{P}$ have distance 2, then there are precisely $q+1$ lines L on p such that $d(p',L) = 1$;*

(v) *the point graph Γ of $(\mathcal{P},\mathfrak{L})$ is connected.*

Then $(\mathcal{P},\mathfrak{L},\in) \cong \left(\left[{V \atop e-1}\right], \left[{V \atop e}\right], \subseteq\right)$ for some vector space V. In particular, Γ is a Grassmann graph. □

9.3.10. Corollary (SPRAGUE [695]). *Let r, q be a positive integers and k, α be real numbers such that $r > q+1 > 2$ and $k > \max(1+\frac{1}{2}(r+1)(rq^2+2rq-2\alpha), (q+1)^2+(2r-1)\alpha)$. Suppose that Γ is a connected graph, regular of degree $r(k-1)$ for which $\lambda(\Gamma) = k-2+\alpha$, $\mu(\Gamma) = (q+1)^2$ and $b_2(\gamma,\delta) > (k-1+\alpha)(r-q-2)$ for all γ, $\delta \in \Gamma$ at distance 2. Then q is a prime power and Γ is the Grassmann graph $\left[{V \atop e}\right]$, where V is a vector space of dimension n over $\mathbf{F}_q$ and $k = \left[{e+1 \atop 1}\right]$, $r = \left[{n-e \atop 1}\right]$.* □

This corollary characterizes the Grassmann graphs $\left[{V \atop e}\right]$ for $\frac{2}{3}n \leqslant e \leqslant n-3$, $(q,e) \neq (2,\frac{2}{3}n)$. The methods used (the 'claw bound' or 'method of grand cliques' of BOSE & LASKAR [107]) are the same as those used by DOWLING [270] for the Johnson graphs. On the other hand, there are many graphs with the same parameters as the Grassmann graph of the lines in $PG(3,2)$, cf. §9.1C.

Problem. Generalize to Grassmann graphs the representation-theoretic approach of the Johnson graphs (using all association scheme parameters and in particular the eigenvalues to obtain information on subgraphs by 'interlacing' and/or graph representations). A first step in this direction is Theorem 4.4.5.

C. Graphs related to Grassmann graphs

Suppose that V is a vector space of dimension $2m+1$ over $\mathbb{F}_q$. Let Γ be the graph whose vertex set is $\left[{V \atop m}\right] \cup \left[{V \atop m+1}\right]$, and such that $\gamma \sim \delta$ for $\gamma,\delta \in \Gamma$ if and only if $\gamma \neq \delta$ and either $\gamma \subset \delta$ or $\delta \subset \gamma$. This graph, discovered by BIGGS & GARDINER [81], will be called a *doubled Grassmann graph*. It is the bipartite double of Δ_m, where Δ is the Grassmann graph $\left[{V \atop m}\right]$. Note that $d(\gamma,\delta) = \dim(\gamma/\gamma \cap \delta) + \dim(\delta/\gamma \cap \delta)$.

9.3.11. Theorem. *A doubled Grassmann graph Γ is bipartite and distance-transitive of diameter $d = 2m+1$ with intersection array given by*

$$b_{2i-1} = b_{2i} = \left[{m+1-i \atop 1}\right]q^i$$

and

$$c_{2i-1} = c_{2i} = \left[{i \atop 1}\right].$$

as listed in Table 6.6, (R3). Its halved graphs are Grassmann graphs with diameter m on V. The graph Γ is not Q-polynomial.

Proof. All statements but the last are direct consequences of the above. For the last one, cf. BANNAI & ITO [33], III.6, Remark 6 (p. 314). □

9.3.12. Remarks. (i) Suppose that V is a $(d+1)$-dimensional vector space over $\mathbb{F}_q$. Let Γ be the graph whose vertex set is $\left[{V \atop 1}\right] \cup \left[{V \atop d}\right]$, adjacency being symmetrized containment, i.e., the incidence graph of the square 2-design formed by points and hyperplanes of the projective space $PG(d,q)$. Then, clearly, Γ is distance-transitive. For $d = 2$ this is a double Grassmann graph; for $d = 3$ a dual polar graph on $[D_3(q)]$ (see §9.4). Characterizations can be deduced from KANTOR [454, 457]; see also §7.7.
(ii) Assume $q > 2$. A $(q-1)$-cover of this graph can be constructed as follows (see also §1.10). Let X be the set of nonzero points of V, and let B be the set of affine hyperplanes of V not containing 0. The incidence graph Γ' of (X,B) is a bipartite graph of rank 6 with $v = 2(q^{d+1}-1)$ vertices and distance distribution diagram

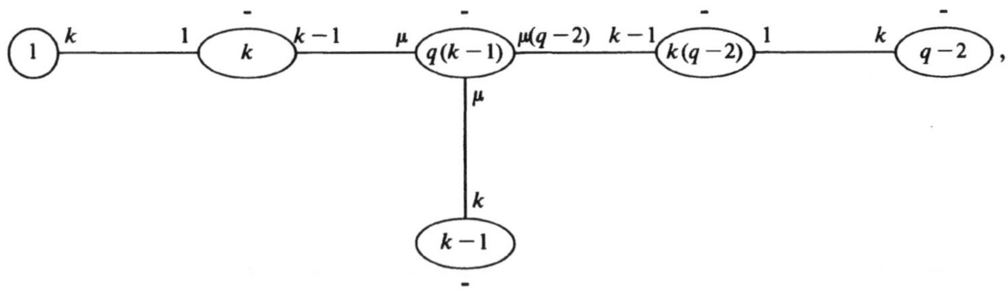

where $\mu = q^{d-1}$ and $k = q^d$, and is therefore not distance-regular. Equivalently, Γ' is the subgraph of the incidence graph of points and hyperplanes in $PG(d+1,q)$ obtained by removing a point p and all hyperplanes on p, and a hyperplane H and all points on H, where $p \notin H$. Another description of Γ' is obtained by taking as vertices the elements x^+, x^- with $x \in \mathbf{F}_q^{d+1} \setminus \{0\}$, and joining two vertices x^α, y^β whenever $\alpha \neq \beta$ and $\sum x_i y_i = 1$. Since $\mathrm{Aut}\,\Gamma'$ is generously transitive, Γ' defines a 5-class association scheme; cf. the remark after Theorem 2.11.6. For a characterization of this graph, and for the construction of other graphs with the same parameters, see Jungnickel & Vedder [451].

(iii) Let X be the set of points V, and let B be the set of affine hyperplanes of V not parallel to a fixed hyperplane of V. Then (X,B) is a symmetric (q,q^{d-1})-net, and its incidence graph Γ'' is a distance-transitive graph with $v = 2q^{d+1}$ vertices, intersection array $\{q^d, q^d - 1, q^d - q^{d-1}, 1;\ 1, q^{d-1}, q^d - 1, q^d\}$ and distance distribution diagram

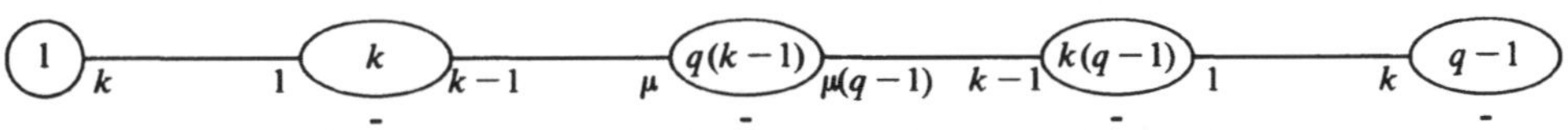

where $\mu = q^{d-1}$ and $k = q^d$. Equivalently, Γ'' is the subgraph of the incidence graph of points and hyperplanes in $PG(d+1,q)$ obtained by removing a point p and all hyperplanes on p, and a hyperplane H and all points on H, where $p \in H$. The graph Γ'' is a q-fold cover of the complete bipartite graph $K_{k,k}$. Another description of Γ'' is obtained by taking as vertices the elements x^+, x^- with $x = (x_0, \cdots, x_d) \in \mathbf{F}_q^{d+1}$, and joining two vertices x^α, y^β whenever $\alpha \neq \beta$ and $x_0 + y_0 = \sum_{i=1}^d x_i y_i$. For a characterization of this graph, see Jungnickel [448].

In the special case $d = 1$, it is easy to reconstruct $PG(2,q)$ from this graph, so that the number of isomorphism classes of distance-regular graphs with parameters $q.K_{q,q}$ equals the number of isomorphism classes of projective planes $PG(2,q)$ with a fixed flag (p,L) (Gardiner [311]); in particular, for $q \leqslant 8$, q a prime power, there is a unique distance-regular graph with intersection array $\{q, q-1, q-1, 1;\ 1, 1, q-1, q\}$, and that graph is distance-transitive.

(iv) Using Proposition 4.2.7 and Corollary 4.2.9, it is trivial to verify that Grassmann graphs of diameter $d \geqslant 2$ do not admit distance-regular antipodal covers. Hemmeter [371] shows that if a non-complete Grassmann graph is the halved graph of a distance-regular graph Γ, then Γ is the (uniquely determined) doubled Grassmann graph.

(v) Erdős-Ko-Rado theorems for Grassmann graphs and related graphs were given by Hsieh [407] and Frankl & Wilson [300].

Problem. If the graph discussed under (iii) is distance-transitive, must it be derived from the Desarguesian projective plane?

9.4 Dual polar graphs

In the sequel q, r are prime powers. Let V be one of the following spaces equipped with a specified form:

$[C_d(q)] = \mathbf{F}_q^{2d}$ with a nondegenerate symplectic form;
$[B_d(q)] = \mathbf{F}_q^{2d+1}$ with a nondegenerate quadratic form;
$[D_d(q)] = \mathbf{F}_q^{2d}$ with a nondegenerate quadratic form of (maximal) Witt index d;
$[{}^2D_{d+1}(q)] = \mathbf{F}_q^{2d+2}$ with a nondegenerate quadratic form of (non-maximal) Witt index d;
$[{}^2A_{2d}(r)] = \mathbf{F}_q^{2d+1}$ with a nondegenerate Hermitean form ($q = r^2$);
$[{}^2A_{2d-1}(r)] = \mathbf{F}_q^{2d}$ with a nondegenerate Hermitean form ($q = r^2$).

Background on these spaces and their forms can be found, e.g., in ARTIN [9], DIEUDONNÉ [264] and HUA & WAN [409]. The spaces $[C_d(q)]$, $[B_d(q)]$, $[D_d(q)]$, $[{}^2D_{d+1}(q)]$, $[{}^2A_{2d}(r)]$, $[{}^2A_{2d-1}(r)]$ are often named $Sp(2d,q)$, $\Omega(2d+1,q)$, $\Omega^+(2d,q)$, $\Omega^-(2d+2,q)$, $U(2d+1,r)$ and $U(2d,r)$, respectively.

A subspace of V is here called *isotropic* whenever the form vanishes completely on this subspace. (In standard terminology this is called totally isotropic, or, in certain cases, totally singular.) Maximal isotropic subspaces have dimension d (in other words, are d-spaces of V). The *dual polar graph* (*on* V) has as vertices the maximal isotropic subspaces; two points γ, δ are adjacent whenever $\dim \gamma \cap \delta = d-1$.

9.4.1. Lemma. *Let k be a natural number, and let e be $1, 1, 0, 2, \frac{3}{2}, \frac{1}{2}$ in the respective cases $[C_d(q)]$, $[B_d(q)]$, $[D_d(q)]$, $[{}^2D_{d+1}(q)]$, $[{}^2A_{2d}(r)]$, $[{}^2A_{2d-1}(r)]$. Then the number of isotropic k-spaces in V is*

$$\begin{bmatrix} d \\ k \end{bmatrix} \prod_{i=0}^{k-1} (q^{d+e-i-1}+1),$$

where, in the last two cases, $q = r^2$.

Proof. For $k = 1$, this is trivial in the symplectic case, and a straightforward computation in the orthogonal and Hermitean cases, see, e.g., ARTIN [9], p. 145-146 and HIRSCHFELD [392], p. 102. Let $k > 1$. The number of isotropic k-spaces of V containing a given isotropic $(k-1)$-space W is

$$\begin{bmatrix} d-k+1 \\ 1 \end{bmatrix} (q^{d+e-k}+1)$$

according to the formula for $k = 1$. By induction, there are $\begin{bmatrix} d \\ k-1 \end{bmatrix} \prod_{i=0}^{k-2} (q^{d+e-i-1}+1)$ such W. As each isotropic k-space contains $\begin{bmatrix} k \\ 1 \end{bmatrix}$ such W, the desired number is

$$\begin{bmatrix} d-k+1 \\ 1 \end{bmatrix} (q^{d+e-k}+1) \begin{bmatrix} d \\ k-1 \end{bmatrix} \prod_{i=0}^{k-2} (q^{d+e-i}+1) / \begin{bmatrix} k \\ 1 \end{bmatrix}.$$

This proves the lemma. □

(These numbers were determined by WAN [769] for the symplectic spaces, by SEGRE [651] (§30), WAN & YANG [771] and BOSE & CHAKRAVARTI [104] for the Hermitean spaces, and by PRIMROSE [599], SEGRE [650], RAY-CHAUDHURI [607], DAI & FENG

[239] and FENG & DAI [284] for the orthogonal spaces.) Later, we shall need a more technical result:

9.4.2. Lemma. *With notation as in the previous lemma, let U be a fixed isotropic m-space. The number of isotropic $(j+k+l)$-spaces W meeting U in a j-space, and $U^\perp$ in a $(j+l)$-space is*

$$q^{l(m-j)+k(2d-m-j-2l+e-1)-k(k-1)/2}\begin{bmatrix}m\\j\end{bmatrix}\begin{bmatrix}m-j\\k\end{bmatrix}\begin{bmatrix}d-m\\l\end{bmatrix}\prod_{i=0}^{l-1}(q^{d+e-m-i-1}+1).$$

Proof. This follows by straightforward computation, but let us present here a somewhat more geometrically oriented proof.

The intersection $Z = U \cap W$ can be chosen in $\begin{bmatrix}m\\j\end{bmatrix}$ ways. Given Z, we are led to consider the same problem in $Z^\perp / Z$ with $d' = d-j$, $m' = m-j$. Thus, we are reduced to the case $j = 0$.

In $U^\perp / U$ there are $\begin{bmatrix}d-m\\l\end{bmatrix}\prod_{i=0}^{l-1}(q^{d-m+e-i-1}+1)$ choices for $(U+(W \cap U^\perp))/U$, and for each such choice, there are q^{lm} possibilities for $Y = W \cap U^\perp$. Given Y, we must find in $Y^\perp / Y$ the number of isotropic k-spaces disjoint from $((Y+U)/Y$ and from) $(Y+U)^\perp / Y$, which is the same problem with $d'' = d-l$, $m'' = m$. Thus, we are reduced to the case $l = 0$.

The subspace $(W+U^\perp)/U^\perp$ can be chosen in $\begin{bmatrix}m\\k\end{bmatrix}$ ways in $V/U^\perp$. Remains to show that the number of choices for W in the $(n-m+k)$-space $W+U^\perp$ is $q^{k(2d-m+e-1)-k(k-1)/2}$. Denote this subspace by $X^\perp$, so that X is an $(m-k)$-space in U. Then we must first choose the subspace $(W+X)/X$ disjoint from $U^\perp / X$ inside $X^\perp / X$, and next choose W in the isotropic space $W+X$. The latter can clearly be done in $q^{k(m-k)}$ ways, and the former is the same problem but with $d''' = d-m+k$, $m''' = k$. Thus, we are reduced to the case $m = k$, and the claim is that the number of isotropic k-spaces disjoint from $U^\perp$, where U is an isotropic k-space, is $q^{k(2d+e-(3k+1)/2)}$.

Now first choose $(W+U)/U$ disjoint from $U^\perp / U$ inside V/U, which can be done in $q^{k(n-2k)}$ ways, then what is left is the number of isotropic k-spaces W, inside $W+U$, disjoint from the k-space U. But this is an affine dual polar space, and its cardinality is the number of symmetric, alternating or Hermitean matrices in the symplectic case (with $n = 2d$, $e = 1$), the orthogonal case (with $n = 2d+e$), and the Hermitean case (with $n = 2d+e-\frac{1}{2}$), respectively, cf. §9.5E. This completes the proof. □

Remark. This shows that polar spaces are partial geometric spaces in the sense of NEUMAIER [551] (i.e., S_n-designs as in NEUMAIER [552]). This, and similar detailed results are given in STANTON [698] (Prop. 4.10), HIRSCHFELD [394], DAI & FENG [239] and WAN, DAI, FENG & YANG [770].

9.4.3. Theorem. *Let Γ be a dual polar graph on a space V, with e as in the previous lemma and $d > 0$. Then Γ has the intersection array given by*

$$b_j = q^{j+e}\begin{bmatrix}d-j\\1\end{bmatrix}, \quad c_j = \begin{bmatrix}j\\1\end{bmatrix} \quad (0 \le j \le d) \tag{5}$$

and has singular line size $s+1 = q^e+1$. Consequently, Γ is a regular near $2d$-gon. Moreover, Γ has classical parameters as given in Table 6.1, (C3) and if $d > 1$ then $\mathrm{Aut}\,\Gamma \cong P\Sigma p(2d,q)$, $P\Gamma O(2d+1,q)$, $P\Gamma O^+(2d,q)$, $P\Gamma O^-(2d+2,q)$, $P\Gamma U(2d+1,r)$, $P\Gamma U(2d,r)$ *in the respective cases* $[C_d(q)]$, $[B_d(q)]$, $[D_d(q)]$, $[^2D_{d+1}(q)]$, $[^2A_{2d}(r)]$,

$[{}^2A_{2d-1}(r)]$, *except that in case* $[D_2(q)]$ *we have* $\mathrm{Aut}(\Gamma) \cong (\mathrm{Sym}(q+1)\times\mathrm{Sym}(q+1)).2$. *The eigenvalues and multiplicities are*

$$\theta_j = q^e\left[{d-j \atop 1}\right] - \left[{j \atop 1}\right],$$

$$f_j = q^j\left[{d \atop j}\right]\frac{1+q^{d+e-2j}}{1+q^{d+e-j}}\prod_{i=1}^{j}\frac{1+q^{d+e-i}}{1+q^{i-e}}\ .$$

Proof. Distance is as in Grassmann graphs: γ, δ are at distance j if and only if $\dim(\gamma \cap \delta) = d-j$. In view of Witt's theorem (cf. ARTIN [9], Theorem 3.9), Γ is distance-transitive.

Taking $\gamma, \delta \in \Gamma$ at distance j, we obtain c_j as the number of $(d-1)$-spaces in δ containing $\gamma \cap \delta$. Since to any such $(d-1)$-space U corresponds the maximal isotropic space $U+(U^\perp \cap \gamma)$ in $\Gamma_{j-1}(\gamma) \cap \Gamma(\delta)$, this yields $c_j = \left[{j \atop j-1}\right] = \left[{j \atop 1}\right]$.

We shall now compute b_j as $b(\gamma,\delta)$ for $\gamma, \delta \in \Gamma$ at distance j. Suppose $\zeta \in \Gamma(\gamma) \cap \Gamma_{j+1}(\delta)$. Then $\dim(\gamma \cap \zeta) = d-1$ and $\dim(\delta \cap \zeta) = d-j-1$, so $\dim(\gamma \cap \delta \cap \zeta) = d-j-1$. On the other hand, if U is a $(d-1)$-space in γ such that $U \cap \delta$ is a $(d-j-1)$-space (there are $\left[{d \atop 1}\right] - \left[{j \atop 1}\right]$ such spaces U), there are q^e maximal isotropic spaces ζ with $\zeta \cap \gamma = U$ as q^e+1 is the number of isotropic 1-spaces of V/U by Lemma 9.4.1. We claim that any such ζ satisfies $\dim(\zeta \cap \delta) = d-j-1$. For, if $z \in \zeta \cap \delta$, then $(\gamma \cap \zeta)+(\gamma \cap \delta)+\mathbf{F}_q z$ is an isotropic space containing γ, so $z \in \gamma$ by maximality of γ. Thus $\zeta \cap \delta = \gamma \cap \zeta \cap \delta = U \cap \delta$ is of dimension $d-j-1$ as claimed. It results that $\zeta \in \Gamma(\gamma) \cap \Gamma_{j+1}(\delta)$ if and only if $\zeta \cap \gamma$ is a $(d-1)$-space and $\zeta \cap \gamma \cap \delta$ is a $(d-j-1)$-space and that $b_j = |\Gamma(\gamma) \cap \Gamma_{j+1}(\delta)| = q^e(\left[{d \atop 1}\right] - \left[{j \atop 1}\right]) = q^{j+e}\left[{d-j \atop 1}\right]$.

Finally, for distinct collinear γ, δ, the singular line $\{\gamma,\delta\}^{\perp\perp}$ consists of the maximal isotropic subspaces containing $\gamma \cap \delta$. Therefore $s+1$ is the number of isotropic 1-spaces of $V/\gamma \cap \delta$, which is q^e+1 as we have seen before. Thus $s+1 = q^e+1$. Since $s+1 = \lambda+2$ we see (by Theorem 6.4.1) that Γ is a regular near $2d$-gon. By now, it is straightforward to verify that Γ has classical parameters $b = q$, $\alpha = 0$ and $\beta = q^e$. It remains to show that $\mathrm{Aut}\,\Gamma$ is as stated. Note that Γ determines the underlying polar space uniquely (cf. CAMERON [172]) so that $\mathrm{Aut}\,\Gamma$ is the full group of automorphisms of this polar space. The result is therefore a consequence of Theorem 8.6 in TITS [747]; see also CHOW [196] and DIEUDONNÉ [263]. □

Remark. $[{}^2A_{2d-1}(r)]$ has two distinct Q-polynomial structures, cf. BANNAI & ITO [33], p. 304, CHIHARA & STANTON [193], and Table 6.1, (C3). The intersection array, eigenvalues and multiplicities of the dual polar graphs were determined in STANTON [698], 5.3-5.5. (In fact Stanton also determined the irreducible constituents of the representations of the Chevalley groups on maximal parabolics other than those corresponding to an end node.)

A. Geometric characterization

In view of the above theorem, any dual polar space is a regular near polygon with classical parameters $(d,q,0,\lambda+1)$. This almost characterizes dual polar spaces.

9.4.4. Theorem. *Suppose that Γ is a distance-regular graph of diameter $d \geqslant 3$ with classical parameters $(d,b,0,\beta)$, where $\lambda \geqslant 1$. If Γ has singular lines of size $\lambda+2$, i.e., if Γ is a regular near 2d-gon, then Γ is a dual polar graph or a Hamming graph.*

Proof. Note that $b = \mu-1$ and $\beta = \lambda+1$. If Γ has singular lines of size $\lambda+2$, then by Theorem 6.4.1 it is a regular near $2d$-gon. Now our conclusion follows from the CAMERON [170] - SHULT & YANUSHKA [673] Theorem and the classification of polar spaces of rank $\geqslant 3$ (TITS [747]), if we show that, in the terminology of [673], Γ has 'quads', and that any point is 'classical' with respect to any 'quad'. But this follows easily from Yanushka's lemma (Proposition 2.5 in [673]) since $\lambda \geqslant 1$ and $\mu > 1$, and from the equalities $c_i = (\mu-1)c_{i-1}+1$ for $i \in \{1,2,\ldots,d\}$, see e.g. BROUWER & WILBRINK [138]. □

Remarks. (i) The dual polar graphs on $[B_d(q)]$ and $[C_d(q)]$ have the same intersection array, but are isomorphic only when q is even.
(ii) By BROUWER & COHEN [132] it suffices to assume only that the near polygon properties hold for points at distance at most 3, i.e., that all singular lines have size $\lambda+2 > 2$, and $a_i = \lambda c_i$, $c_i = (\mu-1)c_{i-1}+1 > 1$ for $i \leqslant 3$.

The preceding theorem covers all dual polar graphs of diameter $\geqslant 3$ with the exception of those with $\lambda = 0$, i.e., the (bipartite) dual polar graphs on $[D_d(q)]$. The following result characterizes these exceptions.

9.4.5. Theorem (CAMERON [172]). *Let $q > 1$ be an integer. Suppose Γ is a bipartite distance-regular graph of diameter $d > 3$ with $c_j = \left[{j \atop 1}\right]$ for $1 \leqslant j \leqslant d$. If, for any three distinct $\gamma_1,\gamma_2,\gamma_3 \in \Gamma$, the set $\{\gamma_1,\gamma_2,\gamma_3\}^{\perp}$ has size 0,1 or $q+1$, then q is a prime power and Γ is the dual polar graph on $[D_d(q)]$.* □

(For $d = 4$ this is Theorem 3.6 in CAMERON & DRAKE [177].)

9.4.6. Remark. The dual polar graph $[D_3(q)]$ is isomorphic to the incidence graph of the square 2-$(\frac{q^4-1}{q-1},\frac{q^3-1}{q-1},\frac{q^2-1}{q-1})$ design formed by the points and planes of $PG(3,q)$. Clearly, the incidence graph of any square 2-design with the same parameters (and there are many of those, cf. JUNGNICKEL [450]) has the same intersection array as the dual polar graphs on $[D_3(q)]$. The dual polar graph $[D_{m+1}(q)]$ has the same intersection array as the extended bipartite double of $[C_m(q)]$, but for odd q these graphs are nonisomorphic (cf. §9.4C).

B. Characterization by parameters

In one case the existence of singular lines of the right size has been proved from the parameters. (For $d = 2$ this is due to CAMERON, GOETHALS & SEIDEL [179].)

9.4.7. Theorem (IVANOV & SHPECTOROV [439]). *Suppose that Γ is a distance-regular graph with the same intersection array as the dual polar graph on $[{}^2A_{2d-1}(r)]$. Then Γ has singular lines of size $r+1$. If $d \geqslant 3$ it follows that Γ is isomorphic to this dual polar graph.*

Proof. Putting $m = (q^d-1)/(q-1)$ we have $k = rm$ and $\lambda+1 = r$. The smallest eigenvalue is $\theta = -m$ with multiplicity $f = k+1-m$ (cf. Corollary 8.4.2 and Theorem 8.4.3, applied with the 'second' Q-polynomial structure). Applying Theorem 4.4.4, we find that $\Gamma(\gamma)$ has eigenvalue λ with multiplicity at least m, so that $\Gamma(\gamma)$ must be a union of m cliques. Now apply Theorem 9.4.4. □

C. Related graphs

The only imprimitive graphs among the dual polar graphs are those on $[D_m(q)]$. Their halved graphs are the *half dual polar spaces*. In accordance with the notation used in the next chapter, we shall also denote these halved graphs by $D_{m,m}(q)$.

9.4.8. Theorem. *Let Γ be a half dual polar graph $D_{m,m}(q)$ obtained by halving the dual polar graph on $[D_m(q)]$. Then Γ is distance transitive with diameter $d = [m/2]$ and with intersection array given by*

$$\left.\begin{aligned} b_j &= q^{4j+1}\begin{bmatrix} m-2j \\ 2 \end{bmatrix} \\ c_j &= \begin{bmatrix} 2j \\ 2 \end{bmatrix} \end{aligned}\right\} \quad (0 \leqslant j \leqslant d). \tag{6}$$

Γ has classical parameters as in Table 6.1, (C4). If $m < 4$, then Γ is complete; otherwise, Γ has singular lines of size $s+1 = q+1$ and $\mathrm{Aut}\,\Gamma \cong P\Gamma\Omega^+(2m,q)$*, a group of index 2 in $P\Gamma O^+(2m,q)$, except when $m = 4$, in which case* $\mathrm{Aut}\,\Gamma \cong P\Gamma O^+(2m,q)$. *$\Gamma$ has eigenvalues and multiplicities given by*

$$\theta_j = q^{2j+1}\begin{bmatrix} m-2j \\ 2 \end{bmatrix} - \frac{q^{2j}-1}{q^2-1},$$

$$f_j = \gamma_j q^j \begin{bmatrix} m \\ j \end{bmatrix} \frac{1+q^{m-2j}}{1+q^{m-j}} \prod_{i=1}^{j} \frac{1+q^{m-i}}{1+q^i},$$

where $\gamma_j = 1$ for $2j < m$, and $\gamma_d = \frac{1}{2}$ if m is even.

Proof. Easy consequence of Theorem 9.4.3 and Proposition 4.2.2, apart from the statement on singular lines, which will follow from the general theory in Chapter 10. For $\mathrm{Aut}\,\Gamma$, cf. CHOW [196] and DIEUDONNÉ [263]. □

The following theorem is a geometric characterization of the half dual polar graphs. Note that the μ-graphs of halved dual polar graphs over $\mathbb{F}_q$ are polar spaces $D_3(q)$.

9.4.9. Theorem (COOPERSTEIN [219]). *Let Γ be a connected non-complete graph in which every singular line has at least three points. Suppose that for any $\gamma, \delta \in \Gamma$ at distance 2 the subgraph $\{\gamma,\delta\}^\perp$ is a nondegenerate polar space of rank 3, and that $\gamma^\perp \cap L^\perp$ is a maximal clique in $\{\gamma,\delta\}^\perp$ whenever L is a singular line on δ contained in $\Gamma_2(\gamma)$. If for any $k > 1$, any $\gamma \in \Gamma$ and any line L contained in $\Gamma_{k-1}(\gamma) \cup \Gamma_k(\gamma)$ we have either that $|L \cap \Gamma_{k-1}(\gamma)| = 1$, or that $L \subseteq \Gamma_k(\gamma)$ and $L^\perp \cap \Gamma_{k-1}(\gamma)$ is a nonempty clique, then Γ is a half dual polar graph.* □

The half dual polar graph $D_{m+1,m+1}(q)$ can be described as the distance 1-or-2 graph of the dual polar graph on $[B_m(q)]$. (Indeed, considering a nondegenerate hyperplane H in $[D_{m+1}(q)]$, we see that H carries a space $[B_m(q)]$, and each maximal totally isotropic subspace M of H (of dimension m) extends uniquely to a maximal totally isotropic subspace M' of the appropriate type in $[D_{m+1}(q)]$ (with dimension $m+1$), and

$\dim M_1' \cap M_2' = m-1$ if and only if $\dim M_1 \cap M_2 \in \{m-1, m-2\}$.)
USTIMENKO [760] observed that also the distance 1-or-2 graph of the dual polar graph on $[C_m(q)]$ is distance-regular. (Indeed, given a distance-regular graph Γ, it can be decided from the intersection array of Γ whether its distance 1-or-2 graph will be distance-regular again (cf. §4.2F). But the dual polar graphs on $[B_m(q)]$ and on $[C_m(q)]$ have the same parameters.) Let us call these graphs *Ustimenko graphs*.

9.4.10. Theorem. *Let Γ be the distance* 1-or-2 *graph of the dual polar graph Δ on $[C_m(q)]$, where q is odd. Then Γ is distance-regular with diameter $d = [m/2]$ and with intersection array given by* (6). *(Cf. Table 6.2, (C4a).) If $d > 1$, then* $\mathrm{Aut}\,\Gamma \cong \mathrm{Aut}\,\Delta \cong P\Sigma p(2m,q)$. *The singular lines have size $q+1$ or 2 (and both occur); in particular, Γ is not distance-transitive.*

Proof. Suppose $d > 1$, i.e., $m \geqslant 4$, and let q be odd. Let γ and δ be two adjacent points of Γ. If $d_\Delta(\gamma,\delta) = 1$, then the singular lines on γ and δ in Γ and Δ coincide, and have size $q+1$. If $d_\Delta(\gamma,\delta) = 2$, then the singular line on γ and δ in Γ is the hyperbolic line $\{\gamma,\delta\}^{\perp\perp}$ on γ and δ in the quad they determine, but this is just $\{\gamma,\delta\}$, since q is odd. Thus, in Γ we can retrieve the graph structure of Δ. Now everything is clear. □

The dual polar graph on $[D_{m+1}(q)]$ can be described as the extended bipartite double of the dual polar graph on $[B_m(q)]$. HEMMETER [373] observed that also the extended bipartite doubles of the dual polar graphs on $[C_m(q)]$ are distance-regular. (Indeed, this follows from Theorem 1.11.2 since $[B_m(q)]$ and $[C_m(q)]$ have the same parameters.) These *Hemmeter graphs* have the same parameters as the dual polar graphs on $[D_{m+1}(q)]$, and their halved graphs are the Ustimenko graphs; it follows that for odd q the Hemmeter graphs are nonisomorphic to the dual polar graphs on $[D_{m+1}(q)]$. From Hemmeter's results it easily follows that when $d \geqslant 4$ any distance-regular graph with a halved graph isomorphic to the distance 1-or-2 graph of $B_d(q)$ or $C_d(q)$ must be the corresponding extended double.

Remarks. (i) M.E. Muzichuk showed that there is no other way to obtain a distance-regular graph by merging classes in the association schemes of type $B_m(q)$ or $C_m(q)$, cf. MUZICHUK [549] and IVANOV, MUZICHUK & USTIMENKO [436].
(ii) Since dual polar spaces are near polygons, it follows from Corollary 4.2.9 that dual polar spaces of diameter $d \geqslant 2$ do not have distance-regular antipodal covers of diameter $2d+1$. Since dual polar spaces are near polygons with quads in which all point-quad relations are classical (see CAMERON [172], SHULT & YANUSHKA [673]), it follows from Proposition 4.2.7 that dual polar spaces of diameter $d \geqslant 3$ do not have distance-regular antipodal covers of diameter $2d$. Dual polar spaces of diameter $d = 2$ are generalized quadrangles, but we do not know precisely which ones have antipodal covers. Half dual polar graphs of diameter $d \geqslant 2$ do not have distance-regular antipodal covers.
(iii) HEMMETER [371] shows that if a dual polar graph of diameter d is the halved graph of a distance-regular graph Γ, then $d \leqslant 2$. Also, that if a half dual polar graph $D_{m,m}(q)$ is the halved graph of a distance-regular graph Γ, then either $m \leqslant 7$, or Γ is isomorphic to the dual polar graph on $[D_m(q)]$.
(iv) Erdős-Ko-Rado theorems for dual polar graphs have been given by STANTON [697].

9.5 Sesquilinear forms graphs

Let F be a field, and let $\mathfrak{M}_{m,n}(F)$ be the collection of all $m \times n$ matrices with entries in F. Since the rank of the sum of two matrices is not larger than the sum of the ranks of these matrices, it follows that $d(M,N) := \mathrm{rk}\,(N-M)$ defines a metric d on $\mathfrak{M}_{m,n}(F)$. We shall see that various subspaces of this metric space are distance-regular graphs (for $F = \mathbb{F}_q$). In §9.5E we discuss Hua's observation (see, for example, HUA [408]) that these spaces of matrices may be regarded as affine Grassmann spaces and affine dual polar spaces.

A. Bilinear forms graphs

Set $V = F^d$ and $W = F^e$. Let B be the vector space (of dimension de over F) of bilinear maps from $V \times W$ to F. (Then B is canonically isomorphic to $(V \otimes W)^*$, the vector space of linear maps from $V \otimes W$ to F.) The null space of f in V is defined as $\{v \in V \,|\, f(v,W) = 0\}$. Defining $f \sim g$ if and only if $\mathrm{rk}\,(f-g) = 1$ (where the rank $\mathrm{rk}\,(f)$ of a bilinear map f is the codimension of each of its null spaces (in V and W)), we get a graph Γ called the *bilinear forms graph (over F with dimensions d and e)*. In a bilinear forms graph, two vertices γ, δ have distance i if and only if $\mathrm{rk}\,(\gamma - \delta) = i$. Note that we might also consider B (and Γ) as the set of $d \times e$ matrices over F (where the distance of two matrices is the rank of their difference). In case $F = \mathbb{F}_q$, this graph is sometimes denoted $H_q(d,e)$, and called the 'q-analogue of the Hamming graph'.

A useful geometric description of the bilinear forms graphs is that of collinearity graphs of so-called *attenuated spaces* (or *affine Grassmann spaces*). Let V be a vector space of dimension $d+e$ over F and W a subspace of V of dimension d. Then the corresponding attenuated space is the collection of subspaces U of V with $U \cap W = 0$, where subspaces U of dimension e are called 'points', and those of dimension $e-1$ 'lines', and incidence is symmetrized inclusion. The identification with the above description of the bilinear forms graph follows by choosing a fixed 'point' U_0; now each 'point' U defines a map $f_U : U_0 \to W$ by $f_U(u_0) + u_0 \in U$ for $u_0 \in U_0$ (i.e., $\{f_U(u_0)\} = W \cap (U - u_0)$), and one easily verifies that $\dim\,(U \cap U') = e - r$ if and only if $\mathrm{rk}\,(f_U - f_{U'}) = r$. But clearly the space $\mathrm{Hom}\,(U_0, W)$ of maps from U_0 into W is isomorphic with $(U_0 \otimes W)^*$. Thus, the bilinear forms graph can be viewed as a subgraph of a Grassmann graph.

Let us write $P\Gamma L(n,F)_m$ for the stabilizer in $P\Gamma L(n,F)$ (in its natural action on $PG(n-1,F)$) of an m-space, and let $G \circ H$ denote a central product of groups G and H, i.e., the quotient of their direct product by a diagonal subgroup of its center.

9.5.1. Theorem (cf. DENG & LI [260]). *Let Γ be the bilinear forms graph over a field F with dimensions d and e. Then Γ is distance-transitive with diameter* $\min(d,e)$. *Assume that d, $e > 1$, and put*

$$G = P\Gamma L(d+e,F)_e \cong F^{de}\,.\,((GL(d,F) \circ GL(e,F))\mathrm{Aut}\,F).$$

If $d \neq e$ then $\mathrm{Aut}\,\Gamma \cong G$, *while if $d = e$ then* $\mathrm{Aut}\,\Gamma \cong G.2$.

Proof. It is clear from the description as attenuated space that the groups given act. (This follows also from the description in terms of matrices: translations of the vector space are automorphisms of the graph, as are left (right-) multiplications by invertible $d \times d$-matrices (resp. $e \times e$-matrices); any field automorphism applied to all coefficients of a matrix leads to a graph automorphism; finally, when $d = e$, we can transpose the matrices.) It easily follows that Γ is distance-transitive with diameter min(d,e). On the other hand, by considering the action of Aut Γ on the maximal cliques of Γ, one may conclude that automorphisms of Γ extend to automorphisms of the Grassmann graph $\left[{d+e \atop e}\right]$ in which Γ in embedded. Thus, Aut Γ cannot be larger. □

9.5.2. Theorem. *Let Γ be the bilinear forms graph over $\mathbf{F}_q$ with dimensions d and e (where $d \leqslant e$). Then Γ is distance-transitive with diameter d, $v = q^{de}$ vertices, and has intersection array given by*

$$\left.\begin{array}{rl} b_j = & q^{2j}(q-1)\left[{d-j \atop 1}\right]\left[{e-j \atop 1}\right] \\ c_j = & q^{j-1}\left[{j \atop 1}\right] \end{array}\right\} \text{ for } 0 \leqslant j \leqslant d.$$

If $d \geqslant 2$, then Γ has singular line size $s+1 = q$. Moreover, Γ has classical parameters as listed in Table 6.1, (C10).

Proof. Clearly, v is the cardinality of $\mathbf{F}_q^{de}$, whence $v = q^{de}$. Let us calculate the parameters k, b_j, c_j and s. The number of forms with distance j to the zero form is found by choosing null spaces of dimensions $d-j$ respectively $e-j$ in V, W and a nondegenerate bilinear form on a j-space, i.e.,

$$k_j = \left[{d \atop d-j}\right]\left[{e \atop e-j}\right](q^j-1)(q^j-q)\cdots(q^j-q^{j-1})$$

and in particular

$$k = \left[{d \atop 1}\right]\left[{e \atop 1}\right](q-1) = \frac{(q^d-1)(q^e-1)}{q-1}.$$

Given a form f of rank j, the number b_j of forms g with $g \sim f$ and rk $(g) = j+1$ equals the number of ways of choosing null spaces for g of codimension 1 in the null spaces of f times the number of ways of enlarging these null spaces to spaces of codimension 1 on which $f = g$ times the number of ways of defining g at a place where $f \neq g$:

$$\begin{aligned} b_j &= \left[{d-j \atop 1}\right]\left[{e-j \atop 1}\right]\left(\left[{j+1 \atop 1}\right]-\left[{j \atop 1}\right]\right)^2\cdot(q-1) \\ &= \frac{(q^d-q^j)(q^e-q^j)}{(q-1)}. \end{aligned}$$

Now, the value c_j is obtained from $c_j = \dfrac{k_{j-1}\cdot b_{j-1}}{k_j}$ (or from a similar argument as used for b_j).

Let us next determine the singular lines. Let f have rank 1. Then $\{0,f\}^\perp$ consists of the forms g of rank 1 that have the same null space as f in at least one of V, W. For $e \geqslant d \geqslant 2$ it follows that $\{0,f\}^{\perp\perp} = \mathbf{F}_q\cdot f$ and we find $s = q-1$.

Finally a straightforward computation shows that Γ has classical parameters as given. □

Remarks. (i) DELSARTE [255] computes the eigenvalues and studies codes and designs in this association scheme.

(ii) Geometric characterizations can be found in HUANG [410, 412]; the case $e = 2$ was treated by THAS & DECLERCK [734]; the case $e = 3$ by RAY-CHAUDHURI & SPRAGUE [609]. In fact, HUANG [412] proves that $H_q(d,e)$ is characterized by its parameters, together with the 'weak 4-vertex condition' (which states that the number of edges in $\gamma^\perp \cap \delta^\perp$ only depends on $d(\gamma,\delta)$), provided $q \geqslant 4$, $e \geqslant 2d \geqslant 6$. CUYPERS [230] has shown that the same conclusion holds if one either drops the weak 4-vertex condition, or replaces the inequalities by $q \geqslant 2$ and $e \geqslant 2d+2 \geqslant 8$.
(iii) Using the geometric description of attenuated spaces, it is not difficult to verify that $C(\gamma,\delta)\setminus\{\gamma,\delta\}$ is connected for all $\gamma,\delta \in \Gamma$, so that Γ does not have antipodal covers of even diameter by Proposition 4.2.7; similarly, Γ does not have antipodal covers of odd diameter by Corollary 4.2.9 (J.T.M. van Bon). HEMMETER [371] shows that non-complete bilinear forms graphs are never halved graphs of bipartite distance-regular graphs.
(iv) An Erdős-Ko-Rado theorem for the bilinear forms graph was given by HUANG [411].

B. Alternating forms graphs

Set $V = F^n$ and let A be the $n(n-1)/2$-dimensional vector space of (bilinear) alternating forms on V. Thus $f \in A$ if and only if f is a bilinear form on V and $f(x,x) = 0$ for all $x \in V$. We define $\mathrm{rk}\,(f) = \dim\,(V/\mathrm{Rad}\,f)$, where $\mathrm{Rad}\,f = \{x \in V \mid f(x,y) = 0 \text{ for all } y \in V\}$. Note that $\mathrm{rk}\,(f)$ takes even values only for $f \in A$. If $f \in A$ and U is a subspace of V, then $f|_U$ denotes the form induced on U by f.
The *alternating forms graph* Γ *on* V is defined on the points of A by $\gamma \in \Gamma(\delta)$ for $\gamma,\delta \in A$ whenever $\mathrm{rk}\,(\gamma-\delta) = 2$; thus two vertices γ, δ have distance i if and only if $\mathrm{rk}\,(\gamma-\delta) = 2i$.

We could also have defined A (and Γ) as the set of all skew symmetric $n \times n$ matrices over F with zero diagonal (where two matrices are adjacent whenever their difference has rank 2).

9.5.3. Theorem (LIU [505]). *Let Γ be the alternating forms graph on the vector space F^n. Then Γ is distance-transitive, with diameter $d = [n/2]$. Assume that $n \geqslant 4$, and let*

$$G \cong F^{n(n-1)/2}.(F^*.P\Gamma L(n,F)) \cong F^{n(n-1)/2}.((\Gamma L(n,F)/\{\pm 1\}) \times (F^*/(F^*)^2))$$

be the group described below. If $n > 4$, then $\mathrm{Aut}\,\Gamma \cong G$, *while if $n = 4$ then* $\mathrm{Aut}\,\Gamma \cong G.2$. *For $n \leqslant 3$ the graph Γ is complete, and* $\mathrm{Aut}\,\Gamma$ *is a symmetric group.*

Proof. Set $V = F^n$. First of all, notice that $\gamma \in \Gamma_j(\delta)$ for $\gamma,\delta \in \Gamma$ if and only if $\mathrm{rk}\,(\gamma-\delta) = 2j$. Thus, Γ has diameter d. Viewing the vertices of Γ as $n \times n$ matrices over F, we see that Γ has a group of automorphisms G acting according to

$$M \mapsto cA^{\mathsf{T}}M^{\sigma}A + N$$

($M, N \in \Gamma$, $A \in GL(V)$, $c \in F^*$, $\sigma \in \mathrm{Aut}\,F$), and it is easy to see that G acts distance-transitively on Γ. For $n = 4$ there is an additional automorphism that can be given explicitly as the automorphism interchanging the 14- and the 23-elements (and of course the 41- and the 32-elements) of the matrices $M \in \Gamma$; viewing Γ as a subgraph of the half dual polar graph $D_{4,4}(F)$ (cf. Proposition 9.5.11 (ii)), we see this automorphism as a graph automorphism of the Dynkin diagram. As Liu shows, this describes the full automorphism group of Γ. □

9.5.4. Lemma. *Let γ, δ be bilinear forms on V. Then:*

(i) $\mathrm{Rad}\,\gamma \cap \mathrm{Rad}\,\delta = \mathrm{Rad}\,\gamma \cap \mathrm{Rad}\,(\gamma-\delta)$.

(ii) $\mathrm{Rad}\,\gamma + \mathrm{Rad}\,\delta = V$ *implies* $\mathrm{Rad}\,\gamma \cap \mathrm{Rad}\,\delta = \mathrm{Rad}\,(\gamma-\delta)$.

Proof. (i) is straightforward.
(ii) Let $x \in \mathrm{Rad}\,(\gamma-\delta)$, $x = c+d$, $c \in \mathrm{Rad}\,\gamma$, $d \in \mathrm{Rad}\,\delta$. Then $\gamma(d,u) = \delta(c,u)$ for all u. But this expression is linear in u and zero on $\mathrm{Rad}\,\gamma + \mathrm{Rad}\,\delta$, i.e., identically zero. So $x, c, d \in \mathrm{Rad}\,\gamma \cap \mathrm{Rad}\,\delta$. □

9.5.5. Lemma. *Let γ, δ be bilinear forms on V with $\mathrm{rk}\,(\gamma) = c$ and $\mathrm{rk}\,(\delta) = d$. Let W be a complement of Rad γ in V and write $U = \mathrm{Rad}\,(\gamma-\delta) \cap W$. Then $c-d \leqslant \mathrm{rk}\,(\gamma-\delta) \leqslant c+d$. Moreover:*

(i) $\mathrm{rk}\,(\gamma-\delta) = c+d$ *if and only if* $V = \mathrm{Rad}\,\gamma + \mathrm{Rad}\,\delta$ *(and* $\mathrm{Rad}\,(\gamma-\delta) = \mathrm{Rad}\,\gamma \cap \mathrm{Rad}\,\delta$*).*

(ii) $\mathrm{rk}\,(\gamma-\delta) = c-d$ *if and only if* $\mathrm{Rad}\,\gamma \subseteq \mathrm{Rad}\,\delta$ *and* $\dim U = d$ *(and* $\gamma|_U$ *is nondegenerate). In this case* δ *is uniquely determined by* $\delta|_U = \gamma|_U$ *and* $\mathrm{Rad}\,\delta = \{v \in V \mid \gamma(v,U) = 0\}$.

Proof. (i) Suppose $\mathrm{rk}\,(\gamma-\delta) = c+d$. Since

$$\dim\,(\mathrm{Rad}\,\gamma \cap \mathrm{Rad}\,\delta) \leqslant \dim\,\mathrm{Rad}\,(\gamma-\delta) = n-c-d\,,$$

we have

$$\dim\,(\mathrm{Rad}\,\gamma + \mathrm{Rad}\,\delta) = (n-c)+(n-d)-\dim\,(\mathrm{Rad}\,\gamma \cap \mathrm{Rad}\,\delta) \geqslant n$$

so that $V = \mathrm{Rad}\,\gamma + \mathrm{Rad}\,\delta$. Conversely, if $V = \mathrm{Rad}\,\gamma + \mathrm{Rad}\,\delta$, then using Lemma 9.5.4 (ii) we find

$$\dim\,\mathrm{Rad}\,(\gamma-\delta) = (n-c)+(n-d)-n = n-c-d$$

and $\mathrm{rk}\,(\gamma-\delta) = c+d$.
(ii) Suppose $\mathrm{rk}\,(\gamma-\delta) = c-d$. Applying (i) (with $\gamma-\delta$, $-\delta$, γ instead of γ, δ, $\gamma-\delta$) we find $\mathrm{Rad}\,\gamma = \mathrm{Rad}\,\delta \cap \mathrm{Rad}\,(\gamma-\delta)$. Moreover, $\dim U = \dim\,(\mathrm{Rad}\,(\gamma-\delta) \cap W) = \dim\,(\mathrm{Rad}\,(\gamma-\delta)\,/\,\mathrm{Rad}\,(\gamma)) = d$. Since $\delta|_U$ is nondegenerate and $\gamma|_U = \delta|_U$ we have $\mathrm{rk}\,(\gamma|_U) = d$. Conversely, if $\mathrm{Rad}\,\gamma \subseteq \mathrm{Rad}\,\delta$ and $\dim U = d$ we have $\mathrm{Rad}\,\gamma = \mathrm{Rad}\,\gamma \cap \mathrm{Rad}\,\delta = \mathrm{Rad}\,(\gamma-\delta) \cap \mathrm{Rad}\,\gamma$ so that $\mathrm{Rad}\,\gamma \subseteq \mathrm{Rad}\,(\gamma-\delta)$. Now $\dim\,\mathrm{Rad}\,(\gamma-\delta) = \dim\,\mathrm{Rad}\,\gamma + \dim U = n-c+d$, so that $\mathrm{rk}\,(\gamma-\delta) = c-d$. Let us check $\mathrm{Rad}\,\delta = \{v \in V \mid \gamma(v,U) = 0\}$. Since the dimensions are equal, we only have to show one inclusion. Let $\gamma(v,U) = 0$. Since, by (i), $V = \mathrm{Rad}\,(\gamma-\delta) + \mathrm{Rad}\,\delta$, we have $\delta(v,V) = \delta(v,\mathrm{Rad}\,(\gamma-\delta)) = \gamma(v,\mathrm{Rad}\,(\gamma-\delta)) = \gamma(v,U+\mathrm{Rad}\,\gamma) = 0$ as desired. □

9.5.6. Theorem. *Let Γ be the alternating forms graph on the vector space $\mathbf{F}_q^n$. Then Γ is distance-transitive with diameter $d = [n/2]$, $v = q^{n(n-1)/2}$ vertices, and intersection array given by*

$$\left.\begin{aligned} b_j &= q^{4j}(q^{n-2j}-1)(q^{n-2j-1}-1)\,/\,(q^2-1) \\ c_j &= q^{2j-2}(q^{2j}-1)\,/\,(q^2-1) \end{aligned}\right\} \quad \textit{for } 0 \leqslant j \leqslant d,$$

and has singular line size $s+1 = q$. Moreover, Γ has classical parameters as listed in

Table 6.1, (C11), and admits a group of automorphisms of the form

$$\mathbb{F}_q^{n(n-1)/2}.((\Gamma L(n,q)/\{\pm 1\})\times 2).$$

Proof. Since Γ is distance-transitive, we may compute b_j, c_j, s as $b(0,\delta), c(0,\delta), s(0,\delta)$ for a single appropriately chosen $\delta \in \Gamma$.

Computation of b_j. Let $\gamma \in \Gamma_j(0)$. By Lemma 9.5.5 (i), any $\delta \in \Gamma_1(0) \cap \Gamma_{j+1}(\gamma)$ leads to an $(n-2)$-space Rad δ of V intersecting Rad γ in an $(n-2j-2)$-space. Conversely, if U is an $(n-2)$-space of V such that $U \cap$ Rad γ is an $(n-2j-2)$-space, then there are $q-1$ forms $\delta \in \Gamma_1(0)$ with $U =$ Rad δ. By Lemma 9.5.5 (ii), the number of such $(n-2)$-spaces is $q^{4j}\left[{n-2j \atop n-2j-2}\right]$, whence we get $b_j = q^{4j}(q^{n-2j}-1)(q^{n-2j-1}-1)/(q^2-1)$ as wanted.

Computation of c_j. Let $\gamma \in \Gamma_j(0)$, and let W be a complement of Rad γ in V. By Lemma 9.5.5 (ii) each $\delta \in \Gamma_1(0) \cap \Gamma_{j-1}(\gamma)$ determines a 2-space U of W. On the other hand, if U is a 2-space of W, then $\delta \in \Gamma$ given by $\delta|_U = \gamma|_U$ and Rad $\delta = \{v \in V \mid \gamma(v,U) = 0\}$ is the unique form in $\Gamma_{j-1}(\gamma) \cap \Gamma_1(0)$ with $\delta|_U = \gamma|_U$. Thus c_j is the number of nonisotropic 2-spaces of W with respect to a nondegenerate alternating form, and since dim $W = 2j$, this number is

$$\left[{2j \atop 2}\right] - \left[{2j \atop 1}\right]\left[{2j-2 \atop 1}\right] / \left[{2 \atop 1}\right] = q^{2j-2}(q^{2j}-1)/(q^2-1).$$

By now it is straightforward to show that Γ has classical parameters.

Finally, we compute s. Let γ be the form in Γ given by $\gamma(x,y) = x_1y_2 - x_2y_1$ for $x = (x_i)_{1 \le i \le n}$, $y = (y_i)_{1 \le i \le n}$ in V. Now

$$\{0,\gamma\}^\perp = \{\delta \in \Gamma_1(0) \mid \dim(\text{Rad } \delta \cap \text{Rad } \gamma) \geq n-3\},$$

so that for any $\delta \in \{0,\gamma\}^\perp \setminus \mathbb{F}_q\gamma$, we have $\delta^\perp \not\subseteq \{0,\gamma\}^\perp$. Consequently, $\{0,\gamma\}^{\perp\perp} \subseteq \mathbb{F}_q\gamma$. But the reverse inclusion clearly holds, too, whence $s = q-1$. □

Remarks. (i) The k_j were found by CARLITZ [185]. We have

$$k_j = q^{j(j-1)}\prod_{i=0}^{2j-1}(q^{n-i}-1) \Big/ \prod_{i=1}^{j}(q^{2i}-1).$$

DELSARTE & GOETHALS [256] compute the eigenvalues and study codes in this association scheme.

(ii) For $n = 4$, the condition rk $A \leq 2$ for an alternating matrix A is equivalent to $a_{12}a_{34} + a_{14}a_{23} + a_{13}a_{42} = 0$, and the alternating forms graph is the (strongly regular) graph with vertex set $\mathbb{F}_q^6$ where two vertices are adjacent when the line joining them hits the hyperplane at infinity in a point of a fixed hyperbolic quadric. In the special case $n = 4$, $q = 2$ we find the complement of the folded halved 8-cube.

(iii) For a discussion of the possible antipodal covers, see §9.5F. HEMMETER [371] shows that non-complete alternating forms graphs are never halved graphs of bipartite distance-regular graphs.

C. Hermitean forms graphs

Set $V = \mathbb{F}_q^n$, where $q = r^2$ with r a prime power, and let H stand for the n^2-dimensional vector space over $\mathbb{F}_r$ of the Hermitean forms on V. Thus $f \in H$ if and only if $f(x,y)$ is linear in y, and $f(y,x) = \overline{f(x,y)}$ for all $x,y \in V$.
The *Hermitean forms graph on V* is the graph whose vertices are the members of H and in which $\gamma, \delta \in H$ are adjacent whenever $\mathrm{rk}\,(\gamma - \delta) = 1$. Here, $\mathrm{rk}\,(\gamma)$ and $\mathrm{Rad}\,\gamma$ have the same meaning as in the previous section. As before we have $d(\gamma,\delta) = \mathrm{rk}\,(\gamma - \delta)$. Note that we could also have defined H as the set of Hermitean $n \times n$ matrices over $\mathbb{F}_q$.

9.5.7. Theorem. *Let Γ be the Hermitean forms graph on the vector space $\mathbb{F}_q^n$, where $q = r^2$ with r a prime power. Then Γ has diameter $d = n$, $v = r^{n^2}$ vertices, is distance-transitive, has intersection array given by*

$$\left.\begin{aligned} b_j &= (q^n - q^j)/(r+1) \\ c_j &= r^{j-1}(r^j - (-1)^j)/(r+1) \end{aligned}\right\} \text{ for } 0 \leqslant j \leqslant d, \tag{7}$$

and has singular line size $s + 1 = r$. Moreover, Γ has classical parameters as listed in Table 6.1, (C12), and admits a group of automorphisms of the form

$$\mathbb{F}_r^{n^2} \cdot (\Gamma L(n,q)/K),$$

where

$$K = \{x \in \mathbb{F}_q \,|\, x^{r+1} = 1\}.$$

Proof. All arguments run parallel to those of the previous section. □

9.5.8. Theorem (IVANOV & SHPEKTOROV [437]). *Let Γ be a distance-regular graph with parameters as in (7), with $q = 4$, $r = 2$. Then Γ is the Hermitean forms graph on $\mathbb{F}_q^n$.* □

Remarks. (i) The Hermitean forms graph on $\mathbb{F}_4^2$ is isomorphic to the folded 5-cube. The Hermitean forms graph on $\mathbb{F}_9^2$ is isomorphic to the coset graph of the truncated ternary Golay code, cf. §11.3C. The Hermitean forms graph on $\mathbb{F}_4^3$ is isomorphic to the coset graph of the doubly truncated binary Golay code, cf. §11.3G.
(ii) For a discussion of the possible antipodal covers, see §9.5F. HEMMETER [371] shows that non-complete Hermitean forms graphs are never halved graphs of bipartite distance-regular graphs.

D. Symmetric bilinear forms graphs

Set $V = \mathbb{F}_q^n$ and let S be the $n(n+1)/2$-dimensional vector space of symmetric bilinear forms on V. We define $\mathrm{rk}\,(f) = \dim\,(V/\mathrm{Rad}\,f)$, where $\mathrm{Rad}\,(f) = \{x \in V \,|\, f(x,y) = 0 \text{ for all } y \in V\}$.
The *symmetric bilinear forms graph Γ on V* is defined on the elements of S by $\gamma \sim \delta$ whenever $\mathrm{rk}\,(\gamma - \delta) = 1$. We could also have defined S (and Γ) as the set of all symmetric $n \times n$ matrices over $\mathbb{F}_q$ (where two matrices are adjacent whenever their difference has rank 1).
The graph Γ has $v = q^{n(n+1)/2}$ vertices, is regular of valency $k = q^n - 1$ and edge-regular with $\lambda = q - 2$; it has singular lines of size q. (Indeed, if each of f, g and $f - g$ has rank 1, then $\mathrm{Rad}\,f = \mathrm{Rad}\,g$, so that the singular line $\{0,f\}^{\perp\perp}$ consists of all multiples of f. The value of k can be seen from the following lemma.)

9.5.9. Lemma. *The number N_i of symmetric bilinear forms of rank i on V is given by $N_{2i-1} = q^{i(i-1)}\left[{n \atop 2i-1}\right]\Pi_i$ and $N_{2i} = q^{i(i+1)}\left[{n \atop 2i}\right]\Pi_i$ (so that $N_{2i-1}+N_{2i} = q^{i(i-1)}\left[{n+1 \atop 2i}\right]\Pi_i$), where $\Pi_i = \prod_{j=1}^{i}(q^{2j-1}-1)$.* □

If f is a symmetric bilinear form of rank i, and $f(a,a) \neq 0$, then $g(x,y) = f(a,x)f(a,y)/f(a,a)$ defines a symmetric bilinear form of rank 1 such that $f-g$ has rank $i-1$. Thus, if q is odd then $d_\Gamma(f,g) = \mathrm{rk}\,(f-g)$, and in particular $k_i = N_i$. (It follows that Γ is not distance-regular in this case: the average value of μ is $kb_1/k_2 = q-q^{-1}$ which is not integral.)
On the other hand, when q is even, alternating forms of rank i do not have neighbours of rank $i-1$ (for, if $f(x,x) = 0$ for all x, and $a \in \mathrm{Rad}\,(f-g) \setminus \mathrm{Rad}\,(f)$, and $f(a,b) \neq 0$, then $g\,|_{\langle a,b\rangle}$ has rank 2).
For $n = 2$ we find a distance-regular graph with intersection array $\{q^2-1, q^2-q, 1; 1, q, q^2-1\}$, isomorphic to the one constructed in Proposition 12.5.1 for $n = 2$; for $n > 2$ the graph Γ is not distance-regular. (In fact, for any q, let Q_f be the quadratic form defined on $V/\mathrm{Rad}\,f$ by $Q_f(x+\mathrm{Rad}\,f) = f(x,x)$, and let n_f be the number of projective points on the quadric Q_f; if $\mathrm{rk}\,f = i$ and f is not alternating then $c_i(0,f) = \left[{i \atop 1}\right] - n_f$. When q is odd, any nondegenerate quadric is of the form Q_f for some f, and c_i takes two values for even i, while $c_i = q^{i-1}$ for odd i. When q is even, Q_f is the square of a linear form and $c_i(0,f) = q^{i-1}$ when f is not alternating.) Diagram for $n = 3$ and q even:

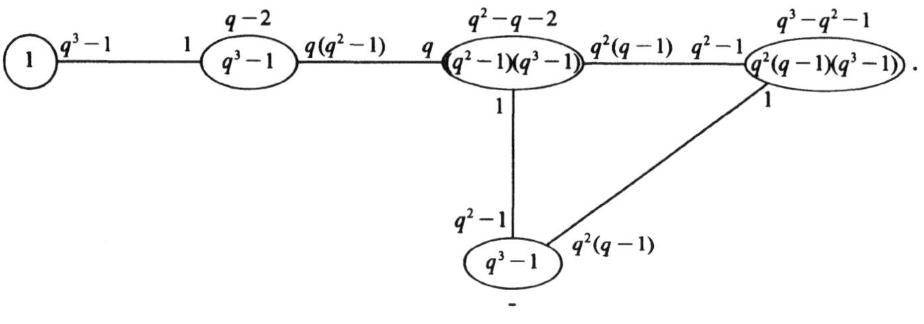

E. Affine subspaces of dual polar spaces

Let Γ be a dual polar graph of diameter d and $\gamma \in \Gamma$. Then often the graph $\Delta = \Gamma_d(\gamma)$ is a forms graph.

9.5.10. Proposition. *Let Γ be the collinearity graph of the dual polar space $[{}^2A_{2d-1}(r)]$ or $[C_d(q)]$ and $\gamma \in \Gamma$. Then the subgraph $\Gamma_d(\gamma)$ is isomorphic to the Hermitean forms graph or the symmetric bilinear forms graph on $\mathbb{F}_q^d$, respectively.*

Proof. Choose the form f such that

$$f(x,y) = x^* \begin{bmatrix} 0 & I \\ -I & 0 \end{bmatrix} y$$

where either f is symplectic and x^* denotes $x^\top$, or f is Hermitean and x^* denotes $\overline{x}^\top$. Let E, F be two complementary maximal totally isotropic subspaces and choose a basis that is a union of bases for E and F. Then the group of linear transformations consisting of the matrices $\begin{bmatrix} I & A \\ 0 & I \end{bmatrix}$ with $A = A^*$ preserves the form, fixes E pointwise, and acts regularly on the maximal totally isotropic subspaces disjoint from E. The map

$$A \mapsto \begin{bmatrix} I & A \\ 0 & I \end{bmatrix} F$$

is an isomorphism from the forms graph to the subgraph $\Gamma_d(E)$ of the dual polar graph. □

9.5.11. Proposition. (i) *Let Γ be the collinearity graph of the dual polar space $[D_d(q)]$, $\gamma \in \Gamma$ and $X = \Gamma_d(\gamma)$. Then X is a coclique in Γ, but the graph induced on X by Γ_2 is isomorphic to the alternating forms graph on $\mathbb{F}_q^d$.*
(ii) *Let Γ be the half dual polar graph $D_{2d,2d}(q)$ and $\gamma \in \Gamma$. Then the subgraph $\Gamma_d(\gamma)$ is isomorphic to the alternating forms graph on $\mathbb{F}_q^{2d}$.*

Proof. The proof is entirely analogous, but this time we choose the form given by the matrix $\begin{bmatrix} 0 & I \\ I & 0 \end{bmatrix}$. Part (ii) follows immediately from part (i). □

These same identifications can also be seen in a more geometric way. Let $d \geqslant 2$. In the dual polar spaces $[^2A_{2d-1}(r)]$, $[C_d(q)]$, $[D_d(q)]$ we have maximal totally isotropic subspaces of dimension d in a vector space of dimension $2d$, so any two disjoint such subspaces span the entire space. (We shall use geometric language, so call subspaces U, V disjoint when they have no projective point in common, i.e., when $U \cap V = 0$ - this cannot cause confusion.)
Now fix a maximal totally isotropic subspace U, and let V, W be maximal totally isotropic subspaces disjoint from U. The we can give V the structure of a polar space by defining for subspaces Z of V:

$$Z^{\perp_W} = ((Z^{\perp} \cap W) + U) \cap V.$$

In this way each maximal totally isotropic subspace W disjoint from U gives rise to a polarity $\perp_W$ of V, and $\perp_W$ will be nondegenerate precisely when $V \cap W = 0$. (In fact, Rad $\perp_W = V \cap W$.) Each polarity $\perp_W$ distinct from the totally degenerate $\perp_V$ is determined by precisely $q-1$ subspaces W (and these do not have common points outside V).

The idea of taking all objects of given type in some geometry of Lie type as far away as possible from a given object, works in other cases, too. For example, if Γ is the graph $E_{7,7}(q)$ (cf. §10.7), and $\gamma \in \Gamma$, then $\Gamma_3(\gamma)$ is the affine E_6 graph on q^{27} vertices discussed in §10.8.
Similarly, the bilinear forms graph is obtained by taking the subgraph of the Grassmann graph $\begin{bmatrix} n \\ k \end{bmatrix}$ consisting of all k-spaces disjoint from a given $(n-k)$-space (cf. the discussion of attenuated spaces above).

The above two propositions were pointed out to us by Kantor. He also asked whether the quadratic forms graphs (see §9.6) are the subgraphs at largest distance from a given point in the Ustimenko graphs, and sometimes this is indeed the case. Let Γ be the symplectic dual polar graph on $[C_m(q)]$, and, for a fixed $\gamma \in \Gamma$, let Δ be the subgraph $\Gamma_m(\gamma)$. When q is odd, $\Gamma_1 \cup \Gamma_2$ induces, on the point set of Δ, the same graph as $\Delta_1 \cup \Delta_2$, and this graph is isomorphic to the quadratic forms graph on $\mathbb{F}_q^m$. Thus, for odd q the quadratic forms graph is a subgraph of the Ustimenko graph, and when both q and m are odd it is a subgraph at largest distance from some point in the Ustimenko graph.

F. Antipodal covers

As an application of the ideas of the previous section, we show that most forms graphs do not admit distance-regular antipodal covers.

9.5.12. Proposition (VAN BON [pers. comm.]). *The alternating forms graphs Γ on $\mathbb{F}_q^n$ with $n \geqslant 4$ do not have distance-regular antipodal covers of even diameter.*

Proof. Note that Γ has diameter $d = [n/2]$. By Proposition 4.2.7 it suffices to show that $C(\gamma,\delta)\setminus\{\gamma,\delta\}$ is connected when $d_\Gamma(\gamma,\delta) = d$. But for arbitrary α, β the graph $C(\alpha,\beta)$ is determined up to isomorphism by $d_\Gamma(\alpha,\beta)$, independent of n, and hence we may suppose $n = 2d$. Let Δ be the collinearity graph of the dual polar space $[D_n(q)]$, then by Proposition 9.5.11 Γ is isomorphic to the graph induced by Δ_2 on $\Delta_n(\infty)$ for some fixed vertex $\infty \in \Delta$. Let ∞, γ, δ correspond to maximal totally isotropic subspaces U, V, W, respectively, then U, V, W are pairwise disjoint. The vertices of $\Gamma_{d-i}(\gamma) \cap \Gamma_i(\delta)$ correspond to maximal totally isotropic subspaces Z with $U \cap Z = 0$, $\dim V \cap Z = 2i$, $\dim W \cap Z = n-2i$. But $Y = V \cap Z$ determines $W \cap Z = W \cap Y^\perp$ and $Z = Y + (W \cap Z)$, and $Z \cap U = 0$ precisely when $Y \cap Y^{\perp_W} = 0$. Thus we find the following description of $C(\gamma,\delta)$:

> *Let γ, δ be two vertices at distance j in an alternating forms graph Γ. Then $C(\gamma,\delta)$ is isomorphic to the graph with as vertices the nondegenerate subspaces Y of a $2j$-space V provided with a nondegenerate symplectic form, where two subspaces Y_1, Y_2 are adjacent when* $\dim(Y_1+Y_2)/(Y_1 \cap Y_2) = 2$. *The vertices of $\Gamma_{j-i}(\gamma) \cap \Gamma_i(\delta)$ correspond to the subspaces Y with* $\dim Y = 2i$.

It follows immediately that $C(\gamma,\delta)\setminus\{\gamma,\delta\}$ is connected; even $\Gamma_{d-1}(\gamma) \cap \Gamma(\delta)$ is connected. □

9.5.13. Proposition. *The alternating forms graphs Γ on $\mathbb{F}_q^n$ with $n \geqslant 4$ do not have distance-regular antipodal covers of odd diameter.*

Proof. This will follow from Proposition 4.2.8 if we show that if $\delta, \epsilon \in \Gamma_d(\gamma)$, $\delta \sim \epsilon$, then $\Gamma(\gamma) \cap \Gamma_{d-1}(\delta) \cap \Gamma_{d-1}(\epsilon) \neq \emptyset$. Translating this to the polar space situation, we see that if n is even, then the size of this set is the number of hyperbolic lines meeting an $(n-2)$-subspace of an n-space with a nondegenerate symplectic form in at least one point. But this is always nonzero. If n is odd, then we may suppose that $\gamma \cap \delta \cap \epsilon = \emptyset$ - otherwise we can divide out this intersection and reduce to the case n even. But now $|\Gamma(\gamma) \cap \Gamma_{d-1}(\delta) \cap \Gamma_{d-1}(\epsilon)|$ is the number of hyperbolic lines in an $(n-2)$-subspace of an n-space with a symplectic form with radical of dimension 1. Also this is nonzero. □

9.5.14. Proposition. *The Hermitean forms graphs Γ of diameter $d \geqslant 3$ do not have distance-regular antipodal covers of diameter $2d$, except in case $d = 3$, $q = 4$, where unique antipodal 2- and 4-covers of diameter 6 exist.*

Proof. Similarly to the proof of Proposition 9.5.12 we find here:

Let γ, δ be two vertices at distance j in a Hermitean forms graph Γ. Then $C(\gamma,\delta)$ is isomorphic to the graph with as vertices the nondegenerate subspaces Y of a j-space V provided with a nondegenerate Hermitean form, where two subspaces Y_1, Y_2 are adjacent when $\dim (Y_1 + Y_2)/(Y_1 \cap Y_2) = 1$. *The vertices of $\Gamma_{j-i}(\gamma) \cap \Gamma_i(\delta)$ correspond to the subspaces Y with* $\dim Y = i$.

It follows that $C(\gamma,\delta)$ is bipartite. For $d \geqslant 3$, the distance two graph of $C(\gamma,\delta) \setminus \{\gamma,\delta\}$ (with $d(\gamma,\delta) = d$) induced on $\Gamma_{d-1}(\gamma) \cap \Gamma(\delta)$ is isomorphic to the graph on the non-isotropic points of a d-space V provided with a nondegenerate Hermitean form, adjacent when the line joining them is hyperbolic. It is easy to see that this graph is connected precisely when $r > 2$. For $d \geqslant 4$, $C(\gamma,\delta) \setminus \{\gamma,\delta\}$ is connected also when $r = 2$. Thus, by Proposition 4.2.7 it follows that Γ does not have antipodal covers of diameter $2d$, except possibly in case $d = 2$, or $(d,r) = (3,2)$. But when $(d,r) = (3,2)$, then such covers do exist, see §11.3H. We have $C(\gamma,\delta) \setminus \{\gamma,\delta\} \cong 4C_6$ for $d(\gamma,\delta) = 3$, and one easily sees that Γ has a unique 4-cover. □

When $d = 2$, then by Corollary 4.2.5 we find that either $r \in \{2,3\}$, or $4r - 3$ is a square, and if we have an antipodal t-cover, then $\sqrt{4r-3} \mid 5(t-1)$. If $r = 2$ then the Hermitean forms graph is the folded 5-cube, and its unique antipodal double cover of diameter 4 is the Wells graph (see §9.2E). If $r = 3$, then the Hermitean forms graph is the coset graph of the truncated ternary Golay code, and has an antipodal 3-cover (see §14). Concerning larger r for which $4r - 3$ is a square (7, 13, 31, $\cdots$), nothing is known.

9.5.15. Proposition. *The Hermitean forms graphs Γ of diameter $d \geqslant 2$ do not have distance-regular antipodal covers of diameter $2d + 1$, except in case $d = 2$, $q = 4$, where we find the 5-cube.*

Proof. Our result will follow from Proposition 4.2.8 if we show that if $\delta, \epsilon \in \Gamma_d(\gamma)$, $\delta \sim \epsilon$, then $\Gamma(\gamma) \cap \Gamma_{d-1}(\delta) \cap \Gamma_{d-1}(\epsilon) \neq \varnothing$. Translating to the polar space situation, we see that the size of this set is the number of nonisotropic points in a hyperplane of a d-space with nondegenerate Hermitean form. For $d \geqslant 3$ this is nonzero and we are done. For $d = 2$, $q = 4$, our graph is isomorphic to the folded 5-cube, and has a unique double cover of diameter 5, the 5-cube. For $d = 2$, $q > 4$ we use the full strength of Proposition 4.2.8. For $\delta \in \Gamma_2(\gamma)$, put $E = \{\delta\} \cup (\Gamma(\delta) \cap \Gamma_2(\gamma))$. We have to show that there is no nontrivial partition of E with the property that if $\epsilon, \epsilon' \in E$ have a common neighbour in $\Gamma(\gamma)$ then they belong to the same part. Embed Γ as $\Delta_2(\infty)$ in the collinearity graph Δ of a generalized quadrangle $GQ(r,r^2)$ with the property that any three pairwise noncollinear points have $r+1$ common neighbours. Put $A = \{\infty,\gamma,\delta\}^\perp$. Then all points of E on some line on δ not meeting A belong to the same part E_0 of the partition. If $\epsilon \in E \setminus E_0$, ϵ on the line $\delta\alpha$, $\alpha \in A$, then for $\beta \in A \setminus \{\alpha\}$ let ζ be the neighbour of ϵ on $\gamma\beta$. Put $B = \{\infty,\delta,\gamma\}^\perp$. Since $r^2 + 1 > 2r + 1$, there are lines on δ not meeting A or B, and a neighbour η of ζ on such a line would belong to E_0, but also to the same part as ϵ, contradiction. Thus $E = E_0$. □

Problem. Complete the classification of covers of Hermitean forms graphs (i.e., treat the case of diameter 2).

9.6 The quadratic forms graphs

The graphs discussed in this section were constructed in EGAWA [274], and this entire section has been taken almost directly from that source.

Set $V = \mathbf{F}_q^n$. A *quadratic form on* V (over $\mathbf{F}_q$) is a map $\gamma : V \to \mathbf{F}_q$ such that

$$\gamma(\lambda x) = \lambda^2 \gamma(x) \text{ for all } \lambda \in \mathbf{F}_q \text{ and } x \in V$$

and such that $B_\gamma : V \times V \to \mathbf{F}_q$ defined by

$$B_\gamma(x,y) = \gamma(x+y) - \gamma(x) - \gamma(y) \quad (x,y \in V)$$

is a symmetric bilinear form (the *symmetric bilinear form associated with* γ). Let Q denote the $n(n+1)/2$-dimensional vector space of all quadratic forms on V. The *radical* of γ, denoted by Rad γ, is defined by

$$\text{Rad}\,\gamma = \{x \in \text{Rad}\,B_\gamma \mid \gamma(x) = 0\} = \{x \in V \mid \gamma(y) = \gamma(x+y) \text{ for all } y \in V\}.$$

We observe that Rad B_γ = Rad γ if q is odd and that dim (Rad B_γ) $\leqslant$ dim (Rad γ) + 1 in general. The *rank* of $\gamma \in Q$, denoted rk (γ), is the number rk (γ) = dim (V / Rad γ). Put $R_i = \{(\gamma,\delta) \in Q^2 \mid \text{rk}(\gamma-\delta) = i\}$ and $R_i(\delta) = \{\gamma \in Q \mid (\gamma,\delta) \in R_i\}$.
The *quadratic forms graph on* V has vertex set Q and γ, $\delta \in Q$ are called adjacent if rk $(\gamma-\delta) \in \{1,2\}$. In this section, we shall denote this graph by Γ.
(Note that the graph Γ' with vertex set Q and where γ, δ are adjacent when rk $(\gamma-\delta) = 1$ is isomorphic to the symmetric bilinear forms graph when q is odd, while Γ' is a union of cliques of size q^n when q is even. For odd q we can define Γ in terms of Γ' by $\Gamma(\gamma) = \Gamma'(\gamma) \cup \Gamma'_2(\gamma)$ and conversely retrieve Γ' from Γ by $\Gamma'(\gamma) = \{\delta \in \Gamma(\gamma) \mid |\{\gamma,\delta\}^{\perp\perp}| = q\}$, so that Aut Γ = Aut Γ'.)

9.6.1. Lemma. *Let* $\gamma, \delta, \zeta \in \Gamma$. *Then*

(i) rk $(\gamma-\delta)$ + rk $(\delta-\zeta) \geqslant$ rk $(\gamma-\zeta)$

(ii) Rad $\gamma \cap$ Rad δ = Rad $\gamma \cap$ Rad $(\gamma-\delta)$.

Proof. Straightforward. □

9.6.2. Lemma. *Let q be even. If* $\gamma, \delta \in \Gamma$ *satisfy* rk $(\gamma+\delta) = 2j-2$, rk $(\delta) = 2$ *and* rk $(\gamma) = 2j-1$, *then* Rad $B_\gamma \cap$ Rad δ = Rad γ.

Proof. Note that, for $\beta \in \Gamma$, the bilinear form B_β is alternating, so that rk (B_β) is even. It follows that, if rk (β) is even, then Rad β = Rad B_β. Suppose $x \in$ Rad $B_\gamma \cap$ Rad $\delta \setminus$ Rad γ. Then $x \in$ Rad $B_{\gamma+\delta} \setminus$ Rad $(\gamma+\delta)$, impossible since rk $(\gamma+\delta)$ is even. Thus Rad $B_\gamma \cap$ Rad $\delta \subseteq$ Rad γ. As rk (δ) is even, rk (δ) = rk (B_δ), so that dim (Rad $B_\gamma \cap$ Rad δ) $\in \{n-2j,\ n-2j+1\}$. But if dim (Rad $B_\gamma \cap$ Rad δ) $= n-2j$, then Rad B_γ + Rad $B_\delta = V$; thus Lemma 9.5.4 (ii) yields

$$n-2j+2 = \dim \text{Rad}\,(\gamma+\delta) = \dim \text{Rad}\,(B_{\gamma+\delta}) =$$
$$= \dim (\text{Rad}\,B_\gamma \cap \text{Rad}\,B_\delta) = n-2j,$$

which is absurd. Consequently, dim (Rad $B_\gamma \cap$ Rad δ) $= n-2j+1$ = dim (Rad γ), whence Rad $B_\gamma \cap$ Rad δ = Rad γ. □

9.6.3. Theorem. *Let Γ be the quadratic forms graph on $\mathbb{F}_q^n$. Then Γ is distance-regular with diameter $d = [(n+1)/2]$, $v = q^{n(n+1)/2}$ vertices, intersection array given by*

$$\left.\begin{array}{rcl} b_j &=& q^{4j}(q^{n-2j+1}-1)(q^{n-2j}-1)/(q^2-1) \\ c_j &=& q^{2j-2}(q^{2j}-1)/(q^2-1) \end{array}\right\} \text{ for } 0 \leqslant j \leqslant d,$$

and singular lines of sizes 2 and q. Moreover, Γ has classical parameters as listed in Table 6.2, (C11a), and admits a group of automorphisms of the form

$$\mathbb{F}_q^{n(n+1)/2}.\mathbb{F}_q^{*}\cdot P\Gamma L(n,q).$$

Proof (sketch) (EGAWA [274]). This case is harder to deal with than the previous ones as Aut Γ is not distance-transitive (see the remark below).

STEP 1 (See ARTIN [9] or DIEUDONNÉ [264]). *The group $G = \mathbb{F}_q^{n(n+1)/2}.\mathbb{F}_q^{*}\cdot P\Gamma L(V)$ acts transitively on Q and preserves the relations R_i $(0 \leqslant i \leqslant n)$. Its origin stabilizer $\mathbb{F}_q^{*}\cdot P\Gamma L(V)$ has precisely two orbits in $R_m(0)$ for even m, and one orbit for odd m $(0 < m \leqslant n)$.*

(When m is even, one has hyperbolic and elliptic quadrics; when m is odd all quadrics look alike, but in case q is also odd, $\Gamma L(V)/\{\pm 1\}$ has two orbits on $\mathbb{F}_q^{*}\gamma$: the multiples $c\gamma$ with c a square, and those with c a nonsquare.)

STEP 2. *Let δ_{2j-1}, δ_{2j}^{+}, δ_{2j}^{-} be representatives of the orbits of G on $R_{2j-1}(0) \cup R_{2j}(0)$. For $l \in \{1,2\}$, $h \in \{2j-3, 2j-2, 2j+1, 2j+2\}$, $\epsilon \in \{-,+\}$ and $\delta \in \{\delta_{2j-1}, \delta_{2j}^{\epsilon}\}$, put $d_{h,l}(\delta) = |R_l(0) \cap R_h(\delta)|$. Then $d_{h,l}(\delta)$ is as given in Table 9.1 below.*

Table 9.1.

	$d_{h,l}(\delta_{2j-1})$		$d_{h,l}(\delta_{2j}^{\epsilon})$	
	$l=1$	$l=2$	$l=1$	$l=2$
$h=2j-3$	0	$\frac{q^{2j-2}(q^{2j-2}-1)}{q^2-1}$	0	0
$h=2j-2$	q^{2j-2}	$q^{2j-2}(q^{2j-2}-1)$	0	$\frac{q^{2j-2}(q^{2j}-1)}{q^2-1}$
$h=2j+1$	0	$\frac{q^{4j}(q^{n-2j}-1)(q^{n-2j+1}-1)}{q^2-1}$	$q^{2j}(q^{n-2j}-1)$	$q^{2j}(q^{n-2j}-1)(q^{2j}-1)$
$h=2j+2$	0	0	0	$\frac{q^{4j+2}(q^{n-2j}-1)(q^{n-2j-1}-1)}{q^2-1}$

All zero entries in Table 9.1 are explained by Lemma 9.6.1 (i). Since the proof of this step consists of numerous small steps, many of which are much alike, we shall restrict ourselves to a single case - this is enough to provide the general flavour. Thus, let q be even and $\delta = \delta_{2j-1}$. We shall determine $d_{2j-3,2}(\delta)$ first. Suppose that $\gamma \in R_2(0) \cap R_{2j-3}(\delta)$. Then rk $(\gamma) = 2$, rk $(B_{\gamma+\delta}) = 2j-4$ and rk $(B_\delta) = 2j-2$, so Rad $B_\delta \subseteq$ Rad B_γ by 9.5.5 (ii). Let W be a complement of Rad B_δ in V. Then by Lemma 9.5.5 (ii) we find a 2-space U of W which is not isotropic with respect to B_γ; and conversely any such U determines B_γ uniquely. Given U, there are q^2 choices for γ such that rk $(\gamma) = 2$ and $B_{\gamma|U} = B_{\delta|U}$. This proves that $d_{2j-3,2}(\delta)$ is q^2 times the number of 2-spaces in a $(2j-2)$-space that are isotropic with respect to $B_{\delta^+_{2j-2}}$. Therefore,

$$d_{2j-3,2}(\delta) = q^2\left(\begin{bmatrix}2j-2\\2\end{bmatrix} - \begin{bmatrix}j-1\\2\end{bmatrix}(q^{j-1}+1)(q^{j-2}+1)\right)$$
$$= q^{2j-2}(q^{2j-2}-1)/(q^2-1)$$

by Lemma 9.3.2, as claimed.

We shall now verify the formula for $d_{2j-2,1}(\delta)$. Suppose $\gamma \in R_1(0) \cap R_{2j-2}(\delta)$. Then $B_\gamma = 0$, and $\text{Rad}\,\gamma+\delta = \text{Rad}\,B_{\gamma+\delta} = \text{Rad}\,B_\delta$. Now $\gamma|_{\text{Rad}\,B_\delta} = \delta|_{\text{Rad}\,B_\delta}$, while γ may be chosen arbitrarily (of rank $\leqslant 1$) on a complement of $\text{Rad}\,B_\delta$. Thus $d_{2j-1,1}(\delta) = q^{2j-2}$.

We continue with $d_{2j-2,2}(\delta)$. Suppose $\gamma \in R_2(0) \cap R_{2j-2}(\delta)$. Then $\text{Rad}\,\gamma \cap \text{Rad}\,B_\delta = \text{Rad}\,\delta$ according to Lemma 9.6.2. Note that there are

$$q^{2j-3}\begin{bmatrix}2j-2\\2j-3\end{bmatrix} = q^{2j-3}\begin{bmatrix}2j-2\\1\end{bmatrix}$$

$(n-2)$-spaces U such that $U \cap \text{Rad}\,B_\delta = \text{Rad}\,\delta$ (cf. Lemma 9.3.2 (ii)). Having fixed $U = \text{Rad}\,\gamma = \text{Rad}\,B_\gamma$, let W be a complement of U in V and define B_γ arbitrarily nonzero on W. This defines B_γ ($q-1$ choices). Now $B_{\gamma+\delta} = B_\gamma + B_\delta$ is determined and $\text{Rad}\,B_{\gamma+\delta} = \text{Rad}\,\delta + \mathbb{F}_q\,x$ for some vector $x \in V\setminus \text{Rad}\,\delta$. (Indeed, $\dim \text{Rad}\,B_{\gamma+\delta} - \dim \text{Rad}\,\delta$ is odd and less than 3 since $\text{Rad}\,B_{\gamma+\delta} \cap \text{Rad}\,B_\gamma \subsetneq \text{Rad}\,B_\delta$, hence is 1.) Let W' be a complement of U in V containing x. It suffices to define γ on W', and since $\text{rk}\,(\gamma+\delta)$ must be even, we have $\gamma(x) = \delta(x)$. This leaves us with q choices for γ (given B_γ). Altogether this means that $d_{2j-2,2}(\delta) = q^{2j-3}\begin{bmatrix}2j-2\\1\end{bmatrix}\cdot(q-1)\cdot q = q^{2j-2}(q^{2j-2}-1)$, as desired.

The final number we shall determine here is $d_{2j+1,2}(\delta)$. For $\gamma \in R_{2j+1}(\delta) \cap R_2(0)$, we have $\text{rk}\,(\gamma) = 2$, $\text{rk}\,(B_{\gamma+\delta}) = 2j$ and $\dim\,(\text{Rad}\,\gamma \cap \text{Rad}\,\delta) = n-2j-1$. There are $q^2(q-1)$ quadratic forms γ of rank 2 with $\text{Rad}\,\gamma \cap \text{Rad}\,\delta$ a given $(n-2j-1)$-space of $\text{Rad}\,\delta$. By Lemma 9.6.1 (ii), all of them satisfy $\text{rk}\,(\gamma+\delta) = 2j+1$. As the number of $(n-2)$-spaces U of V such that $U \cap \text{Rad}\,\delta$ is an $(n-2j-1)$-space is $q^{4j-2}\begin{bmatrix}n-2j+1\\2\end{bmatrix}$ by Lemma 9.3.2 (iii), we get

$$d_{2j+1,2}(\delta) = q^2(q-1)q^{4j-2}\begin{bmatrix}n-2j+1\\2\end{bmatrix}$$
$$= q^{4j}(q^{n-2j+1}-1)(q^{n-2j}-1)(q^2-1),$$

and we are done.

Step 3. From Step 2 we see that $\gamma \in \Gamma_j(\delta)$ if and only if $\text{rk}\,(\gamma-\delta) \in \{2j-1, 2j\}$, so that Γ has diameter $d = [(n+1)/2]$. Now we can read off the values of $b_i(0,\delta)$ and $c_i(0,\delta)$ from Table 9.1 above, and since these do not depend on $\delta \in \Gamma_i(0)$, it follows that Γ is distance-regular. Since the intersection array of Γ coincides with that of an alternating forms graph, we conclude that Γ has classical parameters as described. Finally, it is easily checked that $\{0,\delta\}^{\perp\perp} = \mathbb{F}_q\delta$ for $\delta \in R_1(0)$ and $\{0,\delta\}^{\perp\perp} = \{0,\delta\}$ for $\delta \in R_2(0)$, where $\perp$ is taken with respect to Γ. Thus singular lines have size 2 or q.

□

9.6.4. Proposition. *If $n > 3$, then Γ is not distance-transitive. In particular, Γ is not isomorphic to an alternating forms graph.*

Proof (sketch). For $q > 2$, the statement is obvious since the singular line on 0 and δ has size 2 or q according to whether $\delta \in R_2(0)$ or $\delta \in R_1(0)$. For $q = 2$ a further argument is needed.

If $\gamma, \delta \in \Gamma_1(0)$, denote by $\gamma \star \delta$ the unique $\zeta \in \Gamma$ such that

$$\{0,\gamma,\delta\}^{\perp\perp} = \{0,\gamma,\delta,\zeta\} \text{ if } \gamma \in \Gamma_1(\delta)$$

and

$$\{\zeta\} = M \cap M^{\perp}, \text{ where } M = \Gamma_2(0) \cap \{\gamma,\delta\}^{\perp} \text{ if } \gamma \in \Gamma_2(\delta).$$

Then for $\gamma \in R_2(0)$, there are $\zeta, \eta \in R_2(0)$ with $\zeta + \eta \in R_4(0)$. It follows that $(\zeta\star\gamma)\star\eta \neq \zeta\star(\gamma\star\eta)$. On the other hand, if $\gamma \in R_1(0)$, then $(\zeta\star\gamma)\star\eta = \zeta\star(\gamma\star\eta)$ for all $\zeta, \eta \in \Gamma_1(0)$. Since $n > 3$, the graph Γ is not distance-transitive for $q = 2$. Since, according to Theorem 9.5.3, the alternating forms graphs are distance-transitive, this proves the statements. □

Remarks. For q odd, $n \geqslant 2$, Hua [408] shows that the group given in the statement of the theorem is the full automorphism group of Γ.

For a study of maximal cliques in quadratic forms graphs see Hemmeter [372] (and Hemmeter & Woldar [374, 375]); as a corollary of his investigations, Hemmeter shows there that Γ can be a halved graph only if $n \leqslant 2$.

Chapter 10
Graphs of Coxeter and Lie Type

In this chapter we study Coxeter systems and Tits systems and certain graphs derived from these. In the first few sections finiteness is not assumed. In the later sections almost all known infinite families of distance-transitive graphs are described in this framework. The chapter ends with a determination of all distance-transitive graphs which naturally arise from a Tits system in a finite Chevalley group. Much more information on Tits systems, Chevalley groups and buildings can be found in BOURBAKI [113], CARTER [187, 188] and TITS [747, 753].

10.1 Coxeter systems

A *Coxeter system* (with *Coxeter group W*) is a pair (W,R) consisting of a group W and a subset R of W for which a symmetric matrix (the *Coxeter matrix*) $M = (m_{r,t})_{r,t \in R}$ with entries in $\mathbb{N} \cup \{\infty\}$ and diagonal entries 1 can be found such that

$$\langle R \,|\, (rt)^{m_{r,t}} = 1 \text{ for all } r, t \in R \rangle$$

is a presentation of W. The system is said to be *finite* (or *of spherical type*) whenever the group is finite. The *Coxeter diagram* of (W,R) consists of the pair (R,M). Since M is symmetric this diagram can be viewed as a graph with labelled edges by letting $\{r, t\}$ for two distinct $r, t \in R$ be an edge labelled $m_{r,t}$ whenever $m_{r,t} \geqslant 3$. Labels 3 are usually omitted, and edges with label 4 drawn as double edges. (See Table 10.1 below for examples.)
If J is a subset of R, denote by $W_J = \langle J \rangle$ the subgroup of W generated by J.

10.1.1. Theorem. *Let (W,R) be a Coxeter system.*

(i) *The choice of the matrix $M = (m_{r,t})_{r,t \in R}$ associated with (W,R) is uniquely determined by the fact that $m_{r,t}$ is the order of rt for all $r, t \in R$.*

(ii) *If $I \subseteq R$, then (W_I,I) is a Coxeter system with $W_I \cap R = I$.*

(iii) *The system (W,R) is finite if and only if R is finite and every connected component of its Coxeter diagram appears in Table 10.1 below.*

Proof. See BOURBAKI [113] Ch. V §4.3, Ch. IV §1.8 and Ch. VI §4.1 for (i), (ii) and (iii), respectively. □

By (ii), any Coxeter system (W,R) with diagram (R,M) has a decomposition into a direct product of Coxeter groups whose diagrams correspond to the connected components of (R,M). If the diagram (R,M) is connected, the Coxeter system is called

Table 10.1. *Diagrams of irreducible finite Coxeter systems*

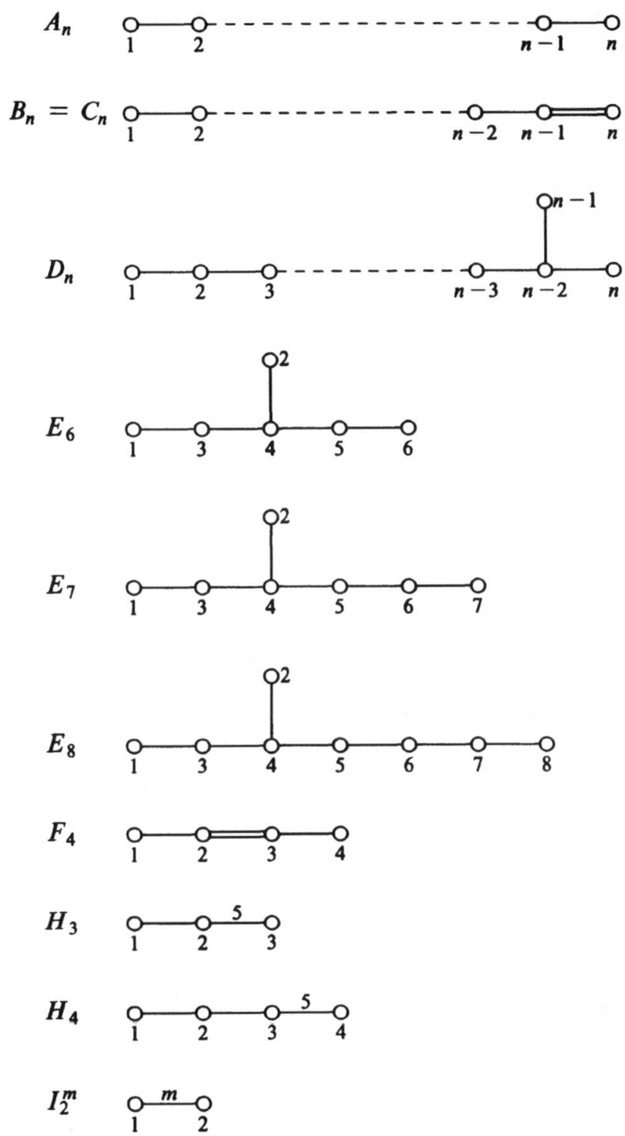

irreducible, otherwise *reducible*. Thus Table 10.1 is a complete list of the finite irreducible Coxeter systems. Note that $I_2^3 = A_2$ and $I_2^4 = B_2$. The diagram I_2^6 will also be referred to as G_2.

The Coxeter group as a reflection group

Although we shall not need this later, let us say something about the geometric interpretation of Coxeter groups. Let V be a vector space over $\mathbb{R}$ with basis $(b_r)_{r \in R}$ indexed by R. Provide V with the inner product (symmetric bilinear form) $(\ ,\ )$ defined by

$$(b_r, b_s) = -\cos(\pi / m_{rs}),$$

so that the vectors b_r become unit vectors. We may define a representation ρ of W in $O(V)$ by

$$(\rho(r))(v) = v - 2(v, b_r)b_r \quad \text{for } r \in R,\ v \in V.$$

Now we see that $\rho(rs)$ restricted to the plane $\langle b_r, b_s \rangle$ is a rotation over an angle $2\pi / m_{rs}$, so that rs has order at least m_{rs}, proving 10.1.1 (i). One may show that ρ is faithful. This gives 10.1.1 (ii). One may also show (cf. BOURBAKI [113], Ch. V, §4.8) that the group W is finite precisely when V is a finite-dimensional Euclidean space, i.e., when R is finite and the matrix $(-\cos(\pi / m_{rs}))_{r,s \in R}$ is positive definite. This leads to a proof of 10.1.1 (iii). (Cf. Theorems 3.2.5 and 3.10.4. See also COXETER [223], COXETER & MOSER [227], GROVE & BENSON [339] and HILLER [390].)

The length function; reduced expressions

Let (W,R) be a Coxeter system. For $w \in W$, the *length* of w is the length $l(w)$, more precisely $l_R(w)$, of a shortest expression of w as a product of elements in R. An expression of $w \in W$ as a product $r_1 \cdots r_p$ of elements of R is called *reduced* when $l(w) = p$. More generally, an expression $w = w_1 \cdots w_q$ is called reduced when $l(w) = l(w_1) + \cdots + l(w_q)$. Define, for subsets I, J of R, the set

$$D_{I,J} = \{w \in W \mid l(iw) = l(wj) > l(w) \text{ for all } i \in I \text{ and } j \in J\}.$$

10.1.2. Theorem. *Let (W,R) be a Coxeter system, and let I, J be subsets of R. Then:*

(i) *Suppose $l(r_1 r_2 \cdots r_p) < p$, where $r_i \in R$ $(1 \leqslant i \leqslant p)$. Then there are indices i, j such that $r_1 r_2 \cdots r_p = r_1 \cdots r_{i-1} r_{i+1} \cdots r_{j-1} r_{j+1} \cdots r_p$. In other words, given a product expression for $w \in W$, one can find a reduced expression by repeatedly cancelling factors in pairs. In particular, $l(r_1 r_2 \cdots r_p) \equiv p \pmod 2$.*

(ii) *For each $w \in W$ there is a subset E_w of R such that $E_w = \{r_1, \cdots, r_p\}$ for each reduced expression $w = r_1 \cdots r_p$ of w as a product of elements of R.*

(iii) *If $w \in W_I$, then the length of w in (W,R) and (W_I, I) is the same: $l_R(w) = l_I(w)$.*

(iv) *The map $\phi_{I,J}$ from $D_{I,J}$ onto the double coset $W_I \backslash W / W_J$ sending $w \in D_{I,J}$ to $W_I w W_J$ is a bijection. Each $w \in W$ has a unique expression $w = dv$ with $d \in D_{\varnothing,J}$ and $v \in W_J$, and a unique expression $w = ud$ with $u \in W_I$ and $d \in D_{I,\varnothing}$, and these expressions are reduced. Thus $D_{I,J} = D_{I,\varnothing} \cap D_{\varnothing,J}$. Each $w \in W$ has a (not necessarily unique) reduced expression $w = udv$ with $d \in D_{I,J}$, $u \in W_I$, $v \in W_J$. In particular, $\phi_{I,J}^{-1}(W_I w W_J)$ is the unique shortest element of $W_I w W_J$.*

(v) *If W is finite, there is a unique longest element w_0 in W. This element is an involution.*

Proof. For (i), (ii), (iv) and (v), see BOURBAKI [113], Ch. IV, Exercice 1.1a., Ch. IV,

§1.8, Proposition 7, Ch. IV, Exercice 1.3 and Ch. IV, Exercice 1.22, respectively. Part (iii) is obvious from (i). □

Note that (i) and (ii) imply results like $W_I \cap W_J = W_{I \cap J}$ and $W_I W_J \cap W_K = W_{I \cap K} W_{J \cap K}$. In the sequel we shall very often use arguments based on (i) and (ii) without explicitly referring to the above theorem.

We close this section with a few technical results needed later.

10.1.3. Proposition. *Let $d \in D_{I,J}$. Then $W_I \cap dW_J d^{-1} = \langle I \cap dJd^{-1} \rangle$.*

Proof. See SOLOMON [692], Lemma 2. □

For $S \subseteq R$, we write $S^\perp := \{r \in R \mid rs = sr$ for all $s \in S\}$, and we write $s^\perp$ instead of $\{s\}^\perp$.

10.1.4. Lemma. *Let $r \in R$ and $S = R \setminus \{r\}$. Let $v, w \in W_S$ satisfy $vr = rw$. Then $v = w \in \langle S \cap r^\perp \rangle$ and hence $r \in E_v^\perp$.*

Proof. Clearly $r \in D_{S,S}$ so that by 10.1.3 we have $v = rwr \in \langle S \cap rSr \rangle$. But if $s = rtr$ with $s, t \in S$, then by 10.1.2 (i) we have $s = t \in S \cap r^\perp$. □

10.1.5. Lemma. *Suppose $S \subseteq R$ and $d \in D_{S,S}$, and let $u, v \in W_S$ satisfy $udv = d$. Then $u \in \langle E_d \rangle \langle E_d^\perp \rangle$.*

Proof. Since E_d and $E_d^\perp$ commute, it suffices to prove $E_u \subseteq E_d \cup E_d^\perp$. Since $u = dv^{-1}d^{-1} \in W_S \cap dW_S d^{-1} = \langle S \cap dSd^{-1} \rangle$ by 10.1.3, any $r \in E_u$ has an expression $r = dsd^{-1}$ for some $s \in S$. If $r \notin E_d$, then $r = s$ by 10.1.2 (i), and hence $r \in E_d^\perp$ by 10.1.4. □

The word problem in Coxeter groups

The following result is a powerful tool when one wants to do explicit computations in Coxeter groups.

10.1.6. Theorem (TITS [745], cf. BOURBAKI [113], Ch. IV §1.5 and Exercice 1.13).
Let (W,R) be a Coxeter system. If $r_1 \cdots r_m = s_1 \cdots s_n$ ($r_i, s_j \in R$), then the expressions (words) $r_1 \cdots r_m$ and $s_1 \cdots s_n$ can be transformed into identical words by a sequence of operations (on each of them) of the following kinds:

(i) *replace a subsequence rr in a word by 1 (the empty sequence) for some $r \in R$.*

(ii) *replace a subsequence $rsr \cdots$ in a word by $srs \cdots$, where both products have $m_{r,s}$ factors, for $r, s \in R$.* □

It follows immediately that if $r_1 \cdots r_m = s_1 \cdots s_m$ ($r_i, s_j \in R$) and these expressions are reduced, then for each i ($1 \leqslant i \leqslant m$) there is a j ($1 \leqslant j \leqslant m$) such that

$$r_1 \cdots \hat{r}_i \cdots r_m = s_1 \cdots \hat{s}_j \cdots s_m.$$

Problem. Consider the diagram

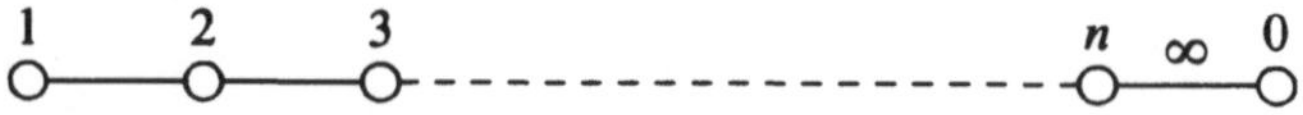

It is easy to see that

$$(012\cdots n)^n = 0w_n0w_{n-1}0\cdots 0w_1$$

where $w_i = n(n-1)\cdots(n-i+2)(n-i+1)(n-i+2)\cdots(n-1)n$ (product of $2i-1$ factors) for $1 \leqslant i \leqslant n$. Show that this element has length $n(n+1)$, and that it has an expression over R with precisely i factors i for any given i $(1 \leqslant i \leqslant n)$ but cannot be written as a product over R with fewer than i factors i.

Bruhat order

For $u, v \in W$ write $u \leqslant v$ if there are reduced expressions $v = r_1\cdots r_n$ and $u = r_{i_1}\cdots r_{i_m}$ with $r_j \in R$ for all j, and $1 \leqslant i_1 < i_2 < \cdots < i_m \leqslant n$. (Note that in view of Theorem 10.1.2 (i) it is not necessary to require that the expression for u be reduced.)

10.1.7. Proposition (VERMA [765], DEODHAR [262]). *$\leqslant$ defines a partial order on W.*

Proof. In order to show transitivity, it suffices to show that if $u \leqslant v$ then for each reduced expression $v = r_1\cdots r_n$ there is a reduced expression $u = r_{i_1}\cdots r_{i_m}$ $(1 \leqslant i_1 < \cdots < i_m \leqslant n)$. But this follows immediately from the result of Theorem 10.1.6 above. □

Clearly, 1 is the smallest element of $(W, \leqslant)$; in case W is finite, the longest element w_0 is also the largest element of $(W, \leqslant)$. If u is covered by v, i.e., if $u < v$ but there is no w such that $u < w < v$, then $l(v) = l(u)+1$.

10.1.8. Lemma. *Let (W,R) be a Coxeter system, and $w \in W$. Put $U := \{u \in W \mid u \leqslant w\}$ and $S := \{s \in R \mid l(sw) < l(w)\}$. Then $SU \subseteq U$.*

Proof. Let $s \in S$ and $u \in U$. Then w has a reduced expression $w = sr_2\cdots r_n$ over R and a reduced expression for u can be obtained by deleting some factors in this product. If the factor s in front was deleted, then also $su \leqslant w$. If it was not, then $su \leqslant u \leqslant w$, and again $su \leqslant w$. □

10.1.9. Proposition. *Let (W,R) be an irreducible Coxeter system. If the center $Z(W)$ of W contains a non-identity element w, then W is finite, and $w = w_0$, the longest element of W.*

Proof. If $r \in R \setminus E_w$ and $s \in E_w$ with $m_{rs} > 2$, then by Lemma 10.1.4 we have $wr \neq rw$. Since we assumed that (W,R) is irreducible, this means that $E_w = R$. From $wr = rw$ for all $r \in R$ it follows by induction on $l(w_2)$ that $w_1w_2 = w_2w_1$ for all reduced expressions $w = w_1w_2$ of w; in particular, w has some reduced expression $w = rv$ for each $r \in R$. In the notation of the previous lemma this means that $S = R$, and we find that $W = U$. □

Lemma 10.2.11 states precisely in which cases $Z(W) = \{1, w_0\}$.

10.2 Coxeter graphs

Let (W,R) be a Coxeter system with nonempty R. Fix $r \in R$ and put $S = R \setminus \{r\}$, $X = W_S$. The graph $\Gamma = \Gamma(W,X,r)$ which has as vertices the left cosets wX ($w \in W$), w_1X and w_2X being adjacent if and only if $w_2^{-1}w_1 \in XrX$, is called a *Coxeter graph*. (Note that we do not make a finiteness assumption.)

In case (W,R) has a diagram named X_n (for some letter X) and r is the node numbered i of this diagram in Table 10.1 or Table 10.2, the graph Γ will be called the graph of Coxeter type $X_{n,i}$ and denoted by $X_{n,i}(1)$ or just $X_{n,i}$.

Table 10.2. *Some infinite Coxeter diagrams*

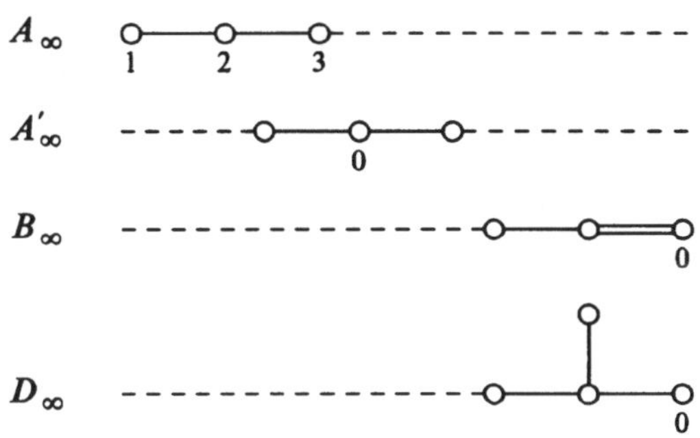

If (W,R) is reducible and C is the connected component of the diagram containing r, then $\Gamma \cong \Gamma(W_C, W_D, r)$ where $D = C \setminus \{r\}$. Consequently, when studying Coxeter graphs there is no loss in generality if one assumes the Coxeter system to be irreducible.

Clearly W acts transitively on Γ by left multiplication; the orbits of the stabilizer (in W) of the vertex X are the (sets of vertices contained in the) double cosets XwX ($w \in W$). In particular, W is a subgroup of $\operatorname{Aut}\Gamma$, and acts edge-transitively on Γ.

The neighbours of a vertex wX are the vertices contained in $wXrX$; hence the points at distance at most k from X are the uX where u has an expression as a product of factors in R with at most k factors equal to r. In particular, Γ is connected. We consider distance 1 and distance 2 in more detail. We write

$$S_j := \{s \in S \mid m_{r,s} = j\}$$

and

$$T := S_2 = S \cap r^\perp.$$

The neighbourhood of a point

Since W acts transitively on Γ, in order to describe the neighbourhood of a vertex, it suffices to do so for the vertex X.

10.2.1. Lemma. *For a Coxeter graph $\Gamma = \Gamma(W,X,r)$ the following hold:*

(i) $X \cap rXr = W_T$.

(ii) *Each neighbour of the vertex X has a unique representation as wrX with $w \in X \cap D_{\varnothing,T}$. In particular, if Γ is finite, then it is regular of valency $k = |X \cap D_{\varnothing,T}|$.*

Proof. (i) This is just a reformulation of 10.1.4.
(ii) A neighbour of X has the form wrX with $w \in X$ and without loss of generality, $w \in X \cap D_{\varnothing,T}$. If $urX = vrX$ with $u,v \in X \cap D_{\varnothing,T}$, then $v^{-1}u \in X \cap rXr = W_T$ so that $uW_T = vW_T$ and consequently $u = v$. □

Since W acts edge-transitively on Γ, in order to describe the common neighbours of two adjacent vertices, it suffices to do so for the adjacent vertices X and rX.

10.2.2. Lemma. *For a Coxeter graph $\Gamma = \Gamma(W,X,r)$, the following hold.*

(i) $X \cap rXrXr = W_T S_3 W_T$.

(ii) *wX is adjacent to both X and rX if and only if $w \in W_T S_3 rX$; in particular, if Γ is finite, then $\lambda(\Gamma) = \sum_{s \in S_3} |W_T \cap D_{\varnothing,s^\perp}|$.*

Proof. (i) It suffices to show that $X \cap rXrXr \cap D_{T,T} = S_3$. Since $s = rsrsr$ for $s \in S_3$, the right hand side is contained in the left hand side. Conversely, suppose $w \in X \cap rXrXr \cap D_{T,T}$. Then rwr is reduced since otherwise $w = rwr \in XrX$, which is absurd. Choose $w_1, w_2 \in X$ such that $w = rw_1rw_2r$, where w_1rw_2 is reduced and $w_1 \in D_{\varnothing,T}$. Then $l(rw_1rw_2) = l(wr) = l(rwr) - 1 = l(w_1rw_2) - 1$ so that rw_1rw_2 is not reduced. By 10.1.2 (i) we must have $wr = rw_1rw_2 = w_1rw_2'w_2''$ where $w_2 = w_2'sw_2''$, $s \in S$ and $w_2'w_2''$ is reduced. Therefore $w_1^{-1}w \in X \cap rXr = W_T$ and since $w, w_1 \in D_{\varnothing,T}$ this implies $w = w_1$, so $w_2'w_2'' = 1$ and consequently $w_2' = w_2'' = 1$. Now we have $w_2 = s$ and $w^{-1}rw = rsr$, and by 10.1.2 (i)-(ii) the expression $w^{-1}rw$ equals an expression of length $l(rsr) = 3$ over $W_{\{r,s\}}$ with at most one r, necessarily srs. It follows that $rsr = srs$, i.e., $s \in S_3$, and that $rws = wsr$, i.e., $ws \in W_T$ and $w \in W_Ts \cap D_{T,T} = \{s\}$.

(ii) The common neighbours of X and rX are the cosets in $XrX \cap rXrX = W_T S_3 rX$ (by (i)), and the formula for λ follows from $W_Ts \cap D_{\varnothing,T} = (W_T \cap D_{\varnothing,s^\perp})s$ for $s \in S_3$. □

10.2.3. Theorem (Cooperstein [218]). *The following statements concerning the Coxeter graph $\Gamma = \Gamma(W,X,r)$ are equivalent.*

(i) *Γ has diameter 1;*

(ii) *$rXr \subseteq X \cup XrX$;*

(iii) *the diagram of (W,R) has a connected component that is a (finite or infinite) path with end node r. (In case (W,R) is irreducible this means that Γ is the graph $A_{n,1}$ with $n \geqslant 1$ or $n = \infty$.)*

Proof. Clearly (i) and (ii) are equivalent, and by 10.2.1 (i) and 10.2.2 (i) these are equivalent to $X \subseteq W_T \cup W_T S_3 W_T$. If this condition holds, then $R = \{r\} \cup T \cup S_3$; moreover $|S_3| \leqslant 1$ for if $s,t \in S_3$, $s \neq t$, then $st \notin W_T \cup W_T S_3 W_T$. If $S_3 = \varnothing$, then the connected component C of the diagram containing r is of type A_1. Otherwise $S_3 = \{s\}$ and $sW_Ts \subseteq W_T \cup W_TsW_T$ and we can argue about (s,T) as we did about (r,S). If C is finite, then after finitely many steps we see that C has type A_n; otherwise we see that C contains a half-infinite path starting with r where all other vertices have degree 2. But then this path must be all of C. Conversely, if C is a path with end node r, then in order to show $rXr \subseteq X \cup XrX$ we have to show that $rwr \in X \cup XrX$ for all $w \in X$; but E_w is finite, so we may assume that C is finite, and now our result follows by induction on $|C|$. □

Recall that we write $\gamma^\perp$ for the set of vertices consisting of γ and all its neighbours, and $A^\perp = \bigcap_{\gamma \in A} \gamma^\perp$. Also recall that singular lines are sets of the form $\{\gamma,\delta\}^{\perp\perp}$ for $\gamma, \delta \in \Gamma$.

10.2.4. Corollary. *The following statements concerning the Coxeter graph $\Gamma = \Gamma(W,X,r)$ are equivalent.*

(i) *Γ has singular lines of size larger than two;*

(ii) *the connected component containing r of the diagram induced on $\{r\} \cup T \cup S_3$ $(= \{s \in R \mid m_{r,s} \leqslant 3\})$ has type A_n (for some $n \geqslant 2$ or $n = \infty$) with end node r;*

(iii) *$S_3 \neq \emptyset$ and $S_3 W_T S_3 \subseteq W_T \cup W_T S_3 W_T$.*

If this is the case, then, for any two adjacent vertices γ, δ, we have $\{\gamma,\delta\}^{\perp\perp} = \{\gamma,\delta\}^\perp$, i.e., the singular lines are precisely the maximal cliques, and Γ can be viewed as a partial linear space with girth at least 4.

Proof. Let $\gamma = X$, $\delta = rX$. Then $\{\gamma,\delta\}^\perp \setminus \{\gamma,\delta\} = W_T S_3 rX$ by 10.2.2(ii). Let $\epsilon \in \{\gamma,\delta\}^\perp \setminus \{\gamma,\delta\}$, so that $\epsilon = usrX$ with $u \in W_T$, $s \in S_3$. Then $\epsilon \in \{\gamma,\delta\}^{\perp\perp}$ if and only if $W_T S_3 rX \subseteq usrX \cup usrXrX$, i.e., if and only if $sW_T S_3 \subseteq rXr \cup rXrXr$. Since $sW_T S_3 \subseteq X$ this condition reduces to $sW_T S_3 \subseteq W_T \cup W_T S_3 W_T$ (cf. 10.2.1(i) and 10.2.2(i)). But there is no $t \in S_3 \setminus \{s\}$, since otherwise we would have $st \notin W_T \cup W_T S_3 W_T$. It follows that either $\{\gamma,\delta\}^{\perp\perp} = \{\gamma,\delta\}$ or $S_3 = \{s\}$ and $sW_T s \subseteq W_T \cup W_T s W_T$ and $\{\gamma,\delta\}^{\perp\perp} = \{\gamma,\delta\}^\perp$. Now apply the above theorem. □

The 2-neighbourhood of a point

The preceding results described what happens at distance one from a given vertex. Let us now widen our horizon and look at distance two. Recall that $\Gamma_j(\gamma)$ is the set of vertices at distance j from γ.

10.2.5. Proposition. *The points in $\Gamma_2(X)$ are precisely those of the form $\delta = wrdrX$, with $w \in X$ and $d \in X \cap D_{T,T}$, $d \neq 1$, $d \notin S_3$; here d is uniquely determined by δ.*

Proof. Clearly the points in $\Gamma_2(X)$ have a representation $\delta = wrdrX$ with $w, d \in X$, and we may suppose $d \in D_{T,T}$. Also $d \notin S_3 \cup \{1\}$ since otherwise δ has distance less than two to X. Finally d is uniquely determined by the fact that rdr is the unique shortest element of $XrdrX$. (Indeed, $rdr \in D_{S,S}$, for if $l(rdrs) < l(rdr)$ for some $s \in S$, then in the expression $rdrs$ we can cancel s and some other factor in R. This other factor is not r since rdr cannot be written with fewer than two r's. But then $l(drs) < l(dr)$ and by 10.1.4 this means that $s \in T$, impossible.) □

For vertices γ, δ at distance two, we shall determine the structure of the subgraphs $\{\gamma,\delta\}^\perp$ of Γ. These 'μ-graphs' turn out to be finite complete graphs K_n $(n \geqslant 1)$ or complete multipartite graphs $K_{n\times 2}$ ($n \geqslant 1$). Without loss of generality we may assume that $\gamma = X$ and $\delta = eX$ with $e \in rXr \cap D_{S,S}$, $e \notin X\langle r\rangle X$.

10.2.6. Theorem. *Let $\gamma = X$, $\delta = eX$ be two vertices with $d(\gamma,\delta) = 2$ and $e \in D_{S,S}$. Write $e = rdr$ (so that $d \in X \cap D_{T,T}$, $d \notin S_3 \cup \{1\}$). Then the following statements hold.*

(i) *$\Gamma(\gamma) \cap \Gamma(\delta) = \langle S \cap eSe^{-1}\rangle rX$; in particular, the stabilizer of γ and δ in Aut Γ is transitive on $\Gamma(\gamma) \cap \Gamma(\delta)$.*

(ii) *$\Gamma(\gamma) \cap \Gamma(\delta)$ consists of a single vertex if and only if $d \notin S_4 \cup S_3 W_T S_3$.*

(iii) *$\Gamma(\gamma) \cap \Gamma(\delta)$ consists of two nonadjacent vertices if and only if $d \in S_4$.*

(iv) *If $d \in S_3 W_T S_3$, then the subgraph $\Gamma(\gamma) \cap \Gamma(\delta)$ is isomorphic to a Coxeter graph of type $A_{n,1}$ ($n \geqslant 1$) or $B_{n,1}$ ($n \geqslant 2$).*

Parts (ii)-(iv) of this theorem will be easy consequences of part (i) and the following lemma.

10.2.7. Lemma. *Let $e = rdr$ be as above, $D = T \cap dTd^{-1}$ and $E = S \cap eSe^{-1}$. Then the connected component K of the diagram $M|_{E \cup \{r\}}$ containing r is of Coxeter type A_n ($n \geqslant 1$), B_n ($n \geqslant 2$) or D_n ($n \geqslant 3$), with 1-endnode r.*
More precisely, we have $E \cap T = D$ and one of the following five cases holds.
1. *$d \notin S_4 \cup S_3 W_T S_3$. Now $E = D$ and K is of type A_1.*
2. *$d \in S_4$. Now $E = \{d\} \cup \{r,d\}^{\perp}$ and K is of type B_2.*
3. *$d = st = ts$, with $s, t \in S_3$, $s \neq t$. Now $E = \{s,t\} \cup \{r,s,t\}^{\perp}$ and K is of type D_3.*
4. *$d = swt$, where $w \in W_T$, $s, t \in S_3$, $s \neq t$, with $w \neq 1$ or $st \neq ts$. Now $E = \{s\} \cup (D \cap s^{\perp})$ and K is of type A_2.*
5. *$d = sws$, where $w \in W_T$, $s \in S_3$. Now $E = \{s\} \cup D$ and K is of one of the types A_n ($n \geqslant 2$), B_n ($n \geqslant 3$) or D_n ($n \geqslant 4$).*

Proof (of the lemma). Let $s \in E$. Then $srdr = rdrt$ for some $t \in S$. If $s \in T$, then $sdr = drt$ so that $d^{-1}sd = t \in T$ (by 10.1.4) and $s \in T \cap dTd^{-1} = D$. Conversely, if $t \in D$, then $t = dsd^{-1}$ with $s \in T$ so that $e^{-1}te = s$ and $t \in E$. This proves $E \cap T = D$.
If $s \notin T$, then consider $drt = rsrdr$. Clearly the expression $rsrdr$ is not reduced, but the expression $srdr$ is reduced, so either $rsrdr = srd$ or $rsrdr = sr\hat{d}r$, where $\hat{d}$ is obtained from d by cancelling one factor in an expression as a product of elements of S. In the former case we have $drt = rsrdr = srd$ so that $dt = sd \in W_T$ (by 10.1.4), but $d \in D_{T,T}$ so $d = s = t$ and $m_{r,d} = 4$. In the latter case we have $drt = rsrdr = sr\hat{d}r$ so that $sd = r\hat{d}rtr \in W_T S_3 W_T$ by 10.2.2(i), and $rsrsr = \hat{d}d^{-1}$ so that $m_{r,s} = 3$ and $sd \in W_T S_3$. This proves 1.
If $d \in S_4$, then by the above $E = \{d\} \cup D$ and $D \subseteq \langle d^{\perp}\rangle$, proving 2.
If $d \in S_3 W_T S_3$, say $d = swt$ with $s, t \in S_3$, $w \in W_T$, then $s \in E$, and conversely the above shows that if $s \in E \setminus T$, then d has such a representation $d = swt$. If s is not uniquely determined by the expression $d = swt$, then $s \neq t$ and $d = tvs$ with $v \in W_T$, but now $tv = wt \in \langle s^{\perp}\rangle$ and since $d \in D_{T,T}$ we have $v = w = 1$ and case 3 occurs. If $s \neq t$, then $D \subseteq \langle s^{\perp}\rangle$ (by 10.1.4) and we are in case 3 or 4. Finally, if $s = t$, then 5 follows by induction on $l(e)$ (apply this lemma to s, d, D instead of r, e, E). □

Proof (of the theorem). Let $wrX \in \Gamma(\gamma) \cap \Gamma(\delta)$, where $w \in X$, $wr \in D_{\varnothing,S}$. Then

$wr = rdrw_1rw_2$ for certain $w_1, w_2 \in X$, and we may assume that w_1 and w_2 have been chosen such that the expression w_1rw_2 is reduced. Now we have $rdrw_1 = wrw_2^{-1}r$ and both sides are reduced (since $rdr \in D_{S,S}$, $wr \in D_{\emptyset,S}$ and $wrw_2^{-1}r$ cannot be written with fewer than two factors r because $d(\gamma,\delta) = 2$) so that

$$l(d)+l(w_1) = l(w)+l(w_2).$$

Also, $(rdr)^{-1}w = w_1rw_2r$ and both sides are reduced, so that

$$l(d)+l(w) = l(w_1)+l(w_2).$$

It follows that $l(d) = l(w_2)$, and since $XrdrX = Xrw_2^{-1}rX$ we have $d = w_2^{-1}$ (using 10.1.2(iv)). Now $w = (rdr)w_1(rdr)^{-1}$ and by 10.1.3 we find the equality part of (i). Clearly, if $s \in S \cap eSe^{-1}$, then $s\gamma = \gamma$ and $s\delta = \delta$, proving the transitivity statement, and hence (i).
The type of $M|_{E \cup \{r\}}$ was determined in the lemma, and suppressing the end node r we find the type of $M|_E$. In the five cases of the lemma, we find that $\Gamma(\gamma) \cap \Gamma(\delta)$ is isomorphic to a single vertex ('type $A_{0,1}$'), two nonadjacent vertices ('type $B_{1,1}$'), a 4-gon (type $B_{2,1}$), an edge (type $A_{1,1}$), and the Coxeter graph $\Gamma(\langle E\rangle,\langle D\rangle,s)$, respectively, and parts (ii)-(iv) of the theorem follow. (Note that the graphs $B_{n,1}$ and $D_{n,1}$ $(n \geqslant 3)$ are isomorphic; both are isomorphic to $K_{n\times 2}$.) □

Remark. If the connected component of (R,M) containing r is of spherical type distinct from H_3 and H_4, then the subgraphs $\Gamma(\gamma) \cap \Gamma(\delta)$ for $d(\gamma,\delta) = 2$ do not have type $A_{n,1}$ $(n \geqslant 1)$.
(Indeed, in case (R,M) is of spherical type, case 4 in the lemma does not occur (since it implies that the diagram is not a tree) and μ-graphs of type $A_{n,1}$ must be derived (via case 5) from case 1. But this can happen only for the graphs $H_{3,1}$, $H_{4,1}$ and $H_{4,2}$.)

Subgraphs from subdiagrams

A subset A of a graph Γ is called *convex* (or *geodetically closed*) if all shortest paths in Γ joining two vertices of A, lie entirely within A.

Let (W,R) be a Coxeter system, $r \in R$, $S = R\setminus\{r\}$, $X = W_S$ and let $\Gamma = \Gamma(W,X,r)$ be the corresponding Coxeter graph. Let $T \subseteq S$ and $\Delta = \Gamma(W_0, Y, r)$, where $W_0 = \langle T \cup \{r\}\rangle$ and $Y = W_T$. Then Δ can be regarded as a convex subgraph of Γ:

10.2.8. Theorem.

(i) *The map $\phi:\Delta\to\Gamma$ defined by $\phi(\delta) = \delta X$ is an isomorphism from Δ onto $\phi[\Delta]$.*

(ii) *$\phi[\Delta]$ is a geodetically closed subgraph of Γ.*

Proof. (i) ϕ is well defined: vertices of Δ have the form uY with $u \in W_0$, and $\phi(uY) = uYX = uX$ is a vertex of Γ; next, ϕ is injective: if $u, v \in W_0$ and $uX = vX$, then $v^{-1}u \in X \cap W_0 = Y$ so that $uY = vY$; finally, ϕ is an isomorphism: for $u, v \in W_0$ we have $v^{-1}u \in XrX$ if and only if $v^{-1}u \in YrY$ (indeed, since $E_{v^{-1}u} \subseteq T \cup \{r\}$, an expression for $v^{-1}u$ in XrX can be reduced to one in YrY).
(ii) Next we want to show that $\phi[\Delta]$ is geodetically closed, i.e., that if $\gamma, \delta \in \phi[\Delta]$ and $\gamma = \gamma_0, \gamma_1, \cdots, \gamma_k = \delta$ is a shortest path in Γ joining γ and δ, then there is a path $\epsilon_0, \epsilon_1, \cdots, \epsilon_k$ in Δ such that $\gamma_j = \phi(\epsilon_j)$ $(0 \leqslant j \leqslant k)$. By transitivity of W_0 on $\phi[\Delta]$ we may assume $\gamma = X$. We write

$$B(J,h) = W_J(rW_J)^h \setminus \bigcup_{i=0}^{h-1} W_J(rW_J)^i \quad (\text{for } J \subseteq S,\ h \geqslant 0),$$

so that $B(S,h)/X = \Gamma_h(X)$; let $B(J,-1) = \varnothing$. From 10.1.2 (i), (ii) we find immediately that $B(S,h) \cap W_0 = B(T,h)$. In particular, $d_\Gamma(\gamma,\delta) = d_\Delta(\gamma,\delta)$.

If we can prove that for some neighbour ϵ_1 of $\epsilon_0 := Y$ in Δ we have $\phi(\epsilon_1) = \gamma_1$, then we are done by induction on $k = d(\gamma,\delta)$. Let $\gamma_j = w_0rw_1r \cdots w_{j-1}rX$ $(0 \leqslant j \leqslant k)$ where $w_i \in X$ $(0 \leqslant i < k)$ and assume $w_0rw_1r \cdots w_{k-1}rw_k \in B(T,k)$ for some $w_k \in X$. Clearly it suffices to prove the following claim.

Claim. *Let $w_0rw_1r \cdots w_{k-1}rw_k \in B(T,k)$, where $w_i \in X$ $(0 \leqslant i \leqslant k)$. Then for all j $(0 \leqslant j \leqslant k)$ we have $w_0rw_1r \cdots w_{j-1}rw_j \in B(T,j)\langle S \cap r^\perp\rangle$.*

We prove this by induction on k and for fixed k induction on j. We may assume $k \geqslant 1$. If $j \geqslant 1$, then by induction (using the case $j = 0$) we have $w_0 = uv$, $u \in B(T,0) = Y$, $v \in \langle S \cap r^\perp\rangle$. Now $w_0rw_1r \cdots rw_j = (ur)(vw_1r \cdots rw_j)$ and $vw_1r \cdots w_{k-1}rw_k = (ur)^{-1}w_0r \cdots w_{k-1}rw_k \in B(S,k-1) \cap W_0 = B(T,k-1)$, so by induction on k we have $vw_1r \cdots rw_j \in B(T,j-1)\langle S \cap r^\perp\rangle$ and $w_0rw_1r \cdots rw_j \in B(T,j)\langle S \cap r^\perp\rangle$ as was to be shown.
So let $j = 0$. Let d be the unique shortest element of XwX for $w = w_0rw_1r \cdots w_{k-1}rw_k$. Then $d \in YwY$ (since $E_d \subseteq E_w \subseteq T \cup \{r\}$ and d is obtained by cancelling factors in $d = uwv$, where $u, v \in X$), and since neither hypothesis nor conclusion of the lemma is affected when we replace w_0 by uw_0 or w_k by w_kv for $u, v \in Y$ we may assume $d = w$. We may also assume that the expression $rw_1r \cdots rw_k$ is reduced, so that

$$l(d) + l(w_0) = k + \sum_{i=1}^{k} l(w_i).$$

If $l(w_0) \leqslant l(w_k)$, then $l(d) \geqslant k + \sum_{i=1}^{k-1} l(w_i) \geqslant l(rw_1r \cdots rw_{k-1}r)$, and since d is the unique shortest element in XdX it follows that $d = rw_1r \cdots rw_{k-1}r$. Now by 10.1.5 we have $w_0 \in \langle E_d\rangle\langle E_d^\perp\rangle$, but

$$\langle E_d\rangle\langle E_d^\perp\rangle \cap X = \langle E_d \cap S\rangle\langle E_d^\perp \cap S\rangle \subseteq \langle T\rangle\langle S \cap r^\perp\rangle$$

and we reach the desired conclusion $w_0 \in Y\langle S \cap r^\perp\rangle$.
On the other hand, if $l(w_0) > l(w_k)$, then writing $e = rw_1r \cdots rw_{k-1}r = w_0^{-1}dw_k^{-1}$ we have $l(w_0e) = l(d) + l(w_k) < l(e) + l(w_0)$ so that w_0e is not reduced. Reduce to obtain $w_0e = w_0'e'$ with $l(w_0') + l(e') = l(w_0e)$ and $l(w_0') < l(w_0)$. By induction on $l(w_0)$ we find from $w_0'e'w_k = w_0ew_k = d \in B(T,k)$ that $w_0'e'r \in B(T,k-1)\langle S \cap r^\perp\rangle$ and, with suitable $u \in \langle S \cap r^\perp\rangle$, that $w_0eru \in B(T,k-1)$. By induction on k it now follows that $w_0 \in Y\langle S \cap r^\perp\rangle$. This proves the claim and hence the theorem. □

Objects and their shadows

We next interpret some fundamental notions of the Buekenhout-Tits geometry of Coxeter systems in terms of subgraphs of Coxeter graphs.

An *object* (of *type* $R \setminus I$ or *cotype* I) is a coset wW_I $(w \in W, I \subseteq R)$. Two objects are *incident* when they have a nonempty intersection. The *shadow* of the object wW_I on the Coxeter graph $\Gamma = \Gamma(W,X,r)$ is the set of vertices incident with it, i.e., the set of vertices contained in wW_IX. As we saw above, this shadow induces a subgraph isomorphic to the Coxeter graph $\Gamma(W_I, W_{I\setminus\{r\}}, r)$ if $r \in I$ (and is a single point when $r \notin I$).

Remark. Let R be finite. We can build a Buekenhout-Tits geometry (cf. §A.9) by taking as objects of the geometry the objects of type $\{r\}$ for all $r \in R$ with incidence as above. The Coxeter diagram of the Coxeter system is a Buekenhout-Tits diagram of this geometry. (Indeed, the maximal flags of the geometry can be identified with the elements $w \in W$ (via $w \leftrightarrow \{wW_{R\setminus\{r\}} \mid r \in R\}$) since $\bigcap_{r \in R} W_{R\setminus\{r\}} = 1$, and, more generally, flags of corank i can be identified with cosets wW_I with $|I| = i$. Thus, the residues of rank 2 and type $\{i,j\}$ are isomorphic to the geometry of points and edges of an m_{ij}-gon. Note that we used that if in a collection of objects any two have nonempty intersection, then the entire collection has nonempty intersection, but this is true only when R is finite.)

We shall see that one can read off from the Coxeter diagram what kinds of cliques occur in the Coxeter graph. Indeed, essentially, cliques can be identified with objects of type $A_{n,1}$. We need some technicalities.

The *r-component* of a subset I of R is the connected component containing r of the subdiagram induced by I, or $\varnothing$ in case $r \notin I$. The subset I is called *r-connected* when it is its own r-component.

10.2.9. Theorem. *Two objects uW_I and vW_J have the same shadow on the Coxeter graph Γ if and only if I and J have the same r-component C, and $u^{-1}v \in W_I\langle S \cap C^\perp\rangle W_J$.*

Proof. 'If' is obvious. Now let $W_IX = wW_JX$, where I and J are r-connected, and $w \in D_{I,J}$. Since $w \in W_IXW_J$ it follows that $w \in X$. Let (for $t \geqslant 1$) $r = r_1, r_2, \cdots, r_t$ be a path in J without repetitions, i.e., the r_i are distinct elements of J, and $r_ir_{i+1} \neq r_{i+1}r_i$ $(1 \leqslant i \leqslant t-1)$. We show that $r_t \in I$ and $w \in \langle r_t^\perp\rangle$. Once we have shown this it will follow by connectedness of J that $J \subseteq I$, and by symmetry that $J = I$; also, that $w \in \langle S \cap J^\perp\rangle$, and we are done.

Consider the element $z = wr_tr_{t-1} \cdots r_1$ of wW_J. Since $wW_J \subseteq W_IX$ and $z \notin X$ (it contains precisely one factor r), there is an $i \in I$ with $l(iz) < l(z)$. But the expressions iw and $wr_tr_{t-1} \cdots r_1$ are reduced, so $iwr_t \cdots r_1 = wr_t \cdots \hat{r}_j \cdots r_1$. The right hand side does not contain a factor r_j, so the left hand side cannot contain precisely one such factor, and we must have $i = r_j$ and by 10.2.1 (i) $wr_t \cdots r_{j+1} \in \langle r_j^\perp\rangle$. But r_{j+1} does not commute with r_j, so $j = t$. Thus $r_t \in I$ and $w \in \langle r_t^\perp\rangle$, as was required. □

10.2.10. Theorem. *Every finite nonempty clique in Γ is the shadow of an object. Γ contains an m-clique (with $m \geqslant 2$) if and only if the diagram (R,M) contains a subdiagram of type A_{m-1} with endnode r. More precisely, any m-clique C in Γ determines a unique such subdiagram (its 'type'), and is the shadow of an object of its type. Cliques are in the same W-orbit if and only if they have the same type. If C, D are cliques of the same type, then there is a $w \in W$ such that $wC = D$ while w fixes $C \cap D$ pointwise.*

Proof. Let C be a finite clique of size $m \geqslant 1$. If $m = 1$, then $C = \{wX\}$ is the shadow of the object $wW_\varnothing$; thus we may assume that $m \geqslant 2$. Choose $\gamma \in C$ and let $C\setminus\{\gamma\}$ be the shadow of wW_I, where I induces the path $r = r_1, r_2, \cdots, r_{m-2}$ (in this order). We may suppose $w = 1$. The vertices of $C\setminus\{\gamma\}$ are the points

$\gamma_i = r_i r_{i-1} \cdots r_1 X$ $(i = 0, \cdots, m-2)$. Let $\gamma = uX$, where $u \in D_{\emptyset,S}$. Let $S_i := S \cap \{r_1, \cdots, r_i\}^{\perp}$ $(i \geqslant 0)$ and $T_0 := \{r\}$, $T_i := \{t \in S_{i-1} \mid (r_i t)^3 = 1\}$ $(i \geqslant 1)$.

Claim. *For each i $(0 \leqslant i \leqslant m-2)$ we have $u = v_i r_i r_{i-1} \cdots r$, where $v_i \in \langle S_i \rangle T_i$.* Indeed, this follows by induction on i. In the case $i = 0$ we claim $u = v_0 \in Xr$, which follows from $u \in XrX$ (since $\gamma \sim \gamma_0$) and $u \in D_{\emptyset,S}$. Next, by induction on j for $0 \leqslant j < i$ we see that $r_{j+1} r_{j+2} \cdots r_i v_i r_i \cdots r_{j+1} \in \langle S_j \rangle T_j \langle S_j \rangle$: for $j = 0$ this says $r_1 r_2 \cdots r_i u \in XrX$, which is true since $\gamma \sim \gamma_i$; and the induction step follows from 10.2.1 (i) since $r_{j+1} \in T_j$ and $T_j \cap S_j = \emptyset$ and (by the induction on i) $v_i r_i = v_{i-1} \in \langle S_{i-1} \rangle T_{i-1}$ so that $E_{v_i} \cap T_j = \emptyset$ for $j < i-1$. Thus $r_i v_i r_i \in \langle S_{i-1} \rangle T_{i-1} \langle S_{i-1} \rangle$. Now $v_i r_i = v_{i-1}$ can be written with a unique factor in T_{i-1} and so can $r_i v_i r_i$, so this unique factor must be r_i and again applying 10.2.1 (i) we find $v_i \in \langle S_i \rangle T_i \langle S_i \rangle$. But since $u \in D_{\emptyset,S}$ we have in fact $v_i \in \langle S_i \rangle T_i$, proving our claim.

Let $v_{m-2} = v r_{m-1}$ with $v \in \langle S_{m-1} \rangle$ and $r_{m-1} \in T_{m-2}$, then $u = v r_{m-1} \cdots r$, and C is the shadow of $v W_{I \cup \{r_{m-1}\}}$. □

Problems. 1. One may ask what conditions should be imposed on a subgraph Δ of a Coxeter graph Γ in order to guarantee that it is the shadow of an object. For example, does this hold when Δ is itself a Coxeter graph? It is not enough to require that Δ be geodetically closed (see the next question).
2. Let γ, δ be vertices of a Coxeter graph Γ. Is the union of all geodesics from γ to δ geodetically closed?

Association scheme and double coset diagram

Recall from Chapter 2 that a (not necessarily symmetric) association scheme is a pair $(X, \mathfrak{R})$, where X is a finite set and $\mathfrak{R} = \{R_0, \cdots, R_s\}$ is a partition of $X \times X$ (the elements of which are called *relations*) satisfying the following requirements: (i) $R_0 = I$, the identity relation; (ii) $R_i^{\mathsf{T}} \in \mathfrak{R}$ for all i; (iii) given $x, y \in X$ with $(y,x) \in R_i$, the number p^i_{jk} of $z \in X$ with $(z,x) \in R_j$ and $(z,y) \in R_k$ does not depend on x and y but only on their relation R_i.

The most important examples of association schemes arise in the situation where a group G acts transitively on a (finite) set X. In this case one takes for $\mathfrak{R}$ the partition of $X \times X$ into G-orbits, and requirements (i)-(iii) are easily verified. Relations can be identified with double cosets: the pair (x,y) is in relation $Gy^{-1}xG$. The number $s+1$ of relations (double cosets) is the *permutation rank* of G on X.

By §2.9, the distance-regular graphs are precisely those graphs Γ for which the distance distribution diagram is linear, i.e., for which the relations $R_i = \{(x,y) \mid d(x,y) = i\}$ $(0 \leqslant i \leqslant \mathrm{diam}(\Gamma))$ define an association scheme. We see that the number of relations is one more than the diameter.

In the present context we find an association scheme on the vertex set of our Coxeter graph Γ (when it is finite), i.e., on the set of cosets wX $(w \in W)$. Since each double coset contains a unique element in $D_{S,S}$, we may label the relations with the elements of $D_{S,S}$. The relation R_u is symmetric (and hence defines an undirected graph) if and only if $u = u^{-1}$ $(u \in D_{S,S})$; in particular R_r is symmetric (and defines Γ). Below we shall sometimes talk about the relation R_w also in cases where X is not finite.

Let us have a look at a nontrivial example, the double coset diagram for $E_{7,2}$.

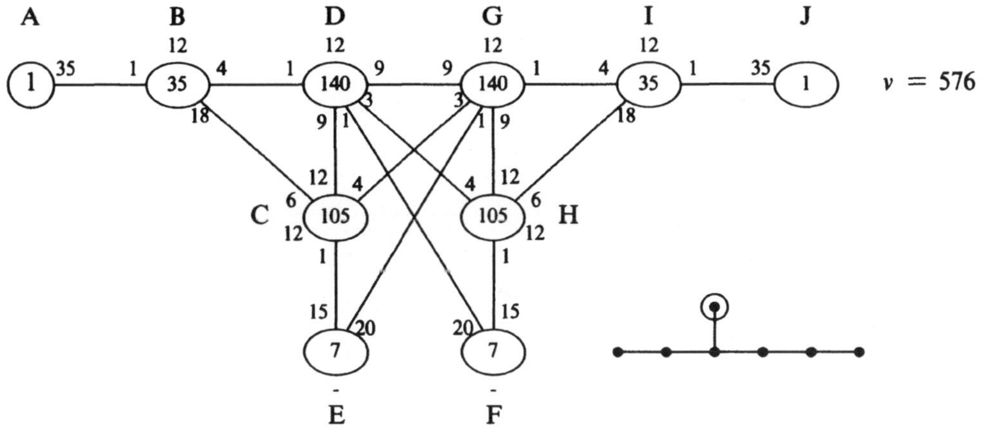

This example shows that in general the stabilizer in $\mathrm{Aut}\,\Gamma$ of two vertices γ, δ at distance $d(\gamma,\delta) = k$ need not be transitive on the set $C_k(\gamma,\delta) = \Gamma(\gamma) \cap \Gamma_{k-1}(\delta)$ as it is for $k = 2$ (cf. 10.2.6 (i)); indeed, we find here for $\gamma = X$, $\delta = gX$ (g in the double coset labelled G) that $C_k(\delta,\gamma)$ has 9 points in the double coset labelled D and 3 points in the double coset labelled C, so that not even the stabilizer of γ in $\mathrm{Aut}\,\Gamma$ is transitive on $C_k(\delta,\gamma)$.

Now that we have this example, let us have a closer look at it.
We have seen that subdiagrams yield geodetically closed subgraphs. Here we recognise subgraphs

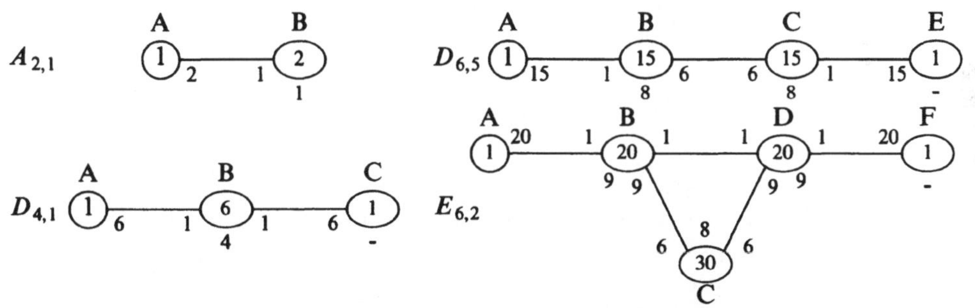

It is a nice and instructive exercise to try and explain the entire diagram above using simple geometric arguments only. Note that $E_{7,2}$ is a 16-cover of K_{36}.

The symmetry of the above diagram is explained by the following result.

10.2.11. Lemma. *Let (W,R) be a finite Coxeter system and w_0 be the longest word in W. Then:*

(i) *$Xw_0X = w_0X$ if and only if $w_0r = rw_0$; if this is the case, the automorphism $\gamma \mapsto w_0\gamma$ of Γ permutes double cosets XwX so that the double coset diagram is symmetric.*

(ii) *w_0 is central in W except in the following cases.*
- A_n ($n \geqslant 2$); here conjugation by w_0 flips the Coxeter diagram and hence w_0

commutes only with the reflection belonging to the middle node (if n is odd).
- D_{2m+1} $(m \geqslant 1)$*; here* w_0 *interchanges the two end nodes labelled* $2m$ *and* $2m+1$ *and commutes with all remaining nodes.*
- E_6*; here* w_0 *flips the diagram and hence commutes only with the two middle nodes.*

Proof. Recall that $l(ww_0) = l(w_0w) = l(w_0) - l(w)$ for all $w \in W$. In particular $l(w_0ww_0^{-1}) = l(w)$ for all $w \in W$ and $w_0Rw_0^{-1} = R$. (Note that $w_0^2 = 1$.) It follows that conjugation with w_0 is an automorphism of the Coxeter diagram. Clearly, $Xw_0X = w_0X$ is equivalent to $Xw_0 = w_0X$, hence to $w_0Xw_0^{-1} = X$, hence to $w_0(R \setminus \{r\})w_0^{-1} = R \setminus \{r\}$, hence to $w_0r = rw_0$. If this is the case, then $\gamma \mapsto w_0\gamma$ sends the double coset XwX to $w_0XwX = Xw_0wX$ hence is an automorphism of the double coset diagram. This settles (i).
The information (ii) can be found in BOURBAKI [113], Ch. VI, §4. □

Product expressions for k and v

If Γ is finite, then clearly it has $v = |W| \;/\; |W_S|$ vertices ($S = R \setminus \{r\}$). The number of vertices in relation i ($i \in D_{S,S}$) to a given vertex is

$$k_i = \frac{|W_SiW_S|}{|W_S|} = \frac{|W_S|}{|iW_Si^{-1} \cap W_S|} = \frac{|W_S|}{|W_{iSi^{-1} \cap S}|}$$

(for: two cosets uiW_S and viW_S with $u, v \in W_S$ are identical if and only if $i^{-1}u^{-1}vi \in W_S$; the last equality follows from 10.1.3).

Table 10.3. *The degrees of the finite irreducible Coxeter systems* (cf. BOURBAKI [113], Planches I-VIII):

Name	Degrees d_i $(1 \leqslant i \leqslant n)$	Coxeter number h
A_n	$2, 3, \cdots, n+1$	$n+1$
B_n	$2, 4, 6, \cdots, 2n$	$2n$
D_n	$2, 4, 6, \cdots, 2n-4, 2n-2, n$	$2n-2$
E_6	2, 5, 6, 8, 9, 12	12
E_7	2, 6, 8, 10, 12, 14, 18	18
E_8	2, 8, 12, 14, 18, 20, 24, 30	30
F_4	2, 6, 8, 12	12
H_3	2, 6, 10	10
H_4	2, 12, 20, 30	30
I_2^m	$2, m$	m

Associated to each finite Coxeter system (W,R) is a sequence $d_1, \cdots, d_n$ $(n = |R|)$ of integers, its *degrees*, satisfying the polynomial identity

$$W(t) := \sum_{w \in W} t^{l(w)} = \prod_{i=1}^{n} \frac{t^{d_i} - 1}{t - 1}$$

(cf. SOLOMON [691]). In particular, $|W| = W(1) = \prod_{i=1}^{n} d_i$, while the longest element w_0 of W has length $\sum_{i=1}^{n} (d_i - 1) = \frac{1}{2}nh$, where h is the *Coxeter number* of (W,R), the largest of the degrees. (The number h is also the order of the 'Coxeter element' $r_1r_2 \cdots r_n$; note that the conjugacy class of $r_1r_2 \cdots r_n$ is independent of the ordering of the r_i.)

Examples. $E_{7,2}$ has $v = \dfrac{2{\cdot}6{\cdot}8{\cdot}10{\cdot}12{\cdot}14{\cdot}18}{2{\cdot}3{\cdot}4{\cdot}5{\cdot}6{\cdot}7} = 2^6.3^2 = 576$ vertices; $H_{3,2}$ has $v = \dfrac{2{\cdot}6{\cdot}10}{2{\cdot}2} = 30$ vertices. Given i we can find k_i. In particular, $k = k_r = \dfrac{|W_S|}{|W_T|}$, so that $E_{7,2}$ has $k = \dfrac{2{\cdot}3{\cdot}4{\cdot}5{\cdot}6{\cdot}7}{2{\cdot}3{\cdot}4{\cdot}2{\cdot}3} = 35$ and $H_{3,2}$ has $k = \dfrac{2{\cdot}2}{1} = 4$.

Incidence graphs

Let (W,R) be a Coxeter system and fix $r, s \in R$. Put $X = W_{R \setminus \{r\}}$, $Y = W_{R \setminus \{s\}}$. The bipartite graph $\Gamma = \Gamma(W,X,Y)$ with as vertices the left cosets of X and Y in W, and adjacencies $xX \sim yY$ whenever $xX \cap yY \neq \emptyset$ is called a *Coxeter incidence graph*. [Thus, this graph is the incidence graph of the objects of types $\{r\}$ and $\{s\}$, as defined above.] In case (W,R) has a diagram named Z_n (for some letter Z) and r, s are the nodes labeled i, j in this diagram in Table 10.1 or 10.2, the graph Γ will be called the incidence graph of Coxeter type $Z_{n,\{i,j\}}$.
The halved graph of Γ on the set of left cosets of X has adjacencies $x_1X \sim x_2X$ whenever $x_2^{-1}x_1 \in XYX$. Thus, this halved graph will be the Coxeter graph $\Gamma(W,X,r)$ precisely when $Y \cap D_{S,S} = \{1,r\}$, where $S = R \setminus \{r\}$. But by 10.2.3 this implies that $(Y, R \setminus \{s\})$ has a diagram with a path with end node r as component. If also the halved graph of Γ on the set of left cosets of Y is the Coxeter graph $\Gamma(W,Y,s)$, then we have one of four possibilities for the component(s) containing r, s of the diagram of (W,R):

$A_{m,1} + A_{n,1}$

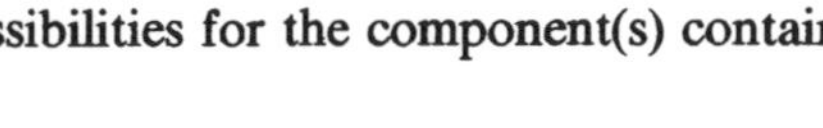

$A_{n,\{1,n\}}$

$A_{n,\{i,i+1\}}$

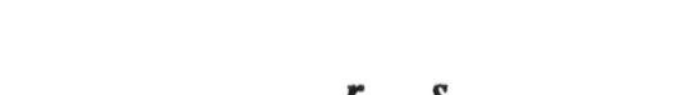

$D_{n,\{n-1,n\}}$

We find in the first case a complete bipartite graph (distance bitransitive of diameter 2), and in the second case a complete bipartite graph minus a 1-factor (distance-transitive of diameter 3, cf. Corollary 1.5.4). In the third case we find for finite i the incidence graph of the i- and $(i+1)$-subsets of some finite or countably infinite set (distance-bitransitive of diameter $\min(2i+1, 2n-2i)$; distance-transitive - namely, the doubled Odd graph - for $n = 2i+1$, cf. §9.1D). In the fourth case we find for finite n the n-cube (distance-transitive of diameter n, cf. §1.12).

10.3 The finite Coxeter graphs; root systems and presentations

Most of the finite Coxeter systems can be described in terms of so-called root systems. We sketch some of this material very briefly and refer to BOURBAKI [113] Ch. VI, for details. We shall also indicate some other descriptions in terms of well-known mathematical structures.

Root systems

From §3.10 we recall the following definition. A (reduced) *root system* is a finite collection Φ of nonzero vectors (called *roots*) spanning some Euclidean space $\mathbb{R}^l$ such that

(i) if $\alpha \in \Phi$, then $\mathbb{R}\alpha \cap \Phi = \{\alpha, -\alpha\}$,

(ii) if $\alpha,\beta \in \Phi$, then $\dfrac{2(\beta,\alpha)}{(\alpha,\alpha)} \in \mathbb{Z}$, and

(iii) if $\alpha,\beta \in \Phi$, then $w_\alpha(\beta) := \beta - \dfrac{2(\beta,\alpha)}{(\alpha,\alpha)}\alpha \in \Phi$.

The w_α are involutions, reflections in the hyperplane perpendicular to α.

The group $W = \langle w_\alpha \,|\, \alpha \in \Phi\rangle$ is called the *Weyl group* of the root system Φ. It is transitive on the set of all roots of any given length. Φ is called *irreducible* when it cannot be nontrivially decomposed as $\Phi_1 \cup \Phi_2$ where all vectors in Φ_1 are perpendicular to all vectors in Φ_2. The $\mathbb{Z}$-linear span $\mathbb{L} = \mathbb{Z}\Phi$ of Φ is a root lattice (cf. §3.10) when all roots have squared length 2.
A subcollection Δ of a root system Φ is called a *fundamental system of roots* when it has the properties

(i) Δ is a basis of $\mathbb{R}^l$,

and

(ii) each root, when written as a linear combination of vectors in Δ, has either only nonnegative or only nonpositive coefficients.

Each irreducible root system Φ contains a fundamental system of roots Δ, and $(W,\{w_\alpha \,|\, \alpha \in \Delta\})$ is a Coxeter system. Each irreducible finite Coxeter system except for H_3, H_4 and I_2^m ($m = 5$ or $m \geqslant 7$) arises in this way. (In fact, the Coxeter system $B_n = C_n$ arises, for $n > 2$, from the two distinct root systems B_n and C_n, see below.) Given Δ there is a unique vector $\tilde{\alpha} \in \Phi$ (called the largest root) such that each of its coordinates is not less than the corresponding coordinate of any other root: if $\tilde{\alpha} = \sum_{\alpha \in \Delta} n_\alpha \alpha$ and for some $\beta \in \Phi$, $\beta = \sum_{\alpha \in \Delta} m_\alpha \alpha$, then $n_\alpha \geqslant m_\alpha$ for all $\alpha \in \Delta$. Put $\tilde{\Delta} = \Delta \cup \{-\tilde{\alpha}\}$. The vectors in $\tilde{\Delta}$ have pairwise nonpositive inner product, i.e., $\tilde{\Delta}$ is an extended fundamental system of roots in the sense of §3.10. The graph $(\tilde{R},\tilde{M})$ where $\tilde{R} = \{w_\alpha \,|\, \alpha \in \tilde{\Delta}\}$ and $\tilde{M} = (m_{rs})_{r,s \in \tilde{R}}$, where m_{rs} is the order of rs, is called the *extended Coxeter diagram.* For irreducible Φ these are given below in Table 10.4.

Suppose that $\tilde{r} := w_{-\tilde{\alpha}}$ has a unique neighbour r in the graph $(\tilde{R},\tilde{M})$ - this holds in all cases except $\tilde{A}_l$. The stabilizer of $\tilde{\alpha}$ in W is precisely $X = \langle R \setminus \{r\}\rangle$ so that the cosets wX ($w \in W$) are in one-to-one correspondence with the roots that have the same length as $\tilde{\alpha}$, i.e., the largest length occurring. Let Φ_0 be this set of longest roots, then the Coxeter graph $\Gamma(W,X,r)$ can be described as the graph on Φ_0 where two roots are adjacent when they make the same angle θ as $\tilde{\alpha}$ and $r\tilde{\alpha}$ - let us call this graph the *root*

Table 10.4. *Extended Coxeter diagrams (the circled node represents $\tilde{r}$)*

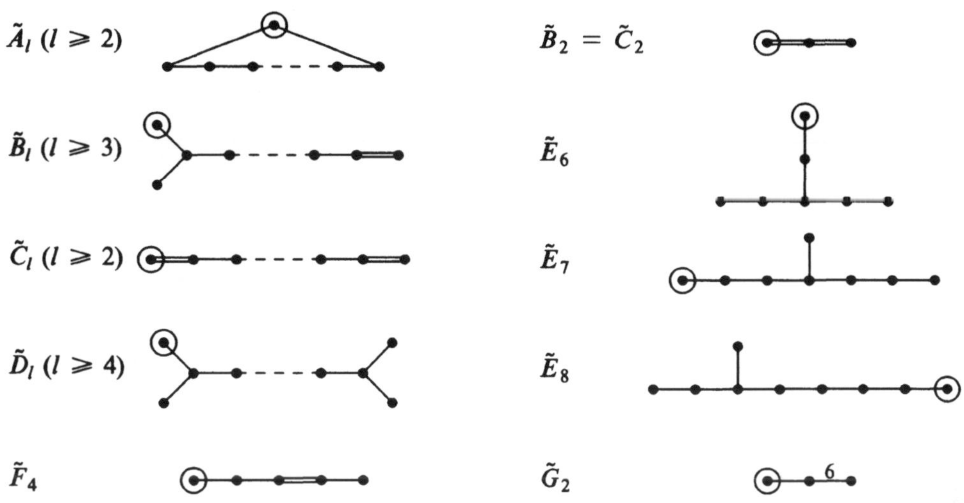

system graph for Φ. The subset Φ_0 is itself a root system, equal to Φ in the cases A_l, D_l, E_6, E_7, E_8; for $\Phi = B_l, C_l, F_4, G_2$ we have $\Phi_0 = D_l, lA_1, D_4, A_2$, respectively. The angle θ equals $\pi/3$ in all cases except C_l where it is $\pi/2$. (See also the remark after Proposition 3.10.1.)

We shall now discuss the individual finite irreducible Coxeter systems.

10.3.1. A_n

Take a fixed set Ω of cardinality $n+1$ (n finite) or $\aleph_0$ ($n = \infty$), say $\{1, 2, \cdots, n+1\}$. The Coxeter group W is the symmetric group on Ω (of order $(n+1)!$ when Ω is finite). A suitable R is the set of the n transpositions (12), (23), $\cdots$, $(n\ n+1)$. Identifying Ω with the standard basis $(e_i)_{1 \leqslant i \leqslant n+1}$ of $\mathbf{R}^{n+1}$, we may take the root system of type A_n to be $\Phi = \{e_i - e_j \mid 1 \leqslant i, j \leqslant n+1; i \neq j\}$.

$A_{n,j}$ is the graph that has the j-subsets of Ω as vertices, two j-subsets being adjacent when their union is a $(j+1)$-set (or, equivalently, when they meet in a $(j-1)$-set). In other words, $A_{n,j}$ is the Johnson graph $J(n+1, j)$ (for $1 \leqslant j \leqslant n$).

10.3.2. B_n and C_n

The Coxeter group W is the semidirect product 2^n:Sym(n), where 2^n denotes the elementary abelian group of this order. Let W act on the set of 2^n vectors of length n and entries ± 1 (in $\mathbf{R}^n$, say) by changing signs and permuting coordinate positions. A suitable R is the set of the $n-1$ transpositions (12), $\cdots$, $(n-1\ n)$ on the coordinate positions and the sign change σ_n changing the sign at coordinate position n. This Coxeter system corresponds to two essentially different root systems: letting

$$\Phi_\alpha = \{\pm e_i \pm e_j \mid 1 \leqslant i < j \leqslant n\} \cup \{\pm \alpha e_i \mid 1 \leqslant i \leqslant n\}$$

we find a root system both for $\alpha=1$ and for $\alpha=2$; the former is called B_n, the latter C_n.

$B_{n,j} = C_{n,j}$ is the graph that has the j-cliques in the complete n-partite graph $K_{n\times 2} = nK_2$ as vertices; two j-cliques are adjacent when their union is a $(j+1)$-clique (for $j < n$) or when they have $n-1$ elements in common (for $j = n$). In particular, $B_{n,n} = C_{n,n}$ is the n-cube.

The graph $B_{n,2}$ is the root system graph of type B_n; it coincides with the root system graph of type D_n; the graph $C_{n,1}$ is the root system graph of type C_n ($n \geqslant 2$).

10.3.3. D_n

The Coxeter group W here is the semidirect product 2^{n-1}:Sym(n), the subgroup of index two of the Coxeter group $W(B_n)$ described above preserving the two classes of the bipartition of the (connected and bipartite) graph $B_{n,n}$. A suitable R is the set of the $n-1$ transpositions (12), $\cdots$, $(n-1\ n)$ on the coordinate positions and the composition of $(n-1\ n)$ and the sign change $\sigma_{n-1}\sigma_n$ changing the signs at coordinate positions $n-1$ and n simultaneously. For Φ we can take $\{\pm e_i \pm e_j \mid 1 \leqslant i < j \leqslant n\}$.

For $j < n-1$, the graph $D_{n,j}$ is isomorphic to the graph $B_{n,j}$. The (isomorphic) graphs $D_{n,n-1}$ and $D_{n,n}$ are derived from the graph $B_{n,n}$ by taking as vertex sets the two classes of the bipartition of $B_{n,n}$, and letting two vertices be adjacent when they have distance two in $B_{n,n}$. (In other words, $D_{n,n}$ has as vertices the binary even weight vectors of length n, two vectors being adjacent when they have Hamming distance two, cf. §3.14, Example (i).) Note that for $\frac{1}{2}n < j < n-1$ the graphs $B_{n,j}$ and $D_{n,j}$ are isomorphic, but that the Coxeter groups of types B_n and D_n have different permutation ranks on these graphs, cf. Proposition 10.4.3.

The graph $D_{n,2}$ is the root system graph of type D_n ($n \geqslant 4$).

10.3.4. E_6

The graph $E_{6,1}$, known as the Schläfli graph, is the complement of the collinearity graph of the unique generalized quadrangle $GQ(2,4)$ (with 3 points on each line and 5 lines on each point - the concurrency graph of the 27 lines on a general cubic surface); it is strongly regular with parameters $(v,k,\lambda,\mu) = (27,16,10,8)$, and is the unique graph that is locally $D_{5,5}$. The vertices of $E_{6,j}$ for $j = 2,3,4,5,6$ are the 6-cliques, edges, triangles, maximal 5-cliques and subgraphs $K_{5\times 2}$, respectively. Two edges are adjacent when they are contained in a triangle, two triangles when their union is a 4-clique, two 6-cliques or two maximal 5-cliques when they have a triangle in common, and two subgraphs $K_{5\times 2}$ when they have a (maximal) 5-clique in common. (The diagram automorphism [conjugation by w_0] is easy to describe: $E_{6,1} \cong E_{6,6}$ by the isomorphism sending a point of the Schläfli graph to its 10 nonneighbours, and similarly $E_{6,3} \cong E_{6,5}$ by sending an edge to the set of 5 points nonadjacent to both of its end points.) The Coxeter group has order $2{\cdot}5{\cdot}6{\cdot}8{\cdot}9{\cdot}12 = 2^7.3^4.5$ and is isomorphic to $G.2$ where $G \cong PSU(4,2) \cong P\Omega^-(6,2) \cong P\Omega(5,3)$; it is the full automorphism group of the generalized quadrangle.

There is also a relatively simple geometric description of $E_{6,2}$: Consider an elliptic quadric Q in $PG(5,3)$ - Q has 112 points and carries the structure of a generalized quadrangle $GQ(3,9)$. Let z be a nonisotropic point, then $Q \cap z^\perp$ is a generalized quadrangle $GQ(3,3)$ on 40 points. Construct a graph on the remaining 72 points by calling points p and q adjacent when $\langle p,q,z\rangle$ is a plane meeting Q in two intersecting lines and the line $\langle p,q\rangle$ is not totally isotropic. This graph is isomorphic to $E_{6,2}$.

The graph $E_{6,2}$ is the root system graph of type E_6.

10.3.5. E_7

The graph $E_{7,7}$, known as the Gosset graph, is the unique distance-regular graph with intersection array $\{27,10,1;1,10,27\}$, and was constructed in Corollary 1.15.3 and in §3.11, Example (i). It is the unique locally Schläfli graph, an antipodal 2-cover of the complete graph K_{28}. Similar to what we saw for E_6, the graphs $E_{7,j}$ for $j<7$ can be described in terms of subgraphs K_i or $K_{6\times 2}$ of $E_{7,7}$ - the description is easily read off from the Coxeter diagram, interpreted as a Buekenhout-Tits diagram. The Coxeter group has order $2\cdot 6\cdot 8\cdot 10\cdot 12\cdot 14\cdot 18 = 2^{10}.3^4.5.7$ and is isomorphic to $G\times 2$, where $G \cong Sp(6,2) \cong P\Omega(7,2)$ is the automorphism group of the regular two-graph on 28 points (or of the 28 bitangents to a general quartic curve in a projective plane).

The graph $E_{7,1}$ is the root system graph of E_7. It has distribution diagram

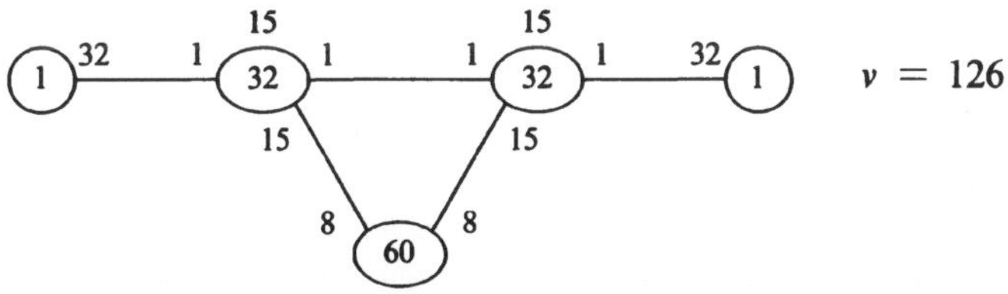

Sometimes a presentation of the root system of type E_7 different from the one given in §3.10 is useful. Take $\mathbb{R}^7$ with standard basis $\{e_i \mid i\in \mathbb{Z}_7\}$ and standard inner product (,). Next, fix a Fano plane structure on $\mathbb{Z}_7$, say such that the lines are $124, 235, \cdots, 713$. Then

$$\Phi := \frac{1}{\sqrt{2}}\{\pm 2e_i,\ \pm e_j \pm e_k \pm e_l \pm e_m \mid i\in \mathbb{Z},\ jklm \text{ is the complement of a line}\}$$

is a root system of type E_7 (with fundamental system $-e_2+e_4-e_5-e_6$, e_1, e_2, $-e_1-e_2-e_3+e_6$, e_3, $-e_3+e_5-e_6-e_7$, e_7).

10.3.6. E_8

$E_{8,8}$ is the root system graph of E_8, described as $E_8(1)$ in §3.11. It is the unique locally $E_{7,7}$ graph; an antipodal 2-cover of the strongly regular graph whose vertices are the nonisotropic points off a hyperbolic quadric in $PG(7,2)$ in which two vertices are adjacent whenever they are nonorthogonal; the parameters of the latter graph are $(v,k,\lambda,\mu) = (120,56,28,24)$. The root system Φ was given in §3.10. A description in terms of the unit Cayley numbers can be found in the ATLAS [216], p. 85. For a different description, consider the division algebra $\mathbb{H}$ of real quaternions with basis $\{1,i,j,k\}$. Set $q=\cos(\pi/5)-\frac{1}{2}i+(\cos(3\pi/5))j = \frac{1}{2}(\tau-i-j+\tau j)$ and $r=\frac{1}{2}(1+i+j+k)$. Let Q be the set of quaternions obtained from 1, q, r by even permutations of the coordinates 1, i, j, k and arbitrary sign changes. Then Q is a subgroup of order 120 of the multiplicative group of quaternions. Let $t:\mathbb{Q}(\tau)\to\mathbb{Q}$ be the trace function given by $t(\alpha+\beta\tau)=\alpha$ $(\alpha,\beta\in\mathbb{Q})$. Let L be the $\mathbb{Z}[\tau]$-linear span of Q

in $\mathbf{Q}(\tau)^4$, viewed as an integral lattice with respect to the quadratic form $q = t \circ N$, where N is the restriction to L of the norm function on $\mathbb{H}$ (thus $N(a_0 + a_1 i + a_2 j + a_3 k) = a_0^2 + a_1^2 + a_2^2 + a_3^2$). Then L has 240 vectors v with $q(v) = 2$, and this set Φ of 240 vectors is a root system of type E_8. (Root systems for E_7 and E_6 are given by $\Phi_7 = \{w \in \Phi \mid (v_0, w) = -1\}$ and $\Phi_6 = \{w \in \Phi_7 \mid (v_1, w) = 0\}$ for fixed $v_0 \in \Phi$ and $v_1 \in \Phi_7$.) The graph $E_{8,8}$ has these vectors as vertices and v, w are adjacent whenever $q(v+w) = 2$. Its double coset diagram was given in §3.11. Similar to what we saw for E_6 and E_7, the graphs $E_{8,j}$ for $j < 8$ can be described in terms of subgraphs K_i or $K_{7\times 2}$ of $E_{8,8}$. The Coxeter group has order 2·8·12·14·18·20·24·30 $= 2^{14}.3^5.5^2.7$ and is isomorphic to $2.P\Omega^+(8,2).2$.

10.3.7. F_4

For $j = 1, 2, 3, 4$ the vertices of $F_{4,j}$ are the 24 points, 96 edges, 96 triangles (2-faces) and 24 octahedra (3-faces), respectively, of the 24-cell. The adjacencies are the obvious ones. The graph $F_{4,1}$ is an antipodal 2-cover of the complete tripartite graph $K_{4,4,4}$ and one of the two graphs that are locally a cube (see Buset [152]; the other is $\overline{3\times 5}$, the complement of the 3×5 grid on 15 vertices). An explicit description follows. Take as vertex set $\{\infty, \infty'\} \cup C \cup C' \cup F$ where C and C' are two copies of a cube and F is the set of (the six) faces of C (or C'); C and C' induce a cube, F is a coclique, ∞ is adjacent to all vertices of C and ∞' to all vertices of C'; a face $f \in F$ is adjacent to the 4 points in C and the 4 points in C' on this face; finally, a vertex $x \in C$ is adjacent to its copy $x' \in C'$. The double coset diagram of this graph was given in §2.9. The Coxeter group is $2.(\mathrm{Alt}(4)\times\mathrm{Alt}(4)).2^2$ of order 2·6·8·12 $= 2^7.3^2$ (Coxeter & Moser [227], §9.6). A root system of type F_4 is gotten by joining $\{(\pm e_1 \pm e_2 \pm e_3 \pm e_4)/2\}$ to the root system of type B_4 given above.

$F_{4,1}$ is the root system graph of type F_4, that is, the root system graph of type D_4. Thus $F_{4,1}$ is isomorphic to $D_{4,2}$ (which provides another simple description of $F_{4,1}$).

10.3.8. H_3

For $j = 1, 2, 3$ the vertices of $H_{3,j}$ are the 12 points, 30 edges and 20 triangles (faces), respectively, of the icosahedron, or, equivalently, the 12 pentagons (faces), 30 edges and 20 points, respectively, of the dodecahedron. The adjacencies are the obvious ones. The double coset diagram of $H_{3,2}$ is

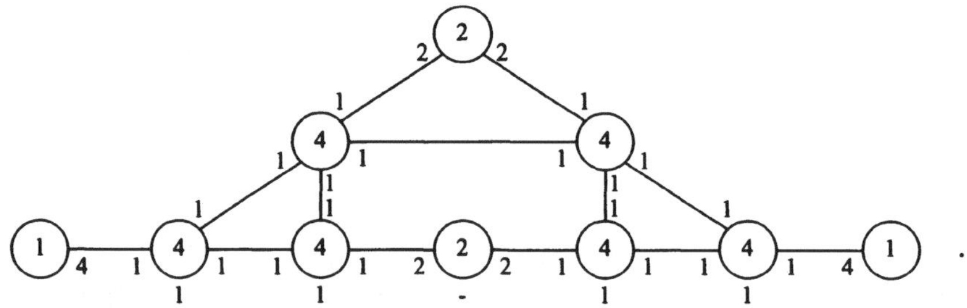

The Coxeter group is $2\times G$, where G is the simple group of order 60: $G \cong \mathrm{Alt}(5) \cong PSL(2,4) \cong PSL(2,5)$.

[Thus, this group is not Sym(5), as stated somewhere in the literature; the identification with $2\times\mathrm{Alt}(5)$ can be seen by taking $R = \{-(12)(34), -(12)(35), -(13)(24)\}$.]

10.3.9. H_4

For $j = 1, 2, 3, 4$ the vertices of $H_{4,j}$ are the 120 points, 720 edges, 1200 triangles (2-faces) and 600 tetrahedra (3-faces) of the 600-cell. The adjacencies are the obvious ones. A description of $H_{4,1}$ follows.
Define a graph with vertex set Q (see §10.3.6 above) by calling two quaternions x, y adjacent when their inner product $\frac{1}{2}(x\bar{y} + y\bar{x})$ equals $\frac{1}{2}\tau$. This graph is isomorphic to $H_{4,1}$. (For more details, see BOURBAKI [113], Ch. VI, Exercice 4.12, and BLOKHUIS et al. [94].) The Coxeter group $2.(\mathrm{Alt}(5)\times\mathrm{Alt}(5)).2 \cong (SL(2,5)\circ SL(2,5)).2$ has order $2\cdot 12\cdot 20\cdot 30 = 2^6.3^2.5^2$.

10.3.10. I_2^m

For $j = 1,2$ the vertices of $I_{2,j}^m$ are the m points and m edges of a regular m-gon, so that both Coxeter graphs are isomorphic to an m-gon. The Coxeter group is the dihedral group of order $2m$.

10.4 Global properties

In this book we are mostly interested in finite graphs (preferably distance-regular). In 10.4.1 we show that this means that we may restrict ourselves to finite Weyl groups: if W is infinite then Γ is not only infinite but usually also has infinite diameter and hence infinite permutation rank. In 10.4.3 we check the finite Coxeter systems one by one and find the diameter and permutation rank in all cases. In 10.4.6 we find all distance-regular Coxeter graphs. Finally, in 10.4.7-11 we investigate which maximal Coxeter subgroups $X = W_S$ lead to multiplicity-free characters 1_X^W.

Finiteness

The following result relates finiteness of W and Γ.

10.4.1. Proposition. *Let (W,R) be an irreducible Coxeter system where W is infinite. Put $S = R\setminus\{r\}$. Then:*

(i) *For each $r \in R$, the graph $\Gamma = \Gamma(W, W_S, r)$ has infinitely many vertices.*

(ii) *Γ has finite diameter d if and only if Γ is the graph of Coxeter type $A_{\infty,d}$.*

Proof. First we assume that R is finite, and prove by induction on $|R|$ that Γ has infinite diameter. If $R = \{r\}$ then W is finite. So there is an $s \in S$ such that $m_{r,s} > 2$. Consider the graph $\Delta_s = \Gamma(W_S, W_{S\setminus\{s\}}, s)$. If it has infinite diameter then there are arbitrarily long reduced words $sw_1sw_2 \cdots sw_{2l}s$ ($w_i \in W_{S\setminus\{s\}}$, $1 \leqslant i \leqslant 2l$) that cannot be written with fewer (than $2l+1$) factors s.
CLAIM. *For such words, the expression $rsw_1sw_2rsw_3 \cdots sw_{2l}rs$ obtained by inserting a factor r in front of every other s is reduced and cannot be written with fewer (than $l+1$) factors r.*
Indeed, according to Theorem 10.1.6, any two words with equal value in W can be transformed into a common reduced form by using only the two replacements $rr \to 1$ (where 1 denotes the empty word) and $rsr \cdots \to srs \cdots$ (where there are $m_{r,s}$ factors on both sides) for arbitrary $r, s \in R$. In our case no transformation involving r other than $rt = tr$ (for $m_{r,t} = 2$) is possible; indeed, the first counterexample cannot start

with $rtr \cdots$ since two factors r are separated by at least two factors s, and it cannot start with trt either because that would mean that before insertion of the factors r our word was not reduced. This proves the claim, and shows that Γ has infinite diameter (when Δ_s has).

If W_S is infinite then one of its direct factors is infinite, and choosing s in that factor we obtain a graph Δ_s that is infinite by inductive assumption, and we are done.

If W_S is finite then Γ has only finitely many points at any given distance from a vertex. But W is infinite, so W / W_S is infinite, and Γ must have infinite diameter.

Now assume that R is infinite and Γ has finite diameter. Considering finite connected subsets F of R containing r we see that the (distance preserving) subgraphs isomorphic to $\Gamma(W_F, W_{F\setminus\{r\}}, r)$ of Γ have infinite diameter (which is impossible) unless W_F is finite. For $n = |F| > 8$ that means that the Coxeter system (W_F, F) has a diagram of type A_n, B_n or D_n. It follows that the type of (W,R) is one of those in Table 10.2. Since $A_{2j-1,j}$ has diameter j, (W,R) cannot have a diagram of type A'_∞. Similarly, the diagram cannot be of type B_∞ or D_∞ since $B_{n,n}$ and $D_{n,n}$ have finite diameter growing with n (namely n and $[\frac{1}{2}n]$, respectively), so that $B_{n,n-j}$ and $D_{n,n-j}$ cannot have diameter bounded in terms of j only (independent of n). (Indeed, if d is an upper bound for this diameter, and s is the node labelled $n-j$ then each word in W can be written with at most d factors s; if r is the node labelled n then between any two factors s we find words that can be written with at most e factors r, where e is the diameter of $B_{j,j}$ or $D_{j,j}$, so that the diameter of $B_{n,n}$ or $D_{n,n}$ would be bounded by $(d+1)e$, a contradiction.) Thus the diagram must be of type A_∞, and it is easy to see that $A_{\infty,d}$ has diameter precisely d. □

10.4.2. Theorem. *For a Coxeter graph $\Gamma = \Gamma(W,X,r)$, the following are equivalent.*

(i) *Γ has finite valencies k_i ($i \in D_{S,S}$),*

(ii) *k is finite,*

(iii) *W_S is finite.*

Proof. Put $X = W_S$, $T = S \cap r^\perp$. Clearly, if X is finite then k is finite and $\Gamma_j(\gamma)$ is finite for all $j \geqslant 0$ and all vertices γ, so that also all valencies k_i ($i \in D_{S,S}$) are finite. Conversely, if all k_i are finite then in particular k is finite. If $r, r_1, \cdots$ is an infinite (simple) path in (R,M) then all vertices $r_j r_{j-1} \cdots r_1 rX$ ($j \geqslant 0$) are distinct neighbours of X, and k is infinite. If $M|_S$ has infinitely many components C_j then choosing $r_j \in C_j \setminus T$ we find infinitely many neighbours $r_j rX$ of X. Thus, if k is finite and X is infinite, then $M|_S$ has only finitely many components, and for some component C and each $s \in C \setminus T$ the graph $\Delta_s = \Gamma(W_S, W_{S\setminus\{s\}}, s)$ has infinite diameter; now let $u_j \in X$ be a word that requires precisely j factors s, then again all $u_j rX$ ($j \geqslant 0$) are distinct neighbours of X, contradiction. □

Diameter and permutation rank

10.4.3. Proposition. *The diameters of the Coxeter graphs with finite diameter and the permutation ranks for the corresponding Coxeter groups are as follows:*

(i) *$A_{n,j}$ has diameter $d = \min(j, n+1-j)$ and permutation rank $d+1$ ($1 \leqslant j \leqslant n$).*

(ii) *$A_{\infty,d}$ has diameter d and permutation rank $d+1$.*

(iii) *$B_{n,n}$ has diameter n and permutation rank $n+1$.*

(iv) *$B_{n,j}$ $(1 \leqslant j \leqslant n-1)$ has diameter $j+1$ and permutation rank $\frac{1}{2}(j+1)(j+2)$ if $j \leqslant \frac{1}{2}n$ and $\frac{1}{2}(n-j+1)(3j-n+2)$ if $\frac{1}{2}n \leqslant j \leqslant n-1$.*

(v) *$D_{n,n} = D_{n,n-1}$ has diameter $d = [n/2]$ and permutation rank $d+1$.*

(vi) *$D_{n,j}$ $(1 \leqslant j \leqslant n-2)$ has diameter $j+1$ and permutation rank $\frac{1}{2}(j+1)(j+2)$ if $j < \frac{1}{2}n$ and $1+\frac{1}{2}(n-j+2)(3j-n+1)$ if $\frac{1}{2}n \leqslant j \leqslant n-2$.*

(vii) *$I_{2,j}^m$ has diameter $d = [m/2]$ and permutation rank $d+1$ $(j = 1, 2)$.*

For E_6, E_7, E_8, F_4, H_3 and H_4, the diameters and permutation ranks are given in Table 10.5 below.

Table 10.5. *Diameters d and permutation ranks r of some Coxeter graphs (each node is labelled d,r for the corresponding Coxeter graph)*

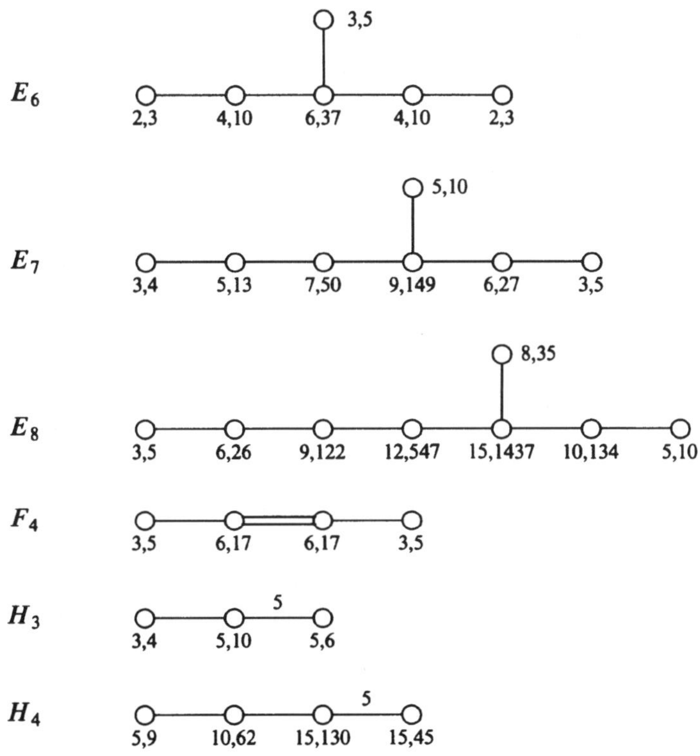

Proof. For the infinite families A_n, B_n and D_n this follows from the descriptions given in §10.3. E.g., for $D_{n,j}$ $(j < n-1)$ the possible relations between the two j-cliques are determined by the number i of classes in $K_{n\times 2}$ they both meet and by the number h $(0 \leqslant h \leqslant i)$ of elements they have in common, except that if $2j = n+i$ there are precisely two possibilities for each h. Thus we find for $j < \frac{1}{2}n$ the permutation rank $\sum_{i=0}^{j}(i+1) = \frac{1}{2}(j+1)(j+2)$ and for $\frac{1}{2}n \leqslant j < n-1$ the permutation rank becomes $2(2j-n+1)+\sum_{i=2j-n+1}^{j}(i+1) = 1+\frac{1}{2}(n-j+2)(3j-n+1)$. For $B_{n,j}$ $(j < n)$ we have similar computations but without the factor 2. For the finitely many cases E_m ($m = 6, 7, 8$), F_4, H_m ($m = 3, 4$) one may ask a computer (cf. BROUWER & COHEN [131]). □

10.4.4. Corollary. *A Coxeter graph has diameter 2 if and only if it has one of the types $A_{n,2}$ ($n \geqslant 3$ or $n = \infty$), $B_{n,1}$ ($n \geqslant 2$), $D_{n,1}$ ($n \geqslant 3$), $D_{5,5}$, $E_{6,1}$ or $I^5_{2,1}$.* □

(These graphs are known as the triangular graphs $T(n+1)$, the complete multipartite graphs $K_{n\times 2}$, the Clebsch graph or half 5-cube, the Schläfli graph and the pentagon, respectively. Note that $B_{2,2} = B_{2,1}$, $D_{4,4} = D_{4,1}$, $D_{n,n-1} = D_{n,n}$ ($n \geqslant 3$), $E_{6,1} = E_{6,6}$, $I^5_{2,2} = I^5_{2,1}$, and $B_{n,1} \cong D_{n,1}$ ($n \geqslant 3$).)

Remark. Concerning the diameters, the following can be said in case W is finite and irreducible. First of all, the diameter d_r of the Coxeter graph $\Gamma = \Gamma(W,X,r)$ equals the distance in that graph between the vertices X and w_0X. (Indeed, each $w \in W$ can be obtained from an arbitrary reduced expression of w_0 over R by omitting factors, since $w \leqslant w_0$ in the Bruhat order. Thus, each w has a product expression containing not more factors r than any product expression of w_0 (or any other element in w_0X).) Next, since the Coxeter diagram is a tree, we can write $R = R' \cup R''$, where R' and R'' induce cocliques in the Coxeter diagram. Now the Coxeter element $c = w'w''$, where $w' = \prod\{s \mid s \in R'\}$ and $w'' = \prod\{s \mid s \in R''\}$, satisfies $w_0 = c^{h/2}$ in case the Coxeter number h is even, and $w_0 = w''c^{(h-1)/2} = c^{(h-1)/2}w'$ in case h is odd (BOURBAKI [113], Ch. V, §6.2, Proposition 2 and Exercice 2), and it follows that $d_r \leqslant [h/2]$ for all r.

Problems. From Proposition 10.4.3 we see that $\max_{r\in R} d_r = [h/2]$; is there a direct proof of this? Is there a proof without case-analysis of the 'linear' behaviour of d_r along unramified simply laced parts of the Coxeter diagram?

Amply regular Coxeter graphs

Below we shall determine all distance-regular Coxeter graphs by showing that the good ones are even distance-transitive, while the bad ones (with the exception of a few graphs of type $H_{m,j}$) are not even amply regular, i.e., do not have constant μ.

10.4.5. Proposition. *Let $\Gamma = \Gamma(W, \langle S\rangle, r)$ be a Coxeter graph.*

(i) *Every two vertices at distance 2 have a unique common neighbour if and only if $S_4 = \emptyset$ and the connected component containing r of the diagram induced on $\{r\} \cup S_3 \cup T$ has type A_n ($n \geqslant 1$ or $n = \infty$) with end node r.*

(ii) *Every two vertices at distance 2 have more than one common neighbour if and only if Γ has one of the types $A_{n,i}$ ($i \geqslant 1$, $n \geqslant i$ or $n = \infty$), $A'_{\infty,0}$, $B_{n,1}$ ($n \geqslant 3$), $B_{n,n}$ ($n \geqslant 2$), $B_{\infty,0}$, $D_{n,1}$ ($n \geqslant 4$), $D_{n,n}$ ($n \geqslant 4$), $D_{\infty,0}$, $E_{6,1}$, $E_{7,7}$ or $H_{3,1}$.*

Proof. (i) By Theorem 10.2.6 (ii), every two vertices at distance two have a unique common neighbour if and only if $d \in X \cap D_{T,T}$, $d \notin S_3 \cup \{1\}$ implies $d \notin S_4 \cup S_3W_TS_3$, that is, if and only if $S_4 \cup S_3W_TS_3 \subseteq W_T \cup W_TS_3W_T$. Now apply Corollary 10.2.4.

(ii) Every two vertices at distance two have more than one common neighbour if and only if $d \in X \cap D_{T,T}$, $d \notin S_3 \cup \{1\}$ implies $d \in S_4 \cup S_3W_TS_3$, that is, if and only if

$$X = W_T S_4 W_T \cup W_T S_3 W_T S_3 W_T \cup W_T S_3 W_T \cup W_T.$$

Suppose first that $S_4 \neq \varnothing$, say $s \in S_4$. If $t \in S_3 \cup S_4$, $t \neq s$, then $d = st$ gives a contradiction, so $S_3 = \varnothing$ and $S_4 = \{s\}$. Also, $sW_{T^s} \subseteq W_T s W_T \cup W_T$ so that $\Gamma(X, W_T, s)$ is a clique, and using 10.2.3 we see that the connected component containing r of the diagram has type B_n ($n \geqslant 2$ or $n = \infty$) with r as the end node incident with the double edge.

Now let $S_4 = \varnothing$. If $S_3 = \varnothing$ then $\Gamma \cong A_{1,1}$ is a single edge. If $|S_3| = 1$ then $S = \{s\} \cup T$, $S_3 = \{s\}$ and $\Delta = \Gamma(X, W_T, s)$ has diameter at most two, and the conclusion follows by Corollary 10.4.4. If $|S_3| > 2$, then $d = t_1 t_2 t_3$ with three distinct $t_j \in S_3$ gives a contradiction. Finally, let $S_3 = \{s,t\}$. Considering $d = sts$ we see that $sts = t$, i.e., $m_{s,t} = 2$. Also, $tsW_{T^s} \subseteq tW_T\langle s\rangle W_T$ so that $\Gamma(\langle\{s\} \cup T\rangle, \langle T\rangle, s)$ is a clique, and similarly for $\Gamma(\langle\{t\} \cup T\rangle, \langle T\rangle, t)$. Using 10.2.3 we see that the diagrams induced on $T \cup \{s\}$ and $T \cup \{t\}$ have connected components containing s (resp. t) of type A_n ($n \geqslant 1$ or $n = \infty$) with end node s (resp. t). If s and t are in the same connected component of the diagram induced on $S = T \cup \{s,t\}$, then we have a subdiagram of type A_m ($m \geqslant 3$) with end nodes s and t. Let $s, r_2, r_3, \cdots, r_{m-1}, t$ be the nodes (in order) of this subdiagram, then $d = sr_2r_3 \cdots r_{m-1}tr_{m-1} \cdots r_3r_2s$ $= tr_{m-1} \cdots r_2sr_2 \cdots r_{m-1}t \notin S_3 W_T S_3$, contradiction. Thus r separates s from t and we find type $A_{n,i}$ or $A'_{\infty,0}$. The converse is clear. □

The finite, non-complete graphs in the conclusion of this proposition are all amply regular; clearly in case (i) (for the graphs $H_{3,2}$, $H_{3,3}$, $H_{4,3}$, $H_{4,4}$ and $I_{2,1}^m$) we have $\mu(\Gamma) = 1$, and the value of $\mu(\Gamma)$ for the finite graphs under (ii) can be read off from Table 10.6 below. Note that $B_{n,1} \cong D_{n,1}$ ($n \geqslant 4$).

Distance-regular Coxeter graphs

Now, let us restrict ourselves to the finite case. In Proposition 10.4.3 we found the diameter and the permutation rank in all cases, and when the permutation rank is just one more than the diameter our graph is distance-transitive. It is conceivable that some Coxeter graph would be distance-regular although its Coxeter group does not act distance-transitively, but in fact this turns out not to be the case.

10.4.6. Theorem. *Let (W,R) be a finite irreducible Coxeter system, $r \in R$ and $\Gamma = \Gamma(W, \langle R \setminus \{r\}\rangle, r)$ the corresponding Coxeter graph. Then Γ is distance-regular if and only if W acts distance-transitively on Γ. This is the case precisely when Γ is one of the graphs $A_{n,j}$ ($1 \leqslant j \leqslant n$), $B_{n,1} \cong D_{n,1}$ ($n \geqslant 3$), $B_{n,n}$ ($n \geqslant 2$), $D_{n,n-1} = D_{n,n}$ ($n \geqslant 3$), $E_{6,1} = E_{6,6}$, $E_{7,7}$, $H_{3,1}$, $H_{3,3}$ or $I_{2,1}^m = I_{2,2}^m$. Some parameters of these graphs are given in Table 10.6 below.*

Proof. We have seen that each of the graphs mentioned is distance-transitive. Conversely, if Γ is distance-regular, then either it has diameter one (and it is $A_{n,1}$), or it has constant μ. If $\mu > 1$ then by Proposition 10.4.5 (ii) Γ is isomorphic to one of $A_{n,j}$, $B_{n,1}$, $B_{n,n}$, $D_{n,n}$, $E_{6,1}$, $E_{7,7}$ or $H_{3,1}$. If $\mu = 1$ and Γ is finite, then by 10.4.5 (i) we have one of $H_{3,2}$, $H_{3,3}$, $H_{4,3}$, $H_{4,4}$ or $I_{2,1}^m$. Remains to exclude $H_{3,2}$ and $H_{4,3}$; but $H_{3,2}$ has nonconstant c_3 ($= 1$ or 2, cf. 10.3.8), and since $H_{4,3}$ contains $H_{3,2}$ as a geodetically closed subgraph, also $H_{4,3}$ has nonconstant c_3. □

Table 10.6. *Distance-regular Coxeter graphs*

Diagram	Graph Γ	d	v	parameters	label
$A_{n-1,j}$	Johnson graph $J(n,j)$	$\min(j,n-j)$	$\binom{n}{j}$	$k = j(n-j)$, $\mu = 4$	(C4)
$B_{n,1}$	$K_{n\times 2}$	2	$2n$	$k = \mu = 2n-2$	-
$B_{n,n}$	n-cube $H(n,2)$	n	2^n	$k = n$, $\mu = 2$	(C8)
$D_{n,n}$	Half n-cube	$[n/2]$	2^{n-1}	$k_i = \binom{n}{2i}$, $\mu = 6$	(C9)
$E_{6,1}$	Schläfli graph, complement of $GQ(2,4)$	2	27	$\{16,5;1,8\}$	-
$E_{7,7}$	Gosset graph	3	56	$\{27,10,1;1,10,27\}$	(C5)
$H_{3,1}$	Icosahedron	3	12	$\{5,2,1;1,2,5\}$	(A3)
$H_{3,3}$	Dodecahedron	5	20	$\{3,2,1,1,1;1,1,1,2,3\}$	(A13)
$I_{2,1}^m$	m-gon ($m \geqslant 5$)	$[m/2]$	m	$k = 2$, $\mu = 1$	(R1)

Multiplicity-free representations

In this subsection we determine all multiplicity-free permutation representations of Coxeter groups on maximal parabolic subgroups, i.e., subgroups of the form $W_{R\setminus\{r\}}$ (up to conjugacy).

10.4.7. Proposition. *Let $r \in R$, $S = R\setminus\{r\}$ and $X = W_S$. The following statements are equivalent.*

(i) *The character 1_X^W is multiplicity-free;*

(ii) *each element of $D_{S,S}$ has order at most 2;*

(iii) *every X-coset distinct from X contains an involution.*

Proof. Since every character of W is real (cf. CARTER [188], §13.2), we get from §2.9 that (i) holds if and only if every orbital of W on W/X is self-paired. This means that $XdX = Xd^{-1}X$ for every $d \in D_{S,S}$. But clearly $d^{-1} \in D_{S,S}$ if and only if $d \in D_{S,S}$, so self-pairedness of the orbital XdX is equivalent to $d = d^{-1}$ (cf. Theorem 10.1.2 (iv)). Thus, we have derived the equivalence of (i) and (ii).

The implication (iii) $\Rightarrow$ (i) is obvious, as (iii) implies that any orbital is self-paired. Suppose (ii). Let $x \in W\setminus X$. Then there are $g \in S$ and $d \in D_{S,S}$ such that $gdX = xX$. By (ii), gdg^{-1} is an involution in xX. Hence (ii) implies (iii). □

10.4.8. Corollary. *Let notation be as above. If $J \subseteq R$ is such that $1_{X\cap Y}^Y$ is not multiplicity-free, where $Y = W_J$, then 1_X^W is not multiplicity-free.*

Proof. Immediate from the equivalence of (i) and (ii) in the above proposition. □

Let us give two easy criteria for showing that an action is not multiplicity-free.

10.4.9. Lemma. *Let the group G have two permutation representations on sets X and Y, respectively, with respective permutation characters π and χ. Then (π,χ) equals the number of orbits of G_x on Y (for any $x \in X$) and also the number of orbits of G_y on X (for any $y \in Y$). In particular, if π and χ are multiplicity-free, and G has permutation rank e on X, then the number of orbits of G_y on X is at most e.*

Proof. Writing $H := G_x$ we find by Frobenius reciprocity $(\pi,\chi) = (1,\chi|_H)$, and by 'Burnside's Lemma' this is the number of orbits of H on Y. □

10.4.10. Lemma. *Let* Γ *be a graph with edge set* E. *If* Γ *has an induced subgraph*

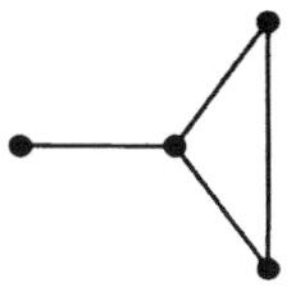

and $G \leqslant \operatorname{Aut}\Gamma$, *then the action of* G *on* E *is not multiplicity-free.*

Proof. One immediately sees that in the corresponding association scheme on E one has $p^h_{ij} \neq p^h_{ji}$ for certain relations h, i, j, so that the centralizer algebra is not commutative. □

[The weaker conclusion that G is not generously transitive on E is even more obvious: the horizontal and vertical edge are in a non-selfpaired relation.]

10.4.11. Theorem. *Suppose* (W,R) *is an irreducible Coxeter group of spherical type* $Y_m = (R,M) \neq H_4$. *The permutation representation of* W *on a maximal parabolic* $X = W_S$, *where* $S = R \setminus \{r\}$ *for some* $r \in R$, *is multiplicity-free if and only* $Y_{m,r}$ *is one of* $A_{n,i}$, $B_{n,i}$, $C_{n,i}$, $D_{n,i}$ *for* $i \leqslant n/2$ *or* $i \in \{n-1,n\}$, $E_{6,1}$, $E_{6,2}$, $E_{6,6}$, $E_{7,1}$, $E_{7,2}$, $E_{7,7}$, $E_{8,1}$, $E_{8,8}$, $F_{4,i}$ $(i = 1,4)$, $H_{3,1}$, $H_{3,3}$, $H_{4,1}$, *or* $I^m_{2,i}$ $(i = 1,2)$.

Proof. We treat the possibilities for M separately.
$M = A_n$. The graph $\Gamma(W,X,r)$ is the Johnson graph of the r-sets in $\{1, \cdots, n+1\}$, and W acts distance-transitively on Γ. Therefore (cf. Proposition 4.1.6 (ii)) the character 1^W_X is multiplicity-free.
$M = D_n$. Set $\Omega = \{\pm 1, \cdots, \pm n\}$. We can view W as the permutation group on Ω consisting of all even permutations preserving the system of imprimitivity $\{\{\pm j\} \mid 1 \leqslant j \leqslant n\}$ (cf. §10.3.3). The elements of R under this identification are $r_1 = (1,2)(-1,-2)$, $r_2 = (2,3)(-2,-3)$, $\cdots$, $r_{n-1} = (n-1,n)(-n+1,-n)$, and $r_n = (n-1,-n)(-n+1,n)$. (By the way, the quotient map $j \mapsto \bar{j} = \{\pm j\}$ induced by the system of imprimitivity yields a morphism $W \to W(A_{n-1}) \cong \operatorname{Sym}(n)$ given by $r_i \mapsto (\bar{i},\overline{i+1})$ $(1 \leqslant i \leqslant n-1)$ and $r_n \mapsto (\bar{n},\overline{n-1})$.)
The graphs $\Gamma(W,X,r)$ with $r = r_n$ and $r = r_{n-1}$ are isomorphic and W acts distance-transitively on them (Theorem 10.4.6), so without loss of generality we shall take $r = r_i$ with $i \leqslant n-2$. Now X is the stabilizer in W of the subset $Y = \{1, \cdots, i\}$ of Ω, and each coset of X consists of all elements of W mapping Y to some fixed i-subset Z of Ω. If $i \leqslant n/2$, then we can find an involution $g \in W$ interchanging Y and Z. Indeed, first of all there is an involution $g_0 \in \operatorname{Sym}(\Omega)$ preserving the system $\{\{\pm j\} \mid 1 \leqslant j \leqslant n\}$ and interchanging Y and Z. For, g_0 can be constructed as follows: it transposes j and $-j$ if $\{j,-j\} \subseteq (Y \setminus Z) \cup (Z \setminus Y)$, interchanges the remaining elements of $Y \setminus Z$ with those of $Z \setminus Y$ (and accordingly those of $(-Y) \setminus Z$ with those of $(-Z) \setminus Y$), and fixes all other elements. If g_0 is an odd permutation, then there is an odd number of $j \in \{1, \cdots, n\}$ such that $\{\pm j\} \subseteq (Y \setminus Z) \cup (Z \setminus Y)$. Thus, if $i \leqslant n/2$, there is an $m \in \Omega \setminus (Y \cup (-Y) \cup Z \cup (-Z))$, and $g = (m,-m)g_0$ is an involution in W interchanging Y and Z. This implies that every X-coset contains an involution, and by Proposition 10.4.7 we are done if $i \leqslant n/2$. On the other hand, if $n/2 < i < n-1$ then we may take $Z = \{-(n-i+1), n-i+2, \cdots, n\}$, and find that no involution in W interchanges Y and Z.
$M = B_n$ or C_n. Since $W(B_n) = W(C_n)$ is the permutation group on Ω consisting of all permutations preserving $\{\{\pm j\} \mid 1 \leqslant j \leqslant n\}$, and the subgroups X are again the

stabilizers of $\{1, \cdots, i\}$, it is immediate from the existence of g_0 above that W is multiplicity-free on W/X for all i ($1 \leqslant i \leqslant n$).

$M = E_n$ ($n = 6, 7, 8$). The permutation representations on the end nodes are studied in BROUWER & COHEN [131], where all distinguished double coset representatives are listed. Specifically, $E_{6,1}$, $E_{6,6}$, and $E_{7,7}$ correspond to graphs on which W acts distance-transitively (Theorem 10.4.6); $E_{6,2}$, $E_{7,1}$ and $E_{8,8}$ correspond to the root groups case of rank 5 (and so are multiplicity-free); all of the distinguished double coset representatives for the cases $E_{7,2}$ and $E_{8,1}$ are involutions (indeed, no two have the same length!); finally, in the case $E_{8,2}$ one finds a noninvolutory element of $D_{S,S}$ (e.g., the element "24534675(243168)57643542", where r_i is abbreviated as i).

Next we consider the unique node r with valency 3 in the Coxeter diagram. We saw above that $D_{5,3}$ does not lead to a multiplicity-free representation, and, by the corollary above, neither does $E_{n,4}$ ($n = 6, 7, 8$).

For the remaining nodes we find, using Lemma 10.4.10 above, that the corresponding representations are not multiplicity-free. Indeed, $E_{n,n-1}$ is the line graph of $E_{n,n}$ ($n = 6, 7, 8$), and $E_{n,n}$ contains the forbidden subgraph since $E_{8,8} \supset E_{7,7} \supset E_{6,6}$, and $E_{6,6}$ is locally $D_{5,5}$, and $D_{5,5}$ contains $K_1 + K_2$ (since it is the halved 5-cube and has $a_2 > 0$). Since $E_{6,5} \cong E_{6,3}$, and in view of the corollary above, this settles the case $M = E_n$ ($n = 6, 7, 8$).

$M = F_4$. If r is one of the end nodes, then by Theorem 10.4.3 the permutation rank is 5, and so the character must be multiplicity-free. On the other hand, $F_{4,2}$ is the line graph of $F_{4,1}$ and $F_{4,1}$ is locally a cube (and hence contains the forbidden subgraph), so that Lemma 10.4.10 yields that $F_{4,2}$ and (by symmetry of the diagram) $F_{4,3}$ do not lead to multiplicity-free permutation representations.

$M = H_3$ or H_4. The graphs $H_{3,1}$ and $H_{3,3}$ are distance-transitive, and the corresponding character is multiplicity-free. The graph $H_{3,2}$ has as vertex set the edges of $H_{3,1}$, and from Lemma 10.4.10 it follows that $H_{3,2}$ does not yield a multiplicity-free representation. Hence neither do $H_{4,2}$ and $H_{4,3}$. From BLOKHUIS e.a. [94] it follows that all orbits of $H_{4,1}$ are self-paired, but $H_{4,4}$ does not lead to a multiplicity-free representation, e.g., because $D_{S,S}$ contains the noninvolutory element "43423(124)32434".

$M = I_2^m$. Here $\Gamma(W,X,r)$ is a polygon, hence distance-transitive, hence multiplicity-free. □

Remarks. (i) The graphs $F_{4,1}$, $B_{4,2}$ and $D_{4,2}$ are isomorphic; the corresponding permutation representations have ranks 5, 6 and 7, respectively, and (hence) are all multiplicity-free. (More generally, if a permutation group G acts multiplicity-freely of rank e, and a subgroup H has permutation rank not more than $e+1$, then H also acts multiplicity-freely.)

(ii) In order to show that $E_{8,2}$ does not yield a multiplicity-free representation, we could also have applied Lemma 10.4.9. Indeed, $E_{8,8}$ has rank 5 and is multiplicity-free, but the stabilizer of a point of $E_{8,2}$ (that is, of an 8-clique in $E_{8,8}$) in $W(E_8)$ (is isomorphic to Sym(8) and) has 7 orbits on the points of $E_{8,8}$. The distance distribution diagram of $E_{8,8}$ around an 8-clique is

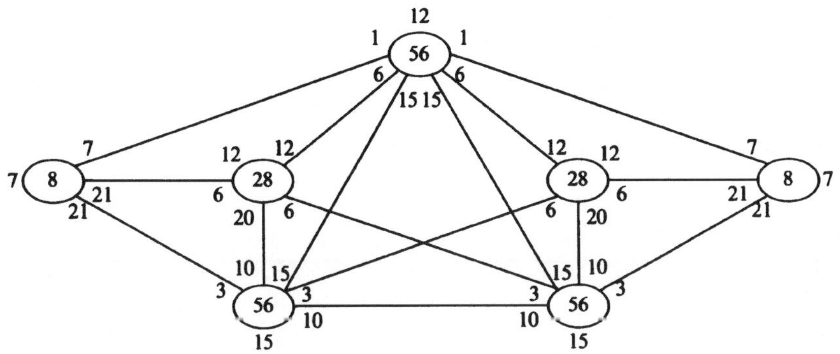

$v = 240.$

10.5 Tits Systems

Let G be a group. A *Tits system* in G is a 4-tuple (B,N,W,R) consisting of two subgroups B, N of G, a group W and a subset R of W such that G is generated by B and N, W is generated by R, the intersection $H = B \cap N$ is a normal subgroup of N with $W = N/H$, and, for any $r \in R$ and $w \in W$, we have

$$rBw \subseteq BwB \cup BrwB \tag{1}$$

and

$$rBr^{-1} \not\subseteq B. \tag{2}$$

Note that the notation Bw makes sense as w is a subset of G; in fact $Bn = Bn'$ for any two $n, n' \in N$ mapping to $w \in W$ under the natural morphism $N \to W$.
The group W is called the *Coxeter group of the Tits system.* If (W,R) is a Coxeter system with diagram X_n, then the Tits system is said to be *of type* X_n. (The above definition is the one given in Tits [744].)

The importance of this definition lies in the two facts that all Chevalley groups have Tits systems (cf. Carter [187], Proposition 8.2.1 and Theorem 13.5.4) and that many properties of these groups can be shown to hold for these groups by use of Tits systems. The second fact may not be surprising to those familiar with the main theorem of Tits [747] which implies that every finite group with a Tits system of rank $\geqslant 3$ is a Chevalley group. (See Fong & Seitz [291] for the case of rank 2.)

As a motivation of this definition, and as a link to the general theme of this book, let us show how Tits systems arise naturally in a geometric context (the special case of buildings of rank 2).
Theorem. *Let Γ be a generalized m-gon of order (s,t), $s, t > 1$, and suppose that G is a group of automorphisms of Γ which is transitive on the set of triples (γ,l,π) consisting of a vertex γ, a line $l \ni \gamma$ and an m-gon π which intersects l in γ and another vertex. Then G has a Tits system of type I_2^m.*

Proof. Let $B := G_{\gamma,l}$ and $N := G_\pi$. Clearly $H := N \cap B \trianglelefteq N$, and H is the pointwise stabilizer of π in G. Since any two point-line flags belong to some m-gon (in the sense that the line of each flag meets the m-gon in the point of the flag and in another point) and since by the hypothesis B is transitive on the set of m-gons on (γ,l) and N is transitive on all flags belonging to π, we see that $\langle B,N\rangle$ is transitive on the set of flags,

whence $G = \langle B,N \rangle$. Put $W = N/H$, and $R = \{g_1 H, g_2 H\}$, where g_1 is an element of $G_{l,\pi}$ moving γ and g_2 an element of $G_{\gamma,\pi}$ moving l. Now W is the dihedral group of order $2m$, R consists of two involutions in this group, and generates W. By assumption on G, the group B is transitive on the points of l different from γ and on the lines on γ different from l. We find $rBr \subseteq BrB \cup B$ for all $r \in R$, and, since $s, t > 1$, $rBr \not\subseteq B$.
Let us prove by induction on $l(w)$ that $rBw \subseteq BwB \cup BrwB$, and, if $l(rw) = l(w)+1$, even $rBw \subseteq BrwB$. If $l(rw) < l(w)$ this follows immediately, so assume $l(rw) = l(w)+1$. We show that $Bw \subseteq rBrwB$. It suffices to show that the set of images of the flag (γ,l) under Bw is contained in the set of images under $rBrw$. But w maps (γ,l) to some flag (γ',l') in π, and the assumption $l(rw) > l(w)$ means that the unique shortest gallery from (γ,l) to (γ',l') is contained in that from $r(\gamma,l)$ to (γ',l'). Now by assumption on G it follows that B is transitive on the set of all shortest galleries starting at (γ,l) of length less than m, and of course so is its conjugate rBr on all shortest galleries starting at $r(\gamma,l)$. This proves our claim, and shows that (B,N,W,R) is a Tits system in G. □

The next theorem justifies the name Coxeter group for W.

10.5.1. Theorem. *Let G be a group with Tits system (B,N,W,R) and let $I, J \subseteq R$. Then*

(i) *The pair (W,R) is a Coxeter system.*

(ii) *The set $G_J = BW_JB$ is a subgroup of G. Conversely, any subgroup of G containing B is of this form.*

(iii) *The map $W_IwW_J \mapsto G_IwG_J$ is a bijection from $W_I \setminus W / W_J$ onto $G_I \setminus G / G_J$. In particular, if for $w, w' \in W$ we have $BwB = Bw'B$, then $w = w'$.*

(iv) *$BrBwB = BrwB$ if and only if $l(rw) > l(w)$; otherwise, $BrBwB = BwB \cup BrwB$.*

(v) *$G_I \cap G_J = G_{I \cap J}$.*

(vi) *If $gBg^{-1} \subseteq G_I$, then $g \in G_I \quad (g \in G)$.*

(vii) *$G_IwG_J = BW_IwW_JB \quad (w \in W)$.*

Proof. Using (1) and (2), we get $B \cup Br^{-1}B = BrBr^{-1}B = (BrBr^{-1}B)^{-1} = B \cup BrB$, from which, by (2), we obtain $Br^{-1}B = BrB$, hence, by (1), $1 \in BrBrB \subseteq BrB \cup Br^2B$, so that r has order 2 for each $r \in R$. Now the present definition of Tits system coincides with that of BOURBAKI [113], Ch. IV, §2.1 and (i) and (iv) are found there in §2.4, and (ii), (iii), (v), (vi) and (vii) in §2.5. □

Remark. $W = N/(B \cap N)$, and R is uniquely determined by $R = \{r \in W \setminus \{1\} \mid B \cup BrB \text{ is a group}\}$ so that, instead of giving the Tits system (B,N,W,R), it suffices to give the *BN-pair* (B,N).

10.5.2. Lemma (COHEN [201]). *Let $r \in R$ and $D \subseteq W$ be such that $rd \in D$ whenever $d \in D$ and $l(rd) < l(d)$. Then $DB \cap BrBDB = (D \cap rD)B$.*

Proof. Let $g \in dB \cap BrBeB$ with $d, e \in D$. If $l(re) > l(e)$ then by 10.5.1 (iv) $BrBeB = BreB$. Since the double cosets BdB and $BreB$ are not disjoint, they coincide, so $d = re$ and $g \in (D \cap rD)B$. If $l(re) < l(e)$ then $re \in D$ and $BrBeB = BeB \cup BreB$ and it follows that $d = re$ or $d = r(re)$ so that again $g \in (D \cap rD)B$. □

10.5.3. Lemma. *Let $V \subseteq W$ and $w \in W_I$ with $W_I V \subseteq V$. Then $wBVB = BVB$.*

Proof. Clearly $wBVB \subseteq BwBVB \subseteq BVB$ and, similarly, $w^{-1}BVB \subseteq BVB$ so that $BVB \subseteq wBVB$. □

Let us write $B_w = B \cap w^{-1}Bw$.

10.5.4. Lemma. *Let $u, v \in W$ with $l(uv) = l(u)+l(v)$. Then*

(i) $BuBvB = BuvB$.

(ii) $Bu^{-1}BuB \cap BvBv^{-1}B = B$.

(iii) $B_{uv} = v^{-1}B_u v \cap B \subseteq B_v$.

(iv) $B = B_u B_{v^{-1}}$.

Proof. (i) follows by induction on $l(u)$ from Theorem 10.5.1 (iv). (ii) Use induction on $l(v)$; the case $l(v) = 0$ is obvious. Take $r \in R$ with $v = wr$ and $l(v) > l(w)$. Then $BvBv^{-1}B = BwBw^{-1}B \cup BwBrBw^{-1}B$; as $Bu^{-1}BuB \cap BwBw^{-1}B = B$ by induction, and $Bu^{-1}BuB \cap BwBrBw^{-1}B = \varnothing$ since $BuBwB \cap BuBwBrB = BuwB \cap BuwrB = \varnothing$, this implies the required identity.
(iii) From (ii) we have $v^{-1}u^{-1}Buv \cap B \subseteq v^{-1}Bv$, and then (iii) is clear.
(iv) Clearly, $B_u B_{v^{-1}} \subseteq B$. Also, $uBv \subseteq BuvB$, so $B \subseteq (u^{-1}Bu)(vBv^{-1})$. But if $b = gh$ with $g \in u^{-1}Bu$ and $h \in vBv^{-1}$, then $g^{-1}b = h \in B$ by (ii), so that $g \in B_u$ and $h \in B_{v^{-1}}$. □

10.5.5. Lemma. *If $w \in D_{J,\varnothing}$, then $w^{-1}G_J w \cap B \subseteq B_w$.*

Proof. Let S_w be the set of all elements of W obtained by omitting factors from a shortest expression of w as a product of elements of R (i.e., $S_w = \{v \in W \mid v \leqslant w\}$ in the notation of 10.1.7). Then, by the exchange condition,

$$W_J w \cap S_w = \{w\},$$

whence $W_J \cap S_w w^{-1} = \{1\}$, so that

$$G_J \cap wBw^{-1} \subseteq BW_JB \cap BwBw^{-1}B \subseteq BW_JB \cap BS_w w^{-1}B = B.$$

Thus, $w^{-1}G_J w \cap B \subseteq w^{-1}Bw \cap B = B_w$. □

10.5.6. Lemma (cf. Richen [612]). *Suppose that W is finite with longest element w_0. Then for each $w \in W$ we have $B_w \cap B_{w_0 w} = B_{w_0}$.*

Proof. Since $l(w_0) = l(w_0 w^{-1})+l(w)$, we find, using 10.5.4 (iii), that $B_w \cap B_{w_0 w} = w^{-1}Bw \cap w^{-1}w_0 B w_0 w \cap B = w^{-1}B_{w_0}w \cap B \subseteq w^{-1}B_{w_0 w^{-1}}w \cap B = B_{w_0}$. □

The association scheme of a Tits system

Let G be a group with Tits system $T = (B,N,W,R)$. A subgroup of G is said to be *parabolic* (with respect to T) if it contains a conjugate of B. It is called a *standard parabolic* (subgroup) of G when it contains B, i.e., when it is of the form G_I for some subset I of R.

As before we find for every $S \subseteq R$ an association scheme on the cosets of a standard parabolic $P = G_S$ in G; relations correspond to double cosets: the pair (xP, yP) is in relation $Py^{-1}xP$, and the following lemma shows that we may index the relations with elements of $D_{S,S}$.

10.5.7. Lemma. *There is a one-to-one correspondence between* $D_{S,S}$ *and* $P \setminus G / P$ *given by* $d \mapsto PdP$ $(d \in D_{S,S})$.

Proof. Clearly any coset PxP can be written as PdP with $d \in D_{S,S}$. If $PdP = PeP$, and $d, e \in D_{S,S}$, then $XdX = XeX$ so that $d = e$. □

Remark. For $d \in D_{S,S}$, the map of the lemma induces a bijective morphism $DC(W,X,d) \to DC(G,P,d)$ of the corresponding double coset graphs, which in general need not be an isomorphism, see 10.6.7.

Below we shall always index the relations on G/P with elements of W (usually of $D_{S,S}$), so that $(xP,yP) \in R_w$ is equivalent to $y^{-1}x \in PwP$. In particular, R_r is the adjacency relation of $\Gamma(G,P,r)$ and R_1 the identity relation. Note that we shall talk about relations R_w also in case G/P is infinite (e.g., in 10.6.3).

Nonexistence results

Since groups with a Tits system of rank 2 (of type I_2^m, say) yield generalized m-gons, it follows from Theorem 6.5.1 that in the finite case we must have $m \in \{3,4,6,8\}$. On the other hand, 'free' constructions of infinite generalized m-gons exist for arbitrary m, and in fact Tits has shown that if (W,R) is a finite Coxeter system, then a group with Tits system (B,N,W,R) exists if and only if the type of (W,R) is not H_3 or H_4 (see also Ronan & Tits [622]).

10.5.8. Theorem.

(i) (Tits [750]) *There is no group with Tits system* (B,N,W,R) *where* (W,R) *has type* H_3 *or* H_4.

(ii) (Feit & Higman [283]) *There is no finite group with Tits system* (B,N,W,R) *where* (W,R) *has type* I_2^m *with* $m = 5$ *or* $m = 7$ *or* $m \geqslant 9$. □

10.6 Graphs of Lie Type

Let G be a group with Tits system (B,N,W,R), and let $r \in R$, $S = R \setminus \{r\}$, $X = W_S$, $P = BXB = G_S$. The graph $\Gamma = \Gamma(G,G_S,r)$ ($= \Gamma(G,G_S,n_r)$ for any representative $n_r \in N$ of $r \in N/H = W$) whose vertices are the left cosets gP $(g \in G)$, g_1P and g_2P being adjacent if and only if $g_2^{-1}g_1 \in PrP$, is called a *graph of Lie type*.

Clearly, G is transitive on the set of vertices and on the set of edges of Γ. Associated with Γ is the Coxeter graph $\Delta := \Gamma(W,X,r)$. When Δ is a Coxeter graph of type $X_{n,i}$ (i.e., when (W,R) is a Coxeter system with diagram X_n for some letter X and r is the node of X_n numbered i in Table 10.1), Γ is called a *graph of Lie type* $X_{n,i}$.

The theory of graphs of Lie type is entirely analogous to that of Coxeter graphs, and most questions on graphs of Lie type can be reduced to the corresponding questions for Coxeter graphs.

Subgraphs from subdiagrams

Just as in the case of Coxeter groups, we find geodetically closed subgraphs from subdiagrams.

10.6.1. Theorem. *Let* $T \subseteq S$, *and* $\Gamma_0 = \Gamma(G_{T \cup \{r\}}, G_T, r)$. *Then:*

(i) *The map* $\phi : \Gamma_0 \to \Gamma$ *defined by* $\phi(\gamma_0) = \gamma_0 P$ *is an isomorphism from* Γ_0 *onto* $\phi[\Gamma_0]$.

(ii) $\phi[\Gamma_0]$ *is a geodetically closed subgraph of* Γ.

Proof. (i) Clearly, ϕ is well defined and preserves adjacency. It is injective by 10.5.1 (v). If we write $Y = W_T, Q = BYB = G_T$ then if $v^{-1}u \in PrP$ for $u,v \in G_{T \cup \{r\}}$ it follows that $v^{-1}u \in PrP \cap BW_{T \cup \{r\}}B = BYrYB = QrQ$.
(ii) We must show that if $\gamma,\delta \in \phi[\Gamma_0]$ and $\gamma = \gamma_0, \gamma_1, \cdots, \gamma_k = \delta$ is a shortest path in Γ joining γ and δ then there is a path $\epsilon_0, \epsilon_1, \cdots, \epsilon_k$ in Γ_0 such that $\gamma_j = \phi(\epsilon_j)$ $(0 \leq j \leq k)$. Without loss of generality, take $\gamma = P$. Now $\delta \in P(rP)^k \setminus P(\langle r \rangle P)^{k-1}$, so if we write $\gamma_j = b_j w_j P$ with $b_j \in B$, $w_j \in W$, then $\gamma_{j+1} \sim \gamma_j$ implies $w_{j+1} \in w_j XrX \cup w_j X$, but $w_{j+1} \in w_j X$ is impossible, since otherwise $\delta \in P(\langle r \rangle P)^{k-1}$. Thus $w_0 X, w_1 X, \cdots, w_k X$ is a shortest path in Δ, with $w_0 = 1$ and $w_k \in W_{T \cup \{r\}}$. By Theorem 10.2.8 we may assume $w_j \in W_{T \cup \{r\}}$ for all j $(0 \leq j \leq k)$, and hence we can take $\epsilon_j = b_j w_j Q$. □

Objects

We can associate a Buekenhout-Tits geometry to a group with a Tits system, just as we did for Coxeter graphs. An *object* (of *type* $R \setminus I$ or *cotype* I) is a coset gG_I $(g \in G, I \subseteq R)$. Two objects are *incident* when they have a nonempty intersection. The *shadow* of the object gG_I on the graph of Lie type $\Gamma = \Gamma(G,P,r)$ is the set of vertices (objects of type $\{r\}$) incident with it, i.e., the set of vertices contained in gG_IP. As we saw above, this shadow induces a subgraph isomorphic to the graph of Lie type $\Gamma(G_I, G_{I \setminus \{r\}}, r)$ if $r \in I$ and is a single point when $r \notin I$.

Lines

A *line* of Γ is a subset of the form $gB\langle r \rangle P / P$ for some $g \in G$, i.e., the shadow of an object of cotype $\{r\}$. (The terminology 'line of Γ' may be misleading: this concept depends on G rather than Γ; see also below under 'Singular lines'.) Let $Q := R \setminus r^{\perp}$. Then the lines are the shadows of objects of type Q. (Indeed, $gG_{r^{\perp}}P = gB\langle r \rangle P$.)

By definition of Γ, two points are adjacent if and only if they are collinear. In particular, lines induce cliques in Γ. The name 'line' is justified by the next proposition.

10.6.2. Proposition. *Two adjacent vertices of* Γ *are on a unique line, and each line contains at least three points.*

Proof. Let the vertices P and rP be on the line $gB\langle r \rangle P / P$. Let $g \in BwB$ with $w \in W$, then $\{w^{-1}, w^{-1}r\} \subseteq X \cup rX$ so that $w^{-1} \in (X \cap rXr) \cup (rX \cap Xr)$. Now $w \in W_T \cup rW_T$ where $T = S \cap r^{\perp}$, so that

$$gB\langle r \rangle P = gB\langle r \rangle XP = B\langle r \rangle P$$

by Lemma 10.5.3. Thus $B\langle r \rangle P$ is the unique line on P and rP. If it contains no other points then $BrP = rP$, i.e., $rBr \subseteq P$, and by Theorem 10.5.1 (vi) this implies $r \in X$, a contradiction. □

10.6.3. Theorem ('Strong gamma space property'). *Let x be a point and L a line. Then either there is a relation R_i ($i \in D_{S,S}$ - cf. Lemma 10.5.7) such that $(x,y) \in R_i$ for all y, or there are two relations R_i, R_j ($i, j \in D_{S,S}$) such that $(x,y_0) \in R_i$ for a unique point $y_0 \in L$, while $(x,y) \in R_j$ for all points $y \in L \setminus \{y_0\}$. In the latter case, $l(j) > l(i)$. In particular, the distance $d(x,y)$ for $y \in L$ takes at most two values, and when it takes two values, then there is a unique point $y_0 \in L$ such that $d(x,L) = d(x,y_0) = d(x,y) - 1$ for $y \in L \setminus \{y_0\}$.*

Proof. Let x be the point gP and L the line $B\langle r\rangle P / P$. The relations of (x,y) for $y \in L$ correspond to the double cosets $P\langle r\rangle BgP$, and if $g \in BwB$ with $w \in W$ then $P\langle r\rangle BgP \subseteq PwP \cup PrwP$. Thus, the only possible relations are R_w and R_{rw}. Since $G_{\{r\}}$ stabilizes L we may assume $l(rw) > l(w)$. Now $(gP,P) \in R_w$ and $(gP,brP) \in R_{rw}$ for all $b \in B$, so that if $R_w \neq R_{rw}$ then relation R_w occurs precisely once. Let $R_w = R_i$, $R_{rw} = R_j$, $i, j \in D_{S,S}$, $i \neq j$. Then w and rw have reduced expressions $w = u_1 i u_2$, $rw = v_1 j v_2$ $(u_1,u_2,v_1,v_2 \in X)$, and since rv_1jv_2 is not reduced we have $w = r(rw) = \widehat{v_1 j v_2}$ where the hat indicates that one factor in an expression as product of elements of R has been cancelled. If this factor occurred in j then we obtain the desired conclusion $l(i) < l(j)$. The cancelled factor cannot have been in v_1. If $w = v_1 j \hat{v_2}$, then $XiX = XjX$ and $i = j$, contrary to our assumption.

Finally, if we have a reduced expression e for rw in which the factor r occurs k times, then since $r(rw)$ is not reduced, the initial factor r in $w = re$ will cancel, so that w has a reduced expression in which the factor r occurs at most k times. In other words: $d(gP,P) \leqslant d(gP,brP)$ for all $b \in B$. □

Singular lines

Often, but not always, the lines of Γ (and, more generally, all objects) can be found from Γ alone. For the lines this is certainly the case when they are the singular lines of Γ, and certainly not in the case of a graph of Lie type A_n ($n \geqslant 2$), since such graphs are complete. In the spherical case we shall see below that the lines can always be found from Γ, except in case A_n ($n \geqslant 2$). Given Γ together with its lines, COHEN [205], §3, shows how to find all remaining objects (in the spherical case).

Let us now determine the singular lines of Γ. By edge transitivity of G we may restrict study to the edge (P,rP).

10.6.4. Lemma *Let $\gamma = P$ and $\delta = rP$ be two adjacent vertices of Γ. Then the singular line on $\{\gamma,\delta\}$ is $\{\gamma,\delta\}^{\perp\perp} = L^{\perp\perp}$, where L is the clique $G_{\{r\}}P / P$ in Γ. If we put $V = W_T S_3 r X$ (with notation as in §10.2) and $K = BVP / P$, then $L^{\perp} \setminus L = K$. Next, either $L = L^{\perp\perp}$ or we have $S_3 = \{s\}$, $sW_Ts \subseteq W_T \cup W_TsW_T$, in which case $L^{\perp} = L^{\perp\perp}$. (Thus, a singular line is either the shadow of an object of cotype $\{r\}$, or of an object of cotype $\{r,s\} \cup T$.) When (W,R) is finite then $L^{\perp} = L^{\perp\perp}$ if and only if when the Lie type of Γ is one of $A_{n,1} = A_{n,n}$ ($n \geqslant 1$), $B_{n,n-1}$ ($n \geqslant 3$), $F_{4,2} = F_{4,3}$.*

Proof. Clearly $L = G_{\{r\}}P / P$ is a clique in Γ containing γ and δ. We claim $\{\gamma,\delta\}^{\perp} \subseteq K \cup L$. Indeed, if $\epsilon \in \{\gamma,\delta\}^{\perp}$, then $\epsilon = xP$ with $x \in PW_{\{r\}}P \cap rPW_{\{r\}}P = BX\langle r\rangle XB \cap rBX\langle r\rangle XB$. Let $D = X\langle r\rangle X$ and apply Lemma 10.5.2 to find $x \in B(D \cap rD)B = B(X \cup rX \cup V)B$ and $\epsilon \in K \cup L$. Thus $L^{\perp} \subseteq \{\gamma,\delta\}^{\perp} \subseteq L \cup K$. In order to show $K = L^{\perp} \setminus L$ we need only prove $K \subseteq L^{\perp} \setminus L$. By Theorem 10.5.1 (vii) we have $K = G_T S_3 rP$ and since both K and L are stabilized by G_T it suffices to show that $S_3 rP \subseteq L^{\perp} \setminus L$. Now $S_3 r \cap \langle r\rangle X = \varnothing$

so $K \cap L = \varnothing$. Finally, when srP (with $s \in S_3$) is a point of S_3rP then we must check that it is adjacent to all points of L, i.e., to P and the points brP, $b \in B$. But this follows immediately from $BrBsrB = BrsrB = BsrsB \subseteq PrP$. Thus $L^{\perp} = \{\gamma,\delta\}^{\perp} = L \cup K$ and $\{\gamma,\delta\}^{\perp\perp} = L^{\perp\perp}$. Suppose some point srP of K is in $L^{\perp\perp}$. Then by (the proof of) 10.2.4 we find $S_3 = \{s\}$, $sW_Ts \subseteq W_T \cup W_TsW_T$ and $L^{\perp} = L^{\perp\perp}$. In the finite case this can happen only in the cases mentioned, or when the Lie type of (W,R) is $H_{3,2}$ or $H_{4,3}$. But by 10.5.8 (i) there is no Tits system with type H_3 or H_4. □

Transitivity properties

10.6.5. Proposition. *G is transitive on the set of incident (point, line) pairs.*

Proof. By definition, G is transitive on the set of lines. But $G_{\{r\}}$ is transitive on the set of points incident with the line $G_{\{r\}}P \,/\, P$. □

10.6.6. Proposition. *The stabilizer in G of a point and a line has at most two orbits on the line. If the point has only one relation to the points of the line then this stabilizer is transitive on the set of points on the line.*

Proof. Let the point be P and the line $wB\langle r\rangle P \,/\, P$ $(w \in W)$, where we may assume that $l(wr) > l(w)$. Now by Lemma 10.5.4 (iv) we have $wBrP = wB_wB_rrP = w(B \cap w^{-1}Bw)rP = (wBw^{-1} \cap B)wrP$ so that $wBw^{-1} \cap B$ stabilizes the point and has two orbits on the line (it also stabilizes wP). □

Relation between a graph of Lie type and the associated Coxeter graph

Let G be a group with Tits system (B,N,W,R) and let $r \in R$. We shall see that many properties of the graph $\Gamma = \Gamma(G,P,r)$ can be read off from the graph $\Delta = \Gamma(W,X,r)$ (where $S = R \setminus \{r\}$, $X = W_S$, $P = BXB = G_S$). In the terminology of Tits [747] this is due to the fact that Γ is the collinearity graph of the building associated with the BN-pair (B,N) of G, while Δ is the collinearity graph of an apartment of this building.

10.6.7. Proposition (cf. Bickel [64]). *The map $\psi : X \setminus W \,/\, X \to P \setminus G \,/\, P$ defined by $\psi(XwX) = PwP$ $(w \in W)$ is an almost-isomorphism between the double coset graphs $\overline{\Delta} := DC(W,X,r)$ and $\overline{\Gamma} := DC(G,P,r)$ (cf. §A.4): it is a bijection of vertex sets, preserving edges, such that all edges of $\overline{\Gamma}$ that are not images of edges of $\overline{\Delta}$ are loops.*
In particular, Γ and Δ have the same diameter, and Γ is distance-transitive if and only if Δ is.

Proof. The map ψ is a well-defined bijection according to 10.5.1 (iii), and clearly preserves edges. Conversely, if PuP and PvP are distinct adjacent vertices in $\overline{\Gamma}$ with $u, v \in W$, then by definition of double coset graph $PvP \subseteq PuPrP$, and by 10.5.1 (iii), (iv), (vii) we find $PvP \subseteq PuXrP$ (since $PvP \neq PuP$) so that XuX and XvX are adjacent in $\overline{\Delta}$. The statements about diameter and distance-transitivity follow from Proposition A.4.2. □

Remark. Bickel [64] was perhaps the first to remark that $\Gamma(G,P,r)$ is distance-transitive if and only if $\Gamma(W,X,r)$ is distance-transitive. He also notes that for the correspondence between the double coset diagrams it is essential that $r \in R$; for example, $\Gamma(W,X,w_0)$ is disconnected when $|R| > 1$ and $Xw_0X = w_0X$ (cf. Lemma 10.2.11), but $\Gamma(G,P,w_0)$ is connected, since P is a maximal subgroup of G.

10.6.8. Proposition.

(i) *The map $\phi : \Delta \rightarrow \Gamma$ defined by $\phi(wX) = wP$ ($w \in W$) is a distance (and relation) preserving embedding. Put $\Sigma = \phi[\Delta]$.*

(ii) *Given points $\gamma_1, \gamma_2 \in \Gamma$ and lines L_1, L_2 with $\gamma_1 \in L_1$, $\gamma_2 \in L_2$, there is a $g \in G$ such that $g\Sigma$ contains γ_1, γ_2 and a second point of each of L_1 and L_2.*

(iii) *A line contains at most two points of Σ.*

(iv) *If vX, wX are adjacent vertices of Δ, both in the double coset XwX, then the line on vP and wP is entirely contained in PwP.*

(v) *Let vX be a point not in the double coset XwX. Then vX has a neighbour in XwX if and only if vP has a neighbour in PwP.*

(vi) *Let $d(X,vX) = 2$ and assume X and vX have precisely one common neighbour in Δ. Then P and vP have precisely one common neighbour in Γ.*

(vii) *Given $m \geqslant 2$, there are $\gamma_1, \gamma_2 \in \Gamma$ with $d(\gamma_1,\gamma_2) = 2$ and $\gamma_1^\perp \cap \gamma_2^\perp$ a polar space of rank m if and only if there are $\delta_1, \delta_2 \in \Delta$ with $d(\delta_1,\delta_2) = 2$ and $\delta_1^\perp \cap \delta_2^\perp$ a complete m-partite graph on $2m$ vertices ($K_{m\times 2}$).*

(viii) *There are $\gamma_1, \gamma_2 \in \Gamma$ with $d(\gamma_1,\gamma_2) = 2$ and $\gamma_1^\perp \cap \gamma_2^\perp$ a coclique (of at least 2 elements) if and only if there are $\delta_1, \delta_2 \in \Delta$ with $d(\delta_1,\delta_2) = 2$ and $\delta_1^\perp \cap \delta_2^\perp$ a coclique (of size 2).*

Proof. (i) Clearly, ϕ is well defined and injective and preserves adjacency. Conversely, if $uP \sim vP$ for $u,v \in W$ then $v^{-1}u \subseteq PrP \cap N = XrX$ so that $uX \sim vX$.

(ii) By Proposition 10.6.5 we may assume $\gamma_i = g_iP$ and $L_i = g_iB\langle r\rangle P$ for certain $g_i \in G$ ($i = 1,2$). Let $g_2^{-1}g_1 \in bwB$ ($b \in B$, $w \in W$). Replacing g_1 by a suitable element of g_1B, we may assume that $g_2^{-1}g_1 \in bw$. Then $g_2b\Sigma$ contains γ_1 and γ_2 and also the points $g_1rP = g_2bwrP$ and g_2brP on L_1 and L_2, respectively.

(iii) If $L = gB\langle r\rangle P\,/\,P$ and $g \in BwB$ ($w \in W$), then $gB\langle r\rangle P \subseteq BwB\langle r\rangle P = B\{w,wr\}P$, so that $\phi^{-1}[L]$ contains at most two points. If it does contain two points in the same double coset XwX, i.e., if $wr \in XwX$, then $gB\langle r\rangle P \subseteq PwP$, proving (iv).

(v) 'Only if' is clear. Suppose vX has no neighbours in XwX, i.e., $vXrX \cap XwX = \varnothing$. Since by assumption also $vX \cap XwX = \varnothing$ we find $vPrP \cap PwP \subseteq B((vX \cup vXrX) \cap XwX)B = \varnothing$, so vP has no neighbour in PwP.

(vi) Suppose vX has a unique neighbour in XwX, say uX. Then $vPrP \cap PwP \subseteq BuP$, assuming $vX \not\subseteq XwX$. In particular, for $w = r$ we find that all common neighbours of X and vX have the form buP with $b \in B$. We may assume $v \in D_{S,S}$, so that $v = rdr$ with $d \in X$. But now rX is a common neighbour of X and vX, so $u = r$. If $brP \neq rP$ then $rbr \notin P$ so that $rbr \in BrB$. Since $brP \sim vP$ we have $rd^{-1}rbr \in PrP$, but this is impossible since $rd^{-1}rbr \in Brd^{-1}BrB = Brd^{-1}rB = Bv^{-1}B$, and $d(X,vX) > 1$. The converse is obvious.

(vii) 'If': Assume that $\delta_1^\perp \cap \delta_2^\perp$ is a complete m-partite graph $K_{m\times 2}$. In view of Theorem 10.2.6 and Lemma 10.2.7 we know a lot about the type. We shall adopt the notation T, E, D, d, e, s, r from there. W.l.o.g. we may take $\gamma_1 = P$ and $\gamma_2 = eP$, where $e = rdr$, $d \in X \cap D_{T,T}$, $d \neq 1$, $d \notin S_3$, $d \in S_3\langle T\rangle S_3$, etc. We establish an isomorphism

$$\gamma_1^{\perp} \cap \gamma_2^{\perp} \to \Gamma(G_E, G_D, S_3)$$

via

$$\gamma = bwrP \mapsto bwG_D \quad (b \in B,\ w \in \langle E \rangle).$$

This map is well defined, for we know that w may be taken right D-reduced (and then is unique), and if $b_1wrP = b_2wrP$, then using Lemma 10.5.5 and Lemma 10.5.4(iii), we find $b_2^{-1}b_1 \in wrPrw^{-1} \cap B \subseteq B'_{wr} \subseteq B'_w$, where $B'_w := B_{w^{-1}}$, so that $b_1wG_D = b_2wG_D$. The inverse map

$$\Gamma(G_E, G_D, S_3) \to \gamma_1^{\perp} \cap \gamma_2^{\perp}$$

can be given as follows:

$$bwG_D \mapsto b_1wrP \quad (b \in B,\ w \in \langle E \rangle),$$

where $b_1 \in B'_{ww_0}$ with $b_1B'_w = bB'_w$, and w_0 is the longest word in $\langle E \cup \{r\} \rangle$. Observe that this map is well defined: first of all, if $bwG_D = b'wG_D$ then by 10.5.5 we find $b^{-1}b' \in B'_w$; next, such a b_1 can be found, as $B = B'_{ww_0}B'_w$ (due to Lemma 10.5.4(iv)); finally, if $b_1wG_D = b_2wG_D$ for $w \in D_{\varnothing,D}$, $b_1, b_2 \in B'_{ww_0}$, then $b_2^{-1}b_1 \in wG_Dw^{-1} \cap B \subseteq B'_w$ by Lemma 10.5.5, so

$$b_2^{-1}b_1 \in B'_w \cap B'_{ww_0} = B'_{w_0} \subseteq B'_{wr}$$

by 10.5.6 and 10.5.4(iii), whence

$$b_1wrP = b_2wrP.$$

Note that $w_0 = ex = rdrx$ for some $x \in \langle E \rangle$ (special property of B_n and D_n). Now the point b_1wrP is obviously collinear to γ_1, and $b_1wrP \subseteq B'_{ww_0}wrP \subseteq ww_0Bw_0^{-1}rP = wexBx^{-1}rP \subseteq eXBx^{-1}rP \subseteq ePrP$, so $b_1wrP \in \gamma_2^{\perp}$. Thus, we have well-defined maps between $\gamma_1^{\perp} \cap \gamma_2^{\perp}$ and $\Gamma(G_E, G_D, S_3)$ which are clearly each other's inverses. It remains to check that these maps preserve collinearity. Suppose $b_1w_1G_D \sim b_2w_2G_D$. Then, w.l.o.g. $b_i \in B'_{w_iw_0}$ $(i = 1, 2)$, and $b_1w_1B \subseteq b_2w_2G_DsG_D$ for some $s \in S_3$, so $b_1w_1rB \subseteq b_2w_2G_DsG_DrB \subseteq b_2w_2G_DsrP = b_2w_2G_DrsrsP \subseteq b_2w_2G_DrPrP = b_2w_2rPrP$, whence b_1w_1rP and b_2w_2rP are collinear. Conversely, if b_1w_1rP and b_2w_2rP are collinear, then $b_1w_1 \in b_2w_2rPrPr \cap G_E \subseteq b_2w_2rG_ErG_Er \cap G_E \subseteq b_2w_2G_DS_3G_D$ (by 10.5.1 and 10.2.2), proving that $b_1w_1G_D$ and $b_2w_2G_D$ are collinear in $\Gamma(G_E, G_D, S_3)$.

(viii) 'If': Without loss of generality $\gamma_1 = P$ and $\gamma_2 = rdrP$ where $d \in S_4$ is as in case 2 of Lemma 10.2.7, and it follows that $\gamma_1^{\perp} \cap \gamma_2^{\perp}$ is a coclique.

(vii) and (viii) 'Only if': Since (vi)-(viii) cover all possibilities for $\delta_1, \delta_2 \in \Delta$ with $d(\delta_1, \delta_2) = 2$ (cf. Lemma 10.2.7), this converse implication follows from the already proved 'If' parts. □

Remarks. (i) Similar (and much more general) statements can be found in Tits [747] 3.9 and 3.10. It is not true in general that if vX has a unique neighbour in XwX then vP has a unique neighbour in PwP, not even when $v, w \in D_{S,S}$ and $l(w) < l(v)$.
(ii) Parts (vi) and (vii) of the proposition (together with Theorem 10.2.6 and the following remark) imply that if W is finite and r satisfies $S_4 = \varnothing$, the μ-graph in the space $\Gamma(G,P,r)$ (with lines) is either a singleton or a nondegenerate polar space. This property has been taken as an axiom for several geometric characterizations, cf. Cooperstein [219], Buekenhout [146], Cohen [200, 205].

10.6.9. Theorem. *Let Γ be a finite graph of (irreducible) Lie type. Then the following statements are equivalent.*

(i) *Γ is distance-regular,*

(ii) *Γ is distance-transitive,*

(iii) *Γ is a graph of Lie type $A_{n,i}$ ($1 \leqslant i \leqslant n$), $B_{n,1} \cong D_{n,1}$ ($n \geqslant 3$), $B_{n,n}$ ($n \geqslant 2$), $D_{n-1,n} = D_{n,n}$ ($n \geqslant 3$), $E_{6,1} = E_{6,6}$, $E_{7,7}$, $I_{2,1}^{6} = I_{2,2}^{6}$ or $I_{2,1}^{8} = I_{2,2}^{8}$.*

Proof. The equivalence of (ii) and (iii) follows from Proposition 10.6.7, Theorem 10.4.6 and Theorem 10.5.8. Clearly, (ii) implies (i). By 10.6.8(vi) and 10.4.5, if Γ is distance-regular (but not complete) then in the associated Coxeter graph either any two vertices at distance two have a unique common neighbour, and we find type I_2^m, or any two vertices at distance two have more than one common neighbour, and we find the remaining types mentioned under (iii). □

Incidence graphs

Let G be a group with Tits system (B,N,W,R), and let $r, s \in R$. Just as we did in the Coxeter case, we can consider bipartite *Lie incidence graphs* $\Gamma = \Gamma(G,P,Q)$, where $P = G_{R\setminus\{r\}}$ and $Q = G_{R\setminus\{s\}}$. Again, there are four cases in which both halved graphs of Γ are the graphs of Lie type $\Gamma(G,P,r)$ and $\Gamma(G,Q,s)$, namely (assuming $|R| = n < \infty$): $A_{m,1} + A_{n,1}$, $A_{n,\{1,n\}}$, $A_{n,\{i,i+1\}}$, and $D_{n,\{n-1,n\}}$. The corresponding graphs are: a complete bipartite graph ($d = 2$), the incidence graph of points and hyperplanes in a projective space (distance-transitive of diameter 3, cf. Remark 9.3.12(i)), the incidence graph of i-spaces and $(i+1)$-spaces in a projective space (distance-bitransitive of diameter $\min(2i+1, 2n-2i)$; distance-transitive - namely, the doubled Grassmann graph - for $n = 2i+1$, cf. §9.3C), and the dual polar graph of type D_n, respectively.

10.7 Chevalley Groups

Let $G = X_l(q)$ be the (ordinary) Chevalley group of type X_l over the finite field $\mathbb{F}_q$, as defined in Carter [187], to which the reader is referred for details. To G we can associate a split saturated Tits system (B,N,W,R), that is a Tits system with the additional properties

(i) *split*: There is a normal subgroup U of B with $B = UH$ and $U \cap H = \{1\}$ (where $H = B \cap N$),

(ii) *saturated*: $\bigcap_{w \in W} {}^{w}B = H$.

Here and below, ${}^{w}A$ stands for wAw^{-1} and A^{w} for $w^{-1}Aw$, if A is a subset of G invariant under conjugation by H. Note that for finite systems, by 10.5.4(iii), we have $\bigcap {}^{w}B = B_{w_0}$. We shall briefly recall some definitions and notation. Start with a finite Coxeter system (W,R) where W is a Weyl group of type X_l. Let Φ be a root system for W. A set of mutually obtuse or perpendicular roots corresponding to the subset R of *fundamental reflections* forms a set of *fundamental roots.* Now, any root $\alpha \in \Phi$ is an integral combination of the fundamental roots such that either all coefficients are nonnegative or all coefficients are nonpositive. In the former case α is called *positive,* notation $\alpha > 0$, in the latter case α is called *negative,* notation $\alpha < 0$.
Now choose a Cartan subgroup H in G, and denote by X_α for $\alpha \in \Phi$ the root

subgroup with respect to α (viewed as a linear character of H). Thus H normalizes each X_α. Next, let N be the normalizer of H in G. Then $W = N/H$ permutes the X_α ($\alpha \in \Phi$) according to ${}^wX_\alpha = X_{w\alpha}$ ($w \in W$). Now $U = \prod_{\alpha>0} X_\alpha$ is a subgroup of G normalized by H, so that $B = UH$ is a subgroup of G with $B \cap N = H$. This explains how B, N, W, R, U occur in G. We need some more subgroups of G. Given $w \in W$, set

$$U_w^- = \prod_{\substack{\alpha>0 \\ w\alpha<0}} X_\alpha.$$

This is a subgroup of U.

Let w_0 be the unique longest element in W. Then w_0 is an involution satisfying

$$U \cap U^{w_0} = \{1\} \text{ and } U_w^- = U \cap U^{w_0 w}.$$

Fundamental for the computation of parameters below is the fact that

$$|U_w^-| = q^{l(w)}$$

for all $w \in W$ (cf. CARTER [187], 8.6).

10.7.1. Lemma. *Let* $r \in R$, $S = R \setminus \{r\}$, $P = G_S = BW_SB$.

(i) *Each coset* xP *has a unique representation* $xP = uw^{-1}P$ *where* $w^{-1} \in D_{\emptyset,S}$ *and* $u \in U_w^-$.

(ii) *If* $u, v \in W$, $l(uv) = l(u)+l(v)$ *then* $(U_u^-)^v \subseteq U_{uv}^-$.

(iii) *Let* $U_w = U \cap w^{-1}Bw$. *Then* $U = U_w.U_w^-$ *and* $U_w \cap U_w^- = 1$. *In particular* $B \cap U^{w_0} = 1$, *i.e.,* $U_{w_0} = 1$. □

Let us quote the parameters of the non-twisted Chevalley groups $X_l(q)$. One has

$$|U| = q^N,\ |H| = \frac{1}{d}(q-1)^l,\ |G| = \frac{q^N}{d}\prod_{i=1}^{l}(q^{d_i}-1) \tag{3}$$

where N is the number of positive roots (half the total number of roots) and the d_i are the *degrees* of the Weyl group W (cf. CARTER [187], 10.2). The parameters N, d, d_i ($1 \leqslant i \leqslant l$) are listed in Table 10.6 below.

Table 10.6. *Parameters for Chevalley groups*

X_l	N	d	$d_1, d_2, \cdots, d_l$	h
A_l	$\frac{1}{2}l(l+1)$	$(l+1, q-1)$	$2, 3, \cdots, l+1$	$l+1$
B_l, C_l	l^2	$(2, q-1)$	$2, 4, \cdots, 2l$	$2l$
D_l	$l(l-1)$	l odd: $(4, q-1)$ l even: $(2, q-1)^2$	$2, 4, \cdots, 2l-2, l$	$2l-2$
E_6	36	$(3, q-1)$	2, 5, 6, 8, 9, 12	12
E_7	63	$(2, q-1)$	2, 6, 8, 10, 12, 14, 18	18
E_8	120	1	2, 8, 12, 14, 18, 20, 24, 30	30
F_4	24	1	2, 6, 8, 12	12
G_2	6	1	2, 6	6

Now, let us consider the twisted case. Let $G = {}^aX_l(q)$ be one of the twisted Chevalley groups ${}^2A_l(q), {}^2D_l(q), {}^3D_4(q), {}^2E_6(q)$ or ${}^2F_4(q)$ as defined in CARTER [187], but with q^a replaced by q. Thus q is a prime power in the first four cases and $q^2 = 2^{2k+1}$ for some $k \in \mathbb{N}$ if $G = {}^2F_4(q)$. Then G is a certain subgroup of a Chevalley group $X_l(q^a)$ consisting of certain elements fixed by an automorphism of order a of this Chevalley group. Again G has a naturally associated Tits system (B,N,W,R) which is split and saturated and has Coxeter group W of type $Y_m = B_{\lceil l/2 \rceil}, B_{l-1}, G_2, F_4, I_2^8$ in the respective cases. The latter type is called the (restricted) Lie type of G. For the parameters one has

$$|U| = q^N,$$
$$|U_w^-| = q^{L(w)},$$
$$|H| = \frac{1}{d}\prod_{i=1}^{l}(q-\epsilon_i)$$

(i.e., $|H| = \frac{1}{d}(q-1)^m(q+1)^{l-m}$, except in the case of ${}^3D_4(q)$, where $|H| = (q-1)^2(q^2+q+1)$), and

$$|G| = \frac{q^N}{d}\prod_{i=1}^{l}(q^{d_i}-\epsilon_i)$$

where $N, d_1, \cdots, d_l$ are as given in Table 10.6 for X_l and $d, \epsilon_1, \cdots, \epsilon_l$ are given in Table 10.7 below and $L(w)$ is the length of w as a word in the Coxeter group of the (non-twisted) Chevalley group $X_l(q)$. For example, we have $|{}^3D_4(q)| = q^{12}(q^8+q^4+1)(q^6-1)(q^2-1)$.

Table 10.7. *Additional parameters for twisted Chevalley groups*

${}^aX_l(q)$	Y_m	d	$\epsilon_1, \epsilon_2, \cdots, \epsilon_l$
${}^2A_l(q)$	$B_{\lceil l/2 \rceil}$	$(l+1, q+1)$	$1, -1, 1, \cdots, (-1)^{(l+1)}$ $(l \geq 2)$
${}^2D_l(q)$	B_{l-1}	$(4, q^l+1)$	$1, 1, 1, \cdots, 1, -1$ $(l \geq 4)$
${}^3D_4(q)$	G_2	1	$1, \omega, 1, \bar{\omega}$ (where $\omega = \exp(2\pi i/3)$)
${}^2E_6(q)$	F_4	$(3, q+1)$	$1, -1, 1, 1, -1, 1$
${}^2F_4(2^{k+\frac{1}{2}})$	I_2^8	1	$1, -1, 1, -1$
${}^2B_2(2^{k+\frac{1}{2}})$	A_1	1	$1, -1$
${}^2G_2(3^{k+\frac{1}{2}})$	A_1	1	$1, -1$

For twisted groups the analogue of the lemma above remains valid.

Graphs of Lie Type from Chevalley groups

Since the Chevalley groups provide examples of groups with a Tits system, they yield examples of graphs of Lie type. If $G = X_l(q)$ or ${}^aX_l(q)$ is an ordinary or twisted Chevalley group with associated Tits system (B,N,W,R) and $r \in R$ then the graph $\Gamma(G, G_{R\setminus\{r\}}, r)$ is denoted by $X_{l,i}(q)$ or ${}^aX_{l,i}(q)$ respectively, where i is the label of r in the Coxeter diagram X_l or Y_m. As a matter of fact, the notation ${}^aX_{l,i}(q)$ does not determine the graph up to isomorphism since in the twisted case it may occur that there is a symmetry of the Coxeter diagram of W mapping r to another element of R that does not correspond to an automorphism of ${}^aX_l(q)$. In these cases, notably ${}^2A_3, {}^2A_4, {}^3D_4, {}^2E_6, {}^2F_4$ we shall take the node numbered 1 to be a member of R

coming from 2, 2, 3, 2, 2 nonadjacent nodes of the Coxeter diagram X_l.

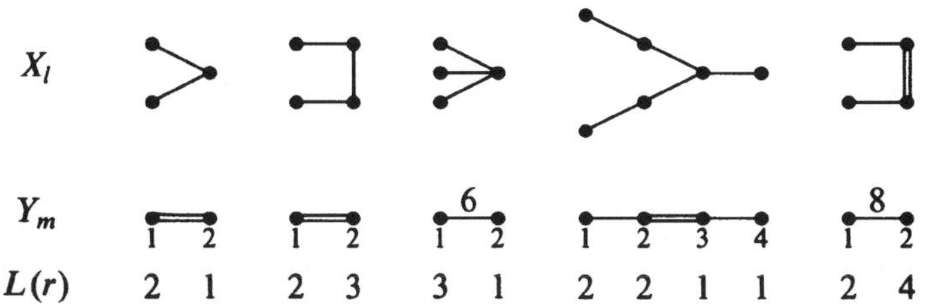

In the twisted case, let $\pi: X_l \to Y_m$ be the projection map. The Weyl group $\overline{W}$ of Y_m is a subgroup of the Weyl group W of X_l, and we shall denote the restriction of the length function of W to $\overline{W}$ by $L(\,)$. For $X_l = (R,M)$ and $Y_m = (\overline{R},\overline{M})$ we may take the elements $\overline{r}$ of $\overline{R}$ to be the longest elements $w_0(\pi^{-1}(\overline{r}))$ of the subgroups $W_{\pi^{-1}(\overline{r})}$ of W (cf. CARTER [187], Proposition 13.1.12); this determines both the diagram $\overline{M}$ and the length function L.
In order to unify notation we shall also write ${}^1X_l(q)$ for $X_l(q)$, and put $L(w) = l(w)$ in the non-twisted case.

10.7.2. Theorem. *Let $G = {}^aX_l(q)$ be an ordinary or twisted finite Chevalley group with Tits system (B,N,W,R), and let $r \in R$. Set $P = G_{R \setminus \{r\}}$. Consider the graph $\Gamma = {}^aX_{l,i}(q) = \Gamma(G,P,r)$. If $|R| = 1$ then Γ is a clique; otherwise, the following three statements are equivalent:*

(i) *Γ is distance-regular*

(ii) *Γ is distance-transitive*

(iii) *${}^aX_{l,i}$ is one of $A_{l,i}$ $(1 \leq i \leq l)$, $B_{l,1}$, $B_{l,l}$, $C_{l,1}$, $C_{l,l}$, $D_{l,1}$, $D_{l,l-1} = D_{l,l}$, $E_{6,1} = E_{6,6}$, $E_{7,7}$, $G_{2,1} = G_{2,2}$, ${}^2A_{l,1}$, ${}^2A_{l,i}$ $(i = \lceil l/2 \rceil)$, ${}^2D_{l,1}$, ${}^2D_{l,l-1}$, ${}^3D_{4,1}$, ${}^3D_{4,2}$, ${}^2F_{4,1}$, ${}^2F_{4,2}$.*

Furthermore, the graphs $A_{l,i}(q)$ are Grassmann graphs $\left[{l+1 \atop i}\right]_q$; the graphs $B_{l,1}(q)$ and $B_{l,l}(q)$ are polar and dual polar graphs in $O_{2l+1}(q)$; the graphs $C_{l,1}(q)$ and $C_{l,l}(q)$ are polar and dual polar graphs in $Sp_{2l}(q)$; the graph $D_{l,1}(q)$ is the polar graph in $O_{2l}^+(q)$; the graph $D_{l,l-1}(q) = D_{l,l}(q)$ is the half dual polar graph in $O_{2l}^+(q)$; the graphs ${}^2A_{l,1}(q)$ and ${}^2A_{l,\lceil l/2 \rceil}(q)$ are polar and dual polar graphs in $U_{l+1}(q^2)$; and the graphs ${}^2D_{l,1}(q)$ and ${}^2D_{l,l-1}(q)$ are polar and dual polar graphs in $O_{2l}^-(q)$. (For these graphs, see §9.4.) *The intersection arrays and singular line sizes $s+1$ of the remaining graphs are as indicated in Table 10.8 below. In particular, ${}^3D_{4,2}(q)$ and $E_{7,7}(q)$ are classical with parameters as given in Table 6.1, (C6) and (C5), respectively.*

Proof. The equivalence of (i)-(iii) is a direct consequence of Theorem 10.6.9. For the singular line size $s+1$ we find by 10.6.4 and 10.7.1 (i): $s = |B\langle r\rangle P/P| - 1 = |BrP/P| = |U_r^-| = q^{L(r)}$. The intersection arrays follow from general formulas derived below. (In the cases of $E_{6,1}$ and $E_{7,7}$ it is a nice exercise to derive them from geometric considerations, cf. 10.7.7-8 below.) Concerning the identifications with classical graphs, these are easily verified given the identification of the corresponding groups (cf. CARTER [187]). □

Remark. The generalized hexagons related to G_2 and ${}^3D_4$ can be constructed directly (i.e., without use of the theory of Chevalley groups) in a fairly elementary way, using

Table 10.8. *Arrays of distance-regular graphs of Lie type*

${}^aX_{l,i}$	q	s	d	Intersection array
$E_{6,1}$	q	q	2	$\{q(q^3+1)\frac{q^8-1}{q-1}, q^7\cdot\frac{q^5-1}{q-1};\ 1, (q^3+1)\frac{q^4-1}{q-1}\}$
$E_{7,7}$	q	q	3	$\{q(q^8+q^4+1)\frac{q^9-1}{q-1}, q^9(q^4+1)\frac{q^5-1}{q-1}, q^{17};$ $1, (q^4+1)\frac{q^5-1}{q-1}, (q^8+q^4+1)\frac{q^9-1}{q-1}\}$
$G_{2,1}$	q	q	3	$\{q(q+1), q^2, q^2;\ 1, 1, q+1\}$
${}^3D_{4,1}$	q	q^3	3	$\{q^3(q+1), q^4, q^4;\ 1, 1, q+1\}$
${}^3D_{4,2}$	q	q	3	$\{q(q^3+1), q^4, q^4;\ 1, 1, q^3+1\}$
${}^2F_{4,1}$	$2^{k+\frac{1}{2}}$	q^2	4	$\{q^2(q^4+1), q^6, q^6, q^6;\ 1, 1, 1, q^4+1\}$
${}^2F_{4,2}$	$2^{k+\frac{1}{2}}$	q^4	4	$\{q^4(q^2+1), q^6, q^6, q^6;\ 1, 1, 1, q^2+1\}$

the Cayley numbers (see SCHELLEKENS [640]). No such construction is known for the generalized octagons related to ${}^2F_4$.
Just like the Chevalley groups of type ${}^2A_n$, B_n, C_n, D_n and ${}^2D_n$ can be constructed using sesquilinear forms, those of types G_2, F_4, E_6, E_7, ${}^3D_4$ and ${}^2E_6$ can be constructed using 3- or 4-linear forms; see ASCHBACHER [15].
The dual polar graphs in $O^+_{2l}(q)$ are found as incidence graphs of Lie type $D_{n,\{n-1,n\}}$.

Parameters

Let $G = {}^aX_l(q)$ be an ordinary or twisted Chevalley group with Tits system (B,N,W,R), and let $r \in R$. Let $S = R\setminus\{r\}$, $X = W_S$ and $P = BXB = G_S$. Let $\Gamma = \Gamma(G,P,r)$. We want to compute the parameters of the graph Γ. To this end, let us write

$$|w| = \begin{cases} l(w) & \text{if } a = 1 \text{ (non-twisted group)} \\ L(w) & \text{if } a > 1 \text{ (L is length in the Weyl group of $X_l(q)$)} \end{cases}$$

so that

$$|U_w^-| = q^{|w|}.$$

(Note that in the twisted case, if $w = r_1 \cdots r_k$ is a reduced expression for w with factors in R, so that $l(w) = k$, then $L(w) = \sum_{i=1}^k L(r_i)$, so that L can be regarded as a weighted length function.)

10.7.3. Proposition. *Let $\Gamma = \Gamma(G,P,r)$, where G is an ordinary or twisted Chevalley group with Tits system (B,N,W,R), P is a maximal parabolic subgroup of G containing B, and $r \in R$ is not contained in P. Then:*

(i) *The number v of vertices of Γ is*

$$v = |G/P| = \sum_{w \in D_{\emptyset,S}} q^{|w|}$$

(ii) *For $i \in D_{S,S}$, the number k_i of vertices of Γ in relation R_i (cf. 10.5.8) to a fixed vertex is $k_i = \sum q^{|w|}$ where the summation runs over $w \in D_{\emptyset,S} \cap W_S i$.* □

Remarks. (i) It follows that k_i is divisible by $q^{|i|}$ (and not by a higher power of q). Also, $k_i = k_{i^{-1}}$.
(ii) Using product formulae for the orders of the groups involved, we find a product formula for ν. In the non-twisted case we have

$$\nu = \frac{\prod_{i=1}^{l}(q^{d_i}-1)}{(q-1)\prod_{i=1}^{l-1}(q^{e_i}-1)}$$

where $d_1, \cdots, d_l$ are the degrees of the Weyl group W, and $e_1, \cdots, e_{l-1}$ are the degrees of the Weyl group $X = W_S$.
In the twisted case, the Weyl group W of type $Y_m = (R,M)$ is a subgroup of a Weyl group $\tilde{W}$ of type $X_l = (\tilde{R},\tilde{M})$, and we have a projection map $\pi : \tilde{R} \to R$ (cf. the diagrams preceding Theorem 10.7.2). Let $c := |\pi^{-1}(r)|$. Then

$$\nu = \frac{\prod_{i=1}^{l}(q^{d_i}-\delta_i)}{(q^c-1)\prod_{i=1}^{l-c}(q^{e_i}-\epsilon_i)}$$

where $d_1, \cdots, d_l$ are the degrees of the Weyl group $\tilde{W}$, $e_1, \cdots, e_{l-c}$ the degrees of its subgroup $\langle \tilde{R} \setminus \pi^{-1}(r)\rangle$, and $\delta_1, \cdots, \delta_l$ and $\epsilon_1, \cdots, \epsilon_{l-c}$ the 'signs' for W and W_S. Here the degrees and signs for a direct product are those for the factors, the degrees of ${}^a(aA_m)$ are those of A_m, each repeated a times, and the signs for ${}^a(aA_m)$ are the a-th roots of unity, repeated m times [so that a component ${}^a(aA_m)$ contributes a factor $\prod_{i=1}^{m}(q^{a(i+1)}-1)$], and the remaining information can be found in Tables 10.6 and 10.7.
Thus, for ${}^2E_{6,1}$ we find $\nu = (q^5+1)(q^9+1)(q^{12}-1)/(q^2-1)$, and for ${}^2E_{6,3}$ we find $\nu = (q^4+1)(q^5+1)(q^9+1)(q^{12}-1)/(q-1)$.
Note that in most cases these formulae for ν reduce to formulae we have seen already (in 9.3.2 for subspaces of the projective space A_l, in 9.4.1 for singular subspaces of the polar spaces ${}^2A_l$, B_l, C_l, D_l and ${}^2D_l$, and in §6.5 for the generalized polygons ${}^2F_4$ and ${}^3D_4$).

10.7.4. Proposition. *Let $i, j, k \in D_{S,S}$. Then, by definition of p^i_{jk}, we have $p^i_{jk} = |PjP/P \cap iPkP/P|$. In the special case $k = r$ we can give a more or less explicit formula:*

$$p^i_{jr} = \sum_{w \in L \cap A \cap E} q^{|w|} + \sum_{w \in Lr \cap Ar \cap E} q^{|w|} + \sum_{w \in Lr \cap A \cap E} q^{|w|}(q-1)$$

where $L = D_{\varnothing,S} \cap Xr$, $A = i^{-1}XjX$, $E = \{w \mid l(iw) > l(iwr)\}$. Note that the third term is present only when $i = j$.

Proof. In order to compute the number of points in $PjP \cap iPkP$, write $iPkP$ as a union of cosets uwP ($u \in U^-_{w^{-1}}$, $w \in D_{\varnothing,S}$) and sum over all w occurring with $w \in XjX$. Let us follow this recipe for $k = r$. Write $XrX = \biguplus_{w \in L} wX$ (where $\biguplus$ denotes disjoint union), so that

$$iPrP = iBXrXB = \biguplus_{w \in L} iBwP$$

and we have to write each $iBwP$ as a union of cosets uwP. If $l(iw) > l(iwr)$ then $iBwP = iU_{w^{-1}}^{-}wP = {}^{i}(U_{w^{-1}}^{-})iwP$ and we have ${}^{i}(U_{w^{-1}}^{-}) \subseteq U_{(iw)^{-1}}^{-}$ (by Lemma 10.7.1 (ii)) as desired. On the other hand, if $l(iw) < l(iwr)$, then write $w = vr$, $v \in X$ and find

$$iBwP = iBvBrP = {}^{i}(U_{v^{-1}}^{-})ivBrP = {}^{i}(U_{v^{-1}}^{-}).[iwP \uplus {}^{iw}(U_r^{-} \setminus \{1\})ivP]$$

where ${}^{i}(U_{v^{-1}}^{-}) \subseteq U_{(iw)^{-1}}^{-}$ and ${}^{i}(U_{v^{-1}}^{-}).{}^{iw}(U_r^{-}) \subseteq U_{(iv)^{-1}}^{-}$. Now each term $iBwP$ with $l(iw) > l(iwr)$ contributes $q^{|w|}$, and each term $iBwP$ with $l(iw) < l(iwr)$ contributes $q^{|wr|}(\delta(iw \in XjX) + (q-1).\delta(iwr \in XjX))$, where $\delta(T)$ for a predicate T denotes 1 if T is true and 0 otherwise. □

10.7.5. Corollary. *Given two vertices γ_1, γ_2 of Γ at mutual distance d, the number $c_d(\gamma_1,\gamma_2)$ of vertices at distance $d-1$ to γ_1 and adjacent to γ_2 is congruent to* 1 (mod q), *and the number $a_d(\gamma_1,\gamma_2)$ of vertices at distance d to γ_1 and adjacent to γ_2 is congruent to* -1 (mod q). *Finally, the valency k is congruent to* 0 (mod q).

Proof. $k = k_r \equiv 0 \pmod q$ was already remarked after 10.7.3. Proposition 10.7.4 shows that

$$p^{i}_{jr} \equiv \delta(ir \in XjX) + (q-1)\cdot\delta(i \in XjX) \pmod q.$$

Thus, all p^{i}_{jr} are congruent 0 (mod q) except p^{i}_{ir} which is congruent -1 (mod q) and p^{i}_{lr} which is congruent 1 (mod q), where l is defined by $ir \in XlX$. Clearly, $d(P,lP) = d(P,iP) - 1$. □

Remarks. As the above corollary shows, Propositions 10.7.3-4 can be used for theoretical purposes. (For example, it is a step in the proof that a thick finite geometry over a Coxeter diagram of spherical type, such that residues of type C_3 are covered by buildings, is itself a building, cf. Brouwer & Cohen [130].) On the other hand, in the cases where W is not very small, the formulae given are not very suitable for hand calculation. They are, however, easily implemented on a computer, and in Brouwer & Cohen [131] the parameters of $X_{l,i}(q)$ for $X_{l,i} = E_{6,1}, E_{6,2}, E_{6,3}, E_{7,1}, E_{7,2}, E_{7,7}, E_{8,1}, E_{8,2}$ and $E_{8,8}$ are listed.
The formulae given are also valid for $q = 1$ and then give the parameters of the corresponding Coxeter graphs.

The following consequence of the above will be needed to determine all distance-transitive graphs on maximal parabolics (cf. §10.9).

10.7.6. Lemma. *Let G, P, r be as in 10.7.2, and suppose that $\Gamma = \Gamma(G,P,r)$ is distance-regular of diameter d. If its distance-i graph Γ_i is distance-regular for some i with $1 < i \leqslant d$, then $i = d = 2$.*

Proof. Suppose $d > 2$. For a thick near $2d$-gon the statement is obvious. Remain the cases ${}^{a}X_{l,i} = A_{l,i}$ $(3 \leqslant i \leqslant l-2)$, $D_{l,l}$ and $E_{7,7}$. In these cases, one checks that $p^{1}_{dd}p^{2}_{dd} > 0$, showing that Γ_d is not distance-regular. Now apply Proposition 4.2.13. □

Computation of the parameters of $E_{7,7}(q)$ - geometric approach

In the previous section we saw how to compute the parameters for the graphs of Lie type from explicit knowledge of the Weyl group. But just as we did in the previous chapter for the classical graphs studied there, we may also derive this information from geometric considerations. In this section we do this for the case of $E_{7,7}$. All the information needed can easily be read off from the E_7 Coxeter diagram (this time interpreted as the diagram of a Buekenhout-Tits geometry) but it would take us too far to explain these things. So, let us instead quote the required information on E_7 from TITS [742], and assume everything that is needed about D_6 and E_6 is known.

Since the E_7 diagram contains a type D_6 diagram, the graph $E_{7,7}(q)$ will contain subgraphs $D_{6,1}(q)$. Let us call such a subgraph a *symplecton.*

[More generally, in the Cooperstein theory of so-called 'parapolar spaces' (cf. COOPERSTEIN [219], BUEKENHOUT [146], COHEN [200], COHEN & COOPERSTEIN [206]) a symplecton is defined as the geodetic closure (as subspace) of a pair of points at distance two, and turns out to be a nondegenerate polar space.]

10.7.7. Lemma.

(i) *$E_{7,7}(q)$ has diameter* 3.

(ii) *Given a symplecton S and a point $x \notin S$, either $x^\perp \cap S$ is a maximal totally isotropic subspace (of one kind) of S, or $x^\perp \cap S$ is a singleton.*

Proof. (ii) is Proposition 7.7′ in TITS [742], and (i) follows directly from (ii). □

Now by 10.2.11 the double coset diagram is symmetric, and by 10.2.5 (since $E_{6,1}(q)$ is distance-transitive of diameter two) we find that there is only one double coset at distance two from a given point, so 10.7.7 (i) implies that $E_{7,7}(q)$ is distance-transitive.

10.7.8. Theorem. *The graph $E_{7,7}(q)$ has the following parameters:*

$$v = (q^5+1)(q^9+1)\frac{q^{14}-1}{q-1}, \quad d = 3,$$

$$k = q\cdot\frac{q^9-1}{q-1}\cdot\frac{q^{12}-1}{q^4-1}, \quad k_2 = q^{10}\cdot\frac{q^9-1}{q-1}\cdot\frac{q^{12}-1}{q^4-1}, \quad k_3 = q^{27},$$

$$a_2 = (q^5-1)(q^4+1)+q^6(1+q^3)\frac{q^8-1}{q-1}, \quad \mu = (q^4+1)\frac{q^5-1}{q-1}.$$

By distance-transitivity all other parameters are determined.

Proof. First of all we find from Remark (ii) following 10.7.3 that

$$v(E_{6,1}) = \frac{q^9-1}{q-1}\cdot\frac{q^{12}-1}{q^4-1} \text{ and } v(E_{7,7}) = (q^5+1)(q^9+1)\frac{q^{14}-1}{q-1}.$$

(Here, for simplicity, we write $v(X_{l,m})$ instead of $v(X_{l,m}(q))$.) The geometry whose points are the lines of $E_{7,7}$ on a fixed point γ and in which two lines are adjacent when their union is a clique, is a geometry of type $E_{6,1}$ and since $E_{7,7}(q)$ has singular line size $q+1$ it follows that $k = q\cdot v(E_{6,1})$.

Next, two vertices at distance two determine a unique symplecton (by geodesic closure), each symplecton has $k_2(D_{6,1}) = q^{10}$ points at distance two from a given point, and each point is in $v(E_{6,1})$ symplecta, so $k_2 = q^{10}\cdot v(E_{6,1})$. Finally $k_3 = v-1-k-k_2$. (Another way of computing k_3 is by observing (10.7.3) that $k_3 = q^{l(w)}$ where w is the shortest word in the appropriate double coset in W.

Without knowledge of W this still suffices since we know that $c_3 \equiv 1 \pmod q$, so that the exponent of q in k_3 is the exponent found in $k_2 b_2$.)
Clearly $\mu = \mu(D_{6,1}) = v(D_{5,1})$. In order to compute a_2, let γ_0 and γ_1 be two points with $d(\gamma_0,\gamma_1) = 2$. Then γ_0 and γ_1 determine a unique symplecton S and the first contribution to a_2 is $a_2(D_{6,1}) = (q-1)\cdot v(D_{5,1}) = (q^5-1)(q^4+1)$. If γ is a vertex outside S, adjacent to γ_1 and at distance two to γ_0, then γ and γ_0 determine a symplecton S', and since any two symplecta on a point have a line in common (this follows since any two symplecta in a geometry of type E_6, i.e., any two shadows of objects of type $E_{6,6}$, have a point in common) we find a common neighbour of γ and γ_1, say δ, on this line. (That the neighbours of γ and γ_1 on this line coincide follows again by geodesic closure.) Now, by 10.7.7 (ii), $\gamma^\perp \cap S$ is a maximal totally isotropic subspace of S so that γ and $S \cap \gamma^\perp$ span a projective 6-space with hyperplane $\gamma^\perp \cap S$ and q^6 points outside S. The numbers of possible maximal totally isotropic subspaces (of the extendible kind) in S on γ_1 is

$$v(D_{5,5}) = (q^4+1)(q^3+1)(q^2+1)(q+1),$$

so that altogether $a_2 = a_2(D_{6,1}) + q^6\cdot v(D_{5,5})$.

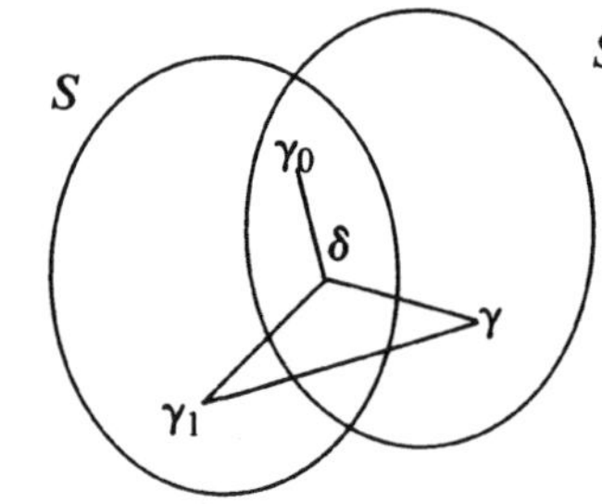

(Alternatively, we might find a_2 from $a_2 = k - b_2 - c_2$ and b_2 from $b_2 k_2 = c_3 k_3$ so that we must find c_3. But since the Coxeter graph of type $E_{7,7}$ has $k_3 = 1$, it follows by 10.6.8 (ii) that $c_3 = k/q$.)
So far, we have found v, $d = 3$, k, k_2, k_3, a_2 and c_2. By distance-regularity, all other parameters follow. □

Remark. The graph $E_{7,7}(q)$ has eigenvalues $\theta_0 = k = q\left[{9 \atop 1}\right](q^{12}-1)/(q^4-1)$, $\theta_1 = q^5\left[{5 \atop 1}\right](q^4+1)-1$, $\theta_2 = q^9 - q^4 - 1$, $\theta_3 = -q^8 - q^4 - 1$.

10.8 The affine E_6 graph

Let Δ be the graph $E_{7,7}(q)$ and let $\infty \in \Delta$. Then the graph $\Delta_3(\infty)$ is called the *affine* E_6*-graph (over* $\mathbb{F}_q$*)*. A direct description may be given as follows.

Let F be a field and denote by $\mathbb{K}_F$, or just $\mathbb{K}$ if no confusion arises, the 27-dimensional vector space over F consisting of ordered triples $x = (x^{(1)}, x^{(2)}, x^{(3)})$ of 3×3 matrices $x^{(i)}$ $(i = 1, 2, 3)$, supplied with the following symmetric cubic form.

$$\mathrm{Dt}(x) = \det\ x^{(1)} + \det\ x^{(2)} + \det\ x^{(3)} - \mathrm{tr}\ x^{(1)}x^{(2)}x^{(3)} \quad (x = (x^{(1)}, x^{(2)}, x^{(3)}) \in \mathbb{K}).$$

Denote by $(\cdot,\cdot,\cdot)$ the linearization of Dt, i.e.,

$$(x,y,z) = \mathrm{Dt}(x+y+z) - \mathrm{Dt}(x+y) - \mathrm{Dt}(x+z) - \mathrm{Dt}(y+z) + \mathrm{Dt}(x) + \mathrm{Dt}(y) + \mathrm{Dt}(z)$$

$(x, y, z \in \mathbb{K})$. Let Γ be the graph whose vertex set is $\mathbb{K}$ and in which x, y are adjacent if and only if $\mathrm{Dt}(x-y, x-y, \mathbb{K}) = 0$. Then Γ is said to be the *affine E_6-graph over F*. The appearance of E_6 in this name is due to the fact that the automorphism group of this graph is a semidirect product of the additive group of $\mathbb{K}$ and (an extension of a central cover of) the group of Lie type E_6 defined over F.

10.8.1. Theorem. *Let Γ be the affine E_6 graph over the finite field F of order q. Then Γ is distance-transitive, has intersection array*

$$\{\frac{(q^{12}-1)(q^9-1)}{q^4-1}, q^8(q^4+1)(q^5-1), q^{16}(q-1);\ 1, q^4(q^4+1), q^8\frac{q^{12}-1}{q^4-1}\},$$

and singular line size $s+1 = q$. Moreover, Γ is classical with parameters as in Table 6.1, (C13), and $\mathrm{Aut}\,\Gamma \cong F^{27}\cdot F^{*}E_6(F)\mathrm{Aut}(F)$.

Proof. The essential part comes down to showing that the group $\mathbb{Z}_{(3,q-1)}E_6(F) = GL(\mathbb{K})_{\mathrm{Dt}}$ has three orbits on the set of projective points of $\mathbb{K}$. (See MARS [518].) These orbits consist of the points Fx ($x \in \mathbb{K}$, $x \neq 0$) with $(x,x,\mathbb{K}) = 0$ (for $Fx \in \Gamma_1(0)$), $\mathrm{Dt}(x) = 0$ and $(x,x,\mathbb{K}) \neq 0$ (for $Fx \in \Gamma_2(0)$), and $\mathrm{Dt}(x) \neq 0$ (for $Fx \in \Gamma_3(0)$). See BUEKENHOUT & COHEN [147].

The parameters of Γ are easily obtained from the discussion of the geometry of E_7 in the previous section. Since each singular line of Δ that meets $\Delta_3(\infty)$ has precisely one point in $\Delta_2(\infty)$, the valency of Γ is $(q-1)/q$ times that of Δ, i.e., $k = (q^8+q^4+1)(q^9-1)$. Similarly $\lambda(\Delta) = q-1+q^2\cdot v(D_{5,5})$, so $\lambda = q-2+q(q-1)\cdot v(D_{5,5})$. It follows that $b_1 = q^8(q^4+1)(q^5-1)$.
If δ is a vertex of Δ, and S a symplecton meeting $\Delta_3(\delta)$, then $\Delta(\delta) \cap S = \{\epsilon\}$ and $\Delta_2(\delta) \cap S = \Delta(\epsilon) \cap S$ for some $\epsilon \in S$. (Indeed, if $\gamma \in \Delta_2(\delta) \cap S$ then γ and δ determine a symplecton S', and $S \cap S'$ is a line L. Now we must have $\Delta(\delta) \cap L = \{\epsilon\}$, and hence $\gamma \in \Delta(\epsilon)$.) It follows that μ is the number of common neighbours of two nonadjacent points in the second subconstituent $D_{6,1}^{(2)}$ of $D_{6,1}$, that is, $\mu = \mu(D_{6,1}) - (q^8-1)/(q-1) = q^8+q^4$. Similar to the computation of $a_2(\Delta)$ in the previous section, we find here $a_2 = a_2(D_{6,1}^{(2)}) + q^5(q-1)\cdot v(D_{5,5}) = q^{16}+q^{13}+q^9-2q^8-2q^4-1$, so that $b_2 = k-\mu-a_2 = q^{17}-q^{16}$. Finally, c_3 follows from $b_2k_2 = c_3k_3$. □

Remark. The subgraph of Γ induced on $\Gamma(0)$ is a $(q-1)$-clique extension of the strongly regular graph of Lie type $E_{6,1}(q)$.

10.9 Distance-transitive representations of Chevalley groups

In this section we determine all multiplicity-free and all distance-regular permutation representations of the finite Chevalley groups on maximal parabolic subgroups.

10.9.1. Proposition. *Let G be a finite group with Tits system* (B,N,W,R). *Let* $r \in R$, $S = R \setminus \{r\}$, $X = W_S$ *and* $P = G_S$. *Then the character* 1_P^G *is multiplicity-free if and only if* 1_X^W *is multiplicity-free.*

Proof. This equivalence follows from the fact (cf. CARTER [188], Corollary 10.11.3) that the centralizer algebras $\mathrm{End}_{\mathbf{C}G}(\mathbf{C}G/P)$ and $\mathrm{End}_{\mathbf{C}W}(\mathbf{C}W/X)$ over $\mathbf{C}$ are isomorphic, and from the general fact that a permutation representation of a group H on a set Ω is multiplicity-free if and only if the corresponding centralizer algebra $\mathrm{End}_{\mathbf{C}H}(\mathbf{C}\Omega)$ is commutative (see §2.9). □

Let us write $X_{l,i}$ for the G-space G/P where G and P are as above, (W,R) is a Coxeter group of spherical type X_l, and r is the node labelled i in its Coxeter diagram (in Table 10.1).

10.9.2. Proposition. *Suppose* (W,R) *is an irreducible Coxeter group of spherical type* $Y_m = (R,M) \neq H_3, H_4$, *and of rank* $m \geqslant 3$. *Let G be a finite group with Tits system* (B,N,W,R). *The permutation representation of G on a maximal parabolic* $P = G_S$, *where* $S = R \setminus \{r\}$ *for some* $r \in R$, *is multiplicity-free if and only* $Y_{m,r}$ *is one of* $A_{n,i}$, $B_{n,i}$, $C_{n,i}$, $D_{n,i}$ *for* $i \leqslant n/2$ *or* $i \in \{n-1, n\}$, $E_{6,1}$, $E_{6,2}$, $E_{6,6}$, $E_{7,1}$, $E_{7,2}$, $E_{7,7}$, $E_{8,1}$, $E_{8,8}$, $F_{4,1}$ *or* $F_{4,4}$.

Proof. By the above lemma, it suffices to consider the permutation representation of W on X, and the conclusion follows from Theorem 10.4.11. □

10.9.3. Lemma. *Let* $Y_{n,m}$ *be one of the indexed types listed in the conclusion of Theorem 10.4.11. If G is a finite group with Tits system* (B,N,W,R) *of type* Y_n *and P is the standard parabolic and r the involution corresponding to m, then* $|PrP/P| < |PwP/P|$ *for every* $w \in D_{S,S} \setminus \{1,r\}$ *(where* $S = R \setminus \{r\}$*), unless* $Y = D$ *and* $1 \leqslant m \leqslant n-2$, *and G is defined over* $\mathbf{F}_2$ *(so that* $D_n(2) \leqslant G \leqslant \mathrm{Aut}\, D_n(2)$*). Here the types* $A_{3,2}$, $D_{4,3}$ *and* $D_{4,4}$ *should be regarded as* $D_{3,1}$, $D_{4,1}$ *and* $D_{4,1}$, *respectively. In the exceptional cases, there is a single* $w \in D_{S,S} \setminus \{1,r\}$ *with* $|PwP/P| \leqslant |PrP/P|$; *it is related to the W-orbital in* Ω *containing the pair* $Y = \{1, \cdots, m\}$, $Z = \{-1, 2, \cdots, m\}$ *in the notation of the proof of 10.4.11.*

Proof. The truth of the assertion in the cases $A_{n,i}$, $B_{n,1}$, $B_{n,n}$, $C_{n,1}$, $C_{n,n}$, $D_{n,1}$, $D_{n,n-1}$, $D_{n,n}$, $E_{6,1}$, $E_{7,7}$ follows from the parameters of the corresponding distance-transitive graphs. The remaining cases for $Y = E$ and $Y = F$ are readily verified using the explicit parameters given in BROUWER & COHEN [131] and COHEN [204], respectively. Thus, the remainder of the proof is concerned with $Y = B$, C or D, and $m \neq 1, n$ (and if $Y = D$, then $m \leqslant n/2$). We shall view $Y_{n,m}(q)$ as the collection of isotropic m-spaces in the geometry $[Y_n(q)]$ of §9.4. We fix an isotropic m-space x and view P as its stabilizer. The P-orbits of isotropic m-spaces are parametrized by the pairs (j,l) in such a way that y belongs to the P-orbit (j,l) if and only if $\dim(x \cap y) = j$ and $\dim(x^{\perp} \cap y) = j + l$. Now r-adjacency occurs for $(j,l) = (m-1, 1)$, so, by Lemma 9.4.2, the suborbit corresponding to (j,l) has size at most $|PrP/P|$ if and only if

$$q\begin{bmatrix} m \\ 1 \end{bmatrix}\begin{bmatrix} n-m \\ 1 \end{bmatrix}(q^{n+e-m-1}+1) \geqslant \tag{*}$$

$$q^{l(m-j)+\frac{1}{2}(m-j-l)(4n-3l-j-3m+2e-1)} \cdot \begin{bmatrix} m \\ j \end{bmatrix}\begin{bmatrix} m-j \\ l \end{bmatrix}\begin{bmatrix} n-m \\ l \end{bmatrix}\prod_{i=0}^{l-1}(q^{n+e-m-i-1}+1).$$

Here, e is as in §9.4. Assume $l \geqslant 1$. Then the right hand side is at least

$$q^{m-j}\begin{bmatrix}m\\j\end{bmatrix}\begin{bmatrix}m-j\\1\end{bmatrix}\begin{bmatrix}n-m\\1\end{bmatrix}(q^{n+e-m-1}+1)$$

and comparison with the left hand side gives $j = m-1$ so that $(j,l) = (m-1,1)$. Therefore we may assume $l = 0$. Now (*) becomes

$$q\begin{bmatrix}m\\1\end{bmatrix}\begin{bmatrix}n-m\\1\end{bmatrix}(q^{n+e-m-1}+1) \geqslant q^{\frac{1}{2}(m-j)(4n-j-3m+2e-1)}\begin{bmatrix}m\\j\end{bmatrix}. \qquad (**)$$

Since the left hand side is smaller than $q^{1+m+n-m+n+e-m}$ and the right hand side has a lower bound $q^{\frac{1}{2}(m-j)(4n+j-3m+2e-1)}$, it follows that

$$2n+e-m+1 > \tfrac{1}{2}(m-j)(4n+2e-2m+2)+\tfrac{1}{2}(m-j)(j-m+1)-2(m-j).$$

Thus

$$(m-j-1)(2n+e-m+1) < \tfrac{1}{2}(m-j)(m-j-1)+2(m-j).$$

Assume now $m-j-1 > 0$. Then

$$2n+e-m+1 < \tfrac{1}{2}(m-j)+2+\frac{2}{m-j-1}.$$

Since $m \leqslant n-1$ and $e \geqslant 0$, this yields

$$2 \leqslant \tfrac{1}{2}(m+j+2) < \frac{2}{m-j-1} \leqslant 2,$$

contradiction. Therefore $l = 0$ and $j = m-1$. The inequality (**) becomes

$$(q^{n-m}-1)(q^{n+e-m-1}+1) \geqslant (q-1)q^{2n-2m+e-1}$$

i.e,

$$(1-\frac{1}{q^{n-m}})(1+\frac{1}{q^{n+e-m-1}}) \geqslant q-1.$$

Since the left hand side is strictly smaller than 2, even smaller than 1 if $e \geqslant 1$, we must have $q = 2$ and $e = 0$. This proves the lemma for the polar space cases. □

We now come to the main goal of this section.

10.9.4. Theorem. *Suppose G is an ordinary or twisted finite Chevalley group with Tits system (B,N,W,R) and P is a maximal parabolic subgroup. If there is a self-paired G-orbital on G/P with representative (P,gP) such that G acts distance-transitively on the graph $\Gamma = \Gamma(G,P,g)$, then Γ is isomorphic to one of the graphs listed in Theorem 10.7.2 (ii).*

Proof. Without loss of generality we may take $P = G_S$, a standard parabolic (i.e., $P \supseteq B$) with $S = R\setminus\{r\}$ for some $r \in R$, and $w = gH \in D_{S,S}$. We have to show that $w = r$. If the rank $|R|$ of the Tits system is $\leqslant 2$, this is obvious (from generalized polygon theory), so we may assume $n := |R| \geqslant 3$. As 1_P^G must be multiplicity-free, we can run down the list of Proposition 10.9.2. Denote by M the type of (W,R). If (M,r) occurs in the list of Theorem 10.7.2 (iii), we are done by Lemma 10.7.6 and Proposition 4.2.14. Also, if (M,r) is one of the rank 5 cases $E_{n,2}$ ($n = 6,7,8$), $F_{4,1}$, $F_{4,4}$, then the diagram has the shape of the second example in §2.9 and is readily seen not to lead to a distance-transitive graph for any other choice of adjacency, cf. COHEN [204]. Thus, it remains to consider the cases $B_{n,i}$ and $C_{n,i}$ for $1 < i < n$, $D_{n,i}$ for $1 < i \leqslant n/2$, $E_{7,2}$ and $E_{8,1}$. In the cases where $M = B_n$, C_n or D_n, we can use polar spaces. Consider the polar space induced on $\Gamma(G,G_{R\setminus\{r_1\}},r_1)$, and denote by Ω

the set of singular subspaces of rank $i-1$. Then as in the previous lemma P may be seen as the stabilizer in G of an element $x \in \Omega$ and the P-orbits on Ω are parametrized by the numbers j, k with $\mathrm{rk}\,(x \cap y) = j-1$ and $\mathrm{rk}\,(x \cap y^\perp) = k-1$ for $y \in \Omega$ (cf. Theorem 10.4.3). The above lemma shows that the P-orbit with $(j,k) = (i-1,i)$ is the smallest nontrivial one or the one but smallest. Suppose it is the smallest. Since this P-orbit corresponds to the G-orbital determined by $r \in D_{S,S}$, Theorem 10.7.2 shows that it does not coincide with $\Gamma_1(x)$. But then, by Proposition 5.1.1 (i), (ii), we must have $Prx = \Gamma_d(x)$, where $d = \mathrm{diam}(\Gamma)$ is known from Theorem 10.4.3. In particular, $d \geqslant 3{\cdot}4/2 = 6$. Now consider $y := gx \in \Gamma_1(x)$. If the parameters of its P-orbit are (j,k), then (via induction by considering the residue of $x \cap y$ and using TITS [747], §7.7 and §7.8) one can find an element $z \in Prx \cap Py$. But this implies $z \in \Gamma_d(x) \cap \Gamma_2(x)$ whence $d \leqslant 3$, which is absurd, as $d \geqslant 6$. In view of the lemma, it follows that $q = 2$, the type is D_n, $i \leqslant n/2$ and $\Gamma(x) = Pgx$ and $\Gamma_2(x) = Prx$, where $y := gx$ satisfies $\mathrm{rk}\,(x \cap y) = \mathrm{rk}\,(x^\perp \cap y) = \mathrm{rk}\,(x) - 1$. But, then an element $z \in \Gamma(y)$ can be found such that $\mathrm{rk}\,(x \cap z) = \mathrm{rk}\,(x) - 2$, and so $z \in \Gamma_2(x)$, a contradiction with $\Gamma_2(x) = Prx$.
Hence Γ is not distance-transitive in the cases $B_{n,i}$ and $C_{n,i}$ with $1 < i < n$ and $D_{n,i}$ with $1 < i \leqslant n/2$.
We are left with the cases $E_{7,2}$ and $E_{8,1}$. From the description of $\Gamma(G,P,r)$ in BROUWER & COHEN [131], we see that this graph is not distance-regular, and that PrP/P is again the smallest nontrivial P-orbit, whence $\Gamma_d(P) = PrP/P$. We also see $d = \mathrm{diam}(\Gamma) = 10$ in both cases. Since $k \leqslant k_{d-1}$ (by 5.1.1 (ii)) we must have that $\Gamma_1(P)$ consists of the second smallest P-orbit on PwP/P ($w \in D_{S,S}$), which occurs (by inspection of [131]) for w = "243542" or w = "243542654376542" if $M = E_7$, and for w = "13425431" if $M = E_8$. In all three cases, we find $rP \subseteq PwPwP$, leading to the contradiction $rP \in \Gamma_d(P) \cap \Gamma_2(P)$. (For instance, in the last case we have $r_1P \subseteq Pr_1r_3r_1P \subseteq Pr_1r_3r_4r_2r_5r_4(r_3r_1r_3)r_4r_5r_2r_4r_3r_1P \subseteq PwBwP$ as $Br_1B \subseteq Br_1Br_1B$.) (Another way to derive this last conclusion is to observe that w actually lies in a Coxeter subgroup of type D_4, D_6 and D_5 in the respective cases, and to argue that $rP \in \Gamma_2(P)$ within a geodetically closed subgraph of Lie type containing P, rP and wP of this type.) □

Chapter 11
Graphs Related to Codes

Let $V = \mathbf{F}_q^n$ be the vector space of n tuples with entries in the finite field $\mathbf{F}_q$ with q elements, and let C be a linear code in V (i.e., a linear subspace of V). We define the *coset graph* $\Gamma(C)$ *of* C by taking as vertices the cosets of C in V, and joining two cosets when they have representatives that differ in one coordinate (i.e., have Hamming distance one). In some cases $\Gamma(C)$ turns out to be distance-regular. In section 11.1 we study this phenomenon in a more general setting. Instead of the vector space V (that is, instead of the Hamming graph $H(n,q)$), we take an arbitrary distance-regular graph Γ, and instead of the partition of V into cosets of C, we take an arbitrary partition Π of Γ. Now there is an obvious concept of quotient graph Γ/Π generalizing that of coset graph, and Theorem 11.1.6 gives a sufficient condition for this quotient graph to be distance-regular. Section 11.1 is the outgrowth of earlier discussions with A.R. Calderbank.

11.1 Completely regular codes

A. Codes in distance-regular graphs

Let Γ be a graph with vertex set X and diameter d. A *code* in Γ is a nonempty subset C of X. The number

$$\delta(C) := \min\{d(x,y) \mid x,\ y \in C,\ x \neq y\}$$

(but $\delta(C) := 2d+1$ if $|C| = 1$) is called the *minimum distance* of C; C is called *e-error correcting* if $\delta(C) \geqslant 2e+1$. The *distance* of $x \in X$ to C is defined as

$$d(x,C) := \min\{d(x,y) \mid y \in C\},$$

and the number

$$t(C) := \max\{d(x,C) \mid x \in X\}$$

is called the *covering radius* of C. Clearly, the covering radius is the smallest number t such that spheres of radius t around the points of C cover X. Minimum distance and covering radius are related by the inequality

$$\delta(C) \leqslant 2t(C)+1; \tag{1}$$

equality holds precisely when the spheres of radius $t(C)$ around the points of C form a partition of X. A code with this property is called *perfect*.

The *outer distribution* of C is the matrix B indexed by $X \times \{0, \cdots, d\}$ with entries

$$B_{xi} := |C \cap \Gamma_i(x)|.$$

A code C is called *s-regular* if, for $d(x,C) = l \leqslant s$, the number B_{xi} only depends on l and i, i.e., if there are numbers e_{li} such that

$$B_{xi} = e_{li} \text{ for all } x \in X \text{ with } d(x,C) = l \tag{2}$$

holds for $l = 0, \cdots, s$ and $i = 0, \cdots, d$. A *completely regular* code is a code C which is $t(C)$-regular, i.e., a code for which B_{xi} only depends on $d(x,C)$ and i. Clearly, for a code C of minimal distance δ, and any $x \in X$ with $d(x,C) = l$, we have

$$B_{xl} > 0,$$
$$B_{xi}B_{xj} = 0 \text{ for } i+j < \delta,\ i \neq j,$$
$$B_{xl} = 1 \text{ if } 2l < \delta,$$
$$B_{xi} = 0 \text{ if } i < \max(l, \delta - l) \text{ and } i \neq l.$$

In particular, a code C is completely regular if and only if the outer distribution matrix has at most (and hence precisely) $t(C)+1$ distinct rows. We refer to the numbers e_{li} as the *outer distribution numbers* of a completely regular code C.

For example, edges in any connected bipartite graph and, more generally, lines in any near polygon (considered as sets of points) are completely regular codes.

It is difficult to make strong statements about codes in general graphs Γ (see e.g. Hammond [361], Smith [685] and Etienne [280] for a few results); to get useful results, an appropriate assumption on Γ is that Γ is distance-regular. In this case, the distance relation defines a metric d-class association scheme on X, and we can use the terminology and results of §2.5. (In fact, many of our results will hold for general association schemes.)

In terms of the inner distribution a of C and its MacWilliams transform aQ, we define the *dual degree* $r(C)$ as the number of indices j with $1 \leqslant j \leqslant d$, $(aQ)_j \neq 0$. (Note that $(aQ)_0 = |C| > 0$.) This is a key parameter, as the following results show.

[Note: the parameter $r(C)$ was called r in 2.5.5. For general codes, it has no direct combinatorial significance. But in the case of linear codes, $r(C)$ is the number of nonzero weights of the dual code (by Theorem 2.10.12), explaining our terminology. Delsarte calls r the 'external distance', which is rather unfortunate; many authors use the term 'external distance' (or sometimes 'true external distance') for the covering radius $t(C)$.]

11.1.1. Theorem (Delsarte [251], pp. 66-68). *Let C be a code in a distance-regular graph, and put $\delta = \delta(C)$, $r = r(C)$, $t = t(C)$. Then:*

(i) *The outer distribution matrix B of C has rank $r+1$. Moreover, each column $B_{\cdot i}$ is a linear combination of the columns $B_{\cdot j}$ with $0 \leqslant j \leqslant r$ and also of the columns $B_{\cdot j}$ with $0 \leqslant j \leqslant r-1$ together with the all-one column* **1**, *in both cases with coefficients depending only on the inner distribution of C.*

(ii) *The covering radius is bounded by the dual degree:*

$$t \leqslant r.$$

Equality holds when C is completely regular.

(iii) *If* $s := \delta - r \geqslant 0$, *then* C *is* s*-regular.*

(iv) *If* C *is* s*-regular with* $s \geqslant r-1$, *and in particular if* $\delta \geqslant 2r-1$, *then* C *is completely regular.*

(v) C *is perfect if and only if* $\delta = 2r+1$; *in particular, any perfect code is completely regular.*

Proof. (i) We have $B^{\mathsf{T}}B = \dfrac{|C|}{|X|}P^{\mathsf{T}}\Delta_{aQ}P$ (see Lemma 2.5.1 (v)), so that $B^{\mathsf{T}}B$ only depends on the inner distribution a of C. Thus, it remains to show that the first $r+1$ columns of B are linearly independent. Here we use that the scheme is metric, so that $P_{ij} = p_j(z_i)$ for certain real z_i and polynomials p_j of degree j. If columns $0, \cdots, r$ of B are linearly dependent, then there are constants α_j, not all zero, such that $\sum_{j=0}^{r}\alpha_j P_{ij} = 0$ for the values of i with $(aQ)_i \neq 0$. But this means that we have a nonzero polynomial of degree at most r with $r+1$ distinct zeros, a contradiction.
Similarly, for the other part of (i), since $B\mathbf{1} = |C|\mathbf{1}$, we have to show that $\mathbf{1}$ is not a linear combination of the first r columns of B. But if it were, then we would find $\sum_{j=0}^{r-1}\alpha_j P_{ij} = \sum_{j=0}^{d} P_{ij} = 0$ for the r values of $i > 0$ with $(aQ)_i \neq 0$, but a nonzero polynomial of degree at most $r-1$ cannot have r zeros, contradiction.
(ii) The first part is a special case of Proposition 2.5.5 (ii) (or, more precisely, of the Remark (ii) following it). For a completely regular code C, the matrix B has only $t+1$ distinct rows, each starting with a different number of zeros, so $r+1 = \mathrm{rk}\, B = t+1$.
(iii) Let $d(x,C) = j$, $j \leqslant s$. Then $B_{xi} = \delta_{ij}$ for $0 \leqslant i \leqslant r-1$, and by (i) the number B_{xi} does not depend on x but only on j, for all i ($0 \leqslant i \leqslant d$). As this argument also works for $j = r$ (if $d(x,C) = r$, then obviously $B_{xi} = 0$ for $0 \leqslant i \leqslant r-1$), (iv) follows immediately.
(v) If $\delta = 2r+1$, then by (1) and part (ii) we have $\delta = 2t+1$, and C is perfect. Conversely, if C is perfect, then $\delta = 2t+1$, and the all-one column $\mathbf{1}$ is the sum of the first $t+1$ columns, so by (i) we have $r \leqslant t$, and hence $r = t$ by (ii). □

A code in a distance-regular graph is called *even* if Γ is bipartite and C is contained in one of the two parts of Γ. The outer distribution matrix B of an even code satisfies $B_{xi} = 0$ if $d(x,C)+i$ is odd. Using this one can strengthen the conclusions of (iii), (iv) to obtain

(iii)′ *If* C *is even and* $s := \delta - r + 1 \geqslant 0$, *then* C *is* s*-regular.*

(iv)′ *If* C *is even and* $\delta \geqslant 2r-2$ *then* C *is completely regular.*

Proof (cf. Roos [626]). Suppose Γ is an imprimitive association scheme, so that we have a set S of relations giving a nontrivial equivalence relation on the point set X of Γ, with equivalence classes $X_1, \cdots, X_m$. Let π_j be the characteristic vector of X_j, and σ the characteristic vector of S. If $C \subseteq X_j$, then π_j is a linear combination of the columns of B, since $B\sigma = |C|\pi_j$.
Now suppose Γ is a bipartite distance-regular graph, and C is contained in a bipartite half of Γ (so that $S = \{0, 2, 4, \cdots\}$). We show that π_0 is not a linear combination of the first $r-1$ columns of B together with $\mathbf{1}$. If it were, then we would have $\sum_{j=0}^{r-2}\alpha_j P_{ij} + \alpha\sum_{j=0}^{d} P_{ij} = \sum_{j\in S} P_{ij}$ for certain real α_j and α, and all $r+1$ values of i with $(aQ)_i \neq 0$. But $\sum_{j=0}^{d} P_{ij}$ is nonzero only for $i = 0$, and $\sum_{j\in S} P_{ij}$ is nonzero only for one nonzero value of i (belonging to the eigenspace with eigenvalue $-k$ for Γ), so we find $r-1$ zeros for the polynomial $\sum_{j=0}^{r-2}\alpha_j P_{ij}$ of degree at most $r-2$, hence $\alpha_0 = \cdots = \alpha_{r-2} = 0$, contradiction. □

For example, the extended Preparata codes have $\delta = 6$ and $r = 4$ and hence are completely regular by (iv)′.

From (1) and part (ii) we find the inequality

$$\delta(C) \leqslant 2r(C) + 1, \tag{3}$$

and, by part (v), equality characterizes perfect codes. Another important class of codes which can be characterized by minimal distance and dual degree are the uniformly packed codes. An e-error correcting code C is called *uniformly packed* if it is not perfect, and there are numbers λ_C, μ_C such that

$$B_{x,e+1} = \begin{cases} \lambda_C & \text{if } d(x,C) = e, \\ \mu_C & \text{if } d(x,C) \geqslant e+1. \end{cases} \tag{4}$$

By 11.1.1 (v), perfect codes also satisfy (4) (for some λ_C and undefined μ_C). Clearly, any completely regular code C with $t(C) = e+1$ is uniformly packed, and, conversely, any uniformly packed code has $t(C) = e+1$. A code in a strongly regular graph is uniformly packed if and only if it is a coclique satisfying the Hoffman bound (Proposition 1.3.2) with equality.

[Concerning the meaning of the term 'uniformly packed': this term was first used by SEMAKOV, ZINOV'EV & ZAITSEV [659] - let us call their concept 'uniformly packed in the narrow sense'; it is equivalent to 'uniformly packed with $\lambda_C = \mu_C$'. A much more general concept was called 'uniformly packed' in BASSALYGO, ZAITSEV & ZINOV'EV [43] - let us call this 'uniformly packed in the wide sense'. Their requirement is that the all-one column **1** is a linear combination of the first $t+1$ columns of B; from the discussion below it will be clear that this is equivalent to $r = t$. They show that codes satisfying this requirement are 0-regular, and derive a version of Lloyd's theorem.]

11.1.2. Corollary (cf. GOETHALS & VAN TILBORG [332]). *A code C in a distance-regular graph is uniformly packed if and only if $\delta(C) \in \{2r(C), 2r(C)-1\}$. In particular, any uniformly packed code is completely regular, with $e_{li} = \delta_{li}$ $(l \leqslant i \leqslant e)$, $e_{e,e+1} = \lambda_C$ and $e_{e+1,e+1} = \mu_C$.*

Proof. If $\delta(C) \in \{2r(C), 2r(C)-1\}$, then C is completely regular by Theorem 11.1.1 (iv), and then $t = r = e+1$ by 11.1.1 (ii), so that C is uniformly packed. Conversely, if C is uniformly packed, then we see that the all-one column **1** is a linear combination of the first $e+2$ columns of B, so that $r \leqslant e+1$ by Theorem 11.1.1 (i), and $\delta \in \{2r, 2r-1\}$. □

In fact we show that the constancy of a few e_{li} is sufficient for complete regularity. Call a code C *s-uniform* if there are constants e_{ii}, $e_{i-1,i}$ $(i = 1, \cdots, s)$ such that $B_{xi} = e_{li}$ for all $i \leqslant s$ and all $x \in \Gamma$ with $d(x,C) = l \in \{i-1, i\}$ (and hence for all x with $d(x,C) = l \geqslant i-1$). E.g., an e-error-correcting code is trivially e-uniform. The next result generalizes an observation in HAMMOND & SMITH [364].

11.1.3. Theorem. *Let C be a code in a distance-regular graph Γ with covering radius t and dual degree r.*

(i) *If C is s-uniform, then there are constants e_{il} $(0 \leqslant i \leqslant s,\ 0 \leqslant l \leqslant t)$ such that $B_{xi} = e_{li}$ for all $i \leqslant s$ and all x with $d(x,C) = l \leqslant t$.*

(ii) *If C is t-uniform, then C is completely regular.*

(iii) *If C is $(r-1)$-uniform, then C is completely regular.*

(iv) *If C is even and $(r-2)$-uniform, then C is completely regular.*

Proof. Let $C_l := \{x \in X \mid d(x,C) = l\}$, and define for $x \in C_l$ the numbers $\hat{c}_l(x) := |x^\perp \cap C_{l-1}|$ and $\hat{a}_l(x) := |x^\perp \cap C_l|$. By induction one sees that $\hat{c}_l(x)$

($l \leqslant s$) and $\hat{a}_l(x)$ ($l < s$) are independent of the choice of $x \in C_l$, because

$$\hat{c}_l(x)\, e_{l-1,l-1} = e_{l,l}\, c_l$$

and

$$(\hat{a}_l(x)+\hat{a}_{l-1}+\cdots+\hat{a}_0)\hat{c}_l\hat{c}_{l-1}\cdots\hat{c}_1 = \\ \big(e_{ll}(a_l+a_{l-1}+\cdots+a_0)+e_{l,l+1}c_{l+1}\big)c_l\cdots c_1.$$

Now we note that the number s_i^j of paths of length i in Γ joining two vertices at distance j depends only on i and j. On the other hand, the number $n_{li}(x)$ of paths of length i in Γ joining $x \in C_l$ to some vertex of C can be expressed in terms of the $\hat{a}_j$, $\hat{c}_j$, and $\hat{b}_j := k - \hat{a}_j - \hat{c}_j$. But $n_{li}(x) = \sum_{i=0}^{j} B_{xj}s_i^j$; so by induction on i we see that, for $i \leqslant s$, the number B_{xi} is independent of $x \in C_l$ for all $l \leqslant i$ (and hence trivially for all $l \leqslant t$). This proves (i). Moreover, if $s = t$ then we can continue the induction for all $i \leqslant t$, which gives (ii).
Now suppose that C is $(r-1)$-uniform. If $t \leqslant r-1$ then C is completely regular by (ii). Otherwise, $t = r$ by 11.1.1 (ii). Let $x \in C_l$. Then (i) implies $B_{xi} = e_{li}$ for $i < r$, and $B_{xt} = |C| - \sum_{i<t} B_{xi} = |C| - \sum_{i<t} e_{li} = e_{lt}$. Hence C is again completely regular, proving (iii).
Finally, if C is even and $(r-2)$-uniform, then the same argument using the additional information that $B_{xi} = 0$ when $l+i$ is odd implies (iv). □

Trivially, codes consisting of one code word only are perfect and completely regular. By Proposition 4.4.6, cliques that meet the Hoffman bound are completely regular. Further trivial completely regular codes arise in distance-regular graphs Γ that are antipodal r-covers (§A.5): If Γ has diameter d, and $e = [(d-1)/2]$, then a (maximal) set C of r vertices at mutual distance d is a completely regular e-error correcting code, with outer distribution numbers $e_{li} = 0$, except for $e_{ii} = 1$, $e_{i,d-i} = r-1$ ($i \leqslant e$) and $e_{e+1,e+1} = r$ (if $d = 2e+2$). In particular, the covering radius is $[d/2]$ so that C is perfect if $d = 2e+1$, and uniformly packed (with parameters $\lambda_C = 0$, $\mu_C = r$) if $d = 2e+2$. For nontrivial examples, see §11.1D, E below.

Remarks. (i) Codes and designs in graphs and association schemes were first studied by BIGGS [69] and DELSARTE [251].

(ii) Let us indicate briefly the relation of the above to the results of BIGGS [74]. A *q-covering* in a graph Γ with vertex set X is a subset C such that $\sum_i B_{xi}q_i = \alpha$ for all $x \in X$, where α is some nonzero constant and $q = (q_0, \cdots, q_d)$ is a given vector of nonnegative numbers. If χ is the characteristic vector of C then this says $\sum_i q_i A_i \chi = \alpha \mathbf{1}$, and it follows by multiplying with $\chi^T E_h$ that $(Pq)_h \cdot (aQ)_h = |X|\alpha\delta_{0h}$. Biggs defines the *nullity* of C as the number of indices h for which $(Pq)_h = 0$; of course this is at least r. Thus, his results follow from the above. Note that for $h = 0$ our relation becomes $\sum q_i k_i = \alpha|X|/|C|$, which is sometimes useful. A *t-design* in the sense of Biggs is a q-covering with $q_i = \binom{d-i}{t}$; with this definition, a classical t-(v,k,λ) design is a (Biggs-) t-design in the Johnson graph (and $\alpha = \lambda q_0$).

Remark (cf. ROOS [633], HEDEN [368], SMITH [684]). Suppose that Γ is an antipodal distance-regular graph and C a completely regular code in Γ. If some point γ at distance i to C has an antipode in C, then the same holds for all points at distance i to C, so the number of such points is at most $|C|.k_d$. For $0 < i < \delta/2$ this is certainly not the case, since $k_i \geqslant k > k_d$. It follows that

If C is a perfect code in an antipodal distance-regular graph Γ, then C is a union of antipodal classes.

Consequently, the image $\bar{C}$ of C in the folded graph $\bar{\Gamma}$ of Γ is a perfect code in $\bar{\Gamma}$.

B. Completely regular partitions and distance-regular quotient graphs

Given a partition Π of the vertex set of a graph Γ (into nonempty sets), we define the *quotient graph* Γ/Π on the classes of Π by calling two classes $C, C' \in \Pi$ adjacent if $C \neq C'$ and Γ contains an edge joining some vertex of C to some vertex of C'. The partition Π is called *regular* if, for any two classes $C, C' \in \Pi$, the number of vertices $y \in C$ adjacent to $x \in C'$ is a constant $e(C',C)$ independent of $x \in C'$.

11.1.4. Lemma. *Let Π be a regular partition of a graph Γ. Then the distance in Γ/Π is given by*

$$d(C',C) = \min\{d(x,y) \mid x \in C', y \in C\}, \tag{5}$$

and we have

$$d(x,C) = d(C',C) \text{ for all } x \in C'.$$

Proof. We proceed by induction, and suppose that, for some $i \geqslant 1$, (5) holds for all $C, C' \in \Pi$ with $d(C',C) = i-1$; this is certainly satisfied for $i = 1$. To complete the induction we choose $C, C' \in \Pi$ at distance i in Γ/Π. Then there is a class $C'' \sim C'$ at distance $i-1$ from C. Since $e(C',C'') > 0$, every vertex $x \in C'$ has a neighbour in C'', so that $d(x,C) \leqslant i = d(C',C)$. But if $x = x_0, x_1, \cdots, x_j = y \in C$ is a shortest path joining x and C, then we find $d(C',C) \leqslant j = d(x,C)$. Therefore $d(x,C) = d(C',C)$, and (5) follows since $\min\{d(x,y) \mid x \in C', y \in C\} = \min\{d(x,C) \mid x \in C'\} = d(C',C)$. □

Certain regular bipartitions (i.e., partitions into two classes) of graphs are discussed in Delsarte [251] (§3.4), Neumaier [556] and Cameron, Delsarte & Goethals [176]. For partitions into many sets it is reasonable to assume more regularity. A regular partition Π is *uniformly regular* if there are constants e_{01} and e_{11} such that

$$e(C',C) = \begin{cases} e_{01} & \text{if } C' = C, \\ e_{11} & \text{if } C' \sim C, \end{cases} \tag{6}$$

i.e., if $e(C',C)$ only depends on the distance of C' and C (since $e(C',C) = 0$ if $d(C',C) > 1$). In other words, Π is uniformly regular if each $C \in \Pi$ is (at least) 1-uniform with the same constants e_{01}, e_{11}. Clearly, $e_{11} > 0$.

11.1.5. Proposition. *Let Γ be a connected regular graph of valency k, and let Π be a uniformly regular partition of Γ. Then:*

(i) *All classes of Π contain the same number c of vertices.*

(ii) *The quotient graph Γ/Π has $\bar{v} = v/c$ vertices, and is regular of valency $\bar{k} = (k - e_{01})/e_{11}$.*

Proof. (i) Let C and C' be adjacent classes with c and c' vertices, respectively. Then the number of edges joining C and C' is $c.e(C,C') = c'.e(C',C)$, whence $c = c'$ by (6). Since Γ is connected, all classes have the same size.
(ii) If a class $C \in \Pi$ has $\bar{k}$ neighbours in Γ/Π, then the number of neighbours of $x \in C$ in Γ is $k = e_{01} + \bar{k}\, e_{11}$, whence $\bar{k} = (k - e_{01})/e_{11}$ is constant. □

An example of a uniformly regular partition is provided by the three 19-cocliques in the construction of the Perkel graph in terms of $\mathrm{Frob}_{19.9}$ (cf. §13.3); here $e_{01} = 0$, $e_{11} = 3$ and $\Gamma/\Pi \cong K_3$.

Most of the partitions that we shall meet later have $e_{01} = 0$, $e_{11} = 1$ (equivalently, all classes are codes with minimum distance $\delta \geqslant 3$); let us call such partitions *congruence partitions*. The graph Γ is called a *cover* of a graph $\bar{\Gamma}$ if there is a mapping $x \mapsto \bar{x}$ from the vertex set of Γ to that of $\bar{\Gamma}$ that preserves adjacencies, and whose restriction to $y^{\perp}$ is injective for each vertex y of Γ. (See also §A.5.) For any such mapping we obtain a congruence partition of Γ as the set of equivalence classes of the relation $\equiv$ defined by $x \equiv y$ when $\bar{x} = \bar{y}$. Conversely, for any congruence partition Π of Γ, the mapping which associates with $x \in \Gamma$ the class $\bar{x}$ of Π containing x, shows that Γ is a cover of Γ / Π. In particular, if Γ is an antipodal covering graph of diameter $d \geqslant 3$, then the partition Π of Γ into maximal cliques of Γ_d is a congruence partition. In the special case that Γ is distance-regular, we know from Proposition 4.2.2 that the quotient graph Γ / Π is also distance-regular. On the other hand, in this case, the classes of Π are completely regular codes in Γ (see the remark after Theorem 11.1.3). We now show that this correspondence is typical.

We call a partition Π *completely regular* if it is regular, and all $C \in \Pi$ are completely regular with the same outer distribution numbers. Lemma 11.1.4 shows that any completely regular partition is uniformly regular.

11.1.6. Theorem. *Let Π be a uniformly regular partition of a distance-regular graph Γ. Then Γ / Π is distance-regular precisely when Π is completely regular.*

Proof. Let Γ and $\bar{\Gamma} = \Gamma / \Pi$ have v and w vertices, respectively, and consider matrices in $\mathbf{R}^{v \times v}$, $\mathbf{R}^{v \times w}$ and $\mathbf{R}^{w \times w}$, indexed by $\Gamma \times \Gamma$, $\Gamma \times \Pi$ and $\Pi \times \Pi$, respectively. For matrices in the Bose-Mesner algebra of Γ we use the notation of §4.1. The class of Π containing $x \in \Gamma$ is denoted by $\bar{x}$. Let K be the $v \times w$-matrix defined by $K_{xC} = 1$ if $x \in C$, and $K_{xC} = 0$ otherwise, and ley A^* be the $w \times w$-matrix with entries $A^*_{CD} = e(C,D)$. Then $AK = KA^*$, and by induction $p(A)K = Kp(A^*)$ for all polynomials $p(\xi)$. In particular, we have

$$A_i K = v_i(A)K = KA_i^*, \tag{7}$$

where $A_i^* = v_i(A^*) \in \mathbf{R}^{w \times w}$. Now $B_{xi}(C) = |C \cap \Gamma_i(x)| = (A_i K)_{xC}$, so that

$$B_{xi}(C) = (A_i^*)_{\bar{x}C} \quad \text{for } x \in \Gamma,\ C \in \Pi,\ i = 0, \cdots, d. \tag{8}$$

We show that (8) is the matrix formulation of the claimed equivalence. Indeed, suppose first that $\bar{\Gamma}$ is distance-regular. Since Π is uniformly regular, A^* belongs to the Bose-Mesner algebra of $\bar{\Gamma}$; therefore the A_i^*, as polynomials in A^*, also belong to it. Hence the entries of A_i^* depend only on the distances of their indices, and (8) shows that all classes C are completely regular with the same outer distribution numbers, i.e., Π is completely regular. Conversely, if Π is completely regular, then (8) shows that

$$[A_i^*]_{CD} = e_{li} \quad \text{if} \quad d(C,D) = l; \tag{9}$$

and since $e_{li} = 0$ for $l > i$ but $e_{ii} > 0$, we see that, for all $i \leqslant t$, the matrices $A_0^*, A_1^*, \cdots, A_i^*$ and the distance-i-matrices $I = \bar{A}_0, \bar{A}_1, \cdots, \bar{A}_i$ of $\bar{\Gamma}$ span the same linear subspace of $\mathbf{R}^{w \times w}$. This implies that the $\bar{A}_i$ are polynomials of degree i in $\bar{A}_1$, so that $\bar{\Gamma}$ is distance-regular. □

Remark. We have $K^{\mathrm{T}}K = cI$, where c is the common size of the classes in Π. It follows that the linear mapping $* : N \mapsto N^* = c^{-1}K^{\mathrm{T}}NK$ from the Bose-Mesner algebra $\mathcal{A}$ of Γ onto the Bose-Mesner algebra of Γ / Π is multiplicative. (Indeed, this map sends A to A^*, and $(NA)^* = c^{-1}K^{\top}NAK = c^{-1}K^{\top}NKA^* = N^*A^*$ for all $N \in \mathcal{Q}$.)

11.1.7. Corollary (cf. RIFÀ & HUGUET [613]). *Let C be a completely regular code (with covering radius t) in a distance-regular graph Γ, and let G be a transitive group of automorphisms of Γ such that for each $g \in G$, either $C^g = C$ or $C^g \cap C = \varnothing$ (i.e., such that C is a block of imprimitivity for the action of G on the vertex set of Γ). Then the graph $\overline{\Gamma}$ defined on the cosets of the setwise stabilizer $H = \{g \in G \mid C^g = C\}$ by $Hg_1 \sim Hg_2$ when there is an edge joining C^{g_1} and C^{g_2} is distance-regular (of diameter t).*

Proof. The set $\Pi = \{C^g \mid g \in G\}$ is a completely regular partition of Γ, and, clearly, $\overline{\Gamma} \cong \Gamma/\Pi$. □

Many interesting examples of this situation arise for completely regular linear codes in Hamming graphs Γ, with an abelian group G of translations acting regularly on the vertices. These examples will be discussed in §11.1D, E.

11.1.8. Theorem. *Let Π be a completely regular partition of a distance-regular graph Γ. If Γ has diameter d, valency k, and intersection numbers a_i, b_i, c_i ($i = 0, \cdots, d$) and if the classes $C \in \Pi$ have dual degree r and outer distribution numbers e_{li} ($l = 0, \cdots, t$; $i = 0, \cdots, d$), then Γ/Π has diameter r, valency $\overline{k} = (k - e_{01})/e_{11}$, and intersection numbers*

$$\overline{c}_i = \frac{e_{ii}}{e_{i-1,i-1}e_{11}}c_i,$$

$$\overline{a}_i = \frac{1}{e_{11}}\left(\frac{e_{i,i+1}}{e_{ii}}c_{i+1} - \frac{e_{i-1,i}}{e_{i-1,i-1}}c_i + a_i - e_{01}\right),$$

$$\overline{b}_i = \overline{k} - \overline{a}_i - \overline{c}_i$$

($i = 1, \cdots, r$).

Proof. Since $A_h^* = v_h(A^*)$, the definition of the polynomials $v_h(\xi)$ in §4.1A implies

$$A^*A_h^* = c_{h+1}A_{h+1}^* + a_hA_h^* + b_{h-1}A_{h-1}^*. \tag{10}$$

By (9) we have $A_i^* = \sum_{l=0}^{i} e_{li}\overline{A}_l$, and substituting this in (10) yields

$$(e_{01}I + e_{11}\overline{A}_1)(\sum_l e_{li}\overline{A}_l) = c_{i+1}\sum_l e_{l,i+1}\overline{A}_l + a_i\sum_l e_{li}\overline{A}_l + b_{i-1}\sum_l e_{l,i-1}\overline{A}_l,$$

that is,

$$e_{11}\overline{A}_1 \cdot \sum_l e_{li}\overline{A}_l = \sum_l (c_{i+1}e_{l,i+1} + a_ie_{li} + b_{i-1}e_{l,i-1} - e_{01}e_{li})\overline{A}_l.$$

A comparison with $\overline{A}_1\overline{A}_i = \overline{c}_{i+1}\overline{A}_{i+1} + \overline{a}_i\overline{A}_i + \overline{b}_{i-1}\overline{A}_{i-1}$ yields

$$e_{11}e_{ii}\overline{c}_{i+1} = c_{i+1}e_{i+1,i+1},$$

$$e_{11}e_{ii}\overline{a}_i + e_{11}e_{i-1,i}\overline{c}_i = c_{i+1}e_{i,i+1} + a_ie_{ii} - e_{01}e_{ii}.$$ □

[An even easier proof is found by comparing with the proof of Theorem 11.1.3. The relation between the parameters computed here and there is $\hat{a}_i = e_{11}\overline{a}_i + e_{01}$, $\hat{b}_i = e_{11}\overline{b}_i$, $\hat{c}_i = e_{11}\overline{c}_i$.]

11.1.9. Proposition. *Let* Π *be a completely regular partition of a distance-regular graph* Γ. *Then the eigenvalues of* $\overline{\Gamma} = \Gamma/\Pi$ *form a subset of* $\{(\theta - e_{01})/e_{11} \mid \theta \in \operatorname{spec}\Gamma\}$. *In particular, if* Π *is a completely regular congruence partition, then* $\operatorname{spec}\overline{\Gamma} \subseteq \operatorname{spec}\Gamma$.

Proof. Let $\overline{\theta}$ be an eigenvalue of $\overline{\Gamma}$, with eigenvector u. Then $A^*u = (e_{01} + e_{11}\overline{\theta})u$ and $AKu = KA^*u = (e_{01} + e_{11}\overline{\theta})Ku$. □

C. Distance-regular graphs with regular abelian automorphism groups

Let X be a finite abelian group. A *translation distance-regular graph* on X is a distance-regular graph Γ with vertex set X such that if $x \sim y$ then $x + z \sim y + z$ for all $x, y, z \in X$. (Equivalently, Γ has a regular abelian automorphism group, which can be identified with its vertex set X.) The association scheme of such a graph Γ is a translation association scheme as defined in §2.10.

Most known translation distance-regular graphs have the additional property that all integers s coprime to $|X|$ are *multipliers*, i.e., that if $x \sim y$ then $sx \sim sy$ for all $x, y \in X$ and all integers s coprime to $|X|$. Exceptions are the Paley graphs with a nonsquare number of vertices v (§1.3).

Clearly, the coset graph of any additive completely regular code in a Hamming graph is a translation distance-regular graph, and we have the following partial converse.

11.1.10. Theorem. *Let* X *be an elementary abelian group of exponent* p, *and let* Γ *be a translation distance-regular graph on* X *of valency* k. *Then we have one of two cases:*

(i) Γ *is the coset graph of a linear completely regular code of length* $n = k/(p-1)$ *and minimum distance* $\delta \geqslant 3$ *over* $\mathbb{F}_p$.

(ii) Γ *has diameter* 2 *and is self-complementary, i.e., is a conference graph; the quadratic residues* mod p *are multipliers, but the quadratic nonresidues* mod p *map* Γ *to its complement.*

Proof. By Theorem 2.10.1, the mapping of $X \times X$ defined for $s \not\equiv 0 \pmod{p}$ by $(x,y) \mapsto (sx,sy)$ permutes the relations of the translation association scheme defined by Γ. Clearly, if R_i is an image of the relation R_1, then $\Gamma_i \cong \Gamma$. Since Γ has at most two P-polynomial structures (Theorem 4.2.12), we find that either all nonzero residues mod p are multipliers, or the quadratic residues are multipliers, and the quadratic nonresidues interchange Γ_1 and Γ_i.
(i) Suppose first that all nonzero residues mod p are multipliers. Then the action of the multiplier group partitions $\Gamma(0)$ into $n := k/(p-1)$ orbits consisting of the nonzero multiples of n particular neighbours $\gamma_1, \cdots, \gamma_n$ of 0. Now let $C := \{x \in \mathbb{F}_p^n \mid \sum_{i=1}^{n} x_i\gamma_i = 0\}$. Clearly, C is a linear code of minimum distance $\delta \geqslant 3$, and the partition of the Hamming graph $H(n,p)$ into cosets of C is a congruence partition. The mapping $x + C \mapsto \sum x_i\gamma_i$ defines an isomorphism of the coset graph of C with Γ. Since a congruence partition is trivially uniformly regular, Theorem 11.1.6 implies that C is completely regular.
(ii) Now suppose that only the quadratic residues mod p are multipliers. Then p is odd and since there exist nonsquares mod p that are the sum of two squares mod p it follows that $i = 2$ and $k = k_2$. If $d \geqslant 3$ then this implies that we have an antipodal 2-cover (by Proposition 5.1.1 (v)), but then $|X|$ is even, impossible. Hence $d = 2$. □

Problem. Are the only graphs under (ii) the Paley graphs with nonsquare v? (One can show that any graph under (ii) necessarily has nonsquare v [Blokhuis & Brouwer, unpublished].)

We note that the same distance-regular graph may be the coset graph of many different completely regular codes. Indeed, BASSALYGO, ZAITSEV & ZINOV'EV [43] remark that if D is a perfect single-error-correcting code of word length m, then the code C of word length $n = md$ obtained by taking as vectors arbitrary concatenations of d code words from D is linear and completely regular, with coset graph $H(d,r)$, where $r = q^m / |D| = m(q-1)+1$. Also BIER [65, 66] constructs many linear codes over $\mathbf{F}_q$ whose duals are completely regular codes C such that the coset graph of C is isomorphic to the Hamming graph $H(n,q)$.

D. Completely regular codes in the Hamming scheme

In view of Corollary 11.1.7 it is of interest to investigate which completely regular codes C exist in the Hamming graph $H(n,q)$. Let us first state some necessary and some sufficient conditions for the most important case where C is linear or at least additive.

11.1.11. Theorem. *Let C be an additive completely regular code in $H(n,q)$ with minimal distance $\delta \geqslant 3$ and covering radius t, and suppose that the dual code $C^\perp$ has weight enumerator*

$$w_{C^\perp}(x) = \sum_{i=0}^{t} m_i x^{w_i} \quad (0 = w_0 < w_1 < \cdots < w_t,\ m_i \neq 0).$$

Then the coset graph $\Gamma(C)$ has diameter $d = t$ and eigenvalues $\theta_i = n(q-1) - qw_i$ with multiplicities $f_i = m_i$ $(i = 0, \cdots, d)$.

Proof. Since $\delta \geqslant 3$, the cosets of C form a congruence partition. Now apply Theorem 2.10.10. The dual scheme has vertex set $C^\perp$ and relations R_i defined by $(x,y) \in R_i$ if and only if $d(x,y) = w_i$. □

(Note that the code $C^\perp$ here is defined as in the discussion following Theorem 2.10.12. It coincides with the dual code for the usual inner product when C is linear.)

11.1.12. Corollary. *Under the hypothesis of Theorem 11.1.11, we have $q^d w_1 \cdots w_d = d!e_{dd}|C^\perp|$, and the number*

$$\frac{q^{d(d-1)/2} \prod_{i>j>0} (w_i - w_j)}{\prod_{i=1}^{d-1} (i!e_{ii})}$$

is an integer dividing $|C^\perp|^d$.

Proof. Apply Proposition 2.7.2 and its dual (Corollary 2.10.11) to the association scheme of $\Gamma(C)$, using $c_i = i$ for the Hamming graph and Theorem 11.1.8 for the parameters of $\Gamma(C)$. □

Special cases of this corollary have been proved by CALDERBANK & GOETHALS [159, 160]. The former paper also contains a transitivity argument for establishing complete regularity of the Kasami codes (see §11.2). Their argument generalizes to

11.1.13. Theorem. *Let C be a linear binary code of dual degree r, and suppose that C has an automorphism group G fixing 0 and permuting the coordinates s-transitively.*

(i) *If $s \geqslant r-1$, then C is completely regular.*

(ii) *If C is even and $s \geqslant r-2$, then C and each shortened code C' are completely regular.*

Proof. The hypothesis implies that C is s-regular, hence the claims about C follow from Theorem 11.1.3. If C is even, then $C^{\perp}$ contains **1**, hence C' has dual degree $\leqslant r-1$ and is still s-regular. So again Theorem 11.1.3 applies. □

The following perfect codes in $H(n,q)$ are known:

(i) the complete code: $C = X$, $e = 0$,

(ii) the trivial code: $|C| = 1$, $e = n$,

(iii) the repetition codes: $|C| = 2$, $q = 2$, $n = 2e+1$,

(iv) the binary Golay code: $|C| = 2^{12}$, $q = 2$, $n = 23$, $e = 3$,

(v) the ternary Golay code: $|C| = 3^6$, $q = 3$, $n = 11$, $e = 2$,

(vi) various single-error-correcting codes: $e = 1$, $|C| = q^{n-m}$, q a prime power, $n = (q^m-1)/(q-1)$.

(For a description of these codes, see, e.g., MacWilliams & Sloane [515] or van Lint [502]. In case (vi) many constructions exist (see Hamming [359], Shapiro & Slotnick [663], Vasil'ev [764], Schönheim [643], Lindström [498], Hergert [377], Bauer, Ganter & Hergert [47], Laborde [470], Mollard [538, 539], Phelps [593]); no complete classification is available; the only linear codes are the Hamming codes.)

On the other hand, it is known that for $e > 2$ (and for $e = 2$, q a prime power) no other perfect codes exist; for arbitrary q, see Bannai [26] (for asymptotic results), Reuvers [611] (for $3 \leqslant e \leqslant 5$), Best [59] (for $e = 7$ and $e \geqslant 9$), and Hong [406] (for $e = 6$ and $e = 8$); for prime powers q, see van Lint [499, 500, 501], Tietäväinen [736, 737], Tietäväinen & Perko [738] and Zinov'ev & Leont'ev [800]. (The case $e \geqslant 2$, $q = 2^{\alpha}3^{\beta}$ is handled in Bassalygo et al. [45].)

Do these perfect codes give rise to completely regular partitions? If $H(n,q)$ can be provided with the structure of a vector space V of dimension n over $\mathbf{F}_q$, and C is a linear code, that is, a subspace of V, then this is certainly the case (the partition is the partition into cosets of C); this works in cases (iii)-(v) and sometimes in case (vi), namely for the Hamming codes. In cases (i)-(ii) we trivially find a completely regular partition. We don't know whether the existence of the partition forces C to be linear in case (vi).
The quotient graphs one finds are: (i) K_1, (ii) the Hamming graph, (iii) the folded Hamming graph, (iv), (v) two graphs treated in §11.3B, E, (vi) a complete graph.

The following uniformly packed codes in $H(n,q)$ are known:

(i) various single-error-correcting codes: $e = 1$, $r = 2$, see, e.g., Table 4.1.1 in van Tilborg [739], and the references below.

[In case these codes are linear, we find strongly regular quotient graphs. In fact there is a 1-1 correspondence between a) strongly regular graphs with a regular elementary abelian group of automorphisms p^a where all nonzero residues mod p are multipliers, b) projective two-weight codes (duals of codes with $e = 1$, $r = 2$), and c) subsets of projective spaces with the property that every hyperplane meets it in one of two given cardinalities. See DELSARTE [250], WOLFMANN [791], CALDERBANK & KANTOR [161] and BROUWER [126].]

(ii) the Preparata codes: $e = 2$, $q = 2$, $n = 2^{2m} - 1$, $|C| = 2^{n+1-4m}$, $\lambda_C + 1 = \mu_C = n/3$, nonlinear.

(iii) binary BCH codes of designed minimum distance 5: $e = 2$, $q = 2$, $n = 2^{2m+1} - 1$, $|C| = 2^{n-4m-2}$, $\lambda_C + 1 = \mu_C = (n-1)/6$ (see §11.2).

(iv) a Hadamard code: $e = 2$, $q = 2$, $n = 11$, $|C| = 24$, $\lambda_C = 2$, $\mu_C = 3$, nonlinear.

(v) the doubly truncated binary Golay code: $e = 2$, $q = 2$, $n = 21$, $|C| = 2^{12}$, $\lambda_C = 1$, $\mu_C = 4$ (see §11.3G).

(vi) the truncated binary Golay code: $e = 2$, $q = 2$, $n = 22$, $|C| = 2^{12}$, $\lambda_C = 0$, $\mu_C = 2$ (see §11.3F).

(vii) the extended binary Golay code: $e = 3$, $q = 2$, $n = 24$, $|C| = 2^{12}$, $\lambda_C = 0$, $\mu_C = 6$ (see §11.3D).

(viii) the extended ternary Golay code: $e = 2$, $q = 3$, $n = 12$, $|C| = 3^6$, $\lambda_C = 0$, $\mu_C = 4$ (see §11.3A).

(ix) the repetition codes: $q = 2$, $n = 2e + 2$, $|C| = 2$, $\lambda_C = 0$, $\mu_C = 2$.

(For codes (ii), (iv) and (iii), see, e.g., VAN LINT [502], §7.4 and §6.6.)

On the other hand, VAN TILBORG [739] shows that other uniformly packed codes with $e \geqslant 4$ do not exist, and also determines the parameters of all uniformly packed binary codes with $e = 3$ and with $e = 1, 2$ and $\mu_C = \lambda_C + 1$. CALDERBANK & GOETHALS [160] and CALDERBANK [158] determine the parameters of all uniformly packed binary linear codes with $e = 2$; their classification rests on Corollary 11.1.12.

Concerning completely regular partitions derived from these codes, in case (iv) no partition exists, since $|C| = 24$ does not divide $|X| = 2^{11}$; partitions of the Hamming graph into Preparata codes are known (cf. BAKER, VAN LINT & WILSON [22]), but we do not know whether any such partition is completely regular. The remaining codes are linear, and the corresponding graphs are discussed in §11.3 for the Golay family, and in §11.2 for the double error-correcting BCH codes. In case (ix) we find folded Hamming cubes.

Many other completely regular codes exist in $H(n,q)$. E.g., the classical graphs (C8)-(C13) give rise to completely regular codes by Theorem 11.1.10, and, as remarked there, sometimes in many ways. A trivial completely regular partition is the one into parallel lines; the corresponding quotient is $H(n-1,q)$.

Another class of completely regular codes, that of the Kasami codes, which contains the uniformly packed binary BCD codes as a subclass, is considered in §11.2. The extended Preparata codes, the doubly truncated binary Golay code, and the truncated ternary Golay code are also completely regular.

Problems. (i) Is the truncation of a completely regular code with only even weights again completely regular? Note that the code obtained by adding a parity check symbol to a completely regular code with not only even weights need not be completely regular. E.g., the extension of code (v) with a parity check bit is not even uniformly packed in the wide sense (BASSALYGO & ZINOV'EV [44]).
(ii) Show (or disprove) that a coset graph with classical parameters (d,b,α,β), where $\alpha+1 = b$, must be one of the known graphs. (These graphs are formally self-dual, with $v = (\beta+1)^d$ vertices; their eigenvalues and multiplicities are given in Corollary 8.4.4. The possible parameters of the code follow from Theorem 11.1.11 and the remark after Theorem 11.1.8.)

E. Completely regular codes in other distance-regular graphs

Only little work has been done on codes in schemes other than the Hamming scheme.

For the Johnson scheme, HAMMOND [362] shows that $J(n,d)$ does not contain non-trivial perfect or nearly perfect codes (as defined, e.g., in HAMMOND [360]; see also GOETHALS & SNOVER [331]) for $e \geqslant 1$ and $n \in \{2d+1, 2d+2\}$. (For the case of perfect codes in $J(2d+1,d)$ this was already shown in BANNAI [25].) On the other hand, ROOS [627] shows that a perfect e-code must satisfy $n \leqslant (d-1)(2e+1)/e$.

For the halved n-cubes, BANNAI [25] shows that no non-trivial perfect codes with $e \geqslant 1$ exist, except that in case $n = 4e+2$ there is a (repetition) code C with $|C| = 2$; also, that the folded halved $2m$-cubes do not have nontrivial perfect codes with $e \geqslant 1$. HAMMOND [362] shows that for $e \geqslant 2$ no non-trivial nearly perfect codes exist in the halved n-cubes when n is odd and $n \geqslant 17$. For the folded n-cubes we find several perfect codes by folding perfect codes in the n-cubes.

For the Odd graphs O_k, HAMMOND & SMITH [363] study perfect codes for $e = 1,2,3$ (many parameter sets seem feasible), and show that arbitrary perfect codes must satisfy $k \geqslant (e^2+4e+2)/2$; SMITH [685] shows that for $e = 4$ no perfect codes exist. Nearly perfect codes in O_k are studied in HAMMOND [360]; he finds that the 2-(11,5,2) biplane is nearly perfect (with $e = 2$) in O_6; it follows that the Steiner system $S(4,5,11)$ is perfect (with $e = 1$) in O_6. In fact, any perfect code with $e = 1$ in O_k is equivalent to a Steiner system $S(k-2,k-1,2k-1)$, so we also have a perfect 1-code in O_4.
[Concerning the line graphs $L(O_k)$, HAMMOND [361] shows that the sets C_x of edges $\gamma\delta$ where $\gamma \cup \delta$ misses the symbol x are perfect codes with $e = 1$, and SMITH [685] shows that no other perfect codes with $e = 1$ occur in this graph.]

CAMERON, THAS & PAYNE [184] show that a perfect 1-code in a generalized hexagon of order (s,s) is the same thing as an *ovoid* (a set of s^3+1 points pairwise at distance 3), and show that the dual of the $G_2(q)$ generalized hexagon does possess ovoids. THAS [731] shows that a perfect 1-code in the dual polar graph on $[C_3(q)]$ is the same as a symplectic spread (consisting of q^3+1 totally isotropic planes), and shows that such spreads exist. THAS [732] carries out a similar construction for $[B_3(q)]$.

The situation in the dual polar schemes and affine schemes was settled in one blow by CHIHARA [192]. She shows that no perfect codes exist in these schemes, except possibly in the dual polar graphs on $[B_N(q)]$ and $[C_N(q)]$, where $N = 2^m-1$. As we just saw, perfect codes do exist in these cases when $N = 3$. She also shows that there are no nontrivial tight designs in these schemes.

BIGGS [72] mentions that in the twelve cubic distance-regular graphs there are only two non-trivial perfect codes: the repetition 1-code in the 3-cube, and a 1-code in the Coxeter graph on 28 vertices (cf. §12.3).

11.2 Graphs from the Kasami codes

Let s, t be powers of 2, $s \geqslant t$, and let $K(s,t)$ be the binary code consisting of all binary words $x \in H(s,2)$, indexed by $F = \mathbb{F}_s$, satisfying

$$\sum_{\alpha \in F} x_\alpha = \sum_{\alpha \in F} x_\alpha \alpha = \sum_{\alpha \in F} x_\alpha \alpha^{t+1} = 0.$$

Results of KASAMI [462, 463] (cf. MACWILLIAMS & SLOANE [515], §15.4) imply that the dual degree of $K(s,t)$ is $r = 4$ precisely when s, t have one of the forms

(i) $s = q^{2j+1}$, $t = q^m$ with $q = 2^i$, $m \leqslant j$, $(m, 2j+1) = 1$,

or

(ii) $s = q^2$, $t = q$ with $q = 2^i$.

We call the codes $K(s,t)$ satisfying (i) or (ii) *extended Kasami codes*. The extended Kasami codes are linear even codes with minimum distance $\delta = 4$, except in case (i) for $q = 2$, where $\delta = 6$. Since t is a power of 2, the group $AGL(1,q)$ acts 2-transitively on the coordinates, and the code words of weight 4 form a 2-$(q, 4, \frac{1}{2}q-1)$ design. Since $r = 4$, Theorem 11.1.13 (ii) implies that the extended Kasami codes are completely regular. Shortening the extended Kasami codes leaves the *Kasami codes* with dual degree $r = 3$; they are also completely regular by Theorem 11.1.13. Their minimum distance is $\delta = 3$, except in case (i) for $q = 2$, where $\delta = 5$. The Kasami codes with minimum distance 5 are uniformly packed; they coincide with the binary BCH-codes of designed minimum distance 5.

11.2.1. Theorem. *The coset graphs of the extended Kasami codes are bipartite distance-regular graphs of diameter 4 with intersection arrays in cases (i) and (ii), respectively,*

$$\{q^{2j+1}, q^{2j+1}-1, q^{2j+1}-q, q^{2j}(q-1)+1;\ 1, q, q^{2j}-1, q^{2j+1}\},\quad v = 2q^{4j+2} \tag{11}$$

$$\{q^2, q^2-1, q^2-q, 1;\ 1, q, q^2-1, q^2\},\quad v = 2q^3. \tag{12}$$

The coset graphs of the Kasami codes are distance-regular of diameter 3, with intersection arrays in cases (i) and (ii), respectively,

$$\{q^{2j+1}-1, q^{2j+1}-q, q^{2j}(q-1)+1;\ 1, q, q^{2j}-1\},\quad v = q^{4j+2} \tag{13}$$

$$\{q^2-1, q^2-q, 1;\ 1, q, q^2-1\},\quad v = q^3. \tag{14}$$

□

Remarks. The association schemes of the Kasami codes are described in CALDERBANK & GOETHALS [159]. The case $q = 2$ of (11) is due to Calderbank [pers. comm.]. The parameters (12) are the same as those for the symmetric net of the points and planes in $AG(3,q)$ minus a parallel class of planes. The graphs with parameters (13) have an eigenvalue -1 and hence give rise to strongly regular graphs.

11.3 Graphs from the Golay codes

We next proceed to discuss graphs obtainable as coset graphs from the Golay codes and some related codes.

Let us first very briefly describe the Golay codes. They may be defined by giving their parameters (DELSARTE & GOETHALS [257]): there are unique codes C with minimum distance δ in $H(n,q)$ with $(n,q,\delta,|C|) = (23,2,7,2^{12})$, $(24,2,8,2^{12})$, $(11,3,5,3^6)$ and $(12,3,6,3^6)$; these are called the binary Golay code, the extended binary Golay code, the ternary Golay code, and the extended ternary Golay code. Here 'unique' means unique up to automorphisms of $H(n,q)$. If we identify $H(n,q)$ with $\mathbf{F}_q^n$ and choose these codes so as to contain 0, then they are linear, and the extended codes are self-dual for the ordinary inner product $(x,y) = \sum x_i y_i$. The non-extended codes are perfect, i.e., $H(n,q)$ is partitioned by the spheres of radius $(\delta-1)/2$ around their code words.

The literature abounds with constructions for these codes. A very direct approach is to give bases for them (considered as linear subspaces of $\mathbf{F}_q^n$). It suffices to do this for the extended Golay codes; then the perfect Golay codes are found by just omitting one fixed coordinate position from the vectors of the extended codes. A basis for the extended binary Golay code is given by the rows of the 12×24 matrix $(I\ J-N)$, where I is the identity matrix of order 12, J the all-one matrix of that order, and N the adjacency matrix of the icosahedron. A basis for the extended ternary Golay code is given by the rows of the 6×12 matrix $\begin{bmatrix} I & \begin{matrix} 0 & \mathbf{1}^{\mathrm{T}} \\ -\mathbf{1} & S \end{matrix} \end{bmatrix}$ where S is the 5×5 circulant matrix with first row $(0\ 1\ -1\ -1\ 1)$ (i.e., the quadratic residue character mod 5).

Assuming that the Golay codes have been chosen so as to contain $\mathbf{0}$, the supports of the words of minimal nonzero weight form the (unique) Steiner systems $S(4,7,23)$, $S(5,8,24)$, $S(4,5,11)$ and $S(5,6,12)$, respectively. (See §A.6 for the definition of a Steiner system.) More details on these codes and Steiner systems, and on their automorphism groups may be found in MACWILLIAMS & SLOANE [515], CONWAY [212] and CONWAY & SLOANE [215], Chapter 10-11.

A. The coset graph of the extended ternary Golay code

11.3.1. Theorem. *If C is the extended ternary Golay code $(q = 3, n = 12, |C| = 3^6)$, then $\Gamma = \Gamma(C)$ is a regular near hexagon on 729 vertices with $(s,t,t_2) = (2,11,1)$; its group of automorphisms $3^6.2.M_{12}$ acts distance-transitively with point stabilizer $2.M_{12}$. The graph Γ is uniquely determined by its intersection array $\{24,22,20;1,2,12\}$. Distance distribution diagram:*

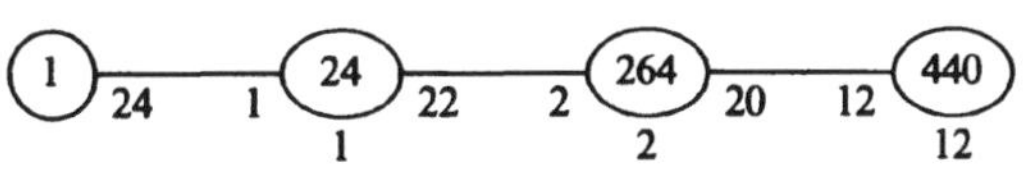

$v = 729.$

The graph Γ has classical parameters, see Table 6.1, (C14) and Table 6.7, (N2).

Proof. Distance in Γ is Hamming distance between cosets; distance-regularity is clear from the parameters of C; distance-transitivity is clear since M_{12} is 3-transitive on the set of 12 coordinate positions. Uniqueness is shown in BROUWER [118]. □

Remark. This near hexagon was first constructed by SHULT & YANUSHKA [673]. Their description is the following: take as vertices the vectors in C and as lines the translates of $\mathbf{F}_3 u$ for $u \in C$, $\mathrm{wt}(u) = 12$.

B. The coset graph of the ternary Golay code

Taking for C the perfect ternary Golay code ($q = 3, n = 11, |C| = 3^6$) we find for $\Gamma(C)$ a strongly regular graph on 243 vertices with intersection array $\{22,20; 1,2\}$, the Berlekamp-van Lint-Seidel graph. (cf. BERLEKAMP, VAN LINT & SEIDEL [54].) An antipodal 3-cover of this graph is constructed in §11.3H.

Problem. Is there a distance-regular graph with $v = 486$ vertices and intersection array $\{45,44,36,5; 1,9,40,45\}$ whose halved graphs are complements of the Berlekamp-van Lint-Seidel graph?

C. The coset graph of the truncated ternary Golay code

Taking for C the truncated ternary Golay code ($q = 3, n = 10, |C| = 3^6$) we find for $\Gamma(C)$ a strongly regular graph on 81 vertices with intersection array $\{20,18; 1,6\}$. This graph is the graph on the points of $\mathbf{F}_3^4$ equipped with a nondegenerate quadratic form of nonmaximal Witt index, where two points are joined when the connecting line has isotropic direction (cf. HUBAUT [413], (C12$^-$)), or, equivalently, the Hermitean forms graph on $\mathbf{F}_9^2$. It admits a group of automorphisms $3^4.2.\mathrm{Sym}(6).2$. An antipodal 3-cover of this graph is constructed in §11.3H.

D. The coset graph of the extended binary Golay code

11.3.2. Theorem. *If C is the extended binary Golay code ($q = 2, n = 24, |C| = 2^{12}$), then $\Gamma(C)$ is a regular thin near octagon on 4096 vertices with $(s,t_2,t_3,t) = (1,1,2,23)$. Its group of automorphisms $2^{12}.M_{24}$ acts distance-transitively with point stabilizer M_{24}. This graph is uniquely determined by its intersection array $\{24,23,22,21; 1,2,3,24\}$. Distance distribution diagram:*

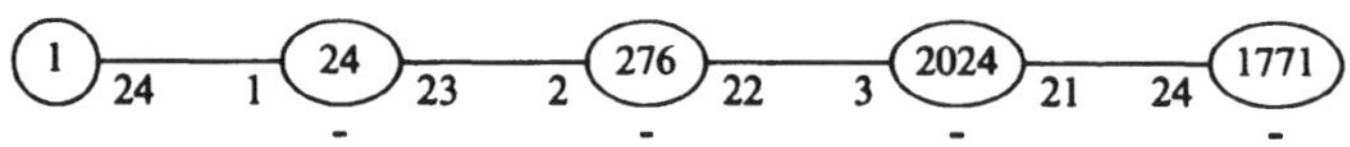

$v = 4096.$

(Cf. Table 6.9, (B7).)

Proof. All statements are clear except for the uniqueness. Uniqueness was shown in BROUWER [123] but the argument as given there was not quite complete. Since we shall use the same type of arguments in uniqueness proofs for other coset graphs, let us give it here in full.

By Proposition 4.3.6, we have a labelling π of our graph Γ with 2^{24}, and putting $C_x := \pi^{-1}(x)$ for $x \in \Gamma$ we find that the set of labels C_x of the vertex x is a code of word length 24, minimum distance at least 8 (Γ is bipartite so C_x has even minimum distance, but $c_3 = 3$ so $\Gamma_2(x_0)$ cannot have points with labels of weight 4, and x_0 cannot have labels of weight 6) and size 4096, hence each C_x is a copy of the extended binary Golay code C. Our aim is to show that all C_x must be translates of each other. (This is where the argument in BROUWER [123] was a bit too short.)

Let $x \sim y$ and let $\psi: C_x \to C_y$ be defined by $d(u, \psi(u)) = 1$. Then ψ can change distances by at most two, but since all distances in C_x are divisible by 4 we must have

$d(\psi(u),\psi(v)) = d(u,v)$ - i.e., ψ preserves distances. We shall identify vectors with subsets of a 24-set (of 'coordinate positions'), and vectors of weight one with elements of this set.

11.3.3. Lemma. *Let C and C' be two copies of the extended binary Golay code in V (i.e., C and C' are isomorphic as metric spaces) and let $\psi: C \to C'$ be a bijection such that $d(u,\psi(u)) = 1$ for all $u \in C$. Then either ψ is a translation by a vector of weight one or there are two coordinate positions e, f such that ψ leaves coordinates in other positions invariant and maps those in positions e, f as follows.*

ef	$\overset{\psi}{\mapsto}$	ef
00		10
10		11
01		00
11		01

Proof. We may suppose that $0 \in C$ and that $\psi(0) = e$. Since the words of weight 8 span C we may also assume that $\psi(u) = u+f$ with $f \neq e$ for some code word u of weight 8. Write $\phi(x) = \psi(x)-x$. If $\mathbf{1}$ is the all-one vector then $\psi(\mathbf{1}) = \mathbf{1}+e$ (since $d(\psi(0),\psi(\mathbf{1})) = 24$). Let $\mathbf{1} = u+v+w$, where u, v, w have weight 8. At most one of the three events $\phi(u) \in u$, $\phi(v) \in v$, $\phi(w) \in w$ can happen since u, v, w have mutual distance 16 and two 7 sets have mutual distance at most 14. If $e \notin u$ then say $e \in v$ and considering the distance to $\mathbf{1}$ we see $f = \phi(u) \in u$ and $\phi(w) = e$ and $\phi(v) = f$. If $e \in u$ then $f \notin u$, say $f \in v$, and $\phi(v) = f$, $\phi(w) = e$. Since the graph that has as vertices the blocks that contain precisely one of e, f, two blocks being adjacent when they are disjoint, is connected (cf. 11.4E below) we see that $\phi(x) = f$ for any such block and that $\phi(x) = e$ for any block containing neither e nor f. Extending this pattern to all of C is straightforward. □

We continue with the proof of 11.2.2. If $x = \pi(u)$, $y = \pi(u+e)$ then either $C_y = C_x+e$ or $C_y = \alpha(C_x)+e$ where $\alpha \in \mathrm{Sym}(24)$ is the transposition (ef). More generally we see that $C_x = \alpha_u(C)+u$ where $\alpha_u \in \mathrm{Sym}(24)$ is a product of at most $\mathrm{wt}(u)$ transpositions and $C = C_{\pi(0)}$. In particular, with $x = x_0$ and $u \in C$ with $\mathrm{wt}(u) = 8$, we find that $\alpha_u \in M_{24}$ fixes at least 8 symbols so either is the identity or is an involution fixing precisely 8 symbols. If the latter occurs then $\alpha_u(u)$ is a block disjoint from u; choose three symbols $\sigma_1, \sigma_2, \sigma_3$ in u, let $\tau = \alpha_u(\sigma_1)$ and let v be a block on $\{\sigma_1,\sigma_2,\sigma_3,\tau\}$ not contained in $u \cup \alpha_u(u)$; now α_v must be the identity but that is impossible. This shows that α_u is the identity for all $u \in C$. If $\alpha_e = (ef)$ then let u be a block on e but not on f, and we find that $\alpha_u \neq 1$. Contradiction. This shows that indeed each C_x is a coset of C as desired. □

E. The coset graph of the binary Golay code

11.3.4. Theorem. *If C is the (perfect) binary Golay code ($q = 2, n = 23, |C| = 2^{12}$), then $\Gamma(C)$ is a regular thin near heptagon on 2048 vertices with $(s,t_2,t_3,t) = (1,1,2,22)$. Its group of automorphisms $2^{11}.M_{23}$ acts distance-transitively with point stabilizer M_{23}. This graph is uniquely determined by its intersection array $\{23,22,21;1,2,3\}$. Distance distribution diagram:*

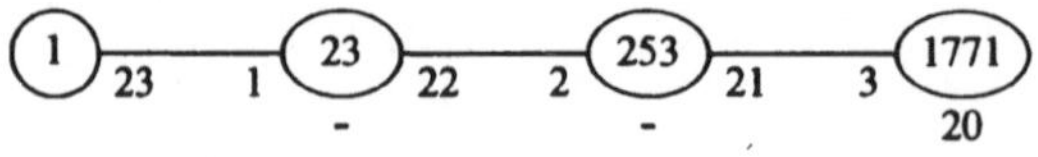

$v = 2048.$

(Cf. Table 6.7, (N4).)

Proof. Again everything is immediately clear except for uniqueness. By Proposition 4.3.6 we find for the vertices of a graph Γ with the above intersection array a labelling such that the collection C_x of labels of vertex x is a code of size 2^{12} and minimum distance at least 7, that is, each C_x is isomorphic to the perfect binary Golay code C.

Let $x \sim y$ and let $\psi: C_x \to C_y$ be defined by $d(u, \psi(u)) = 1$. Since all distances in C are congruent to -1 or 0 (mod 4) and ψ changes distances by 0 or ± 2 it follows that ψ preserves distances. The analogue of the lemma in the previous section is true again (where the proof uses decompositions $\mathbf{1} = u + v + w$ with $\mathbf{1}$ the 23-set and u, v, w two 8-sets and a 7-set). Defining α_u as before we see that, if $u \in C$, $\mathrm{wt}(u) = 7$, then α_u is the identity (since α_u has at least 9 fixed points and $\alpha_u \in M_{23}$). Now it follows as before that each α_u is the identity and each C_x is a coset of C. □

Remarks. The graph Γ has spectrum $23^1\, 7^{506}\, (-1)^{1288}\, (-9)^{253}$ so (by Proposition 4.2.17) $\overline{\Gamma}_3 = \Gamma_1 \cup \Gamma_2$ is strongly regular (with $(v,k,\lambda,\mu) = (2048,276,44,36)$).

Since Γ is a near heptagon, also Γ_2 is distance-regular (and then distance-transitive) with intersection array $\{253,210,3;1,30,231\}$, distance distribution diagram

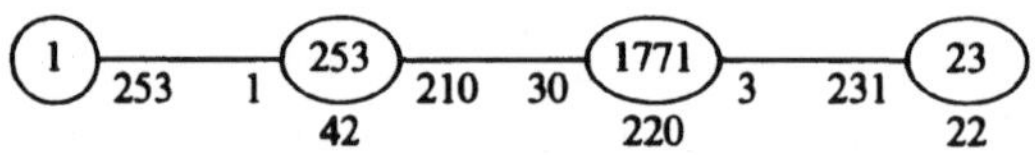

$v = 2048,$

and spectrum $253^1\, 29^{253}\, 13^{506}\, (-11)^{1288}$. (Cf. Table 6.8, (D8).) For the same reason, Γ has a distance-regular bipartite double Δ on 4096 vertices (cf. §1.11); it is distance-transitive with distance distribution diagram

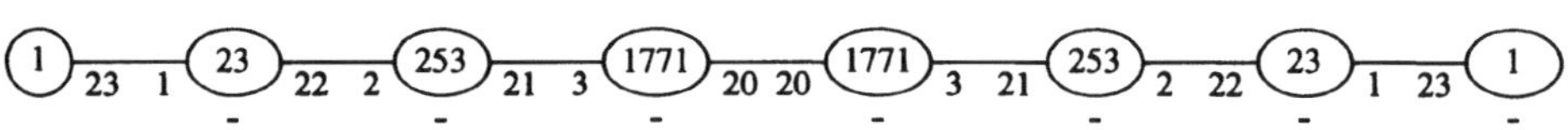

and automorphism group $2^{11}.M_{23} \times 2$. (Cf. Table 6.9, (B13).) Of course both Γ_2 and Δ are determined by their intersection array.

F. The coset graph of the truncated binary Golay code

11.3.5. Theorem. *If C is a truncation of the binary Golay code ($q = 2, n = 22, |C| = 2^{12}$), then $\Gamma = \Gamma(C)$ is a regular thin near heptagon on* 1024 *vertices with $(s,t_2,t_3,t) = (1,1,5,21)$. Its group of automorphisms $2^{10}.M_{22}.2$ acts distance transitively with point stabilizer $M_{22}.2$. Γ has intersection array* $\{22,21,20;1,2,6\}$ *and distance distribution diagram*

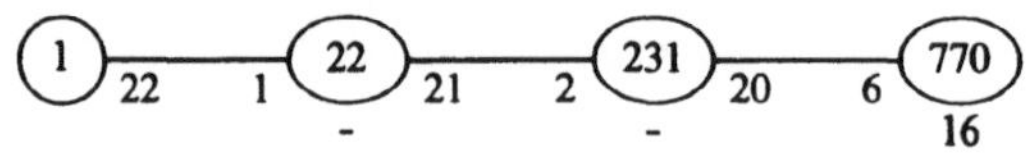

$v = 1024.$

It has spectrum $22^1\, 6^{330}\, (-2)^{616}\, (-10)^{77}$. (Cf. Table 6.7, (N3).) □

It is unknown whether Γ is uniquely determined by its intersection array. Γ_2 is distance-transitive with distance distribution diagram

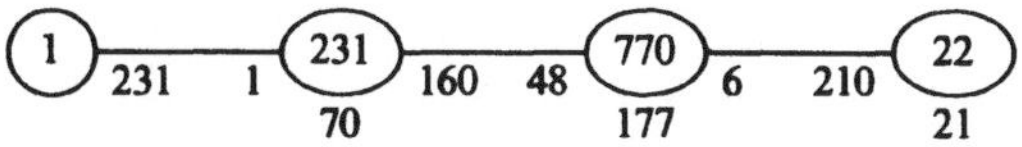

$v = 1024$

and spectrum $231^1\,39^{77}\,7^{330}\,(-9)^{616}$. (Cf. Table 6.8, (D7).) The bipartite double of Γ is distance-transitive on 2048 vertices with distance distribution diagram

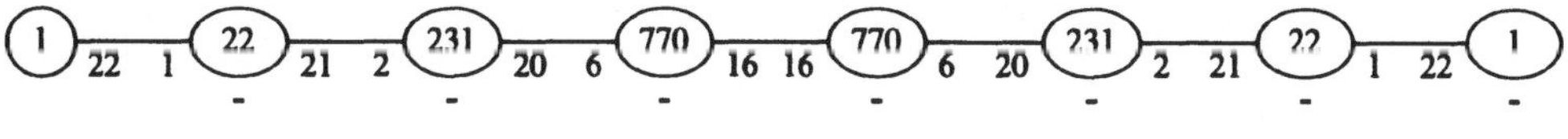

and automorphism group $(2^{10}.M_{22}.2)\times 2$ (cf. Table 6.9, (B12)). An antipodal 2-cover of Γ is constructed in §11.3H.

G. The coset graph of the doubly truncated binary Golay code and the graph of the unitals in $PG(2,4)$

11.3.6. Theorem. *If C is a double truncation of the binary Golay code ($q = 2, n = 21, |C| = 2^{12}$), then $\Gamma = \Gamma(C)$ is a distance-transitive graph on 512 vertices with intersection array $\{21,20,16; 1,2,12\}$; its automorphism group is $2^9.P\Gamma L(3,4)$. The distance distribution diagram is*

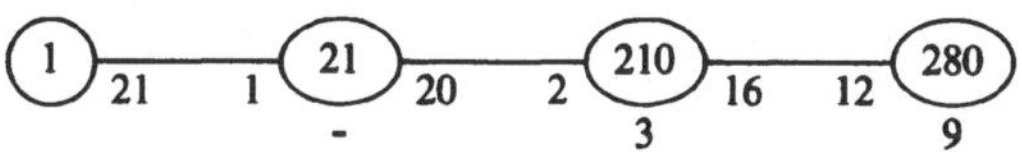

$v = 512.$

(Cf. Table 6.1, (C12) for $r = 2$.) It has spectrum $21^1\,(-11)^{21}\,5^{210}\,(-3)^{280}$. Γ is uniquely determined by its intersection array.

Proof. Everything except uniqueness is straightforward. Uniqueness is due to IVANOV & SHPECTOROV [437]. Let us briefly sketch the proof. The eigenvalue $\theta = -11$ has multiplicity 21 and standard sequence $u = (1, -\frac{11}{21}, \frac{5}{21}, -\frac{3}{21})$, and by Proposition 4.4.1 and scaling we find a representation $\gamma \mapsto \bar{\gamma}$ of Γ in $\mathbf{R}^{21}$ such that $(\bar{\gamma},\bar{\delta}) = 21u_{d(\gamma,\delta)}$ for all γ, $\delta \in \Gamma$. Fix a vertex α. We may assume that $\bar{\alpha} = -\mathbf{1}$, where $\mathbf{1}$ is the vector with all coordinates equal to 1. The 21 vectors $\bar{\beta}$ for $\beta \sim \alpha$ are linearly independent and hence span $\mathbf{R}^{21}$. We may assume that $\bar{\beta_i} = \frac{1}{3}\mathbf{1} + 4e_i$ $(1 \leqslant i \leqslant 21)$, where e_i is the i-th unit vector, and $\beta_1, \cdots, \beta_{21}$ is an arbitrary numbering of $\Gamma(\alpha)$. Let $\gamma \in \Gamma_2(\alpha)$. Then $\bar{\gamma}_i = -\frac{7}{3}, \frac{5}{3}, -\frac{1}{3}$ for $d(\gamma,\beta_i) = 1$, 2, 3, respectively. Similarly, if $\delta \in \Gamma_3(\alpha)$, then $\bar{\delta}_i = 1$, -1 for $d(\delta,\beta_i) = 2$, 3, respectively. For $\gamma \in \Gamma_2(\alpha)$, put $l(\gamma) = \Gamma(\alpha) \cap \Gamma_{\leqslant 2}(\gamma)$. If $\epsilon \in \Gamma(\gamma) \cap \Gamma_2(\alpha)$, then by considering the inner product $(\bar{\gamma},\bar{\epsilon})$ one finds that $l(\gamma) = l(\epsilon)$. It follows that $\Gamma_2(\alpha)$ is the disjoint union of 21 Petersen graphs, and that any two vertices at distance 2 determine a subgraph isomorphic to the folded 5-cube. The sets $l(\gamma)$ give $\Gamma(\alpha)$ the structure of a projective plane $PG(2,4)$. By considering the inner product $(\bar{\gamma},\bar{\delta})$ for $\delta \in \Gamma(\gamma) \cap \Gamma_3(\alpha)$ we find that $\Gamma_2(\gamma) \cap \Gamma(\alpha) \subset \Gamma_3(\delta) \cap \Gamma(\alpha)$. Now the 280 sets $\Gamma_3(\delta) \cap \Gamma(\alpha)$, for $\delta \in \Gamma_3(\alpha)$, are the 280 unitals (affine subplanes $AG(2,3)$) of $PG(2,4)$, and by computing $(\bar{\delta}_1,\bar{\delta}_2)$ we see that adjacent vertices in $\Gamma_3(\alpha)$ correspond to unitals meeting in one point. Now Γ is uniquely determined. □

Remark. Γ has classical parameters $(d,b,\alpha,\beta) = (3,-2,-3,7)$, and hence has the same parameters as the Hermitean forms graph with $d = 3$ and $q = 4$. And indeed it is easy to see directly that these are isomorphic: writing down the 21 rank one 3×3 Hermitean matrices over $\mathbf{F}_4$ in a 9×21 matrix we obtain a parity check matrix for C. Antipodal 2- and 4-covers of Γ are constructed in §11.3H.

11.3.7. Theorem. *For $\gamma \in \Gamma$, the subgraph $\Delta = \Gamma_3(\gamma)$ of Γ is distance-transitive on* 280 *vertices with intersection array* $\{9,8,6,3;1,1,3,8\}$ *and distance distribution diagram*

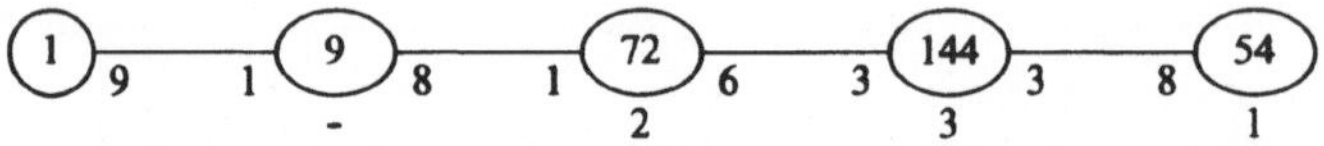

$v = 280.$

(Cf. Table 6.8, (D11).) It has spectrum $9^1\,(-5)^{20}\,4^{64}\,(-3)^{90}\,1^{105}$. *The group* $\mathrm{Aut}\,\Delta$ *is* $\mathrm{Aut}(L_3(4)) \cong P\Gamma L(3,4).2$ *with point stabilizer* $3^2{:}2S_4 \times 2$. □

For Δ we have the three descriptions: (i) the rank 3 Hermitean 3×3 matrices over $\mathbf{F}_4$, where adjacent matrices differ by a rank 1 matrix; (ii) subplanes $AG(2,3)$ of $PG(2,4)$, where two subplanes are adjacent whenever they have precisely one vertex in common (and have a common tangent at that vertex). (These two descriptions are derived from the two descriptions of Γ, but are easily seen to be equivalent directly.) Finally (iii): words of weight 9 in C (a double truncation of the binary Golay code), where adjacent code words have Hamming distance 16. In the second description, the other relations are: two affine subplanes have distance 3 whenever they have precisely three (noncollinear) points in common, and distance 2 (resp. 4) whenever they have five vertices in common, carrying one (resp. two) collinear triples. Note that $P\Gamma L(3,4)$ has two orbits on pairs of subplanes having three noncollinear points in common, so that $P\Gamma L(3,4)$ acts rank 6. But $PGL(3,4).2$, where the 2 represents the graph automorphism (interchange points and lines), acts distance transitively.
Any two vertices at distance two determine a Petersen subgraph; in description (ii), the Petersen graphs correspond to the antiflags (nonincident (point,line) pairs in $PG(2,4)$).

This graph was first described in Biggs & Gardiner [81] (and mentioned in Biggs [75]). For a detailed study of this graph, see Tchuda [712]. Lambeck [473] has shown that Δ is uniquely determined by its parameters. For antipodal 2- and 4-covers of Γ, see the following section.

H. Variations on the theme of coset graph - some antipodal covers

Let C be a linear code over $\mathbf{F}_q$ with minimum distance $\delta \geqslant 3$, and let C_0 be a subcode of C with $\dim C/C_0 = t$. Then the coset graph $\Gamma(C_0)$ is a q^t-cover of the coset graph $\Gamma(C)$ (with covering map $x + C_0 \mapsto x + C$).

In case C is the t-fold truncation of a linear code C_1, and C_0 is the t-fold shortening of C_1 (i.e., C is obtained from C_1 by deleting t coordinate positions, while C_0 is obtained from C_1 by deleting these t coordinate positions and only retaining the code words that had only zeros at these positions), then $\Gamma(C_0)$ is isomorphic to the partial graph of $\Gamma(C_1)$ where two cosets of C_1 are adjacent when they have representatives that differ in only one coordinate position, but not one of the t chosen positions.

We find the following graphs $\Gamma(C_0)$.

	Table	v	intersection array	q	n	dim C_0	r
a)	(A14)	2048	$\{22,21,20,3,2,1;1,2,3,20,21,22\}$	2	22	11	2
b)	(A15)	1024	$\{21,20,16,6,2,1;1,2,6,16,20,21\}$	2	21	11	2
c)	(A16)	2048	$\{21,20,16,9,2,1;1,2,3,16,20,21\}$	2	21	10	4
d)	(A17)	243	$\{20,18,4,1;1,2,18,20\}$	3	10	5	3
e)	(A18)	729	$\{22,20,18,2,1;1,2,9,20,22\}$	3	11	5	3

Here n denotes the word length of C_0 and $\Gamma = \Gamma(C_0)$ is an antipodal r-cover. The code C_0 is obtained in cases a), d), and e) by once shortening the binary Golay code, the ternary Golay code and the extended ternary Golay code, respectively. In case b) C_0 is obtained by taking all code words in the binary Golay code that start with 00 or 11 and deleting these two coordinate positions. In case c) C_0 is obtained by taking all code words in the extended binary Golay code that start with 000 or 111 and deleting these three coordinate positions (this is a code of index 4 in the doubly truncated binary Golay code). The graphs (A14) and (A15) have extended bipartite doubles (B13) and (B14), respectively.

The distance-regular graphs in cases a), b) and c) are uniquely determined by their intersection array. Indeed, in cases b) and c) we have antipodal covers of known graphs, and we see that cubes lift to cubes, so that each 3-claw in one of these graphs determines a unique 3-cube. Now by Proposition 4.3.6 we find a map $\pi : 2^k \to \Gamma$ preserving distances ≤ 2. Since the composition of π with the quotient map is known, we need only consider in what way one of the codes C involved can be partitioned into 2, 2, 4 subcodes (respectively) such that any vector from one subcode has distance at least 6 to the vectors of the other subcode(s). We may assume that $0 \in C$. In case c) we see that both the code C and the subcode containing 0 have 21 words of weight 5, and since in C the words of weight 5 span a subcode C_0 of dimension 10, it follows that our partition of C is the partition into the four cosets of C_0. This shows uniqueness in case c). By the same argument, in case b) the partition must be into the two cosets of a linear code C_1 that contains C_0 as index 2 subcode. This can be done in three (isomorphic) ways, but given $\pi^{-1}(\gamma)$, and $\delta \sim \gamma$, only one choice is possible for $\pi^{-1}(\delta)$. This shows uniqueness in case b). In case a), the graph $\Delta = \Gamma_1 \cup \Gamma_6$ is distance-regular; it is the coset graph of the binary Golay code. From this one obtains uniqueness.

There is another feasible parameter set that looks as if it belongs to this section: $\{21,20,16,8,2,1;1,2,4,16,20,21\}$ is the intersection array for an antipodal 3-cover of the coset graph of the doubly truncated binary Golay code; but this graph has $a_2 = 3$, and we learn from Corollary 4.3.15 that $\frac{1}{6}b_1b_2 = \frac{1}{6}\cdot 20\cdot 16$ must be an integer, which is not the case, so no such graphs exist.

Problems. Graph d) is an antipodal 3-cover of the graph constructed in §11.3C; but also the 2- and 6-covers have feasible parameters ($v = 162$, $\{20,18,3,1;1,3,18,20\}$ and $v = 486$, $\{20,18,5,1;1,1,18,20\}$). Do such graphs exist? Are graphs d) and e) uniquely determined by their parameters?

11.4 Graphs related to the Witt designs

Instead of looking at coset graphs we can also look at graphs defined on the code words of fixed weight in a code. For the Golay codes this leads to further distance-regular graphs. We shall first be concerned with graphs derived from the Steiner systems $S(5,8,24)$ and $S(5,6,12)$. (For the definition of Steiner system, see §A.6. The blocks of the two Steiner systems mentioned may be found as the supports of the code words of minimal nonzero weight in the extended binary and ternary Golay codes, respectively.)

A. The Witt graph associated to M_{24}

The *large Witt graph* is the graph whose vertices are the 759 blocks of a Steiner system $(\Sigma,\mathfrak{B})$ with parameters $S(5,8,24)$, in which two blocks are adjacent whenever they are disjoint.

11.4.1. Theorem. *The large Witt graph* Γ *is distance-transitive with* 759 *vertices, intersection array* $\{30,28,24;1,3,15\}$, *and full automorphism group* M_{24}. *It has singular lines of size* $s+1=3$ *and is a regular near hexagon with parameters* $(s,t,t_2)=(2,14,2)$. *Distance distribution diagram:*

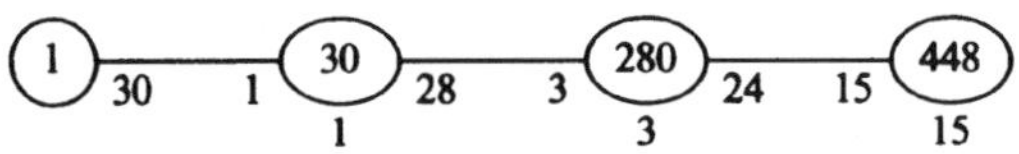

$v = 759.$

Furthermore, Γ *has classical parameters as given in Table 6.1, (C7). Any distance-regular graph with this intersection array is isomorphic to* Γ.

Proof. From the parameters of $S(5,8,24)$ one sees that two blocks have distance 0, 1, 2, 3 in Γ whenever they meet 8, 0, 4, 2 points, respectively. Also distance-regularity can be inferred from the parameters.

(Indeed, two blocks meet in 0, 2, 4 or 8 points since the extended binary Golay code C is self-dual and has minimum weight 8; the singular lines are triples of pairwise disjoint blocks - their existence follows from the fact that C contains the all-1 vector. If F_0 is a 4-subset of Σ then for any $x \notin F_0$ the set $F_0 \cup \{x\}$ is in a unique block, so F_0 determines 5 blocks $F_0 \cup F_i$ $(1 \leqslant i \leqslant 5)$, and $F_i \cup F_j$ is a block for all $i \neq j$ (by linearity of C); this set of 15 blocks induces a subgeometry isomorphic to the generalized quadrangle $GQ(2,2)$, called a *quad*. One easily checks that quads are geodetically closed (so that $|B \cap C| = 4$ implies $d(B,C) = 2$ and $\mu(B,C) = 3$). If a block B meets one block of a singular line in 2 points, then it meets the three blocks in 2, 2, 4 points (and it follows that $|B \cap C| = 2$ implies $d(B,C) = 3$ and $\mu = 3$, $c_3 = t+1 = 15$). We see that Γ is a near hexagon (a result due to Shult & Yanushka [673]) and hence that $a_i = c_i$ $(i = 1,2,3)$.)

The stabilizer of a block in M_{24} is $2^4.GL(4,2)$ acting in the natural way on the 16 points outside the block (3-transitively); the 30 neighbouring blocks are the 30 hyperplanes in this affine 4-space; the 280 blocks at distance two yield the 140 (4-point) planes (each occurs twice). The pointwise stabilizer (centralizer) of a block is 2^4, and thus the stabilizer of a block induces $GL(4,2) \cong \mathrm{Alt}(8)$ on the block. Now it easily follows that M_{24} acts distance-transitively. For much more information on this situation, cf. Conway [212].

Uniqueness of the regular near hexagon with parameters $(s,t,t_2) = (2,14,2)$ has been shown in BROUWER [120]. But if Γ is any distance-regular graph with intersection array $\{30,28,24; 1,3,15\}$ then Γ has singular lines of size 3 (as $\lambda = 1$); together with its singular lines, Γ is a near polygon since $a_i = (s-1)c_i$ for all i (cf. Theorem 6.4.1). □

Remarks. (i) (cf. RONAN & SMITH [621]) The lines and quads on a given point form the points and lines of a $PG(3,2)$, so we find the Buekenhout diagram (with 'ghost' node)

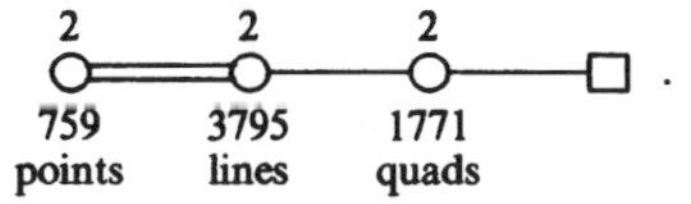

(ii) For $\gamma, \delta \in \Gamma$, $d(\gamma,\delta) = 3$, the subgraph $C(\gamma,\delta) \setminus \{\gamma,\delta\}$ is isomorphic to the collinearity graph of the generalized quadrangle $GQ(2,2)$, and, in particular, is connected. It follows (by 4.2.7 and 4.2.9) that Γ does not have distance-regular antipodal covers.

B. The truncated Witt graph associated to M_{23}

Fix a symbol $\sigma \in \Sigma$, and let $\mathcal{B}'$ be the collection of 506 blocks in $\mathcal{B}$ missing σ. Then $(\Sigma \setminus \{\sigma\}, \mathcal{B}')$ is a 4-(23,8,4) design such that any two distinct blocks meet in 0, 2 or 4 symbols, and it is the unique such design (TONCHEV [754]).

11.4.2. Theorem. *In the large Witt graph Γ described above $\mathcal{B}'$ induces a distance-transitive graph Δ with 506 vertices, with intersection array $\{15,14,12; 1,1,9\}$, and full automorphism group M_{23}. Distance distribution diagram:*

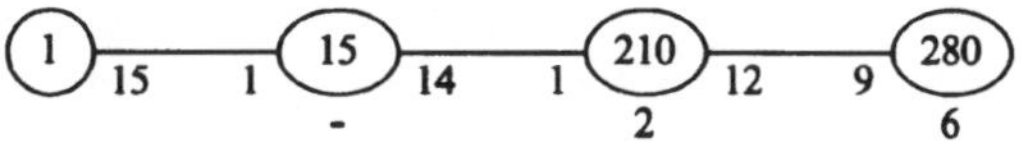

$v = 506.$

Δ has classical parameters, see Table 6.1, (C7a). It has spectrum (with Q-polynomial ordering of the eigenspaces) $15^1\,(-8)^{22}\,4^{230}\,(-3)^{253}$. Any distance-regular graph with this intersection array is isomorphic to Δ.

Proof. Throwing away from Γ the vertices that are blocks containing σ, we lose one point from each line and an ovoid from each quad (so that in Δ any two vertices at distance 2 determine a unique geodetically closed Petersen subgraph). This shows $\lambda = 0$, $\mu = 1$, $a_2 = 2$. The stabilizer in M_{24} of the subset σ is M_{23}. If we also stabilize a block B in $\mathcal{B}'$ then we are left with $PGL(2,4)$ acting on the 15 symbols outside $B \cup \{\sigma\}$ as on the points of $PG(3,2)$ and acting on B as Alt(8); transitivity on $\Delta_1(B)$ and $\Delta_2(B)$ is clear and on $\Gamma_3(B) \cap \Delta$ not difficult to check. Thus distances in Γ and Δ agree and Δ is distance-transitive. In order to show that Aut Δ cannot be larger than M_{23}, observe that Δ contains precisely $1771 = \binom{23}{3}$ Petersen graphs, labelled in a natural way with triples from a 23-set; if we call a symbol τ incident with a block (or a Petersen graph) if τ is not in the block (or not in the label of the Petersen graph) then we find a geometry with Buekenhout diagram

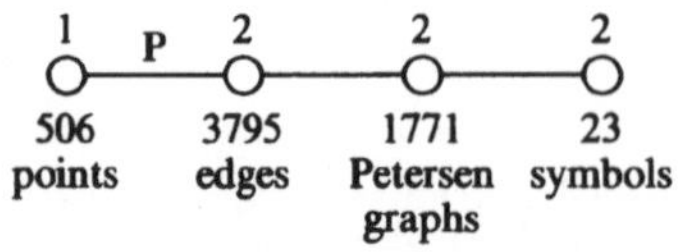

(that is, the 'ghost node' we had before has been filled in). We can recognize the symbols locally, but a collection of 231 Petersen graphs that agrees in each point with a symbol necessarily is a symbol, and hence Aut Δ permutes the set of symbols. If an automorphism of Δ fixes all symbols then it also fixes all points and hence this automorphism is the identity. Thus Aut $\Delta \cong M_{23}$.
Uniqueness has been shown in BROUWER & LAMBECK [133]. □

Remark. There are feasible intersection arrays, viz. $\{15,14,12,6,1,1;1,1,3,12,14,15\}$ and $\{15,14,12,8,1,1;1,1,1,12,14,15\}$, that would belong to antipodal 3- and 9-covers of Δ; it is unknown whether such graphs exist.

C. The doubly truncated Witt graph associated to M_{22}

Fix two symbols $\sigma,\sigma' \in \Sigma$, and let $\mathfrak{B}''$ be the collection of 330 blocks missing both σ and σ'. Then $(\Sigma \setminus \{\sigma,\sigma'\},\mathfrak{B}'')$, is a 3-(22,8,12) design such that any two distinct blocks meet in 0, 2 or 4 symbols, and it is the unique such design (TONCHEV [754]).

11.4.3. Theorem. *In the large Witt graph* Γ *described above,* $\mathfrak{B}''$ *induces a graph* E *with* 330 *vertices that is distance-transitive with intersection array* $\{7,6,4,4;1,1,1,6\}$, *and conversely, any distance-regular graph with this intersection array is isomorphic to* E. *We have* Aut E $\cong M_{22}.2$, *and* M_{22} *already acts distance-transitively. The point stabilizers are* $2^3{:}L_3(2)\times 2$ *and* $2^3{:}L_3(2)$, *respectively. Distance distribution diagram:*

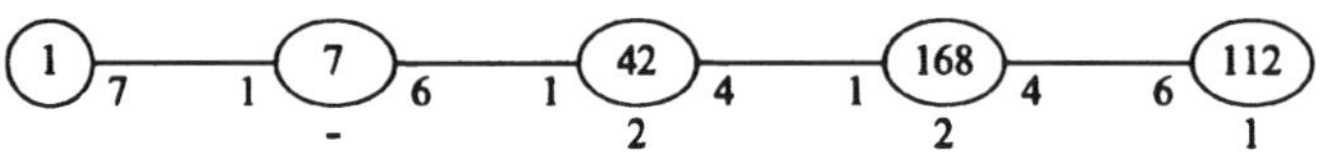

$v = 330.$

E *is not Q-polynomial. (Cf. Table 6.8, (D10).)*

Proof. We have the following correspondence between block intersections and graph distances in E.

$d(B,C)$	0	1	2	3	4
$\lvert B \cap C\rvert$	8	0	4	2	4.

For $|B \cap C| = 4$ we have $d(B,C) = 2$ or 4 depending on whether or not there is an octad containing $(B \cap C) \cup \{\sigma,\sigma'\}$.
Each quad in Γ loses two ovoids; if these coincide then we have a Petersen graph in E, otherwise we have $3K_2$. Thus, any two vertices at distance 2 in E determine a unique Petersen graph and we have $\lambda = 0$, $\mu = 1$, $a_2 = 2$. In the case that a quad Q is reduced to $3K_2$, the vertices from different components of the $3K_2$ have distance (at least) 4 in E (since Q is geodetically closed and the projection of an edge in $\Gamma_1(Q)$ is an edge in Q), and assuming the description of the relation $d = 4$, it is clear that $a_4 = 1$. Finally, if $|B \cap C| = 2$ and C has more than one neighbour in $E_2(B)$, then C together with two such neighbours (say D,D') determines a Petersen graph P contained in a unique quad Q. Since P does not contain quadrangles we have $d(B,Q) = 2$, and the quads $Q(B,D)$ and $Q(B,D')$ correspond to disjoint lines on the

local $PG(3,2)$ at B. But $Q(B,D)$ and $Q(B,D')$ must have survived as Petersen graphs in E, that is, they correspond to lines in the plane determined by the symbol σ' in the local $PG(3,2)$ at B, contradiction. Now since there are 168 edges between $E_2(B)$ and $E_3(B)$ we must have $c_3 = 1$ and the description of the distance relations is as claimed above. Finally, $a_3 = 2$ since it is at least 2 because of Petersen graphs meeting both $E_1(B)$ and $E_3(B)$, but it cannot be more because of the number of edges between $E_3(B)$ and $E_4(B)$.) This shows that E is distance-regular. Uniqueness is shown by first finding the 231 Petersen graphs and identifying the structure of the collection of Petersen graphs as the 3-class association scheme of M_{22} acting on $\binom{22}{2}$ (see BROUWER [127]), and from the uniqueness proof it follows that $\mathrm{Aut}\,E = \mathrm{Aut}\,S(3,6,22) = \mathrm{Aut}\,M_{22} = M_{22}.2$. Concerning distance-transitivity: it is clear from the description that M_{22} acts transitively, and looking at the permutation character we see that the action is rank 5. □

Remarks. (i) This graph was described in BIGGS & GARDINER [81] and in BIGGS [74]. The suborbit sizes (1+7+42+168+112) were given incorrectly in FISCHER & MCKAY [287] but correctly by IVANOV, KLIN & FARADJEV [435]. The Buekenhout geometry we find here is a residue of the one we found in the previous section:

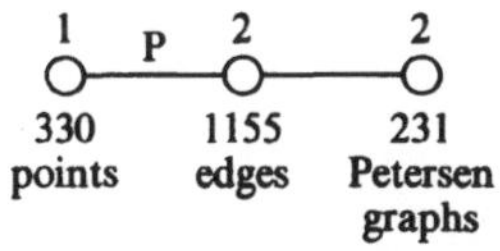

(ii) E is a commuting involutions graph in the wide sense of §13.2. More precisely, the vertex set of E can be identified with the involutions in $\mathrm{Aut}\,M_{22} \setminus M_{22}$ of ATLAS [216] type 2B. Two involutions x, y in E are adjacent if and only if their product xy lies in $O_2(C(x))$. We have:

order xy	2	2	4	3
$d_\Gamma(x,y)$	1	2	3	4.

(iii) If we try to continue the sequence Γ, Δ, E and look at the graph induced on the 210 blocks missing three fixed symbols we find the disjoint union of 21 Petersen graphs.
If we fix $\sigma, \sigma' \in \Sigma$ and let $\mathcal{B}'''$ be the set of 77 blocks that contain both σ and σ', and join blocks when they meet only in $\{\sigma,\sigma'\}$, we obtain the unique (cf. BROUWER [124]) strongly regular graph with parameters $(v,k,\lambda,\mu) = (77,16,0,4)$. Its automorphism group is $M_{22}.2$ acting distance-transitively with point stabilizer $L(3,4).2$. It is the second subconstituent of the Higman-Sims graph, cf. §13.2.

D. The Ivanov-Ivanov-Faradjev graph

The parameters of distance-regular antipodal 2- and 3-covers (of diameter 8) of the graph E constructed above pass all usual feasibility criteria. IVANOV, IVANOV & FARADJEV [282] constructed an antipodal 3-cover, and BROUWER [127] showed that there is no antipodal 2-cover, and that the antipodal 3-cover is unique. Thus there exists a unique distance-regular graph $\tilde{E}$ with intersection array $\{7,6,4,4,4,1,1,1;1,1,1,2,4,4,6,7\}$ on 990 vertices (cf. Table 6.10, (A19).) It is distance-transitive with automorphism group $3{\cdot}M_{22}.2$. Distance distribution diagram:

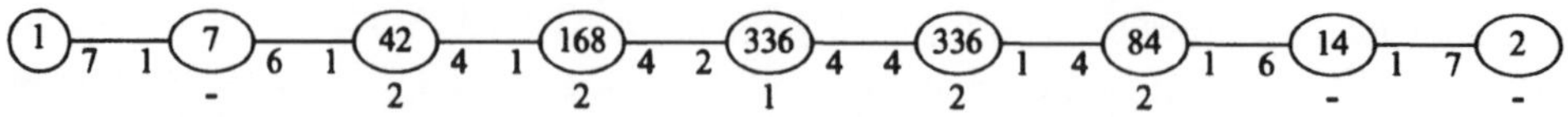

This graph is an example of a distance-regular graph whose sequence $(a_i)_i$ is non-unimodal.

E. Higman's symmetric design

Fix two symbols $\sigma, \tau \in \Sigma$ and let Z be the (bipartite) graph that has as vertices the blocks in $\mathcal{B}$ that contain precisely one of the symbols σ, τ, and where two blocks are adjacent whenever they are disjoint. Z is not distance-regular; its distance distribution diagram is:

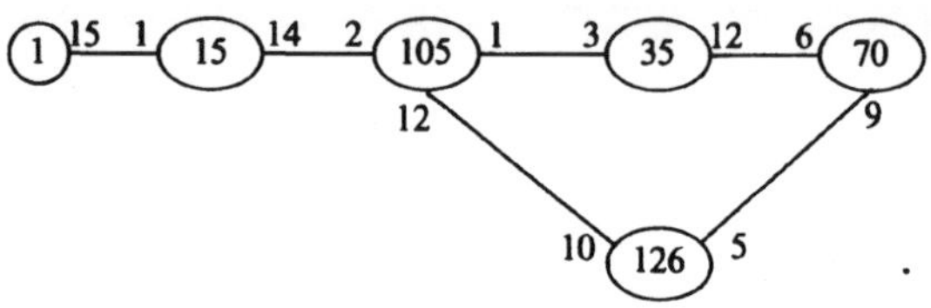

(It was used in §11.3D that this graph is connected.) Next define Z' to be the graph on the same vertex set, where now two blocks are adjacent whenever they meet in 0 or 4 symbols in $\Sigma \setminus \{\sigma,\tau\}$. Z' has distance distribution diagram

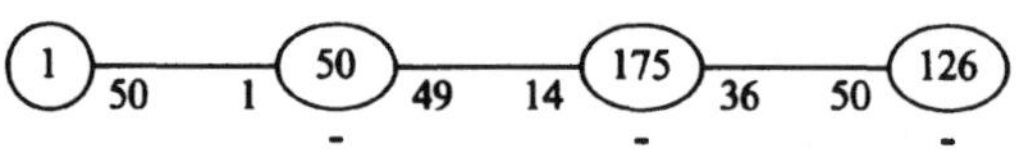

$v = 352$

and is distance-transitive. We have $\mathrm{Aut}\, Z' \cong \mathrm{Aut}\, HS \cong HS.2$. Z' is the incidence graph of G. Higman's symmetric design (a square 2-(176,50,14) design), cf. HIGMAN [388] and SMITH [687].

The halved graph of Z is strongly regular with parameters $(v,k,\lambda,\mu) = (176,105,68,54)$; its complement is strongly regular with parameters $(v,k,\lambda,\mu) = (176,70,18,34)$. These graphs are in the switching class of (complementary) regular two-graphs. Thus, we find Taylor graphs with intersection arrays $\{175,72,1;1,72,175\}$ and $\{175,102,1;1,102,175\}$ on 352 vertices. The latter graph, with distance distribution diagram

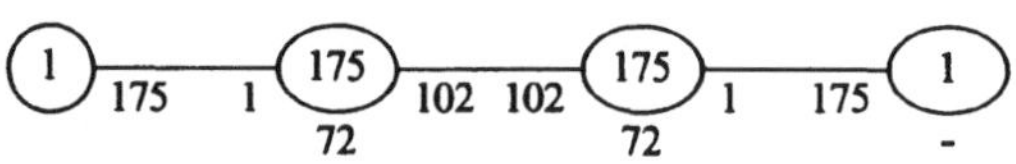

$v = 352,$

is locally the distance-2 graph of the line graph of the Hoffman-Singleton graph (cf. §13.1).

An involution in M_{24} fixing 8 points of Σ and interchanging σ and τ yields a polarity of Higman's 2-design with all points absolute. Thus, this design has a symmetric incidence matrix N with all-one diagonal, and $N - I$ is the adjacency matrix of a strongly regular graph with parameters $(v,k,\lambda,\mu) = (176,49,12,14)$ admitting Sym(8) as a group of automorphisms; cf. BROUWER [119].

F. The Leonard graph - $M_{12}.2$ over $PGL(2,11)$

11.4.4. Theorem. *There is a unique distance-regular graph* Γ *with* 288 *vertices and intersection array* $\{12,11,10,7;1,2,5,12\}$. Γ *is bipartite, and its automorphism group* $M_{12}.2$ *has two orbits on the point set of* Γ *(being transitive on each bipartite half).* Γ *has distance distribution diagram*

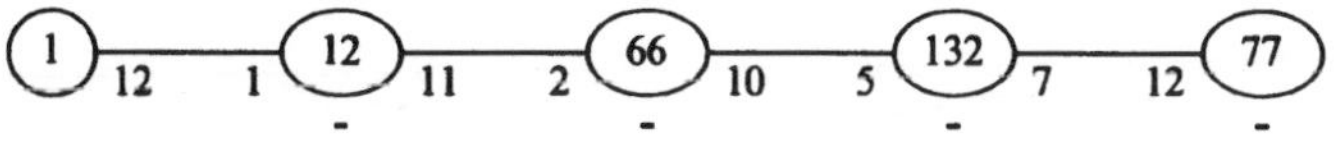

$v = 288$

and is listed under Table 6.9, (B6). □

Γ may be constructed as follows: Look at the Steiner system $S(5,8,24)$ and let D_1 and D_2 be two complementary dodecads, so that there are 132 blocks with 6 points in D_1 and 2 points in D_2 (say of type 6+2), 132 blocks of type 2+6 and 495 blocks of type 4+4. Let Δ be the graph whose vertices are the ordered pairs $(d_1,d_2) \in D_1 \times D_2$; call two such pairs (d_1,d_2), (e_1,e_2) nonadjacent whenever either $d_1 = e_1$ or $d_2 = e_2$ or there is a block B in the Steiner system with $B \cap D_1 = \{d_1,e_1\}$ and $B \cap D_2 \supset \{d_2,e_2\}$. Then Δ is a strongly regular graph with parameters $(v,k,\lambda,\mu) = (144,66,30,30)$. This graph has 144 12-cliques, and the design formed by the vertices and 12-cliques of Δ is a partial 2-geometry; its incidence graph Γ is distance-regular with parameters as given above.
The two components of Γ_2 are nonisomorphic; they are defined by the two distinct (up to conjugacy) subgroups isomorphic to $PGL(2,11)$ of $M_{12}.2$. The subgroups in one conjugacy class meet M_{12} in a maximal subgroup $PSL(2,11)$ - this yields a rank 4 presentation on 144 points; the subgroups in the other conjugacy class meet M_{12} in a subgroup $PSL(2,11)$ contained in M_{11} - this yields the imprimitive presentation Δ described above.
Γ (viewed as a pair of dual semi-biplanes) was first constructed by LEONARD [484]. Uniqueness was proved by BROUWER [121] (and most of the statements above can be found there), and also, under certain strong transitivity assumptions on $\mathrm{Aut}(\Gamma)$, by LEEMANS [482]. BUEKENHOUT [145] discusses the present situation in §10, describing two semibiplanes associated with $PSL(2,11)$ in M_{12}. For a discussion of the subgroup structure of M_{12} and $M_{12}.2$, see CONWAY [212], and the ATLAS [216].

The Hadamard association scheme

A connection with Hadamard matrices was exhibited by IVANOV & CHUVAEVA [433]. They showed that one may define a 4-class association scheme on the set X of the $16m^2$ positions of a Hadamard matrix $H = (h_{ij})$ of order $4m$ with the relations R_0: identity, R_1: in the same row, R_2: in the same column, R_3, R_4: the positions (i,j) and (k,l) where $i \neq k$ and $j \neq l$ are in relation R_3 when $\epsilon := h_{ij}h_{il}h_{kj}h_{kl} = 1$, and in relation R_4 when $\epsilon = -1$. (The valencies are $k_0 = 1$, $k_1 = k_2 = 4m-1$, $k_3 = 8m^2-6m+1$, $k_4 = 8m^2-2m$; all p^i_{jk} are given in [433].) Merging classes we may obtain strongly regular graphs with parameters $v = 16m^2$, $k = 8m^2-2m$, $\lambda = \mu = 4m^2-2m$, namely (X,R_4) and $(X,R_1 \cup R_3)$ and $(X,R_2 \cup R_3)$.
In the case of the (unique) Hadamard matrix of order 12, the graph (X,R_4) turns out to be the graph Δ constructed above.

G. Antipodal 2-covers of the Gewirtz graph

The Gewirtz graph is the unique strongly regular graph with parameters $(v,k,\lambda,\mu) = (56,10,0,2)$ (cf. GEWIRTZ [324,323]). In terms of the Steiner system $S(3,6,22)$ one can give the following description: Vertices are the $77-21 = 56$ blocks not containing a fixed symbol; two blocks are adjacent whenever they are disjoint.

11.4.5. Proposition. *The only distance-regular antipodal cover of the Gewirtz graph is its bipartite double.*

Proof. Apart from the bipartite double (Table 6.9, (B9)), the parameters allow two antipodal covers, with intersection arrays $\{10,9,1,1;1,1,9,10\}$ and $\{10,9,4,2,1;1,2,4,9,10\}$. The former is a 2-cover without 4-cycles, but one easily checks that the Gewirtz graph Γ has no such 2-covers - equivalently, one may verify that the edges of Γ cannot be labelled with 0 and 1 in such a way that each 4-cycle contains an odd number of edges labelled 1.
Let us now consider the other intersection array. If $\gamma, \delta \in \Gamma$ with $d(\gamma,\delta) = 2$, then $\{\epsilon \mid d(\gamma,\epsilon)+d(\delta,\epsilon) = 3\}$ induces a subgraph isomorphic to $2C_8$. But if Δ is a graph with intersection array $\{10,9,4,2,1;1,2,4,9,10\}$, and $\gamma, \delta \in \Delta$ with $d(\gamma,\delta) = 2$, then $\{\epsilon \mid d(\gamma,\epsilon)+d(\delta,\epsilon) = 3\}$ induces a subgraph isomorphic to C_8; it follows that for fixed γ and for $\delta \in \Delta_2(\gamma)$ we know the partition of the 8 neighbours of $\phi(\delta)$ in $\Gamma_2(\phi(\gamma))$ into two 4-sets $\phi(\Delta(\delta) \cap \Delta_2(\gamma))$ and $\phi(\Delta(\delta) \cap \Delta_3(\gamma))$, where $\phi:\Delta \rightarrow \Gamma$ is the projection.
A geometric description showing this partition is the following: Let B, C be two disjoint blocks of $S(3,6,22)$. The graph $\Gamma_2(\gamma)$ (for some $\gamma \in \Gamma$) is isomorphic to the graph that has as vertices the 45 blocks meeting both B and C, where disjoint blocks are adjacent. But if we call the pairs in B 'Points', and the pairs in C 'Lines', then these 45 blocks define 45 incidences between Points and Lines, and one easily checks that this defines the generalized quadrangle $GQ(2,2)$. Thus, our graph has as vertices the flags of $GQ(2,2)$, where $(p,L) \sim (q,M)$ whenever $p \neq q$, $L \neq M$ and either $p \in M$ or $q \in L$. (In the representation as 45 blocks we can also recognize the 10 vertices of $\Gamma(\gamma)$: these are the 10 symbols not in $B \cup C$, and adjacency is containment.) The partition of the 8 neighbours of (p,L) into 4+4 is the partition into 4 flags (q,M) with $p \in M$ and 4 flags (q,M) with $q \in L$.
Now one easily verifies that $\Gamma_2(\gamma)$ does not have a partial graph of valency 4 such that we have one of the two distinguished sets of 4 neighbours at each vertex. Consequently, for $\delta \in \Delta$, the subgraph $\Delta_2(\delta)$ does not exist. Thus, the Gewirtz graph has no non-bipartite antipodal covers. □

Remark. A similar question is whether one can direct the edges in Γ in such a way that if γ, $\delta \in \Gamma$, $d(\gamma,\delta) = 2$, then among the two paths of length two from γ to δ precisely one path has two edges directed in the same direction (yielding a directed path from γ to δ or from δ to γ). This would provide a new elliptic semiplane $GD[11,1,2;112]$, cf. §1.10 and BAKER [21]; however, it is impossible to direct the edges of Γ in such a way.

H. The regular two-graph on 276 vertices and the McLaughlin graph

Let $(X,\mathfrak{B})$ be the Steiner system $S(4,7,23)$, so that $|X| = 23$ and $|\mathfrak{B}| = 253$. Construct a graph Γ with vertex set $Y = X \cup \mathfrak{B}$ by letting a point $x \in X$ be adjacent to the blocks $B \in \mathfrak{B}$ that do not contain it, and letting two blocks B, B' be adjacent when they meet in a single point.
The two-graph $(Y,\mathfrak{D})$ that contains Γ in its switching class (cf. §1.5), is regular, and

has full automorphism group Co_3, acting 2-transitively on Y with point stabilizer $McL.2$. GOETHALS & SEIDEL [329] show that it is the unique regular two-graph on 276 vertices (up to complementation).
The Taylor graph Δ corresponding to $(Y, \mathcal{D})$ (cf. Theorem 1.5.3) is distance-transitive with intersection array $\{275,162,1;1,162,275\}$ and distance distribution diagram

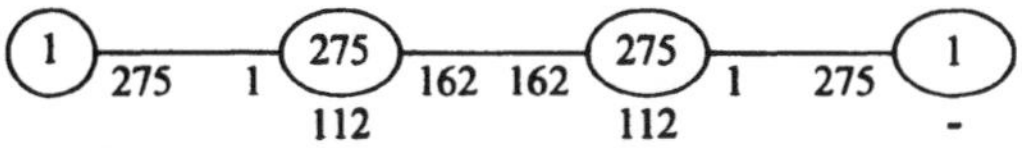

$v = 552.$

$\mathrm{Aut}\,\Delta \cong Co_3 \times 2$. Corresponding to the complementary two-graph $(Y, \overline{\mathcal{D}})$ we find a distance-transitive graph with (the same automorphism group and) intersection array $\{275,112,1;1,112,275\}$.
The derived designs $(Y \setminus \{\infty\}, \mathcal{D}_\infty)$ (for $\infty \in Y$), that is, the local graphs Δ_δ (for $\delta \in \Delta$), are strongly regular graphs with parameters $(v,k,\lambda,\mu) = (275,112,30,56)$. But, up to isomorphism, there is a unique such graph (indeed, adding an isolated point we find a graph in the switching class of the unique regular two-graph on 276 points); it was first constructed in MCLAUGHLIN [531], and is known as the *McLaughlin graph*. Its automorphism group is $McL.2$, with point stabilizer $U_4(3).2$. The graph is locally $GQ(3,9)$. The subconstituents $\Lambda(\lambda)$ and $\Lambda_2(\lambda)$ of the McLaughlin graph Λ are again strongly regular, with parameters $(v,k,\lambda,\mu) = (112,30,2,10)$ and $(162,105,72,60)$. Both are uniquely determined by their parameters (CAMERON, GOETHALS & SEIDEL [179]). See also BROUWER & VAN LINT [134].

11.5 The van Lint-Schrijver partial geometry

Let C be the code $\mathbf{F}_3\mathbf{1}$ of dimension one in $\mathbf{F}_3^6$. The coset graph $\Gamma(C)$ is tripartite, a tripartition being given by $\{A_i \mid i \in \mathbf{F}_3\}$, where $A_i = \{x + C \mid (x,\mathbf{1}) = i\}$. Taking any two of the parts, e.g. A_{-1} and A_1, we find that $\Gamma(C)$ induces on their union the incidence graph Δ of a self-dual partial geometry with parameters $pg(6,6,2)$. Δ is a regular thin near octagon with intersection array $\{6,5,5,4;1,1,2,6\}$ and distance distribution diagram

$v = 162.$

Δ is distance-transitive; its full group of automorphisms is $3^4.(\mathrm{Sym}(6)\times 2)$. See VAN LINT & SCHRIJVER [503], CAMERON & VAN LINT [182] and SMITS & VAN VROONHOVEN [689].

Chapter 12
Graphs Related to Classical Geometries

Graphs on isotropic points of a classical geometry (i.e., a geometry related to a semilinear or quadratic form) have been studied explicitly in Chapter 9 and implicitly in the context of parabolic representations of groups of Lie type. The nonisotropic points usually fall into a few orbits of the isometry group. The permutation rank of these orbits depends on the cardinality of the underlying field. We show that only in a few cases the related graphs are distance-regular. In the last three sections we construct several infinite families of antipodal covers of complete graphs (starting from affine instead of projective points), and an infinite family of partial geometries yielding bipartite distance-regular graphs of diameter 4 (starting from complete arcs in a projective plane).

Warning: in this chapter $x^{\perp}$ denotes the set of projective points with inner product zero with x (with respect to some form), and not the set consisting of x and its neighbours in the graph under consideration.

12.1 The even orthogonal case; the Doro graph

Let q be a power of 2 and $n \geqslant 3$. Equip the vector space $V = \mathbf{F}_q^n$ with a nondegenerate quadratic form Q. Let B be the bilinear form associated to Q defined by $B(u,v) = Q(u+v) - Q(u) - Q(v)$. The form Q defines a nondegenerate quadric (also denoted by Q) in the projective space $PG(n-1,q)$ of linear subspaces of V. When n is odd, there is a (unique) projective point $N = \langle u_0 \rangle$ such that $u_0 \in V^{\perp}$, i.e., $B(u_0,\cdot)$ is identically zero. Consider the set X of all projective points not on Q and (for n odd) distinct from the *nucleus* N, i.e.,

$$X = \{\langle u \rangle \,|\, Q(u) \neq 0 \text{ and } B(u,v) \neq 0 \text{ for some } v \in V\}.$$

Define four relations R_i ($0 \leqslant i \leqslant 3$) on X:

$$\begin{aligned} R_0 &= \{(x,x) \,|\, x \in X\}, \text{ the identity relation;} \\ R_1 &= \{(x,y) \,|\, x+y \text{ is a hyperbolic line (secant) }\}; \\ R_2 &= \{(x,y) \,|\, x+y \text{ is an elliptic line (exterior line, non-secant)}\}, \\ R_3 &= \{(x,y) \,|\, x+y \text{ is a tangent }\}. \end{aligned}$$

12.1.1. Theorem. $(X, \{R_0,R_1,R_2,R_3\})$ *is an association scheme with* 3 *classes.*

Proof (sketch). First let $n = 3$ so that Q is an oval in a projective plane of even order. (The computations below are valid for any oval, not just for conics.) One easily checks that $(X, \{R_0,R_1,R_2,R_3\})$ is an association scheme (with $v = q^2-1$) by explicitly computing the parameters:

$$(p^i_{0j})_{ij} = \begin{bmatrix} 1\,0\,0\,0 \\ 0\,1\,0\,0 \\ 0\,0\,1\,0 \\ 0\,0\,0\,1 \end{bmatrix}, \quad (p^i_{1j})_{ij} = \begin{bmatrix} 0 & \frac{1}{2}q(q-2) & 0 & 0 \\ 1 & \frac{1}{4}(q-2)^2 & \frac{1}{4}q(q-2) & \frac{1}{2}(q-4) \\ 0 & \frac{1}{4}(q-2)^2 & \frac{1}{4}q(q-2) & \frac{1}{2}(q-2) \\ 0 & \frac{1}{4}q(q-4) & \frac{1}{4}q^2 & 0 \end{bmatrix},$$

$$(p^i_{2j})_{ij} = \begin{bmatrix} 0 & 0 & \frac{1}{2}q^2 & 0 \\ 0 & \frac{1}{4}q(q-2) & \frac{1}{4}q^2 & \frac{1}{2}q \\ 1 & \frac{1}{4}q(q-2) & \frac{1}{4}q^2 & \frac{1}{2}q-1 \\ 0 & \frac{1}{4}q^2 & \frac{1}{4}q^2 & 0 \end{bmatrix}, \quad (p^i_{3j})_{ij} = \begin{bmatrix} 0 & 0 & 0 & q-2 \\ 0 & \frac{1}{2}(q-4) & \frac{1}{2}q & 0 \\ 0 & \frac{1}{2}(q-2) & \frac{1}{2}q-1 & 0 \\ 1 & 0 & 0 & q-3 \end{bmatrix}.$$

By way of example, let us compute p^3_{11}. If $x, y \in X$ with $l = x+y$ a tangent then each of x, y is on $\frac{1}{2}q$ secants, and a point z of intersection of a secant on x and a secant on y contributes to $p^3_{11}(x,y)$ unless $z \in Q$. Consequently $p^3_{11} = \frac{1}{2}q \cdot \frac{1}{2}q - q = \frac{1}{4}q(q-4)$.

In the general case (i.e., $n \geqslant 3$) one finds $p^k_{ij}(x,y)$ by summing the contributions of all planes Π on the line $x+y$ (being careful to distinguish the various kinds of tangent planes and the planes in which x or y is the nucleus). □

Remark. Note that in general the group $P\Omega^{\pm}(n,q)$ does not act as a rank 4 group, so that the proof has to be a geometric one. Not surprisingly, the conclusion of the theorem also holds in case Q is an oval (not necessarily a conic) in $PG(2,q)$, or an ovoid (not necessarily an elliptic quadric) in $PG(3,q)$.
The scheme is imprimitive for $n = 3$: R_3 is an equivalence relation.

12.1.2. Proposition. *Let $q > 2$. Then (X,R_j) is a distance-regular graph if and only if $j = 1$ and either $(n,q) = (3,4)$ or $(n,q) = (4,4)$, Q an elliptic quadric.*
In the former case, (X,R_1) is the line graph of the Petersen graph, with $v = 15$ and intersection array $\{4,2,1;1,1,4\}$; it is distance-transitive with group Sym(5).
In the latter case (X,R_1) is a distance-transitive graph with $v = 68$, intersection array $\{12,10,3;1,3,8\}$ and automorphism group $P\Gamma L(2,16)$.

Proof. In view of the previous theorem, (X,R_1) will be distance-regular if and only if $p^3_{11} = 0$. If $n = 3$, then $p^3_{11} = \frac{1}{4}q(q-4) = 0$ for $q = 4$ and if $n = 4$ and Q has Witt index 1, then $p^3_{11} = \frac{1}{4}q(q-2)(q-4) = 0$ for $q = 4$. For $(n,q) = (3,4)$ all claims are immediate. For $(n,q) = (4,4)$, GORDON & LEVINGSTON [336] showed distance-transitivity for the group $P\Sigma L(2,16)$ in the latter case (and uniqueness of this graph as a distance-transitive graph with given intersection array). □

Remarks. (i) When $q = 2$, a line has only three points, so relation R_1 is empty, and (X,R_3) is strongly regular. (E.g., if $n = 4$ and Q is an elliptic quadric then (X,R_3) is the Petersen graph.)
(ii) The graph (X,R_1) with $(n,q) = (4,4)$ was first constructed by S. Doro, cf. BUEKENHOUT & ROWLINSON [149] and is known as the *Doro graph*. Another construction of the Doro graph is the following. Let Q be a nondegenerate quadratic form with Witt index 1 on $V = \mathbf{F}_4^4$. Let $X = \{x \in V \mid Q(x) = 1\}$. For $x, y \in X$, let

$x \sim y$ if and only if $x + y \in X$. This defines a graph Γ on 68 vertices, isomorphic to the Doro graph.
For each vertex x of Γ the graph $\Gamma_3(x)$ is isomorphic to the line graph of the Petersen graph since $\Gamma_3(x) = x^\perp \cap X$.

Both graphs in the above proposition are determined by their intersection array. This is trivial for the line graph of the Petersen graph; for the Doro graph this is shown by the following theorem.

12.1.3. Theorem. *There exists a unique distance-regular graph on* Γ *on* 68 *points with intersection array* $\{12,10,3; 1,3,8\}$ *and distance distribution diagram*

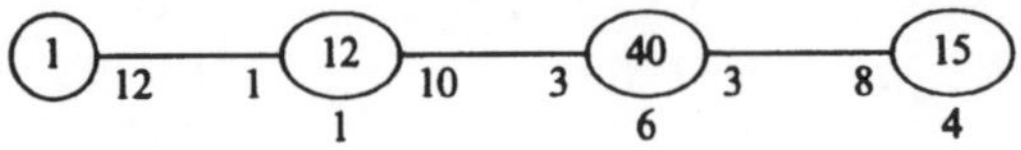

$v = 68.$

The graph is distance-transitive with $\mathrm{Aut}\,\Gamma \cong P\Gamma L(2,4^2) \cong P\Gamma O^-(4,4)$, *and point stabilizer* $(2\times \mathrm{Alt}(5)).2$. *(Cf. Table 6.8, (D6).)*

Proof of uniqueness. We proceed in 15 steps.

STEP 1. Γ *is the point graph of a partial linear space* $(X, \mathfrak{L}_H)$ *with lines of size 3.*
Indeed, $\lambda = 1$. We shall call the elements of $\mathfrak{L}_H$ *hyperbolic lines* (or just lines).

STEP 2. *For no* $x \in X$, $l \in \mathfrak{L}_H$ *do we have* $l \subset \Gamma_2(x)$.
Let $l = \{a_1,a_2,a_3\} \subset \Gamma_2(x)$. The three 3-sets $M(x,a_j) = \Gamma(x) \cap \Gamma(a_j)$ $(j = 1,2,3)$ are pairwise disjoint, and disjoint from the 3-set $\Gamma(x) \cap \Gamma_3(a_i)$ for any fixed i. Consequently $\Gamma(x) \cap \Gamma_3(a_i) = S$ is independent of i. If $s \in S$, and t is the third point on the line xs then t cannot be in $M(x,a_j)$ for some j (otherwise $d(s,a_j) < 3$) so $t \in S$. Consequently $|S|$ is even, a contradiction since $|S| = 3$.

STEP 3. *Let the type of* $l = \{a_1,a_2,a_3\}$ *with respect to* x *be the (unordered) vector* $(d(x,a_i))_i$. *There are* 6 *lines of type* (011), 60 *lines of type* (122), 60 *lines of type* (223) *and* 10 *lines of type* (333) *with respect to any point* x. *In particular* $\Gamma_3(x)$ *is a subspace of* $(X, \mathfrak{L}_H)$.

For $A \subset X$, let us write $\Gamma_i(A) = \bigcap_{a \in A} \Gamma_i(a)$.

STEP 4. *Let* $l \in \mathfrak{L}_H$. *Then* $\Gamma_3(l)$ *is a* Γ_2*-clique of size* 5.
Indeed, $|\Gamma_3(l)| = p^1_{33} = 5$. If x, $y \in \Gamma_3(l)$ and $d(x,y) = 3$ then $l \subset \Gamma_3(\{x,y\})$, but $|\Gamma_3(\{x,y\})| = p^3_{33} = 2$ so this is impossible. Suppose $x \sim y$. Then there is a line m on x, y, and since $\Gamma_3(l) = \bigcap_{z \in l} \Gamma_3(z)$, we have $m \subset \Gamma_3(l)$. But $|\Gamma(l)| = 30$, $|\Gamma_2(l)| = 30$ and $\Gamma(m) \subset \Gamma_2(l) \cup \Gamma_3(l)$ so that $|\Gamma(m) \cap \Gamma_2(l)| \geqslant 28$ (because there is at most one more line in $\Gamma_3(l)$). On the other hand, if $p \in m$ then $\Gamma_3(p)$ has 3 points on l, 6 points in $\Gamma(l)$ and 6 points in $\Gamma_2(l)$, and clearly $\Gamma(m) \cap \Gamma_3(p) = \varnothing$, so $|\Gamma(m) \cap \Gamma_2(l)| \leqslant 24$. Contradiction.

Let us call the Γ_2-cliques of the form $\Gamma_3(l)$ for $l \in \mathfrak{L}_H$, *elliptic lines* and let $\mathfrak{L}_E$ be the set of all elliptic lines.

Step 5. *$(X, \mathfrak{L}_E)$ is a partial linear space with point graph Γ_2.*
Since $p^2_{33} = 3$, two points on an elliptic line $\Gamma_3(l)$ determine l uniquely. Given a point x, there are 10 lines l in $\Gamma_3(x)$, and these determine 10 elliptic lines on x. Since $|\Gamma_2(x)| = 40$ any pair $\{x,y\}$ with $d(x,y) = 2$ is on an elliptic line.

Step 6. *Let $d(x,y) = 3$ and $\Gamma_3(\{x,y\}) = \{z,w\}$. Then $d(z,w) = 3$, so that $\{x,y,z,w\}$ is a Γ_3-clique.*
If $w \sim z$ then $\Gamma_3(\{x,y\})$ would contain the third point of the line wz impossible. If $d(w,z) = 2$ then $\{w,z\}$ is in an elliptic line $\Gamma_3(l)$ for some $l \in \mathfrak{L}_H$, and $x,y \in l$, impossible.

Let us call the Γ_3-cliques of size 4 *tangents* and let $\mathfrak{L}_T$ be the set of all tangents.

Step 7. *$(X, \mathfrak{L}_T)$ is a partial linear space with point graph Γ_3.*

Step 8. *$\Gamma_3(x)$ is isomorphic to the line graph of the Petersen graph. Distances measured in the subgraph $\Gamma_3(x)$ are the same as those measured in Γ. The set $x^\perp := \Gamma_3(x) \cup \{x\}$ is a subspace of $(X, \mathfrak{L})$, where $\mathfrak{L} = \mathfrak{L}_H \cup \mathfrak{L}_E \cup \mathfrak{L}_T$.*

It is clear that $\Gamma_3(x) \cup \{x\}$ is linearly closed for hyperbolic lines (by Step 3) and tangents (by Step 6). But if $d(y,z) = 2$ for $y,z \in \Gamma_3(x)$, then y,z determine the elliptic line $\Gamma_3(l)$ and $x \in \Gamma_3(y) \cap \Gamma_3(z) = l$ so that $\Gamma_3(l) \subset \Gamma_3(x)$. Now $\Gamma_3(x)$ is a linear space with 15 points, 6 5-lines and 15 3-lines which is necessarily the generalized quadrangle of order (2,2) together with its six ovoids. One parallel class of 3-lines comes from the tangents and the remaining 10 3-lines define a Petersen graph.

Step 9. *Let $x \sim y$. Then $\Gamma_3(x) \cap \Gamma_2(y)$ is isomorphic to $2C_5$, the disjoint union of two 5-circuits; Γ_3 induces on this 10-set a matching joining the two 5-circuits in such a way that $\Gamma_1 \cup \Gamma_3$ induces the Petersen graph. In particular, $\Gamma_3(x) \cap \Gamma_2(y)$ does not contain three points with pairwise distance 3 in Γ.*
We have $\Gamma_3(x) \cap \Gamma_2(y) = \Gamma_3(x) \setminus \Gamma_3(\{x,y\})$ but $\Gamma_3(\{x,y\})$ is an elliptic line. Now $\Gamma_1 \cup \Gamma_3$ induces on $\Gamma_3(x)$ the generalized quadrangle of order (2,2), and removing an ovoid leaves the Petersen graph.

Step 10. *Let $d(x,y) = 2$. Then $\Gamma_3(x) \cap \Gamma_3(y)$ is a hyperbolic line and $\Gamma_3(x) \cap \Gamma_1(y)$ is a Γ_2-clique of size 3, a projective line that is not totally isotropic in the projective space $PG(3,2)$ (with symplectic form) whose totally isotropic lines are the $(\Gamma_1 \cup \Gamma_3)$-cliques of size 3 in $\Gamma_3(x)$. Each such projective line (there are 20) occurs as $\Gamma_3(x) \cap \Gamma_1(y)$ for precisely two points $y \in \Gamma_2(x)$.*
The first statement is immediate from Step 5. Since $\Gamma_3(x)$ is linearly closed, $\Gamma_3(x) \cap \Gamma_1(y)$ is a Γ_2-clique (of size $p^2_{13} = 3$). Let $u,v \in \Gamma_3(x) \cap \Gamma_1(y)$. The third point w in $\Gamma_3(x) \cap \Gamma_1(y)$ is uniquely determined in $\Gamma_3(x)$ by the requirements that $d(u,w) = d(v,w) = 2$ and that $\Gamma_{\geq 2}(u) \cap \Gamma_{\geq 2}(v) \cap \Gamma_{\geq 2}(w) \cap \Gamma_3(x)$ contains a triangle (namely the hyperbolic line $\Gamma_3(x) \cap \Gamma_3(y)$). Each set $\Gamma_3(x) \cap \Gamma_1(y)$ occurs for at most two points y (since $\mu = 3$) but there are 20 such sets and 40 points $y \in \Gamma_2(x)$.

STEP 11. *There are* 136 *elliptic lines;* 10 *of type* (02222), 6 *of type* (33333), 60 *of type* (11223) *and* 60 *of type* (22223) *with respect to any point* x. *If* $d(u,v) = 3$, *then* u *is on* 4 *elliptic lines of type* (22223) *with respect to* v.
There are 10 elliptic lines on each point, so $10 \cdot 68 / 5 = 136$ in all. In $\Gamma_3(x)$ there are 2 elliptic lines on each point, so $2 \cdot 15 / 5 = 6$ in all. The remaining $136 - 6 - 10 = 120 = (10-2).15$ elliptic lines each meet $\Gamma_3(x)$ in a single point (by Step 8). Given $y \sim x$ there are 5 vertices $z \in \Gamma_3(x) \cap \Gamma_3(y)$ and for each z two lines $l \in \mathfrak{L}_H$, $z \in l \subset \Gamma_3(y)$; now $\Gamma_3(l)$ is an elliptic line on y of type (11223) with respect to x, as we can see in $\Gamma_3(z)$. This accounts for all elliptic lines meeting $\Gamma(x)$. Thus, the remaining 60 must be of type (22223). Finally, if $d(u,v) = 3$ then the number of elliptic lines of type (11223) on u is $\frac{1}{2} p^3_{12} = 4$. The remaining $10-2-4 = 4$ are of type (22223).

STEP 12. *Let* $l,m \in \mathfrak{L}_T \cup \mathfrak{L}_H$ *be two intersecting lines. Then there is a unique subspace (plane)* $p^\perp$ *of* $(X,\mathfrak{L})$ *containing* l *and* m. *The same holds when* $l \in \mathfrak{L}_H$ *and* $m \in \mathfrak{L}_E$.

STEP 13. *Let* $T \in \mathfrak{L}_T$ *and* $E \in \mathfrak{L}_E$ *be two lines with a common point* x, *and assume that* $E \not\subseteq \Gamma_3(y)$ *for* $y \in T$. *Then there is a unique subspace* π *of* $(X,\mathfrak{L})$ *with* $T \cup E \subset \pi$ *and* $|\pi| = 20$. π *is embeddable in a projective plane* $PG(2,4)$.
Note that $T \cup E$ is not contained in a plane $p^\perp$ (otherwise $p \in T$ and $E \subset \Gamma_3(p)$). Let $y \in T \setminus \{x\}$ and $z \in E \setminus \{x\}$. Then $d(y,z) = 2$. (For: if $d(y,z) = 3$ then $E \subset \Gamma_3(y)$ contrary to our assumption; if $d(y,z) = 1$ then the plane $p^\perp$ containing T and yz also contains E, contradiction.) Thus there is an elliptic line E_{yz} on y and z and clearly $E_{yz} \not\subseteq \Gamma_3(u)$ for $u \in T$. Varying $z \in E \setminus \{x\}$ we find four lines E_{yz}; let $\pi = T \cup \bigcup_z E_{yz}$, then $|\pi| = 20$. We have to show that π is linearly closed.
First of all, if $p \in T$ and $q \in \pi \setminus T$, then $d(p,q) = 2$. (Otherwise there would be a plane $a^\perp$ containing the lines T and pq, and hence also the line E_{yz} containing q, and also the line E, contradiction.)
Let us show that no two points in π are adjacent. If $u,v \in \pi$ and $u \sim v$ then $u,v \notin T$ so that $T \subset \Gamma_2(\{u,v\})$. Let w be the third point of line uv. Then w has distance 1 or 3 to each point of T (by Step 2) and distance 1 to at most one point of T, so that $|T \cap \Gamma_3(w)| \geqslant 3$. But this contradicts Step 6.
Next assume $u,v \in \pi$ and $d(u,v) = 3$. If $u,v \in T$ then the tangent uv is T and is contained in π. So let $u,v \notin T$ and let $S \in \mathfrak{L}_T$ be the tangent on u and v.
Let $S = \{u,v,w,t\}$ and assume $w \notin \pi$. The four lines E_{yz} are the only lines on y not in a plane $p^\perp$ together with T. Thus the lines yw and T are in plane $p^\perp$, where clearly $p \in T$. In the generalized quadrangle on $\Gamma_3(w)$ we have a line $\{u,v,t\}$ and $p \not\sim u,v$ so $p \sim t$. Since $p \sim t$ we must have $t \notin \pi$ and we find in the same way a point $q \in T$ with $t \in q^\perp$ and $q \sim w$. Now $\Gamma_3(w) \cap \Gamma_3(q)$ is an elliptic line, but it contains p,t and $p \sim t$ contradiction. This shows that π is closed for tangents.
Finally let $u,v \in \pi$ with $d(u,v) = 2$. Let F be the elliptic line on u,v. If $F \cap T = \varnothing$ then by Step 11 each point of T has distance 3 to at least one point of F. If two points $p,q \in T$ have distance 3 to the same point $w \in F$ then $w \in T$, impossible. So some point in T has distance 3 to a point in $F \cap \pi$, impossible.
So assume $F \cap T = \{u\}$. Let $w \in F \setminus \pi$. As before we find $p \in T$ such that $p^\perp$ contains the lines T and yw. But now $p^\perp \cap F \supset \{u,w\}$, so $F \subset p^\perp$, but $d(p,v) = 2$, contradiction. This shows, that π is linearly closed. Now π is a linear space on 20 points with lines of sizes 4,5 only, and in view of Step 11 and the fact that y is in one 4-line and four 5-lines one immediately sees that each point of π is in precisely one 4-line, so that π is a projective plane $PG(2,4)$ minus a point.

STEP 14. *It is possible to define a collection of planes* Π *such that the design* $(\mathfrak{L}_x,\Pi_x)$ *of lines and planes on* x *is a projective plane* $PG(2,4)$ *for each* $x \in X$. *Of these planes,* 68 *have* 16 *points and are of the form* $x^\perp$. *The remaining* 17 *planes have* 20 *points and contain elliptic lines and tangents only. Each plane is embeddable in* $PG(2,4)$. *Any two planes meet in a line.*
Given $x \in X$ there are 5 tangents, 6 hyperbolic lines and 10 elliptic lines on x. For each $z \in \{x\} \cup \Gamma_3(x)$ we find a plane $z^\perp$ on x, and two such planes meet in a line. Thus locally at x we find a partial projective plane with 21 points and 16 lines of size 5. Given two lines on x, they already determine a unique plane when at least one is hyperbolic or when both are tangents. Otherwise Step 13 provides us with the plane we need. If π is a 'tangent' plane then $z^\perp \cap \pi$ cannot be more than a line since $z^\perp$ is linearly closed, and 3 non-collinear points generate π. Thus $(\mathfrak{L}_x,\Pi_x)$ is a partial linear space with 21 points and 21 lines of size 5, i.e., $PG(2,4)$. Counting the total number of pairs of planes and the number of such pairs meeting in a line we find that any two planes meet in a line.

STEP 15. *It is possible to define incidences between the points of a set* Y *of size* 17 *(disjoint from* X*) and the lines in* $\mathfrak{L}$ *in such a way that each line in* $\mathfrak{L}$ *is incident with* 5 *points of* $X \cup Y$, *and each* $\pi \in \Pi$ *becomes a projective plane.*
For each 20-point plane π, let p_π be a new point (incident with π and the tangents in π), and let Y be the set of these p_π. Given a line l and a point p_π let p_π be incident to l if and only if either l is a tangent in π or l is a hyperbolic line such that there is a tangent t in π such that l and t are disjoint and coplanar.
Let us first show that if s,t are tangents in π, and l is a hyperbolic line disjoint and coplanar with t, then l is also disjoint and coplanar with s. Indeed, l cannot meet s since otherwise a plane containing l and t must contain $l \cap s$ and t and hence coincide with π, impossible. There are 16 lines l that do not meet π (all hyperbolic) and 16 elliptic lines in π, and if E is an elliptic line in π then $\Gamma_3(E) \cap \pi = \emptyset$. Thus for each line l with $l \cap \pi = \emptyset$ the set $E = \Gamma_3(l)$ is an elliptic line in π. Now let $p \in E \cap s$. Then $s, l \subset p^\perp$ so that s, l are coplanar.
Next we show that each line is incident with 5 points in $X \cup Y$. Indeed, this is trivial for $l \in \mathfrak{L}_E \cup \mathfrak{L}_T$, so let $l \in \mathfrak{L}_H$. If p_π is incident with l then $\Gamma_3(l) \subset \pi$, but the elliptic line is in precisely two 20-point planes π.
Finally we show that each plane $\pi \in \Pi$ is projective. Indeed this is trivial for the planes with 20 points in X, so let $\pi = z^\perp$. Now π has 16 points in X and contains 5 tangents (the lines on z) giving 5 points in Y. Each hyperbolic line in π is disjoint from (and coplanar with) two tangents so is incident with 2 of these points, so also in this case is π extended to a projective plane.

Now we have reconstructed $PG(3,4)$ and find that X is the complement of an ovoid Y. But any ovoid in $PG(3,4)$ is an elliptic quadric, by BARLOTTI [42], cf. HIRSCHFELD [393], Theorem 16.1.7, so this ends the proof of uniqueness. □

Remark. Γ_3 is a commuting involutions graph (cf. §13.2): for $\gamma \in \Gamma$, the pointwise stabilizer of $\{\gamma\} \cup \Gamma_3(\gamma)$ in $P\Gamma L(2,16)$ is an involution σ_γ, and we have $\gamma \sim \delta$ if and only if $\langle \sigma_\gamma, \sigma_\delta \rangle \cong \mathrm{Sym}(3)$, and $d(\gamma,\delta) = 3$ if and only if σ_γ and σ_δ commute.

12.2 The odd orthogonal case

Now let Q be a nondegenerate quadric in $PG(d,q)$ with q odd. Consider the set X of all points x with $Q(x)$ a nonzero square, with relations R_i as before. Let us investigate whether $(X,\{R_0,R_1,R_2,R_3\})$ is an association scheme.
For $d = 2$, we only find an association scheme when q is small, but in any case p_{ij}^k is always constant when $3 \in \{i,j,k\}$.

[Indeed, for $d = 2$ we find

$$v = \tfrac{1}{2}q(q+\epsilon),$$

$$k_0 = 1,\ k_1 = \tfrac{1}{4}(q-\epsilon)(q-3),\ k_2 = \tfrac{1}{4}(q-\epsilon)(q-1),\ k_3 = (\epsilon+1)(q-1),$$

where $\epsilon = -1$ if X is the set of interior points, and $\epsilon = 1$ if X is the set of exterior points of Q. In the former case R_3 is empty, and p_{ij}^k is undefined for $k = 3$ and zero for $3 \in \{i,j\}$. In the latter case we have

$$p_{13}^1 = q-5,\ p_{23}^1 = q-1,\ p_{33}^1 = 4,$$

$$p_{13}^2 = q-3,\ p_{23}^2 = q-3,\ p_{33}^2 = 4,$$

$$p_{11}^3 = \tfrac{1}{8}(q-3)(q-5),\ p_{12}^3 = \tfrac{1}{8}(q-1)(q-3),\ p_{22}^3 = \tfrac{1}{8}(q-1)(q-3),$$

$$p_{13}^3 = \tfrac{1}{2}(q-3),\ p_{23}^3 = \tfrac{1}{2}(q-1),\ p_{33}^3 = q-1.$$

Thus, in both cases we know the p_{ij}^k involving relation R_3, and we find an association scheme as soon as we find that p_{11}^1 and p_{11}^2 are constant. However, this is only the case when q is small.
For $q = 3$ relation R_1 is empty. Thus, we find on the interior points the complete graph $(X,R_2) \cong K_3$ and on the exterior points the graph $(X,R_2) \cong 3K_2$. Note that here $R_2 = \perp$, where $\perp$ is the relation of being orthogonal.
For $q = 5$ we have $R_1 = \perp$, which enables us to compute p_{11}^1 and p_{11}^2. Indeed, for general q we have for the exterior points $p_{\perp\perp}^1 = 1$, $p_{\perp\perp}^2 = 0$, and for the interior points $p_{\perp\perp}^1 = 0$, $p_{\perp\perp}^2 = 1$. Thus, we find for $q = 5$ on the interior points the Petersen graph (X,R_1) and on the exterior points the line graph of the Petersen graph (X,R_2).
The case $q = 7$ is considered in the next section.
Problem. For which q is p_{11}^1 constant? For which q is p_{11}^2 constant?]

For $d = 3$ however, we always find an association scheme.

12.2.1. Theorem. *Let X be the set of nonisotropic points x in $PG(3,q)$ for which $Q(x)$ is a square. Then $(X,\{R_0,R_1,R_2,R_3\})$ is an association scheme.*

Proof. Put $\epsilon = 1$ if Q is hyperbolic and $\epsilon = -1$ if Q is elliptic. We easily find

$$v = \tfrac{1}{2}q(q^2-\epsilon),$$

$$k_0 = 1,\ k_1 = \tfrac{1}{4}q(q+\epsilon)(q-3),\ k_2 = \tfrac{1}{4}q(q-1)(q-\epsilon),\ k_3 = q^2-1.$$

Using our results for $d = 2$, we find all $p_{ij}^k(x,y)$ with $3 \in \{i,j,k\}$ by summing the contributions of the planes on the line xy. For example, in the elliptic case

$$p_{11}^3 = \tfrac{1}{8}q(q-3)(q-5),\ p_{21}^3 = \tfrac{1}{8}q(q-1)(q-3).$$

Thus, we need only show that p_{11}^1 and p_{11}^2 are constant. This requires some computations using the values of the quadratic form Q. Thus, we have to work in the 4-dimensional vector space V over $F := \mathbb{F}_q$ rather than in the projective space PV. Define an inner product on V by $(u,v) := (Q(u+v)-Q(u)-Q(v))/2$, for $u, v \in V$.

Let '$a = \square$' denote that a is a nonzero square in F.
Lemma. *Let Q_1 be a nondegenerate quadratic form on F^2, and let $\eta = 1$ if it is hyperbolic, and $\eta = -1$ if it is elliptic. Then for $c \in F$ we have:*

$$\#\{(x,y) \in F^2 \mid Q_1(x,y) = c\} = \begin{cases} q+\eta(q-1) & \text{if } c = 0 \\ q-\eta & \text{if } c \neq 0. \end{cases} \qquad \square$$

Lemma. $\#\{x \in F \mid x^2 - 1 = \square\} = \frac{1}{2}(q-3)$. $\square$

Lemma. *Let u, $v \in V$. The subspace $\langle u,v\rangle$ of V is a secant (hyperbolic line) if and only if $(u,v)^2 - Q(u)Q(v) = \square$.* $\square$

Let u, $v \in V$ with $\langle u\rangle, \langle v\rangle \in X$, i.e., $Q(u) = \square$, $Q(v) = \square$, and suppose that $\langle u,v\rangle$ is nondegenerate. Count the number of vectors $w \in V$ such that both $\langle u,w\rangle$ and $\langle v,w\rangle$ are secants. To this end, decompose V as the orthogonal direct sum $V = V_1 \perp V_2$, where $V_1 = \langle u,v\rangle$. Then Q decomposes accordingly as $Q = Q_1 + Q_2$. Let η be the character of Q_2, i.e., $\eta = 1$ if V_2 is a hyperbolic line, and $\eta = -1$ if it is an elliptic line. Then the character of Q_1 is $\epsilon\eta$. Write $w = w_1 + w_2$ with $w_i \in V_i$, and put

$$a := Q_1(w_1) + Q_2(w_2), \quad b := (w_1,u)^2 - aQ_1(u), \quad c := (w_1,v)^2 - aQ_1(v).$$

Then the requirements on w are $a = \square$, $b = \square$, $c = \square$.
Given a, there are (according to the second lemma above) $\frac{1}{2}(q-3)$ choices for each of (w_1,u) and (w_1,v), so $\frac{1}{4}(q-3)^2$ choices for w_1.
If $a = Q_1(w_1)$, there are (according to the first lemma above) $q+\eta(q-1)$ choices for w_2, and otherwise $q-\eta$ choices. But the former case can occur only when V_1 is a hyperbolic line; indeed, if $a = Q_1(w_1)$, then both requirements $b = \square$ and $c = \square$ are equivalent to the requirement that $V_1 = \langle u,w_1\rangle = \langle v,w_1\rangle$ is a secant. If this is indeed the case, then the remaining requirement for w_1 is that $Q(w_1) = \square$, $\langle w_1\rangle \neq \langle u\rangle, \langle v\rangle$, and we find $\frac{1}{2}(q-1)-2$ choices for $\langle w_1\rangle$, and $\frac{1}{2}(q-1)(q-5)$ for w_1. Altogether, we find

$$\tfrac{1}{2}(q-1)\cdot\tfrac{1}{4}(q-3)^2\cdot(q-\eta) \;+\; \eta q\cdot\frac{1+\epsilon\eta}{2}\cdot\tfrac{1}{2}(q-1)(q-5)$$

choices for w. This number is $(q-1)p^1_{11}$ if $\epsilon\eta = 1$, and $(q-1)p^2_{11}$ if $\epsilon\eta = -1$. Thus we proved:

$$p^1_{11} = \tfrac{1}{8}(q-3)^2(q-\epsilon) + \tfrac{1}{2}\epsilon q(q-5), \quad p^2_{11} = \tfrac{1}{8}(q-3)^2(q+\epsilon).$$

This shows that we have an association scheme. $\square$

For $q = 3$, relation R_1 is empty and (X,R_2) is the (strongly regular) point graph of the generalized quadrangle $GQ(2,2)$. (This illustrates the sporadic isomorphisms $P\Omega^-(4,3) \cong PSL(2,9) \cong \mathrm{Alt}(6)$.)

For $q = 5$, $p^3_{11} = 0$, and we shall see below that (X,R_1) is distance-regular of diameter 3. Note that $(x,y) \in R_1$ if and only if $x \perp y$; it is this fact that makes the case $q = 5$ special.

For $q > 5$ none of the R_i defines a distance-regular graph.

12.2.2. Proposition (Hall [345]). *There is a unique distance-regular graph Γ on 65 vertices with intersection array $\{10,6,4; 1,2,5\}$ and distance distribution diagram*

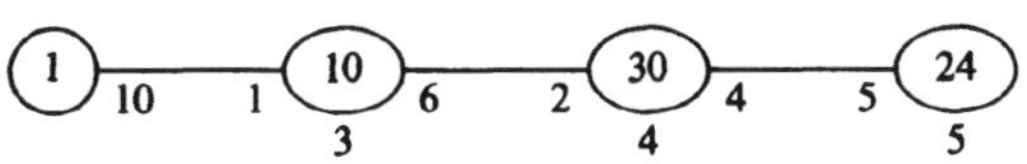

$v = 65.$

(Cf. Table 6.8, (D4).) We have $\mathrm{Aut}\,\Gamma \cong P\Sigma L(2,25)$, *and already* $PSL(2,25)$ *acts distance-transitively. The point stabilizers in these groups are* $\mathrm{Sym}(5)\times 2$ *and* $\mathrm{Sym}(5)$, *respectively. For each* $\gamma \in \Gamma$, *the graph induced on* $\Gamma(\gamma)$ *is isomorphic to the Petersen graph.* □

S. Doro (cf. Buekenhout & Rowlinson [149]) and, independently, Gordon & Levingston [336] and A.A. Ivanov [429] found this graph; Gordon & Levingston showed the uniqueness of this graph as a distance-transitive graph with given parameters, and determined its group of automorphisms; Hall [345] found this graph in his classification of locally Petersen graphs, and showed uniqueness as a distance-regular graph, cf. Theorem 1.16.5. Γ can be described as a commuting involutions graph (cf. §13.2): for $\gamma \in \Gamma$, the pointwise stabilizer of $\gamma^{\perp}$ in $P\Sigma L(2,25)$ is an involution, and two vertices are adjacent precisely when the corresponding involutions commute.

12.3 The Coxeter graph for $PSL(2,7)$

Let Γ be the graph on the antiflags of the Fano plane (P,L) (i.e., $\gamma \in \Gamma$ if and only if $\gamma = (p,l)$ for some line $l \in L$ and some point $p \in P$ not incident with l) in which $\gamma = (p,l)$ and $\delta = (q,m)$ are adjacent whenever $P = l \cup m \cup \{p,q\}$. The graph thus obtained is called the *Coxeter graph* (not to be confused with the Coxeter graphs of §10.2).

One may describe the relations for two vertices $\gamma = (p,l)$, $\delta = (q,m)$ in Γ as follows: we have $d(\gamma,\delta) = 2$ if and only if $p \in m$ and $q \in l$; $d(\gamma,\delta) = 3$ if and only if ($p \in m$ and $q \notin l$) or ($q \in l$ and $p \notin m$); $d(\gamma,\delta) = 4$ if and only if $\gamma \neq \delta$ and ($p = q$ or $l = m$).

This graph can be regarded as a graph on nonisotropic points; indeed, Γ is isomorphic to the graph on the exterior points for a conic in $PG(2,7)$, where two points are adjacent if and only if they are orthogonal; then two points have distance 3 if and only if they are on a common tangent. (It follows that each tangent yields a perfect 1-error-correcting code in this graph.)

See Biggs & Smith [84] and Coxeter [225] for other descriptions of this graph.

12.3.1. Theorem. *The Coxeter graph is the unique distance-regular graph with intersection array* $\{3,2,2,1;1,1,1,2\}$. *It has 28 vertices, is distance-transitive with automorphism group* $PGL(2,7)$ *and point stabilizer* D_{12}, *the dihedral group of order 12. It has distance distribution diagram*

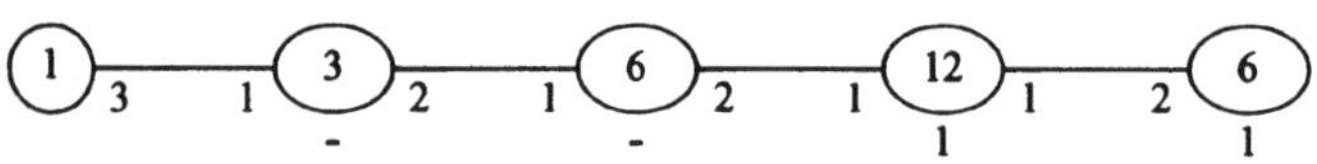

$v = 28$

(cf. Table 6.8, (D9)).

Proof. Γ is so small that showing uniqueness given the intersection array $\{3,2,2,1;1,1,1,2\}$ is relatively straightforward: fix a vertex ∞ and draw the tree T on 10 points induced by $\Gamma_{\leqslant 2}(\infty)$. Now continue drawing, labelling each vertex γ with the 10-tuple $(d(\gamma,\tau))_{\tau \in T}$. One may complete the drawing without loss of generality, without the need to backtrack or distinguish cases.

From the description given it is clear that the automorphism group contains

$PGL(3,2).2 \cong PGL(2,7)$ of order 336, but if one fixes two vertices at distance 3 then one fixes a heptagon, and also the 7 points at distance one to the heptagon and also all other points, so the group cannot have order larger than $28 \cdot 12 = 336$. Clearly the group $PGL(3,2).2$ acts distance-transitively. □

Remarks. (i) Equivalent to the first description given is the following: Γ is the flag graph of the unique biplane 2-(7,4,2), where two flags (p,B) and (q,C) are adjacent if and only if $B \cap C = \{p,q\}$. But (p,B) is uniquely determined by $B \setminus \{p\}$, and we find that Γ is the subgraph of the Odd graph O_4 obtained by deleting 7 triples forming a Fano plane (cf. JEURISSEN [444]).

(ii) BIGGS [70] shows that the Petersen graph and the Coxeter graph are the only two primitive distance-transitive trivalent graphs that are not Hamiltonian. (See also TUTTE [758].)

(iii) The graph Γ_3 is isomorphic to the triangular graph $T(8)$.

(iv) Note that $PSL(2,7)$ acts imprimitively: there are two systems of imprimitivity consisting of the seven sets $(p,*)$ and $(*,l)$.

12.4 The unitary case

For a square prime power $r = q^2$, equip $V = \mathbb{F}_r^3$ with (a nondegenerate Hermitean form, e.g.,) the standard unitary inner product

$$(x,y) = \sum_{i=1}^{3} x_i y_i^q \qquad (x,y \in V).$$

The *unitary nonisotropics graph* on V is the graph Γ whose vertex set is $\{\langle v\rangle \mid v \in V,\ (v,v) \neq 0\}$, the set of nonisotropic points of PV, the projective space associated with V; two vertices $\langle v\rangle$ and $\langle w\rangle$ are adjacent if and only if $(v,w) = 0$.

12.4.1. Theorem (GORDON & LEVINGSTON [336]). *Let Γ be the unitary nonisotropics graph on V. Then Γ has $v = q^2(q^2-q+1)$ vertices. If $q = 2$ then $\Gamma \cong 4K_3$, the disjoint union of four triangles. If $q > 2$ then Γ is connected and distance-regular with intersection array $\{q^2-q,\ q^2-q-2,\ q+1;\ 1,\ 1,\ q^2-2q\}$. (Cf. Table 6.8, (D1).) Γ has singular lines of size 3.*

Proof. Let X be the set of nonisotropic points and U the set of isotropic points in PV. For $p = \langle v\rangle \in PV$ write $p^\perp = \{\langle w\rangle \mid (v,w) = 0\}$. According to Lemmas 9.3.2 and 9.4.1 we have $|X \cup U| = q^4+q^2+1$ and $|U| = q^3+1$ so that $|X| = q^2(q^2-q+1)$.
Let x be a nonisotropic point. Then $x^\perp$ is nondegenerate, hence contains $q+1$ isotropic and q^2-q nonisotropic points. Thus Γ has valency $k = q^2-q$. The subgraph $\Gamma(x)$ is the disjoint union of $\frac{1}{2}q(q-1)$ edges, so $\lambda = 1$, $b_1 = k-2 = q^2-q-2$ and singular lines have size 3. Now let $x, y \in X$ be two distinct and nonadjacent vertices of Γ. Then the set of common neighbours of x and y is $x^\perp \cap y^\perp \cap X$, so that x and y have precisely one common neighbour whenever the point $x^\perp \cap y^\perp$ is nonisotropic (i.e., whenever the line $x+y$ is nondegenerate), and distance at least 3 otherwise. In the latter case

$$\Gamma(y) \cap \Gamma_2(x) = (X \cap y^\perp) \setminus \bigcup \{u^\perp \cap y^\perp \mid u \in x^\perp \cap U,\ u \neq x^\perp \cap y^\perp\},$$

so that $|\Gamma(y) \cap \Gamma_2(x)| = (q^2-q)-q = q^2-2q$ so that if $q > 2$, then Γ has diameter 3 and $c_3 = q^2-2q$. When $d(x,y) = 2$ and $q > 2$ then, since $\Gamma_3(x)$ is the set of nonisotropic points on the $q+1$ lines $x+u$ for $u \in x^\perp \cap U$, it follows that $\Gamma(y) \cap \Gamma_3(x)$ is the collection of intersection points of these lines with the line $y^\perp$ and it follows that $b_2 = q+1$. This proves that Γ has the stated parameters. □

Remarks. (i) The group $P\Gamma U(3,q^2)$ acts as an automorphism group of Γ. Already $PGU(3,q^2)$ is transitive on vertices, edges and pairs of vertices at distance 3, but it is distance-transitive only for $q = 3$. The group $P\Gamma U(3,q^2)$ is also distance-transitive for $q = 4$ (but not for larger q). BUEKENHOUT & ROWLINSON [149] prove that this graph (for $q = 4$) is the unique distance-transitive graph with intersection array $\{12,10,5;1,1,8\}$.
(ii) In the above theorem for $q > 2$, we have $k = b_2+c_3-1$ so that Γ_3 is strongly regular (by Proposition 4.2.17). In fact, it is the block graph of the unital (Steiner system $S(2,q+1,q^3+1)$) on the isotropic points. See also the remarks following 1.17.8.
(iii) If $q = 3$, then Γ is a generalized hexagon of order (2,2). It is known that there are precisely two isomorphism classes of generalized hexagons of order (2,2), dual to each other (see COHEN & TITS [207]). The two types can be distinguished by looking at $\Gamma_3(\gamma)$ for an arbitrary point γ. In systems of one kind $\Gamma_3(\gamma)$ is connected for all γ, in systems of the other kind $\Gamma_3(\gamma)$ is disconnected for all γ. Here we have a system of the latter kind: $\Gamma_3(\gamma)$ has two components of 16 points each, with distance distribution diagram

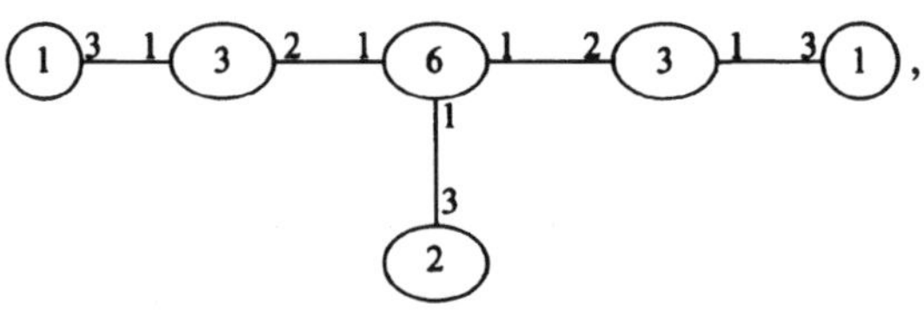

i.e., we have double covers of the cube 2^3 (cf. Remark 9.3.12 (ii)). [In fact there is a unique graph with the above distribution diagram, namely the unique double cover of 2^3 without 4-gons (cf. Theorem 9.2.10); it was first constructed by FOSTER [292].]
If $y \in \Gamma_3(x)$ and $u = (x+y) \cap U$ then the antipode $\bar{y}$ of y in $\Gamma_3(x)$ is determined by the requirement: $\{x,u,y,\bar{y}\}$ is a Baer subline. For $x = 001$ the two components satisfy $X^2Y^2 = \pm\alpha^2 Z^4 \neq 0$, where α is a primitive element of $\mathbb{F}_9$.
Γ is a commuting involutions graph, cf. §13.2.

12.4.2. Proposition. *The full automorphism group of* Γ *is* $P\Gamma U(3,q^2)$.

Proof (as communicated by H.A. Wilbrink). We can recognize the tangent lines in our graph: they correspond to sets of q^2 vertices, pairwise at distance 3; but if A is any set of q^2 vertices, pairwise on a tangent but not all on one tangent, then any tangent contains at most $q+1$ points of A, and we can find four points that determine six distinct tangents. But O'NAN [579], p. 507, shows that the unital does not contain configurations consisting of four blocks that pairwise intersect in six distinct points, and dualizing we see that no such sets A exist. This means that we can reconstruct the unital from the graph: points correspond to tangents, i.e., to q^2-cliques of Γ_3, and blocks correspond to nonisotropic points, i.e., to vertices of Γ. A point is on a block if and only if the corresponding q^2-clique of Γ_3 contains the corresponding vertex. Now O'NAN [579] proves that the unital has full automorphism group $P\Gamma U(3,q^2)$, so that Aut Γ is a subgroup of this group. By Remark (i), we are done. □

12.5 Antipodal covers of complete graphs

In this section we discuss a few 'geometric' constructions of antipodal covers of complete graphs.

First of all we give a 'symplectic cover' of the complete graph on q^{n+1} points if n is even.

12.5.1. Proposition. *Let V be a vector space over $\mathbb{F} = \mathbb{F}_q$ of even dimension n, and let B be a nondegenerate symplectic form on V. Then the graph Γ whose vertices are the pairs (α,x) with $\alpha \in \mathbb{F}$ and $x \in V$, and in which the vertices (α,x) and (β,y) are adjacent if and only if $B(x,y) = \alpha - \beta$ and $x \neq y$, is distance-regular with $v = q^{n+1}$ vertices and intersection array*

$$\{q^n - 1, q^n - q^{n-1}, 1; 1, q^{n-1}, q^n - 1\},$$

an antipodal q-cover of the complete graph on q^n vertices. See Table 6.10, (A1).

Proof. Straightforward. □

Remarks. (i) This was a 'vector space' description; an equivalent 'projective' description was given by Somma [693]: Let q be a prime power, and consider a symplectic polarity on $PG(n+1,q)$, where n is even. Let H be a hyperplane, and $N = H^{\perp}$. Construct a graph Γ by taking as point set $PG(n+1,q) \setminus H$, and letting $x \sim y$ when the line xy is totally isotropic. Then Γ is distance-regular, isomorphic to the graph Γ constructed in the proposition. (Note that $d(x,y) = 3$ precisely when the line xy passes through N.)

(ii) For q a power of two an equivalent construction was already given by Thas [731]: Let Q be a nonsingular quadric in $PG(n,q)$, and let N be the nucleus of Q (i.e., the point such that each line on it is a tangent to Q). Embed $PG(n,q)$ as a hyperplane H in $PG(n+1,q)$, and define a graph Γ with vertex set $PG(n+1,q) \setminus H$ by letting $x \sim y$ when the line xy meets Q. Then Γ is distance-regular, isomorphic to the graph Γ constructed in the proposition.

(iii) For $n = 2$, q even, the graphs found are the symmetric bilinear forms graphs on $V = \mathbb{F}_q^2$. In this case, the construction may be generalized a little: under (ii) we may take for Q an arbitrary oval instead of a conic. This is a particular case of Proposition 12.5.2 below.

(iv) Hensel [376] generalized this construction as follows. Let A be a subgroup of the additive group of $\mathbb{F}$. Then the graph Γ whose vertices are the pairs $(\alpha+A,x)$ with $\alpha \in \mathbb{F}$ and $x \in V$ and in which the vertices $(\alpha+A,x)$ and $(\beta+A,y)$ are adjacent if and only if $x \neq y$ and $B(x,y) \in \alpha-\beta+A$ is distance-regular with $v = rq^n$ vertices, where $|A| = q/r$, and intersection array

$$\{q^n - 1, (r-1)q^n/r, 1; 1, q^n/r, q^n - 1\}.$$

A *spread* in some point-line geometry is a collection of lines partitioning the point set.

12.5.2. Proposition (Brouwer [125]). *Let $(X,\mathfrak{L})$ be a generalized quadrangle of order (s,t) (where $t > 1$) with a spread $\mathfrak{S}$. Then the collinearity graph Γ of $(X,\mathfrak{L}\setminus\mathfrak{S})$ is distance-regular of diameter 3, with $v = (s+1)(st+1)$ vertices, and intersection array $\{st, s(t-1), 1; 1, t-1, st\}$, an antipodal $(s+1)$-cover of the complete graph K_{st+1}.*

More generally, given a strongly regular graph Δ *with parameters* $(v,k,\lambda,\mu) = ((s+1)(st+1), s(t+1), s-1, t+1)$ *such that there is a partition* $\mathcal{S}$ *of its point set into* $(s+1)$*-cliques, the partial graph* Γ *obtained by deleting the edges contained in the members of* $\mathcal{S}$ *is distance-regular of diameter 3 with intersection array as given above. Conversely, any graph* Γ *with these parameters arises in this way.* □

As an application one can show that there are precisely two nonisomorphic distance-regular graphs with intersection array $\{8,6,1;1,3,8\}$, since there is a unique strongly regular graph with parameters (27,10,1,5), namely the collinearity graph of the unique generalized quadrangle of order (2,4), and this generalized quadrangle has precisely two nonisomorphic spreads (BROUWER & WILBRINK [137]). Part of this proposition was rediscovered by FISHER [288].

Known generalized quadrangles with spreads have orders $(s,t) = (s,1)$, $(1,t)$, (q,q), (q,q^2), $(q-1,q+1)$ and $(r+1,r-1)$ ($t \geqslant 2$, q a prime power, $r = 2^i > 2$), cf. PAYNE & THAS [588], §3.4.

Another 'symplectic cover' is the following.

12.5.3. Proposition. *Let* $q = rm+1$ *be a prime power, where* $r > 1$ *and either* m *is even or* q *is a power of 2. Let* V *be a vector space of dimension 2 over* $\mathbb{F} = \mathbb{F}_q$ *provided with a nondegenerate symplectic form* B*. Let* K *be the subgroup of the multiplicative group* $\mathbb{F}^*$ *of index* r*, and let* $b \in \mathbb{F}^*$*. Then the graph* Γ *with vertex set* $\{Kv \mid v \in V \setminus \{0\}\}$ *where* $Ku \sim Kv$ *if and only if* $B(u,v) \in bK$ *is distance-regular of diameter 3 with* $r(q+1)$ *vertices and intersection array* $\{q,\ q-m-1,\ 1;\ 1,\ m,\ q\}$*, an antipodal* r*-cover of the complete graph* K_{q+1}*. (Cf. Table 6.10, (A3).)* □

Remarks. (i) In case we take a vector space V of even dimension more than 2, we find an antipodal r-cover of the strongly regular symplectic graph; this cover has diameter 3 but is not distance-regular.

(ii) Γ does not depend on the choice of b (up to isomorphism), and is isomorphic to the graph constructed in §12.7A; in fact, if we put $B((x_1,x_2),(y_1,y_2)) = x_1y_2 - x_2y_1$ then $(x,i) \mapsto (a^{-i}x, a^{-i}b)$, $(\infty,i) \mapsto (a^i,0)$ constitutes an isomorphism from that graph onto Γ. Letting b vary, we obtain the colouring of $K_{r(q+1)}$ considered in MATHON [521].

(iii) Aut Γ contains the subgroup of $GL(2,q)$ consisting of the matrices with determinant in K, modulo $\{kI \mid k \in K\}$, that is, $L_2(q).(K/K^2)$. [More precisely, if q is even, $L_2(q)$; if q and r are odd, $L_2(q).2$; if q is odd and r is even, $2{\times}L_2(q)$.] This group acts edge-transitively. (Already $L_2(q)$ does.) Choosing $b = 1$, we see that Aut Γ also contains the field automorphisms. If there are enough of these, i.e., if r is prime and p (= char $\mathbb{F}$) is a primitive element (mod r), then Γ is distance-transitive. Examples are $(q,r,v) = (16,3,51)$, $(16,5,85)$, $(64,3,195)$, $(81,5,410)$, $(256,5,1285)$. (See also §7.6C.) For each $\gamma \in \Gamma$, the local graph $\Gamma(\gamma)$ is isomorphic to the graph with vertex set $\mathbb{F}_q$, where two vertices are adjacent when they differ by an element of K. In particular, for $q = 16$, $r = 3$ we see that Γ is locally a folded 5-cube. In fact, according to BÖINCK [95], it is the only such graph.

(iv) There is a unique distance-regular graph with intersection array $\{7,4,1;1,2,7\}$ on 24 vertices; this graph is not distance-transitive (its full group is $PSL(2,7).2$, acting transitively with rank 6, with point stabilizer D_{14}).

(v) For $m = 1$, q a power of 2, we find graphs with intersection array $\{q, q-2, 1;\ 1, 1, q\}$; these were found earlier in §1.17, Example e), cf. Proposition 1.16.3.

(vi) When $K \cup \{0\}$ is a subfield of $\mathbb{F}$, say $K = \mathbb{F}_t^*$, where $q = t^e$, then the vertex set of Γ can be naturally identified with the projective space $PG(2e-1,t)$, and there are $e-1$ symplectic forms B_i such that two points x, y are adjacent or antipodal when $B_i(x,y) = 0$ for all i. In particular, in case $e = 2$, $\Gamma \cup \Gamma_3$ is the symplectic graph in $PG(3,q)$, and Γ itself is obtained by removing the lines of a symplectic spread. In other words, we have an example of the construction described in Proposition 12.5.2.

12.6 Thin near octagons from Denniston's complete arcs in projective planes of even order

A *complete n-arc* in a projective plane of order q is a nonempty proper subset X of its point set such that each line meets this set in either 0 or n points. (For $n = 2$ such a set is known as a *hyperoval.*) Clearly, we must have $|X| = 1+(q+1)(n-1)$ and $n \mid q$.

[Note that in this area of geometry some authors use the word 'complete' for sets maximal under inclusion in some collection of sets, while they use 'maximal' for such sets attaining some upper bound for their cardinality. We adhere to the established meaning of 'maximal'.]

DENNISTON [261] constructs complete n-arcs in the Desarguesian plane $PG(2,q)$ when q is a power of 2, and $n \mid q$.

[As follows: let $Q(x,y) = x^2 + axy + y^2$ be an irreducible quadratic form over $\mathbb{F}_q$, and let H be a subgroup of order n of the additive group of $\mathbb{F}_q$. Then $X = \{(x,y,1) \mid Q(x,y) \in H\}$ is a complete n-arc in $PG(2,q)$, the union of $n-1$ conics and their common nucleus $(0,0,1)$.]

On the other hand, for $q > 3$ no complete 3-arcs exist (THAS [730]), and, more generally, for odd $q > n$ no complete n-arcs are known.

Complete n-arcs occur in a variety of combinatorial contexts; e.g., they provide examples of Steiner systems $S(2,n,qn-q+n)$. (Indeed, of very special such Steiner systems: each exterior point gives a parallel class, and each exterior line gives a resolution, so these Steiner systems are resolvable in many ways; for $(n,q) = (4,8)$ we find the so-called 'Ree unital' on 28 points.) Our interest here stems from the fact that the complement of a complete n-arc (together with its secants) is a partial geometry with $(q+1)(q+1-q/n)$ points and parameters $pg(q+1-n,\ q+1-q/n,\ q+1-n-q/n)$. In particular, when $q = n^2$, we find a partial 1-geometry with nexus $q+1-2n$, and its incidence graph is distance-regular of diameter 4.

12.6.1. Proposition. *Let n be a power of 2. Then the incidence graph of the partial geometry with parameters* $pg(n^2-n+1, n^2-n+1, (n-1)^2)$ *constructed above is distance-regular with* $v = 2(n^2+1)(n^2-n+1)$ *vertices and intersection array* $\{n^2-n+1, n(n-1), n(n-1), n;\ 1, 1, (n-1)^2, n^2-n+1\}$. *(Cf. Table 6.9, (B3).)* □

12.7 Antipodal covers of complete graphs from pseudocyclic association schemes

Let $(X, \mathfrak{R})$ be an r-class association scheme, let $I = \{1, \cdots, r\}$, and suppose that $(I, \oplus)$ is a commutative quasigroup (i.e., $i \oplus j = j \oplus i$, and any two of i, j and $i \oplus j$ uniquely determine the third).
Then we can construct a graph Γ with vertex set $(X \cup \{\infty\}) \times I$, letting $(\infty,i) \sim (x,j)$ when $i = j$, and $(x,i) \sim (y,j)$ when $(x,y) \in R_{i \oplus j}$.
Clearly, Γ is an r-cover of the complete graph K_{n+1}, where $n = |X|$ - projection onto the first coordinate provides the covering map. In particular, Γ is regular of valency n. (Conversely, in case $r = 2$, we have seen in §1.5 that every distance-regular 2-cover of K_{n+1} arises in this way.)

12.7.1. Proposition. *The graph Γ is distance-regular of diameter 3 (an antipodal r-cover of the complete graph K_{n+1}), if and only if either*

(a) *there is a number μ such that $n_1 = n_2 = \cdots = n_r = \mu$,*
$\sum_i p^l_{ii} = \mu - 1$ $(l \in I)$ *and* $\sum_i p^l_{i\oplus j, i\oplus k} = \mu$ $(j,k,l \in I,\ j \neq k)$,

or

(b) *there is a number μ and an element $a \in I$ such that $i \oplus i = a$ $(i \in I)$, $n_i = \mu$ $(i \in I,\ i \neq a)$, $\sum_i p^l_{ii} = n_l - 1$ $(l \in I)$ and $\sum_i p^l_{i\oplus j, i\oplus k} = n_{l\oplus(j\oplus k)}$ $(j,k,l \in I,$ $j \neq k)$.*

Proof. The common neighbours of (∞,i) and (x,j) are the vertices (y,i) with $(x,y) \in R_{i\oplus j}$, and there are $n_{i\oplus j}$ of these. Thus, if Γ is to be distance-regular, we must have $\lambda(\Gamma) = n_{i\oplus i}$ and $\mu(\Gamma) = n_{i\oplus j}$ for all $i,j \in I$, $i \neq j$. The common neighbours of (x,j) and (y,j) are (∞,j) and the vertices (z,i) with $(x,z), (y,z) \in R_{i\oplus j}$, and there are $\sum_i p^l_{ii}$ of these, if $(x,y) \in R_l$. Finally, (x,j) and (y,k), where $(x,y) \in R_l$ and $j \neq k$, have $\sum_i p^l_{i\oplus j, i\oplus k}$ common neighbours. □

12.7.2. Corollary (MATHON [520]). *Suppose that $(X, \mathfrak{R})$ is pseudocyclic, and that $p^{i\oplus l}_{i\oplus j, i\oplus k} = p^l_{jk}$ for all i, j, k, $l \in I$. Then Γ is distance-regular with intersection array*

$$\{r\mu + 1, (r-1)\mu, 1;\ 1, \mu, r\mu + 1\}$$

where $\mu = (|X| - 1)/r$. Its spectrum is

$$(r\mu + 1)^1\ (-1)^{r\mu+1}\ (\pm\sqrt{r\mu + 1})^{(r-1)(r\mu+2)/2}.$$

Proof. The conditions under (a) of the proposition above are fulfilled (cf. Proposition 2.2.7), so Γ is distance-regular. Computation of the parameters is straightforward. □

12.7.3. Corollary. *If $r = 3$, then Γ is distance-regular if and only if $(X, \mathfrak{R})$ is pseudocyclic.*

Proof. There is no commutative quasigroup of order 3 with constant $i \oplus i$, so we must be in case (a) of the proposition above. The first two requirements there are equivalent to the requirement that $(X, \mathfrak{R})$ must be pseudocyclic (cf. Proposition 2.2.7). Moreover, the following lemma states that the third condition is always fulfilled. □

12.7.4. Lemma (MATHON [520]). *A 3-class pseudocyclic association scheme has parameters*

$$(p^1_{ij})_{ij} = \begin{bmatrix} 0 & 1 & 0 & 0 \\ 1 & a-1 & b & c \\ 0 & b & c & a \\ 0 & c & a & b \end{bmatrix}, \quad (p^2_{ij})_{ij} = \begin{bmatrix} 0 & 0 & 1 & 0 \\ 0 & b & c & a \\ 1 & c & a-1 & b \\ 0 & a & b & c \end{bmatrix}, \quad (p^3_{ij})_{ij} = \begin{bmatrix} 0 & 0 & 0 & 1 \\ 0 & c & a & b \\ 0 & a & b & c \\ 1 & b & c & a-1 \end{bmatrix}$$

for certain constants a, b, c, and $a+b+c = n_1 = n_2 = n_3 = m$, say. Moreover, $a^2+b^2+c^2-ab-ac-bc = a$. We have for the eigenmatrices P, Q

$$P = Q = \begin{bmatrix} 1 & m & m & m \\ 1 & \theta_1 & \theta_2 & \theta_3 \\ 1 & \theta_2 & \theta_3 & \theta_1 \\ 1 & \theta_3 & \theta_1 & \theta_2 \end{bmatrix}$$

where the θ_i are the three roots of $\theta^3+\theta^2-m\theta+bc-a^2 = 0$. This latter polynomial is reducible over $\mathbf{Q}$ if and only if $b = c$, and in that case we have $\theta_1 = 2a-2b-1$, $\theta_2 = \theta_3 = b-a$, and each of the relations R_1, R_2, R_3 defines a strongly regular graph.

Proof. Straightforward. □

In view of the corollaries above, it is of interest to see which pseudocyclic association schemes are known.

A. Cyclotomic schemes

Let $q = r\mu+1$ be a prime power, and let a be a primitive root in $\mathbf{F}_q$. Let $K \subset \mathbf{F}_q{}^*$ be the subgroup of the r-th powers. Consider the *cyclotomic scheme* on $X = \mathbf{F}_q$ defined by $R_i = \{(x,y) \mid y-x \in a^iK\}$ $(1 \leqslant i \leqslant r)$ (cf. §2.10). This association scheme is a translation scheme; it is symmetric when $(q-1)/r$ is even, or when q is a power of 2. Let $(I, \oplus)$ be the abelian group defined on $I = \{1, 2, \cdots, r\}$ by $i \oplus j = i+j$ (mod r). Then $p^{i\oplus l}_{i\oplus j\; i\oplus k} = p^l_{jk}$ $(i,j,k,l \in I)$ and by Corollary 12.7.2 this leads to distance-regular graphs with intersection array $\{q, (r-1)\mu, 1; 1, \mu, q\}$, see Table 6.10, (A3). For another description of these graphs, see §12.5.

12.7.5. Proposition. *Let $(X, \mathfrak{R})$ be a d-class cyclotomic scheme on $X = \mathbf{F}_p$, where p is prime and $d \geqslant 2$. Then its group of automorphisms (relation preserving permutations of X) consists of the maps $x \mapsto bx+c$, where b is a d-th power and $c \in \mathbf{F}_p$.*

Proof. Let ϕ be an automorphism of $(X, \mathfrak{R})$ and consider the graph $\{(x, \phi(x)) \mid x \in X\}$ of ϕ. Since ϕ preserves relations we find that the slope $(\phi(y)-\phi(x))/(y-x)$ is a d-th power for all x, $y \in X$, $x \neq y$, and hence this slope takes not more than $(p-1)/2$ values. By RÉDEI [610], p.237 (see also LOVÁSZ & SCHRIJVER [510]), it follows that ϕ is linear, as was to be proved. □

Remark. For $d = 2$ this result is due to CARLITZ [186]. More generally, MCCONNEL [525] shows that if q is a prime power, and $d \mid (q-1)$, then any permutation ϕ of $\mathbf{F}_q$ such that $(\phi(y)-\phi(x))/(y-x)$ is a d-th power for all x, $y \in \mathbf{F}_q$, $x \neq y$, must be semilinear, i.e., of the form $x \mapsto bx^\sigma+c$, where b is a d-th power, and σ a field automorphism. (See also BRUEN & LEVINGER [141].)

B. The Mathon and Hollmann schemes on 28 points, and conics in $PG(2,q)$

There exist precisely two pseudocyclic 3-class schemes on 28 points with parameters as in Corollary 12.7.2 for $t = 9$ (HOLLMANN [404]). These lead to distance-regular graphs with intersection array $\{28,18,1;1,9,28\}$ on 87 vertices, see Table 6.10, (A4).

The first of these two schemes (MATHON [520]) is the special case $q = 8$ of the following construction: Let $q > 4$ be a power of 2 and let $\mathcal{O}$ be a conic in $PG(2,q)$. Consider the action of $PO(3,q)$ on the $\frac{1}{2}q(q-1)$ exterior lines of $\mathcal{O}$. (Equivalently, consider the action of $PSL(2,q)$ (by conjugation) on its cyclic subgroups of order $q+1$.) This action is generously transitive of rank $\frac{1}{2}q$, and we find a $(\frac{1}{2}q-1)$-class pseudocyclic association scheme with valencies $q+1$.

The second of the two schemes on 28 points arises from the first by a switching construction, see HOLLMANN [404].

C. The Hollmann scheme on 496 points

Enlarging the group $PO(3,q)$ used above yields an association scheme with fewer classes. In particular, when $q = 2^p$, where p is prime, the action of $P\Sigma O(3,q)$ on the exterior lines of $\mathcal{O}$ yields a $(\frac{1}{2}q-1)/p$-class pseudocyclic association scheme with valencies $p(q+1)$. Thus, for $q = 32$, HOLLMANN [403], p. 135, constructs a pseudocyclic 3-class scheme on 496 points with parameters as in Lemma 12.7.4 with $(a,b,c) = (57,58,50)$. Consequently, we find a distance-regular graph with intersection array $\{496,330,1;1,165,496\}$ on 1491 vertices, see Table 6.10, (A5).

Chapter 13
Sporadic Graphs

The preceding four chapters were concerned with graphs belonging to an infinite series of distance-regular graphs or constructed in a uniform way. A handful of (known) distance-regular graphs remain. These are the subject of the present chapter.

13.1 Graphs related to the Hoffman - Singleton graph

There exists a unique strongly regular graph Δ with parameters $(v,k,\lambda,\mu) = (50,7,0,1)$, the largest known Moore graph (HOFFMAN & SINGLETON [401]; for uniqueness, see also JAMES [440]). Its automorphism group is $P\Sigma U(3,5^2)$ acting distance transitively (i.e., as a rank 3 group) with point stabilizer isomorphic to Sym(7) and edge stabilizer to Aut(Sym(6)) = Sym(6).2 (BENSON & LOSEY [51]).

We give four constructions of Δ.

A. The following construction is due to N. Robertson. Let P_j ($j \in \mathbf{Z}_5$) be five pentagons (so that vertex i of P_j is adjacent to the vertices $i\pm1$ of P_j ($i \in \mathbf{Z}_5$)), and let Q_l ($l \in \mathbf{Z}_5$) be five pentagrams (so that vertex i of Q_l is adjacent to the vertices $i\pm2$ of Q_l ($i \in \mathbf{Z}_5$)). Now join vertex i of P_j to vertex $i + jl$ of Q_l for all $i, j, l \in \mathbf{Z}_5$. This yields the Hoffman-Singleton graph.
(This construction shows that the point set of the Hoffman-Singleton graph can be partitioned into two halves that each induce $5C_5$. The union of any pentagon and any pentagram induces the Petersen graph. By taking three pentagons and three pentagrams, we find a (5,5)-cage, see §6.9.)

B. Starting from the unique Steiner system $S(3,6,22)$ (with 22 points and 77 blocks), construct a graph on 100 vertices with vertex set $\{\Omega\} \cup X \cup \mathfrak{B}$, where X is the set of points and $\mathfrak{B}$ is the set of blocks of the Steiner system and Ω is a new symbol. Let Ω be adjacent to the elements of X, let $x \in X$ be adjacent to $B \in \mathfrak{B}$ if and only if $x \in B$, let X be a coclique and let $B \sim B'$ for $B, B' \in \mathfrak{B}$ if and only if $B \cap B' = \varnothing$. This defines a strongly regular graph H with parameters $(v,k,\lambda,\mu) = (100,22,0,6)$ called the *Higman-Sims graph*. Aut $H \cong$ HS.2, where HS is the sporadic simple group of Higman and Sims. Now regard the Steiner system $S(3,6,22)$ as the derived system (at ∞) from $S(4,7,23)$, and let $B_0 \subset X$ be a block of the latter system (not containing ∞). Then

$$\{\{\Omega\} \cup B_0 \cup \{B \in \mathfrak{B} |\ |B \cap B_0| = 1\},\ X \backslash B_0 \cup \{B \in \mathfrak{B} |\ |B \cap B_0| = 3\}\}$$

is a partition of the point set of H into two halves, and the induced subgraph on each of these two halves is isomorphic to the Hoffman-Singleton graph.

C. There are 30 distinct Steiner triple systems $STS(7)$ (i.e., $S(2,3,7)$, i.e., $PG(2,2)$, the

Fano plane) on a given 7-set Σ. Two such systems have 0, 1, 3 or 7 triples in common, and the collection of 30 systems decomposes into two sets X_1, X_2 of 15 systems such that two distinct systems both in X_j have precisely one triple in common ($j = 1,2$). Thus we find a projective space $PG(3,2)$ with point set X_1, line set $\mathfrak{L} = \binom{7}{3}$, and set of planes X_2 (with the obvious incidence relations).
Now define a graph Δ with point set $X_1 \cup \mathfrak{L}$ where X_1 is a coclique, $l, l' \in \mathfrak{L}$ are adjacent if and only if l, l' are disjoint triples, and $x \in X_1$ is adjacent to $l \in \mathfrak{L}$ if and only if in this $PG(3,2)$ the point x is on the line l (i.e., if and only if the triple l is a line of the Steiner triple system x). Then Δ is isomorphic to the Hoffman-Singleton graph.
(This construction is contained in the previous, where we find the 15-coclique $X_1 = X \setminus B_0$ on the second half.)

[That the Higman-Sims graph can be partitioned into two Hoffman-Singleton graphs is suggested by SIMS [675]; the identification of the set of lines in $PG(3,2)$ and the triples in a 7-set was worked out in detail by HALL [346]; the decomposition of the 30 Fano planes on a 7-set into two collections of size 15 was noticed by LINDNER & ROSA [497]. Most of the results in this section, and connections to some other strongly regular graphs, are discussed in BROUWER & VAN LINT [134], §10.]

Incidentally, from this construction we find a Buekenhout-Tits geometry with diagram

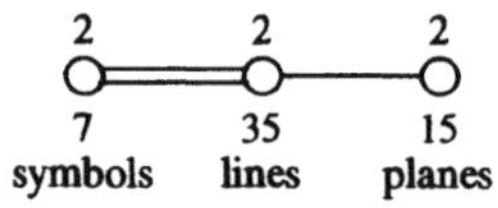

with set of objects $\Sigma \cup \mathfrak{L} \cup X_1$ and obvious incidences (each symbol is incident with each plane); see NEUMAIER [560].

D. Starting from the Sylvester graph on 36 points discussed in §13.1A below, the uniqueness proof given there shows how to extend it to the Hoffman-Singleton graph.

13.1.1. Theorem. *Let Δ be the Hoffman-Singleton graph.*

(i) *The line graph Γ of Δ is distance-transitive with $v = 175$ vertices, intersection array $\{12,6,5;1,1,4\}$ and distance distribution diagram*

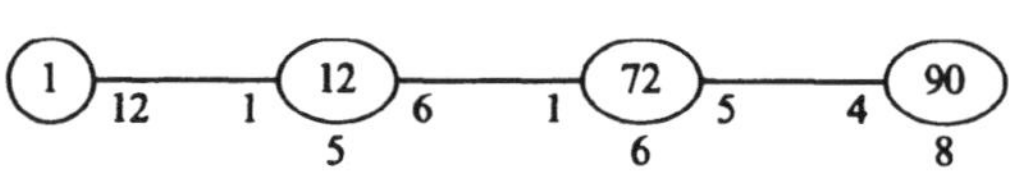

$v = 175$;

$\mathrm{Aut}\,\Gamma \cong P\Sigma U(3,5^2)$ *with point stabilizer* Sym(6).2 *(cf. Table 6.8, (D5)). The intersection array determines Γ uniquely. Γ_2 is strongly regular with parameters $(v,k,\lambda,\mu) = (175,72,20,36)$ and is the point graph of a partial geometry* pg(5,18,2).

(ii) *Let $\delta \in \Delta$. The subgraph Γ' induced on $\Delta_2(\delta)$ is an antipodal 6-cover of K_7, the complete graph on 7 points. Γ' is distance-transitive with $v = 42$ vertices, intersection array $\{6,5,1;1,1,6\}$ and distance distribution diagram*

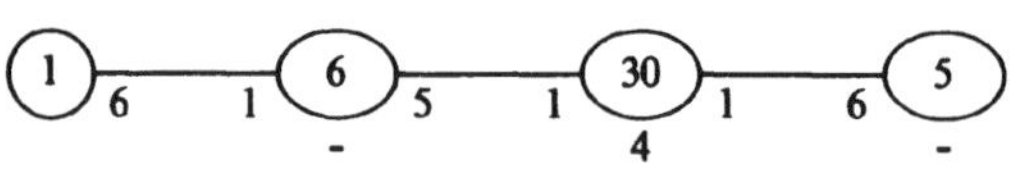

$v = 42$;

$\mathrm{Aut}\,\Gamma' \cong$ Sym(7) *with point stabilizer* Sym(5). *The intersection array determines the graph uniquely. (Cf. Table 6.10, (A6).)*

(iii) *Let C be a 15-coclique of Δ. The graph Γ'' induced on the complement of C is isomorphic to the Odd graph O_4 of disjoint triples in a 7-set. (Cf. Table 6.6, (R2).) Γ'' is distance-transitive with $v = 35$ vertices, intersection array $\{4,3,3;1,1,2\}$ and*

Aut $\Gamma'' \cong$ Sym(7) *with point stabilizer* Sym(3) $\times$ Sym(4). *The intersection array determines* Γ'' *uniquely.*

(iv) *The graph* Γ''' *with as vertex set the* 100 15-*cocliques in* Δ, *two cocliques being adjacent when they have* 8 *points in common, is a distance-transitive bipartite graph with* $v = 100$ *vertices, intersection array* $\{15,14,10,3;1,5,12,15\}$ *and distance distribution diagram*

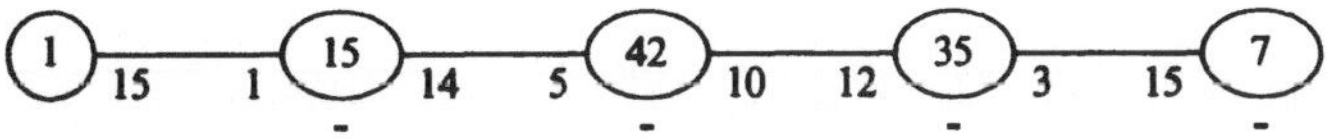

$v = 100;$

Aut $\Gamma''' \cong P\Sigma U(3,5^2)$ *with point stabilizer* Alt(7). *The intersection array determines* Γ''' *uniquely.*

(v) *The bipartite double* Γ'''' *of* Δ *is a distance-transitive bipartite graph with* $v = 100$ *vertices, intersection array* $\{7,6,6,1,1;1,1,6,6,7\}$ *and distance distribution diagram*

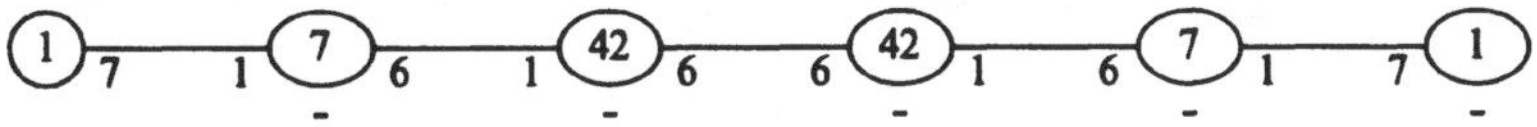

$v = 100;$

a regular thin near decagon (cf. Table 6.9, (B8)). We have Aut $\Gamma'''' \cong P\Sigma U(3,5^2) \times 2$ *with point stabilizer* Sym(7). *Again the intersection array determines the graph uniquely.*

Proof. (i) Claims about the parameters follow directly from the parameters of Δ. Since $c_3(a_3+a_2-a_1) = b_1a_2$ it follows from Proposition 4.2.17 that Γ_2 is strongly regular. A partial geometry pg(5,18,2) with point graph Γ_2 was constructed by Haemers [341]. Aut Γ = Aut $\Delta \cong P\Sigma U(3,5^2)$ acts as a rank 4 group on Γ. Uniqueness follows by observing that, since $\mu = 1$, we must have $\Gamma(\gamma) \cong 2K_6$ for each vertex γ, so that each vertex is in two 7-cliques and there are 50 7-cliques altogether. The graph on the cliques (with cliques adjacent when they meet) has the parameters of, and hence is, the Hoffman-Singleton graph.
(ii) The parameters follow easily from those of Δ. From the intersection array one finds that there is a (unique) partition into seven sets of size 6 such that two distinct vertices from the same set have distance 3. Consequently, there is a unique way of embedding a graph with this intersection array as $\Delta_2(\delta)$ in Δ. It follows that the graph has automorphism group Sym(7).
(iii) The largest cocliques in Δ have size 15, and all such cocliques are equivalent under Aut Δ. (There are 100 of them.) The structure of the graph is clear from construction C above. Uniqueness was shown by Gewirtz [323] (and is of course a special case of Moon [543]; uniqueness may also be shown by constructing an embedding of the graph in the Hoffman-Singleton graph).
(iv) See Brouwer & van Lint [134], p. 108, or Neumaier [560].
(v) Uniqueness follows easily: Γ_4'''' is the disjoint union of two Hoffman-Singleton graphs, and Γ_5'''' correlates the labellings of these two graphs. □

Remarks. (i) Two edges in Δ have distance 1, 2 or 3 in Γ if they intersect, form disjoint edges in a pentagon, or form opposite edges in some hexagon, respectively.
(ii) Let $\gamma \in \Gamma'$. Then Γ'_2 induces a distance-regular graph on 36 vertices with intersection array $\{5,4,2;1,1,4\}$, see §13.1A.
(iii) From construction B above one sees that Γ''' is a partial graph of the Higman-Sims graph. In fact, $\Gamma_1''' \cup \Gamma_4'''$ is isomorphic to the Higman-Sims graph. But this

means that there exists a graph Θ on 3×50 vertices with vertex set $X = X_1 \cup X_2 \cup X_3$ such that X_i induces a Hoffman-Singleton graph and $X \setminus X_i$ induces a Higman-Sims graph for $i = 1, 2, 3$.
(iv) The second subconstituent of the Higman-Sims graph is a strongly regular graph on 77 points; it is the unique strongly regular graph with parameters $(v,k,\lambda,\mu) = (77,16,0,4)$ (see also §11.4C).
(v) We may construct two geometries related to the Hoffman-Singleton graph with Buekenhout-Tits diagrams

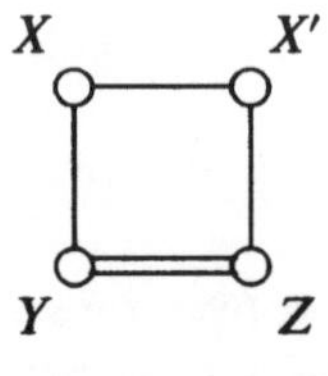

and

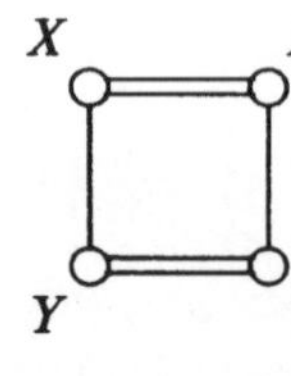

as follows.
For the first geometry, let $\{X,Y,Z\}$ be a system of three linked Hoffman-Singleton graphs so that the union of any two of them induces a Higman-Sims graph (cf. Remark (iii) above). Let X' be a copy of X. Incidence between X and X' or X and Y or Y and Z or X' and Z is adjacency in this graph, but incidence between X and Z or X' and Y is nonadjacency in this graph.
For the second geometry, take a partition $\{X,Y\}$ of the Higman-Sims graph H into two Hoffman-Singleton graphs (cf. construction B above), and let X', Y' be copies of X, Y. Incidence between X and X' or X and Y or X' and Y' or Y and Y' is adjacency in H, but incidence between X and Y' or X' and Y is nonadjacency in H.
These geometries were constructed by NEUMAIER [560] and WESTER [780]. Uniqueness assuming the rank 3 residues was shown by HEISS [369].
(vi) Γ_2 is the local graph of a Taylor graph, cf. §11.4E.

A. Sylvester's double six graph

13.1.2. Theorem. *There exists a unique distance-regular graph* Γ *with* $v = 36$ *vertices, intersection array* $\{5,4,2;1,1,4\}$ *and distance distribution diagram*

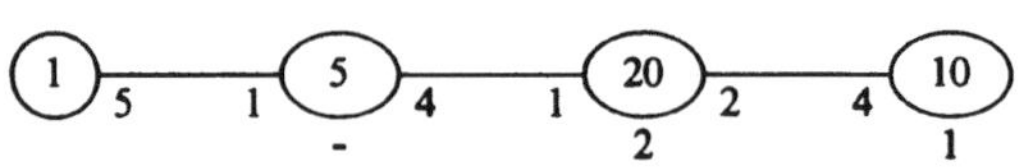

$v = 36.$

It is distance-transitive with automorphism group $\mathrm{Aut}(S_6)$ *and point stabilizer* $AGL(1,5)\times\mathbf{Z}_2$. *(Cf. Table 6.8, (D2).) The graph* Γ_3 *is the lattice graph* $L_2(6)$ *(the* 6×6*-grid).*

Proof. (i). Existence. We indicate several constructions of distance-regular graphs with the indicated intersection array.

A. Let Δ be the Hoffman-Singleton graph, and let γ, δ be two adjacent vertices of Δ. One easily verifies that the graph Γ induced on $\Delta_2(\gamma) \cap \Delta_2(\delta)$ has the required parameters (BIGGS [75]).

B. Let Γ be the graph that has as point set the $\mathrm{Aut}(S_6)$-conjugacy class of 36 involutions contained in $\mathrm{Aut}(S_6)\setminus S_6$, where two involutions are adjacent whenever they commute. [Now distance 1, 2, 3 in Γ corresponds to a product of order 2, 5, 4, respectively.]

Equivalently, but formulated in a more geometric language:

C. Let Γ be the graph that has as vertices the ordered pairs (O,S) of an ovoid and a spread in the (unique) generalized quadrangle of order (2,2), where $(O,S) \sim (O',S')$ if and only if $O \neq O', S \neq S'$ and the unique point in $O \cap O'$ is on the unique line in $S \cap S'$. In this representation (O,S) and (O',S') have distance three if and only if $O = O'$ or $S = S'$ (but not both).

[The automorphism group of the generalized quadrangle is S_6; the involutions in $\mathrm{Aut}(S_6) \setminus S_6$ are the polarities of the generalized quadrangle, but it is easily seen that the absolute points (resp. lines) under a polarity of a generalized quadrangle form a ovoid (resp. spread); conversely one may check that given a pair (O,S), there is a unique polarity of the generalized quadrangle with O (resp. S) as set of absolute points (resp. lines).]

(ii) Uniqueness.

Since $k = b_2 + c_3 - 1$ it follows that Γ_3 is strongly regular (by Proposition 4.2.17) and one finds that Γ_3 has the parameters of the lattice graph $L_2(6)$. But for $m \neq 4$ the lattice graph $L_2(m)$ is uniquely determined by its parameters (SHRIKHANDE [668], cf. §9.2B) so $\Gamma_3 \cong L_2(6)$. Invent 14 new symbols γ, γ_i $(1 \leqslant i \leqslant 6)$, δ, δ_i $(1 \leqslant i \leqslant 6)$; join γ to δ and to each γ_i and δ to each δ_i $(1 \leqslant i \leqslant 6)$, and join each of the points γ_i, δ_i to the 6 points of a Γ_3-clique, different points to different cliques, such that the cliques belonging to the γ_i (resp. δ_i) form a parallel class (partition) of the point set of Γ. Call the graph thus obtained Δ. One easily verifies that Δ is strongly regular with the parameters of the Hoffman-Singleton graph, but this graph is known to be unique (HOFFMAN & SINGLETON [401]), and has an edge-transitive group of automorphisms, so that Γ is uniquely determined (as in construction A above). □

Remark. Bagchi & Sastry [pers. comm. J. Saxl] remarked that the permutation representation of $Sp_4(q)$ on cosets of $Sz(q)$ (with $q = 2^{2h+1}$) is multiplicity-free of rank $q+3$. But $Sp_4(2) \cong \mathrm{Sym}(6)$ and $Sz(2) \cong 5{:}4$, so for $q = 2$ we find the graph Γ. (And indeed, Sym(6) acts rank 5 on Γ.)

13.2 Commuting involutions graphs and Fischer spaces

We may obtain graphs (called *commuting involutions graphs*) from a group G by taking for the vertex set a conjugacy class of involutions in that group and for adjacency commuting. Clearly, such graphs have the property that for each $\gamma \in \Gamma$ there is an involution in $\mathrm{Aut}\,\Gamma$ fixing γ and each of its neighbours, but no other vertices. Conversely, if a graph Γ has a transitive group of automorphisms G such that for each $\gamma \in \Gamma$ there is a unique involution σ_γ in G (pointwise) fixing precisely $\gamma^\perp$, then the involutions σ_γ form a conjugacy class, and σ_γ commutes with σ_δ if and only if $\gamma \sim \delta$ (for: if $\sigma = \sigma_\gamma$, $\tau = \sigma_\delta$, then $\tau\sigma\tau^{-1}$ is the involution fixing $\tau(\gamma)^\perp$, so that $\tau\sigma\tau^{-1} = \sigma$ if and only if $\gamma \in \delta^\perp$).

If Γ is the commuting involutions graph of a conjugacy class K of involutions in G, then G acts on Γ by conjugation, and $G_\gamma = C_G(\gamma)$ for $\gamma \in K$. A commuting involutions graph need not be edge-transitive; for example, if we let Γ be the commuting involutions graph corresponding to the class $2B$ of involutions in $M_{22}.2$ in the notation of the ATLAS [216], then Γ is the distance 1-or-2 graph of the doubly truncated Witt graph (on 330 vertices, see §11.4C), and not edge-transitive.

We have already met several (families of) commuting involutions graphs. The complement of the triangular graph, $\overline{T(n)}$, is the commuting graph of the class of transpositions in the symmetric group Sym (n). Also the Hall-Janko near octagon on 315 vertices described in §13.6, and the Sylvester graph on 36 vertices, described in §13.1A, are commuting involutions graphs.

The unitary nonisotropics graph Γ on 63 vertices (one of the two generalized hexagons of order (2,2), see §12.4) is a commuting involutions graph.

[In fact $U_3(3) = G_2(2)'$ has a single conjugacy class of involutions; each involution fixes 7 points (some set $\gamma^\perp$ for $\gamma \in \Gamma$) and 15 lines (all lines meeting $\gamma^\perp$); $U_3(3)$ acts rank 4 (i.e., distance transitively) on the points and rank 5 on the lines; $G_2(2)$ also acts rank 4 on the lines.]

Also the generalized hexagon of order (2,8) derived from ${}^3D_4(2)$ (cf. §10.7) is a commuting involutions graph.

[Involutions from conjugacy class 2A fix 19 points ($\gamma^\perp$ for some point γ) and 153 lines (all lines meeting $\gamma^\perp$).]

Somewhat similarly we find for the generalized octagon of order (2,4) derived from ${}^2F_4(2)$ that involutions from conjugacy class 2A fix 91 points ($\Gamma_{\leq 2}(\gamma)$ for some point γ) and 45 lines (all lines meeting $\gamma^\perp$) - so no commuting involution graph in this case.

The three locally Petersen graphs (cf. Theorem 1.16.5) are commuting involution graphs; apart from $\overline{T(7)}$ we have the graph on 65 vertices described in §12.2 (with automorphism group $P\Sigma L(2,25)$) and a 3-cover of $\overline{T(7)}$ (with automorphism group 3·Sym (7)) described below.

The case of diameter 2 is well known in group theory. If G possesses a conjugacy class D of involutions (generating G) the product of any two of which has order 1, 2 or p, then D is called a *class of p-transpositions* of G. It is natural to ask when the permutation representation of G on D has rank 3, the 3 orbitals being parametrized by the numbers 1, 2 and p (so that the commuting involutions graph is strongly regular).
If p is a power of 2, then Baer's Theorem (Baer [18], cf. Gorenstein [337], Ch. 3, Th. 8.2) yields that G is a 2-group, and no primitive rank 3 permutation representation arises.
If $p = 2m$ is even for $m \geq 3$, then $(xy)^p = 1$ for $x, y \in D$ implies $(x x^y)^m = 1$, so $m = 1$ or 2, contradiction.
Hence p is odd. The case $p = 3$ is dealt with in the fundamental work by Fischer [285]. Apart from the infinite series Sym (n), $Sp(2m, 2)$, $O^\pm(2m, 2)$, $PSU(n, 2^2)$ and $P\Omega^\pm(n, 3)$, three sporadic groups Fi_{22}, Fi_{23} and Fi_{24} appear. The corresponding strongly regular graphs have parameters $(v,k,\lambda,\mu) = (3510,693,180,126)$, $(31671,3510,693,351)$, $(306936,31671,3510,3240)$.

[Fi_{22} and Fi_{23} each have another rank three representation; these give rise to strongly regular graphs with parameters $(v,k,\lambda,\mu) = (14080,3159,918)$ and $(137632,28431,6030)$.]

Later, Aschbacher [11, 12] completely answered the above question, and more generally determined all D where for $x, y \in D$ the product xy either has odd order, or order 1 or 2, and where G is primitive on D. Timmesfeld [740, 741] studied the more general case where the product xy of two elements $x, y \in D$ either has odd order, or order 1, 2 or 4, where in case xy has order 4 we also have $(xy)^2 \in D$. Roughly speaking, he finds all Chevalley groups (twisted or not) over finite fields of characteristic 2, except for ${}^2F_4(q)$, and the Fischer groups, $P\Omega^\pm(n, 5)$, $L(2,q)\,\mathrm{wr}\,\mathrm{Sym}(n)$ (q a power of 2), and also Alt (6) and HJ.

Sometimes pairs of commuting involutions from D fall into several G-orbitals; choosing one orbital for adjacency, one obtains 'commuting involutions graphs in the wide sense'. For example, the doubly truncated Witt graph (cf. §11.4C) on 330 vertices is a commuting involutions graph in the wide sense for $M_{22}.2$. VAN BON [99] determines all primitive distance-transitive commuting involution graphs in the wide sense for groups G with a normal sporadic simple subgroup, and finds the graphs belonging to M_{22}, to HJ (on 315 vertices, cf. §13.6), and to Fi_{22}, Fi_{23}, Fi_{24}'.

Five antipodal 3-covers

S.D. SMITH [688] studies the situation where one has the following:

> Z is a group of order 3, normal in a finite group H, and H/Z is generated by a class of 3-transpositions; $H' = C_H(Z)$ and is of index 2 in H, with H'/Z simple.

It turns out that if we put $G = H/Z$, then G' is one of Alt(6), Alt(7), $P\Omega^-(6,3)$, $P\Omega(7,3)$, Fi_{24}'. The group G, being a Fischer group, acts rank 3 on a class D of 3-transpositions, so that we find a strongly regular graph with certain parameters (v,k,λ,μ) with vertex set D, where two elements of D are adjacent whenever they commute. In the five cases mentioned, the parameters are

G'	v	k	λ	μ
Alt(6)	15	6	1	3
Alt(7)	21	10	3	6
$P\Omega^-(6,3)$	126	45	12	18
$P\Omega(7,3)$	378	117	36	36
Fi_{24}'	306936	31671	3510	3240.

Each element of D has three preimages in H, and H acts rank 5 on this set of $3v$ preimages. Thus, with the same definition for adjacency (adjacent $\equiv$ commuting), we find a distance-transitive graph on $3v$ vertices with intersection array $\{k, k-\lambda-1, \frac{2}{3}\mu, 1; 1, \frac{1}{3}\mu, k-\lambda-1, k\}$, an antipodal 3-cover of the strongly regular graph found earlier. Below we shall give explicit constructions of the first four graphs in 'geometric' or 'combinatorial' terms. NORTON [572] explicitly constructs the last two graphs.

A. The Foster graph for 3·Sym(6).2 and the hexacode

Let X be the complement of a hyperoval O (i.e., a set O of 6 points, no three on a line) in the projective plane Π of order 4, and let $\mathfrak{L}$ be the set of 15 secants of O. Then $(X,\mathfrak{L})$ is a generalized quadrangle with parameters (2,2). In order to construct a 3-cover of this generalized quadrangle, let us consider the situation over $\mathbf{F}_2$ instead of $\mathbf{F}_4$.

The points and lines of Π are 1-spaces and 2-spaces in a vector space V of dimension 3 over $\mathbf{F}_4$; let us write V_0 for the same set of vectors but now considered as a vector space over $\mathbf{F}_2$ (so that dim $V_0 = 6$). In V_0, the points and lines of Π are 2-spaces and 4-spaces and hence they are lines and projective 3-spaces in the projective space $\Pi_0 = PV_0$.

Now define a 3-cover $(Y,\mathfrak{M})$ of $(X,\mathfrak{L})$ as follows: Let $\pi:\Pi_0 \to \Pi$ be the map sending a point x to the point $\mathbf{F}_4 x$. Put $Y = \pi^{-1}X$, and let $\mathfrak{M}$ be the set of lines M of Π_0 such that $\pi M = L \cap X$ for some line $L \in \mathfrak{L}$.

13.2.1. Theorem. *The partial linear space $(Y, \mathfrak{M})$ is a 3-cover of $(X, \mathfrak{L})$. The collinearity graph Γ of $(Y, \mathfrak{M})$ is distance-transitive with $v = 45$ vertices, intersection array $\{6,4,2,1;1,1,4,6\}$ and distance distribution diagram*

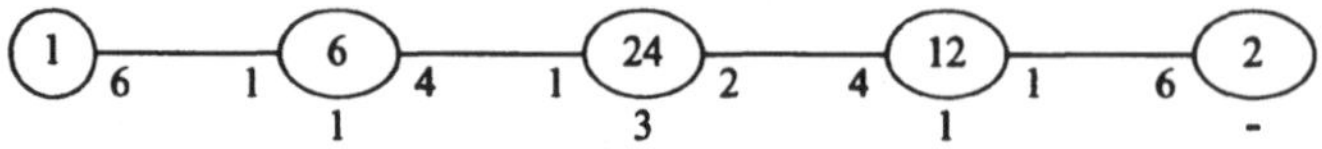

$v = 45$;

it has spectrum $6^1\ 3^{12}\ 1^9\ (-2)^{18}\ (-3)^5$; $\mathrm{Aut}\,\Gamma \cong 3{\cdot}\mathrm{Sym}(6)$ *with point stabilizer* $\mathrm{Sym}(4)\times 2$. *(Cf. Table 6.10, (A7).)*
The incidence graph Δ of $(Y, \mathfrak{M})$ is distance-transitive with $v = 90$ vertices, intersection array $\{3,2,2,2,2,1,1,1;1,1,1,1,2,2,2,3\}$ and distance distribution diagram

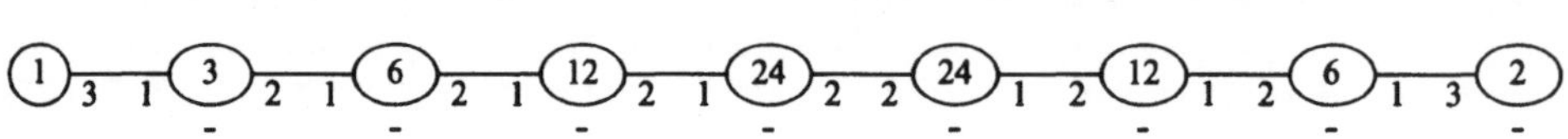

;

it has spectrum $\pm 3^1\ \pm\sqrt{6}^{12}\ \pm 2^9\ \pm 1^{18}\ 0^{10}$; $\mathrm{Aut}\,\Delta \cong 3{\cdot}\mathrm{Sym}(6).2$ *with point stabilizer* $\mathrm{Sym}(4)\times 2$. *(Cf. Table 6.9, (B14).) Both graphs are uniquely determined by their intersection arrays.*

Proof. If $L \in \mathfrak{L}$, then the projective 3-space L_0 meets Y in a regulus (3×3 grid) containing 3 lines x_0 obtained from points $x \in L \cap X$, and 3 lines in $\mathfrak{M}$. Now the only nonobvious fact we have to verify is that $(Y, \mathfrak{M})$ is connected. But if it were not, it would have a component Z isomorphic to $(X, \mathfrak{L})$, meeting each fiber $\pi^{-1}x$ ($x \in X$) in a single point. If K is a line of Π not meeting O, then $K_0 \cap Z$ is a 5-coclique that spans K_0, so $K_0 \subseteq \langle Z\rangle$. It follows that $Y \subseteq \langle Z\rangle$, so that $\langle Z\rangle = \Pi_0$. But the universal projective embedding of $(X, \mathfrak{L})$ has affine dimension 5, so Z cannot span Π_0, contradiction. Concerning uniqueness, cf. ITO [428]. □

The graph Δ is known as the *Foster graph*, cf. FOSTER [293], SMITH [680]. Δ is the unique 3-cover of the incidence graph of $GQ(2,2)$ without 8-gons.

Choosing coordinates, we can make this construction more explicit. An elegant way of doing so is provided by the *hexacode* (cf. the ATLAS [216], p. 4). The hexacode $\mathcal{H}$ is the linear code over $\mathbf{F}_4 = \{0,1,\omega,\bar\omega\}$ of dimension 3 and word length 6 spanned by the vectors (001111), ($0101\omega\bar\omega$) and ($1001\bar\omega\omega$). Viewed projectively, we see a projective plane Π of order 4, and the set O of the six projective points without zero coordinate is a hyperoval. Thus, we can construct $(Y, \mathfrak{M})$ as before. The code $\mathcal{H}$ is self-dual for the standard Hermitean inner product, so this inner product does not provide us with any structure; but if we change our mind and regard $0,1,\omega,\bar\omega$ as elements of $\mathbf{C}$ (of course with $\omega^2+\omega+1 = 0$) and use the Hermitean inner product $(x,y) = \sum_{i=1}^{6} \bar{x}_i y_i$, then the collinearity graph Γ of $(Y, \mathfrak{M})$ can be described as the graph on the set $\mathcal{H}_0$ of the 45 code words of $\mathcal{H}$ with precisely two nonzero coordinates, where two code words are adjacent when their inner product equals 2.

The hexacode provides us with another distance-regular graph. Indeed, the set of 18 code words without zero coordinate can be regarded as the set of blocks of a resolvable transversal design $RT[6,2;3]$, and the point-block incidence graph of this design is distance-regular with intersection array $\{6,5,4,1;1,2,5,6\}$ and distance distribution

diagram

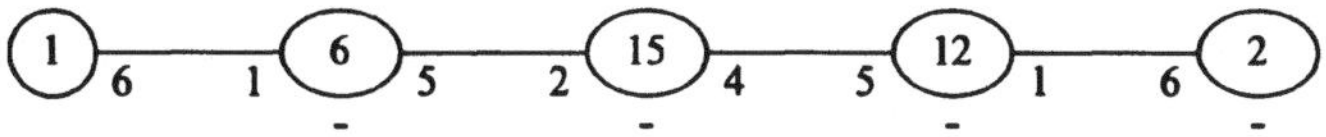

$v = 36$.

13.2.2. Theorem. *There is a unique distance-regular graph with intersection array* $\{6,5,4,1;1,2,5,6\}$. *It is distance-transitive with automorphism group* 3·Sym(6).2 *and point stabilizer* Sym(5).

Proof. Straightforward. □

B. The Conway-Smith graph for 3·Sym(7)

Continuing the discussion from the previous section, if we adjoin to the set $\mathcal{K}_0$ the 18 vectors $2\epsilon e_i$, where e_i is a unit vector $(1 \leqslant i \leqslant 6)$ and $\epsilon \in \{1,\omega,\bar{\omega}\}$, then, with the same definition for adjacency, we find a graph E on 63 vertices.

13.2.3. Theorem. *The graph* E *is distance-transitive with* $v = 63$ *vertices, intersection array* $\{10,6,4,1;1,2,6,10\}$ *and distance distribution diagram*

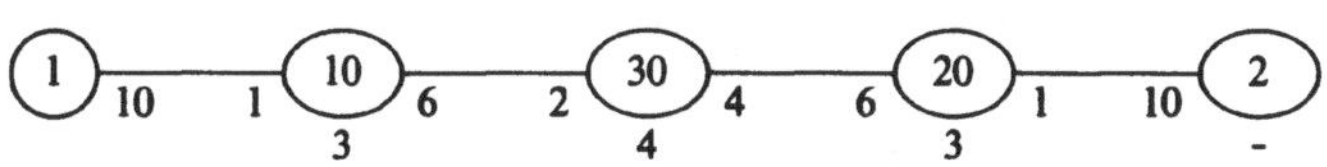

$v = 63$;

Aut E $\cong$ 3·Sym(7) *with point stabilizer* Sym(5)×2. *(Cf. Table 6.10, (A8).)* E *is locally Petersen, and an antipodal* 3*-cover of the graph* $\overline{T(7)}$. *This graph is uniquely determined by its intersection array.*

Proof. Uniqueness was shown by HALL [345]. Everything else is straightforward. □

Remarks. In both Γ (of the previous subsection) and E, two vectors have distance two precisely when they are orthogonal. The graphs Γ and E can be defined in terms of each other: The graph E has 7 subgraphs isomorphic to Γ (permuted by Sym(7)). Conversely, given Γ which is an antipodal 3-cover of $\overline{T(6)}$, we find 6 subgraphs that are 3-covers of $\overline{T(5)}$, and upon inspection these turn out to be disconnected, the disjoint union of three Petersen graphs; inventing 18 new points with these 18 Petersen graphs as neighbourhoods, we reconstruct E. SMITH [688] gives an explicit description of the group 3·Sym(7), due to Conway.

C. The locally *GQ*(4,2) graph on 3×126 points

In BLOKHUIS & BROUWER [89] it is shown as part of a classification of certain graphs that are locally a generalized quadrangle, that there exists a unique amply regular graph Γ with $\mu = 6$ that is locally $GQ(4,2)$. It is distance-regular with intersection array $\{45,32,12,1;1,6,32,45\}$. (Cf. Table 6.10, (A9).)

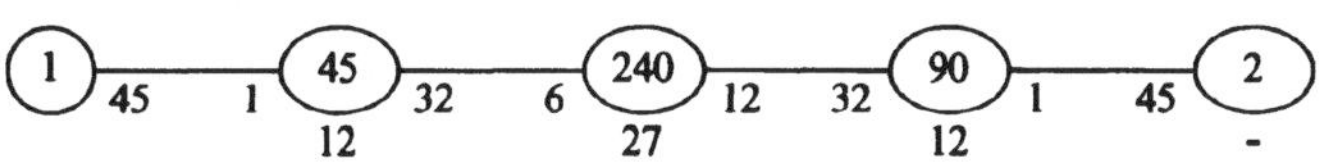

$v = 378$

Let Δ be the strongly regular graph on the 126 nonisotropic points of one kind in

$PG(5,3)$ provided with a quadratic form of non-maximal Witt index, where two points are adjacent whenever they are orthogonal. (The uniqueness of Δ as a Zara graph with $(v,K,e) = (126,6,2)$ is shown in BLOKHUIS & BROUWER [91].) In Δ, the μ-graphs are isomorphic to $3K_{3,3}$. We may construct an antipodal 3-cover Γ of Δ as follows.

Consider $PG(3,4)$ with a nondegenerate Hermitean form; we find 45 isotropic and 40 nonisotropic points. Let X be the set of isotropic points and B the set of 40 orthogonal bases (4 pairwise orthogonal nonisotropic points). Construct a graph Γ with vertex set $\{\infty_1, \infty_2, \infty_3\} \cup X_1 \cup X_2 \cup X_3 \cup \tilde{B}$, where X_j $(j = 1,2,3)$ are three copies of X, and $\tilde{B}$ is the set of 240 objects $(\{pq,rs\}, \{pr,qs\}, \{ps,qr\})$ for $\{p,q,r,s\} \in B$: the vectors of length 3 with as entries the three possible partitions of a basis into two pairs. We may identify a partition $\{pq,rs\}$ of a basis $\{p,q,r,s\}$ with a $K_{3,3}$ subgraph of X: the two lines pq and rs meet X in three points each, and these six points induce a $K_{3,3}$ subgraph in the generalized quadrangle on X. Let ∞_j be adjacent to the elements of X_j $(j = 1,2,3)$, and let two vertices of X_j be adjacent when they are orthogonal. Let $x \in X_j$ be adjacent to $b \in \tilde{B}$ when the point x lies in the $K_{3,3}$ determined by b_j. Finally, let $b, c \in \tilde{B}$ be adjacent when for some i $(1 \leqslant i \leqslant 3)$ the $K_{3,3}$'s determined by b_i and c_i have a point in common. This defines the graph Γ, and it is not difficult to show that it has all the properties claimed.

D. The $3.O_7(3)$ graph on 3×378 points

This graph is most easily described in terms of a system of complex vectors found in the ATLAS [216] (p. 108). Let ω be a primitive cube root of unity, and let V be the lattice $\mathbf{Z}[\omega]^{28}$ provided with the Hermitean form $(v,w) = \sum \bar{v}_i w_i$. Construct a graph Γ with $3 \times \binom{28}{2} = 1134$ vertices from the regular two-graph $(X, \mathfrak{D})$ on 28 vertices by taking as vertices the vectors $\eta(u(x,y) - u)$ in the hyperplane $\mathbf{1}^{\perp}$ of V, where η is a cube root of unity, $u = \frac{10+2\sqrt{-3}}{28}\mathbf{1}$ and for each unordered pair $\{x,y\}$ of points in X the vector $u(x,y)$ is determined by $u(x,y)_x = u(x,y)_y = \sqrt{-3}$ and $u(x,y)_z = 1$ if $\{x,y,z\}$ is a coherent triple, $u(x,y)_z = 0$ otherwise. Join two vectors v and w when $(v,w) = 4$.

13.2.4. Theorem. *The graph Γ is distance-transitive with $v = 1134$ vertices, intersection array $\{117,80,24,1; 1,12,80,117\}$ and distance distribution diagram*

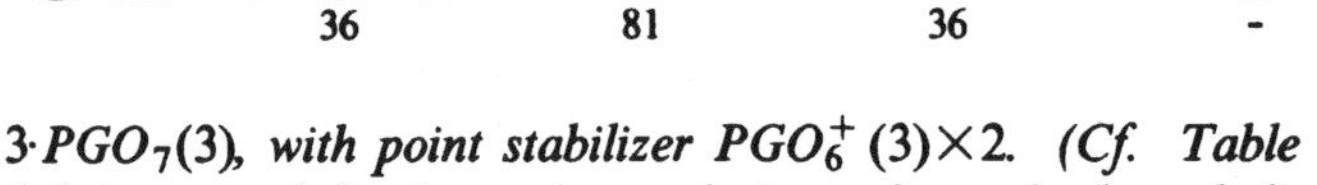

$v = 1134;$

$\mathrm{Aut}\,\Gamma \cong 3 \cdot PGO_7(3)$, with point stabilizer $PGO_6^+(3) \times 2$. (Cf. Table 6.10, (A10).) Γ is an antipodal 3-cover of the (strongly regular) graph on the hyperbolic points in $PG(6,3)$ provided with a nondegenerate quadric, where points are adjacent when they are orthogonal.

Proof. Straightforward. □

Remark. Distance and inner product are related by

$d(v,w)$	0	1	2	3	4
(v,w)	24	4	0	$4\omega, 4\bar{\omega}$	$24\omega, 24\bar{\omega}$.

13.3 The Perkel graph for $L(2,19)$

Let Γ be one of the two conjugacy classes in $G = PSL(2,19)$ of subgroups isomorphic to Alt(5), see HUPPERT [416]. This class is turned into a graph, called the *Perkel graph* (cf. NEUMANN [565, 567], BIGGS & GARDINER [81], PERKEL [589, 590]), by letting $\gamma, \delta \in \Gamma$ be adjacent whenever $\gamma \cap \delta$ contains an element of order 5. Since $PGL(2,19)$ interchanges the two conjugacy classes of G of subgroups isomorphic to Alt(5), the graph is uniquely defined up to isomorphism.

13.3.1. Theorem. *The Perkel graph is distance-transitive with intersection array* $\{6,5,2;1,1,3\}$ *and distance distribution diagram*

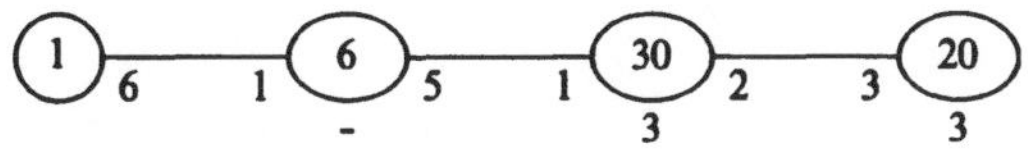

$v = 57.$

(Cf. Table 6.8, (D3).) It has spectrum 6^1 $(\frac{3\pm\sqrt{5}}{2})^{18}$ $(-3)^{20}$. *Its full automorphism group is* $PSL(2,19)$ *with point stabilizer* Alt(5).

Proof. By construction, G is transitive on Γ. Fix $\gamma \in \Gamma$. Since γ is a maximal subgroup of G and G is simple, and we have $N_G(\gamma) = \gamma$, so that $v = |G/\gamma| = 57$. Moreover, if x is a 5-group of γ, then $N_G(x) \cong D_{10} \times \mathbf{Z}_2$, the direct product of a dihedral group of order 10 and a cyclic group of order 2, while $N_\gamma(x) \cong D_{10}$. It follows that γ contains $|\gamma/N_\gamma(x)| = 6$ groups of order 5 and that each group of order 5 is in $|N_G(x)/N_\gamma(x)| = 2$ vertices of Γ. Thus, $k = 6$ and each subgroup of γ isomorphic to D_{10} belongs to a unique member of $\Gamma \setminus \{\gamma\}$. Hence, $\delta \in \Gamma(\gamma)$ if and only if $|\gamma \cap \delta| = 10$. Now let x_1, x_2 be two subgroups of γ isomorphic to D_{10}. If δ_i is the unique member of $\Gamma \setminus \{\gamma\}$ containing x_i for $i = 1, 2$, then $\delta_1 \cap \delta_2 \supseteq x_1 \cap x_2 \cong \mathbf{Z}_2$. Now $|\delta_1 \cap \delta_2| = 2$ for: any subgroup of order 4 or 6 is contained in a unique member of Γ and the induced graph on the five vertices of Γ containing an involution is a 5-circuit (it is a graph of valency 2 as the involution is contained in 5 dihedral subgroups of order 10 and the 5-group centralizing the involution acts nontrivially on the 5 vertices), so that $\delta_1 \cap \delta_2$ contains no 5-group. Consequently, $\delta \in \Gamma_2(\gamma)$ if and only if $\delta \cap \gamma \cong \mathbf{Z}_2$, and $\Gamma_2(\gamma)$ is a γ-orbit of size $|\gamma/\gamma \cap \delta| = 30$ whence $k_2 = 30$. Moreover, if x_1 is a subgroup of γ of order 5 and δ_1 is as above, then there are 5 other subgroups of γ of order 5, each accounting for a single neighbour of γ at distance 2 to δ_1. Therefore, $b_1 = 5$ and $c_2 = 1$. Since $\Gamma_{\leq 2}(\gamma)$ contains $1 + 6 + 30 \geq \frac{1}{2}v$ points, Γ is connected.

Next, let u be a subgroup of γ of order 3. Since $N_G(u) \cong \mathrm{Dih}_{18}$ and $N_\gamma(u) \cong \mathrm{Dih}_6$, the subgroup u is contained in exactly three members of Γ. If $\delta \in \Gamma \setminus \{\gamma\}$ is such that $\gamma \cap \delta$ contains u, then $\gamma \cap \delta = u$, so $|\gamma/(\gamma \cap \delta)| = 20 = v - |\Gamma_{\leq 2}(\gamma)|$. Therefore, Γ is distance-transitive, and $k_3 = 20$. As u does not fix an involution in γ, the number c_3 must be a (nonzero) multiple of $|u| = 3$. If $c_3 = 6$, then $b_2 = c_3k_3/k_2 = 4$, so $a_2 = k - b_2 - c_2 = 1$ and each path of length two would be contained in a single 5-circuit and $\Gamma_2(\gamma)$ would consist of 15 disjoint edges, whereas each path of length two with extremities in $\Gamma_3(\gamma)$ would lead to a pentagon with an edge in $\Gamma_2(\gamma)$. As each edge is then in exactly five 5-circuits, there can be at most $15 \times (5-1) = 60$ paths of length two with extremities in $\Gamma_3(\gamma)$, a contradiction with their total number $k_3c_3(b_2-1)/2 = 180$. Therefore $c_3 = 3$ and $b_2 = 2$, and the intersection array is determined. This proves the theorem. □

Let us give a few other descriptions of the same graph.

(i) First of all an explicit description in terms of $\mathbf{Z}_{19}$: take vertices α_i, β_i and γ_i for $i \in \mathbf{Z}_{19}$, and join α_i to $\beta_{i+2}, \beta_{i+5}, \beta_{i+7}, \gamma_{i+5}, \gamma_{i-4}, \gamma_{i-8}$; also join β_i to $\gamma_{i+7}, \gamma_{i-4}, \gamma_{i-5}$ ($i \in \mathbf{Z}_{19}$). This yields the Perkel graph, and shows that it is a union of three cocliques.

(ii) Next a description in $PG(2,19)$. According to HUPPERT [416], Satz II.8.13, $L_2(q)$ has a subgroup isomorphic to Alt(5) if and only if $q \equiv 0$, 1 or 4 (mod 5). For q not a power of 2 or 5, this can be seen geometrically as follows. Let $\tau = (1+\sqrt{5})/2$. Put $S := \{(0,1,\pm\tau), (\pm\tau,0,1), (1,\pm\tau,0)\}$. Then $S \cup -S$ considered as a subset of $\mathbf{R}^3$ is the set of vertices of an icosahedron, and the Alt(5)×2 acting on this icosahedron consists of orthogonal matrices with entries in $\mathbf{Z}[\tau,\frac{1}{2}]$. (For example, $\frac{1}{2}\begin{bmatrix} \tau^{-1} & -\tau & 1 \\ \tau & 1 & \tau^{-1} \\ -1 & \tau^{-1} & \tau \end{bmatrix}$ is an element of order 5 fixing $(0,1,\tau)$; note that $\tau^{-1} = \tau - 1$.) By our choice of q, all this can be carried out over $\mathbf{F}_q$. Thus, we find a 6-set S inside $PG(2,q)$, and an Alt(5) stabilizing S inside the $O_3(q)$ ($\cong L_2(q)$) that preserves the standard inner product on $PG(2,q)$. Let us call 6-subsets S of $PG(2,q)$ with stabilizer isomorphic to Alt(5) *icosahedrals*. Each Alt(5) in $O_3(q)$ can have at most one orbit of size 6 (since each involution in Alt(5) fixes 2 points in any such orbit), and thus there is a one-to-one correspondence between the $O_3(q)$-orbits of icosahedrals and the $O_3(q)$-conjugacy classes of Alt(5)'s.

All points of an icosahedral are exterior when $q \equiv 1$ (mod 5), and interior when $q \equiv 4$ (mod 5). In the former case, each exterior point is on $(q-1)/10$ icosahedrals from each orbit; in the latter case, each interior point is on $(q+1)/10$ icosahedrals from each orbit. Each orbit of icosahedrals has size $(q-1)q(q+1)/120$. The 15 lines meeting an icosahedral in two points are all hyperbolic if $q \equiv 1$ (mod 4), and elliptic if $q \equiv 3$ (mod 4).

In particular, for $q = 19$ we find two conjugacy classes of Alt(5)'s in $O_3(q)$, corresponding to the two orbits of icosahedrals represented by the above set S and its image under $\begin{bmatrix} 1 & 0 & 0 \\ 0 & 0 & 1 \\ 0 & 1 & 0 \end{bmatrix}$. Each icosahedral contains elliptic (or 'interior' or 'minus'-) points only, and each of the 171 elliptic points lies on two icosahedrals (from one orbit). The graph on the icosahedrals (from one orbit) where two icosahedrals are adjacent when they have a point in common (that is, when the corresponding Alt(5)'s have a D_{10} in common) turns out to be the Perkel graph.

(iii) The final description is based upon the observation that, in the Perkel graph Γ, the subgraphs $\Gamma_3(\gamma)$ are isomorphic to the dodecahedron. Recall that a Petrie polygon is a closed sequence of edges, no three successive ones in one face. We shall call a subpath of length l of a Petrie polygon a Petrie l-path. Fix an orientation of the dodecahedron. The vertices of our graph Γ will be: (i) a symbol ∞; (ii) the six Petrie 10-gons; (iii) the 30 Petrie 3-paths that are positively oriented, that is, of which the first two edges follow the direction of the face they are in (and the last two edges do not); (iv) the 20 vertices of the dodecahedron. Adjacencies are as follows: ∞ is adjacent to the six 10-gons; a 10-gon is adjacent to the five positively oriented 3-paths in it; a 3-path is adjacent to its two extremal vertices; two vertices of the dodecahedron are adjacent when joined by an edge of the dodecahedron. Finally, a Petrie 3-path $\pi = e'e\,e''$ is adjacent to three other 3-paths: let $\bar{e}$ be the edge opposite to e; then the first neighbour of π is the unique 3-path $\bar{\pi} = f'\bar{e}f''$ that is not in the same Petrie 10-gon as π; the remaining two neighbours of π are the two 3-paths distinct from $\bar{\pi}$ that contain f' resp. f''. This again yields the Perkel graph.

Remarks. An Alt(5) from the other conjugacy class has orbits of sizes $5+10+12+30$ on Γ, where the orbit of size 10 induces a Petersen graph, and the orbit of size 5 is a clique in Γ_3. (And conversely, each of the 57 Petersen subgraphs and each of the 57 5-cliques in Γ_3 arises in this way.) The distribution diagram for this partition follows.

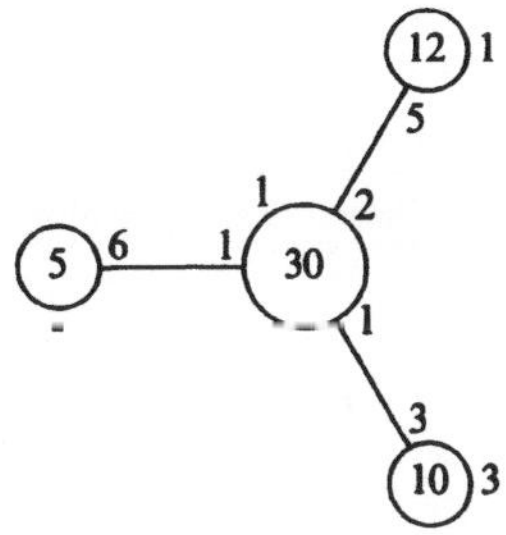

The graph Δ defined on the 2·57 subgroups of $PGL(2,19)$ isomorphic to Alt(5) by calling two such subgroups adjacent when they intersect in an Alt(4), is bipartite and has distance distribution diagram

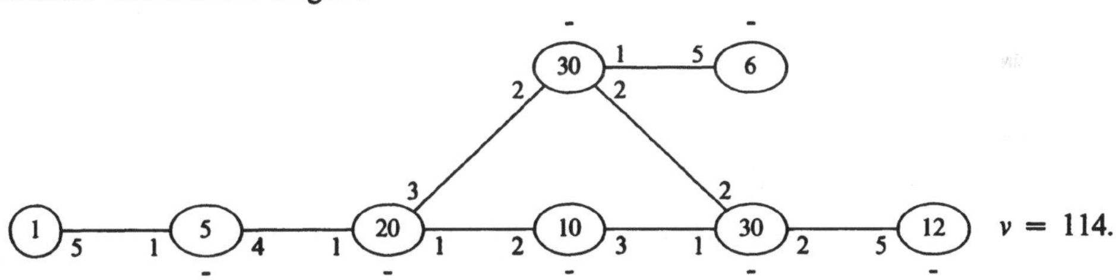

The distance-2 graph of the Perkel graph is a halved graph of Δ.
For related material, see COXETER [224], COXETER & WEISS [229] and VANDEN CRUYCE [762].

Problem. Is this graph uniquely determined by its intersection array? Neumaier showed uniqueness assuming a Frobenius group. FISHER [290] characterized this graph under certain additional assumptions by the fact that the design with as points the vertices of Γ and as blocks the 1-spheres $\Gamma(\gamma)$ for $\gamma \in \Gamma$ is a semi-symmetric design that is 'almost nexus': any point x is collinear to $e = 3$ points of any block not containing it, except one.

13.4 The Biggs-Smith graph for $L(2,17)$

13.4.1. Theorem. *There is a unique distance-regular graph Γ with intersection array $\{3,2,2,2,1,1,1;1,1,1,1,1,1,3\}$ and distance distribution diagram*

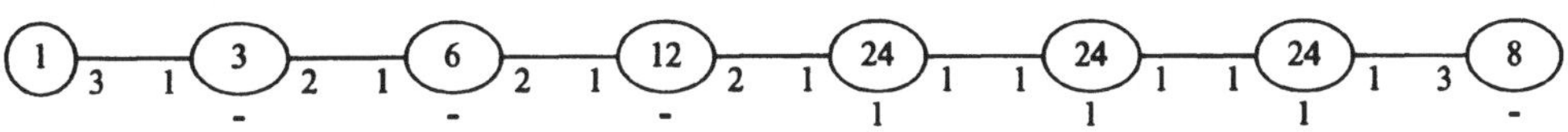

(cf. Table 6.8, (D14)). It has 102 *vertices and full automorphism group $PSL(2,17)$, acting distance transitively with point stabilizer* Sym(4). *It has diameter 7, girth 9 and does not contain 10-cycles. Its spectrum is* 3^1, 2^{18}, 0^{17}, $(\frac{1\pm\sqrt{17}}{2})^9$, θ_j^{16} $(j=1,2,3)$ *where the θ_j are the three roots of $\theta^3+3\theta^2-3=0$.*

Proof. A graph with these parameters can be constructed by taking as set of vertices a conjugacy class of subgroups of $PSL(2,17)$ isomorphic to Sym(4), and joining two vertices if these subgroups intersect in a group isomorphic to the dihedral group of order 8. In order to show uniqueness, fix a vertex ∞ and label each vertex γ of the graph with the ordered 6-tuple $(d(\gamma,\gamma_i))_{i=1}^{6}$, where the γ_i are the six vertices in $\Gamma_2(\infty)$. Since the γ_i fall into three pairs at mutual distance 2, the ordered 6-tuple will be written as an ordered triple of ordered pairs. It is easy to see what labels will occur: e.g., if $d(\gamma,\infty) = 4$ then γ has distances 3, 4 and 5 to the three neighbours of ∞ and distances 2,4; 3,5; 5,6 to the points γ_i, only the ordering remains to be determined. We shall see that not all $3!.2^3 = 48$ orderings occur, but only half of them. This motivates the following description of a graph Δ. Take as vertex set the following collection of ordered triples of ordered pairs; we list representatives for the orbits of Sym(4), where Sym(4) acts by arbitrarily permuting the triple and interchanging the elements of an even number of pairs (like its action on the edge set of K_4, the pairs corresponding to disjoint edges):

#	representative
1	(22,22,22)
3	(11,33,33)
6	(02,44,44)
12	(13,45,45)
24	(24,56,35)
24	(34,67,46)
24	(45,66′,57)
8	$(56,56,56)_i$ $(i = 1,2)$

(Here odd permutations of a triple in the last orbit change the index i.) Join two 6-tuples whenever the corresponding 6 elements are adjacent, where for $0 \leqslant i \leqslant 6$, i is adjacent to $i+1$ and for $i = 4,5,6$ also i adjacent to i. The 6′ behaves as 6 but serves to define the adjacency (45,66′,57) $\sim$ (45,75,6′6) (and (45,66′,57) $\not\sim$ (54,57,6′6), (54,75,66′),(45,57,66′)). Concerning $(56,56,56)_1$, it is adjacent to the three cyclic shifts of (45,66′,57), while $(56,56,56)_2$ is adjacent to the three cyclic shifts of (45,57,66′). This defines a graph Δ, regular of valency 3.

We claim that for an arbitrary graph Γ with the above intersection array and an arbitrary vertex ∞ the labelling necessarily is as described. The 6′ is defined by: if $d(\gamma,\infty) = 6$ and $d(\gamma,\gamma_i) = 6$ then write 6′ as i-coordinate of the label of γ whenever $d(\delta,\gamma_i) = 7$, for $\gamma \sim \delta \in \Gamma_5(\infty)$. Now for any γ with $d(\gamma,\infty) \geqslant 3$ and any δ with $\gamma \sim \delta$ we see that if γ has one of the allowed labels then also δ has. Since $\Gamma_{\geqslant 3}(\infty)$ is connected, this shows that only the labels indicated occur. Given the label of a vertex γ with $3 \leqslant d(\gamma,\infty) \leqslant 7$, the labels of its three neighbours are uniquely determined. Consequently, if a vertex δ had the same label as γ then walking along a geodesic from γ (resp. δ) to ∞ we would find distinct vertices γ' and δ' with the same label, both adjacent to a vertex ϵ, but this is impossible. Therefore, each of the labels indicated occurs precisely once. Finally we consider $\Gamma_7(\infty)$. Since Γ has girth 9, the paths

(45,66′,57) $\sim$ (34,67,46) $\sim$ (24,56,35) $\sim$ (35,56,24) $\sim$ (46,67,34) $\sim$ (57,66′,45)

and

(45,66′,57)$\sim$(34,67,46)$\sim$(24,56,35)$\sim$(13,45,45)$\sim$(24,35,56)$\sim$(34,46,67)$\sim$(45,57,66′)

show that if (45,66′,57) $\sim$ $(56,56,56)_1$ then $(56,56,56)_2$ $\sim$ (57,66′,45), (45,57,66′). By symmetry this determines all adjacencies for points of $\Gamma_7(\infty)$. This shows that any graph Γ with the above intersection array has the structure of Δ around any vertex , so there is a unique such graph and it has a transitive group. But the description given above shows that the stabilizer of a vertex ∞ in Γ is isomorphic to Sym(4) and acts transitively on $\Gamma_j(\infty)$, $0 \leqslant j \leqslant 7$. So Γ is a distance-transitive graph. □

Remark. See Brouwer [129]. Uniqueness of this graph was also proved by Boshier, cf. Biggs, Boshier & Shawe-Taylor [80].

We give several other descriptions of this graph.

(i) (Biggs & Smith [84])

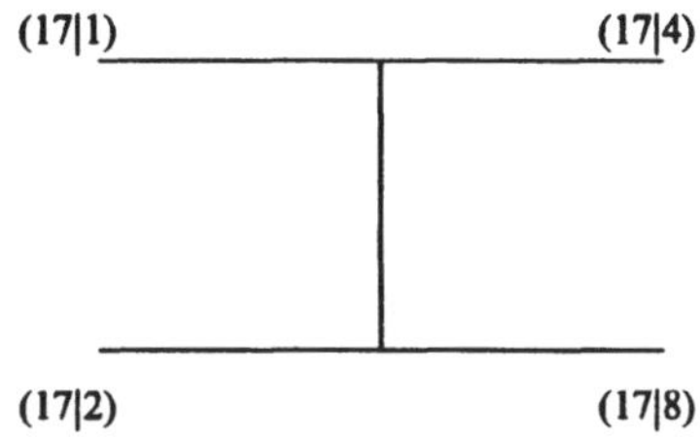

This picture describes the following graph: take 17 copies of this 'letter H', so that we have the $6 \cdot 17 = 102$ vertices. The middle vertices have valency 3 already; join the outer vertices in four 17-gons, where the vertices labelled $(17 \mid j)$ are in a 17-gon with vertices $\gamma_0^{(j)}, \ldots, \gamma_{16}^{(j)}$, where $\gamma_i^{(j)}$ is joined to $\gamma_{i \pm j}^{(j)}$ (indices mod 17).

(ii) Let C be the extended binary quadratic residue code of length 18. (For definitions, cf. van Lint [502]; for an explicit description, see Berlekamp, Conway & Guy [53], Vol. 2, p. 435.) It has weight enumerator

$$1 + 102X^6 + 153X^8 + 153X^{10} + 102X^{12} + X^{18}.$$

Now Γ_2 is isomorphic to the graph whose vertices are the 102 vectors of weight 6 and where two vectors are adjacent whenever they have Hamming distance 12. Note that Γ can be recovered form Γ_2 - in fact, for a vertex x the three Γ-neighbours form a Γ_2-triangle, the unique triangle at distance 4 from x in Γ_2. The correspondence between graph distance d_Γ and Hamming distance d_H is:

d_Γ	0	1	2	3	4	5	6	7
d_H	0	8	12	8	6	10	8	6

Since the full automorphism group of C is $PSL(2,17)$, the same holds for Γ.

(iii) Consider the projective plane $PG(2,17)$ provided with a nondegenerate quadratic form (conic). There are 204 orthogonal bases consisting of three exterior points, and $PGO(3,17)$ permutes them transitively. Under $O_3(17) \cong L_2(17)$ this set of bases splits into two orbits of size 102. If we call two bases adjacent when they have a point in common, then each basis is adjacent to three others in the same orbit, and to six in the other orbit. The graph induced on each orbit is the Biggs-Smith graph.

13.5 The Livingstone graph for J_1

Consider the first Janko group $G = J_1$ of order 175560 - see JANKO [441]. It has a single conjugacy class of subgroups isomorphic to $PSL(2,11)$ and a single conjugacy class of involutions. The centralizer $C_G(z)$ of an involution z has the shape $\mathrm{Alt}(5)\times\mathbf{Z}_2$. Take a subgroup L of G isomorphic to $PSL(2,11)$ and an involution i of G such that $C_G(i) \cap L \cong \mathrm{Alt}(5)$.
Then $\Gamma = \Gamma(G,L,i)$ (cf. §A.3) will be called the *Livingstone graph* (*associated with the first Janko group*), since the permutation representation of G on Γ was first discussed in LIVINGSTONE [507].

13.5.1. Theorem. *The Livingstone graph* Γ *is distance-transitive on* 266 *vertices with intersection array* $\{11,10,6,1;1,1,5,11\}$ *and distance distribution diagram*

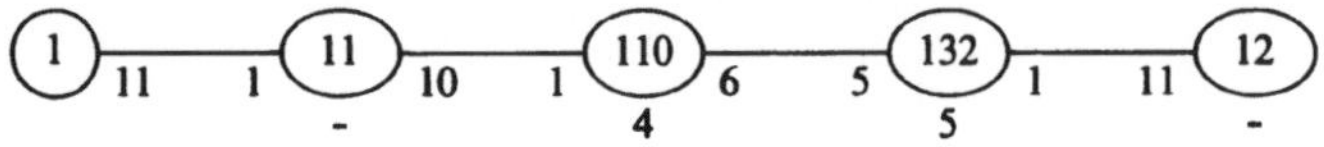

$v = 266.$

(Cf. Table 6.8, (D12).) $\mathrm{Aut}\,\Gamma \cong J_1$ *acts transitively on geodesics of any given length. The point stabilizer is* $L(2,11)$. *Its spectrum is* 11^1, 4^{76}, 1^{77}, $(\frac{-7\pm\sqrt{5}}{2})^{56}$. *If* $\gamma \in \Gamma$ *then any two vertices in* $\Gamma_4(\gamma)$ *have mutual distance* 3.

Proof. There are two conjugacy classes of subgroups isomorphic to Alt(5) in G, one of which is Γ. Two subgroups of L isomorphic to Alt(5) are conjugate in L if and only if they are conjugate in G, see JANKO [441]. Thus, there is a set S of 11 involutions j in G with $C_G(j) \cap L \cong \mathrm{Alt}(5)$, and these are all conjugate to i by L. We shall quote the description of the 266 cosets of L in G from CONWAY [212]. First of all, there is a permutation group isomorphism $\pi : S \to T$, where T is the L-orbit of the involution

$$\pi_i = (0,\infty)(1,2)(4,8)(5,10)(9,7)(3,6)$$

in the symmetric group on the projective line $\Omega = \mathbf{F}_{11} \cup \{\infty\}$, such that $\pi(i) = \pi_i$. (Here L acts on the symmetric group by conjugation.) For $j \in S$, we shall write $\pi_j = \pi(j)$. Note that for any two points $u,v \in \Omega$ there is a unique member of T interchanging them. If $i,j,k \in S$ are distinct, then $\pi_i\pi_j\pi_k$ has either no fixed point or exactly two fixed points in Ω. In the latter case, denote by $[i,j,k]$ the unique fixed point u such that the unique $h \in S$ for which $v = u\pi_h$ is the second fixed point, interchanges $u\pi_i$ and $v\pi_k$ (so that $v = [k,j,i]$). Now, let L_x for $x \in \Omega$ be the coset $Lijkh$ where $i,j,k,h \in S$ are distinct and satisfy $x = [i,j,k] = [h,i,j]$. Then L_x is well defined and the following rules for the multiplication of cosets by involution from S hold (see CONWAY [212], pp. 222-223) with $l,j,k,h \in S$ distinct.

$$\begin{cases} Ljl = Lj, & \\ Ljk = Ljl & \text{if } \pi_l\pi_j\pi_k \text{ has no fixed point in } \Omega, \\ Ljk = L_xh & \text{if } x = [l,j,k] = [h,l,j], \\ L_xjk = L_yh & \text{if } x = [h,k,j] \text{ and } y = [j,k,h], \\ L_xjk = Lhl & \text{if } x\pi_j = [l,j,k] = [h,l,j]. \end{cases} \tag{1}$$

The L-orbits of cosets with representative L, Li, Lij, L_xi, L_x for $j \in S$, $j \neq i$ and $x \in \Omega$ clearly have sizes at most 1, 11, 110, 132, 12, respectively. But their union is invariant under right multiplication by elements of $L \cup S$, so this must be the complete collection of 266 cosets of L in G, and the numbers above are the exact sizes of the L-orbits. By now, it is straightforward to establish from (1) that Γ is distance-transitive with the given intersection array. □

The distribution diagram of Γ_4 is interesting since it is a tree:

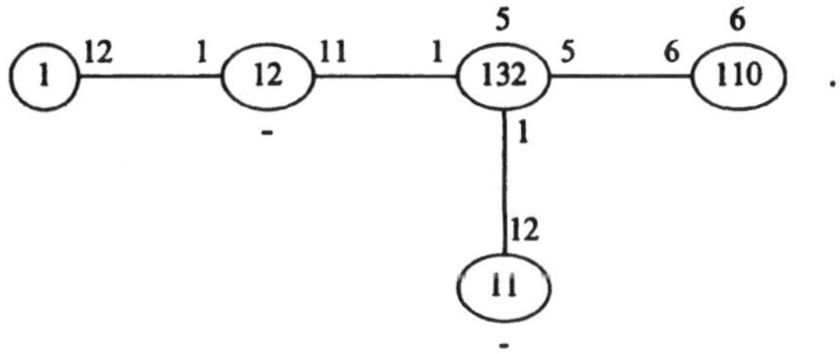

The distribution diagram of Γ around an edge is as shown below,

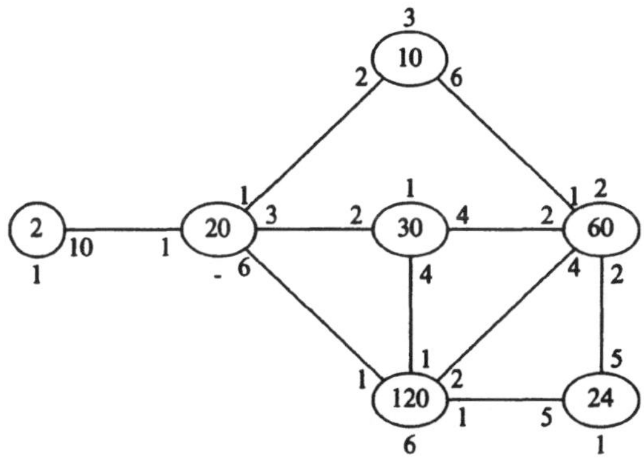

and this is also the orbit diagram: the edge stabilizer $2\times\mathrm{Alt}(5)$ has 7 orbits on vertices.

The group G has two orbits on pentagons (of sizes 2926 and 8778, with stabilizers $D_6\times D_{10}$ and $2\times D_{10}$, respectively), and if we call the pentagons in the small orbit 'special' then we find a (flag-transitive) geometry with Buekenhout diagram

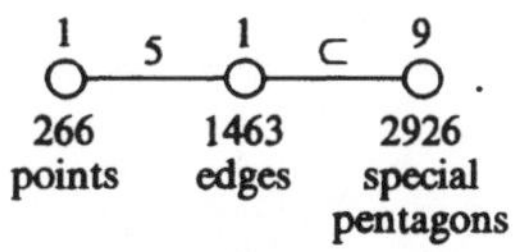

G has a single conjugacy class of involutions, of size 1463, with centralizer $2\times\mathrm{Alt}(5)$. Each involution interchanges the end points of a unique edge. Each involution fixes 10 vertices of Γ, and Γ induces a Petersen graph on such a set of 10 vertices. Each of the 1463 Petersen subgraphs arises in this way, and G_P induces $\mathrm{Alt}(5)$ on each Petersen subgraph P of Γ. Under $\mathrm{Alt}(5)$, the 12 pentagons of P fall into two orbits: 6 special and 6 non-special pentagons. Each special pentagon lies in three Petersen graphs, and each non-special pentagon in one. We find a (flag-transitive) geometry with Buekenhout diagram:

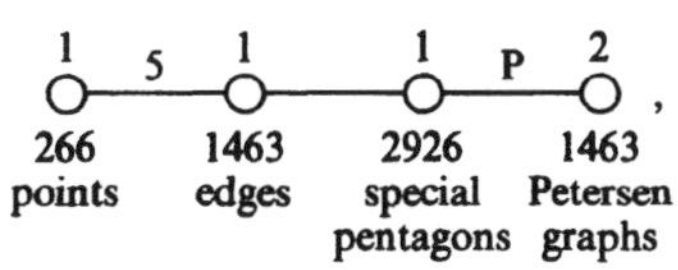

where P denotes the geometry of vertices and edges of the Petersen graph.

Also interesting are the 1045 subgraphs $7K_2$ consisting of 7 edges, pairwise in a Petersen graph. The stabilizer of such a subgraph is a maximal subgroup of G of shape $2^3{:}7{:}3$, and the 7 edges and the 7 Petersen graphs that meet the $7K_2$ in 3 edges form a Fano plane. The distribution diagram around such a subgraph is as shown below; again this is also the orbit diagram.

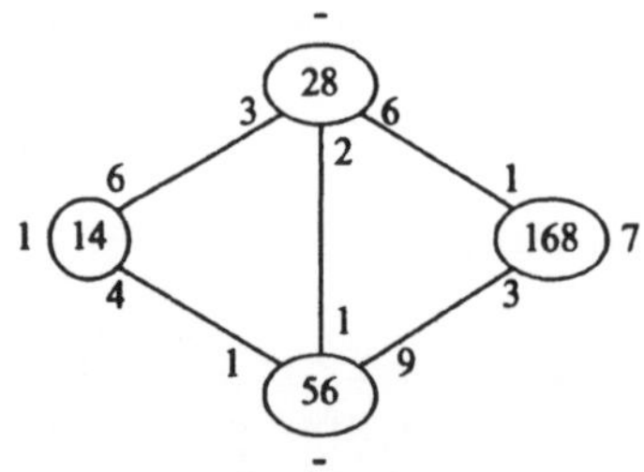

Problem. Is the Livingstone graph uniquely determined by its intersection array?

Remarks. The character table and the maximal subgroups of J_1 can be found in McKay [529] and Fischer & McKay [287], corrected in Ivanov, Klin & Faradjev [435]. The parameters of the rank 5 permutation representation of degree 266 are given by D.G. Higman [380]. Details on the geometry of pentagons and Petersen subgraphs can be found in Perkel [590], see also Ivanov [432], §6. An explicit description of the 266 cosets of L in G is given in Conway [212], cf. also the Atlas [216]. The intersections of maximal subgroups of J_1 are given in Komissartschik & Tsaranov [467].

13.6 The near octagon associated with the Hall-Janko group

13.6.1. Theorem. *There exists a unique distance-regular graph* Γ *with intersection array* $\{10,8,8,2;1,1,4,5\}$. *It has* $v = 315$ *vertices and is distance-transitive with* $\mathrm{Aut}\,\Gamma \cong \mathrm{Aut}(HJ)$. *Its distance distribution diagram is*

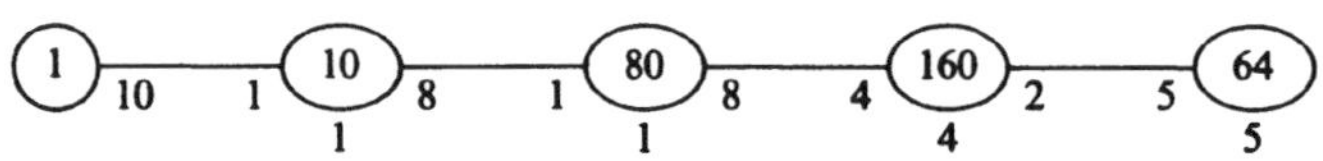

$v = 315.$

(Cf. Table 6.7, (N5).) The point stabilizer is $2_-^{1+4}.S_5$. *The graph* Γ *is a regular near octagon with line size three.*

Proof. The Hall-Janko simple group HJ of order 604800 (cf. Janko [442] and Hall & Wales [357, 358]) has a (unique) conjugacy class V_0 of 315 involutions whose centralizers contain Sylow 2-subgroups. Let Γ be the graph whose vertices are these 315 involutions, two involutions being adjacent when they commute, or, equivalently, when their product is again an element of V_0. Then Γ is distance-regular with intersection array $\{10,8,8,2;1,1,4,5\}$. The group HJ has index 2 in its automorphism group (cf. Tits [746]), and the two irreducible constituents of degree 14 in the permutation character of HJ on V_0 (cf. the Atlas [216]) are fused in $HJ.2$. Thus, HJ acts rank 6, and $HJ.2$ acts distance-transitively on Γ. Since $\lambda = 1$ it follows from Theorem 6.4.1 that the maximal cliques are the lines of a regular near octagon with line size three. Uniqueness is due to Cohen & Tits [207]. □

It is possible (cf. COHEN [199]) to present this near octagon without assuming any previous knowledge on HJ. In order to do so we exploit the quaternionic root system given in COHEN [198] (see also TITS [752] and WILSON [788]). Let $\mathbf{H} = \mathbf{R}1+\mathbf{R}i+\mathbf{R}j+\mathbf{R}k$ be the skew field of real quaternions whose multiplication is determined by $ij = k = -ji$ and $i^2 = j^2 = k^2 = -1$. The multiplicative subgroup $Q = \{\pm 1, \pm i, \pm j, \pm k\}$ is the well-known quaternion group. Set $\tau = (1+\sqrt{5})/2$ and $\zeta = (-1-i-j-k)/2$. (Then $\zeta^3 = 1$ and $i\zeta = \zeta k$, etc.)

The projective plane P over $\mathbf{H}$ is represented by homogeneous coordinates (i.e., left cosets) $[x] = x\mathbf{H}$ for $x \in \mathbf{H}^3 \setminus \{0\}$. Here $\mathbf{H}^3$ stands for the right vector space over $\mathbf{H}$ of dimension 3. Let A be the collineation group of P of order 1152 generated by the permutations of the homogeneous coordinates, scalar multiplication by ζ (acting on the left) and the transformations

$$\begin{bmatrix} x_1 \\ x_2 \\ x_3 \end{bmatrix} \mapsto \begin{bmatrix} px_1 \\ qx_2 \\ rx_3 \end{bmatrix} \quad \text{for } p,q,r \in Q \text{ with } pqr \in \{\pm 1\}$$

Write

$$V_1 = A\begin{bmatrix} 1 \\ 0 \\ 0 \end{bmatrix} \cup A\begin{bmatrix} 1 \\ 1 \\ 0 \end{bmatrix} \cup A\begin{bmatrix} 1 \\ \zeta^2+\tau \\ 1 \end{bmatrix} \cup A\begin{bmatrix} 1 \\ \zeta(1-\tau) \\ \zeta^2\tau \end{bmatrix}.$$

Then V_1 consists of four A-orbits having lengths 3, 24, 192, 96 respectively, so $|V_1| = 315$. Furthermore, the cosines of the angles between elements $[x], [y]$ of V_1 (i.e., $\sqrt{\frac{(x\,|\,y)(y\,|\,x)}{(x\,|\,x)(y\,|\,y)}}$, where $(x\,|\,y)$ stands for the standard unitary inner product on $\mathbf{H}^3$ of x and y linear in y) attain a limited number of values. The first A-orbit of V_1 consists of an *orthogonal triple,* i.e. a triple in which the cosine of the angle of each pair is 0. Denote by L_1 the set of all orthogonal triples of V_1.

13.6.2. Proposition. *Let* (V_1,L_1) *be as above. Then the point graph* Γ *of* (V_1,L_1) *is a regular near octagon with intersection array* $\{10,8,8,2;\,1,1,4,5\}$.

Proof. In view of Theorem 6.4.1, it suffices to establish that Γ is distance-regular with intersection array $\{10,8,8,2;\,1,1,4,5\}$. Consider the collineation s_X of P for $X = [x] \in V_1$ defined by

$$s_X(y) = y - 2x(x\,|\,y)/(x\,|\,x) \quad (y \in \mathbf{H}^3).$$

Each of the s_X $(X \in V_1)$ stabilizes V_1 and leaves the angles invariant, whence $s_X \in \mathrm{Aut}(\Gamma)$. Let H be the subgroup of $\mathrm{Aut}(\Gamma)$ generated by the s_X $(X \in V_1)$ and A. It is straightforward that H acts transitively of rank 6 on V_1, the orbitals being indexed by the angles. In fact, the distances and the angles of two points $X, Y \in V_1$ are related to the order of $s_X s_Y$ as indicated in the table

distance	0	1	2	3	4
cosine of angle	1	0	$1/\sqrt{2}$	$1/2$	$\tau/2, (\tau-1)/2$
order of $s_X s_Y$	1	2	4	3	5

Verification of distance-regularity is thus reduced to the straightforward check whether the intersection number c_4 for a pair of points from the orbital indexed by angle $\tau/2$ coincides with the number c_4 for a pair from the $(\tau-1)/2$-orbital. □

Remarks. (i) The group H constructed in the proof above is in fact HJ.
(ii) Γ_2 is the second subconstituent $\Delta_2(\infty)$ for a vertex ∞ of the strongly regular graph Δ with parameters $(v,k,\lambda,\mu) = (416,100,36,20)$ and $\mathrm{Aut}\,\Delta \cong G_2(4).2$. For each of the 100 vertices $\delta \in \Delta(\infty)$ we find a subset $\delta^\perp \cap \Delta_2(\infty)$ of size 63 of the vertex set of Γ, inducing the unitary nonisotropics graph ($G_2(2)$ generalized hexagon) in Γ. Also the first subconstituent $\Delta(\infty)$ of Δ is strongly regular; it has parameters $(v,k,\lambda,\mu) = (100,36,14,12)$.
(iii) $\Gamma_4(\gamma)$ has distance distribution diagram

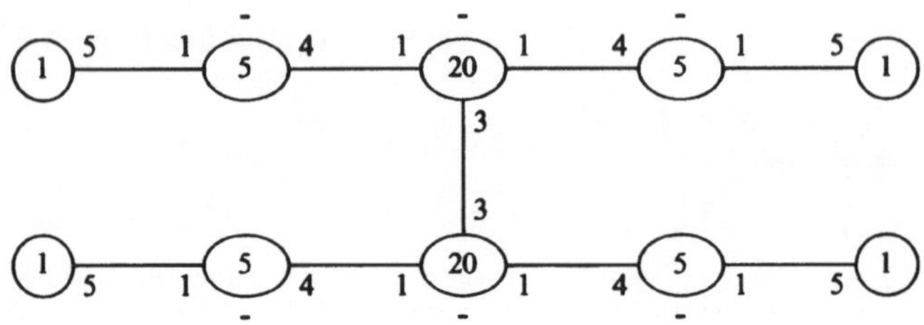

$v = 64;$

it is the unique double cover of the 5-cube without quadrangles (cf. Cohen & Tits [207], p. 23); it is also a double cover of Wells' graph, and, of course, a fourfold cover of the folded 5-cube, cf. §9.2E.
(iv) There is a feasible intersection array $\{12,8,6,4;1,1,2,9\}$ for a graph on 525 vertices, and one might think that this is realized by the line graph of (V_1,L_1). However, this line graph is not distance-regular (it has distance distribution diagram

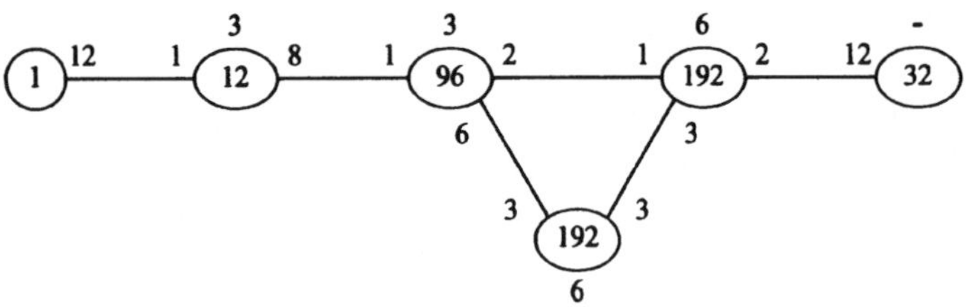

)

and in fact no distance-regular graph with the intersection array mentioned does exist (Cohen [199]).
[Indeed, if Γ is a distance-regular graph with $\mu = 1$, $c_3 = 2$, and $a_2 = \lambda + (\frac{k}{\lambda+1} - 1) \leqslant 2\lambda$ then Γ has singular lines of size $\lambda+2$, and any singular line meeting both $\Gamma_2(\gamma)$ and $\Gamma_3(\gamma)$ has two points in $\Gamma_2(\gamma)$; thus, if $\delta \in \Gamma_3(\gamma)$, then the four points in $C(\gamma,\delta) \setminus \{\gamma,\delta\}$ form a quadrangle, contradiction.]

13.7 The Patterson graph for Suz

By Suz we denote the sporadic simple group of Suzuki of order $2^{13}.3^7.5^2.7.11.13$.

13.7.1. Theorem. *Let Γ be the graph whose vertices are the 22880 centers (of order 3) of the Sylow 3-groups of* Suz *and in which two vertices are adjacent whenever they generate an abelian subgroup of order 3^2. Then Γ is distance-transitive with intersection array $\{280,243,144,10;1,8,90,280\}$, spectrum 280^1, 80^{364}, 20^{5940}, -8^{15795}, -28^{780}, and distance distribution diagram*

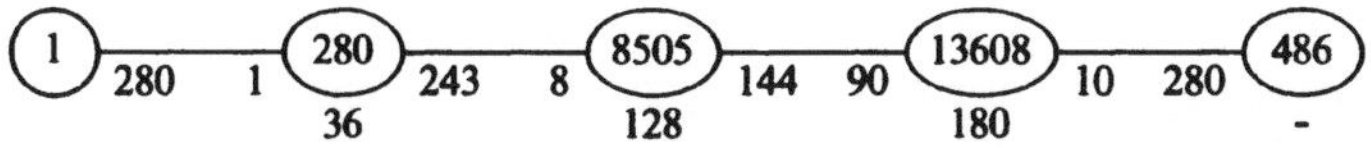

$v = 22880.$

(Cf. Table 6.8, (D13).) The automorphism group of Γ *is* Suz.2 *with point stabilizer* $3{\cdot}PSU(4,3).2^2$. *Moreover,* $q^4_{11} = q^4_{12} = q^4_{14} = 0$. Γ *is locally* $GQ(9,3)$.

Proof. Let γ, δ be two vertices of Γ. Then, by STELLMACHER [701] and PATTERSON & WONG [583], we have one of the following relations between γ and δ.

notation	$\langle\gamma,\delta\rangle$	$C_{\mathrm{Suz}}(\langle\gamma,\delta\rangle)$	$\mathrm{Suz}_{\gamma,\delta}$	$N_{\mathrm{Suz}}(\langle\gamma,\delta\rangle)$
A	$\gamma = \delta$	$3{\cdot}PSU(4,3)$	$3{\cdot}PSU(4,3).2$	$3{\cdot}PSU(4,3).2$
B	$\mathbf{Z}_3\times\mathbf{Z}_3$	$3^{2+4}2\mathrm{Alt}_4$	$3^{2+4}{:}2(\mathrm{Alt}_4\times 2^2)$	$3^{2+4}{:}2(\mathrm{Alt}_4\times 2^2).2$
C	$SL(2,3)$	$2^{1+4}P\Omega^+(4,2)$ $= 2^{1+4}\mathrm{Sym}_3\times\mathrm{Sym}_3$	$(2^{1+4}\mathrm{Sym}_3\times\mathrm{Sym}_3).2$	
D	Alt_5	Alt_6	$(2\times\mathrm{Alt}_6).2$	$(\mathrm{Alt}_5\times\mathrm{Alt}_6).2$
E	Alt_4	$L(3,4)$	$L(3,4).2$	$(\mathrm{Alt}_4\times L(3,4)).2$

From line A, we see that Γ has $v = |\mathrm{Suz}|\,/\,|3{\cdot}PSU(4,3).2| = 22880$ vertices. Furthermore, there are 1, 280, 8505, 13608, 486 vertices δ in relation A, B, C, D, E to γ, respectively, while each of these relations represents a suborbit of Suz. By definition $\delta \sim \gamma$ if and only if γ, δ are in relation B. The action of the stabilizer $3{\cdot}PSU(4,3).2$ on the 280 points of $\Gamma(\gamma)$ is as on the points of the generalized quadrangle of order (9,3), and two vertices of $\Gamma(\gamma)$ are adjacent whenever they are collinear in the generalized quadrangle. In particular, we obtain $a_1 = 36$ and $b_1 = 243$. If $\delta \in \Gamma_2(\gamma)$, then there is an $\epsilon \in \Gamma(\delta) \cap \Gamma(\gamma)$ such that δ, γ are noncommuting subgroups of $\mathrm{Suz}_\epsilon \cong 3{\cdot}PSU(4,3).2$. Thus $\langle\gamma,\delta\rangle \cong SL(2,3)$, whence γ, δ are in relation C, and $k_2 = 8505$. Consequently, $c_2 = kb_1\,/\,k_2 = 280{\cdot}243\,/\,8505 = 8$. In fact, since $\Gamma(\delta) \cap \Gamma(\gamma)$ is a graph of valency 4 without triangles on 8 vertices, it is $\cong K_{4,4}$.

Now $L(3,4)$ is transitive on the point set of the $PSU(4,3)$ generalized quadrangle. Thus vertices in relation E to γ are adjacent to vertices in only one relation, say X, with respect to γ. (Clearly $X = C$ or D.) On the other hand, according to WILSON [787], there is an embedding of Alt(6) in Suz with the property that the subgroups of order 3 fixing 3 letters correspond to points of Γ. Taking γ, δ, ϵ to correspond to (123), (456), (124), respectively, we see that γ, δ (resp. γ, ϵ and δ, ϵ) generate a subgroup isomorphic to $\mathbf{Z}_3^2$ (resp. Alt(4) and Alt(5)) so that they are in relation B (resp. E and D). It follows that $X = D$, whence ϵ has distance 4 and by connectedness, δ has distance 3 to γ. This shows that Γ is distance-regular, with $k_3 = 13608$, $k_4 = 486$ and $c_4 = 280$, $a_4 = 0$. Now $b_3 = k_4c_4\,/\,k_3 = 10$. It remains to determine c_3, b_2 and a_2. As Alt(6) has 3 orbits on the 280 points of the $PSU(4,3)$ generalized quadrangle, with sizes 10, 90, 180, the stabilizer of γ and $\delta \in \Gamma_3(\gamma)$ has 3 orbits of the same sizes. The orbit of size 10 being accounted for by $\Gamma(\delta) \cap \Gamma_4(\gamma)$, we see that $c_3 \in \{90,180\}$, so that $b_2 \in \{144,288\}$. But $b_2 = 288$ is clearly impossible. Consequently $b_2 = 144$, $a_2 = k - b_2 - c_2 = 128$, and $c_3 = 90$. The computation of the spectrum is straightforward. □

Remarks. (i) In the process of the proof it has become clear that Γ, when viewed as space with thin lines, is a nondegenerate space such that $\gamma^{\perp}/\gamma = GQ(9,3)$, $\{\gamma,\delta\}^{\perp} = GQ(1,3)$ when $d(\gamma,\delta) = 2$, and $|\{\gamma,\delta,\epsilon\}^{\perp}| = 2$ for each $\epsilon, \delta \in \Gamma_2(\gamma)$ with $\epsilon \in \Gamma(\delta)$. (As a consequence, any two vertices at distance 2 determine a unique geodetically closed subgraph $\cong K_{4,4,4}$.) This shows that thickness is necessary in the hypothesis of the Main Theorem of COHEN [203].

(ii) The maximal cliques have size 11 and stabilizer $3^5 M_{11}$. (They are all conjugate.) The distribution diagram with respect to a clique is as follows

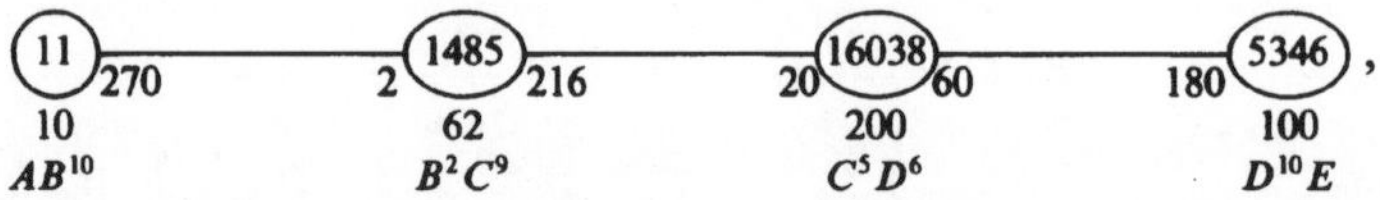

where the expressions underneath describe the relation of a vertex in the suborbit with respect to the 11 points in the maximal clique which is stabilized. In particular, a maximal clique is a completely regular code in Γ.

(iii) The graph appears in Patterson's thesis and in STELLMACHER [701]. More details can be found in PATTERSON & WONG [583]; the intersection array is mentioned in RONAN [619]. See also WILSON [787].

In Chapter 7 we have 'indicated' that all primitive distance-transitive graphs will be known before too long. In this light we wonder

Problem. Is it possible to determine all primitive distance-regular graphs with $d \geqslant 3$, or at least those with large d?

The tables in the next chapter may give an idea about our present knowledge regarding distance-regular graphs on at most 4096 points.

Chapter 14
Tables of Parameters for Distance Regular Graphs

This chapter contains tables of parameters for primitive distance-regular graphs with diameter 3 on at most 1024 vertices, for non-bipartite distance-regular graphs with diameter 4 on at most 4096 vertices, and for arbitrary distance-regular graphs of diameter at least 5 on at most 4096 vertices. In each category the parameter sets are ordered by k (not v). We only list intersection arrays that pass all feasibility criteria known to us. We do not give any information on the polygons (e.g., these have many P- and Q-polynomial structures).

Concerning the remarks about $p_{ij}^k = 0$ or $q_{ij}^k = 0$, we only list the non-obvious ones; for example, for bipartite graphs one has $p_{ij}^k = 0$ whenever $i+j+k$ is odd, and, similarly, for antipodal graphs $q_{ij}^k = 0$ when precisely two of the eigenspaces i, j and k belong to the quotient, and for antipodal 2-covers also when none of them belongs to the quotient (cf. §2.4). (Note that, in our ordering of eigenspaces, eigenspace i belongs to the quotient precisely when i is even.) For Q-polynomial graphs we indicate the ordering of the eigenspaces, unless it is $0, 1, 2, \cdots, d$. Since graphs with classical parameters are Q-polynomial (with ordering of eigenvalues $\theta_i = b^{-i}b_i - \left[{i \atop 1}\right]_b$, $i = 0, \cdots, d$, cf. Corollary 8.4.2), we suppress this latter information except when there are several Q-polynomial orderings. Of course, Q-polynomiality (with ordering $0', 1', \cdots, d'$) implies $q_{i'j'}^{h'} = 0$ for $h > i+j$ or $h < |i-j|$, and this is not listed separately. Note that '(generalized) Odd graph' implies $p_{ij}^h = 0$ for odd $i+j+h < 2d$, while 'distance 2 graph of generalized Odd graph' implies $p_{ij}^h = 0$ for the corresponding triples i, j, h and for triples whose corresponding triple violates the triangle inequality. (For the correspondence in distances, see §4.2D.)
The distance 1-or-2 graph $\Gamma_1 \cup \Gamma_2$ of Γ is denoted here by $\Gamma_{1,2}$.

Large diameter

Since for large diameter the tables are short, we start with those.
First the graphs with diameter $d \geqslant 9$ and valency $k > 2$ and $v \leqslant 4096$. Only the Doubled Odd graphs and the d-cubes occur; these are both antipodal and bipartite.

d	v	k	description
13	3432	7	Doubled Odd graph on 13 points
12	4096	12	Hamming 12-cube
11	924	6	Doubled Odd graph on 11 points
11	2048	11	Hamming 11-cube
10	1024	10	Hamming 10-cube
9	252	5	Doubled Odd graph on 9 points
9	512	9	Hamming 9-cube

(The n-cube has $v = 2^n$ vertices, diameter $d = n$, intersection array $\{n, n-1, \cdots, 1; 1, 2, \cdots, n\}$ and eigenvalues $n - 2j$ with multiplicity $\binom{n}{j}$ $(0 \leqslant j \leqslant n)$. It is self-dual, so that $p_{ij}^l = q_{ij}^l$ for all i, j, l. In particular, it is Q-polynomial.

The Doubled Odd graph on $2m+1$ points has $v = 2\binom{2m+1}{m}$ vertices, diameter $d = 2m+1$, intersection array $\{m+1,m,m,\cdots,1,1;1,1,\cdots,m,m,m+1\}$ and eigenvalues $\pm(m+1-j)$ with multiplicity $\binom{2m+1}{j}-\binom{2m+1}{j-1}$ $(0 \leq j \leq m)$.)

Next the three graphs with $d = 8$, $k > 2$, $v \leq 4096$. All are antipodal; the Foster graph and the Hamming cube are also bipartite. The first line for each graph gives $v = k_0 + k_1 + \cdots + k_d$ followed by the intersection array, the second line the spectrum, and the remaining lines give the name of the graph and various other remarks. The symbol preceding v is ! for graphs uniquely determined by their intersection array, + for intersection arrays such that all corresponding graphs are known (and there are at least two), ? for intersection arrays for which no graph is known, and a blank in the remaining cases.

Diameter 8

! $v = 90 = 1+3+6+12+24+24+12+6+2$ $\{3,2,2,2,2,1,1,1; 1,1,1,1,2,2,2,3\}$
3^1 2.449^{12} 2^9 1^{18} 0^{10} -1^{18} -2^9 -2.449^{12} -3^1
Foster graph (see §13.2A), bipartite, antipodal 3-cover of the incidence graph of $GQ(2,2)$

! $v = 990 = 1+7+42+168+336+336+84+14+2$ $\{7,6,4,4,4,1,1,1; 1,1,1,2,4,4,6,7\}$
7^1 5^{42} 4^{55} 2.372^{210} 1^{154} 0^{198} -3^{99} -3.372^{210} -4^{21}
Ivanov-Ivanov-Faradjev graph (see §11.4D), antipodal 3-cover of the M_{22}-graph on 330 vertices, $a_1 = a_7 = 0$, equality in absolute bound for $(i,j) = (8,8)$

! $v = 256 = 1+8+28+56+70+56+28+8+1$ $\{8,7,6,5,4,3,2,1; 1,2,3,4,5,6,7,8\}$
8^1 6^8 4^{28} 2^{56} 0^{70} -2^{56} -4^{28} -6^8 -8^1
Hamming 8-cube, bipartite, antipodal 2-cover; $\Gamma_{1,2}$ is distance-regular; classical $(b = 1)$; 2nd Q-polynomial ordering: 0 7 2 5 4 3 6 1 8; $q^6_{66} = 0$

Next the graphs with $d = 7$, $k > 2$, $v \leq 4096$. We separate them into the three categories 'primitive', 'antipodal but not bipartite', and 'both antipodal and bipartite'. Except for the Biggs-Smith graph, the Hamming graph 3^7 and the Johnson graph $J(14,7)$, all are bipartite antipodal 2-covers.

Diameter 7 - Primitive graphs

! $v = 102 = 1+3+6+12+24+24+24+8$ $\{3,2,2,2,1,1,1; 1,1,1,1,1,1,3\}$
3^1 2.562^9 2^{18} 0.879^{16} 0^{17} -1.347^{16} -1.562^9 -2.532^{16}
Biggs-Smith graph, cf. §13.5;
$q^3_{11} = q^5_{11} = q^6_{11} = q^7_{11} = q^6_{12} = q^6_{16} = q^6_{36} = q^6_{56} = q^7_{66} = q^7_{77} = 0$;
$a_1 = a_2 = a_3 = a_7 = p^3_{22} = p^5_{25} = p^7_{27} = p^7_{36} = p^7_{47} = p^7_{67} = 0$; equality in absolute bound for $(i,j) = (1,1), (6,6)$

! $v = 2187 = 1+14+84+280+560+672+448+128$ $\{14,12,10,8,6,4,2; 1,2,3,4,5,6,7\}$
14^1 11^{14} 8^{84} 5^{280} 2^{560} -1^{672} -4^{448} -7^{128}
Hamming graph 3^7; classical $(b = 1)$

Diameter 7 - Antipodal (but not bipartite) graphs

! $v = 3432 = 1+49+441+1225+1225+441+49+1$ $\{49,36,25,16,9,4,1; 1,4,9,16,25,36,49\}$
49^1 35^{13} 23^{77} 13^{273} 5^{637} -1^{1001} -5^{1001} -7^{429}
Johnson graph J(14,7); classical $(b = 1)$

Diameter 7 - Bipartite antipodal 2-covers

Here we meet the first parameter set for which existence has not been settled (on 352 vertices), and also the first parameter set for which uniqueness is unknown (on 2048 vertices).

! $v = 70 = 1+4+12+18+18+12+4+1$ $\{4,3,3,2,2,1,1;1,1,2,2,3,3,4\}$
$4^1\ 3^6\ 2^{14}\ 1^{14}\ -1^{14}\ -2^{14}\ -3^6\ -4^1$ $q_{11}^4 = q_{13}^6 = q_{46}^6 = 0$
Doubled Odd graph on 7 points; $\Gamma_{1,2}$ is distance-regular; equality in absolute bound for $(i,j) = (1,1), (6,6)$

! $v = 128 = 1+7+21+35+35+21+7+1$ $\{7,6,5,4,3,2,1;1,2,3,4,5,6,7\}$
$7^1\ 5^7\ 3^{21}\ 1^{35}\ -1^{35}\ -3^{21}\ -5^7\ -7^1$ $q_{16}^6 = q_{55}^6 = q_{66}^6 = 0$
Hamming 7-cube; $\Gamma_{1,2}$ is distance-regular; classical ($b = 1$)

? $v = 352 = 1+7+42+126+126+42+7+1$ $\{7,6,6,5,2,1,1;1,1,2,5,6,6,7\}$
$7^1\ 4^{32}\ 3^{66}\ 1^{77}\ -1^{77}\ -3^{66}\ -4^{32}\ -7^1$
2-cover of unknown near heptagon, cf. §4.2D

$v = 2048 = 1+22+231+770+770+231+22+1$ $\{22,21,20,16,6,2,1;1,2,6,16,20,21,22\}$
$22^1\ 10^{77}\ 6^{330}\ 2^{616}\ -2^{616}\ -6^{330}\ -10^{77}\ -22^1$ $q_{11}^6 = q_{66}^6 = 0$
Double coset graph of truncated Golay code, cf. §11.3F

! $v = 4096 = 1+23+253+1771+1771+253+23+1$ $\{23,22,21,20,3,2,1;1,2,3,20,21,22,23\}$
$23^1\ 9^{253}\ 7^{506}\ 1^{1288}\ -1^{1288}\ -7^{506}\ -9^{253}\ -23^1$ $q_{11}^6 = q_{66}^6 = 0$
Double coset graph of binary Golay code, cf. §11.3E

Then the graphs with $d = 6$, $k > 2$, $v \leqslant 4096$. We separate them into the four categories 'primitive', '(only) antipodal', '(only) bipartite', and 'both antipodal and bipartite'. The primitive graphs are all well known.

Diameter 6 - Primitive graphs

! $v = 189 = 1+4+8+16+32+64+64$ $\{4,2,2,2,2,2;1,1,1,1,1,2\}$
$4^1\ 3.449^{21}\ 2.414^{27}\ 1^{28}\ -0.414^{27}\ -1.449^{21}\ -2^{64}$ $p_{22}^2 = p_{23}^3 = p_{24}^4 = p_{33}^4 = 0$
Generalized dodecagon of order (2,1)

$v = 1456 = 1+6+18+54+162+486+729$ $\{6,3,3,3,3,3;1,1,1,1,1,2\}$
$6^1\ 5^{104}\ 3.732^{168}\ 2^{182}\ 0.268^{168}\ -1^{104}\ -2^{729}$ $p_{22}^2 = p_{23}^3 = p_{24}^4 = p_{33}^4 = 0$
Generalized dodecagon of order (3,1)

! $v = 1716 = 1+7+42+126+315+525+700$ $\{7,6,6,5,5,4;1,1,2,2,3,3\}$
$7^1\ 5^{65}\ 3^{429}\ 1^{429}\ -2^{572}\ -4^{208}\ -6^{12}$
Odd graph on 13 points; $\Gamma_{1,2}$ is distance-regular; Q-polynomial ordering: 0 6 1 5 2 4 3; equality in absolute bound for $(i,j) = (6,6)$

! $v = 729 = 1+12+60+160+240+192+64$ $\{12,10,8,6,4,2;1,2,3,4,5,6\}$
$12^1\ 9^{12}\ 6^{60}\ 3^{160}\ 0^{240}\ -3^{192}\ -6^{64}$
Hamming graph 3^6; classical ($b = 1$)

! $v = 4096 = 1+13+78+286+715+1287+1716$ $\{13,12,11,10,9,8;1,2,3,4,5,6\}$
$13^1\ 9^{78}\ 5^{715}\ 1^{1716}\ -3^{1287}\ -7^{286}\ -11^{13}$ $q_{44}^6 = q_{45}^5 = q_{55}^5 = q_{55}^6 = q_{66}^6 = 0$
Folded 13-cube; generalized Odd graph; $\Gamma_{1,2}$ is distance-regular; Q-polynomial; 2nd Q-polynomial ordering: 0 6 1 5 2 4 3

+ $v = 4096 = 1+18+135+540+1215+1458+729$ $\{18,15,12,9,6,3;1,2,3,4,5,6\}$
$18^1\ 14^{18}\ 10^{135}\ 6^{540}\ 2^{1215}\ -2^{1458}\ -6^{729}$
Hamming (or Doob) graph 4^6; classical ($b = 1$)

! $v = 1716 = 1+42+315+700+525+126+7$ $\{42,30,20,12,6,2;1,4,9,16,25,36\}$
$42^1\ 29^{12}\ 18^{65}\ 9^{208}\ 2^{429}\ -3^{572}\ -6^{429}$
Johnson graph J(13,6); dist. 2 in generalized Odd graph; classical ($b = 1$); equality in absolute bound for $(i,j) = (1,1)$

! $v = 3003 = 1+48+420+1120+1050+336+28$ $\{48,35,24,15,8,3;1,4,9,16,25,36\}$
$48^1\ 34^{13}\ 22^{77}\ 12^{273}\ 4^{637}\ -2^{1001}\ -6^{1001}$ $p_{36}^6 = p_{45}^6 = p_{46}^6 = p_{55}^5 = p_{55}^6 = p_{56}^6 = p_{66}^6 = 0$
Johnson graph J(14,6); classical ($b = 1$); equality in absolute bound for $(i,j) = (1,1)$

! $v = 4096 = 1+78+715+1716+1287+286+13$ $\{78,55,36,21,10,3;1,6,15,28,45,66\}$
$78^1\ 54^{13}\ 34^{78}\ 18^{286}\ 6^{715}\ -2^{1287}\ -6^{1716}$ $q_{11}^1 = q_{13}^3 = q_{15}^5 = q_{33}^3 = q_{33}^5 = 0$
Halved 13-cube; dist. 2 in generalized Odd graph; classical ($b = 1$); 2nd Q-polynomial ordering: 0 2 4 6 5 3 1

Diameter 6 - Antipodal (but not bipartite) graphs

? $v = 1518 = 1+15+210+840+420+30+2$ $\{15,14,12,6,1,1;1,1,3,12,14,15\}$

$15^1\ 6^{184}\ 4^{230}\ 1^{506}\ -3^{253}\ -5^{322}\ -8^{22}$ $q^6_{46} = 0;\ a_1 = a_5 = 0$

equality in absolute bound for $(i,j) = (6,6)$

! $v = 1024 = 1+21+210+560+210+21+1$ $\{21,20,16,6,2,1;1,2,6,16,20,21\}$

$21^1\ 9^{56}\ 5^{210}\ 1^{336}\ -3^{280}\ -7^{120}\ -11^{21}$ $q^6_{11} = q^6_{46} = q^6_{55} = q^6_{66} = 0;\ a_1 = a_5 = 0$

Coset graph of the once shortened and once truncated binary Golay code, cf. §11.3H; $\Gamma_{1,2}$ is distance-regular

! $v = 2048 = 1+21+210+1120+630+63+3$ $\{21,20,16,9,2,1;1,2,3,16,20,21\}$

$21^1\ 9^{168}\ 5^{210}\ 1^{1008}\ -3^{280}\ -7^{360}\ -11^{21}$ $q^5_{11} = q^6_{11} = q^6_{46} = q^6_{55} = q^6_{66} = 0;\ a_1 = a_5 = 0$

Coset graph of a subcode of the doubly truncated binary Golay code, cf. §11.3H

! $v = 2048 = 1+22+231+1540+231+22+1$ $\{22,21,20,3,2,1;1,2,3,20,21,22\}$

$22^1\ 8^{176}\ 6^{330}\ 0^{672}\ -2^{616}\ -8^{176}\ -10^{77}$ $q^6_{11} = q^6_{55} = q^6_{66} = 0;\ a_1 = a_2 = a_4 = a_5 = p^5_{24} = 0$

coset graph of the shortened binary Golay code, cf. §11.3H; $\Gamma_{1,2}$ is distance-regular

! $v = 924 = 1+36+225+400+225+36+1$ $\{36,25,16,9,4,1;1,4,9,16,25,36\}$

$36^1\ 24^{11}\ 14^{54}\ 6^{154}\ 0^{275}\ -4^{297}\ -6^{132}$

Johnson graph J(12,6); classical ($b = 1$)

! $v = 2048 = 1+66+495+924+495+66+1$ $\{66,45,28,15,6,1;1,6,15,28,45,66\}$

$66^1\ 44^{12}\ 26^{66}\ 12^{220}\ 2^{495}\ -4^{792}\ -6^{462}$

Halved 12-cube; classical ($b = 1$)

Diameter 6 - Bipartite (but not antipodal) graphs

! $v = 126 = 1+3+6+12+24+48+32$ $\{3,2,2,2,2,2;1,1,1,1,1,3\}$

$3^1\ 2.449^{21}\ 1.414^{27}\ 0^{28}\ -1.414^{27}\ -2.449^{21}\ -3^1$

Generalized dodecagon of order (1,2)

$v = 728 = 1+4+12+36+108+324+243$ $\{4,3,3,3,3,3;1,1,1,1,1,4\}$

$4^1\ 3^{104}\ 1.732^{168}\ 0^{182}\ -1.732^{168}\ -3^{104}\ -4^1$

Generalized dodecagon of order (1,3)

$v = 2730 = 1+5+20+80+320+1280+1024$ $\{5,4,4,4,4,4;1,1,1,1,1,5\}$

$5^1\ 3.464^{350}\ 2^{650}\ 0^{728}\ -2^{650}\ -3.464^{350}\ -5^1$

Generalized dodecagon of order (1,4)

? $v = 686 = 1+7+42+126+210+210+90$ $\{7,6,6,5,4,3;1,1,2,3,4,7\}$

$7^1\ 4.583^{42}\ 2.646^{210}\ 0^{180}\ -2.646^{210}\ -4.583^{42}\ -7^1$ $q^3_{11} = q^5_{13} = q^5_{35} = 0$

! $v = 2048 = 1+12+66+220+495+792+462$ $\{12,11,10,9,8,7;1,2,3,4,5,12\}$

$12^1\ 8^{66}\ 4^{495}\ 0^{924}\ -4^{495}\ -8^{66}\ -12^1$ $q^5_{35} = q^5_{44} = q^5_{45} = q^5_{55} = 0$

Folded 12-cube; Q-polynomial

Diameter 6 - Graphs that are both bipartite and antipodal

Note that by Corollary 5.8.3 the only bipartite antipodal 2-covers of diameter 6 are the 12-gon and the 6-cube. No other examples are known; we conjecture that no other graphs exist. In fact, by similar arguments one finds that any antipodal r-cover must satisfy $(r-1)\mu^2(\mu+2) = k\mu + k - \mu$, and that no feasible array exists for $3 \leqslant r \leqslant 13$. For $r = 14$ one finds the array $\{70,69,68,65,2,1;1,2,5,68,69,70\}$ with $v = 14\times 4832$ but the folded graph corresponds to a square 2-(2416,70,2) design and this fails Bruck-Chowla-Ryser. The smallest r for which parameters exist that also survive Bruck-Chowla-Ryser is $r = 42$, where $\mu = 9$, $c_3 = 87$, $k = 3654$ is feasible.

! $v = 64 = 1+6+15+20+15+6+1$ $6^1\ 4^6\ 2^{15}\ 0^{20}\ -2^{15}\ -4^6\ -6^1$ $\{6,5,4,3,2,1;1,2,3,4,5,6\}$

Hamming 6-cube; $\Gamma_{1,2}$ is distance-regular; classical ($b = 1$); 2nd Q-polynomial ordering: 0 5 2 3 4 1 6

Next the graphs with $d = 5$. The primitive graphs here all belong to the known infinite series.

Diameter 5 - Primitive graphs

! $v = 462 = 1+6+30+75+150+200$ $6^1\ 4^{44}\ 2^{165}\ -1^{132}\ -3^{110}\ -5^{10}$ $\{6,5,5,4,4; 1,1,2,2,3\}$
Odd graph on 11 points; $\Gamma_{1,2}$ is strongly regular; Q-polynomial ordering: 0 5 1 4 2 3; equality in absolute bound for $(i,j) = (5,5)$

! $v = 243 = 1+10+40+80+80+32$ $10^1\ 7^{10}\ 4^{40}\ 1^{80}\ -2^{80}\ -5^{32}$ $\{10,8,6,4,2; 1,2,3,4,5\}$
Hamming graph 3^5; classical ($b = 1$)

! $v = 1024 = 1+11+55+165+330+462$ $11^1\ 7^{55}\ 3^{330}\ -1^{462}\ -5^{165}\ -9^{11}$ $\{11,10,9,8,7; 1,2,3,4,5\}$
Folded 11-cube; $q^4_{44} = q^5_{44} = q^5_{55} = 0$; generalized Odd graph; $\Gamma_{1,2}$ is strongly regular; Q-polynomial; 2nd Q-polynomial ordering: 0 5 1 4 2 3

+ $v = 1024 = 1+15+90+270+405+243$ $15^1\ 11^{15}\ 7^{90}\ 3^{270}\ -1^{405}\ -5^{243}$ $\{15,12,9,6,3; 1,2,3,4,5\}$
Hamming (or Doob) graph 4^5; classical ($b = 1$)

! $v = 3125 = 1+20+160+640+1280+1024$ $20^1\ 15^{20}\ 10^{160}\ 5^{640}\ 0^{1280}\ -5^{1024}$ $\{20,16,12,8,4; 1,2,3,4,5\}$
Hamming graph 5^5; classical ($b = 1$)

! $v = 462 = 1+30+150+200+75+6$ $30^1\ 19^{10}\ 10^{44}\ 3^{110}\ -2^{165}\ -5^{132}$ $\{30,20,12,6,2; 1,4,9,16,25\}$
Johnson graph J(11,5); dist. 2 in generalized Odd graph; classical ($b = 1$); equality in absolute bound for $(i,j) = (1,1)$

! $v = 792 = 1+35+210+350+175+21$ $35^1\ 23^{11}\ 13^{54}\ 5^{154}\ -1^{275}\ -5^{297}$ $\{35,24,15,8,3; 1,4,9,16,25\}$
Johnson graph J(12,5); $p^5_{35} = p^5_{44} = p^5_{45} = p^5_{55} = 0$; classical ($b = 1$); equality in absolute bound for $(i,j) = (1,1)$

! $v = 1287 = 1+40+280+560+350+56$ $40^1\ 27^{12}\ 16^{65}\ 7^{208}\ 0^{429}\ -5^{572}$ $\{40,28,18,10,4; 1,4,9,16,25\}$
Johnson graph J(13,5); $p^5_{45} = p^5_{55} = 0$; classical ($b = 1$); equality in absolute bound for $(i,j) = (1,1)$

! $v = 2002 = 1+45+360+840+630+126$ $45^1\ 31^{13}\ 19^{77}\ 9^{273}\ 1^{637}\ -5^{1001}$ $\{45,32,21,12,5; 1,4,9,16,25\}$
Johnson graph J(14,5); $p^5_{55} = 0$; classical ($b = 1$); equality in absolute bound for $(i,j) = (1,1)$

! $v = 3003 = 1+50+450+1200+1050+252$ $\{50,36,24,14,6; 1,4,9,16,25\}$
$50^1\ 35^{14}\ 22^{90}\ 11^{350}\ 2^{910}\ -5^{1638}$
Johnson graph J(15,5); classical ($b = 1$); equality in absolute bound for $(i,j) = (1,1)$

! $v = 1024 = 1+55+330+462+165+11$ $\{55,36,21,10,3; 1,6,15,28,45\}$
$55^1\ 35^{11}\ 19^{55}\ 7^{165}\ -1^{330}\ -5^{462}$ $q^1_{11} = q^3_{13} = q^3_{33} = 0$
Halved 11-cube; dist. 2 in generalized Odd graph; classical ($b = 1$); 2nd Q-polynomial ordering: 0 2 4 5 3 1

Diameter 5 - Antipodal (but not bipartite) graphs

There is an infinite series of feasible intersection arrays

$$\{2\mu^2+\mu, 2\mu^2+\mu-1, \mu^2, \mu, 1; 1, \mu, \mu^2, 2\mu^2+\mu-1, 2\mu^2+\mu\}$$

for graphs with $v = 4\mu^2(2\mu+3)$ vertices and spectrum

$$k^1\ \theta^k\ \mu^{(2\mu-1)(2\mu+1)(2\mu+3)/3}\ 0^{2\mu(\mu+1)(2\mu-1)}\ (-2\mu)^{2\mu(\mu+1)(2\mu+1)/3}\ (-\theta)^k$$

where $k = \mu(2\mu+1)$ and $\theta = \mu\sqrt{2\mu+3}$. The first of these is the dodecahedron. The second is $\{10,9,4,2,1; 1,2,4,9,10\}$ and we saw in §11.4G that no such graph exists. For the remaining cases existence is undecided.

! $v = 20 = 1+3+6+6+3+1$ $3^1\ 2.236^3\ 1^5\ 0^4\ -2^4\ -2.236^3$ $\{3,2,1,1,1; 1,1,1,2,3\}$
Dodecahedron; $q^4_{11} = q^5_{45} = 0$; $a_1 = a_4 = 0$; equality in absolute bound for $(i,j) = (1,1), (3,3), (4,4), (5,1), (5,5)$

? $v = 324 = 1+21+140+140+21+1$ $21^1\ 9^{21}\ 3^{105}\ 0^{120}\ -6^{56}\ -9^{21}$ $\{21,20,9,3,1; 1,3,9,20,21\}$
$q^4_{11} = q^5_{45} = 0$; $a_1 = a_4 = 0$

$v = 729 = 1+22+220+440+44+2$ $\{22,20,18,2,1; 1,2,9,20,22\}$
$22^1\ 7^{132}\ 4^{132}\ -2^{330}\ -5^{110}\ -11^{24}$ $q^5_{11} = q^5_{35} = q^5_{45} = 0$
coset graph of the shortened extended ternary Golay code, cf. §11.3H

! $v = 252 = 1+25+100+100+25+1$ $25^1\ 15^9\ 7^{35}\ 1^{75}\ -3^{90}\ -5^{42}$ $\{25,16,9,4,1; 1,4,9,16,25\}$
Johnson graph J(10,5); classical ($b = 1$)

? $v = 704 = 1+36+315+315+36+1$ $\{36,35,16,4,1; 1,4,16,35,36\}$
$36^1\ 13.266^{36}\ 4^{231}\ 0^{280}\ -8^{120}\ -13.266^{36}$ $q^4_{11} = q^5_{45} = 0$; $a_1 = a_4 = 0$

! $v = 512 = 1+45+210+210+45+1$ $45^1\ 27^{10}\ 13^{45}\ 3^{120}\ -3^{210}\ -5^{126}$ $\{45,28,15,6,1; 1,6,15,28,45\}$
Halved 10-cube; classical ($b = 1$)

? $v = 2352 = 1+50+1125+1125+50+1$ $\{50,45,24,2,1; 1,2,24,45,50\}$
$50^1\ 14^{175}\ 8^{500}\ 0^{756}\ -6^{675}\ -10^{245}$

? $v = 3528 = 1+50+1125+2250+100+2$ $\{50,45,32,2,1; 1,2,16,45,50\}$
$50^1\ 14^{350}\ 8^{500}\ 0^{1512}\ -6^{675}\ -10^{490}$

? $v = 1300 = 1+55+594+594+55+1$ $\{55,54,25,5,1;1,5,25,54,55\}$
$55^1\ 18.028^{55}\ 5^{429}\ 0^{540}\ -10^{220}\ -18.028^{55}$ $q_{11}^4 = q_{45}^5 = 0;\ a_1 = a_4 = 0$

? $v = 2160 = 1+78+1001+1001+78+1$ $\{78,77,36,6,1;1,6,36,77,78\}$
$78^1\ 23.238^{78}\ 6^{715}\ 0^{924}\ -12^{364}\ -23.238^{78}$ $q_{11}^4 = q_{45}^5 = 0;\ a_1 = a_4 = 0$

? $v = 2912 = 1+105+1350+1350+105+1$ $\{105,90,49,7,1;1,7,49,90,105\}$
$105^1\ 35^{78}\ 14^{480}\ 0^{1248}\ -7^{975}\ -21^{130}$ $q_{12}^5 = 0$

? $v = 3332 = 1+105+1560+1560+105+1$ $\{105,104,49,7,1;1,7,49,104,105\}$
$105^1\ 28.862^{105}\ 7^{1105}\ 0^{1456}\ -14^{560}\ -28.862^{105}$ $q_{11}^4 = q_{45}^5 = 0;\ a_1 = a_4 = 0$

Diameter 5 - Bipartite (but not antipodal) graphs

$v = 310 = 1+7+42+84+112+64$ $7^1\ 4.243^{30}\ 2^{124}\ -2^{124}\ -4.243^{30}\ -7^1$ $\{7,6,6,4,4;1,1,3,3,7\}$
Double Grassmannian (q = 2); $q_{11}^3 = q_{12}^4 = q_{34}^4 = 0$

? $v = 650 = 1+9+72+288+252+28$ $9^1\ 4^{168}\ 1^{156}\ -1^{156}\ -4^{168}\ -9^1$ $\{9,8,8,7,1;1,1,2,8,9\}$
$p_{25}^5 = 0$

! $v = 512 = 1+10+45+120+210+126$ $10^1\ 6^{45}\ 2^{210}\ -2^{210}\ -6^{45}\ -10^1$ $\{10,9,8,7,6;1,2,3,4,10\}$
Folded 10-cube; $q_{34}^4 = q_{44}^4 = 0$; Q-polynomial

$v = 2420 = 1+13+156+468+1053+729$ $\{13,12,12,9,9;1,1,4,4,13\}$
$13^1\ 6.928^{120}\ 3^{1089}\ -3^{1089}\ -6.928^{120}\ -13^1$ $q_{11}^3 = q_{12}^4 = q_{34}^4 = 0$
Double Grassmannian (q = 3)

? $v = 3500 = 1+55+594+1485+1155+210$ $\{55,54,50,35,10;1,5,20,45,55\}$
Q-polynomial $55^1\ 20^{132}\ 5^{1617}\ -5^{1617}\ -20^{132}\ -55^1$ $q_{34}^4 = q_{44}^4 = 0$

Diameter 5 - Graphs that are both antipodal and bipartite

There is an infinite series of feasible intersection arrays

$$\{2\mu^2+\mu, 2\mu^2+\mu-1, 2\mu^2, \mu, 1;\ 1, \mu, 2\mu^2, 2\mu^2+\mu-1, 2\mu^2+\mu\}.$$

The first of these is the Desargues graph, the bipartite double of the Petersen graph. The second is the bipartite double of the Gewirtz graph. All others are unknown.

! $v = 20 = 1+3+6+6+3+1$ $3^1\ 2^4\ 1^5\ -1^5\ -2^4\ -3^1$ $\{3,2,2,1,1;1,1,2,2,3\}$
Doubled Odd graph on 5 points; Desargues graph; $\Gamma_{1,2}$ is strongly regular; equality in absolute bound for $(i,j) = (1,1), (4,4)$

! $v = 32 = 1+5+10+10+5+1$ $5^1\ 3^5\ 1^{10}\ -1^{10}\ -3^5\ -5^1$ $\{5,4,3,2,1;1,2,3,4,5\}$
Hamming 5-cube; $q_{44}^4 = 0$; $\Gamma_{1,2}$ is strongly regular; classical ($b = 1$)

! $v = 100 = 1+7+42+42+7+1$ $7^1\ 3^{21}\ 2^{28}\ -2^{28}\ -3^{21}\ -7^1$ $\{7,6,6,1,1;1,1,6,6,7\}$
Double Hoffman-Singleton, cf. §13.1

! $v = 112 = 1+10+45+45+10+1$ $10^1\ 4^{20}\ 2^{35}\ -2^{35}\ -4^{20}\ -10^1$ $\{10,9,8,2,1;1,2,8,9,10\}$
Double Gewirtz

! $v = 154 = 1+16+60+60+16+1$ $16^1\ 6^{21}\ 2^{55}\ -2^{55}\ -6^{21}\ -16^1$ $\{16,15,12,4,1;1,4,12,15,16\}$
Double 77-graph

? $v = 324 = 1+21+140+140+21+1$ $21^1\ 6^{56}\ 3^{105}\ -3^{105}\ -6^{56}\ -21^1$ $\{21,20,18,3,1;1,3,18,20,21\}$

! $v = 200 = 1+22+77+77+22+1$ $22^1\ 8^{22}\ 2^{77}\ -2^{77}\ -8^{22}\ -22^1$ $\{22,21,16,6,1;1,6,16,21,22\}$
Double Higman-Sims; $q_{44}^4 = 0$; Q-polynomial

? $v = 352 = 1+25+150+150+25+1$ $25^1\ 7^{55}\ 3^{120}\ -3^{120}\ -7^{55}\ -25^1$ $\{25,24,21,4,1;1,4,21,24,25\}$

? $v = 704 = 1+26+325+325+26+1$ $\{26,25,24,2,1;1,2,24,25,26\}$
$26^1\ 6^{143}\ 4^{208}\ -4^{208}\ -6^{143}\ -26^1$

? $v = 420 = 1+33+176+176+33+1$ $33^1\ 9^{55}\ 3^{154}\ -3^{154}\ -9^{55}\ -33^1$ $\{33,32,27,6,1;1,6,27,32,33\}$

? $v = 704 = 1+36+315+315+36+1$ $\{36,35,32,4,1;1,4,32,35,36\}$
$36^1\ 8^{120}\ 4^{231}\ -4^{231}\ -8^{120}\ -36^1$

? $v = 1408 = 1+37+666+666+37+1$ $\{37,36,35,2,1;1,2,35,36,37\}$
$37^1\ 7^{296}\ 5^{407}\ -5^{407}\ -7^{296}\ -37^1$

? $v = 532 = 1+45+220+220+45+1$ $\{45,44,36,9,1;1,9,36,44,45\}$
$45^1\ 12^{56}\ 3^{209}\ -3^{209}\ -12^{56}\ -45^1$

? $v = 784 = 1+46+345+345+46+1$ $\{46,45,40,6,1;1,6,40,45,46\}$
$46^1\ 10^{115}\ 4^{276}\ -4^{276}\ -10^{115}\ -46^1$

? $v = 1276 = 1+49+588+588+49+1$ $\{49,48,45,4,1;1,4,45,48,49\}$
$49^1\ 9^{231}\ 5^{406}\ -5^{406}\ -9^{231}\ -49^1$

? $v = 2552 = 1+50+1225+1225+50+1$ $\{50,49,48,2,1;1,2,48,49,50\}$
$50^1\ 8^{550}\ 6^{725}\ -6^{725}\ -8^{550}\ -50^1$

? $v = 1300 = 1+55+594+594+55+1$ $\{55,54,50,5,1;1,5,50,54,55\}$
$55^1\ 10^{220}\ 5^{429}\ -5^{429}\ -10^{220}\ -55^1$

? $v = 648 = 1+57+266+266+57+1$ $\{57,56,45,12,1;1,12,45,56,57\}$
Q-polynomial $57^1\ 15^{57}\ 3^{266}\ -3^{266}\ -15^{57}\ -57^1$ $q^4_{44} = 0$

? $v = 2146 = 1+64+1008+1008+64+1$ $\{64,63,60,4,1;1,4,60,63,64\}$
$64^1\ 10^{406}\ 6^{666}\ -6^{666}\ -10^{406}\ -64^1$

? $v = 1104 = 1+76+475+475+76+1$ $\{76,75,64,12,1;1,12,64,75,76\}$
$76^1\ 16^{114}\ 4^{437}\ -4^{437}\ -16^{114}\ -76^1$

? $v = 2160 = 1+78+1001+1001+78+1$ $\{78,77,72,6,1;1,6,72,77,78\}$
$78^1\ 12^{364}\ 6^{715}\ -6^{715}\ -12^{364}\ -78^1$

? $v = 3404 = 1+81+1620+1620+81+1$ $\{81,80,77,4,1;1,4,77,80,81\}$
$81^1\ 11^{666}\ 7^{1035}\ -7^{1035}\ -11^{666}\ -81^1$

? $v = 1600 = 1+85+714+714+85+1$ $\{85,84,75,10,1;1,10,75,84,85\}$
$85^1\ 15^{204}\ 5^{595}\ -5^{595}\ -15^{204}\ -85^1$

? $v = 1334 = 1+96+570+570+96+1$ $\{96,95,80,16,1;1,16,80,95,96\}$
$96^1\ 20^{115}\ 4^{551}\ -4^{551}\ -20^{115}\ -96^1$

? $v = 3300 = 1+97+1552+1552+97+1$ $\{97,96,91,6,1;1,6,91,96,97\}$
$97^1\ 13^{582}\ 7^{1067}\ -7^{1067}\ -13^{582}\ -97^1$

? $v = 2356 = 1+99+1078+1078+99+1$ $\{99,98,90,9,1;1,9,90,98,99\}$
$99^1\ 15^{341}\ 6^{836}\ -6^{836}\ -15^{341}\ -99^1$

? $v = 3332 = 1+105+1560+1560+105+1$ $\{105,104,98,7,1;1,7,98,104,105\}$
$105^1\ 14^{560}\ 7^{1105}\ -7^{1105}\ -14^{560}\ -105^1$

? $v = 1568 = 1+116+667+667+116+1$ $\{116,115,96,20,1;1,20,96,115,116\}$
Q-polynomial $116^1\ 24^{116}\ 4^{667}\ -4^{667}\ -24^{116}\ -116^1$ $q^4_{44} = 0$

? $v = 2916 = 1+141+1316+1316+141+1$ $\{141,140,126,15,1;1,15,126,140,141\}$
$141^1\ 21^{329}\ 6^{1128}\ -6^{1128}\ -21^{329}\ -141^1$

? $v = 2380 = 1+145+1044+1044+145+1$ $\{145,144,125,20,1;1,20,125,144,145\}$
$145^1\ 25^{203}\ 5^{986}\ -5^{986}\ -25^{203}\ -145^1$

? $v = 4004 = 1+161+1840+1840+161+1$ $\{161,160,147,14,1;1,14,147,160,161\}$
$161^1\ 21^{506}\ 7^{1495}\ -7^{1495}\ -21^{506}\ -161^1$

? $v = 2788 = 1+175+1218+1218+175+1$ $\{175,174,150,25,1;1,25,150,174,175\}$
$175^1\ 30^{204}\ 5^{1189}\ -5^{1189}\ -30^{204}\ -175^1$

? $v = 3200 = 1+205+1394+1394+205+1$ $\{205,204,175,30,1;1,30,175,204,205\}$
Q-polynomial $205^1\ 35^{205}\ 5^{1394}\ -5^{1394}\ -35^{205}\ -205^1$ $q^4_{44} = 0$

Diameter 4 - Primitive graphs

There is an infinite series of feasible formally self-dual intersection arrays

$$\{\mu(2\mu+1), (\mu-1)(2\mu+1), \mu^2, \mu;\ 1, \mu, \mu(\mu-1), \mu(2\mu+1)\}$$

with $v = 8\mu^2(\mu+1)$ and spectrum

$$k^1\ \mu(1+\sqrt{2\mu+2})^{k_1}\ \mu^{k_2}\ -\mu^{k_3}\ \mu(1-\sqrt{2\mu+2})^{k_4}.$$

These graphs have two P-polynomial and two Q-polynomial structures. For $\mu \leqslant 3$ no such graph exists; for $\mu > 3$ none is known.
Irrational eigenvalues occur also for the generalized octagons of order $(s, 1)$; their spectrum is

$$2s^1\ (s-1+\sqrt{2s})^{\frac{1}{2}s(s+1)^2}\ (s-1)^{s(s^2+1)}\ (s-1-\sqrt{2s})^{\frac{1}{2}s(s+1)^2}\ (-2)^{s^4}.$$

! $v = 28 = 1+3+6+12+6$ $3^1\ 2^8\ 0.414^6\ -1^7\ -2.414^6$ $\{3,2,2,1;1,1,1,2\}$
Coxeter graph, cf. §12.3; $q^3_{22} = q^4_{34} = 0$; $a_1 = a_2 = p^4_{24} = 0$; equality in absolute bound for $(i,j) = (2,2), (3,3), (4,4)$

! $v = 45 = 1+4+8+16+16$ $4^1\ 3^9\ 1^{10}\ -1^9\ -2^{16}$ $\{4,2,2,2;1,1,1,2\}$
Generalized octagon of order (2,1); $p^2_{22} = 0$; equality in absolute bound for $(i,j) = (1,1), (3,3)$

? $v = 176 = 1+5+20+60+90$ $5^1\ 3.236^{32}\ 1^{66}\ -1.236^{32}\ -3^{45}$ $\{5,4,3,3;1,1,1,2\}$
$a_1 = 0$

! $v = 126 = 1+5+20+40+60$ $5^1\ 3^{27}\ 1^{42}\ -2^{48}\ -4^8$ $\{5,4,4,3;1,1,2,2\}$
Odd graph on 9 points; $\Gamma_{1,2}$ is strongly regular; Q-polynomial ordering: 0 4 1 3 2; equality in absolute bound for $(i,j) = (4,4)$

! $v = 160 = 1+6+18+54+81$ $6^1\ 4.449^{24}\ 2^{30}\ -0.449^{24}\ -2^{81}$ $\{6,3,3,3;1,1,1,2\}$
Generalized octagon of order (3,1); $p^2_{22} = 0$

! $v = 330 = 1+7+42+168+112$ $7^1\ 4^{55}\ 1^{154}\ -3^{99}\ -4^{21}$ $\{7,6,4,4;1,1,1,6\}$
M_{22} graph, cf. §11.4C; $q^4_{34} = 0$; $a_1 = 0$; equality in absolute bound for $(i,j) = (4,4)$

! $v = 425 = 1+8+32+128+256$ $8^1\ 5.828^{50}\ 3^{68}\ 0.172^{50}\ -2^{256}$ $\{8,4,4,4;1,1,1,2\}$
Generalized octagon of order (4,1); $p^2_{22} = 0$

! $v = 81 = 1+8+24+32+16$ $8^1\ 5^8\ 2^{24}\ -1^{32}\ -4^{16}$ $\{8,6,4,2;1,2,3,4\}$
Hamming graph 3^4; classical ($b = 1$)

! $v = 280 = 1+9+72+144+54$ $9^1\ 4^{64}\ 1^{105}\ -3^{90}\ -5^{20}$ $\{9,8,6,3;1,1,3,8\}$
Unital graph, cf. §11.3G; $q_{34}^4 = 0$; $a_1 = 0$

! $v = 256 = 1+9+36+84+126$ $9^1\ 5^{36}\ 1^{126}\ -3^{84}\ -7^9$ $\{9,8,7,6;1,2,3,4\}$
Folded 9-cube; $q_{33}^4 = q_{44}^4 = 0$; generalized Odd graph; $\Gamma_{1,2}$ is strongly regular; Q-polynomial; 2nd Q-polynomial ordering: 0 4 1 3 2

$v = 936 = 1+10+50+250+625$ $10^1\ 7.162^{90}\ 4^{130}\ 0.838^{90}\ -2^{625}$ $\{10,5,5,5;1,1,1,2\}$
Generalized octagon of order (5,1); $p_{22}^2 = 0$

! $v = 315 = 1+10+80+160+64$ $10^1\ 5^{36}\ 3^{90}\ -2^{160}\ -5^{28}$ $\{10,8,8,2;1,1,4,5\}$
Near octagon; J_2 graph, cf. §13.1B; $q_{12}^4 = 0$

$v = 1755 = 1+10+80+640+1024$ $10^1\ 5^{351}\ 1^{650}\ -3^{675}\ -5^{78}$ $\{10,8,8,8;1,1,1,5\}$
Generalized octagon of order (2,4); $q_{44}^4 = 0$

$v = 266 = 1+11+110+132+12$ $11^1\ 4^{76}\ 1^{77}\ -2.382^{56}\ -4.618^{56}$ $\{11,10,6,1;1,1,5,11\}$
Livingstone graph, cf. §13.6; $a_1 = a_4 = p_{24}^4 = p_{44}^4 = 0$

? $v = 1813 = 1+12+72+432+1296$ $12^1\ 8.464^{147}\ 5^{222}\ 1.536^{147}\ -2^{1296}$ $\{12,6,6,6;1,1,1,2\}$
Generalized octagon of order (6,1); $q_{23}^3 = 0$; $p_{22}^2 = 0$

$v = 2925 = 1+12+96+768+2048$ $12^1\ 7^{351}\ 3^{650}\ -1^{675}\ -3^{1248}$ $\{12,8,8,8;1,1,1,3\}$
Generalized octagon of order (4,2)

+ $v = 256 = 1+12+54+108+81$ $12^1\ 8^{12}\ 4^{54}\ 0^{108}\ -4^{81}$ $\{12,9,6,3;1,2,3,4\}$
Hamming (or Doob) graph 4^4; classical ($b = 1$)

$v = 3200 = 1+14+98+686+2401$ $14^1\ 9.742^{224}\ 6^{350}\ 2.258^{224}\ -2^{2401}$ $\{14,7,7,7;1,1,1,2\}$
Generalized octagon of order (7,1); $p_{22}^2 = 0$

! $v = 625 = 1+16+96+256+256$ $16^1\ 11^{16}\ 6^{96}\ 1^{256}\ -4^{256}$ $\{16,12,8,4;1,2,3,4\}$
Hamming graph 5^4; classical ($b = 1$)

! $v = 126 = 1+20+60+40+5$ $20^1\ 11^8\ 4^{27}\ -1^{48}\ -4^{42}$ $\{20,12,6,2;1,4,9,16\}$
Johnson graph J(9,4); dist. 2 in generalized Odd graph; classical ($b = 1$); equality in absolute bound for $(i,j) = (1,1)$

! $v = 1296 = 1+20+150+500+625$ $20^1\ 14^{20}\ 8^{150}\ 2^{500}\ -4^{625}$ $\{20,15,10,5;1,2,3,4\}$
Hamming graph 6^4; classical ($b = 1$)

? $v = 1300 = 1+21+378+756+144$ $21^1\ 8^{168}\ 3^{468}\ -4^{624}\ -7^{39}$ $\{21,18,12,4;1,1,6,21\}$
$q_{34}^4 = 0$; $a_4 = 0$

! $v = 210 = 1+24+90+80+15$ $24^1\ 14^9\ 6^{35}\ 0^{75}\ -4^{90}$ $\{24,15,8,3;1,4,9,16\}$
Johnson graph J(10,4); $p_{34}^4 = p_{44}^4 = 0$; classical ($b = 1$); equality in absolute bound for $(i,j) = (1,1)$

! $v = 2401 = 1+24+216+864+1296$ $24^1\ 17^{24}\ 10^{216}\ 3^{864}\ -4^{1296}$ $\{24,18,12,6;1,2,3,4\}$
Hamming graph 7^4; classical ($b = 1$)

? $v = 2576 = 1+25+600+1800+150$ $25^1\ 7^{552}\ 2^{896}\ -3^{552}\ -7^{575}$ $\{25,24,18,2;1,1,6,24\}$
$a_1 = 0$

! $v = 330 = 1+28+126+140+35$ $28^1\ 17^{10}\ 8^{44}\ 1^{110}\ -4^{165}$ $\{28,18,10,4;1,4,9,16\}$
Johnson graph J(11,4); $p_{44}^4 = 0$; classical ($b = 1$); equality in absolute bound for $(i,j) = (1,1)$

! $v = 4096 = 1+28+294+1372+2401$ $28^1\ 20^{28}\ 12^{294}\ 4^{1372}\ -4^{2401}$ $\{28,21,14,7;1,2,3,4\}$
Hamming graph 8^4; classical ($b = 1$)

! $v = 2295 = 1+30+280+960+1024$ $30^1\ 13^{135}\ 3^{1190}\ -5^{918}\ -15^{51}$ $\{30,28,24,16;1,3,7,15\}$
dual polar graph $B/C_4(2)$; $q_{34}^4 = q_{44}^4 = 0$; $\Gamma_{1,2}$ is strongly regular; classical ($b = 2$); equality in absolute bound for $(i,j) = (4,4)$

! $v = 495 = 1+32+168+224+70$ $32^1\ 20^{11}\ 10^{54}\ 2^{154}\ -4^{275}$ $\{32,21,12,5;1,4,9,16\}$
Johnson graph J(12,4); classical ($b = 1$); equality in absolute bound for $(i,j) = (1,1)$

! $v = 256 = 1+36+126+84+9$ $36^1\ 20^9\ 8^{36}\ 0^{84}\ -4^{126}$ $\{36,21,10,3;1,6,15,28\}$
Halved 9-cube; $q_{11}^1 = q_{13}^3 = 0$; dist. 2 in generalized Odd graph; classical ($b = 1$); 2nd Q-polynomial ordering: 0 2 4 3 1

! $v = 715 = 1+36+216+336+126$ $36^1\ 23^{12}\ 12^{65}\ 3^{208}\ -4^{429}$ $\{36,24,14,6;1,4,9,16\}$
Johnson graph J(13,4); classical ($b = 1$); equality in absolute bound for $(i,j) = (1,1)$

? $v = 640 = 1+36+243+324+36$ $36^1\ 16.649^{36}\ 4^{243}\ -4^{324}\ -8.649^{36}$ $\{36,27,16,4;1,4,12,36\}$
$a_4 = p_{34}^4 = 0$; Q-polynomial; 2nd Q-polynomial ordering: 0 4 2 3 1

? $v = 768 = 1+39+312+390+26$ $39^1\ 15^{52}\ 7^{117}\ -1^{468}\ -9^{130}$ $\{39,32,20,2;1,4,16,30\}$
$q_{11}^4 = q_{12}^2 = q_{12}^4 = 0$; $p_{34}^4 = 0$

! $v = 1001 = 1+40+270+480+210$ $40^1\ 26^{13}\ 14^{77}\ 4^{273}\ -4^{637}$ $\{40,27,16,7;1,4,9,16\}$
Johnson graph J(14,4); classical ($b = 1$); equality in absolute bound for $(i,j) = (1,1)$

! $v = 1365 = 1+44+330+660+330$ $44^1\ 29^{14}\ 16^{90}\ 5^{350}\ -4^{910}$ $\{44,30,18,8;1,4,9,16\}$
Johnson graph J(15,4); classical ($b = 1$); equality in absolute bound for $(i,j) = (1,1)$

! $v = 1820 = 1+48+396+880+495$ $48^1\ 32^{15}\ 18^{104}\ 6^{440}\ -4^{1260}$ $\{48,33,20,9;1,4,9,16\}$
Johnson graph J(16,4); classical ($b = 1$); equality in absolute bound for $(i,j) = (1,1)$

! $v = 2380 = 1+52+468+1144+715$ $52^1\ 35^{16}\ 20^{119}\ 7^{544}\ -4^{1700}$ $\{52,36,22,10;1,4,9,16\}$
Johnson graph J(17,4); classical ($b = 1$); equality in absolute bound for $(i,j) = (1,1)$

? $v = 2916 = 1+53+530+1484+848$ $\quad 53^1\ 29^{53}\ 11^{530}\ -1^{1484}\ -7^{848}$ $\quad \{53,40,28,16;1,4,10,28\}$
Q-polynomial

? $v = 1200 = 1+55+484+605+55$ $\quad 55^1\ 22.321^{55}\ 5^{484}\ -5^{605}\ -12.321^{55}$ $\quad \{55,44,25,5;1,5,20,55\}$
$a_4 = p^4_{34} = 0$; Q-polynomial; 2nd Q-polynomial ordering: 0 4 2 3 1

! $v = 3060 = 1+56+546+1456+1001$ $\quad 56^1\ 38^{17}\ 22^{135}\ 8^{663}\ -4^{2244}$ $\quad \{56,39,24,11;1,4,9,16\}$
Johnson graph J(18,4); classical ($b = 1$); equality in absolute bound for $(i,j) = (1,1)$

! $v = 3876 = 1+60+630+1820+1365$ $\quad 60^1\ 41^{18}\ 24^{152}\ 9^{798}\ -4^{2907}$ $\quad \{60,42,26,12;1,4,9,16\}$
Johnson graph J(19,4); classical ($b = 1$); equality in absolute bound for $(i,j) = (1,1)$

? $v = 2016 = 1+78+845+1014+78$ $\quad 78^1\ 28.450^{78}\ 6^{845}\ -6^{1014}\ -16.450^{78}$ $\quad \{78,65,36,6;1,6,30,78\}$
$a_4 = p^4_{34} = 0$; Q-polynomial; 2nd Q-polynomial ordering: 0 4 2 3 1

? $v = 3136 = 1+105+1350+1575+105$ $\quad 105^1\ 35^{105}\ 7^{1350}\ -7^{1575}\ -21^{105}$ $\quad \{105,90,49,7;1,7,42,105\}$
$a_4 = p^4_{34} = 0$; Q-polynomial; 2nd Q-polynomial ordering: 0 4 2 3 1

Diameter 4 - Antipodal (but not bipartite) graphs

For all $t \geqslant 1$ and each divisor $r > 1$ of $t(t+1)$ we have a feasible array $\{k, k-\mu, (r-1)\mu, 1; 1, \mu, k-1, k\}$ where $k = t(t^2+3t+1)$ and $r\mu = t(t+1)$. For $t = 1$, $r = 2$ we find the Wells graph (cf. §9.2E); no other examples are known. For $t = 2$ the examples must be covers of the Higman-Sims graph. By BAGCHI [19], no examples exist for $t = 3$. As we saw in Corollary 4.2.5 (and the remark after Proposition 5.6.1) all eigenvalues are integral when $\lambda \neq 0$; for $\lambda = 0$ we find eigenvalues $\pm\sqrt{k}$.

! $v = 32 = 1+5+20+5+1$ $\quad 5^1\ 2.236^8\ 1^{10}\ -2.236^8\ -3^5$ $\quad \{5,4,1,1;1,1,4,5\}$
Wells graph, cf. §9.2E; $q^4_{44} = 0$; $a_1 = a_3 = 0$

! $v = 45 = 1+6+24+12+2$ $\quad 6^1\ 3^{12}\ 1^9\ -2^{18}\ -3^5$ $\quad \{6,4,2,1;1,1,4,6\}$
3.Sym(6).2, cf. §13.2A; equality in absolute bound for $(i,j) = (4,4)$

! $v = 63 = 1+10+30+20+2$ $\quad 10^1\ 5^{12}\ 1^{14}\ -2^{30}\ -4^6$ $\quad \{10,6,4,1;1,2,6,10\}$
Locally Petersen graph; 3.Sym(7), cf. §13.2B; $q^4_{11} = 0$; no 4-cycles; equality in absolute bound for $(i,j) = (4,4)$

! $v = 70 = 1+16+36+16+1$ $\quad 16^1\ 8^7\ 2^{20}\ -2^{28}\ -4^{14}$ $\quad \{16,9,4,1;1,4,9,16\}$
Johnson graph J(8,4); classical ($b = 1$)

? $v = 162 = 1+20+120+20+1$ $\quad 20^1\ 5^{36}\ 2^{60}\ -4^{45}\ -7^{20}$ $\quad \{20,18,3,1;1,3,18,20\}$
$\Gamma_{1,2}$ is strongly regular

$v = 243 = 1+20+180+40+2$ $\quad 20^1\ 5^{72}\ 2^{60}\ -4^{90}\ -7^{20}$ $\quad \{20,18,4,1;1,2,18,20\}$
coset graph of shortened ternary Golay code, cf. §11.3H

? $v = 486 = 1+20+360+100+5$ $\quad 20^1\ 5^{180}\ 2^{60}\ -4^{225}\ -7^{20}$ $\quad \{20,18,5,1;1,1,18,20\}$

? $v = 486 = 1+21+420+42+2$ $\quad 21^1\ 4.583^{162}\ 3^{105}\ -4.583^{162}\ -6^{56}$ $\quad \{21,20,2,1;1,1,20,21\}$
$a_1 = a_3 = 0$

? $v = 200 = 1+22+154+22+1$ $\quad 22^1\ 4.690^{50}\ 2^{77}\ -4.690^{50}\ -8^{22}$ $\quad \{22,21,3,1;1,3,21,22\}$
$q^4_{44} = 0$; $a_1 = a_3 = 0$

? $v = 300 = 1+22+231+44+2$ $\quad 22^1\ 4.690^{100}\ 2^{77}\ -4.690^{100}\ -8^{22}$ $\quad \{22,21,4,1;1,2,21,22\}$
$q^4_{44} = 0$; $a_1 = a_3 = 0$

? $v = 600 = 1+22+462+110+5$ $\quad 22^1\ 4.690^{250}\ 2^{77}\ -4.690^{250}\ -8^{22}$ $\quad \{22,21,5,1;1,1,21,22\}$
$q^4_{44} = 0$; $a_1 = a_3 = 0$

? $v = 352 = 1+25+300+25+1$ $\quad 25^1\ 5^{88}\ 3^{120}\ -5^{88}\ -7^{55}$ $\quad \{25,24,2,1;1,2,24,25\}$
$a_1 = a_3 = 0$; $\Gamma_{1,2}$ is strongly regular

? $v = 704 = 1+25+600+75+3$ $\quad 25^1\ 5^{264}\ 3^{120}\ -5^{264}\ -7^{55}$ $\quad \{25,24,3,1;1,1,24,25\}$
$a_1 = a_3 = 0$

? $v = 704 = 1+26+650+26+1$ $\quad 26^1\ 5.099^{176}\ 4^{208}\ -5.099^{176}\ -6^{143}$ $\quad \{26,25,1,1;1,1,25,26\}$
$a_1 = a_3 = 0$

? $v = 264 = 1+27+180+54+2$ $\quad 27^1\ 9^{44}\ 3^{55}\ -3^{132}\ -6^{32}$ $\quad \{27,20,6,1;1,3,20,27\}$

! $v = 128 = 1+28+70+28+1$ $\quad 28^1\ 14^8\ 4^{28}\ -2^{56}\ -4^{35}$ $\quad \{28,15,6,1;1,6,15,28\}$
Halved 8-cube; classical ($b = 1$)

? $v = 210 = 1+32+144+32+1$ $\quad 32^1\ 8^{35}\ 2^{84}\ -4^{70}\ -10^{20}$ $\quad \{32,27,6,1;1,6,27,32\}$
$\Gamma_{1,2}$ is strongly regular

? $v = 315 = 1+32+216+64+2$ $\quad 32^1\ 8^{70}\ 2^{84}\ -4^{140}\ -10^{20}$ $\quad \{32,27,8,1;1,4,27,32\}$

? $v = 420 = 1+32+288+96+3$ $\quad 32^1\ 8^{105}\ 2^{84}\ -4^{210}\ -10^{20}$ $\quad \{32,27,9,1;1,3,27,32\}$

? $v = 630 = 1+32+432+160+5$ $\quad 32^1\ 8^{175}\ 2^{84}\ -4^{350}\ -10^{20}$ $\quad \{32,27,10,1;1,2,27,32\}$

? $v = 420 = 1+33+352+33+1$ $\quad 33^1\ 5.745^{105}\ 3^{154}\ -5.745^{105}\ -9^{55}$ $\quad \{33,32,3,1;1,3,32,33\}$
$a_1 = a_3 = 0$

? $v = 630 = 1+33+528+66+2$ $\quad 33^1\ 5.745^{210}\ 3^{154}\ -5.745^{210}\ -9^{55}$ $\quad \{33,32,4,1;1,2,32,33\}$
$a_1 = a_3 = 0$

? $v = 1260 = 1+33+1056+165+5$ $\quad 33^1\ 5.745^{525}\ 3^{154}\ -5.745^{525}\ -9^{55}$ $\quad \{33,32,5,1;1,1,32,33\}$
$a_1 = a_3 = 0$

? $v = 704 = 1+36+630+36+1$ $36^1\ 6^{176}\ 4^{231}\ -6^{176}\ -8^{120}$ $\{36,35,2,1;1,2,35,36\}$
$a_1 = a_3 = 0$; $\Gamma_{1,2}$ is strongly regular
? $v = 1408 = 1+36+1260+108+3$ $36^1\ 6^{528}\ 4^{231}\ -6^{528}\ -8^{120}$ $\{36,35,3,1;1,1,35,36\}$
$a_1 = a_3 = 0$
? $v = 1408 = 1+37+1332+37+1$ $37^1\ 6.083^{352}\ 5^{407}\ -6.083^{352}\ -7^{296}$ $\{37,36,1,1;1,1,36,37\}$
$a_1 = a_3 = 0$
? $v = 252 = 1+45+160+45+1$ $45^1\ 15^{21}\ 3^{90}\ -3^{105}\ -9^{35}$ $\{45,32,9,1;1,9,32,45\}$
Q-polynomial
$v = 378 = 1+45+240+90+2$ $45^1\ 15^{42}\ 3^{90}\ -3^{210}\ -9^{35}$ $\{45,32,12,1;1,6,32,45\}$
$3.O_6^-(3)$, cf. §13.2C; $q_{11}^4 = 0$
? $v = 756 = 1+45+480+225+5$ $45^1\ 15^{105}\ 3^{90}\ -3^{525}\ -9^{35}$ $\{45,32,15,1;1,3,32,45\}$
$q_{11}^4 = 0$
? $v = 392 = 1+45+300+45+1$ $45^1\ 9^{70}\ 3^{150}\ -5^{126}\ -11^{45}$ $\{45,40,6,1;1,6,40,45\}$
$\Gamma_{1,2}$ is strongly regular
? $v = 588 = 1+45+450+90+2$ $45^1\ 9^{140}\ 3^{150}\ -5^{252}\ -11^{45}$ $\{45,40,8,1;1,4,40,45\}$
? $v = 784 = 1+45+600+135+3$ $45^1\ 9^{210}\ 3^{150}\ -5^{378}\ -11^{45}$ $\{45,40,9,1;1,3,40,45\}$
? $v = 1176 = 1+45+900+225+5$ $45^1\ 9^{350}\ 3^{150}\ -5^{630}\ -11^{45}$ $\{45,40,10,1;1,2,40,45\}$
? $v = 2352 = 1+45+1800+495+11$ $45^1\ 9^{770}\ 3^{150}\ -5^{1386}\ -11^{45}$ $\{45,40,11,1;1,1,40,45\}$
? $v = 798 = 1+45+660+90+2$ $45^1\ 6.708^{266}\ 3^{209}\ -6.708^{266}\ -12^{56}$ $\{45,44,6,1;1,3,44,45\}$
$a_1 = a_3 = 0$
? $v = 2394 = 1+45+1980+360+8$ $45^1\ 6.708^{1064}\ 3^{209}\ -6.708^{1064}\ -12^{56}$ $\{45,44,8,1;1,1,44,45\}$
$a_1 = a_3 = 0$
? $v = 784 = 1+46+690+46+1$ $46^1\ 6.782^{196}\ 4^{276}\ -6.782^{196}\ -10^{115}$ $\{46,45,3,1;1,3,45,46\}$
$a_1 = a_3 = 0$
? $v = 1176 = 1+46+1035+92+2$ $46^1\ 6.782^{392}\ 4^{276}\ -6.782^{392}\ -10^{115}$ $\{46,45,4,1;1,2,45,46\}$
$a_1 = a_3 = 0$
? $v = 2352 = 1+46+2070+230+5$ $46^1\ 6.782^{980}\ 4^{276}\ -6.782^{980}\ -10^{115}$ $\{46,45,5,1;1,1,45,46\}$
$a_1 = a_3 = 0$
? $v = 1276 = 1+49+1176+49+1$ $49^1\ 7^{319}\ 5^{406}\ -7^{319}\ -9^{231}$ $\{49,48,2,1;1,2,48,49\}$
$a_1 = a_3 = 0$; $\Gamma_{1,2}$ is strongly regular
? $v = 2552 = 1+49+2352+147+3$ $49^1\ 7^{957}\ 5^{406}\ -7^{957}\ -9^{231}$ $\{49,48,3,1;1,1,48,49\}$
$a_1 = a_3 = 0$
? $v = 2552 = 1+50+2450+50+1$ $50^1\ 7.071^{638}\ 6^{725}\ -7.071^{638}\ -8^{550}$ $\{50,49,1,1;1,1,49,50\}$
$a_1 = a_3 = 0$
? $v = 650 = 1+54+540+54+1$ $54^1\ 9^{130}\ 4^{234}\ -6^{195}\ -11^{90}$ $\{54,50,5,1;1,5,50,54\}$
$\Gamma_{1,2}$ is strongly regular
? $v = 1625 = 1+54+1350+216+4$ $54^1\ 9^{520}\ 4^{234}\ -6^{780}\ -11^{90}$ $\{54,50,8,1;1,2,50,54\}$
? $v = 3250 = 1+55+2970+220+4$ $55^1\ 7.416^{1300}\ 5^{429}\ -7.416^{1300}\ -10^{220}$ $\{55,54,4,1;1,1,54,55\}$
$a_1 = a_3 = 0$
? $v = 324 = 1+56+210+56+1$ $56^1\ 14^{36}\ 2^{140}\ -4^{126}\ -16^{21}$ $\{56,45,12,1;1,12,45,56\}$
$q_{44}^4 = 0$; $\Gamma_{1,2}$ is strongly regular; Q-polynomial
? $v = 486 = 1+56+315+112+2$ $56^1\ 14^{72}\ 2^{140}\ -4^{252}\ -16^{21}$ $\{56,45,16,1;1,8,45,56\}$
$q_{11}^4 = q_{44}^4 = 0$
? $v = 648 = 1+56+420+168+3$ $56^1\ 14^{108}\ 2^{140}\ -4^{378}\ -16^{21}$ $\{56,45,18,1;1,6,45,56\}$
$q_{11}^4 = q_{44}^4 = 0$
? $v = 972 = 1+56+630+280+5$ $56^1\ 14^{180}\ 2^{140}\ -4^{630}\ -16^{21}$ $\{56,45,20,1;1,4,45,56\}$
$q_{11}^4 = q_{44}^4 = 0$
? $v = 1296 = 1+56+840+392+7$ $56^1\ 14^{252}\ 2^{140}\ -4^{882}\ -16^{21}$ $\{56,45,21,1;1,3,45,56\}$
$q_{11}^4 = q_{44}^4 = 0$
? $v = 552 = 1+75+400+75+1$ $75^1\ 15^{69}\ 3^{230}\ -5^{207}\ -17^{45}$ $\{75,64,12,1;1,12,64,75\}$
$\Gamma_{1,2}$ is strongly regular
? $v = 828 = 1+75+600+150+2$ $75^1\ 15^{138}\ 3^{230}\ -5^{414}\ -17^{45}$ $\{75,64,16,1;1,8,64,75\}$
? $v = 1104 = 1+75+800+225+3$ $75^1\ 15^{207}\ 3^{230}\ -5^{621}\ -17^{45}$ $\{75,64,18,1;1,6,64,75\}$
? $v = 1656 = 1+75+1200+375+5$ $75^1\ 15^{345}\ 3^{230}\ -5^{1035}\ -17^{45}$ $\{75,64,20,1;1,4,64,75\}$
? $v = 2208 = 1+75+1600+525+7$ $75^1\ 15^{483}\ 3^{230}\ -5^{1449}\ -17^{45}$ $\{75,64,21,1;1,3,64,75\}$
? $v = 3312 = 1+75+2400+825+11$ $75^1\ 15^{759}\ 3^{230}\ -5^{2277}\ -17^{45}$ $\{75,64,22,1;1,2,64,75\}$
? $v = 1104 = 1+76+950+76+1$ $76^1\ 8.718^{276}\ 4^{437}\ -8.718^{276}\ -16^{114}$ $\{76,75,6,1;1,6,75,76\}$
$a_1 = a_3 = 0$
? $v = 1656 = 1+76+1425+152+2$ $76^1\ 8.718^{552}\ 4^{437}\ -8.718^{552}\ -16^{114}$ $\{76,75,8,1;1,4,75,76\}$
$a_1 = a_3 = 0$
? $v = 2208 = 1+76+1900+228+3$ $76^1\ 8.718^{828}\ 4^{437}\ -8.718^{828}\ -16^{114}$ $\{76,75,9,1;1,3,75,76\}$
$a_1 = a_3 = 0$
? $v = 3312 = 1+76+2850+380+5$ $76^1\ 8.718^{1380}\ 4^{437}\ -8.718^{1380}\ -16^{114}$ $\{76,75,10,1;1,2,75,76\}$
$a_1 = a_3 = 0$

? $v = 1080 = 1+77+924+77+1$ $77^1\ 11^{210}\ 5^{385}\ -7^{330}\ -13^{154}$ $\{77,72,6,1;1,6,72,77\}$
$\Gamma_{1,2}$ is strongly regular

? $v = 1620 = 1+77+1386+154+2$ $77^1\ 11^{420}\ 5^{385}\ -7^{660}\ -13^{154}$ $\{77,72,8,1;1,4,72,77\}$

? $v = 2160 = 1+77+1848+231+3$ $77^1\ 11^{630}\ 5^{385}\ -7^{990}\ -13^{154}$ $\{77,72,9,1;1,3,72,77\}$

? $v = 3240 = 1+77+2772+385+5$ $77^1\ 11^{1050}\ 5^{385}\ -7^{1650}\ -13^{154}$ $\{77,72,10,1;1,2,72,77\}$

? $v = 2160 = 1+78+2002+78+1$ $78^1\ 8.832^{540}\ 6^{715}\ -8.832^{540}\ -12^{364}$ $\{78,77,3,1;1,3,77,78\}$
$a_1 = a_3 = 0$

? $v = 3240 = 1+78+3003+156+2$ $78^1\ 8.832^{1080}\ 6^{715}\ -8.832^{1080}\ -12^{364}$ $\{78,77,4,1;1,2,77,78\}$
$a_1 = a_3 = 0$

? $v = 750 = 1+81+504+162+2$ $81^1\ 27^{50}\ 6^{144}\ -3^{450}\ -9^{105}$ $\{81,56,18,1;1,9,56,81\}$
$q^4_{11} = 0$

? $v = 2250 = 1+81+1512+648+8$ $81^1\ 27^{200}\ 6^{144}\ -3^{1800}\ -9^{105}$ $\{81,56,24,1;1,3,56,81\}$
$q^4_{11} = 0$

? $v = 3404 = 1+81+3240+81+1$ $81^1\ 9^{851}\ 7^{1035}\ -9^{851}\ -11^{666}$ $\{81,80,2,1;1,2,80,81\}$
$a_1 = a_3 = 0$; $\Gamma_{1,2}$ is strongly regular

? $v = 800 = 1+84+630+84+1$ $84^1\ 14^{120}\ 4^{315}\ -6^{280}\ -16^{84}$ $\{84,75,10,1;1,10,75,84\}$
$\Gamma_{1,2}$ is strongly regular

? $v = 1600 = 1+84+1260+252+3$ $84^1\ 14^{360}\ 4^{315}\ -6^{840}\ -16^{84}$ $\{84,75,15,1;1,5,75,84\}$

? $v = 2000 = 1+84+1575+336+4$ $84^1\ 14^{480}\ 4^{315}\ -6^{1120}\ -16^{84}$ $\{84,75,16,1;1,4,75,84\}$

? $v = 4000 = 1+84+3150+756+9$ $84^1\ 14^{1080}\ 4^{315}\ -6^{2520}\ -16^{84}$ $\{84,75,18,1;1,2,75,84\}$

? $v = 1600 = 1+85+1428+85+1$ $85^1\ 9.220^{400}\ 5^{595}\ -9.220^{400}\ -15^{204}$ $\{85,84,5,1;1,5,84,85\}$
$a_1 = a_3 = 0$

? $v = 4000 = 1+85+3570+340+4$ $85^1\ 9.220^{1600}\ 5^{595}\ -9.220^{1600}\ -15^{204}$ $\{85,84,8,1;1,2,84,85\}$
$a_1 = a_3 = 0$

? $v = 644 = 1+96+450+96+1$ $96^1\ 24^{46}\ 4^{252}\ -4^{276}\ -16^{69}$ $\{96,75,16,1;1,16,75,96\}$
Q-polynomial

? $v = 1288 = 1+96+900+288+3$ $96^1\ 24^{138}\ 4^{252}\ -4^{828}\ -16^{69}$ $\{96,75,24,1;1,8,75,96\}$
$q^4_{11} = 0$

? $v = 2576 = 1+96+1800+672+7$ $96^1\ 24^{322}\ 4^{252}\ -4^{1932}\ -16^{69}$ $\{96,75,28,1;1,4,75,96\}$
$q^4_{11} = 0$

? $v = 1650 = 1+96+1456+96+1$ $96^1\ 12^{330}\ 6^{572}\ -8^{495}\ -14^{252}$ $\{96,91,6,1;1,6,91,96\}$
$\Gamma_{1,2}$ is strongly regular

? $v = 2475 = 1+96+2184+192+2$ $96^1\ 12^{660}\ 6^{572}\ -8^{990}\ -14^{252}$ $\{96,91,8,1;1,4,91,96\}$

? $v = 3300 = 1+96+2912+288+3$ $96^1\ 12^{990}\ 6^{572}\ -8^{1485}\ -14^{252}$ $\{96,91,9,1;1,3,91,96\}$

? $v = 3300 = 1+97+3104+97+1$ $97^1\ 9.849^{825}\ 7^{1067}\ -9.849^{825}\ -13^{582}$ $\{97,96,3,1;1,3,96,97\}$
$a_1 = a_3 = 0$

? $v = 3534 = 1+99+3234+198+2$ $99^1\ 9.950^{1178}\ 6^{836}\ -9.950^{1178}\ -15^{341}$ $\{99,98,6,1;1,3,98,99\}$
$a_1 = a_3 = 0$

? $v = 2736 = 1+104+2106+520+5$ $104^1\ 26^{304}\ 8^{247}\ -4^{1976}\ -10^{208}$ $\{104,81,20,1;1,4,81,104\}$

? $v = 2625 = 1+114+2280+228+2$ $114^1\ 19^{420}\ 9^{475}\ -6^{1330}\ -11^{399}$ $\{114,100,10,1;1,5,100,114\}$

? $v = 784 = 1+115+552+115+1$ $115^1\ 23^{70}\ 3^{345}\ -5^{322}\ -25^{46}$ $\{115,96,20,1;1,20,96,115\}$
$q^4_{44} = 0$; $\Gamma_{1,2}$ is strongly regular; Q-polynomial

? $v = 1568 = 1+115+1104+345+3$ $115^1\ 23^{210}\ 3^{345}\ -5^{966}\ -25^{46}$ $\{115,96,30,1;1,10,96,115\}$
$q^4_{11} = q^4_{44} = 0$

? $v = 1960 = 1+115+1380+460+4$ $115^1\ 23^{280}\ 3^{345}\ -5^{1288}\ -25^{46}$ $\{115,96,32,1;1,8,96,115\}$
$q^4_{11} = q^4_{44} = 0$

? $v = 3136 = 1+115+2208+805+7$ $115^1\ 23^{490}\ 3^{345}\ -5^{2254}\ -25^{46}$ $\{115,96,35,1;1,5,96,115\}$
$q^4_{11} = q^4_{44} = 0$

? $v = 3920 = 1+115+2760+1035+9$ $115^1\ 23^{630}\ 3^{345}\ -5^{2898}\ -25^{46}$ $\{115,96,36,1;1,4,96,115\}$
$q^4_{11} = q^4_{44} = 0$

? $v = 1568 = 1+116+1334+116+1$ $\{116,115,10,1;1,10,115,116\}$
$116^1\ 10.770^{392}\ 4^{667}\ -10.770^{392}\ -24^{116}$ $q^4_{44} = 0$; $a_1 = a_3 = 0$

? $v = 3136 = 1+116+2668+348+3$ $\{116,115,15,1;1,5,115,116\}$
$116^1\ 10.770^{1176}\ 4^{667}\ -10.770^{1176}\ -24^{116}$ $q^4_{44} = 0$; $a_1 = a_3 = 0$

? $v = 3920 = 1+116+3335+464+4$ $\{116,115,16,1;1,4,115,116\}$
$116^1\ 10.770^{1568}\ 4^{667}\ -10.770^{1568}\ -24^{116}$ $q^4_{44} = 0$; $a_1 = a_3 = 0$

? $v = 756 = 1+117+520+117+1$ $117^1\ 39^{27}\ 9^{182}\ -3^{351}\ -9^{195}$ $\{117,80,18,1;1,18,80,117\}$
Q-polynomial

$v = 1134 = 1+117+780+234+2$ $117^1\ 39^{54}\ 9^{182}\ -3^{702}\ -9^{195}$ $\{117,80,24,1;1,12,80,117\}$
$3.O_7(3)$, cf. §13.2D; $q^4_{11} = 0$

? $v = 1512 = 1+117+1040+351+3$ $117^1\ 39^{81}\ 9^{182}\ -3^{1053}\ -9^{195}$ $\{117,80,27,1;1,9,80,117\}$
$q^4_{11} = 0$

? $v = 2268 = 1+117+1560+585+5$ $117^1\ 39^{135}\ 9^{182}\ -3^{1755}\ -9^{195}$ $\{117,80,30,1;1,6,80,117\}$
$q^4_{11} = 0$

? $v = 3402 = 1+117+2340+936+8$	$117^1\ 39^{216}\ 9^{182}\ -3^{2808}\ -9^{195}$	$\{117,80,32,1;1,4,80,117\}$
$q_{11}^4 = 0$		
? $v = 2420 = 1+117+2184+117+1$	$117^1\ 13^{495}\ 7^{819}\ -9^{715}\ -15^{390}$	$\{117,112,6,1;1,6,112,117\}$
$\Gamma_{1,2}$ is strongly regular		
? $v = 3630 = 1+117+3276+234+2$	$117^1\ 13^{990}\ 7^{819}\ -9^{1430}\ -15^{390}$	$\{117,112,8,1;1,4,112,117\}$
? $v = 2100 = 1+128+1584+384+3$	$128^1\ 32^{175}\ 8^{308}\ -4^{1400}\ -12^{216}$	$\{128,99,24,1;1,8,99,128\}$
? $v = 2432 = 1+135+2160+135+1$	$135^1\ 15^{456}\ 7^{855}\ -9^{760}\ -17^{360}$	$\{135,128,8,1;1,8,128,135\}$
$\Gamma_{1,2}$ is strongly regular		
? $v = 1458 = 1+140+1176+140+1$	$140^1\ 20^{189}\ 5^{588}\ -7^{540}\ -22^{140}$	$\{140,126,15,1;1,15,126,140\}$
$\Gamma_{1,2}$ is strongly regular		
? $v = 2187 = 1+140+1764+280+2$	$140^1\ 20^{378}\ 5^{588}\ -7^{1080}\ -22^{140}$	$\{140,126,20,1;1,10,126,140\}$
? $v = 3645 = 1+140+2940+560+4$	$140^1\ 20^{756}\ 5^{588}\ -7^{2160}\ -22^{140}$	$\{140,126,24,1;1,6,126,140\}$
? $v = 1190 = 1+144+900+144+1$	$144^1\ 24^{119}\ 4^{510}\ -6^{476}\ -26^{84}$	$\{144,125,20,1;1,20,125,144\}$
$\Gamma_{1,2}$ is strongly regular		
? $v = 2380 = 1+144+1800+432+3$	$144^1\ 24^{357}\ 4^{510}\ -6^{1428}\ -26^{84}$	$\{144,125,30,1;1,10,125,144\}$
? $v = 2975 = 1+144+2250+576+4$	$144^1\ 24^{476}\ 4^{510}\ -6^{1904}\ -26^{84}$	$\{144,125,32,1;1,8,125,144\}$
? $v = 2380 = 1+145+2088+145+1$		$\{145,144,10,1;1,10,144,145\}$
	$145^1\ 12.042^{595}\ 5^{986}\ -12.042^{595}\ -25^{203}$	$a_1 = a_3 = 0$
? $v = 2002 = 1+160+1680+160+1$	$160^1\ 20^{286}\ 6^{780}\ -8^{715}\ -22^{220}$	$\{160,147,14,1;1,14,147,160\}$
$\Gamma_{1,2}$ is strongly regular		
? $v = 4004 = 1+160+3360+480+3$	$160^1\ 20^{858}\ 6^{780}\ -8^{2145}\ -22^{220}$	$\{160,147,21,1;1,7,147,160\}$
? $v = 4004 = 1+161+3680+161+1$		$\{161,160,7,1;1,7,160,161\}$
	$161^1\ 12.689^{1001}\ 7^{1495}\ -12.689^{1001}\ -21^{506}$	$a_1 = a_3 = 0$
? $v = 3402 = 1+170+3060+170+1$	$170^1\ 17^{630}\ 8^{1190}\ -10^{1071}\ -19^{510}$	$\{170,162,9,1;1,9,162,170\}$
$\Gamma_{1,2}$ is strongly regular		
? $v = 3952 = 1+175+3600+175+1$	$175^1\ 35^{247}\ 23^{455}\ -5^{1729}\ -7^{1520}$	$\{175,144,7,1;1,7,144,175\}$
? $v = 1360 = 1+175+1008+175+1$	$175^1\ 35^{85}\ 5^{560}\ -5^{595}\ -25^{119}$	$\{175,144,25,1;1,25,144,175\}$
Q-polynomial		
? $v = 3400 = 1+175+2520+700+4$	$175^1\ 35^{340}\ 5^{560}\ -5^{2380}\ -25^{119}$	$\{175,144,40,1;1,10,144,175\}$
$q_{11}^4 = 0$		
? $v = 1344 = 1+176+990+176+1$	$176^1\ 44^{56}\ 8^{440}\ -4^{616}\ -16^{231}$	$\{176,135,24,1;1,24,135,176\}$
Q-polynomial		
? $v = 2016 = 1+176+1485+352+2$	$176^1\ 44^{112}\ 8^{440}\ -4^{1232}\ -16^{231}$	$\{176,135,32,1;1,16,135,176\}$
$q_{11}^4 = 0$		
? $v = 2688 = 1+176+1980+528+3$	$176^1\ 44^{168}\ 8^{440}\ -4^{1848}\ -16^{231}$	$\{176,135,36,1;1,12,135,176\}$
$q_{11}^4 = 0$		
? $v = 4032 = 1+176+2970+880+5$	$176^1\ 44^{280}\ 8^{440}\ -4^{3080}\ -16^{231}$	$\{176,135,40,1;1,8,135,176\}$
$q_{11}^4 = 0$		
? $v = 1276 = 1+189+896+189+1$	$189^1\ 63^{29}\ 15^{231}\ -3^{609}\ -9^{406}$	$\{189,128,27,1;1,27,128,189\}$
Q-polynomial		
? $v = 1914 = 1+189+1344+378+2$	$189^1\ 63^{58}\ 15^{231}\ -3^{1218}\ -9^{406}$	$\{189,128,36,1;1,18,128,189\}$
$q_{11}^4 = 0$		
? $v = 3828 = 1+189+2688+945+5$	$189^1\ 63^{145}\ 15^{231}\ -3^{3045}\ -9^{406}$	$\{189,128,45,1;1,9,128,189\}$
$q_{11}^4 = 0$		
? $v = 3552 = 1+200+3150+200+1$	$200^1\ 20^{592}\ 8^{1295}\ -10^{1184}\ -22^{480}$	$\{200,189,12,1;1,12,189,200\}$
$\Gamma_{1,2}$ is strongly regular		
? $v = 1600 = 1+204+1190+204+1$	$204^1\ 34^{120}\ 4^{714}\ -6^{680}\ -36^{85}$	$\{204,175,30,1;1,30,175,204\}$
$q_{44}^4 = 0$; $\Gamma_{1,2}$ is strongly regular; Q-polynomial		
? $v = 2400 = 1+204+1785+408+2$	$204^1\ 34^{240}\ 4^{714}\ -6^{1360}\ -36^{85}$	$\{204,175,40,1;1,20,175,204\}$
$q_{11}^4 = q_{44}^4 = 0$		
? $v = 3200 = 1+204+2380+612+3$	$204^1\ 34^{360}\ 4^{714}\ -6^{2040}\ -36^{85}$	$\{204,175,45,1;1,15,175,204\}$
$q_{11}^4 = q_{44}^4 = 0$		
? $v = 4000 = 1+204+2975+816+4$	$204^1\ 34^{480}\ 4^{714}\ -6^{2720}\ -36^{85}$	$\{204,175,48,1;1,12,175,204\}$
$q_{11}^4 = q_{44}^4 = 0$		
? $v = 3200 = 1+205+2788+205+1$		$\{205,204,15,1;1,15,204,205\}$
	$205^1\ 14.318^{800}\ 5^{1394}\ -14.318^{800}\ -35^{205}$	$q_{44}^4 = 0$; $a_1 = a_3 = 0$
? $v = 2450 = 1+216+2016+216+1$	$216^1\ 27^{280}\ 6^{1008}\ -8^{945}\ -29^{216}$	$\{216,196,21,1;1,21,196,216\}$
$\Gamma_{1,2}$ is strongly regular		
? $v = 3675 = 1+216+3024+432+2$	$216^1\ 27^{560}\ 6^{1008}\ -8^{1890}\ -29^{216}$	$\{216,196,28,1;1,14,196,216\}$
? $v = 3825 = 1+234+3120+468+2$	$234^1\ 39^{340}\ 9^{884}\ -6^{2210}\ -21^{390}$	$\{234,200,30,1;1,15,200,234\}$
? $v = 3432 = 1+245+2940+245+1$	$245^1\ 35^{286}\ 11^{1078}\ -7^{1430}\ -19^{637}$	$\{245,216,18,1;1,18,216,245\}$
? $v = 2256 = 1+245+1764+245+1$	$245^1\ 35^{188}\ 5^{987}\ -7^{940}\ -37^{140}$	$\{245,216,30,1;1,30,216,245\}$
$\Gamma_{1,2}$ is strongly regular		
? $v = 3384 = 1+245+2646+490+2$	$245^1\ 35^{376}\ 5^{987}\ -7^{1880}\ -37^{140}$	$\{245,216,40,1;1,20,216,245\}$

? $v = 4032 = 1+260+3510+260+1$ $260^1\ 26^{560}\ 8^{1560}\ -10^{1456}\ -28^{455}$ $\{260,243,18,1;1,18,243,260\}$
$\Gamma_{1,2}$ is strongly regular
? $v = 1800 = 1+261+1276+261+1$ $261^1\ 87^{30}\ 21^{261}\ -3^{870}\ -9^{638}$ $\{261,176,36,1;1,36,176,261\}$
Q-polynomial
? $v = 3600 = 1+261+2552+783+3$ $261^1\ 87^{90}\ 21^{261}\ -3^{2610}\ -9^{638}$ $\{261,176,54,1;1,18,176,261\}$
$q_{11}^4 = 0$
? $v = 3500 = 1+264+2970+264+1$ $264^1\ 44^{210}\ 14^{924}\ -6^{1540}\ -16^{825}$ $\{264,225,20,1;1,20,225,264\}$
? $v = 2538 = 1+288+1960+288+1$ $288^1\ 48^{141}\ 6^{1080}\ -6^{1128}\ -36^{188}$ $\{288,245,36,1;1,36,245,288\}$
Q-polynomial
? $v = 3807 = 1+288+2940+576+2$ $288^1\ 48^{282}\ 6^{1080}\ -6^{2256}\ -36^{188}$ $\{288,245,48,1;1,24,245,288\}$
$q_{11}^4 = 0$
? $v = 4050 = 1+308+3432+308+1$ $308^1\ 77^{100}\ 38^{308}\ -4^{1925}\ -7^{1716}$ $\{308,234,21,1;1,21,234,308\}$
? $v = 3872 = 1+315+3240+315+1$ $315^1\ 35^{396}\ 7^{1620}\ -9^{1540}\ -37^{315}$ $\{315,288,28,1;1,28,288,315\}$
$\Gamma_{1,2}$ is strongly regular
? $v = 3234 = 1+320+2592+320+1$ $320^1\ 80^{77}\ 26^{440}\ -4^{1540}\ -10^{1176}$ $\{320,243,30,1;1,30,243,320\}$
? $v = 2916 = 1+329+2256+329+1$ $329^1\ 47^{189}\ 5^{1316}\ -7^{1269}\ -49^{141}$ $\{329,288,42,1;1,42,288,329\}$
$q_{44}^4 = 0$; $\Gamma_{1,2}$ is strongly regular; Q-polynomial
? $v = 2816 = 1+336+2142+336+1$ $336^1\ 84^{64}\ 16^{693}\ -4^{1344}\ -16^{714}$ $\{336,255,40,1;1,40,255,336\}$
Q-polynomial
? $v = 3906 = 1+384+3136+384+1$ $384^1\ 48^{279}\ 6^{1736}\ -8^{1674}\ -50^{216}$ $\{384,343,42,1;1,42,343,384\}$
$\Gamma_{1,2}$ is strongly regular
? $v = 4050 = 1+414+3220+414+1$ $414^1\ 69^{162}\ 9^{1610}\ -6^{1863}\ -36^{414}$ $\{414,350,45,1;1,45,350,414\}$
Q-polynomial
? $v = 3564 = 1+416+2730+416+1$ $416^1\ 104^{66}\ 20^{780}\ -4^{1716}\ -16^{1001}$ $\{416,315,48,1;1,48,315,416\}$
Q-polynomial

Diameter 4 - graphs that are both antipodal and bipartite

These graphs have intersection array

$$v = 2m^2\mu \quad \{m\mu, m\mu-1, (m-1)\mu, 1;\ 1, \mu, m\mu-1, m\mu\}$$

and spectrum

$$k^1\ \sqrt{k}^{(m-1)k}\ 0^{2k-2}\ (-\sqrt{k})^{(m-1)k}\ (-k)^1 \quad \text{where } k = m\mu.$$

They are precisely the incidence graphs of symmetric (m,μ)-nets.
For $m = 2$ these graphs are Hadamard graphs, which gives the additional restriction that μ must be even. Hadamard graphs have two Q-polynomial orderings 1234 and 3214 and satisfy $q_{11}^1 = q_{33}^3 = 0$, i.e., are Q-bipartite.
For $\mu = 1$ these graphs are the incidence graphs of affine planes of order m from which a parallel class of lines has been removed, hence the additional restriction that m is a sum of two squares. For $m = \mu+1$, μ a prime power $\equiv 1 \pmod 4$, these graphs can be used to construct square 2-(m^3-m+1, m^2, m) designs (RAJKUNDLIA [602]), and the Bruck-Ryser conditions imply that m must be a sum of two squares.

Diameter 3 - Primitive graphs

There is an infinite sequence of formally self-dual classical graphs with $b = 2$, $\alpha = 1$, $\beta = n-1$, $v = n^3$ $(n > 5)$ realized when n is a power of 2 by a bilinear forms graph.

! $v = 21 = 1+4+8+8$ $4^1\ 2.414^6\ -0.414^6\ -2^8$ $\{4,2,2;1,1,2\}$
Generalized hexagon of order (2,1); equality in absolute bound for $(i,j) = (1,1), (2,2)$
! $v = 35 = 1+4+12+18$ $4^1\ 2^{14}\ -1^{14}\ -3^6$ $\{4,3,3;1,1,2\}$
Odd graph on 7 points; Γ_3 is strongly regular; Q-polynomial ordering: 0 3 1 2; equality in absolute bound for $(i,j) = (3,3)$
! $v = 36 = 1+5+20+10$ $5^1\ 2^{16}\ -1^{10}\ -3^9$ $\{5,4,2;1,1,4\}$
Sylvester graph, cf. §13.1A; Sym(6).2; $q_{33}^3 = 0$; $a_1 = 0$; Γ_3 is strongly regular
? $v = 56 = 1+5+20+30$ $5^1\ 2.414^{20}\ -0.414^{20}\ -3^{15}$ $\{5,4,3;1,1,2\}$ $a_1 = 0$
! $v = 52 = 1+6+18+27$ $6^1\ 3.732^{12}\ 0.268^{12}\ -2^{27}$ $\{6,3,3;1,1,2\}$
Generalized hexagon of order (3,1)
! $v = 27 = 1+6+12+8$ $6^1\ 3^6\ 0^{12}\ -3^8$ $\{6,4,2;1,2,3\}$
Hamming graph 3^3; classical $(b = 1)$

$+\,v = 63 = 1+6+24+32$ $\quad 6^1\ 3^{21}\ -1^{27}\ -3^{14}$ $\quad \{6,4,4;1,1,3\}$
Generalized hexagon of order (2,2); Γ_3 is strongly regular

$v = 57 = 1+6+30+20$ $\quad 6^1\ 2.618^{18}\ 0.382^{18}\ -3^{20}$ $\quad \{6,5,2;1,1,3\}$ $\quad a_1 = 0$
Perkel graph, cf. §13.3

$!\,v = 64 = 1+7+21+35$ $\quad 7^1\ 3^{21}\ -1^{35}\ -5^7$ $\quad \{7,6,5;1,2,3\}$ $\quad q^3_{33} = 0$
Folded 7-cube; generalized Odd graph; Γ_3 is strongly regular; Q-polynomial; 2nd Q-polynomial ordering: 0 3 1 2

$?\,v = 176 = 1+7+42+126$ $\quad 7^1\ 3^{66}\ -1^{77}\ -4^{32}$ $\quad \{7,6,6;1,1,2\}$
generalized Odd graph; Γ_3 is strongly regular

$!\,v = 105 = 1+8+32+64$ $\quad 8^1\ 5^{20}\ 1^{20}\ -2^{64}$ $\quad \{8,4,4;1,1,2\}$
Generalized hexagon of order (4,1); Γ_2 is strongly regular

$?\,v = 135 = 1+8+56+70$ $\quad 8^1\ 3^{54}\ -1^{50}\ -4^{30}$ $\quad \{8,7,5;1,1,4\}$ $\quad a_1 = 0$
Γ_3 is strongly regular

$+\,v = 64 = 1+9+27+27$ $\quad 9^1\ 5^9\ 1^{27}\ -3^{27}$ $\quad \{9,6,3;1,2,3\}$
Hamming (or Doob) graph 4^3; Γ_2 is strongly regular; classical ($b = 1$)

$!\,v = 186 = 1+10+50+125$ $\quad 10^1\ 6.236^{30}\ 1.764^{30}\ -2^{125}$ $\quad \{10,5,5;1,1,2\}$
Generalized hexagon of order (5,1)

$!\,v = 65 = 1+10+30+24$ $\quad 10^1\ 5^{13}\ 0^{26}\ -3^{25}$ $\quad \{10,6,4;1,2,5\}$
Locally Petersen graph, cf. §12.2; no 4-cycles

$?\,v = 231 = 1+10+80+140$ $\quad 10^1\ 4^{77}\ -1^{98}\ -4^{55}$ $\quad \{10,8,7;1,1,4\}$
Γ_3 is strongly regular

$?\,v = 210 = 1+11+110+88$ $\quad 11^1\ 4^{55}\ 1^{77}\ -4^{77}$ $\quad \{11,10,4;1,1,5\}$ $\quad a_1 = 0$
Γ_2 is strongly regular

$!\,v = 35 = 1+12+18+4$ $\quad 12^1\ 5^6\ 0^{14}\ -3^{14}$ $\quad \{12,6,2;1,4,9\}$
Johnson graph J(7,3); dist. 2 in generalized Odd graph; Γ_2 is strongly regular; classical ($b = 1$); equality in absolute bound for $(i,j) = (1,1)$

$!\,v = 175 = 1+12+72+90$ $\quad 12^1\ 7^{28}\ 2^{21}\ -2^{125}$ $\quad \{12,6,5;1,1,4\}$
Edges of Hoffman-Singleton graph; Γ_2 is strongly regular

$!\,v = 125 = 1+12+48+64$ $\quad 12^1\ 7^{12}\ 2^{48}\ -3^{64}$ $\quad \{12,8,4;1,2,3\}$
Hamming graph 5^3; classical ($b = 1$)

$v = 364 = 1+12+108+243$ $\quad 12^1\ 5^{104}\ -1^{168}\ -4^{91}$ $\quad \{12,9,9;1,1,4\}$
Generalized hexagon of order (3,3); Γ_3 is strongly regular

$?\,v = 88 = 1+12+60+15$ $\quad 12^1\ 4^{22}\ 1^{32}\ -4^{33}$ $\quad \{12,10,2;1,2,8\}$

$!\,v = 68 = 1+12+40+15$ $\quad 12^1\ 4^{17}\ 0^{34}\ -5^{16}$ $\quad \{12,10,3;1,3,8\}$
Doro graph, cf. §12.1

$v = 208 = 1+12+120+75$ $\quad 12^1\ 4^{78}\ -1^{64}\ -4^{65}$ $\quad \{12,10,5;1,1,8\}$
Unitary nonisotropics graph, cf. §12.4; Γ_3 is strongly regular

$?\,v = 144 = 1+13+65+65$ $\quad 13^1\ 5^{39}\ -1^{78}\ -5^{26}$ $\quad \{13,10,7;1,2,7\}$
Γ_3 is strongly regular

$!\,v = 456 = 1+14+98+343$ $\quad 14^1\ 8.646^{56}\ 3.354^{56}\ -2^{343}$ $\quad \{14,7,7;1,1,2\}$ $\quad q^2_{22} = 0$
Generalized hexagon of order (7,1)

$!\,v = 135 = 1+14+56+64$ $\quad 14^1\ 5^{35}\ -1^{84}\ -7^{15}$ $\quad \{14,12,8;1,3,7\}$ $\quad q^3_{33} = 0$
dual polar graph $B/C_3(2)$; Γ_3 is strongly regular; classical ($b = 2$); equality in absolute bound for $(i,j) = (3,3)$

$?\,v = 855 = 1+14+168+672$ $\quad 14^1\ 5^{266}\ -1^{399}\ -5^{189}$ $\quad \{14,12,12;1,1,3\}$
Γ_3 is strongly regular

$!\,v = 56 = 1+15+30+10$ $\quad 15^1\ 7^7\ 1^{20}\ -3^{28}$ $\quad \{15,8,3;1,4,9\}$ $\quad p^3_{33} = 0$
Johnson graph J(8,3); classical ($b = 1$); equality in absolute bound for $(i,j) = (1,1)$

$!\,v = 216 = 1+15+75+125$ $\quad 15^1\ 9^{15}\ 3^{75}\ -3^{125}$ $\quad \{15,10,5;1,2,3\}$
Hamming graph 6^3; classical ($b = 1$)

$?\,v = 160 = 1+15+90+54$ $\quad 15^1\ 5^{48}\ -1^{75}\ -5^{36}$ $\quad \{15,12,6;1,2,10\}$
Γ_3 is strongly regular

$!\,v = 506 = 1+15+210+280$ $\quad 15^1\ 4^{230}\ -3^{253}\ -8^{22}$ $\quad \{15,14,12;1,1,9\}$ $\quad a_1 = 0$
Residual of S(5,8,24), cf. §11.4B; classical ($b = -2$); equality in absolute bound for $(i,j) = (3,3)$

$!\,v = 657 = 1+16+128+512$ $\quad 16^1\ 9.828^{72}\ 4.172^{72}\ -2^{512}$ $\quad \{16,8,8;1,1,2\}$
Generalized hexagon of order (8,1)

$?\,v = 324 = 1+17+136+170$ $\quad 17^1\ 5^{102}\ -1^{170}\ -7^{51}$ $\quad \{17,16,10;1,2,8\}$
$q^3_{33} = 0$; $a_1 = 0$; Γ_3 is strongly regular

$v = 910 = 1+18+162+729$ $\quad 18^1\ 11^{90}\ 5^{90}\ -2^{729}$ $\quad \{18,9,9;1,1,2\}$
Generalized hexagon of order (9,1)

$!\,v = 84 = 1+18+45+20$ $\quad 18^1\ 9^8\ 2^{27}\ -3^{48}$ $\quad \{18,10,4;1,4,9\}$
Johnson graph J(9,3); classical ($b = 1$); equality in absolute bound for $(i,j) = (1,1)$

$!\,v = 343 = 1+18+108+216$ $\quad 18^1\ 11^{18}\ 4^{108}\ -3^{216}$ $\quad \{18,12,6;1,2,3\}$
Hamming graph 7^3; classical ($b = 1$)

? $v = 190 = 1+18+126+45$ $18^1\ 5.623^{57}\ -1^{75}\ -4.623^{57}$ $\{18,14,5;1,2,14\}$
Γ_3 is strongly regular

? $v = 532 = 1+18+270+243$ $18^1\ 5.623^{171}\ -1^{189}\ -4.623^{171}$ $\{18,15,9;1,1,10\}$
Γ_3 is strongly regular

! $v = 819 = 1+18+288+512$ $18^1\ 5^{324}\ -3^{468}\ -9^{26}$ $\{18,16,16;1,1,9\}$
Generalized hexagon of order (2,8); classical ($b = -2$); equality in absolute bound for $(i,j) = (3,3)$

? $v = 96 = 1+19+57+19$ $19^1\ 7^{19}\ -1^{57}\ -5^{19}$ $\{19,12,5;1,4,15\}$
Γ_3 is strongly regular; Γ_2 is strongly regular; Q-polynomial

? $v = 324 = 1+19+152+152$ $19^1\ 7^{57}\ 1^{152}\ -5^{114}$ $\{19,16,8;1,2,8\}$
Γ_2 is strongly regular

? $v = 441 = 1+20+320+100$ $20^1\ 6^{144}\ -1^{100}\ -4^{196}$ $\{20,16,5;1,1,16\}$
Γ_3 is strongly regular

? $v = 525 = 1+20+360+144$ $20^1\ 5^{210}\ -1^{125}\ -5^{189}$ $\{20,18,6;1,1,15\}$
Γ_3 is strongly regular

? $v = 189 = 1+20+120+48$ $20^1\ 5^{63}\ -1^{90}\ -7^{35}$ $\{20,18,6;1,3,15\}$
Γ_3 is strongly regular

! $v = 64 = 1+21+35+7$ $21^1\ 9^7\ 1^{21}\ -3^{35}$ $\{21,10,3;1,6,15\}$ $q^1_{11} = 0$
Halved 7-cube; dist. 2 in generalized Odd graph; Γ_2 is strongly regular; classical ($b = 1$); 2nd Q-polynomial ordering: 0 2 3 1

! $v = 120 = 1+21+63+35$ $21^1\ 11^9\ 3^{35}\ -3^{75}$ $\{21,12,5;1,4,9\}$
Johnson graph J(10,3); Γ_2 is strongly regular; classical ($b = 1$); equality in absolute bound for $(i,j) = (1,1)$

! $v = 512 = 1+21+147+343$ $21^1\ 13^{21}\ 5^{147}\ -3^{343}$ $\{21,14,7;1,2,3\}$
Hamming graph 8^3; classical ($b = 1$)

? $v = 154 = 1+21+84+48$ $21^1\ 7^{33}\ -1^{98}\ -7^{22}$ $\{21,16,8;1,4,14\}$
Γ_3 is strongly regular; classical ($b = 2$)

? $v = 330 = 1+21+168+140$ $21^1\ 7.325^{77}\ -1^{175}\ -5.325^{77}$ $\{21,16,10;1,2,12\}$
Γ_3 is strongly regular

? $v = 792 = 1+21+420+350$ $21^1\ 5^{315}\ -1^{252}\ -6^{224}$ $\{21,20,10;1,1,12\}$ $a_1 = 0$
Γ_3 is strongly regular

! $v = 512 = 1+21+210+280$ $21^1\ 5^{210}\ -3^{280}\ -11^{21}$ $\{21,20,16;1,2,12\}$
Coset graph of doubly truncated binary Golay code, cf. §11.3G; $q^3_{33} = 0$; $a_1 = 0$; classical ($b = -2$)

? $v = 243 = 1+22+176+44$ $22^1\ 7^{66}\ -2^{132}\ -5^{44}$ $\{22,16,5;1,2,20\}$
Γ_2 is strongly regular

? $v = 320 = 1+22+231+66$ $22^1\ 6^{55}\ 2^{154}\ -6^{110}$ $\{22,21,4;1,2,14\}$ $a_1 = 0$
Γ_2 is strongly regular

$v = 1024 = 1+22+231+770$ $22^1\ 6^{330}\ -2^{616}\ -10^{77}$ $\{22,21,20;1,2,6\}$ $q^3_{33} = 0$
Coset graph of truncated binary Golay code, cf. §11.3F; generalized Odd graph

! $v = 165 = 1+24+84+56$ $24^1\ 13^{10}\ 4^{44}\ -3^{110}$ $\{24,14,6;1,4,9\}$
Johnson graph J(11,3); classical ($b = 1$); equality in absolute bound for $(i,j) = (1,1)$

! $v = 729 = 1+24+192+512$ $24^1\ 15^{24}\ 6^{192}\ -3^{512}$ $\{24,16,8;1,2,3\}$
Hamming graph 9^3; classical ($b = 1$)

? $v = 700 = 1+24+432+243$ $24^1\ 8^{175}\ -1^{224}\ -4^{300}$ $\{24,18,9;1,1,16\}$
Γ_3 is strongly regular

? $v = 221 = 1+24+168+28$ $24^1\ 6^{52}\ 1.606^{84}\ -5.606^{84}$ $\{24,21,3;1,3,18\}$

! $v = 729 = 1+24+264+440$ $24^1\ 6^{264}\ -3^{440}\ -12^{24}$ $\{24,22,20;1,2,12\}$
Near hexagon; Extended ternary Golay code; classical ($b = -2$)

? $v = 256 = 1+25+200+30$ $25^1\ 5^{80}\ 1^{100}\ -7^{75}$ $\{25,24,3;1,3,20\}$ $a_1 = 0$

? $v = 896 = 1+25+600+270$ $25^1\ 5^{420}\ -3^{300}\ -7^{175}$ $\{25,24,9;1,1,20\}$ $a_1 = 0$

? $v = 729 = 1+26+208+494$ $26^1\ 8^{156}\ -1^{494}\ -10^{78}$ $\{26,24,19;1,3,8\}$ $q^3_{33} = 0$
Γ_3 is strongly regular

! $v = 220 = 1+27+108+84$ $27^1\ 15^{11}\ 5^{54}\ -3^{154}$ $\{27,16,7;1,4,9\}$
Johnson graph J(12,3); classical ($b = 1$); equality in absolute bound for $(i,j) = (1,1)$

! $v = 1000 = 1+27+243+729$ $27^1\ 17^{27}\ 7^{243}\ -3^{729}$ $\{27,18,9;1,2,3\}$
Hamming graph 10^3; classical ($b = 1$)

? $v = 208 = 1+27+135+45$ $27^1\ 8.211^{45}\ -1^{117}\ -6.211^{45}$ $\{27,20,7;1,4,21\}$
Γ_3 is strongly regular

? $v = 448 = 1+27+270+150$ $27^1\ 9^{96}\ -1^{216}\ -5^{135}$ $\{27,20,10;1,2,18\}$
Γ_3 is strongly regular

$v = 765 = 1+28+224+512$ $28^1\ 11^{84}\ 1^{476}\ -7^{204}$ $\{28,24,16;1,3,7\}$
dual polar graph ${}^2D_4(2)$; classical ($b = 2$)

? $v = 378 = 1+29+232+116$ $29^1\ 11^{29}\ 3.243^{174}\ -5.243^{174}$ $\{29,24,7;1,3,14\}$

! $v = 286 = 1+30+135+120$ $30^1\ 17^{12}\ 6^{65}\ -3^{208}$ $\{30,18,8;1,4,9\}$
Johnson graph J(13,3); classical ($b = 1$); equality in absolute bound for $(i,j) = (1,1)$
? $v = 350 = 1+30+220+99$ $30^1\ 10^{63}\ 0^{154}\ -5^{132}$ $\{30,22,9;1,3,20\}$
? $v = 486 = 1+30+260+195$ $30^1\ 12^{30}\ 3^{260}\ -6^{195}$ $\{30,26,9;1,3,12\}$
Q-polynomial
? $v = 486 = 1+30+420+35$ $30^1\ 6^{135}\ 3^{140}\ -6^{210}$ $\{30,28,2;1,2,24\}$
? $v = 416 = 1+30+280+105$ $30^1\ 6^{156}\ -2^{195}\ -9^{64}$ $\{30,28,9;1,3,24\}$
! $v = 759 = 1+30+280+448$ $30^1\ 7^{252}\ -3^{483}\ -15^{23}$ $\{30,28,24;1,3,15\}$
Near hexagon; S(5,8,24); classical ($b = -2$); equality in absolute bound for $(i,j) = (3,3)$
? $v = 280 = 1+31+186+62$ $31^1\ 8.746^{62}\ -1^{155}\ -6.746^{62}$ $\{31,24,8;1,4,24\}$
Γ_3 is strongly regular
? $v = 1024 = 1+31+465+527$ $31^1\ 7^{310}\ -1^{527}\ -9^{186}$ $\{31,30,17;1,2,15\}$ $a_1 = 0$
Γ_3 is strongly regular
? $v = 625 = 1+32+448+144$ $32^1\ 7^{224}\ -3^{320}\ -8^{80}$ $\{32,28,9;1,2,28\}$
! $v = 364 = 1+33+165+165$ $33^1\ 19^{13}\ 7^{77}\ -3^{273}$ $\{33,20,9;1,4,9\}$
Johnson graph J(14,3); classical ($b = 1$); equality in absolute bound for $(i,j) = (1,1)$
? $v = 1024 = 1+33+495+495$ $33^1\ 9^{198}\ 1^{495}\ -7^{330}$ $\{33,30,15;1,2,15\}$
Γ_2 is strongly regular
? $v = 216 = 1+35+140+40$ $35^1\ 11^{35}\ -1^{140}\ -7^{40}$ $\{35,24,8;1,6,28\}$
Γ_3 is strongly regular; Γ_2 is strongly regular; classical ($b = 2$)
? $v = 624 = 1+35+490+98$ $35^1\ 9^{168}\ -1^{182}\ -5^{273}$ $\{35,28,6;1,2,30\}$
Γ_3 is strongly regular
? $v = 624 = 1+35+525+63$ $35^1\ 9^{90}\ 5^{182}\ -5^{351}$ $\{35,30,3;1,2,25\}$
Γ_2 is strongly regular
? $v = 486 = 1+35+350+100$ $35^1\ 8^{140}\ -1^{210}\ -7^{135}$ $\{35,30,8;1,3,28\}$
Γ_3 is strongly regular
? $v = 756 = 1+35+560+160$ $35^1\ 7^{270}\ -1^{245}\ -7^{240}$ $\{35,32,8;1,2,28\}$
Γ_3 is strongly regular
! $v = 455 = 1+36+198+220$ $36^1\ 21^{14}\ 8^{90}\ -3^{350}$ $\{36,22,10;1,4,9\}$
Johnson graph J(15,3); classical ($b = 1$); equality in absolute bound for $(i,j) = (1,1)$
? $v = 352 = 1+36+225+90$ $36^1\ 14^{32}\ 4^{99}\ -4^{220}$ $\{36,25,8;1,4,20\}$
Γ_2 is strongly regular
$v = 462 = 1+36+225+200$ $36^1\ 14^{54}\ 0^{275}\ -6^{132}$ $\{36,25,16;1,4,18\}$
Folded Johnson graph J(12,6); Q-polynomial
? $v = 625 = 1+36+504+84$ $36^1\ 11^{84}\ 6^{120}\ -4^{420}$ $\{36,28,4;1,2,24\}$
Γ_2 is strongly regular
? $v = 847 = 1+36+360+450$ $36^1\ 14^{63}\ 3^{420}\ -6^{363}$ $\{36,30,15;1,3,12\}$
! $v = 560 = 1+39+234+286$ $39^1\ 23^{15}\ 9^{104}\ -3^{440}$ $\{39,24,11;1,4,9\}$
Johnson graph J(16,3); classical ($b = 1$); equality in absolute bound for $(i,j) = (1,1)$
? $v = 300 = 1+39+234+26$ $39^1\ 9^{78}\ -1^{117}\ -6^{104}$ $\{39,30,4;1,5,36\}$
Γ_3 is strongly regular; Γ_2 is strongly regular
? $v = 456 = 1+39+312+104$ $\{39,32,10;1,4,30\}$
Γ_3 is strongly regular $39^1\ 9.718^{104}\ -1^{247}\ -7.718^{104}$
? $v = 560 = 1+39+312+208$ $39^1\ 11^{104}\ -1^{364}\ -9^{91}$ $\{39,32,16;1,4,24\}$
Γ_3 is strongly regular
? $v = 525 = 1+40+440+44$ $40^1\ 10^{84}\ 5^{132}\ -5^{308}$ $\{40,33,3;1,3,30\}$
? $v = 250 = 1+40+165+44$ $40^1\ 10^{40}\ 0^{165}\ -10^{44}$ $\{40,33,8;1,8,30\}$
Q-polynomial
? $v = 861 = 1+40+440+380$ $\{40,33,19;1,3,22\}$
Γ_3 is strongly regular $40^1\ 11.311^{164}\ -1^{532}\ -8.311^{164}$
! $v = 680 = 1+42+273+364$ $42^1\ 25^{16}\ 10^{119}\ -3^{544}$ $\{42,26,12;1,4,9\}$
Johnson graph J(17,3); classical ($b = 1$); equality in absolute bound for $(i,j) = (1,1)$
? $v = 176 = 1+42+126+7$ $42^1\ 9^{32}\ 2^{66}\ -6^{77}$ $\{42,30,2;1,10,36\}$
dist. 2 in generalized Odd graph; Γ_2 is strongly regular
? $v = 343 = 1+42+210+90$ $42^1\ 14^{42}\ 0^{210}\ -7^{90}$ $\{42,30,12;1,6,28\}$
classical ($b = 2$)
? $v = 351 = 1+42+224+84$ $42^1\ 15^{26}\ 3^{168}\ -6^{156}$ $\{42,32,9;1,6,24\}$
Γ_2 is strongly regular; Q-polynomial
? $v = 795 = 1+42+560+192$ $42^1\ 7^{318}\ -3^{371}\ -11^{105}$ $\{42,40,12;1,3,35\}$
! $v = 891 = 1+42+336+512$ $42^1\ 9^{252}\ -3^{616}\ -21^{22}$ $\{42,40,32;1,5,21\}$ $q^3_{33} = 0$
dual polar graph ${}^2A_6(2)$; classical ($b = 4$); classical ($b = -2$); equality in absolute bound for $(i,j) = (3,3)$
? $v = 560 = 1+43+387+129$ $\{43,36,11;1,4,33\}$
Γ_3 is strongly regular $43^1\ 10.165^{129}\ -1^{301}\ -8.165^{129}$

? $v = 540 = 1+44+440+55$ $44^1\ 14^{88}\ -1^{176}\ -4^{275}$ $\{44,30,5;1,3,40\}$
Γ_3 is strongly regular; Γ_2 is strongly regular

? $v = 375 = 1+44+264+66$ $44^1\ 14^{55}\ -1^{220}\ -6^{99}$ $\{44,30,9;1,5,36\}$
Γ_3 is strongly regular; Γ_2 is strongly regular

? $v = 375 = 1+44+308+22$ $44^1\ 9^{110}\ -1^{110}\ -6^{154}$ $\{44,35,3;1,5,42\}$
Γ_3 is strongly regular; Γ_2 is strongly regular

? $v = 243 = 1+44+176+22$ $44^1\ 8^{66}\ -1^{132}\ -10^{44}$ $\{44,36,5;1,9,40\}$
Γ_3 is strongly regular; Γ_2 is strongly regular

? $v = 765 = 1+44+528+192$ $44^1\ 11^{170}\ -1^{374}\ -7^{220}$ $\{44,36,12;1,3,33\}$
Γ_3 is strongly regular

? $v = 525 = 1+44+352+128$ $44^1\ 9^{132}\ -1^{308}\ -11^{84}$ $\{44,40,12;1,5,33\}$
Γ_3 is strongly regular

? $v = 342 = 1+44+264+33$ $44^1\ 6^{132}\ -1^{152}\ -12^{57}$ $\{44,42,5;1,7,40\}$
Γ_3 is strongly regular

! $v = 816 = 1+45+315+455$ $45^1\ 27^{17}\ 11^{135}\ -3^{663}$ $\{45,28,13;1,4,9\}$
Johnson graph J(18,3); classical ($b = 1$); equality in absolute bound for $(i,j) = (1,1)$

? $v = 896 = 1+45+675+175$ $45^1\ 17^{90}\ 9^{70}\ -3^{735}$ $\{45,30,7;1,2,27\}$
Γ_2 is strongly regular

? $v = 896 = 1+45+600+250$ $45^1\ 9^{280}\ -3^{525}\ -11^{90}$ $\{45,40,15;1,3,36\}$

? $v = 990 = 1+46+483+460$ $46^1\ 13^{120}\ 1^{644}\ -10^{225}$ $\{46,42,20;1,4,21\}$
Γ_2 is strongly regular

! $v = 969 = 1+48+360+560$ $48^1\ 29^{18}\ 12^{152}\ -3^{798}$ $\{48,30,14;1,4,9\}$
Johnson graph J(19,3); classical ($b = 1$); equality in absolute bound for $(i,j) = (1,1)$

? $v = 343 = 1+48+240+54$ $48^1\ 13^{56}\ -1^{216}\ -8^{70}$ $\{48,35,9;1,7,40\}$
Γ_3 is strongly regular; Γ_2 is strongly regular

? $v = 540 = 1+49+441+49$ $49^1\ 13^{105}\ -1^{189}\ -5^{245}$ $\{49,36,5;1,4,45\}$
Γ_3 is strongly regular; Γ_2 is strongly regular

? $v = 400 = 1+49+294+56$ $49^1\ 13^{70}\ -1^{224}\ -7^{105}$ $\{49,36,8;1,6,42\}$
Γ_3 is strongly regular; Γ_2 is strongly regular

$v = 512 = 1+49+294+168$ $49^1\ 17^{49}\ 1^{294}\ -7^{168}$ $\{49,36,16;1,6,28\}$
Bilinear forms graph $H_2(3,3)$; classical ($b = 2$)

? $v = 750 = 1+49+392+308$ $49^1\ 14^{112}\ -1^{539}\ -11^{98}$ $\{49,40,22;1,5,28\}$
Γ_3 is strongly regular; Q-polynomial

? $v = 546 = 1+50+440+55$ $50^1\ 10^{105}\ 2.623^{220}\ -7.623^{220}$ $\{50,44,5;1,5,40\}$

? $v = 576 = 1+50+375+150$ $50^1\ 14^{50}\ 2^{375}\ -10^{150}$ $\{50,45,12;1,6,30\}$
Γ_2 is strongly regular; Q-polynomial

? $v = 800 = 1+51+561+187$ $51^1\ 11^{187}\ -1^{425}\ -9^{187}$ $\{51,44,13;1,4,39\}$
Γ_3 is strongly regular

? $v = 800 = 1+51+612+136$ $51^1\ 11^{102}\ 3^{425}\ -9^{272}$ $\{51,48,8;1,4,36\}$
Γ_2 is strongly regular

? $v = 768 = 1+52+455+260$ $52^1\ 20^{65}\ 4^{182}\ -4^{520}$ $\{52,35,16;1,4,28\}$
Γ_2 is strongly regular

? $v = 625 = 1+52+364+208$ $52^1\ 17^{52}\ 2^{364}\ -8^{208}$ $\{52,42,16;1,6,28\}$
Γ_2 is strongly regular; Q-polynomial

? $v = 550 = 1+54+432+63$ $54^1\ 14^{99}\ -1^{252}\ -6^{198}$ $\{54,40,7;1,5,48\}$
Γ_3 is strongly regular; Γ_2 is strongly regular

? $v = 1000 = 1+54+756+189$ $54^1\ 14^{189}\ -1^{432}\ -6^{378}$ $\{54,42,11;1,3,44\}$
Γ_3 is strongly regular

? $v = 672 = 1+55+495+121$ $55^1\ 19^{77}\ -1^{363}\ -5^{231}$ $\{55,36,11;1,4,45\}$
Γ_3 is strongly regular; Γ_2 is strongly regular

? $v = 896 = 1+55+660+180$ $55^1\ 11^{224}\ -1^{440}\ -9^{231}$ $\{55,48,12;1,4,44\}$
Γ_3 is strongly regular

? $v = 616 = 1+55+440+120$ $55^1\ 11^{140}\ -1^{363}\ -11^{112}$ $\{55,48,12;1,6,44\}$
Γ_3 is strongly regular

? $v = 936 = 1+55+660+220$ $\{55,48,14;1,4,42\}$
Γ_3 is strongly regular $55^1\ 11.392^{220}\ -1^{495}\ -9.392^{220}$

? $v = 756 = 1+55+550+150$ $55^1\ 10^{200}\ -1^{405}\ -11^{150}$ $\{55,50,12;1,5,44\}$
Γ_3 is strongly regular

? $v = 855 = 1+56+672+126$ $56^1\ 20^{95}\ -1^{360}\ -4^{399}$ $\{56,36,9;1,3,48\}$
Γ_3 is strongly regular; Γ_2 is strongly regular

? $v = 729 = 1+56+392+280$ $56^1\ 20^{56}\ 2^{392}\ -7^{280}$ $\{56,42,20;1,6,28\}$
Γ_2 is strongly regular; classical ($b = 2$)

? $v = 855 = 1+56+672+126$ $56^1\ 11^{224}\ -1^{360}\ -8^{270}$ $\{56,48,9;1,4,48\}$
Γ_3 is strongly regular

? $v = 768 = 1+59+531+177$ $59^1\ 11^{177}\ -1^{472}\ -13^{118}$ $\{59,54,15;1,6,45\}$
Γ_3 is strongly regular
? $v = 670 = 1+60+420+189$ $60^1\ 20^{67}\ 0^{402}\ -7^{200}$ $\{60,42,18;1,6,40\}$
? $v = 322 = 1+60+225+36$ $60^1\ 14^{45}\ 0^{207}\ -10^{69}$ $\{60,45,8;1,12,50\}$ $p^3_{33} = 0$
Γ_2 is strongly regular; Q-polynomial
? $v = 438 = 1+60+312+65$ $60^1\ 12^{73}\ 0^{292}\ -13^{72}$ $\{60,52,10;1,10,48\}$
? $v = 900 = 1+62+558+279$ $62^1\ 14^{155}\ -1^{620}\ -13^{124}$ $\{62,54,21;1,6,42\}$
Γ_3 is strongly regular
? $v = 430 = 1+63+294+72$ $63^1\ 20^{42}\ 0^{258}\ -7^{129}$ $\{63,42,12;1,9,49\}$
Γ_2 is strongly regular; classical ($b = 2$)
? $v = 512 = 1+63+378+70$ $63^1\ 15^{84}\ -1^{315}\ -9^{112}$ $\{63,48,10;1,8,54\}$
Γ_3 is strongly regular; Γ_2 is strongly regular
? $v = 1000 = 1+63+504+432$ $63^1\ 23^{63}\ 3^{504}\ -7^{432}$ $\{63,48,24;1,6,28\}$
classical ($b = 2$)
? $v = 784 = 1+63+630+90$ $63^1\ 14^{160}\ -1^{343}\ -7^{280}$ $\{63,50,8;1,5,56\}$
Γ_3 is strongly regular
? $v = 924 = 1+65+715+143$ $65^1\ 21^{105}\ -1^{455}\ -5^{363}$ $\{65,44,11;1,4,55\}$
Γ_3 is strongly regular; Γ_2 is strongly regular
? $v = 768 = 1+65+585+117$ $65^1\ 17^{65}\ 5^{312}\ -7^{390}$ $\{65,54,9;1,6,45\}$
Γ_2 is strongly regular
? $v = 924 = 1+65+585+273$ $\{65,54,21;1,6,45\}$
Γ_3 is strongly regular $65^1\ 15.748^{143}\ -1^{637}\ -11.748^{143}$
? $v = 864 = 1+65+728+70$ $65^1\ 13^{135}\ 5^{273}\ -7^{455}$ $\{65,56,5;1,5,52\}$
$v = 1024 = 1+66+495+462$ $66^1\ 26^{66}\ 2^{495}\ -6^{462}$ $\{66,45,28;1,6,30\}$
Folded half-cube; Γ_2 is strongly regular; Q-polynomial
? $v = 825 = 1+68+612+144$ $68^1\ 17^{110}\ 2^{340}\ -7^{374}$ $\{68,54,12;1,6,51\}$
? $v = 392 = 1+69+276+46$ $69^1\ 13^{69}\ -1^{276}\ -15^{46}$ $\{69,56,10;1,14,60\}$
$q^3_{33} = 0$; $p^3_{33} = 0$; Γ_3 is strongly regular; Γ_2 is strongly regular; Q-polynomial
? $v = 598 = 1+72+405+120$ $72^1\ 26^{45}\ 0^{345}\ -6^{207}$ $\{72,45,16;1,8,54\}$
Γ_2 is strongly regular; Q-polynomial
? $v = 783 = 1+72+630+80$ $72^1\ 9^{232}\ 0^{406}\ -15^{144}$ $\{72,70,8;1,8,63\}$
? $v = 630 = 1+74+444+111$ $74^1\ 20^{74}\ -1^{444}\ -10^{111}$ $\{74,54,15;1,9,60\}$
Γ_3 is strongly regular; Γ_2 is strongly regular; Q-polynomial
? $v = 1000 = 1+74+814+111$ $74^1\ 19^{148}\ -1^{444}\ -6^{407}$ $\{74,55,9;1,5,66\}$
Γ_3 is strongly regular; Γ_2 is strongly regular
? $v = 630 = 1+74+518+37$ $74^1\ 11^{185}\ -1^{259}\ -10^{185}$ $\{74,63,5;1,9,70\}$
Γ_3 is strongly regular; Γ_2 is strongly regular
? $v = 756 = 1+75+600+80$ $75^1\ 15^{105}\ 3^{350}\ -9^{300}$ $\{75,64,8;1,8,60\}$
? $v = 540 = 1+77+385+77$ $77^1\ 17^{77}\ -1^{385}\ -13^{77}$ $\{77,60,13;1,12,65\}$
Γ_3 is strongly regular; Γ_2 is strongly regular; Q-polynomial
? $v = 378 = 1+78+260+39$ $78^1\ 24^{26}\ 3^{156}\ -6^{195}$ $\{78,50,9;1,15,60\}$ $p^3_{33} = 0$
Γ_2 is strongly regular; Q-polynomial
? $v = 945 = 1+80+720+144$ $80^1\ 26^{80}\ 5^{144}\ -4^{720}$ $\{80,54,12;1,6,60\}$
? $v = 729 = 1+80+560+88$ $80^1\ 17^{120}\ -1^{440}\ -10^{168}$ $\{80,63,11;1,9,70\}$
Γ_3 is strongly regular; Γ_2 is strongly regular
? $v = 585 = 1+80+420+84$ $80^1\ 20^{52}\ 2^{350}\ -10^{182}$ $\{80,63,12;1,12,60\}$
Q-polynomial
? $v = 621 = 1+80+480+60$ $80^1\ 11^{160}\ -1^{368}\ -16^{92}$ $\{80,72,9;1,12,72\}$
Γ_3 is strongly regular
? $v = 800 = 1+85+459+255$ $85^1\ 35^{34}\ 5^{255}\ -5^{510}$ $\{85,54,25;1,10,45\}$
Γ_2 is strongly regular; Q-polynomial
? $v = 726 = 1+87+522+116$ $87^1\ 21^{87}\ -1^{522}\ -12^{116}$ $\{87,66,16;1,11,72\}$
Γ_3 is strongly regular; Γ_2 is strongly regular; Q-polynomial
? $v = 616 = 1+90+450+75$ $90^1\ 27^{48}\ 2^{252}\ -6^{315}$ $\{90,60,12;1,12,72\}$
Γ_2 is strongly regular
? $v = 676 = 1+90+540+45$ $90^1\ 12^{180}\ -1^{360}\ -14^{135}$ $\{90,78,7;1,13,84\}$
Γ_3 is strongly regular; Γ_2 is strongly regular
? $v = 1000 = 1+99+792+108$ $99^1\ 19^{165}\ -1^{594}\ -11^{240}$ $\{99,80,12;1,10,88\}$
Γ_3 is strongly regular; Γ_2 is strongly regular
? $v = 729 = 1+104+572+52$ $104^1\ 32^{52}\ 5^{104}\ -4^{572}$ $\{104,66,8;1,12,88\}$
Γ_2 is strongly regular; Q-polynomial ordering: 0 2 3 1
? $v = 952 = 1+105+630+216$ $105^1\ 35^{51}\ 3^{441}\ -7^{459}$ $\{105,72,24;1,12,70\}$
classical ($b = 2$)
? $v = 672 = 1+110+495+66$ $110^1\ 26^{55}\ 2^{385}\ -10^{231}$ $\{110,81,12;1,18,90\}$ $p^3_{33} = 0$
Γ_2 is strongly regular; Q-polynomial

? $v = 1000 = 1+111+814+74$ $\quad 111^1\ 21^{148}\ 1^{444}\ -9^{407}$ $\quad \{111,88,9;1,12,99\}$
Γ_2 is strongly regular

? $v = 704 = 1+111+555+37$ $\quad \{111,90,7;1,18,105\}$ $\quad p^3_{33} = 0$
Γ_3 is strongly regular $\quad 111^1\ 15.832^{148}\ -1^{407}\ -13.832^{148}$

? $v = 750 = 1+112+539+98$ $\quad 112^1\ 32^{49}\ 2^{392}\ -8^{308}$ $\quad \{112,77,16;1,16,88\}$
Γ_2 is strongly regular; Q-polynomial

? $v = 960 = 1+119+714+126$ $\quad 119^1\ 23^{119}\ -1^{714}\ -17^{126}$ $\quad \{119,96,18;1,16,102\}$
Γ_3 is strongly regular; Γ_2 is strongly regular; Q-polynomial

? $v = 800 = 1+119+595+85$ $\quad 119^1\ 19^{119}\ -1^{595}\ -21^{85}$ $\quad \{119,100,15;1,20,105\}$
$q^3_{33} = 0$; $p^3_{33} = 0$; Γ_3 is strongly regular; Γ_2 is strongly regular; Q-polynomial

? $v = 900 = 1+145+609+145$ $\quad 145^1\ 55^{29}\ 5^{261}\ -5^{609}$ $\quad \{145,84,25;1,20,105\}$
Γ_2 is strongly regular; Q-polynomial

? $v = 870 = 1+154+616+99$ $\quad 154^1\ 49^{29}\ 9^{231}\ -6^{609}$ $\quad \{154,100,18;1,25,112\}$ $\quad p^3_{33} = 0$
Q-polynomial

? $v = 900 = 1+174+638+87$ $\quad 174^1\ 54^{29}\ 9^{232}\ -6^{638}$ $\quad \{174,110,18;1,30,132\}$ $\quad p^3_{33} = 0$
Γ_2 is strongly regular; Q-polynomial

$v = 1024 = 1+231+770+22$ $\quad 231^1\ 39^{77}\ 7^{330}\ -9^{616}$ $\quad \{231,160,6;1,48,210\}$ $\quad q^1_{11} = 0$
dist. 2 in generalized Odd graph

Diameter 3 - antipodal (but not bipartite) graphs

These graphs have intersection array

$$v = r(k+1) \quad \{k, \mu(r-1), 1;\ 1, \mu, k\} \quad (\mu < k-1)$$

and spectrum

$$k^1\ n^f\ (-1)^k\ (-m)^g$$

where n and $-m$ are the solutions of the equation $\theta^2 + (\mu-\lambda)\theta - k = 0$, and

$$f = \frac{m}{m+n}(r-1)(k+1), \quad g = \frac{n}{m+n}(r-1)(k+1).$$

If $\lambda \neq \mu$, then the eigenvalues are integral, and we may express the parameters in terms of m, n and r:

$$k = mn, \quad \mu = (m-1)(n+1)/r, \quad \lambda = \mu + n - m.$$

The integrality of the multiplicities gives the divisibility condition $(m+n)\,|\,(r-1)m(m^2-1)$; but equality is excluded by the absolute bound. The Krein condition $q^3_{33} \geqslant 0$ gives the further restriction $m \leqslant n^2$ when $r > 2$.
When $r = 2$ we have the Taylor graphs; they have two Q-polynomial orderings 123 and 321, and also $q^1_{11} = q^3_{33} = 0$, i.e., Taylor graphs are Q-bipartite. Moreover, Taylor graphs have two P-polynomial orderings 123 and 213, i.e., for Taylor graphs Γ, also Γ_2 is distance regular (with the same parameters as Γ). When $\mu = 1$ we have the graphs characterized in Proposition 1.17.3 (which leads to some parameter restrictions).
If $\lambda = \mu$, then we get the family

$$v = r(r\mu+2), \quad \{r\mu+1, (r-1)\mu, 1;\ 1, \mu, r\mu+1\}$$

with spectrum

$$k^1\ \sqrt{k}^f\ (-1)^k\ (-\sqrt{k})^f \quad \text{where } k = r\mu+1,\ f = \tbinom{r}{2}\mu + r - 1.$$

Lemma 4.3.1 gives the restriction $(r-1)\mu \equiv 0 \pmod 2$; further restrictions come from Hasse-Minkowski theory (Proposition 1.10.5). Examples exist whenever $r\mu+1$ is a prime power, and in a few other cases.

Gardiner [315] shows as a consequence of the feasibility conditions that if $\lambda = 0$ then either $r = 2$ or $r - 2 > \sqrt{\mu}$.

Diameter 3 - bipartite (but not antipodal) graphs

These graphs have intersection array

$$v = 2w = 2\left(1 + \frac{k(k-1)}{\mu}\right), \quad \{k, k-1, k-\mu; 1, \mu, k\} \quad (\mu < k-1)$$

and spectrum

$$k^1 \ \sqrt{k-\mu}^{\,w-1} \ (-\sqrt{k-\mu})^{w-1} \ (-k)^1 \ ;$$

they are just the incidence graphs of nontrivial square 2-(w,k,μ) designs. Apart from the restriction $\mu \mid k(k-1)$ we have a further restriction from Hasse-Minkowski theory (Theorem 1.10.4). Each graph has two Q-polynomial orderings 123 and 213 and Γ_3 is again distance-regular.

Diameter 3 - graphs that are both antipodal and bipartite

These graphs have intersection array

$$v = 2(k+1), \quad \{k, k-1, 1; 1, k-1, k\}$$

and spectrum

$$k^1 \ 1^k \ (-1)^k \ (-k)^1 .$$

They are realized uniquely by the complement of the $2\times(k+1)$-grid. (The P- and Q-polynomial structures are unique; for the other relations the scheme is disconnected.)

Appendix

Here we collect some standard material on graphs, groups, designs and the like. Most of the definitions given here are also given in the main text.

A.1 Graphs

A *graph* is a pair $\Gamma = (V,E)$ consisting of a set V, referred to as the *vertex set* of Γ and a set E of 2-subsets of V, referred to as the *edge set* of Γ. That is, our graphs are undirected, without loops or multiple edges. [*Directed graphs*, that is, pairs (V,E) where E is a subset of $V \times V$, will occur occasionally.] Two graphs (V,E) and (W,F) are *isomorphic* when there is a bijection $f: V \to W$ such that $f[E] = F$. Usually, we shall not distinguish between isomorphic graphs. Unless stated otherwise (as in Chapter 10), all graphs in this book are finite. A graph (V,E) is called *empty* when $V = \emptyset$. Elements of V, E are called *vertices* (or *points*), and *edges*, respectively. A 2-subset of V not in E is called a *nonedge* of Γ. We shall often write $\gamma\delta$ for the 2-subset $\{\gamma,\delta\}$. The *complement* of Γ (often denoted $\overline{\Gamma}$) is the graph with vertex set V whose edges are all the nonedges of Γ. Let $\Gamma = (V,E)$ be a graph, and let X be a subset of V. The *subgraph of* Γ *induced on* X is the graph with vertex set X whose edges are the edges of Γ contained in X. By abuse of notation, we generally denote this subgraph by X. A *partial graph* of Γ is a graph $\Delta = (V,F)$, where $F \subset E$. When two graphs $\Gamma = (V,E)$ and $\Delta = (V,F)$ have the same vertex set, then $\Gamma \cup \Delta$ denotes the graph $(V,E \cup F)$.

We write $\gamma \in \Gamma$ if γ is a vertex of Γ and $\gamma \sim \delta$ (or sometimes $\gamma\Gamma\delta$) if $\gamma,\delta \in \Gamma$ and $\{\gamma,\delta\}$ is an edge of Γ; in this case γ and δ are called *neighbours* or *joined*. The cardinality $|\Gamma|$ of the vertex set of Γ is denoted by $v(\Gamma)$ or simply v. A *path of length i joining* two vertices $\gamma, \delta \in \Gamma$ is a sequence $\gamma_0, \gamma_1, \cdots, \gamma_i$ of vertices such that $\gamma = \gamma_0 \sim \gamma_1 \sim \cdots \sim \gamma_i = \delta$. Being joined by a path is an equivalence relation. Its equivalence classes are called the (*connected*) *components* of Γ. The graph is called *connected* if it consists of one connected component. (In particular, we do not regard the empty graph as connected - it has no components.) [A directed graph is called *strongly connected* when for any two vertices γ, δ there is a sequence of vertices $\gamma = \gamma_0, \gamma_1, \cdots, \gamma_i = \delta$ such that $(\gamma_{j-1},\gamma_j) \in E$ for $j = 1, \cdots, i$.] The *distance* $d(\gamma,\delta)$ of two vertices γ, δ is the length of a shortest path (*geodesic*) from γ to δ in Γ. If there is no such path, the distance is defined to be ∞. A subset C of Γ is called *convex* or *geodetically closed* if it contains all geodesics between any two of its vertices. The *diameter* of Γ, denoted by d (or $d(\Gamma)$, or $\mathrm{diam}(\Gamma)$), is the maximal distance occurring in Γ. The *girth* of Γ, denoted by g, is the length of a shortest circuit (subgraph of valency 2) in Γ. A 1-*factor* of Γ is a partition of the vertex set of Γ into edges. A μ-*graph* in Γ is the subgraph induced by the set of common neighbours of two vertices at distance 2.

Given $\delta \in \Gamma$, we write $\Gamma_i(\delta)$ for the set of vertices γ with $d(\gamma,\delta) = i$, and $\Gamma_{\leqslant i}(\delta)$ for the set of vertices γ with $d(\gamma,\delta) \leqslant i$. In particular, $\Gamma(\delta) = \Gamma_1(\delta)$ denotes the set of neighbours of δ; $\Gamma(\delta)$ is called a local graph of Γ. The graph Γ is called *locally* $\mathfrak{X}$ for some graph property or class of graphs $\mathfrak{X}$ when $\Gamma(\gamma)$ has this property or belongs to this class for all $\gamma \in \Gamma$ (e.g., we talk about locally connected, locally complete multipartite, locally triangular, locally Petersen, etc.). The *valency* or *degree* $k(\gamma)$ of a vertex γ is the cardinality of $\Gamma(\gamma)$. A graph is called *regular* (with *valency* k) if each vertex has the same valency k. A *cubic* graph is a regular graph of valency 3.

A graph is called *complete* (or a *clique*) when any two of its vertices are adjacent. The complete graph on n vertices is denoted by K_n. A *coclique* is a graph in which no two vertices are adjacent. A *polygon* is a connected graph of valency 2. An s-*claw* is a subgraph consisting of a vertex of valency s and s vertices of valency 1. A graph Γ is the *disjoint union* of graphs $\Gamma^1, \cdots, \Gamma^s$, written $\Gamma = \Gamma^1 + \cdots + \Gamma^s$, if $\Gamma^1, \cdots, \Gamma^s$ are subgraphs of Γ such that the vertex sets of Γ^i partition that of Γ, and the edge sets of Γ^i partition that of Γ. The disjoint union of n graphs isomorphic to a graph Γ is written as $n\Gamma$. A graph is called *bipartite* (with *parts* or *classes* M, N) if its vertex set can be partitioned into two cocliques (M and N). We write $K_{m,n}$ for the complete bipartite graph with parts an m-set M and an n-set N, where each vertex in M is adjacent to each member of N. A graph is the *complete union* of graphs $\Gamma^1, \cdots, \Gamma^t$, written $\Gamma = \Gamma^1 \oplus \cdots \oplus \Gamma^t$, if $\Gamma^1, \cdots, \Gamma^t$ are subgraphs of Γ such that the vertex sets of the Γ^i partition the vertex set of Γ and the edges of Γ are precisely those of the Γ^i and those of the form $\gamma\delta$ with γ,δ in different Γ^i. A *complete multipartite graph* $K_{m_1,m_2,\cdots,m_t}$ is the complete union of cocliques of size $m_1, \cdots, m_t$; we shall write $K_{t\times m}$ for the complete union of t cocliques of size m. (Thus, $K_{2\times m}$ is the same as $K_{m,m}$.) A bipartite graph with parts M, N is called *semiregular* if there are constants m, n such that each vertex of M has valency m and each vertex of N has valency n. A graph is called *strongly regular* when it is regular and there are constants λ, μ such that any two adjacent (non-adjacent) vertices have λ (resp. μ) common neighbours. A first (second) *subconstituent* of a strongly regular graph is a subgraph $\Gamma(\infty)$ (resp. $\Gamma_2(\infty)$) for some $\infty \in \Gamma$.

For $\gamma, \delta \in \Gamma$ at distance j, we write $k_j(\gamma) := |\Gamma_j(\gamma)|$, $c_j(\gamma,\delta) := |\Gamma_{j-1}(\gamma) \cap \Gamma(\delta)|$, $a_j(\gamma,\delta) := |\Gamma_j(\gamma) \cap \Gamma(\delta)|$, and $b_j(\gamma,\delta) := |\Gamma_{j+1}(\gamma) \cap \Gamma(\delta)|$. We put $k(\gamma) = k_1(\gamma)$, $\lambda(\gamma,\delta) = a_1(\gamma,\delta)$ when $\gamma \sim \delta$, and $\mu(\gamma,\delta) = c_2(\gamma,\delta)$ when $d(\gamma,\delta) = 2$. In case $k(\gamma)$ does not depend on the vertex $\gamma \in \Gamma$ (i.e., when Γ is regular) we write $k = k(\Gamma) = k(\gamma)$, and similarly for the other parameters. The graph Γ is called *distance-regular* when $a_j(\gamma,\delta)$, $b_j(\gamma,\delta)$ and $c_j(\gamma,\delta)$ are independent of the vertices $\gamma, \delta \in \Gamma$ at distance j.

For non-complete distance-regular graphs Γ we define the *numerical girth* g as follows. Suppose $(c_1,a_1,b_1) = (c_2,a_2,b_2) = \cdots = (c_{i-1},a_{i-1},b_{i-1}) \neq (c_i,a_i,b_i)$. Then the numerical girth g of Γ is $2i$ if $c_i > 1$ and $2i+1$ otherwise. [Thus, the numerical girth of a distance-regular graph is always at least 4. In case $\lambda = 0$, the numerical girth equals the girth. In case Γ has singular lines of size $\lambda+2$, i.e., in case Γ has geometric girth larger than 3 (cf. §A.8), the numerical girth equals the geometric girth. Often this is the case, but, e.g., the Shrikhande graph has geometric girth 3 and numerical girth 4. In any case the geometric girth is not larger than the numerical girth.]

The *adjacency matrix* of Γ is the $v \times v$-matrix $A = A(\Gamma)$ indexed by the vertices of Γ, whose entries $a_{\gamma\delta}$ are given by $a_{\gamma\delta} = 1$ if $\gamma \sim \delta$, and $a_{\gamma\delta} = 0$ otherwise. The matrix A is symmetric, $A = A^{\top}$. Consequently, the eigenvalues of the adjacency matrix are real; they are called the *eigenvalues* of Γ. If Γ is regular of valency k, its adjacency matrix A satisfies the equations $AJ = kJ$ and $A\mathbf{1} = k\mathbf{1}$; in particular, k is an eigenvalue of Γ. Here, and throughout the book, J stands for the matrix all of whose entries are equal to one (its dimensions are usually clear from the context or specified), and $\mathbf{1}$ denotes a vector all of whose entries are one.

For any graph property $\mathcal{P}$ we say that a graph Γ has property *co-*$\mathcal{P}$ when the complementary graph $\overline{\Gamma}$ satisfies $\mathcal{P}$. Thus, Γ is coconnected whenever $\overline{\Gamma}$ is connected, the cocomponents of Γ are the components of $\overline{\Gamma}$, a coclique in Γ is a clique in $\overline{\Gamma}$, etc.

A.2 Permutation groups

A *permutation group* G is a group of permutations of a set X. Let us write gx for the image of x under g. An *orbit* of G is a set $Gx = \{gx \mid g \in G\}$; clearly, the orbits partition X. We write G_x for the stabilizer in G of x, that is, for the subgroup $\{g \in G \mid gx = x\}$. The orbits of G_x on X (or on $X \setminus \{x\}$) are called *suborbits* of G. An *orbital* of G is an orbit of G in its natural action on $X \times X$ (given by $g(x,y) = (gx,gy)$).

A.3 Automorphisms

An *automorphism* of a graph Γ is a permutation g of the vertex set of Γ such that, for all $\gamma, \delta \in \Gamma$, we have $\gamma \sim \delta$ if and only if $g\gamma \sim g\delta$. The group of all automorphisms of Γ is denoted by $\operatorname{Aut}\Gamma$. A subgroup G of $\operatorname{Aut}\Gamma$ is called *edge-transitive* if it is transitive on the edge set. In this case, the graph Γ is also called *edge-transitive.* A *flag* of a graph is an ordered pair (vertex,edge), and G (or Γ) is called *flag-transitive* when G is transitive on the set of flags, or, equivalently, when G is transitive on ordered pairs of adjacent vertices. Clearly, flag transitivity implies edge transitivity, but the two concepts are not equivalent, not even for regular graphs. (For example, the generalized 12-gon of order (1,2) is a connected cubic bipartite graph with automorphism group acting edge-transitively but not flag-transitively, indeed, not even vertex transitively. Examples of vertex- and edge- but not flag-transitive graphs have been given by Bouwer [114] and Holt [405].) Flag-transitive graphs can be described entirely in terms of groups, as follows.

Let G be a finite group, H a subgroup of G, and $x \in G \setminus H$ be an element satisfying $HxH = Hx^{-1}H$. Then we denote by $\Gamma(G,H,x)$ the graph whose vertex set is the set $G/H = \{gH \mid g \in G\}$ and in which g_1H and g_2H are adjacent whenever $g_2^{-1}g_1 \in HxH$. We shall refer to this graph as the *coset graph* of G on H (with respect to x). The group G acts on the graph $\Gamma(G,H,x)$ by left multiplication, and is flag-transitive. The stabilizer G_γ of a vertex $\gamma = gH$ is the group gHg^{-1}. Moreover, $\Gamma(G,H,x)$ is connected if and only if G is generated by H and x. Conversely we have:

A.3.1. Proposition. *Let Γ be a graph without isolated vertices, and let G be a group of automorphisms of Γ that acts flag-transitively. If $\gamma \sim \delta$, then there is an element $x \in G$ with $x\gamma = \delta$ such that Γ is isomorphic to $\Gamma(G,G_\gamma,x)$.*

Proof. The existence of x follows from $\gamma \sim \delta$, $\delta \sim \gamma$ and flag transitivity. It is easy to verify that the map $\phi: \Gamma \rightarrow \Gamma(G,G_\gamma,x)$ given by $\phi(\zeta) = gG_\gamma$ whenever $g \in G$ with

$g\gamma = \zeta$ is well defined and an isomorphism. □

Remark. More generally, we might for $X \subset G$ with $HXH = HX^{-1}H$ define $\Gamma(G,H,X)$ as the graph with vertex set G/H where $g_1H \sim g_2H$ whenever $g_2^{-1}g_1 \in HXH$. Now the appropriate generalization of Proposition A.3.1 is the result of SABIDUSSI [634] that says that any vertex-transitive graph is isomorphic to $\Gamma(G,H,X)$ for suitable G, H, X.

A.4 Regular partitions, distribution diagrams and double coset graphs

On numerous occasions one wants to picture the structure of some graph in relation to a fixed vertex or subset. We shall use various such pictures (diagrams), and the formal status of these diagrams depends a bit on the context. In order to cover all these cases at once, let us introduce a rather general version of the concept of distribution diagram.

Let $\Gamma = (V,E)$ be a graph. A partition $\mathcal{P}$ of V is called *regular* when for $A, B \in \mathcal{P}$ and $\alpha \in A$, the number $e_{AB} = |\Gamma(\alpha) \cap B|$ does not depend on the choice of $\alpha \in A$.

The partition into singletons is always regular; the partition $\{V\}$ is regular precisely when Γ is regular. For any group G of automorphisms of Γ, the partition of V into G-orbits is regular.

Regular partitions give information about the spectrum: the eigenvalues of the matrix $(e_{AB})_{A,B \in \mathcal{P}}$ are also eigenvalues of Γ (with at least the same multiplicity). Much more information can be found in GODSIL & MCKAY [326] (who, following SCHWENK [645], call such partitions *equitable*).

The *distribution diagram* of Γ with respect to a regular partition $\mathcal{P}$ consists of a number of balloons b_A, one for each element $A \in \mathcal{P}$, and a number of lines l_{AB} ($= l_{BA}$) joining the two balloons b_A and b_B, one for each pair $\{A,B\}$ for which $e_{AB} \neq 0$ (then also $e_{BA} \neq 0$). Lines l_{AA} are usually not drawn. This diagram is provided with numbers as follows: in the balloon b_A we write $|A|$, and at the A-end of l_{AB} we write e_{AB}. The number e_{AA} is just written next to b_A; sometimes, when $e_{AA} = 0$, we write '-'.

If we forget the numerical labels, take the balloons as vertices and the lines as edges, then we find the *distribution graph*.

Example. The structure of the Petersen graph around an edge is shown by

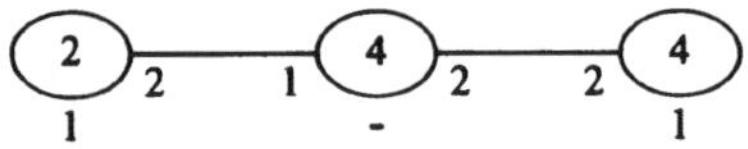

and if we consider the vertices of the edge separately, we get

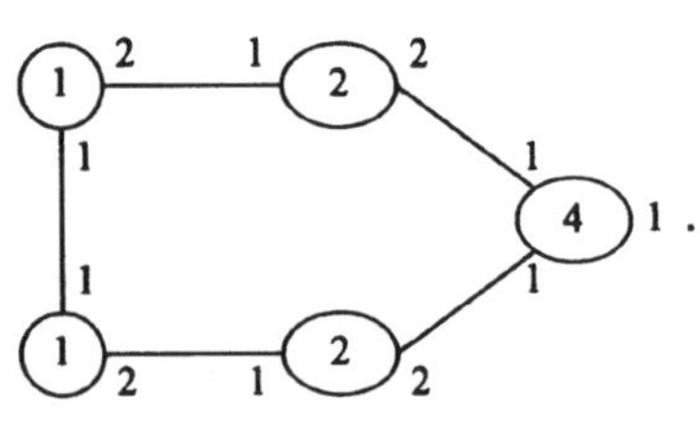

As suggested by the example, we want to determine $\mathcal{P}$ by giving a few of its elements. And indeed this can be done in a canonical way:

A.4.1. Proposition. *Let $\mathbb{S}$ be an arbitrary partition of the vertex set V of a graph Γ. Then there is a unique coarsest partition $\mathcal{P}$ of V finer than $\mathbb{S}$, that is regular in the above sense.* □

Thus, we have associated a distribution diagram with an arbitrary partition of V; the 'distribution diagram around C' for a subset C of V is that associated with $\{C, V\setminus C\}$; when unqualified, 'distribution diagram' refers to the distribution diagram around γ (that is, around $\{\gamma\}$) for some vertex γ.

Graphs for which the distribution diagram around each vertex is linear (this is stronger than saying that the distribution graph is a path) were studied in §4.1G - they turned out to be either distance-regular or distance biregular.

Given a graph $\Gamma = \Gamma(G,H,r)$, the associated *double coset graph* $\Delta = DC(G,H,r)$ is the distribution graph of Γ with respect to the partition $\{HgH \mid g \in G\}$ of the vertex set of Γ. (Note that this partition is regular.) Thus, Δ has vertex set $H\setminus G/H = \{HgH \mid g \in G\}$ and Hg_1H is adjacent to Hg_2H in Δ if and only if $Hg_1H \subseteq Hg_2HrH$. The number of distinct cosets HgH is the *rank* of the permutation representation of G on (the vertex set of) Γ. The vertex $H = H1H$ plays a distinguished role in Δ. Its unique neighbour is the double coset HrH. The double coset graph of Γ reflects certain properties of Γ itself:

A.4.2. Proposition. *Let $\Gamma = \Gamma(G,H,r)$ be a coset graph of G on H, and let $\Delta = DC(G,H,r)$ be the associated double coset graph. Then, for arbitrary $g_1, g_2 \in G$:*

(i) *g_1H and g_2H have distance i in Γ if and only if H and $Hg_2^{-1}g_1H$ have distance i in Δ.*

(ii) *The graph Γ is connected if and only if Δ is connected; in this case, Γ and Δ have the same diameter. This holds if and only if $G = \langle H,r\rangle$.*

(iii) *G (in its action by left multiplication) acts distance-transitively on Γ if and only if Δ is a path, possibly with loops added.*

Proof. Part (i) follows by induction on i, and (ii) is an obvious consequence of (i). (Alternatively, observe that by Lemma 11.1.4 we have $d(g_1H,g_2H) = d(H,g_2^{-1}g_1H) = d(H,Hg_2^{-1}g_1H)$.) Part (iii) is a direct translation of the definition of distance-transitivity. □

A.5 Primitivity

For any connected graph Γ of diameter d we denote by Γ_i $(i = 1, \cdots, d)$ the graphs whose vertices are those of Γ and whose edges are the 2-subsets of points at mutual distance i in Γ. In particular, $\Gamma_1 = \Gamma$. We call a graph Γ *primitive* if it is connected and all Γ_i $(i = 1, \cdots, d)$ are connected, and *imprimitive* otherwise. These notions are closely related to the corresponding concepts for permutation groups. A permutation group G on a set X is called *primitive* if the only G-invariant equivalence relations $\equiv$ on X are the two trivial relations defined by

$$\gamma \equiv \delta \text{ if and only if } \gamma = \delta,$$

and by

$$\gamma \equiv \delta \text{ for all } \gamma, \delta \in X;$$

G is called *imprimitive* otherwise (cf. WIELANDT [782]).

A.5.1. Proposition. *Let* Γ *be a connected graph, and let* G *be a subgroup of* Aut Γ. *If* G *is primitive then* Γ *is primitive.*

Proof. We observe that G preserves the distance between points of Γ, hence G preserves adjacency in each Γ_i. Therefore the relation $\equiv$ given by

$$\gamma \equiv \delta \text{ if there is a path in } \Gamma_i \text{ between } \gamma \text{ and } \delta,$$

is a G-invariant equivalence relation whose equivalence classes are the connected components of Γ_i. This immediately implies the Proposition. □

Every connected bipartite graph Γ is imprimitive since Γ_2 has precisely two connected components Γ^+ and Γ^-, say. The graphs Γ^+ and Γ^- are called the *halved graphs* of Γ.

An *antipodal* graph is a connected graph Γ of diameter $d > 1$ for which Γ_d is a disjoint union of cliques. Such a graph is imprimitive by definition. In this case, the *folded graph* of Γ is defined as the graph $\overline{\Gamma}$ whose vertices are the maximal cliques of Γ_d, adjacent if their union contains an edge of Γ. If, in addition, each $\gamma \in \Gamma$ has the same valency as its image under folding (so that, in particular, $d \geqslant 3$), then Γ is called an *antipodal covering graph* of $\overline{\Gamma}$. (Note that the folding map will be a local isomorphism precisely when $d > 3$.) If, moreover, all maximal cliques of Γ_d have the same size r then Γ is also called an *antipodal r-cover* (*double cover* if $r = 2$, *triple cover* if $r = 3$) of $\overline{\Gamma}$. (More general covers of graphs occur only in §11.1.)

A.6 Designs

A *design* is an ordered pair $(X, \mathfrak{B})$ with *point set* X and *set of blocks* $\mathfrak{B}$ such that $\mathfrak{B}$ is a collection of subsets of X. Occasionally, we shall use the slightly more general definition of a design as an ordered triple $(X, \mathfrak{B}, I)$, where X and $\mathfrak{B}$ are sets and I is a subset of $X \times \mathfrak{B}$, called the *incidence relation.* In this notation, the design $(X, \mathfrak{B})$ is written $(X, \mathfrak{B}, \in)$. The extended notation is necessary when several blocks may be incident with the same set of points, and is also useful when the elements of $\mathfrak{B}$ are given in a representation other than as subsets of X. The *incidence matrix* of $(X, \mathfrak{B})$ is the 0-1 matrix with rows indexed by X and columns by $\mathfrak{B}$ and (x,B)-entry 1 when $x \in B$ and 0 otherwise.

The *point graph* of the design $(X, \mathfrak{B}, I)$ is the graph whose vertex set is X and in which two points are adjacent whenever there is a block containing both. The *dual* of the design $(X, \mathfrak{B}, I)$ is the design $(\mathfrak{B}, X, I')$, where $I' = \{(B,x) \mid (x,B) \in X \times \mathfrak{B}\}$. A design isomorphic to its dual is called *self-dual.* The *block graph* of a design is the point graph of its dual. The *incidence graph* of a design $(X, \mathfrak{B}, I)$ is the bipartite graph with vertex set $X \cup \mathfrak{B}$ and edge set $\{\{x, \mathfrak{B}\} \mid (x, \mathfrak{B}) \in I\}$.

A *t-(v,k,λ) design* is a design $(X, \mathfrak{B})$ with $|X| = v$, $|B| = k$ for each $B \in \mathfrak{B}$, and such that for each subset T of X with $|T| = t$, there are precisely λ blocks containing T. Such a design is also called a *t-design* when v,k,λ are not explicitly given.

One can show by counting that a t-design is also an i-design for $0 \leqslant i \leqslant t$. In fact, a t-(v,k,λ) design is an i-(v,k,λ_i) design with $\lambda_i = \lambda\binom{v-i}{t-i}/\binom{k-i}{t-i}$ for $0 \leqslant i \leqslant t$. Note

that $(k-i)\lambda_i = (v-i)\lambda_{i+1}$ for $0 \leqslant i \leqslant t$.
Given a design $(X,\mathfrak{B})$ and a point $x \in X$, the *derived design* at x is the design $(X \setminus \{x\}, \{B \setminus \{x\} \mid x \in B \in \mathfrak{B}\})$, and the *residual design* at x is the design $(X \setminus \{x\}, \{B \mid x \notin B \in \mathfrak{B}\})$. If $(X,\mathfrak{B})$ is a t-(v,k,λ) design, then each derived design is a $(t-1)$-$(v-1,k-1,\lambda)$ design and each residual design is a $(t-1)$-$(v-1,k,\lambda_{t-1}-\lambda_t)$ design.
A *Steiner system* $S(t,k,v)$ is a t-$(v,k,1)$ design. A *Steiner triple system* $STS(v)$ is a Steiner system $S(2,3,v)$.
A *balanced incomplete block design* (BIBD) is a 2-design. The parameters λ_1 and λ_0 are usually denoted by r (replication number) and b (number of blocks).
A *square* (or *symmetric*) 2-design is a 2-(v,k,λ) design with just as many points as blocks, i.e., $b = v$, whence $k = r$ and $k(k-1) = (v-1)\lambda$. When $\lambda = 2$, it is also called a *biplane*.
A *finite projective plane of order n* is a Steiner system $S(2,n+1,n^2+n+1)$, i.e., a square 2-$(n^2+n+1,n+1,1)$ design. A *finite affine plane of order n* is a Steiner system $S(2,n,n^2)$. Note that we do not require projective and affine planes to be Desarguesian (i.e., coordinatized by a field). Arbitrary projective and affine planes of order n are denoted by $PG(2,n)$ and $AG(2,n)$, respectively. For $d > 2$, $PG(d,q)$ will denote the d-dimensional *projective space* over $\mathbf{F}_q$, i.e., the lattice of subspaces of a $(d+1)$-dimensional vector space over the field $\mathbf{F}_q$. On the set of projective points (i.e., the set of 1-spaces), we find a Steiner system $S(2,q+1,v)$ with $v = (q^{d+1}-1)/(q-1)$ if we take the lines as blocks, and a square 2-$(v, (q^d-1)/(q-1), (q^{d-1}-1)/(q-1))$ design if we take the hyperplanes as blocks. Similarly, $AG(d,q)$ is the d-dimensional *affine space* over $\mathbf{F}_q$.
A *divisible design* of index λ is a triple $(X,\mathcal{G},\mathfrak{B})$, where X is the set of 'points', $\mathcal{G}$ is a partition of X (into 'classes' or 'parts'), and $\mathfrak{B}$ is a collection of subsets of X (called 'blocks') such that each block meets every class in at most one point, and any two points of X that belong to different classes are contained in λ blocks. [More general definitions exist, but the above suffices for our purposes.]
If all blocks have size k and all classes have size m and the total number of points is v, then such a design is called a $GD[k,\lambda,m;v]$ design.

[In the statistical literature this concept is known as a *group-divisible design*, and the classes are called *groups*. Mathematicians wanted to avoid the word 'group' since that already has a different meaning. Now 'divisible design' is fairly standard, but there are still several names around for the classes (like 'part' or 'level' or 'groop'). Concerning the denotation: out of old habit we write *GD*, but *DD* would be more logical.]

A *transversal design* $TD[k,\lambda;n]$ is a divisible design $GD[k,\lambda,n;kn]$ (i.e., every block meets every class precisely once). When $\lambda = 1$, it is usually omitted from the notation. A $TD[k;n]$ is equivalent to an *orthogonal array* $OA(n,k)$ (that is, a $k \times n^2$ array with entries in $\{1,2,\cdots,n\}$ such that in any $2 \times n^2$ submatrix all possible columns occur precisely once). It is also equivalent (when two classes are singled out) to a set of $k-2$ mutually orthogonal Latin squares of order n.
A design is called *resolvable* when its set of blocks can be partitioned into *parallel classes*, that is, into partitions of its point set. (For example, affine planes are resolvable.) We use notations like $RGD[k,\lambda,n;v]$ and $RT[k,\lambda;n]$ for resolvable divisible designs and resolvable transversal designs.
A *partial linear space* is a design $(X,\mathfrak{L})$ in which the blocks are called *lines* such that the lines have size at least 2 and two distinct points are joined by at most one line. A *triangle* of $(X,\mathfrak{L})$ is a clique of the point graph of size 3 which is not contained in a line. The *girth* g of $(X,\mathfrak{L})$ is half of the girth of its incidence graph, i.e., g equals 3 if there is a triangle; otherwise, it is the minimal number $\geqslant 4$ such that the point graph

contains a g-gon. The point graph of a partial linear space is usually referred to as its *collinearity graph* and its block graph as the *line graph.* Any graph Γ can be viewed as a partial linear space all of whose lines have size 2 (namely the edges of Γ). Thus, line graphs of graphs are defined. (We write $L(\Gamma)$ for the line graph of Γ.) The line graph of K_n is the *triangular graph* $T(n)$, and the line graph of $K_{m,n}$ is the *grid* $m \times n$.
A *partial geometry* $pg(K,R,T)$ (as introduced by BOSE [103]) is a partial linear space in which each line has size K, each point is in R lines, and each point outside a line l is in T lines meeting l. The point graph of a partial geometry is strongly regular.

A.7 Codes

Let Q be a set and n a natural number. A *code* C of *(word) length* n over the *alphabet* Q is a subset of Q^n. The code is called *binary* (*ternary*) when $Q = \mathbb{F}_2$ (resp. $Q = \mathbb{F}_3$). The elements of C are called *code words* (and those of Q^n *vectors*). The *Hamming distance* $d_H(u,v)$ between two code words (vectors) u and v is the number of coordinate positions where they differ. The *minimum distance* $\delta(C)$ of C is defined as $\min d_H(u,v)$ for $u, v \in C$, $u \neq v$. Now let 0 be a distinguished element of Q. The *weight* $\mathrm{wt}(u)$ of a code word (vector) u is its number of nonzero coordinates; the *weight enumerator* of a code C is the polynomial $w_C(x) = \sum_i a_i x^i$, where a_i is the number of code words of weight i. (Thus, $w_C(1) = |C|$.) The *support* of a code word (vector) u is the set $\{i \mid u_i \neq 0\}$. The code C is called *e-error correcting* when the spheres $S_e(u) = \{v \mid d_H(u,v) \leqslant e\}$ with $u \in C$ are mutually disjoint. It is *perfect* if these spheres partition Q^n. If C is a code of word length n, then a *truncation* of C is a code of word length $n-1$ obtained by deleting a fixed coordinate position (i.e., by projecting into Q^{n-1}); a *shortening* of C is a code of word length $n-1$ obtained by deleting a fixed coordinate position and only retaining the code words that were 0 at that position. (Thus, a shortening of C is a subcode of a truncation of C; shortening does not diminish minimum distance and truncating preserves cardinality when $\delta(C) > 1$. Note that in coding theory one often uses 'puncture' instead of 'truncate'.) Conversely, in the case of a binary code C of word length n, the *extended code* is the code of word length $n+1$ obtained by adding a *parity check bit* to the words of C, i.e., adding an extra coordinate so as to make the weight even. (Thus, C is a truncation of its extension.)
Often one takes for Q a finite field $\mathbb{F}_q$. Then $V := Q^n$ has a natural structure of vector space over Q (with distinguished basis) and we can use vector space concepts. In particular, C is called *linear* when it is a linear subspace of V. The vector space V also has a natural inner product given by $(u,v) := \sum u_i v_i$. We say that a code C has *parity check matrix* H when $H^\perp = C$ (where we regard the matrix as a set of (row)vectors).

A.8 Singular subspaces

Let Γ be a graph. If γ is a vertex of Γ, we define $\gamma^\perp = \Gamma_{\leqslant 1}(\gamma)$, and, if X is a set of vertices, we define $X^\perp = \bigcap_{\gamma \in X} \gamma^\perp$. If Γ is edge-regular, then Γ can be viewed as the point graph of a partial linear space whose points are the vertices of Γ and whose lines are the subsets $\{\gamma,\delta\}^{\perp\perp}$ for $\gamma,\delta \in \Gamma$ with $\gamma \sim \delta$, cf. §1.14. These lines will be called *singular lines*. The girth of this partial linear space is called the *geometric girth* of Γ. The set of all singular lines is written as $\mathfrak{L}(\Gamma)$, or just $\mathfrak{L}$ when Γ is clear from the

context. If $\gamma \in \Gamma$, and L is a singular line, the *distance* $d(\gamma,L)$ between γ and L is the minimum over all distances $d(\gamma,\delta)$ for $\delta \in L$: a point $\delta \in L$ achieving this minimum is called a point on L *nearest to* γ. A clique Δ of Γ is called a *singular subspace* if $|L \cap \Delta| \in \{0,1,|L|\}$ for all $L \in \mathfrak{L}$.
Two points $\gamma, \delta \in \Gamma$ are called *neighbour equivalent* if $\gamma^\perp = \delta^\perp$; the equivalence class of γ is denoted by $\bar{\gamma}$. If $\{\gamma,\delta\}$ is an edge of Γ, then the union of the two equivalence classes containing γ, δ, respectively, is a clique. Hence it makes sense to define the *reduced graph* $\bar{\Gamma}$ *of* Γ, the quotient of Γ with respect to equivalence by calling two classes $\bar{\gamma}$, $\bar{\delta}$ adjacent if they are distinct and their union is a clique. Γ is called *reduced* when all equivalence classes have size 1. The above quotient $\bar{\Gamma}$ is reduced. [This equivalence relation and construction of $\bar{\Gamma}$ from Γ are a special case of the contraction defined by ASCHBACHER [13].] The equivalence class consisting of all points γ for which $\gamma^\perp$ is the whole vertex set of Γ is called the *radical* of Γ and denoted Rad Γ. More generally, given a subset X of vertices of Γ, the *radical of* X in Γ is the set $X \cap X^\perp$. Thus, a singular subspace is a subspace contained in its own radical. If the radical of X is empty, we say that X is *nondegenerate*.
The *residue* res (X) of a singular subspace X of Γ is the quotient graph of $X^\perp \setminus X$ with respect to equivalence on $X^\perp$. By $\pi_X : X^\perp \setminus X \to \text{res}\,(X)$ we denote the natural quotient map. In particular, a point of the residue res (γ) of a point γ (viewed as the subspace $\{\gamma\}$) is a singular line of Γ containing γ.

A.8.1. Proposition. *Let X be a clique of the graph Γ.*

(i) *If C is a singular subspace of Γ containing X, then its image $\pi_X(C)$ in* res (X) *is a singular subspace.*

(ii) *If L is a singular line of Γ contained in $X^\perp$, then $\pi_X(L)$ is either a point or a singular line of* res (X) *of size at least $|L|$.*

A connected graph Γ (together with its collection $\mathfrak{L}$ of singular lines) is called a *polar space* (cf. BUEKENHOUT & SHULT [150]) if $\gamma^\perp \cap L \neq \emptyset$ for any $\gamma \in \Gamma$ and $L \in \mathfrak{L}$. The *rank* of a polar space Γ is the maximal number k such that there is a chain $\emptyset = \Delta_0 \subset \Delta_1 \subset \cdots \subset \Delta_k$ of distinct singular subspaces in Γ. A polar space of rank two (i.e., a graph where all maximal cliques are singular lines) is called a *generalized quadrangle.*

A.9 Geometries

A *(Buekenhout-Tits) geometry* is a graph Γ together with a fixed partition T of its vertex set into cocliques. The *objects* of the geometry are the vertices of Γ; the *type* $\tau(\gamma)$ of an object γ is the element of T containing it. Two objects are *incident* when they coincide or are adjacent. A *flag* is a set of pairwise incident objects, i.e., a clique in Γ.

An automorphism of the geometry is an automorphism of the graph Γ preserving types. The geometry is called *flag-transitive* when for any $T_0 \subseteq T$ its automorphism group acts transitively on the flags of type T_0.

If F is a flag, then its *residue* Γ_F is the subgraph $F^\perp \setminus F$ of Γ together with the partition induced by $T \setminus \tau^{-1}(F)$. The *rank* of the geometry is the number $|T|$ of types. The *corank* of a flag F is the rank $|T \setminus \tau^{-1}(F)|$ of its residue. A *point-line geometry* is a rank 2 geometry where the two types of objects are called points and lines; the *dual* of a point-line geometry is the same geometry, but with the labels point and line interchanged.

A *Buekenhout-Tits diagram* is a labelled directed graph $\mathfrak{D}$ where each edge (i,j) is labelled with a class of rank 2 geometries $\mathfrak{D}_{ij}$. A geometry *has / belongs to / satisfies* a Buekenhout-Tits diagram $\mathfrak{D}$ when there is a bijection d from the type set T of the geometry to the vertex set of $\mathfrak{D}$ such that for $s, t \in T$ each residue of type $\{s,t\}$ (i.e., each residue Γ_F with $\tau(F) = T \setminus \{s,t\}$) is isomorphic to a point-line geometry in $\mathfrak{D}_{ij}$, where $i = d(s)$, $j = d(t)$ and s corresponds to the point type. (Thus, one might assume at the outset that $\mathfrak{D}_{ji}$ is the class of duals of geometries in $\mathfrak{D}_{ij}$. See also BUEKENHOUT [143].)

When drawing diagrams, we use the following conventions:

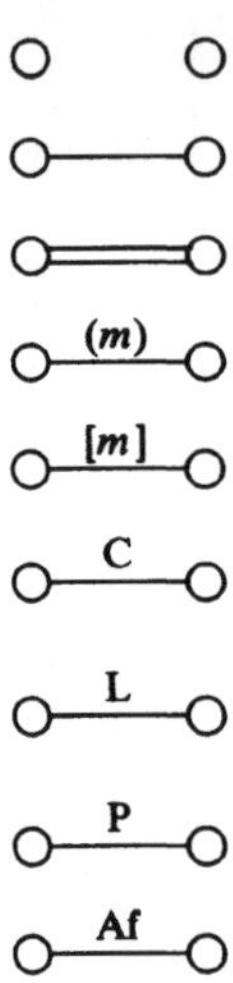

no edge: generalized digon - each point is incident with each line

single edge: (generalized) projective plane

double edge: generalized quadrangle

edge labelled (m): generalized m-gon

edge labelled $[m]$: arbitrary geometry of girth m

edge labelled C: geometry of the points and unordered pairs of points in some set (with obvious incidence)

edge labelled L: geometry of the points and lines in some linear space (point-line geometry where any two points are joined by a unique line)

edge labelled P: points and edges of the Petersen graph

edge labelled Af: points and lines of an affine plane

Example: the geometry of points, lines and planes of a 3-dimensional affine space has diagrams

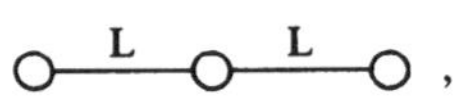

,

and (somewhat more precisely)

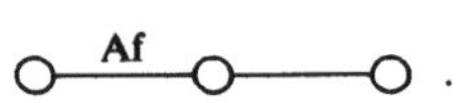

.

A.10 Miscellaneous notation

N, **Z**, **Q**, **R**, **C** and $\mathbf{F}_q$ denote the set of natural numbers, the ring of integers, the field of rational numbers, the real field, the complex field, and the finite field with q elements (also known as $GF(q)$), respectively; **H** is the skew field of real quaternions; $\mathbf{Q}_p$ the field of p-adic numbers. $\mathrm{Sym}(n)$ and $\mathrm{Alt}(n)$ denote the symmetric and alternating group on n symbols, respectively. (Sometimes these are written Sym_n and Alt_n or S_n and A_n for brevity.) Dih_{2n} denotes the dihedral group of order $2n$, $\mathrm{Frob}_{n.k}$ the Frobenius group of order nk with cyclic normal subgroup of order n, Q_8 the quaternion group of order 8, $\mathbf{Z}_n$ the cyclic group of order n, $[n]$ some group of order n (of

unspecified structure), and $A.B$ a group with a normal subgroup A and quotient by A isomorphic to B. For unexplained group denotations, maybe the **ATLAS** [216] or GORENSTEIN [338] will help.
When $a = (a_1, \cdots, a_n)$, we sometimes write Δ_a or diag $(a_1, \cdots, a_n)$ for the matrix A of order n with $A_{ii} = a_i$ $(1 \leqslant i \leqslant n)$ and $A_{ij} = 0$ for $i \neq j$.
The golden ratio is denoted by τ $(= (1+\sqrt{5})/2)$.
The cardinality (or size) of a set X is denoted by $|X|$ or $\#X$.
The expressions $\lfloor x \rfloor$ (or $[x]$) and $\lceil x \rceil$ (with $x \in \mathbf{R}$) denote the largest integer not larger than x, and the smallest integer not smaller than x, respectively.
Finally, the end (or absence) of a proof is denoted by $\square$

References

(After each reference we mention in square brackets the page(s) where the paper is quoted.)

1. Adel'son-Vel'skii, G. M., B. Ju. Veĭsfeĭler, A. A. Leman, and I. A. Faradžev, *Example of a graph without a transitive automorphism group*, Soviet Math. Dokl. **10** (1969) 440-441. (Russian original: Dokl. Akad. Nauk SSSR **185** (1969) 975-976.) [136]
2. Agaian, S. S., *Hadamard matrices and their applications*, Lecture Notes in Math. 1168, Springer, Berlin (1985). [20]
3. Aigner, M., *On the tetrahedral graph*, Pacific J. Math. **25** (1968) 219-228. [258]
4. Aigner, M., *A characterization problem in graph theory*, J. Combin. Th. **6** (1969) 45-55. [258]
5. Aigner, M., *The uniqueness of the cubic lattice graph*, J. Combin. Th. **6** (1969) 282-297. [262]
6. Aldred, R. E. L., *Distance-regular antipodal covering graphs*, Master's thesis, Univ. of Melbourne (Dec. 1986). [189]
7. Alegre, I., M. A. Fiol, and J. L. A. Yebra, *Some large graphs with given degree and diameter*, J. Graph Th. **10** (1986) 219-224. [207]
8. Armanios, C., *A new 5-valent distance transitive graph*, Ars Combin. **19A** (1985) 77-85. MR 86k:05093; Zbl 565.05036 [266]
9. Artin, E., *Geometric Algebra*, Interscience, New York (1957). [x, 274, 276, 291]
10. Aschbacher, M., *The non-existence of rank three permutation groups of degree 3250 and subdegree 57*, J. Algebra **19** (1971) 538-540. [149, 207]
11. Aschbacher, M., *On finite groups generated by odd transpositions, I*, Math. Z. **127** (1972) 45-56. [396]
12. Aschbacher, M., *On finite groups generated by odd transpositions, II, III, IV*, J. Algebra **26** (1973) 451-459, 460-478, 479-491. [396]
13. Aschbacher, M., *A homomorphism theorem for finite graphs*, Proc. AMS **54** (1976) 468-470. [441]
14. Aschbacher, M., *Finite group theory*, Cambridge studies in advanced mathematics 10, Cambridge Univ. Press, Cambridge (1986). [214]
15. Aschbacher, M., *Some multilinear forms with large isometry groups*, Geom. Dedic. **25** (1988) 417-465. [336]
16. Askey, R. and J. Wilson, *A set of orthogonal polynomials that generalize the Racah coefficients on* $6-j$ *symbols*, SIAM J. Math. Anal. **10** (1979) 1008-1016. [246]
17. Assmus, E. F., jr. and H. F. Mattson, jr., *New 5-designs*, J. Combin. Th. **6** (1969) 122-151. MR 42#7528; Zbl 179, p. 29 [60]
18. Baer, R., *Engelsche Elemente noethersche Gruppen*, Math. Ann. **133** (1957) 256-270. [396]
19. Bagchi, Bh., *No extendable biplane of order nine*, J. Comb. Th. (A) **49** (1988) 1-12. [421]
20. Baker, H. F., *A locus with 25920 linear self-transformations*, Cambridge Tracts in Math. and Mathematical Physics 39, Cambridge Univ. Press, London (1946). [30]
21. Baker, R. D., *An elliptic semiplane*, J. Combin. Th. (A) **25** (1978) 193-195. [24, 210, 372]
22. Baker, R. D., J. H. van Lint, and R. M. Wilson, *On the Preparata and Goethals codes*, IEEE Trans. Inf. Theory **IT-29** (1983) 342-345. Zbl 521.94015 [356]
23. Balaban, A. T., *A trivalent graph of girth ten*, J. Combin. Th. (B) **12** (1972) 1-5. [210]
24. Balaban, A. T., *Trivalent graphs of girth nine and eleven, and relationships among cages*, Rev. Roum. Math. Pures et Appl. **18** (1973) 1033-1043. [210]
25. Bannai, E., *Codes in bipartite distance-regular graphs*, J. London Math. Soc. (2) **16** (1977) 197-202. Zbl 365.94015 [357]
26. Bannai, E., *On perfect codes in the Hamming scheme* $H(n,q)$ *with* q *arbitrary*, J. Combin. Th. (A) **23** (1977) 56-67. Zbl 356.94008 [355]

27. Bannai, E. and E. Bannai, *How many P-polynomial structures can an association scheme have?*, European J. Combin. **1** (1980) 289-298. MR 82a:05073; Zbl 458.94047 [58, 145, 146]
28. Bannai, E., P.J. Cameron, and J. Kahn, *Nonexistence of certain distance-transitive digraphs*, J. Combin. Th. (B) **31** (1981) 105-110. [232]
29. Bannai, E. and T. Ito, *On finite Moore graphs*, J. Fac. Sci. Univ. Tokyo, Sect IA **20** (1973) 191-208. Zbl 275.05121 [207]
30. Bannai, E. and T. Ito, *On the spectra of certain distance-regular graphs*, J. Combin. Th. (B) **27** (1979) 274-293. MR 81f:05133; Zbl 427.15005 [208]
31. Bannai, E. and T. Ito, *On the spectra of certain distance-regular graphs, II*, Quart. J. Math. Oxford (2) **32** (1981) 389-411. MR 83c:05089; Zbl 475.05059 [208]
32. Bannai, E. and T. Ito, *Regular graphs with excess one*, Discrete Math. **37** (1981) 147-158. [207, 210]
33. Bannai, E. and T. Ito, *Algebraic Combinatorics I: Association Schemes*, Benjamin-Cummings Lecture Note Ser. 58, The Benjamin/Cummings Publishing Company, Inc., London (1984). Zbl 555.05019 [43, 47, 58, 145, 147, 235, 240, 241, 247, 260, 272, 276]
34. Bannai, E. and T. Ito, *Current researches on algebraic combinatorics*, Graphs Combin. **2** (1986) 287-308. [viii, ix]
35. Bannai, E. and T. Ito, *On distance-regular graphs with fixed valency*, Graphs Combin. **3** (1987) 95-109. [viii, 163, 184, 185]
36. Bannai, E. and T. Ito, *On distance-regular graphs with fixed valency, III*, J. Algebra **107** (1987) 43-52. [184, 185]
37. Bannai, E. and T. Ito, *The study of distance-regular graphs from the algebraic (i.e., character theoretic) viewpoint*, Proc. Symp. Pure Math. **47** (1987) 343-349. [184, 225]
38. Bannai, E. and T. Ito, *On distance-regular graphs with fixed valency, II*, Graphs and Combinatorics **4** (1988) 219-228. [184, 221]
39. Bannai, E. and T. Ito, *Algebraic Combinatorics II*, to appear. [136]
40. Bannai, E. and T. Ito, *On distance-regular graphs with fixed valency, IV*, in preparation. [184]
41. Bannai, E. and T. Ito, *On distance-regular graphs with fixed valency, V*, in preparation. [184]
42. Barlotti, A., *Un' estensione del teorema di Segre-Kustaanheimo*, Boll. Un. Mat. Ital. (3) **10** (1955) 498-506. [379]
43. Bassalygo, L.A., G.V. Zaitsev, and V.A. Zinov'ev, *Uniformly packed codes*, Problemy Peredachi Informatsii **10** (1974) 9-14. (Russian. English transl. Problems Inform. Transm. **10** (1974) 6-9.) MR 49#4647; Zbl 317.94016 [348, 354]
44. Bassalygo, L.A. and V.A. Zinov'ev, *Remark on uniformly packed codes*, Probl. Pered. Inform. **13** (3) (1977) 22-25. (Russian. English transl. Probl. Inform. Transm. **13** (1977) 178-180.) MR 58#15664; Zbl 403.94016 [357]
45. Bassalygo, L.A., V.A. Zinov'ev, V.K. Leont'ev, and N.I. Fel'dman, *Nonexistence of perfect codes for some composite alphabets*, Probl. Pered. Inform. **11** (1975) 3-13. (Russian. English transl. in Problems Inform. Transm. **11** (1975) 181-189.) MR 58#9760; Zbl 326.94008 [355]
46. Batten, L.M., *Combinatorics of finite geometries*, Cambridge University Press, Cambridge (1986). [x]
47. Bauer, H., B. Ganter, and F. Hergert, *Algebraic techniques for nonlinear codes*, Combinatorica **3** (1983) 21-33. Zbl 525.94011 [355]
48. Belevitch, V., *Theory of 2n-terminal networks with application to conference telephony*, Elect. Commun. **27** (1950) 231-244. [10]
49. Benson, C.T., *Minimal regular graphs of girth eight and twelve*, Canad. J. Math. **18** (1966) 1091-1094. [209]
50. Benson, C.T., *On the structure of generalized quadrangles*, J. Algebra **15** (1970) 443-454. [204]

51. Benson, C.T. and N.E. Losey, *On a graph of Hoffman and Singleton*, J. Combin. Th. (B) **11** (1971) 67-79. [391]
52. Bercov, R.D., *On permutation groups with regular subgroup*, Canad. Math. Bull. **17** (1974) 359-361. Correction *ibid.* **17** (1974) 623. MR 51#8222 [65, 68]
53. Berlekamp, E.R., J.H. Conway, and R.K. Guy, *Winning Ways*, Academic Press, London (1982). [405]
54. Berlekamp, E.R., J.H. van Lint, and J.J. Seidel, *A strongly regular graph derived from the perfect ternary Golay code,* pp. 25-30 in: *A survey of combinatorial theory, Symp. Colorado State Univ., 1971* (J.N. Srivastava e.a., eds.), North Holland, 1973. Zbl 258.05129 [360]
55. Berman, A. and R.J. Plemmons, *Nonnegative matrices in the mathematical sciences*, Acad. Press, New York (1979). [81]
56. Berman, G., *Families of generalized weighing matrices*, Canad. J. Math. **30** (1978) 1016-1028. [19]
57. Bermond, J.-C., C. Delorme, and G. Farhi, *Large graphs with given degree and diameter, II*, J. Combin. Th. (B) **36** (1984) 32-48. [207]
58. Bermond, J.-C., C. Delorme, and J.-J. Quisquater, *Strategies for interconnection networks: some methods from graph theory*, J. of Parallel and Distributed Computing **3** (1986) 433-449. [207]
59. Best, M.R., *A contribution to the nonexistence of perfect codes*, Ph.D. thesis, Univ. of Amsterdam, 1982, Math. Centr., Amsterdam (1983). MR 85m:94034; Zbl 526.94014 [355]
60. Best, M.R. and A.E. Brouwer, *The triply shortened binary Hamming code is optimal*, Discrete Math. **17** (1977) 235-245. MR 55#14400; Zbl 356.94009 [55]
61. Best, M.R., A.E. Brouwer, F.J. MacWilliams, A.M. Odlyzko, and N.J.A. Sloane, *Bounds for binary codes of length less than 25*, IEEE Trans. Inform. Theory **IT-24** (1978) 81-93. MR 57#19066; Zbl 369.94011 (Russian transl. in: Kibern. Sb. Nov. Ser. **17** (1980) 28-59. Zbl 456.94015) [55]
62. Beth, Th., D. Jungnickel, and H. Lenz, *Design Theory*, Bibl. Institut - Wissenschaftsverlag, Mannheim (1985). MR 86j:05026 [17, 19]
63. Bichara, A. and G. Tallini, *On a characterization of Grassmann space representing the h-dimensional subspaces in a projective space*, Ann. Discrete Math. **18** (1983) 113-132. [270]
64. Bickel, Th.F., *Circles in (B,N)-groups*, Math. Z. **130** (1973) 249-253. MR 48#384 [329]
65. Bier, Th., *Hyperplane codes*, Graphs Combin. **1** (1985) 207-212. Zbl 606.94006 [354]
66. Bier, Th., *A family of nonbinary linear codes*, Discrete Math. **65** (1987) 47-51. [354]
67. Biggs, N.L., *Finite groups of automorphisms*, London Math. Soc. Lecture Note Ser. 6, Cambridge Univ. Press, London (1971). [132]
68. Biggs, N.L., *Intersection matrices for linear graphs,* pp. 15-23 in: *Combinatorial Mathematics and its applications (Proc. Oxford, 7-10 July 1969)* (D.J.A. Welsh, ed.), Acad. Press, London, 1971. [128, 131, 132]
69. Biggs, N., *Perfect codes in graphs*, J. Combin. Th. (B) **15** (1973) 289-296. Zbl 266.94009 (256.94009) [349]
70. Biggs, N., *Three remarkable graphs*, Canad. J. Math. **25** (1973) 397-411. Zbl 256.05114 [383]
71. Biggs, N.L., *Algebraic Graph Theory*, Cambridge Tracts in Math. 67, Cambridge University Press, Cambridge (1974). Zbl 284.05101 [1, 25, 82, 128, 133]
72. Biggs, N.L., *Perfect codes and distance-transitive graphs,* pp. 1-8 in: *Combinatorics, Proc. British Combinatorial Conf. Aberystwyth 1973*, London Math. Soc. Lecture Note Ser. 13 (T.P. McDonough & V.C. Mavron, eds.), Cambridge University Press, 1974. Zbl 298.94013 [viii, 358]
73. Biggs, N.L., *The symmetry of line graphs*, Utilitas Math. **5** (1974) 113-121. Zbl 279.05116 [x, 148]
74. Biggs, N.L., *Designs, factors and codes in graphs*, Quart. J. Math. Oxford (2) **26** (1975) 113-119. [349, 369]

75. Biggs, N. L., *Automorphic graphs and the Krein condition*, Geom. Dedicata **5** (1976) 117-127. MR 57#12288; Zbl 333.05108 [47, 50, 133, 178, 364, 394]
76. Biggs, N., *Girth, valency and excess*, Lin. Alg. Appl. **31** (1980) 55-59. [210]
77. Biggs, N. L., *Distance-regular graphs with diameter three,* pp. 69-80 in: *Algebraic and Geometric Combinatorics*, Ann. Discrete Math. 15 (E. Mendelsohn, ed.), 1982. Zbl 506.05057 [37, 130]
78. Biggs, N. L., *Excess in vertex-transitive graphs*, Bull. London Math. Soc. **14** (1982) 52-54. [viii, 210]
79. Biggs, N. L., *Graphs with large girth*, Ars Combinatoria **25C** (1988) 73-80. [210]
80. Biggs, N. L., A. G. Boshier, and J. Shawe-Taylor, *Cubic distance-regular graphs*, J. London Math. Soc. (2) **33** (1986) 385-394. Zbl 591.05045 (576.05039) [191, 192, 221, 405]
81. Biggs, N. L. and A. Gardiner, *The classification of distance transitive graphs*, unpublished manuscript (1974). [25, 141, 143, 259, 266, 272, 364, 369, 401]
82. Biggs, N. L. and M. J. Hoare, *A trivalent graph with 58 vertices and girth 9*, Discrete Math. **30** (1980) 299-301. [210]
83. Biggs, N. and T. Ito, *Graphs with even girth and small excess*, Math. Proc. Cambridge Philos. Soc. **88** (1980) 1-10. [210]
84. Biggs, N. L. and D. H. Smith, *On trivalent graphs*, Bull. London Math. Soc. **3** (1971) 155-158. [ix, 221, 382, 405]
85. Biondi, P., *A characterization of the Grassmann space representing the 2-dimensional subspaces of a projective space*, preprint (1986). [270]
86. Biondi, P., *On finite Grassmann spaces*, preprint (1986). [270]
87. Blass, A., F. Harary, and Z. Miller, *Which trees are link graphs?*, J. Combin. Theory (B) **29** (1980) 277-292. MR 82h:05016 [5]
88. Blokhuis, A., *Few-distance sets*, Reidel, Dordrecht (1984). Thesis (T.H. Eindhoven, 1983) = C.W.I. Tract 7 (Amsterdam, 1984) [89]
89. Blokhuis, A. and A. E. Brouwer, *Graphs that are locally a generalized quadrangle, with complete bipartite μ-graphs*, unpublished manuscript (1983). [399]
90. Blokhuis, A. and A. E. Brouwer, *Locally 4-by-4 grid graphs*, Math. Centr. Report PM-R8401, Amsterdam (Jan. 1984). [257, 258]
91. Blokhuis, A. and A. E. Brouwer, *Uniqueness of a Zara graph on 126 points and non-existence of a completely regular two-graph on 288 points,* pp. 6-19 in: *Papers dedicated to J.J. Seidel* (P.J. de Doelder, J. de Graaf & J.H. van Lint, eds.), Eindhoven Univ. Techn. report 84-WSK-03, Aug. 1984. Zbl 554.05049 [400]
92. Blokhuis, A. and A. E. Brouwer, *Amply regular graphs with* $(k,\lambda,\mu) = (10,3,2)$ *are locally Petersen*, manuscript (1986). [37]
93. Blokhuis, A. and A. E. Brouwer, *Geodetic graphs of diameter two*, Geom. Dedicata **25** (1988) 527-533. [38]
94. Blokhuis, A., A. E. Brouwer, D. Buset, and A. M. Cohen, *The locally icosahedral graphs,* pp. 19-22 in: *Finite geometries, Proc. Winnipeg 1984*, Lecture Notes in Pure and Applied Math. 103 (C.A. Baker & L.M. Batten, eds.), Marcel Dekker, New York, 1985. MR 87k:05133; Zbl 587.05059 [5, 36, 315, 322]
95. Böinck, F., *Stageverslag over grafen die lokaal een gegeven graaf zijn*, Manuscript (1988). (Dutch) [386]
96. Bollobás, B., *Graph Theory, an introductory course*, Graduate Texts in Math. 63, Springer, New York (1969). [75]
97. Bollobás, B., *Combinatorics*, Cambridge Univ. Press, Cambridge (1986). MR 88g:05001 [181]
98. Bon, J. T. M. van, *Distance regular antipodal covers of Johnson and Hamming graphs*, CWI Report PM-N8701, Amsterdam (Nov. 1987). [143, 171, 266]
99. Bon, J. van, *On distance-transitive graphs and involutions*, preprint (1988). [397]

100. Bon, J. T. M. van, and A. E. Brouwer, *The distance-regular antipodal covers of classical distance-regular graphs,* pp. 141-166 in: *Colloq. Math. Soc. János Bolyai, Proc. Eger 1987,* 1988. [212]
101. Bon, J. T. M. van, and A. M. Cohen, *Linear groups and distance-transitive graphs*, CWI Report PM-R8805, Amsterdam (1988). [230]
102. Bon, J. T. M. van, A. M. Cohen, and H. Cuypers, *Graphs related to Held's simple group*, CWI Report PM-R8707, Amsterdam (Nov. 1987). [230]
103. Bose, R. C., *Strongly regular graphs, partial geometries, and partially balanced designs*, Pacific J. Math. **13** (1963) 389-419. [440]
104. Bose, R. C. and I. M. Chakravarti, *Hermitian varieties in a finite projective space* $PG(N,q^2)$, Canad. J. Math. **18** (1966) 1161-1182. [274]
105. Bose, R. C. and W. S. Connor, *Combinatorial properties of group divisible incomplete block designs*, Ann. Math. Statist. **23** (1952) 367-383. [23]
106. Bose, R. C. and T. A. Dowling, *A generalization of Moore graphs of diameter two*, J. Combin. Th. (B) **11** (1971) 213-226. [4, 6, 207]
107. Bose, R. C. and R. Laskar, *A characterization of tetrahedral graphs*, J. Combin. Th. **3** (1967) 366-385. [258, 272]
108. Bose, R. C. and R. Laskar, *Eigenvalues of the adjacency matrix of tetrahedral graphs*, Aequationes Math. **4** (1970) 37-43. (Inst. Statist. Mimeo Series 571, Univ. of North Carolina, 1968) [258]
109. Bose, R. C. and D. M. Mesner, *On linear associative algebras corresponding to association schemes of partially balanced designs*, Ann. Math. Statist. **30** (1959) 21-38. [vii, 45]
110. Bose, R. C. and K. R. Nair, *Partially balanced incomplete block designs*, Sankhyā **4** (1939) 337-372. [vii]
111. Bose, R. C. and T. Shimamoto, *Classification and analysis of partially balanced incomplete block designs with two associate classes*, J. Amer. Statist. Assoc. **47** (1952) 151-184. [vii, 43]
112. Boshier, A. and K. Nomura, *A remark on the intersection arrays of distance-regular graphs*, J. Combin. Th. (B) **44** (1988) 147-153. Zbl 595.05048 [191]
113. Bourbaki, N., *Groupes et algèbres de Lie, Chap. IV, V, VI*, Hermann, Paris (1968). [84, 98, 294, 296, 297, 308, 310, 315, 318, 324]
114. Bouwer, I. Z., *Vertex and edge transitive, but not 1-transitive graphs*, Canad. Math. Bull. **13** (1970) 231-237. [435]
115. Bridges, W. G., D. M. Finucane, and R. A. Mena, *On graphs with unique 2-paths for unjoined vertices*, J. Comb. Inf. Syst. Sci. **2** (1977) 12-19. MR 57#16130; Zbl 418.05036 [38]
116. Bridges, W. G. and R. A. Mena, *Rational circulants with rational spectra and cyclic strongly regular graphs*, Ars Combin. **8** (1979) 143-161. Zbl 443.05046 [68]
117. Bridges, W. G. and S. Toueg, *On the impossibility of directed Moore graphs*, J. Comb. Th. (B) **29** (1980) 339-341. Zbl (388.05019) **444**.05042 [207]
118. Brouwer, A. E., *The uniqueness of the near hexagon on 729 points*, Combinatorica **2** (1981) 333-340. MR 85c:51026; Zbl 509.05028 [359]
119. Brouwer, A. E., *Polarities of G. Higman's symmetric design and a strongly regular graph on 176 vertices*, Aequationes Math. **25** (1982) 77-82. MR 85e:05089; Zbl 518.05039 [370]
120. Brouwer, A. E., *The uniqueness of the near hexagon on 759 points,* pp. 47-60 in: *Finite Geometries, T.G. Ostrom Conf. Pullman 1981*, Lecture Notes in Pure and Applied Math. 82 (N.L. Johnson, M.J. Kallaher & C.T. Long, eds.), Marcel Dekker, New York, 1982. MR 84d:51021; Zbl 505.05020 (= Math. Centr. report ZW154, Mar. 1981. MR 82j:05038; Zbl 448.05020) [367]
121. Brouwer, A. E., *On the uniqueness of a regular thin near octagon on 288 vertices (or the semibiplane belonging to the Mathieu group* M_{12}*)*, Math. Centre Report ZW196, Amsterdam (Jul. 1983). MR 85g:05046; Zbl 547.51007 [371]
122. Brouwer, A. E., *A note on the uniqueness of the Johnson scheme*, Math. Centr. report ZN105, Amsterdam (Sep. 1983). [258]

123. Brouwer, A. E., *On the uniqueness of a certain thin near octagon (or partial 2-geometry, or parallelism) derived from the binary Golay code*, IEEE Trans. Inform. Theory **IT-29** (1983) 370-371. MR 85e:05150; Zbl 505.94014 [153, 264, 360]
124. Brouwer, A. E., *The uniqueness of the strongly regular graph on 77 points*, J. Graph Th. **7** (1983) 455-461. MR 86b:05062; Zbl 523.05021 [369]
125. Brouwer, A. E., *Distance regular graphs of diameter 3 and strongly regular graphs*, Discrete Math. **49** (1984) 101-103. MR 86a:05094; Zbl 538.05024 [150, 385]
126. Brouwer, A. E., *Some new two-weight codes and strongly regular graphs*, Discrete Appl. Math. **10** (1985) 111-114. MR 86d:05023; Zbl 561.94008 [356]
127. Brouwer, A. E., *Uniqueness and nonexistence of some graphs related to M_{22}*, Graphs Combin. **2** (1986) 21-29. Zbl 592.05027 [155, 156, 369]
128. Brouwer, A. E., *A remark on association schemes with two P-polynomial structures*, Preprint (Fall 1988). [147]
129. Brouwer, A. E., *Uniqueness of the Biggs-Smith graph*, in: *Graph theory in memory of G.A. Dirac, Proc. Sandbjerg 1985* (L.D. Andersen e.a., eds.), North Holland, 1988. [405]
130. Brouwer, A. E. and A. M. Cohen, *Some remarks on Tits geometries*, Indag. Math. **45** (1983) 393-402. (With an appendix by J. Tits.) MR 85c:51023; Zbl 541.51011 [338]
131. Brouwer, A. E. and A. M. Cohen, *Computation of some parameters of Lie geometries*, pp. 1-48 in: *Algorithms in Combinatorial Design Theory*, Ann. Discrete Math. 26 (C.J. & M.J. Colbourn, eds.), Math. Centr. report ZW198/83, Amsterdam, Sep. 1983, North Holland, Amsterdam, 1985. MR 87h:51015; Zbl 575.51008 [318, 322, 338, 342, 344]
132. Brouwer, A. E. and A. M. Cohen, *Local recognition of Tits geometries of classical type*, Geom. Dedicata **20** (1986) 181-199. MR 87m:51006; Zbl 585.51010 [153, 155, 277]
133. Brouwer, A. E. and E. W. Lambeck, *An inequality on the parameters of distance regular graphs and the uniqueness of a graph associated to M_{23}*, Ann. Discrete Math. **34** (1987) 100-106. [176, 368]
134. Brouwer, A. E. and J. H. van Lint, *Strongly regular graphs and partial geometries*, pp. 85-122 in: *Enumeration and Design - Proc. Silver Jubilee Conf. on Combinatorics, Waterloo, 1982* (D.M. Jackson & S.A. Vanstone, eds.), Academic Press, Toronto, 1984. MR 87c:05033; Zbl 555.05016 [11, 36, 193, 373, 392, 393]
135. Brouwer, A. E. and A. Neumaier, *Determination of the graphs with largest eigenvalue between 2 and* $\sqrt{2+\sqrt{5}}$, preprint (August 1986). (= *The graphs with ...*, Lin. Alg. Appl., to appear) [84]
136. Brouwer, A. E. and A. Neumaier, *A remark on partial linear spaces of girth 5 with an application to strongly regular graphs*, Combinatorica **8** (1988) 57-61. (= *Strongly regular graphs where mu equals two and lambda is large*, Math. Centrum Report ZW151/81, Amsterdam, Mar. 1981. MR 82k:05032; Zbl 448.05021) [6]
137. Brouwer, A. E. and H. A. Wilbrink, *Ovoids and fans in the generalized quadrangle GQ*(4,2), Math. Centr. Report ZN102 (Oct. 1981). [386]
138. Brouwer, A. E. and H. A. Wilbrink, *The structure of near polygons with quads*, Geom. Dedicata **14** (1983) 145-176. MR 85b:05045; Zbl 521.51013 [206, 277]
139. Brown, W. G., *On the non-existence of a type of regular graphs of girth 5*, Canad. J. Math. **19** (1967) 644-648. [210]
140. Bruck, R. H. and H. J. Ryser, *The nonexistence of certain finite projective planes*, Canad. J. Math. **1** (1949) 88-93. MR 10, p. 319; Zbl 37, p. 375 [23]
141. Bruen, A. and B. Levinger, *A theorem on permutations of a finite field*, Canad. J. Math. **25** (1973) 1060-1065. MR 48#8452 [389]
142. Buczak, J. M. J., *Some topics in group theory*, Ph.D. Thesis, Oxford (1980). [231]
143. Buekenhout, F., *Diagrams for geometries and groups*, J. Combin. Th. (A) **27** (1979) 121-151. [36, 442]
144. Buekenhout, F., *The basic diagram of a geometry*, pp. 1-29 in: *Geometries and groups (Proceedings Berlin 1981)*, Lecture Notes in Math. 893 (M. Aigner & D. Jungnickel, eds.), Springer, Berlin, 1981. [36]

145. Buekenhout, F., *Geometries for the Mathieu group* M_{12}, pp. 74-85 in: *Combinatorial Theory*, Lecture Notes in Math. 969 (D. Jungnickel & K. Vedder, eds.), Springer, Berlin, 1982. [371]
146. Buekenhout, F., *Cooperstein's theory*, Simon Stevin **57** (1983) 125-140. [331, 339]
147. Buekenhout, F. and A. M. Cohen, *Diagram Geometry*, in preparation. [341]
148. Buekenhout, F. and X. Hubaut, *Locally polar spaces and related rank 3 groups*, J. Algebra **45** (1977) 391-434. [228, 258]
149. Buekenhout, F. and P. Rowlinson, *The uniqueness of certain automorphic graphs*, Geom. Dedicata **11** (1981) 443-446. Zbl 469.05036 [375, 382, 384]
150. Buekenhout, F. and E. E. Shult, *On the foundations of polar geometry*, Geom. Dedicata **3** (1974) 155-170. [441]
151. Burnside, W., *Theory of groups of finite order (2nd ed.)*, Cambridge Univ. Press, London (1911). (Reprinted Chelsea, New York, 1958) [66]
152. Buset, D., *Graphs which are locally a cube*, Discrete Math. **46** (1983) 221-226. [314]
153. Bussemaker, F. C., D. M. Cvetković, and J. J. Seidel, *Graphs related to exceptional root systems,* pp. 185-191 in: *Combinatorics*, Colloquia Mathematica Soc. János Bolyai 18 (A. Hajnal & V.T. Sós, eds.), North Holland, Amsterdam, 1978. (Details in: Report TH Eindhoven 76-WSK-05, 91 pp., 1976.) [103, 104, 106, 107]
154. Bussemaker, F. C., R. A. Mathon, and J. J. Seidel, *Tables of two-graphs,* pp. 70-112 in: *Combinatorics and Graph Theory, Proc. Calcutta 1980,* Lecture Notes in Math. 885 (S.B. Rao, ed.), Springer, 1981. Full details in Rep. 79-WSK-05, Technische Hogeschool Eindhoven (1979). [16, 37, 95, 259]
155. Bussemaker, F. C. and A. Neumaier, *Exceptional graphs with smallest eigenvalue* -2 *and related problems*, preprint (1987). [106, 110, 120, 164, 198, 259, 265]
156. Bussemaker, F. C., S. Čobeljić, D. M. Cvetković, and J. J. Seidel, *Computer investigation of cubic graphs*, T.H. Report 76-WSK-01, Techn. Univ. Eindhoven (Jan. 1976). [82]
157. Butson, A. T., *Generalized Hadamard matrices*, Proc. Amer. Math. Soc. **13** (1962) 894-898. [19]
158. Calderbank, A. R., *Nonexistence of a uniformly packed* [70,58,5] *code*, IEEE Trans. Inform. Theory **IT-32** (1986) 828-833. [356]
159. Calderbank, A. R. and J.-M. Goethals, *Three-weight codes and association schemes*, Philips J. Res. **39** (1984) 143-152. MR 86e:94026 [354, 358]
160. Calderbank, A. R. and J.-M. Goethals, *On a pair of dual subschemes of the Hamming scheme* $H_n(q)$, Europ. J. Combin. **6** (1985) 133-147. Zbl 579.05021 [354, 356]
161. Calderbank, A. R. and W. M. Kantor, *The geometry of two-weight codes*, Bull. London Math. Soc. **18** (1986) 97-122. [356]
162. Cameron, P. J., *Bounding the rank of certain permutation groups*, Math. Z. **124** (1972) 343-352. [64]
163. Cameron, P. J., *Suborbits in transitive permutation groups,* pp. 98-129 in: *Combinatorics (Proc. of the 1974 Conf. held at Nijenrode Castle, Breukelen)*, Math. Centre Tracts 57 (M. Hall, jr. & J.H. van Lint, eds.), 1974. MR 51#5718 [231]
164. Cameron, P. J., *Partial Quadrangles*, Quart. J. Math. Oxford (2) **26** (1975) 61-73. [199]
165. Cameron, P. J., *Parallelisms of complete designs*, London Math. Soc. Lecture Note Ser. 23, Cambridge University Press, Cambridge (1976). MR 54#7269; Zbl 333.05007 [153]
166. Cameron, P. J., *A note on triple transitive graphs*, J. London Math. Soc. **15** (1977) 197-198. [231]
167. Cameron, P. J., *Strongly regular graphs,* pp. 337-360 in: *Selected topics in graph theory* (L.W. Beineke & R.J. Wilson, eds.), Acad. Press, 1978. Zbl 468.05064 [11]
168. Cameron, P. J., *6-transitive graphs*, J. Combin. Th. (B) **28** (1980) 168-179. Zbl 451.05038 [231]
169. Cameron, P. J., *Finite permutation groups and finite simple groups*, Bull. London Math. Soc. **13** (1981) 1-22. Zbl 463.20003 (522.20004) [229]

170. Cameron, P. J., *Flat embeddings of near 2n-gons,* pp. 61-71 in: *Finite Geometries and designs, Proc. Second Isle of Thorns Conference 1980,* London Math. Soc. Lecture Note Ser. 49 (P.J. Cameron, J.W.P. Hirschfeld & D.R. Hughes, eds.), Cambridge University Press, Cambridge, 1981. MR 83g:51014; Zbl 469.51011 [277]
171. Cameron, P. J., *Orbits of permutation groups on unordered sets, II,* J. London Math. Soc. (2) **23** (1981) 249-264. MR 82k:20005 [32]
172. Cameron, P. J., *Dual polar spaces,* Geom. Dedicata **12** (1982) 75-85. Zbl 473.51002 [276, 277, 279]
173. Cameron, P. J., *There are only finitely many finite distance-transitive graphs of given valency greater than two,* Combinatorica **2** (1982) 9-13. MR 83k:05050; Zbl 507.05037 [220]
174. Cameron, P. J., *Automorphism groups of graphs,* pp. 89-127 in: *Selected topics in Graph theory 2* (L.W. Beineke & R.J. Wilson, eds.), Academic Press, New York, 1983. Zbl 536.05037 [231]
175. Cameron, P. J., *Infinite versions of some topics in finite geometry,* pp. 13-20 in: *Geometrical combinatorics, Conf. Milton Keynes, 1984,* Research Notes in Math. 114 (F.C. Holroyd & R.J. Wilson, eds.), Pitman, 1984. [33]
176. Cameron, P. J., Ph. Delsarte, and J.-M. Goethals, *Hemisystems, orthogonal configurations, and dissipative conference matrices,* Philips J. Res. **34** (1979) 147-162. Zbl 447.05020 [350]
177. Cameron, P. J. and D. A. Drake, *Partial λ-geometries of small nexus,* pp. 19-29 in: *Combinatorial math., optimal designs and their appl., Proc. Fort Collins 1978,* Ann. Discrete Math. 6 (J. Srivastava, ed.), North-Holland, Amsterdam, 1980. Zbl 448.05015 [18, 277]
178. Cameron, P. J., J.-M. Goethals, and J. J. Seidel, *The Krein condition, spherical designs, Norton algebras and permutation groups,* Kon. Nederl. Akad. Wet. Proc. Ser. A **81** (1978) 196-206. Zbl 408.05016 (399.20001) [51, 78]
179. Cameron, P. J., J.-M. Goethals, and J. J. Seidel, *Strongly regular graphs having strongly regular subconstituents,* J. Algebra **55** (1978) 257-280. Zbl 444.05045 [viii, 277, 373]
180. Cameron, P. J., J.-M. Goethals, J. J. Seidel, and E. E. Shult, *Line graphs, root systems and elliptic geometry,* J. Algebra **43** (1976) 305-327. Zbl 337.05142 [x, 102, 106, 107]
181. Cameron, P. J. and J. H. van Lint, *Graphs, Codes and Designs,* London Math. Soc. Lecture Notes 43, Cambridge Univ. Press, Cambridge (1980). Zbl 427.05001 [x, 102, 106, 107]
182. Cameron, P. J. and J. H. van Lint, *On the partial geometry* $pg(6,6,2)$, J. Combin. Th. (A) **32** (1982) 252-255. Zbl 484.05016 [373]
183. Cameron, P. J., C. E. Praeger, J. Saxl, and G. M. Seitz, *On the Sims conjecture and distance-transitive graphs,* Bull. London Math. Soc. **15** (1983) 499-506. Zbl 536.20003 [214, 220]
184. Cameron, P. J., J. A. Thas, and S. E. Payne, *Polarities of generalized hexagons and perfect codes,* Geom. Dedicata **5** (1976) 525-528. MR 58#5289 [357]
185. Carlitz, L., *Representations by skew forms in a finite field,* Arch. Math. **5** (1954) 19-31. [284]
186. Carlitz, L., *A theorem on permutations in a finite field,* Proc. Amer. Math. Soc. **11** (1960) 456-459. MR 22#8005; 22-2547 [x, 389]
187. Carter, R. W., *Simple Groups of Lie type,* Wiley, New York (1972). [294, 323, 332-335]
188. Carter, R. W., *Finite groups of Lie type: conjugacy classes and complex characters,* Wiley, Chichester (1985). [294, 320, 342]
189. Chang, G. J., *Existence and non-existence of association schemes,* Bull. Inst. Mat. Acad. Sinica **14** (1986) 257-270. Zbl 609.05023 [53]
190. Chang, L. C., *The uniqueness and non-uniqueness of the triangular association scheme,* Sci. Record Peking Math. (New Ser.) **3** (1959) 604-613. [106, 258]
191. Chang, L. C., *Association schemes of partially balanced block designs with parameters* $v = 28$, $n_1 = 12$, $n_2 = 15$ *and* $p_{11}^2 = 4$, Sci. Record Peking Math. (New Ser.) **4** (1960) 12-18. [106, 258]
192. Chihara, L., *On the zeros of the Askey-Wilson polynomials with applications to coding theory,* SIAM J. Math. Anal. **18** (1987) 191-207. [357]

193. Chihara, L. and D. Stanton, *Association schemes and quadratic transformations for orthogonal polynomials*, Graphs Combin. **2** (1986) 101-112. [276]
194. Chima, J., *Near n-gons with thin lines*, Ph.D. Dissertation, Kansas State University (1982). [156]
195. Chima, J. and E. E. Shult, *Regular thin near polygons*, preprint (1981?). [153]
196. Chow, W.-L., *On the geometry of algebraic homogeneous spaces*, Ann. of Math. (2) **50** (1949) 32-67. MR 10, p. 396 [268, 276, 278]
197. Chowla, S. and H. J. Ryser, *Combinatorial Problems*, Canad. J. Math. **2** (1950) 93-99. [23]
198. Cohen, A. M., *Finite quaternionic reflection groups*, J. Algebra **64** (1980) 293-324. MR 81g:20093 [409]
199. Cohen, A. M., *Geometries originating from certain distance-regular graphs*, pp. 81-87 in: *Finite Geometries and Designs, Proc. Second Isle of Thorns Conference 1980*, London Math. Soc. Lecture Note Ser. 49 (P.J. Cameron, J.W.P. Hirschfeld & D.R. Hughes, eds.), Cambridge University Press, Cambridge, 1981. MR 82k:05059; Zbl 472.05030 (436.05010) [409, 410]
200. Cohen, A. M., *An axiom system for metasymplectic spaces*, Geom. Dedicata **12** (1982) 417-433. MR 84c:51011 [331, 339]
201. Cohen, A. M., *Semisimple Lie groups from a geometric viewpoint*, pp. 41-77 in: *The structure of real semisimple Lie groups*, MC syllabus 49 (T.H. Koornwinder, ed.), Amsterdam, 1982. [324]
202. Cohen, A. M., *Exceptional presentations of three generalized hexagons of order 2*, J. Combin. Th. (A) **35** (1983) 79-88. MR 84f:51038 [204]
203. Cohen, A. M., *On a theorem of Cooperstein*, European J. Combin. **4** (1983) 107-126. MR 84j:51011 [271, 412]
204. Cohen, A. M., *Points and lines in metasymplectic spaces*, pp. 193-196 in: *Combinatorics '81 (Proc. Rome 1981)*, Ann. Discrete Math. 18, 1983. [342, 343]
205. Cohen, A. M., *Point-line characterizations of buildings*, pp. 191-206 in: *Buildings and the geometry of diagrams, Proc. Como 1984*, Lecture Notes in Math. 1181 (L.A. Rosati, ed.), Springer, Berlin, 1986. [328, 331]
206. Cohen, A. M. and B. N. Cooperstein, *A characterization of some geometries of exceptional Lie type*, Geom. Dedicata **15** (1983) 73-105. [339]
207. Cohen, A. M. and J. Tits, *On generalized hexagons and a near octagon whose lines have three points*, European J. Combin. **6** (1985) 13-27. Zbl 565.05014 [204, 267, 384, 408, 410]
208. Cohen, A. M. and H. Zantema, *A computation concerning doubly transitive permutation groups*, J. reine angew. Math. **347** (1984) 196-211. [229]
209. Cohn, H., *Advanced number theory*, Dover, New York (1962). [97]
210. Connor, W. S., jr., *Some relations among the blocks of symmetrical group divisible designs*, Ann. Math. Statist. **23** (1952) 602-609. MR 14, p. 442 [22, 263]
211. Connor, W. S., jr., *The uniqueness of the triangular association scheme*, Ann. Math. Statist. **29** (1958) 262-266. MR 20#3620 [258]
212. Conway, J. H., *Three lectures on exceptional groups*, pp. 215-247 in: *Finite Simple Groups* (M.B. Powell & G. Higman, eds.), Academic Press, 1971. [111, 359, 366, 371, 406, 408]
213. Conway, J. H. and N. J. A. Sloane, *Laminated lattices*, Ann. of Math. **116** (1982) 593-620. [111]
214. Conway, J. H. and N. J. A. Sloane, *Complex and integral laminated lattices*, Trans. Amer. Math. Soc. **280** (1983) 463-490. [x, 111]
215. Conway, J. H. and N. J. A. Sloane, *Sphere Packings, Lattices and Groups*, Springer, New York (1988). [55, 359]
216. Conway, J. H., R. A. Wilson, R. T. Curtis, S. P. Norton, and R. P. Parker, *Atlas of finite groups*, Clarendon Press, Oxford (1985). [313, 369, 371, 395, 398, 400, 408, 443]
217. Cook, C. R., *A note on the exceptional graph of the cubic lattice graph characterization*, J. Combin. Th. (B) **14** (1973) 132-136. [262]

218. Cooperstein, B. N., *Some geometries associated with parabolic representations of groups of Lie type*, Canad. J. Math. **28** (1976) 1021-1031. [300]
219. Cooperstein, B. N., *A characterization of some Lie incidence structures*, Geom. Dedicata **6** (1977) 205-258. [271, 278, 331, 339]
220. Courant, R. and D. Hilbert, *Methoden der Mathematischen Physik*, Springer Verlag, Berlin (1924). [85]
221. Coxeter, H. S. M., *Self-dual configurations and regular graphs*, Bull. Amer. Math. Soc. **56** (1950) 413-455. [221]
222. Coxeter, H. S. M., *Extreme forms*, Canad. J. Math. **3** (1951) 391-441. [103]
223. Coxeter, H. S. M., *Regular polytopes*, Dover, New York (1973). [296]
224. Coxeter, H. S. M., *Ten toroids and fifty-seven hemi-dodecahedra*, Geom. Dedicata **13** (1982) 87-99. [403]
225. Coxeter, H. S. M., *My graph*, Proc. London Math. Soc. **46** (1983) 117-136. [382]
226. Coxeter, H. S. M., R. Frucht, and D. L. Powers, *Zero-symmetric graphs*, Academic Press, New York (1981). [231]
227. Coxeter, H. S. M. and W. O. J. Moser, *Generators and Relations for Discrete Groups*, Springer, Berlin (1980). [296, 314]
228. Coxeter, H. S. M., *Regular honeycombs in hyperbolic space,* pp. 155-169 in: *Proc. Intern. Congress Mathematicians, Amsterdam 1954*, vol. 3. [5]
229. Coxeter, H. S. M. and A. I. Weiss, *Twisted honeycombs* $\{3,5,3\}_t$ *and their groups*, Geom. Dedicata **17** (1984) 169-179. [403]
230. Cuypers, H., *Two remarks on Huang's characterization of the bilinear forms graphs*, Preprint, Univ. Utrecht (1988). [282]
231. Cvetković, D. M., *New characterizations of the cubic lattice graph*, Publ. Inst. Math. Beograd (Nouvelle Série) **10** (1970) 195-198. [262, 263]
232. Cvetković, D. and M. Doob, *Root systems, forbidden subgraphs, and spectral characterizations of line graphs,* pp. 69-99 in: 4th *Yugoslav Seminar on Graph Theory, Novi Sad, 1983* (D.M. Cvetković, I. Gutman, T. Pisansky & R. Tošić, eds.), Novi Sad, 1984. [106]
233. Cvetković, D. M., M. Doob, and I. Gutman, *On graphs whose spectral radius does not exceed* $\sqrt{2+\sqrt{5}}$, Ars Combin. **14** (1982) 225-239. [84]
234. Cvetković, D., M. Doob, I. Gutman, and A. Torgašev, *Recent results in the theory of graph spectra*, Annals of Discrete Math. 36, North Holland, Amsterdam (1988). [82]
235. Cvetković, D. M., M. Doob, and H. Sachs, *Spectra of graphs*, V.E.B. Deutscher Verlag der Wissenschaften, Berlin (1979). (also: Academic Press, New York, 1980) [82, 258, 262]
236. Cvetković, D. M. and I. Gutman, *On spectral structure of graphs having the maximal eigenvalue not greater than two*, Publ. Inst. Math. Beograd (Nouvelle Série) **18** (1975) 39-45. [84]
237. Cvetković, D. and M. Petrić, *A table of connected graphs on six vertices*, Discrete Math. **50** (1984) 37-49. [82]
238. Cvetković, D. and Z. Radosavljević, *A construction of the 68 connected, regular graphs, non-isomorphic but cospectral to line graphs,* pp. 101-123 in: 4th *Yugoslav Seminar on Graph Theory, Novi Sad, 1983* (D.M. Cvetković, I. Gutman, T. Pisansky & R. Tošić, eds.), Novi Sad, 1984. [106]
239. Dai, Z.-d. and X.-n. Feng, *Studies in finite geometries and the construction of incomplete block designs. IV. Some 'Anzahl' theorems in orthogonal geometry over finite fields of characteristic* $\neq 2$, Acta Math. Sinica **15** (1965) 545-558. MR 32#6313d (Chinese. English transl. in Chinese Math. **7** (1965) 265-280.) [274, 275]
240. Damerell, R. M., *On Moore graphs*, Proc. Cambridge Philos. Soc. **74** (1973) 227-236. MR 47#6553 [207]
241. Damerell, R. M., *Distance-transitive and distance-regular digraphs*, J. Combin. Th. (B) **31** (1981) 46-53. [58, 232]
242. Damerell, R. M., *On Moore geometries II*, Math. Proc. Cambridge Philos. Soc. **90** (1981) 33-40. MR 82j:05034b [207]

243. Damerell, R. M. and M. A. Georgiacodis, *On Moore geometries, I*, J. London Math. Soc. (2) **23** (1981) 1-9. MR 82j:05034a [207]
244. Damerell, R. M. and M. A. Georgiacodis, *On the maximum diameter of a class of distance-regular graphs*, Bull. London Math. Soc. **13** (1981) 316-322. MR 82k:05075; Zbl 457.05055 [208]
245. Dawson, J. E., *A construction for generalized Hadamard matrices $GH(4q, EA(q))$*, J. Statist. Planning and Inference **11** (1985) 103-110. MR 86f:05035 [19]
246. Delorme, C., *Régularité métrique forte*, Rapport de Recherche No. 156, Univ. Paris Sud, Orsay (Dec. 1983). [20, 138]
247. Delorme, Ch., *Grands graphes de degré et diamètre donnés*, Europ. J. Combin. **6** (1985) 291-302. [207]
248. Delorme, Ch. and G. Farhi, *Large graphs with given degree and diameter - part I*, IEEE Trans. Comput. **33** (1984) 857-860. [207]
249. Delsarte, P., *Bounds for unrestricted codes, by linear programming*, Philips Res. Repts. **27** (1972) 272-289. [viii, 55]
250. Delsarte, Ph., *Weights of linear codes and strongly regular normed spaces*, Discrete Math. **3** (1972) 47-64. [356]
251. Delsarte, Ph., *An algebraic approach to the association schemes of coding theory*, Philips Research Reports Suppl. **10** (1973). [43, 48, 54-56, 58, 66, 70-72, 128, 136, 346, 349, 350]
252. Delsarte, Ph., *Association schemes and t-designs in regular semilattices*, J. Combin. Th. (A) **20** (1976) 230-243. [60]
253. Delsarte, Ph., *Properties and applications of the recurrence $F(i+1,k+1,n+1) = q^{k+1}F(i,k+1,n) - q^kF(i,k,n)$*, SIAM J. Appl. Math. **31** (1976) 262-270. [251]
254. Delsarte, Ph., *Pairs of vectors in the space of an association scheme*, Philips Res. Reports **32** (1977) 373-411. [54, 55, 61, 163]
255. Delsarte, Ph., *Bilinear forms over a finite field with applications to coding theory*, J. Combin. Th. (A) **25** (1978) 226-241. [281]
256. Delsarte, Ph. and J.-M. Goethals, *Alternating bilinear forms over $GF(q)$*, J. Combin. Th. (A) **19** (1975) 26-50. [284]
257. Delsarte, Ph. and J.-M. Goethals, *Unrestricted codes with the Golay parameters are unique*, Discrete Math. **12** (1975) 211-224. Zbl 307.94013 [55, 359]
258. Delsarte, Ph., J.-M. Goethals, and J. J. Seidel, *Spherical codes and designs*, Geom. Dedicata **6** (1977) 363-388. Zbl 376.05015 [73, 77, 78, 90, 132]
259. Dembowski, P., *Finite Geometries*, Ergebnisse der Math. und ihre Grenzgebiete **44**, Springer, Berlin (1968). MR 38#1597; Zbl 159, p. 500 [24]
260. Deng, Sh.-t. and Q. Li, *On the affine geometry of algebraic homogeneous spaces*, Acta Math. Sinica **15** (1965) 651-663. (Chinese) [280]
261. Denniston, R. H. F., *Some maximal arcs in finite projective planes*, J. Combin. Th. **6** (1969) 317-319. [387]
262. Deodhar, V. V., *Some characterizations of Bruhat ordering on a Coxeter group and determination of the relative Möbius function*, Invent. Math. **39** (1977) 187-198. [298]
263. Dieudonné, J., *Algebraic homogeneous spaces over fields of characteristic two*, Proc. Amer. Math. Soc. **2** (1951) 295-304. [276, 278]
264. Dieudonné, J., *La géometrie des groupes classiques*, Springer, Berlin (1963). [274, 291]
265. Dixmier, S. and F. Zara, *Etude d'un quadrangle généralisé autour de deux de ses points non liés*, preprint (1976). [30]
266. Doob, M., *On graph products and association schemes*, Utilitas Math. **1** (1972) 291-302. Zbl 315.05130 [26, 27, 262]
267. Doob, M., *Pseudocyclic graphs*, pp. 107-114 in: *Proc. 6th Yugoslav Semin. Graph Theory, Dubrovnik 1985* (R. Tošić, D.M. Acketa & V. Petrović, eds.), Novi Sad, 1986. [130]
268. Doob, M. and D. M. Cvetković, *On spectral characterizations and embeddings of graphs*, Lin. Alg. Appl. **27** (1979) 17-26. [109]

269. Dowling, T. A., *Note on 'A characterization of cubic lattice graphs'*, J. Combin. Th. **5** (1968) 425-426. [262]

270. Dowling, T. A., *A characterization of the T_m-graph*, J. Combin. Th. **6** (1969) 251-263. [257, 258, 272]

271. Drake, D. A., *Partial λ-geometries and generalized Hadamard matrices over groups*, Canad. J. Math. **31** (1979) 617-627. [18-20]

272. Duval, A. M., *A directed graph version of strongly regular graphs*, J. Combin. Th. (A) **47** (1988) 71-100. [232]

273. Egawa, Y., *Characterization of H(n,q) by the parameters*, J. Combin. Th. (A) **31** (1981) 108-125. MR 82k:05092; Zbl 472.05056 [262, 263]

274. Egawa, Y., *Association schemes of quadratic forms*, J. Combin. Th. (A) **38** (1985) 1-14. Zbl 564.05014 [290, 291]

275. Enomoto, H., *Characterization of families of finite permutation groups by the subdegrees, II*, J. Fac. Sci. Univ. Tokyo Sect. IA Math. **20** (1973) 1-11. [4, 133, 262]

276. Enomoto, H. and R. A. Mena, *Distance-regular digraphs of girth 4*, J. Combin. Th. (B) **43** (1987) 293-302. MR 88j:05031 [232]

277. Erdős, P., S. Fajtlowicz, and A. J. Hoffman, *Maximum degree in graphs of diameter 2*, Networks **10** (1980) 87-90. [207]

278. Erdős, P., Ch. Ko, and R. Rado, *Intersection theorems for systems of finite sets*, Quart. J. Math. Oxford (2) **12** (1961) 313-320. [261]

279. Erdős, P. and H. Sachs, *Reguläre Graphen gegebener Taillenweite mit minimaler Knotenzahl*, Wiss. Z. Uni. Halle (Math. Nat.) **12** (1963) 251-257. MR 29#2797 [209]

280. Etienne, G., *Perfect codes and regular partitions in graphs and groups*, European J. Combin. **8** (1987) 139-144. [ix, 346]

281. Fan, P. S., *The Thompson-Wielandt theorem*, Proc. Amer. Math. Soc. **97** (1986) 590-592. Zbl 597.20003 [217]

282. Faradjev, I. A., A. A. Ivanov, and A. V. Ivanov, *Distance-transitive graphs of valency 5, 6 and 7*, European J. Combin. **7** (1986) 303-319. Zbl 608.05041 (Дистанционно-транзитивные графы степени 5, 6 и 7, Žurnal Vičisl. Mat. i Mat. Fiz. 24 (1984) 1704-1718, MR 86d:05057, Zbl 573.05033; *Distance-transitive graphs of valency 5, 6 and 7*, USSR Comput. Maths. Math. Phys. 24 (1984) 67-76, Zbl 592.05026.) [176, 184, 223, 369]

283. Feit, W. and G. Higman, *The non-existence of certain generalized polygons*, J. Algebra **1** (1964) 114-131. [200, 326]

284. Feng, X.-n. and Z.-d. Dai, *Studies in finite geometries and the construction of incomplete block designs. V. Some 'Anzahl' theorems in orthogonal geometry over finite fields of characteristic* 2, Acta Math. Sinica **15** (1965) 664-682. MR 32#6313e (Chinese. English transl. in Chinese Math. **7** (1965) 392-410.) [275]

285. Fischer, B., *Finite groups generated by 3-transpositions*, Invent. Math. **13** (1966) 232-246. [396]

286. Fischer, E., *Über quadratische Formen mit reellen Koeffizienten*, Monatsh. f. Math. u. Phys. **16** (1905) 234-249. [85]

287. Fischer, J. and J. McKay, *The Nonabelian Simple Groups G, $|G| < 10^6$ - Maximal Subgroups*, Mathematics of Computation **32** (1978) 1293-1302. [369, 408]

288. Fisher, P. H., *Some infinite families of antipodal distance-regular graphs with diameter three*, Report 1986/11, Dept. of Math., Univ. of Western Australia (June 1986). [386]

289. Fisher, P. H., *A note on distance-regular graphs*, Report 1987/15, Dept. of Math., Univ. of Western Australia (March 1987). [179]

290. Fisher, P. H., *On almost nexus semi-symmetric designs*, Geom. Dedicata **22** (1987) 211-224. MR 88c:05036 [403]

291. Fong, P. and G. M. Seitz, *Groups with a BN-pair of rank 2, I, II*, Invent. Math. **21** (1973) 1-57. and **24** (1974) 191-239. [323]

292. Foster, R. M., *Geometrical circuits of electrical networks*, Trans. Amer. Inst. Elec. Engrs **51** (1932) 309-317. [384]

293. Foster, R. M., *A census of trivalent symmetric graphs*, unpublished paper presented at the University of Waterloo (1966). [398]
294. Foulser, D. A., *Solvable primitive permutation groups of low rank*, Trans. Amer. Math. Soc. **143** (1969) 1-54. [229]
295. Foulser, D. A. and M. J. Kallaher, *Solvable, flag-transitive rank 3 collineation groups*, Geom. Dedicata **7** (1978) 111-130. [229]
296. Fraïssé, R., *Sur l'extension aux relations de quelques propriétés d'ordres*, Ann. Sci. École Norm. Sup. (3) **71** (1954) 363-388. MR 16, p. 1006 [232]
297. Frame, J. S., *The double cosets of a finite group*, Bull. Amer. Math. Soc. **47** (1941) 458-467. [46]
298. Frankl, P., *The Erdős-Ko-Rado theorem is true for n = ckt,* pp. 365-375 in: *Combinatorics, Proc. 5th Hungarian Colloq. Keszthely 1976*, Colloq. Math. Soc. János Bolyai 18 (A. Hajnal & Vera T. Sós, eds.), North Holland, Amsterdam, 1978. [261]
299. Frankl, P. and Z. Füredi, *An exact result for 3-graphs*, Discrete Math. **50** (1984) 323-328. [16]
300. Frankl, P. and R. M. Wilson, *The Erdős-Ko-Rado theorem for vector spaces*, J. Combin. Th. (A) **41** (1986) 228-236. [273]
301. Freudenthal, H., *Une étude de quelques quadrangles généralisés*, Ann. Mat. Pura Appl. (4) **102** (1975) 109-133. [30]
302. Friedman, H. D., *On the impossibility of certain Moore graphs*, J. Combin. Th. (B) **10** (1971) 245-252. [207]
303. Frobenius, G., *Ueber Matrizen aus positiven Elementen*, Sitz. Ber. Preuss. Akad. Wiss. (Berlin) (1908) 471-476, (1909) 514-518. [80]
304. Frobenius, G., *Ueber Matrizen aus nicht negativen Elementen*, Sitz. Ber. Preuss. Akad. Wiss. (Berlin) (1912) 456-477. [80]
305. Fuglister, F. J., *On finite Moore geometries*, J. Combin. Th. (A) **23** (1977) 187-197. MR 56#158 [207]
306. Fuglister, F. J., *The nonexistence of Moore geometries of diameter 4*, Discrete Math. **45** (1983) 229-238. MR 85e:05039 [207]
307. Fuglister, F. J., *On generalized Moore geometries I, II*, Discrete Math. **67** (1987) 249-258 and 259-269. [208]
308. Gantmacher, F. R., *Applications of the Theory of Matrices*, Interscience, New York (1959). [45, 81, 86]
309. Gardiner, A., *Arc transitivity in graphs*, Quart. J. Math. Oxford (2) **24** (1973) 399-407. [218]
310. Gardiner, A., *Antipodal covering graphs*, J. Combin. Th. (B) **16** (1974) 255-273. [141]
311. Gardiner, A., *Imprimitive distance-regular graphs and projective planes*, J. Combin. Th. (B) **16** (1974) 274-281. Zbl 275.05120 (267.05112) [273]
312. Gardiner, A., *On trivalent graphs*, J. London Math. Soc. (2) **10** (1975) 507-512. [221]
313. Gardiner, A., *Redrawing distance-regular graphs*, unpublished manuscript (1980?). [58, 140, 145, 146]
314. Gardiner, A., *When is an array realised by a distance-regular graph?,* pp. 209-219 in: *Algebraic Methods in Graph Theory, Szeged, 1978, Vol. 1*, Colloq. Math. Soc. János Bolyai 25 (L. Lovász & Vera T. Sós, eds.), North Holland Publ. Co., Amsterdam, 1981. MR 83d:05080; Zbl 473.05044 [173, 174, 179]
315. Gardiner, A., *Antipodal graphs of diameter three*, Lin. Alg. Appl. **46** (1982) 215-219. Zbl 497.05053 [431]
316. Gardiner, A., *Classifying distance-transitive graphs,* pp. 67-88 in: *Combinatorial Math. IX, Brisbane 1981*, Lecture Notes in Math. 952 (E.J. Billington, S. Oates-Williams & A.P. Street, eds.), Springer Verlag, 1982. MR 84a:05036; Zbl 494.05026 [223]
317. Gardiner, A., *An elementary classification of distance-transitive graphs of valency four*, Ars Combin. **19A** (1985) 129-141. Zbl 564.05031 [222]

318. Gardiner, A. and C. E. Praeger, *Distance-transitive graphs of valency six*, Ars Combin. **21A** (1986) 195-210. Zbl 598.05038 [223]
319. Gardiner, A. and Ch. E. Praeger, *Distance-transitive graphs of valency five*, Proc. Edinburgh Math. Soc. (2) **30** (1987) 73-81. Zbl 608.05042 [223]
320. Georgiacodis, M. A., *On the impossibility of certain distance-regular graphs*, Serdica **9** (1983) 12-17. Zbl 523.05044 [208]
321. Georgiacodis, M. A., *An application of Newton polygons*, Serdica **10** (1984) 429-432. Zbl 579.05035 [209]
322. Geramita, A. V. and J. Seberry, *Orthogonal designs*, Lecture Notes Pure Appl. Math. 45, Dekker, New York (1979). [20]
323. Gewirtz, A., *Graphs with maximal even girth*, Canad. J. Math. **21** (1969) 915-934. [372, 393]
324. Gewirtz, A., *The uniqueness of* $g(2,2,10,56)$, Trans. New York Acad. Sci. **31** (1969) 656-675. [372]
325. Godsil, C. D., *Bounding the diameter of distance-regular graphs*, Combinatorica **8** (1988) 333-343. [162, 163, 171, 172]
326. Godsil, C. D. and B. D. McKay, *Feasibility conditions for the existence of walk-regular graphs*, Lin. Alg. Appl. **30** (1980) 51-61. [436]
327. Godsil, Chr. D. and J. Shawe-Taylor, *Distance-regularised graphs are distance-regular or distance-biregular*, J. Combin. Th. (B) **43** (1987) 14-24. MR 88j:05037; Zbl 616.05041 [138]
328. Goethals, J.-M., *The extended Nadler code is unique*, IEEE Trans. Inform. Theory **23** (1977) 132-135. [55]
329. Goethals, J.-M. and J. J. Seidel, *The regular two-graph on 276 vertices*, Discrete Math. **12** (1975) 143-158. [112, 373]
330. Goethals, J.-M. and J. J. Seidel, *Spherical designs*, 1979. [73, 77, 78]
331. Goethals, J.-M. and S. L. Snover, *Nearly perfect binary codes*, Discrete Math. 3 (1972) 65-88. [357]
332. Goethals, J.-M. and H. C. A. van Tilborg, *Uniformly packed codes*, Philips Research Reports **30** (1975) 9-36. Zbl 444.94025 [348]
333. Goldschmidt, D., *Automorphisms of trivalent graphs*, Ann. of Math. **111** (1980) 377-406. [222]
334. Gómez, J. and M. A. Fiol, *Dense compound graphs*, Ars Combin. **20A** (1985) 211-237. [207]
335. Goodman, F., P. de la Harpe, and V. Jones, *Chapter 1: Matrices over natural numbers: Values of the norms, classification, and variations,* in: *Dynkin diagrams and towers of algebras*, Report, Univ. de Genève, June 1986. [84]
336. Gordon, L. M. and R. Levingston, *The construction of some automorphic graphs*, Geom. Dedicata **10** (1981) 261-267. MR 82g:05048; Zbl 469.05035 [375, 382, 383]
337. Gorenstein, D., *Finite Groups*, Chelsea, New York (1968). [x, 396]
338. Gorenstein, D., *Finite Groups*, 2nd ed., Chelsea, New York (1980). [x, 137, 214, 443]
339. Grove, L. C. and C. T. Benson, *Finite reflection groups*, Graduate texts in mathematics 99, Springer, New York (1985). [296]
340. Haemers, W. H., *Eigenvalue techniques in design and graph theory*, Reidel, Dordrecht (1980). Thesis (T.H. Eindhoven, 1979) = Math. Centr. Tract 121 (Amsterdam, 1980) [41, 85, 86, 161, 204]
341. Haemers, W. H., *A new partial geometry constructed from the Hoffman-Singleton graph,* pp. 119-127 in: *Finite Geometries and designs, Proc. Second Isle of Thorns Conference 1980,* London Math. Soc. Lecture Note Ser. 49 (P.J. Cameron, J.W.P. Hirschfeld & D.R. Hughes, eds.), Cambridge University Press, Cambridge, 1981. [393]
342. Haemers, W. H. and C. Roos, *An inequality for generalized hexagons*, Geom. Dedicata **10** (1981) 219-222. Zbl 463.51012 [201]

343. Haemers, W. H. and C. Weug, *Linear programming bounds for codes and designs*, Technical Note N96, M.B.L.E. Brussels (1974). [55]

344. Hale, M. P., jr. and E. E. Shult, *Equiangular lines, the graph extension theorem, and transfer in triply transitive groups*, Math. Z. **135** (1974) 111-123. [225]

345. Hall, J. I., *Locally Petersen graphs*, J. Graph Th. **4** (1980) 173-187. MR 81h:05076 [35-37, 381, 382, 399]

346. Hall, J. I., *On identifying $PG(3,2)$ and the complete 3-design on seven points*, Ann. Discr. Math. **7** (1980) 131-141. [392]

347. Hall, J. I., *Graphs, geometry, 3-transpositions and symplectic $\mathbf{F}_2$-transvection groups*, preprint (1985). [258]

348. Hall, J. I., *Graphs with constant link and small degree or order*, J. Graph Th. **8** (1985) 419-444. [5]

349. Hall, J. I., *A local characterization of the Johnson scheme*, Combinatorica **7** (1987) 77-85. MR 88g:05035 [5, 258]

350. Hall, J. I., *Classifying copolar spaces and graphs*, Preprint. [28]

351. Hall, J. I. and E. E. Shult, *Locally cotriangular graphs*, Geom. Dedicata **18** (1985) 113-159. Zbl 578.05058 [5]

352. Hall, M., jr., *Uniqueness of the projective plane with 57 points*, Proc. Amer. Math. Soc. **4** (1953) 912-916. MR 15, p. 460. Correction: ibid. **5** (1954) 994-997. Zbl 56, p. 385 [204]

353. Hall, M., jr., *Hadamard matrices of order 16*, JPL Research Summary No 36-10, vol. 1 (1961). [20]

354. Hall, M., jr., *Note on the Mathieu group M_{12}*, Arch. Math. **13** (1962) 334-340. [227]

355. Hall, M., jr., *Hadamard matrices of order 20*, JPL Technical Report No 32-761. [20]

356. Hall, M., jr., J. D. Swift, and R. Walker, *Uniqueness of the projective plane of order eight*, Math. Tables Aids Comput. **10** (1956) 186-194. MR 18, p. 816; Zbl 73, p. 365 [204]

357. Hall, M., jr. and D. Wales, *The simple group of order 604,800*, J. Algebra **9** (1968) 417-450. [408]

358. Hall, M., jr. and D. Wales, *The simple group of order 604,800,* pp. 79-90 in: *The theory of finite groups* (R. Brauer & C.-H. Sah, eds.), Benjamin, 1969. [408]

359. Hamming, R. W., *Error detecting and error correcting codes*, Bell Syst. Tech. J. **29** (1950) 147-160. [355]

360. Hammond, P., *Nearly perfect codes in distance-regular graphs*, Discrete Math. **14** (1976) 41-56. Zbl 339.94004 [357]

361. Hammond, P., *q-coverings and completely regular codes in distance-regular graphs*, Quart. J. Math. Oxford (2) **29** (1978) 23-30. Zbl 381.05051 [346, 357]

362. Hammond, P., *On the non-existence of perfect and nearly perfect codes*, Discrete Math. **39** (1982) 105-109. Zbl 493.05059 [357]

363. Hammond, P. and D. H. Smith, *Perfect codes in the graphs O_k*, J. Combin. Th. (B) **19** (1975) 239-255. Zbl 308.94005 (282.94007) [357]

364. Hammond, P. and D. H. Smith, *An analogue of Lloyd's theorem for completely regular codes,* pp. 261-267 in: *Proc. 5th British Combin. Conf. Aberdeen 1975*, Congr. Numer. 15 (C.St.J.A. Nash-Williams & J. Sheehan, eds.), Utilitas Math., Winnipeg, 1976. Zbl 327.94013 [348]

365. Hasse, H., *Zahlentheorie*, Akademie Verlag, Berlin (1963). [68]

366. Heawood, P. J., *Map-colour theorem*, Quart. J. Math. Oxford Ser. **24** (1890) 332-338. [209, 221]

367. Hedayat, A. and W. D. Wallis, *Hadamard matrices and their applications*, Ann. Statist. **6** (1978) 1184-1238. [20]

368. Heden, O., *Perfect codes in antipodal distance-transitive graphs*, Math. Scand. **35** (1974) 29-37. Zbl 295.94016 [349]

369. Heiss, S., *Two sporadic geometries related to the Hoffman-Singelton [sic] graph*, Preprint (1988). [394]

370. Hemmeter, J., *Halved graphs, Johnson and Hamming graphs*, Utilitas Math. **25** (1984) 115-118. Zbl 562.05045 [260, 265]
371. Hemmeter, J., *Distance-regular graphs and halved graphs*, European J. Combin. **7** (1986) 119-129. Zbl 606.05041 [260, 265, 273, 279, 282, 284, 285]
372. Hemmeter, J., *The large cliques in the graph of quadratic forms*, Eur. J. Comb. **9** (1988) 395-410. [293]
373. Hemmeter, J., *A new family of distance-regular graphs*, manuscript (1988). [279]
374. Hemmeter, J. and A. Woldar, *Classification of the maximal cliques of size* $\geqslant q+4$ *in the quadratic forms graph in odd characteristic*, preprint (Eur. J. Comb., submitted) (1988). [293]
375. Hemmeter, J. and A. Woldar, *On the maximal cliques of the quadratic forms graph in even characteristic*, preprint (Eur. J. Comb., submitted) (1989). [293]
376. Hensel, A. D., *Antipodal distance regular graphs*, Master's thesis, Univ. of Waterloo (1988). [385]
377. Hergert, F., *Beiträge zur Theorie nichtlinearer Fehler-korrigierender Codes*, Diplomarbeit, Techn. Hochschule Darmstadt (1980). [355]
378. Hering, C., *Transitive linear groups and linear groups which contain irreducible subgroups of prime order*, Geom. Dedicata **2** (1974) 425-460. [229]
379. Hering, C., *Transitive linear groups and linear groups which contain irreducible subgroups of prime order, II*, J. Algebra **93** (1985) 151-164. [viii, 229]
380. Higman, D. G., *Intersection matrices for finite permutation groups*, J. Algebra **6** (1967) 22-42. [408]
381. Higman, D. G., *Partial geometries, generalized quadrangles and strongly regular graphs*, pp. 263-293 in: *Atti del Convegno di Geometria, Combinatoria e sue Applicazioni (Univ. degli Studi di Perugia, Perugia, 1970)* (A. Barlotti, ed.), Perugia, 1971. MR 51#2945 [201]
382. Higman, D. G., *Invariant relations, coherent configurations and generalized polygons*, pp. 27-43 in: *Combinatorics*, Math. Centre Tracts 57 (M. Hall & J.H. van Lint, eds.), Amsterdam, 1974. (Reidel, Dordrecht) MR 52#150 [44, 201]
383. Higman, D. G., *Coherent configurations, Part I: Ordinary representation theory*, Geom. Dedicata **4** (1975) 1-32. [44]
384. Higman, D. G., *Coherent configurations, Part II: Weights*, Geom. Dedicata **5** (1976) 413-424. Zbl 353.05009 [44]
385. Higman, D. G., *Admissible graphs*, pp. 211-222 in: *Finite Geometries, Proc. Pullman 1981*, Lecture Notes in Pure and Applied Math. 82 (N.L. Johnson, M.J. Kallaher & C.T. Long, eds.), Marcel Dekker, New York, 1982. [29]
386. Higman, D. G., *Coherent algebras*, Lin. Alg. Appl. **93** (1987) 209-239. [44, 57]
387. Higman, D. G., *Admissible graphs and gamma and delta spaces*, Preprint. [28]
388. Higman, G., *On the simple group of D.G. Higman and C.C. Sims*, Illinois J. Math. **13** (1969) 74-80. [370]
389. Hilano, T. and K. Nomura, *Distance degree regular graphs*, J. Comb Th. (B) **37** (1984) 96-100. Zbl 567.05028 [139]
390. Hiller, H., *Geometry of Coxeter groups*, Research Notes in Math., Pitman, New York (1982). [296]
391. Hine, T. C. and V. C. Mavron, *Translations of symmetric and complete nets*, Math. Z. **182** (1983) 237-244. [19]
392. Hirschfeld, J. W. P., *Projective Geometries over Finite Fields*, Oxford Mathematical Monographs, Clarendon Press, Oxford (1979). [274]
393. Hirschfeld, J. W. P., *Finite projective spaces of three dimensions*, Oxford Mathematical Monographs, Clarendon Press, Oxford (1985). [30, 379]
394. Hirschfeld, J. W. P., *Quadrics over finite fields*, pp. 53-87 in: *Proc. Rome 1983*, 1986. [275]
395. Hoffman, A. J., *On the uniqueness of the triangular association scheme*, Ann. Math. Statist. **31** (1960) 492-497. [258]

396. Hoffman, A. J., *On the duals of symmetric partially balanced incomplete block designs*, Ann. Math. Statist. **34** (1963) 528-531. [22, 263]
397. Hoffman, A. J., *On the polynomial of a graph*, Amer. Math. Monthly **70** (1963) 30-36. [263]
398. Hoffman, A. J., *On eigenvalues and colourings of graphs,* pp. 79-91 in: *Graph theory and its applications* (B. Harris, ed.), Acad. Press, New York, 1970. [10, 92, 106, 160]
399. Hoffman, A. J., *On graphs whose least eigenvalue exceeds* $-1-\sqrt{2}$, Lin. Alg. Appl. **16** (1977) 153-166. [110]
400. Hoffman, A. J., *On limit points of the least eigenvalue of a graph*, Ars Combin. **3** (1977) 3-14. [111]
401. Hoffman, A. J. and R. R. Singleton, *On Moore graphs with diameters 2 and 3*, IBM J. Res. Develop. **4** (1960) 497-504. [207, 209, 391, 395]
402. Hoffman, A. J. and J. H. Smith, *On the spectral radii of topologically equivalent graphs,* pp. 273-281 in: *Recent advances in graph theory, Proc. Symp. held in Prague, 1974*, Academia Praha, Prague, 1975. [84]
403. Hollmann, H. D. L., *Association schemes*, Master's thesis, Eindhoven University of Technology (1982). [48, 390]
404. Hollmann, H., *Pseudocyclic 3-class association schemes on 28 points*, Discrete Math. **52** (1984) 209-224. Zbl 559.05019 [390]
405. Holt, D. F., *A graph which is edge transitive but not arc transitive*, J. Graph Th. **5** (1981) 201-204. [435]
406. Hong, Y., *On the nonexistence of unknown perfect 6- and 8-codes in Hamming schemes* $H(n,q)$ *with q arbitrary*, Osaka J. Math. **21** (1984) 687-700. MR 86b:94016; Zbl 551.94011 [355]
407. Hsieh, W. N., *Intersection theorems for systems of finite vector spaces*, Discrete Math. **12** (1975) 1-16. [273]
408. Hua, L.-K., *Geometry of symmetric matrices over any field with characteristic other than two*, Ann. Math. **50** (1949) 8-31. [280, 293]
409. Hua, L.-g. and Zh.-x. Wan, *Tien hsing chün (The classical groups)*, Shanghai Science and Technology Press (1963). (Chinese) MR 34#2715; erratum 34, p. 1786 [274]
410. Huang, T., *Some results on the association schemes of bilinear forms*, Ph.D. Thesis, Ohio State Univ. (1985). [282]
411. Huang, T., *An analogue of the Erdős-Ko-Rado theorem for distance-regular graphs of bilinear forms*, Discrete Math. **64** (1987) 191-198. [282]
412. Huang, T., *A characterization of the association schemes of bilinear forms*, European J. Combin. **8** (1987) 159-173. [282]
413. Hubaut, X., *Strongly regular graphs*, Discrete Math. **13** (1975) 357-381. [11, 193, 360]
414. Hughes, D. R., *On designs,* pp. 43-67 in: *Geometries and groups (Proceedings Berlin 1981)*, Lecture Notes in Math. 893 (M. Aigner & D. Jungnickel, eds.), Springer, Berlin, 1981. [17]
415. Hughes, D. R. and F. C. Piper, *Design theory*, Cambridge Univ. Press (1985). [17]
416. Huppert, B., *Endliche Gruppen, I*, Springer Verlag, Berlin (1967). [68, 401, 402]
417. Huppert, B. and N. Blackburn, *Finite groups III*, Grundlehren der Math. Wiss. 243, Springer, Berlin (1982). [215]
418. Hussain, Q. M., *On the totality of the solutions for the symmetrical incomplete block designs:* $\lambda=2$, $k=5$ *or* 6, Sankhyā **7** (1945) 204-208. [264]
419. Imrich, W., *Explicit construction of graphs without small cycles*, Combinatorica **4** (1984) 53-59. [210]
420. Inglis, N. F. J., *Multiplicity-free permutation characters, distance-transitive graphs and classical groups*, Ph.D. Thesis, Univ. of Cambridge (1988). [230]
421. Inglis, N. F. J., M. W. Liebeck, and J. Saxl, *Multiplicity-free permutation representations of finite linear groups*, Math. Z. **192** (1986) 329-337. Zbl 598.20006 (578.20006) [230]

422. Isaacs, I. M., *Character Theory of Finite Groups*, Academic Press, New York (1976). [64, 231]
423. Ito, N., *Hadamard matrices with 'doubly transitive' automorphism groups*, Arch. Math. **35** (1980) 100-111. [228]
424. Ito, N., *Hadamard graphs I, II*, Graphs Combin. **1** (1985) 57-64 and 331-337. [265]
425. Ito, N. and H. Kimura, *Studies on Hadamard matrices with '2-transitive' automorphism groups*, J. Math. Soc. Japan **36** (1984) 63-73. Zbl 557.05020 [228]
426. Ito, N. and J. S. Leon, *An Hadamard matrix of order 36*, J. Combin. Th. (A) **34** (1983) 244-247. [227]
427. Ito, T., *On a graph of O'Keefe and Wong*, J. Graph Th. **5** (1981) 87-94. [210]
428. Ito, T., *Bipartite distance-regular graphs of valency three*, Lin. Alg. Appl. **46** (1982) 195-213. MR 83j:05046; Zbl 501.05052 [221, 398]
429. Ivanov, A. A., *Construction by the computer of some new automorphic graphs,* pp. 144-146 in: *Aerofisika i prikladnaya matematika (Aerophysics and applied mathematics)*, MFTI, Moscow, 1981. (Russian) [382]
430. Ivanov, A. A., Ограничение диаметра дистанционно-регулярного графа, Dokl. Akad. Nauk SSSR **271** (1983) 789-792. MR 84m:05045; Zbl 573.05040 (Bounding the diameter of a distance regular graph, Soviet Math. Doklady **28** (1983) 149-152. Zbl 552.05034) [ix, 184, 189]
431. Ivanov, A. A., *Distance-transitive representations of the symmetric groups*, J. Combin. Th. (B) **41** (1986) 255-274. MR 87m:20008; Zbl 605.05020 [230, 260]
432. Ivanov, A. A., *On 2-transitive graphs of girth 5*, European J. Combin. **8** (1987) 393-420. [155, 408]
433. Ivanov, A. A. and I. V. Chuvaeva, Действие группы M_{12} на матрицах Адамара (Action of the group M_{12} on Hadamard matrices), pp. 159-169 in: Исследования по алгебраической теории комбинаторных объектов *(Investigations in the Algebraic Theory of Combinatorial Objects)* (M.H. Klin & I.A. Faradjev, eds.), Institute for System Studies, Moscow, 1985. (Russian) MR 88k:20033 [371]
434. Ivanov, A. A. and A. V. Ivanov, *Distance-transitive graphs of valency k,* $8 \leqslant k \leqslant 13$, pp. 112-145 in: *Algebraic, Extremal and Metric Combinatorics, 1986*, Cambridge Univ. Press, Cambridge, 1988. [223-225]
435. Ivanov, A. A., M. H. Klin, and I. A. Faradjev, Примитивные представления неабелевых простых групп порядка меньше 10^6, Часть I *(The primitive representations of the nonabelian simple groups of order less than* 10^6*, Part I)*, Preprint, The Institute for System Studies, Moscow (1982). (Russian) [369, 408]
436. Ivanov, A. A., M. E. Muzichuk, and V. A. Ustimenko, *On a new family of (P&Q)-polynomial schemes*, preprint (1988). [261, 279]
437. Ivanov, A. A. and S. V. Shpectorov, *Characterization of the association schemes of Hermitian forms over* $GF(2^2)$, Preprint (1987). (Geom. Dedicata, submitted) [160, 285, 363]
438. Ivanov, A. A. and S. V. Shpectorov, *Geometries for sporadic groups related to the Petersen graph*, Comm. Algebra **16** (1988) 925-953. [36]
439. Ivanov, A. A. and S. V. Shpectorov, *The association schemes of dual polar spaces of type* ${}^2A_{2d-1}(p^f)$ *are characterized by their parameters if* $d \geqslant 3$, Lin. Alg. Appl. (to appear). [277]
440. James, L. O., *A combinatorial proof that the Moore (7,2) graph is unique*, Utilitas Math. **5** (1974) 79-84. [391]
441. Janko, Z., *A new finite simple group with abelian 2-Sylow subgroups and its characterization*, J. Algebra **3** (1966) 147-186. [406]
442. Janko, Z., *Some new simple groups of finite order,* pp. 63-64 in: *The theory of finite groups* (R. Brauer & C.-H. Sah, eds.), Benjamin, 1969. [408]
443. Janko, Z. and T. van Trung, *Answers on two questions about semi-symmetric designs*, J. Combin. Th. (A) **37** (1984) 357-358. [18]
444. Jeurissen, R. H., *His graph in theirs*, Ars Combinatorica **25C** (1988) 31-36. [383]

445. Johnson, P. and E. Shult, *Gamma spaces which are locally polar spaces,* pp. 243-249 in: *Finite Geometries, Proc. Pullman 1981*, Lecture Notes in Pure and Applied Math. 82 (N.L. Johnson, M.J. Kallaher & C.T. Long, eds.), Marcel Dekker, New York, 1982. Zbl 514.51001 [5]
446. Johnson, P. and E. Shult, *Local characterizations of polar spaces*, Geom. Dedic. **28** (1988) 127-151. [5]
447. Jungnickel, D., *On difference matrices, resolvable transversal designs and generalized Hadamard matrices*, Math. Z. **167** (1979) 49-60. [19]
448. Jungnickel, D., *Transitive symmetric nets*, Arch. Math. **36** (1981) 92-96. [19, 225, 273]
449. Jungnickel, D., *Symmetric translation nets*, J. reine angew. Math. **235** (1982) 215-230. [19]
450. Jungnickel, D., *The number of designs with classical parameters grows exponentially*, Geom. Dedicata **16** (1984) 167-178. [277]
451. Jungnickel, D. and K. Vedder, *Square divisible designs with* $k = (n+1)\mu$, Arch. Math. **43** (1984) 275-284. [273]
452. Kantor, W.M., *Automorphism groups of designs*, Math. Z. **109** (1969) 246-252. [227]
453. Kantor, W.M., *Automorphisms of Hadamard matrices*, J. Combin. Th. **6** (1969) 279-281. [227]
454. Kantor, W.M., *2-Transitive designs,* pp. 44-97 in: *Combinatorics (Part 3), Proc. of the Nijenrode Conf.*, Mathematical Centre Tracts 57 (M. Hall, jr. & J.H. van Lint, eds.), 1974. (= Reidel, Dordrecht, 1975, pp. 365-418) [272]
455. Kantor, W.M., *Symplectic groups, symmetric designs and line ovals*, J. Algebra **33** (1975) 43-58. [226]
456. Kantor, W.M., *Moore geometries and rank 3 groups having* $\mu = 1$, Quart. J. Math. Oxford (2) **28** (1977) 309-328. MR 57#6153 [38, 40, 207]
457. Kantor, W.M., *Classification of 2-transitive symmetric designs*, Graphs Combin. **1** (1985) 165-166. [226, 272]
458. Kantor, W.M., *Homogeneous designs and geometric lattices*, J. Combin. Th. (A) **38** (1985) 66-74. [229]
459. Kantor, W.M., *Generalized polygons, SCABs and GABs,* pp. 79-158 in: *Buildings and the Geometry of Diagrams, Proceedings Como 1984*, Lecture Notes in Math. 1181 (L.A. Rosati, ed.), Springer Verlag, Berlin, 1986. [204]
460. Kantor, W.M. and R.A. Liebler, *The rank three representations of the finite classical groups*, Trans. Amer. Math. Soc. **271** (1982) 1-71. [229]
461. Kárteszi, F., *Piani finiti ciclici come risoluzioni di un certo problema di minimo*, Boll. Un. Mat. Ital. (3) **15** (1960) 522-528. [209, 210]
462. Kasami, T., *Weight distributions of Bose-Chaudhuri-Hocquenghem codes,* pp. 335-357 in: *Proc. Conf. Combin. Math. & Appl.* (R.C. Bose & T.A. Dowling, eds.), Univ. of North Carolina Press, Chapel Hill, N.C., 1969. [358]
463. Kasami, T., *The weight enumerators for several classes of subcodes of the 2nd order binary Reed-Muller codes*, Info. & Control **18** (1971) 369-394. [358]
464. Kelly, J.B., *A characteristic property of quadratic residues*, Proc. Amer. Math. Soc. **5** (1954) 38-46. [68]
465. Kilmoyer, R. and L. Solomon, *On the theorem of Feit-Higman*, J. Combin. Th. (A) **15** (1973) 310-322. [201]
466. Knapp, W., *On the point stabilizers in a primitive permutation group*, Math. Z. **133** (1973) 137-168. Zbl 253.20004 [217]
467. Komissartschik, E.A. and S.V. Tsaranov, *Intersections of maximal subgroups in simple groups of order less than* 10^6, Comm. Algebra **14** (1986) 1623-1678. [408]
468. Kovács, P., *The non-existence of certain regular graphs of girth 5*, J. Combin. Th. (B) **30** (1981) 282-284. [ix, x, 210]
469. Kumar, V., S.B. Rao, and N.M. Singhi, *Graphs with eigenvalues at least* -2, Lin. Alg. Appl. **46** (1982) 27-42. [110]

470. Laborde, J.-M., *Une nouvelle famille de codes binaires, parfaits, non linéaires*, C.R. Acad. Sci. Paris Sér. I Math. **297** (1983) 67-70. MR 84j:94031; Zbl 538.94021 [355]

471. Lam, C. W. H., *Distance-transitive digraphs*, Discrete Math. **29** (1980) 265-274. [232]

472. Lambeck, E. W., *A remark on the intersection arrays of distance-regular graphs and the distance-regular graphs of diameter $d = 3i - 1$ with $b_i = 1$ and $k > 2$*, manuscript (Fall 1988). [172]

473. Lambeck, E. W., *The uniqueness of a distance-regular graph on 280 vertices*, preprint (1988). [364]

474. Lander, E. S., *Symmetric designs: an algebraic approach*, London Math. Soc. Lect. Note Series 74, Cambridge Univ. Press, Cambridge (1983). [17]

475. Larman, D. G., A. A. Rogers, and J. J. Seidel, *On 2-distance sets in Euclidean space*, Bull. London Math. Soc. **9** (1977) 261-267. [89]

476. Laskar, R., *A characterization of cubic lattice graphs*, J. Combin. Th. **3** (1967) 386-401. [262]

477. Laskar, R., *Eigenvalues of the adjacency matrix of cubic lattice graphs*, Pacific J. Math. **29** (1969) 623-629. [262]

478. Launey, W. de, *On the nonexistence of generalized Hadamard matrices*, J. Statist. Planning and Inference **10** (1984) 385-396. [19]

479. Launey, W. de, *A survey of generalised Hadamard matrices and difference matrices $D(k, \lambda; G)$ with large k*, Utilitas Math. **30** (1986) 5-29. [19]

480. Lee, H.-J., *A note on geodetic graphs of diameter two and their relation to orthogonal Latin squares*, J. Combin. Th. (B) **22** (1977) 165-167. [38, 42]

481. Leech, J., *Notes on sphere packings*, Canad. J. Math. **19** (1967) 251-267. [viii, 111]

482. Leemans, H., *On partial two-geometries of small nexus*, preprint (1983). [371]

483. Lemmens, P. W. H. and J. J. Seidel, *Equiangular lines*, J. Algebra **24** (1973) 494-512. [83, 111, 112]

484. Leonard, D. A., *Thesis*, Ohio State University (1979). [371]

485. Leonard, D. A., *Orthogonal polynomials, duality and association schemes*, SIAM J. Math. Anal. **13** (1982) 656-663. Zbl 495.33006 [58, 235, 237, 241, 245, 246]

486. Leonard, D. A., *Metric, co-metric association schemes*, pp. 277-282 in: *Combinatorics, graph theory and computing, Proc. 15th Southeast. Conf., La. State Univ. 1984*, Congr. Numerantium 44, 1984. Zbl 559.05018 [58, 240]

487. Leonard, D. A., *Parameters of association schemes that are both P- and Q-polynomial*, J. Combin. Th. (A) **36** (1984) 355-363. Zbl 533.05016 [58, 240]

488. Leonard, D. A., *Non-symmetric, metric, cometric association schemes*, preprint (1988). [58, 232]

489. Leonard, D. A., *Non-symmetric, metric, cometric association schemes are self-dual*, manuscript (1988). [58, 232]

490. Liebeck, M. W., *Permutation modules for rank 3 symplectic and orthogonal groups*, J. Algebra **92** (1985) 9-15. [229]

491. Liebeck, M. W., *The affine permutation groups of rank three*, Proc. London Math. Soc. (3) **54** (1987) 477-516. [229]

492. Liebeck, M. W. and Ch. E. Praeger, and J. Saxl, *Distance transitive graphs with symmetric or alternating automorphism group*, Bull. Austral. Math. Soc. **35** (1987) 1-25. Zbl 601.05024 [230]

493. Liebeck, M. W. and Ch. E. Praeger, and J. Saxl, *On the 2-closures of finite permutation groups*, preprint. [230]

494. Liebeck, M. W. and J. Saxl, *The finite primitive permutation groups of rank three*, Bull. London Math. Soc. **18** (1986) 165-172. [229]

495. Liebler, R. A., *On the uniqueness of the tetrahedral association schemes*, J. Combin. Th. (B) **22** (1977) 246-262. Zbl 358.05011 [258]

496. Liebler, R. A. and R. A. Mena, *Certain distance-regular digraphs and related rings of characteristic 4*, J. Combin. Th. (A) **47** (1988) 111-123. [x, 232]

497. Lindner, C.C. and A. Rosa, *Construction of large sets of almost disjoint Steiner triple systems*, Canad. J. Math. **27** (1975) 256-260. [392]
498. Lindström, B., *On group and nongroup perfect codes in q symbols*, Math. Scand. **25** (1969) 149-158. [355]
499. Lint, J.H. van, *Nonexistence theorems for perfect error-correcting codes,* pp. 89-95 in: *Computers in Algebra and Number Theory, Symp. New York 1970*, SIAM-AMS Proc. IV (G. Birkhoff & M. Hall, jr., eds.), AMS, Providence, 1971. [355]
500. Lint, J.H. van, *Recent results on perfect codes and related topics,* pp. 158-178 in: *Combinatorics (Part 1), Proc. of the Nijenrode Conf.*, Mathematical Centre Tracts 55 (M. Hall, jr. & J.H. van Lint, eds.), 1974. (Also published by Reidel, Dordrecht, 1975.) [355]
501. Lint, J.H. van, *A survey of perfect codes*, Rocky Mountain J. Math. **5** (1975) 199-224. [355]
502. Lint, J.H. van, *Introduction to coding theory*, Graduate Texts in Math. 86, Springer, New York (1982). Zbl 485.94015 [355, 356, 405]
503. Lint, J.H. van, and A. Schrijver, *Construction of strongly regular graphs, two-weight codes and partial geometries by finite fields*, Combinatorica **1** (1981) 63-73. [373]
504. Lint, J.H. van, and J.J. Seidel, *Equilateral point sets in elliptic geometry*, Proc. Kon. Ned. Acad. Wet., Ser. A **69** (1966) 335-348. (= Indag. Math. **28**) [10, 95]
505. Liu, M.-l., *Geometry of alternate matrices*, Acta Math. Sinica **16** (1966) 104-135. (Chinese) [282]
506. Livingston, M., *An ordered version of the Erdős-Ko-Rado theorem*, J. Combin. Th. (A) **26** (1979) 162-165. [265]
507. Livingstone, D., *On a permutation representation of the Janko group*, J. Algebra **6** (1967) 43-55. [406]
508. Lloyd, S.P., *Binary block coding*, Bell System Tech. J. **36** (1957) 517-535. [56]
509. Lo Re, P.M. and D. Olanda, *Grassmann spaces*, J. Geometry **17** (1981) 50-60. [270]
510. Lovász, L. and A. Schrijver, *Remarks on a theorem of Rédei*, Studia Scient. Math. Hungar. **16** (1981) 449-454. Zbl 535.51009 [389]
511. Lubotzky, A., R. Phillips, and P. Sarnak, *Ramanujan graphs*, to appear. [210]
512. MacInnes, C.R., *Finite planes with less than eight points on a line*, Amer. Math. Monthly **14** (1907) 171-174. [204]
513. Macpherson, H.D., *Infinite distance transitive graphs of finite valency*, Combinatorica **2** (1982) 63-69. MR 84g:05075; Zbl 492.05036 [128, 220, 232]
514. MacWilliams, F.J., *A theorem on the distribution of weights in a systematic code*, Bell Syst. Tech. J. **42** (1963) 79-94. [71]
515. MacWilliams, F.J. and N.J.A. Sloane, *The Theory of Error-Correcting Codes*, North Holland Publ. Co., Amsterdam (1977). [55, 56, 128, 136, 355, 358, 359]
516. Marcus, M. and H. Minc, *A survey of matrix theory and matrix inequalities*, Allyn and Bacon, Boston (1964). [81]
517. Margulis, G.A., *Explicit constructions of graphs without short cycles and low density codes*, Combinatorica **2** (1982) 71-78. [210]
518. Mars, J.G.M., *Les nombres de Tamagawa de certains groupes exceptionels*, Bull. Soc. Math. France **94** (1966) 97-140. [341]
519. Martinov, N.J., Транзитивные-по-расстоянию графы *(Distance-transitive graphs)*, Serdica **1** (1975) 381-385. Zbl 438.05034 [140]
520. Mathon, R., *3-class association schemes,* pp. 123-155 in: *Proc. Conf. Alg. Aspects Comb., Sem. Toronto 1975*, Congressus Numer. 13, Toronto, 1975. Zbl 326.05023 [44, 48, 388-390]
521. Mathon, R., *Lower bounds for Ramsey numbers and association schemes*, J. Combin. Th. (B) **42** (1987) 122-127. Zbl 606.05047 [386]
522. Mathon, R., *A new elliptic semiplane* $S(135,12,3)$, {lecture, London, 1987} (in preparation). [24]

523. Mavron, V. C., *Translations and parallel classes of lines in affine designs*, J. Combin. Th. (A) **22** (1977) 322-330. [19]
524. Mavron, V. C., *A characterization of some symmetric substructures of projective and affine geometries*, Arch. Math. **36** (1981) 281-288. [19]
525. McConnel, R., *Pseudo-ordered polynomials over a finite field*, Acta Arith. **8** (1963) 127-151. MR 29#2244 [389]
526. McEliece, R. J., E. R. Rodemich, H. Rumsey, jr., and L. R. Welch, *New upper bounds on the rate of a code via the Delsarte-MacWilliams inequalities*, IEEE Trans. Inform. Theory **IT-23** (1977) 157-166. [55]
527. McFeat, R. B. and A. Neumaier, *Some tuple system constructions with applications to resolvable designs*, J. Combin. Th. (A) **36** (1984) 92-100. [19]
528. McGee, W. F., *A minimal cubic graph of girth seven*, Canad. Math. Bull. **3** (1960) 149-152. MR 23#A1550 [209]
529. McKay, J., *The non-abelian simple groups G, $|G| < 10^6$ - character tables*, Comm. Algebra **7** (1979) 1407-1445. [408]
530. McKay, J., *Graphs, Singularities and Finite Groups,* pp. 183-186 in: *Proc. Conf. Finite Groups Santa Cruz 1979*, AMS Proc. Symp. Pure Math. 37 (B. Cooperstein & G. Mason, eds.), Providence, 1980. [73]
531. McLaughlin, J., *A simple group of order 898128000,* pp. 109-111 in: *The theory of finite groups* (R. Brauer & C.-H. Sah, eds.), Benjamin, 1969. [373]
532. Melone, N. and D. Olanda, *A characteristic property of the Grassmann manifold representing the lines of a projective space*, European J. Combin. **5** (1984) 323-330. Zbl 563.51019 [270]
533. Meredith, G. H. J., *Triple transitive graphs*, J. London Math. Soc. (2) **13** (1976) 249-257. MR 53#10642 [231]
534. Mesner, D. M., *An investigation of certain combinatorial properties of partially balanced incomplete block experimental designs and association schemes, with a detailed study of designs of Latin square and related types*, unpublished doctoral thesis, Michigan State Univ. (1956). [262]
535. Miller, G. A., H. F. Blichfeldt, and L. E. Dickson, *Theory and applications of finite groups*, Dover reprint, New York (1961). (Part III.) [103]
536. Mohar, B. and M. Omladič, *Divisors and the spectrum of infinite graphs*, Lin. Alg. Appl. **91** (1987) 99-106. Zbl 615.05041 [128]
537. Mohar, B. and J. Shawe-Taylor, *Distance-biregular graphs with 2-valent vertices and distance-regular line graphs*, J. Combin. Th. (B) **38** (1985) 193-203. Zbl 565.05057 (549.05056) [138, 148]
538. Mollard, M., *Une nouvelle famille de 3-codes parfaits sur $GF(q)$*, Discrete Math. **49** (1984) 209-212. MR 85k:94049; Zbl 549.94028 [355]
539. Mollard, M., *A generalized parity function and its use in the construction of perfect codes*, SIAM J. Alg. Discr. Meth. **7** (1986) 113-115. Zbl 585.94015 [355]
540. Moon, A., unpublished (1982). [258]
541. Moon, A., *An analog of the Erdős-Ko-Rado theorem for the Hamming schemes $H(n,q)$*, J. Combin. Th. (A) **32** (1982) 386-390. [265]
542. Moon, A., *Characterization of the graphs of the Johnson schemes $G(3k,k)$ and $G(3k+1,k)$*, J. Combin. Th. (B) **33** (1982) 213-221. MR 84f:05077a; Zbl 506.05053 [258]
543. Moon, A., *Characterization of the Odd graphs O_k by parameters*, Discrete Math. **42** (1982) 91-97. MR 83m:05114; Zbl 494.05035 [257, 258, 260, 393]
544. Moon, A., *On the uniqueness of the graphs $G(n,k)$ of the Johnson schemes*, J. Combin. Th. (B) **33** (1982) 256-264. MR 84f:05077b; Zbl 506.05054 [258]
545. Moon, A., *The graphs $G(n,k)$ of the Johnson schemes are unique for $n \geqslant 20$*, J. Combin. Th. (B) **37** (1984) 173-188. Zbl 557.05049 [258]
546. Moore, E. F., personal communication referred to in Hoffman & Singleton (1960). [206]
547. Mulder, H. M., *$(0,\lambda)$-graphs and n-cubes*, Discrete Math. **28** (1979) 179-188. [20, 28]

548. Munemasa, A., *An analogue of t-designs in the association schemes of alternating bilinear forms*, Graphs Combin. **2** (1986) 259-267. [60]
549. Muzichuk, M. E., *V-rings of permutation groups with invariant metric*, Ph.D. Thesis, Kiev (1987). (Russian) [279]
550. Neumaier, A., *Distances, graphs and designs*, European J. Combin. **1** (1980) 163-174. Zbl 435.05015 [89, 136]
551. Neumaier, A., *Combinatorial configurations in terms of distances*, Memorandum 81-09 (Wiskunde), Eindhoven Univ. Technol. (1981). [275]
552. Neumaier, A., *Distance matrices and n-dimensional designs*, European J. Combin. **2** (1981) 165-172. Zbl 467.05017 [246, 275]
553. Neumaier, A., *Distance matrices, dimension, and conference graphs*, Proc. Kon. Ned. Acad. Wet., Ser. A **84** (1981) 385-391. (= Indag. Math. **43**) [89]
554. Neumaier, A., *New inequalities for the parameters of an association scheme,* pp. 365-367 in: *Combinatorics and Graph Theory, Proc. Calcutta 1980*, Lecture Notes in Math. 885 (S.B. Rao, ed.), Springer, Berlin, 1981. Zbl 479.05047 [50]
555. Neumaier, A., *On a class of edge-regular graphs,* pp. 223-230 in: *Geometries and groups (Proceedings Berlin 1981)*, Lecture Notes in Math. 893 (M. Aigner & D. Jungnickel, eds.), Springer, Berlin, 1981. MR 83e:05068; Zbl 499.05035 [viii, 7, 16]
556. Neumaier, A., *Regular cliques in graphs and special 1½-designs,* pp. 244-259 in: *Finite Geometries and designs, Proc. Second Isle of Thorns Conference 1980*, London Math. Soc. Lecture Note Ser. 49 (P.J. Cameron, J.W.P. Hirschfeld & D.R. Hughes, eds.), Cambridge University Press, Cambridge, 1981. [18, 183, 350]
557. Neumaier, A., *Completely regular twographs*, Arch. Math. **38** (1982) 378-384. [16, 227]
558. Neumaier, A., *Rectagraphs, diagrams and Suzuki's sporadic simple group,* pp. 305-318 in: *Algebraic and Geometric Combinatorics*, Ann. Discrete Math. 15 (E. Mendelsohn, ed.), 1982. [28]
559. Neumaier, A., *On norm three vectors in integral Euclidean lattices I*, Math. Z. **183** (1983) 565-574. [111, 168]
560. Neumaier, A., *Some sporadic geometries related to PG(3,2)*, Arch. Math. **42** (1984) 89-96. [392-394]
561. Neumaier, A., *Characterization of a class of distance regular graphs*, J. reine angew. Math. **357** (1985) 182-192. Zbl 552.05042 [27, 122, 154, 258, 264, 267]
562. Neumaier, A., *Some relations between roots, holes and pillars*, European J. Combin. **8** (1987) 29-33. MR 88c:05045 [111]
563. Neumaier, A., *Duality in coherent configurations*, Combinatorica (to appear). [71]
564. Neumaier, A., *Graph representations, two distance sets, and equiangular lines*, Lin. Alg. Appl. (to appear). [89, 112]
565. Neumann, P. M., *Primitive permutation groups of degree* $3p$, Preprint (1970?). [401]
566. Neumann, P. M., *Generosity and characters of multiply transitive permutation groups*, Proc. London Math. Soc. (3) **31** (1975) 457-481. [63]
567. Neumann, P. M., *Permutationsgruppen von Primzahlgrad und verwandte Themen*, Vorlesungen Math. Inst. Giessen, Heft 5 (1977). [401]
568. Neumann, P. M., *A lemma that is not Burnside's*, Math. Scientist **4** (1979) 133-141. MR 81g:01012; Zbl 409.20001 [64]
569. Nomura, K., *An inequality between intersection numbers of a distance-regular graph*, J. Combin. Th. (B) **43** (1987) 358-359. Zbl 618.05028 (592.05046) [176]
570. Nomura, K., *On local structure of a distance-regular graph of Hamming type*, preprint (1988). [154]
571. Nomura, K., *Distance-regular graphs with valency four and girth three*, J. Combin. Th. (B) (to appear). Zbl 607.05055 [192, 225]
572. Norton, S. P., *On the group* Fi_{24}, Geom. Dedicata **25** (1988) 483-501. [397]
573. Numata, M., *On the graphical characterization of the projective space over a finite field*, J. Combin. Th. (B) **38** (1985) 143-155. [271]

574. Obreschkoff, N., *Verteilung und Berechnung der Nullstellen reeller Polynome*, VEB Deutscher Verlag der Wissenschaften, Berlin (1963). [47]

575. Odlyzko, A.M. and N.J.A. Sloane, *New upper bounds on the number of unit spheres that can touch a unit sphere in n dimensions*, J. Combin. Th. (A) **26** (1979) 210-214. [55]

576. O'Keefe, M. and P.K. Wong, *A smallest graph of girth 5 and valency 6*, J. Combin. Th. (B) **26** (1979) 145-149. [210]

577. O'Keefe, M. and P.K. Wong, *A smallest graph of girth 10 and valency 3*, J. Combin. Th. (B) **29** (1980) 91-105. [210]

578. O'Keefe, M. and P.K. Wong, *The smallest graph of girth 6 and valency 7*, J. Graph Th. **5** (1981) 79-85. [210]

579. O'Nan, M.E., *Automorphisms of unitary block designs*, J. Algebra **20** (1972) 495-511. [384]

580. Ore, O., *Theory of graphs*, Amer. Math. Soc., Providence, R.I. (1962). [38]

581. Ott, U., *Eine Bemerkung über Polaritäten eines verallgemeinerten Hexagons*, Geom. Dedic. **11** (1981) 341-345. [205]

582. Paley, R.E.A.C., *On orthogonal matrices*, J. Math. Phys. Mass. Inst. Tech. **12** (1933) 311-320. [10, 227]

583. Patterson, N.J. and S.K. Wong, *A characterization of the Suzuki sporadic simple group of order 448,345,497,600*, J. Algebra **39** (1976) 277-286. [411, 412]

584. Paulus, A.J.L., *Conference matrices and graphs of order 26*, Technische Hogeschool Eindhoven, report WSK 73/06, 89 pp., Eindhoven (1973). [37]

585. Payne, S.E., *Symmetric representations of nondegenerate generalized n-gons*, Proc. Amer. Math. Soc. **19** (1968) 1321-1326. [204]

586. Payne, S.E., *All generalized quadrangles of order 3 are known*, J. Combin. Th. (A) **18** (1975) 203-206. [204]

587. Payne, S.E., *Generalized quadrangles of order 4, I, II*, J. Combin. Th. (A) **22** (1977) 267-279 and 280-288. [204]

588. Payne, S. and J.A. Thas, *Finite generalized quadrangles*, Pitman, New York (1985). [30, 200, 204, 386]

589. Perkel, M., *Bounding the valency of polygonal graphs with odd girth*, Canad. J. Math. **31** (1979) 1307-1321. [401]

590. Perkel, M., *A characterization of J_1 in terms of its geometry*, Geom. Dedicata **9** (1980) 291-298. [401, 408]

591. Perron, O., *Zur Theorie der Matrizen*, Math. Ann. **64** (1907) 248-263. [80]

592. Petersen, J., *Sur le théorème de Tait*, L'Intermédiare des Mathématiciens **5** (1898) 225-227. [ix, 1, 209]

593. Phelps, K.T., *A combinatorial construction of perfect codes*, SIAM J. Alg. Discr. Meth. **4** (1983) 398-403. MR 85e:94024; Zbl 549.94029 [355]

594. Plesken, W. and M. Pohst, *Constructing integral lattices with prescribed minimum, I*, Math. Comput. **45** (1985) 209-221, 505-516. [111]

595. Powers, D.L., *Eigenvectors of distance-regular graphs*, SIAM J. Matrix Anal. Appl. **9** (1988) 399-407. [163]

596. Praeger, C.E., *Symmetric graphs and the classification of the finite simple groups*, pp. 99-110 in: *Groups, Proc. Conf. on combinatorial group theory, Kyoungju, Korea, 1983*, Lect. Notes Math. 1098 (A.C. Kim & B.H. Neumann, eds.), 1984. Zbl 558.05031 [231]

597. Praeger, Ch.E., J. Saxl, and K. Yokoyama, *Distance transitive graphs and finite simple groups*, Proc. London Math. Soc. (3) **55** (1987) 1-21. MR 88c:05062, Zbl 619.05026 [229]

598. Praeger, C.E. and M.-Y. Xu, *On symmetric graphs of twice prime valency*, Eur J. Comb. (to appear). [231]

599. Primrose, E.J.F., *Quadrics in finite geometries*, Proc. Camb. Phil. Soc. **47** (1951) 299-304. MR 12, p. 850 [274]

600. Quirin, W.L., *The circle number of orbital graphs*, J. Combin. Th. **10** (1971) 93-94. [vii, 133]

601. Raghavarao, D., *Constructions and combinatorial problems in designs of experiments*, J. Wiley & Sons, New York (1971). [22, 23]
602. Rajkundlia, D. P., *Some techniques for constructing new infinite families of balanced incomplete block designs*, Thesis, Dept. of Math. and Statist., Queen's Univ. Kingston, Ontario, Canada (1978). [425]
603. Rajkundlia, D. P., *Some techniques for constructing infinite families of BIBDs*, Discrete Math. **44** (1983) 61-96. [23]
604. Rao, S. B., D. K. Ray-Chaudhuri, and N. M. Singhi, *On imprimitive association schemes,* pp. 273-291 in: *Combinatorics and applications - Proc. of the Seminar on Combinatorics and Applications in honour of Prof. S.S. Shrikhande, Calcutta, 1982*, Indian Statist. Inst. (K.S. Vijayan & N.M. Singhi, eds.), Calcutta, 1984. MR 88c:05037 [51, 52]
605. Rao, S. B., N. M. Singhi, and K. S. Vijayan, *The minimal forbidden subgraphs for generalized line graphs,* pp. 459-472 in: *Proc. Intern. Symp. in Combinatorics (Calcutta, 1980)*, Lecture Notes in Math. 885 (S.B. Rao, ed.), Springer, Berlin, 1981. [110]
606. Ray-Chaudhuri, D. K., *Application of the geometry of quadrics for constructing PBIB designs*, Ann. Math. Statist. **33** (1962) 1175-1186. MR 28#683 [110]
607. Ray-Chaudhuri, D. K., *Some results on quadrics in finite projective geometry based on Galois fields*, Canad. J. Math. **14** (1962) 129-138. MR 24#A3550 [274]
608. Ray-Chaudhuri, D. K. and A. P. Sprague, *Characterization of projective incidence structures*, Geom. Dedicata **5** (1976) 361-376. [271]
609. Ray-Chaudhuri, D. K. and A. P. Sprague, *A combinatorial characterization of attenuated spaces*, Utilitas Math. **15** (1979) 3-29. [282]
610. Rédei, L., *Lückenhafte Polynome über endlichen Körpern*, Birkhäuser Verlag, Basel (1970). MR 45#3366 (English translation: Lacunary polynomials over finite fields, North Holland, Amsterdam, 1973, MR 50#4548) [389]
611. Reuvers, H. F. H., *Some non-existence theorems for perfect codes over arbitrary alphabets*, Ph.D. thesis, Techn. Univ. Eindhoven (1977). Zbl 352.94015 [355]
612. Richen, F. A., *Modular representations of split BN pairs*, Trans. Amer. Math. Soc. **140** (1969) 435-460. [325]
613. Rifà, J. and Ll. Huguet, *Characterization of completely regular codes through P-polynomial association schemes,* pp. 157-167 in: *Proc. 4th Conf. Applicable Algebra and Error Correcting Codes, Karlsruhe 1986*, Lecture Notes in Computer Sci. 307, Springer, Berlin, 1988. [352]
614. Robertson, N., *The smallest graph of girth 5 and valency 4*, Bull. Amer. Math. Soc. **70** (1964) 824-825. [209]
615. Rolland, P. T., *On the uniqueness of the tetrahedral association scheme*, Ph.D. Thesis, City University of New York (1976). [258]
616. Ronan, M. A., *A geometric characterization of Moufang hexagons*, Invent. Math. **57** (1980) 227-262. Zbl 429.51002 [204]
617. Ronan, M. A., *A note on the* ${}^3D_4(q)$ *generalized hexagons*, J. Combin. Th. (A) **29** (1980) 249-250. Zbl 444.05028 [204]
618. Ronan, M. A., *A combinatorial characterization of the dual Moufang hexagons*, Geom. Dedicata **11** (1981) 61-67. Zbl 485.51014 (477.51010) [204]
619. Ronan, M. A., *Coverings of certain finite geometries,* pp. 316-331 in: *Finite Geometries and designs, Proc. Second Isle of Thorns Conference 1980*, London Math. Soc. Lecture Note Ser. 49 (P.J. Cameron, J.W.P. Hirschfeld & D.R. Hughes, eds.), Cambridge University Press, Cambridge, 1981. [412]
620. Ronan, M. A., *Extending locally truncated buildings and chamber systems*, Proc. London Math. Soc. (3) **53** (1986) 385-406. [153]
621. Ronan, M. A. and S. D. Smith, *2-Local geometries for some sporadic groups,* pp. 283-289 in: *Proceedings of Symposia in Pure Mathematics, Volume 37*, 1980. [367]
622. Ronan, M. and J. Tits, *Building buildings*, Math. Ann. **278** (1987) 291-306. [326]

623. Rooij, A. C. M. van, and H. S. Wilf, *The interchange graph of a finite graph*, Acta Math. Acad. Sci. Hungar. **16** (1965) 263-269. MR 33#3959 [110]

624. Roos, C., *An alternative proof of the Feit-Higman theorem on generalized polygons*, Delft Progr. Rep. **5** (1980) 67-77. Zbl 452.05033 [201, 203]

625. Roos, C., *Metric schemes in which one of the (nontrivial) multiplicities equals one*, manuscript (1980). [163]

626. Roos, C., *Some results on t-constant codes*, Delft Progr. Rep. **5** (1980) 3-13. [347]

627. Roos, C., *A note on the existence of perfect constant weight codes*, preprint (1983?). [357]

628. Roos, C. and A. J. van Zanten, *On the existence of certain distance-regular graphs*, J. Combin. Th. (B) **33** (1982) 197-212. MR 84g:05101; Zbl 502.05056 (488.05055: *The existence of ...*) (An earlier version was published as *On a class of Distance-Regular Graphs*, Delft Progr. Rep. 7 (1982) 26-49. MR 84j:05089a; Zbl 496.05030) [182, 208]

629. Roos, C. and A. J. van Zanten, *On the existence of certain generalized Moore geometries, Part I*, Discrete Math. **51** (1984) 179-190. Zbl 547.05021 (An earlier version was published as *On a Class of Distance-Regular Graphs whose Members have Small Diameters*, Delft Progr. Rep. **7** (1982) 50-67. MR 84j:05089b; Zbl 504.05034 {The preprint was called: *On a class of distance-regular graphs whose members cannot have large diameter (Part I).*} Corrected in *Erratum on 'On a Class of Distance Regular Graphs whose Members have Small Diameters'*, Delft Progr. Rep. **8** (1983) 292-297. MR 84j:05089d; Zbl 519.05046) [208]

630. Roos, C. and A. J. van Zanten, *On the existence of certain generalized Moore geometries, Part II*, Discrete Math. **51** (1984) 277-282. Zbl 547.05022 (An earlier version was published as *On a Class of Distance Regular Graphs whose Members have Small Diameters (Part II)*, Delft Progr. Rep. **8** (1983) 3-16. MR 84j:05089c; Zbl 512.05041) [208]

631. Roos, C. and A. J. van Zanten, *On the existence of certain generalized Moore geometries, Part III*, Discrete Math. **58** (1986) 275-283. Zbl 594.05020 [208]

632. Roos, C., A. J. van Zanten, and M. J. Coster, *On the existence of certain generalized Moore geometries, Part IV*, Discrete Math. **62** (1986) 139-144. Zbl 613.05016 [208]

633. Roos, J. E., *An algebraic study of group and nongroup error-correcting codes*, Inform. & Control **8** (1965) 195-214. [349]

634. Sabidussi, G., *Vertex-transitive graphs*, Monatsh. Math. **68** (1964) 426-438. [436]

635. Sachs, H., *Regular graphs with given girth and restricted circuits*, J. London Math. Soc. **38** (1963) 423-429. [209]

636. Sauer, N., *Extremaleigenschaften regulärer Graphen gegebener Taillenweite, I, II*, Österreich. Akad. Wiss. Math. Natur. Kl. S.-B. II **176** (1967) 9-25; **176** (1967) 27-43. [209]

637. Sawade, K., *A Hadamard matrix of order 268*, Graphs Combin. **1** (1985) 185-187. [20]

638. Saxl, J., *On multiplicity-free permutation representations*, pp. 337-353 in: *Finite Geometries and Designs (Proc. Isle of Thorns, 1980)*, London Math. Soc. Lecture Note Ser. 49 (P.J. Cameron, J.W.P. Hirschfeld & D. Hughes, eds.), Cambridge University Press, Cambridge, 1981. [230]

639. Scapellato, R., *Geodetic graphs of diameter two and some related structures*, J. Combin. Th. (B) **41** (1986) 218-229. [38]

640. Schellekens, G. L., *On a hexagonic structure, I, II*, Indag. Math. **24** (1962) 201-217, 218-234. (= Proc. Kon. Ned. Akad. Wet. **A65**) [336]

641. Schellenberg, P. J., *A computer construction for balanced orthogonal matrices*, pp. 513-522 in: *Proc. 6th S.E. Conf. Comb., Gr. Th. & Comp. (Boca Raton 1975)*. [24]

642. Schläfli, L., *An attempt to determine the twenty-seven lines upon a surface of the third order, and to divide such surfaces into species in reference to the reality of the lines upon the surface*, Quart. J. Pure Appl. Math. **2** (1858) 55-65, 110-120. (Gesammelte mathematische Abhandlungen, Birkhäuser, Basel, 1953, Vol. II, 198-218.) [30]

643. Schönheim, J., *On linear and nonlinear single-error-correcting q-nary perfect codes*, Information and Control **12** (1968) 23-26. [355]

644. Schur, I., *Zur Theorie der einfach transitiven Permutationsgruppen*, Sitzungsberichte d. Preussischen Akademie d. Wissenschaften, Physikalisch-Mathematische Klasse (1933) 598-623. [66]
645. Schwenk, A. J., *Computing the characteristic polynomial of a graph,* pp. 153-172 in: *Graphs and Combinatorics, Proc. at George Washington Univ., 1973*, Lecture Notes in Math. 406 (R.A. Bari & F. Harary, eds.), Springer, Berlin, 1974. [436]
646. Schwenk, A. J. and R. J. Wilson, *Eigenvalues of graphs,* pp. 307-336 in: *Selected topics in graph theory* (L.W. Beineke & R.J. Wilson, eds.), Acad. Press, New York, 1978. [82]
647. Scott, L. L., jr., *A condition on Higman's parameters*, Notices Amer. Math. Soc. **20** (1973) A-97. (701-20-45) [49]
648. Scott, L. L., *Some properties of character products*, J. Algebra **45** (1977) 259-265. [49, 50, 137]
649. Seberry, J., *A construction for generalized Hadamard matrices*, J. Statist. Planning and Inference **4** (1980) 365-368. [19]
650. Segre, B., *Le geometrie di Galois*, Ann. Mat. Pura Appl. (4) **48** (1959) 1-96. [274]
651. Segre, B., *Forme e geometrie hermitiane con particolare riguardo al caso finito*, Ann. Mat. Pura Appl. (4) **70** (1965) 1-201. MR 35#4802 [274]
652. Seidel, J. J., *Strongly regular graphs of L_2 type and of triangular type*, Indag. Math. **29** (1967) 188-196. (= Proc. Kon. Ned. Akad. Wet. Ser. A **70**) [106]
653. Seidel, J. J., *Strongly regular graphs with* $(-1,1,0)$ *adjacency matrix having eigenvalue 3*, Lin. Alg. Appl. **1** (1968) 281-298. [30, 109]
654. Seidel, J. J., *Graphs and two-graphs,* pp. 125-143 in: *Proc. 5th Southeastern Conf. on Combinatorics, Graph Theory, and Computing (Boca Raton, 1974)*, Congressus Numerantium 10 (F. Hoffman et al., eds.), Utilitas Math., Winnipeg, 1974. [16, 106]
655. Seidel, J. J., *A survey of two-graphs,* pp. 481-511 in: *Coll. Intern. Teorie Combin. Acc. Naz. Lincei, Roma 1973*, Atti dei convegni Lincei 17, Rome, 1976. [16, 94]
656. Seidel, J. J., *Strongly regular graphs,* pp. 157-180 in: *Surveys in Combinatorics, Proc. 7th British Combinatorial Conference*, London Math. Soc. Lecture Note Ser. 38 (B. Bollobás, ed.), Cambridge University Press, Cambridge, 1979. [11, 48-50]
657. Seidel, J. J., *Graphs and two distance sets,* pp. 90-98 in: *Combinatorial Mathematics VIII*, Lecture Note in Math. 884 (K.L. McAvaney, ed.), Springer, Berlin, 1981. [89]
658. Seidel, J. J. and D. E. Taylor, *Two-graphs, a second survey,* pp. 689-711 in: *Algebraic Methods in Graph Theory, Proc. Intern. Colloq. Szeged 1978*, Coll. Math. Soc. J. Bolyai 25 (L. Lovász & Vera T. Sós, eds.), North Holland Publ. Co., Amsterdam, 1981. [16, 95]
659. Semakov, N. V., V. A. Zinov'ev, and G. V. Zaitsev, *Uniformly packed codes*, Problemy Peredachi Informatsii **7** (1971) 38-50. (Russian; translation in: Problems of Information Transmission **7** (1971) 30-39.) [348]
660. Seneta, E., *Non-negative matrices and Markov chains*, Springer Series in Statistics, Springer, New York (1981). [81]
661. Serre, J.-P., *Cours d'arithmétique*, Presses Universitaires de France, Paris (1970). (English translation: A course in arithmetic, Springer, New York - Heidelberg - Berlin, 1973) [22]
662. Shad, S. A., *Regular thin near n-gons and balanced incomplete block designs*, Arabian J. Sci. Engrg. **9** (1984) 251-260. Zbl 545.05023 [19, 20]
663. Shapiro, H. S. and D. L. Slotnick, *On the mathematical theory of error correcting codes*, IBM J. Res. Develop. **3** (1959) 25-37. [355]
664. Shawe-Taylor, J., *Automorphism groups of primitive distance-bitransitive graphs are almost simple*, European J. Combin. **8** (1987) 187-197. MR88i:05098 [139]
665. Shearer, J., *On the distribution of the maximum eigenvalue of graphs*, preprint (1986). [84]
666. Shrikhande, S. S., *The impossibility of certain symmetrical balanced incomplete block designs*, Ann. Math. Statist. **21** (1950) 106-111. [23]
667. Shrikhande, S. S., *On a characterization of the triangular association scheme*, Ann. Math. Statist. **30** (1959) 39-47. [258]

668. Shrikhande, S. S., *The uniqueness of the L_2 association scheme*, Ann. Math. Statist. **30** (1959) 781-798. [25, 105, 258, 262, 395]
669. Shrikhande, S. S., *Affine resolvable balanced incomplete block designs: a survey*, Aequationes Math. **14** (1976) 251-269. [19]
670. Shult, E. E., *Characterizations of certain classes of graphs*, J. Combin. Th. (B) **13** (1972) 142-167. [30]
671. Shult, E. E., *The graph extension theorem*, Proc. Amer. Math. Soc. **33** (1972) 278-284. [16, 225]
672. Shult, E. E., *Characterizations of the Lie incidence geometries*, pp. 157-186 in: *Surveys in Combinatorics (Invited papers for the Ninth British Combin. Conf.)*, London Math. Soc. Lecture Note Ser. 82 (E. Keith Lloyd, ed.), Cambridge Univ. Press, 1983. [153]
673. Shult, E. E. and A. Yanushka, *Near n-gons and line systems*, Geom. Dedicata **9** (1980) 1-72. [111, 155, 199, 277, 279, 360, 366]
674. Sidel'nikov, V. M., *New bounds for densest packings of spheres in n-dimensional Euclidean space*, Mat. Sbornik **95** (1974) 148-158. (In Russian. English translation in Math. USSR Sbornik **24** (1974) 147-157.) [ix, 77]
675. Sims, Ch. C., *On the isomorphism of two groups of order 44,352,000*, pp. 101-108 in: *The theory of finite groups* (R. Brauer & C.-H. Sah, eds.), Benjamin, 1969. [ix, 392]
676. Singleton, R. R., *On minimal graphs of maximal even girth*, J. Combin. Th. **1** (1966) 306-322. [ix, 208]
677. Singleton, R. R., *There is no irregular Moore graph*, Amer. Math. Monthly **75** (1968) 42-43. [206]
678. Sloane, N. J. A., *An introduction to association schemes and coding theory*, pp. 225-260 in: *Theory and Application of Special Functions, Proc. Sem. Madison 1975* (R.A. Askey, ed.), Academic Press, New York, 1975. Zbl 333.94001 [55]
679. Smith, D. H., *Highly symmetrical graphs of low valency*, Ph.D. thesis, Univ. of Southampton (1971). [221, 222]
680. Smith, D. H., *Primitive and imprimitive graphs*, Quart. J. Math. Oxford (2) **22** (1971) 551-557. [133, 140, 398]
681. Smith, D. H., *On tetravalent graphs*, J. London Math. Soc. (2) **6** (1973) 659-662. Zbl 257.05118 [222]
682. Smith, D. H., *Distance-transitive graphs of valency four*, J. London Math. Soc. (2) **8** (1974) 377-384. MR 52#202b; Zbl 285.05109 [222, 260]
683. Smith, D. H., *On bipartite tetravalent graphs*, Discrete Math. **10** (1974) 167-172. MR 50#4406; Zbl 288.05117 [180, 221, 222]
684. Smith, D. H., *An improved version of Lloyd's theorem*, Discrete Math. **15** (1976) 175-184. [349]
685. Smith, D. H., *Perfect codes in the graphs O_k and $L(O_k)$*, Glasgow Math. J. **21** (1980) 169-172. Zbl 432.05048 [346, 357]
686. Smith, J. H., *Some properties of the spectrum of a graph*, pp. 403-406 in: *Combinatorial Structures and their applications (Proc. Conf. Calgary, 1969)* (R. Guy et al., eds.), Gordon and Breach, New York, 1970. [83]
687. Smith, M. S., *On the isomorphism of two simple groups of order 44,352,000*, J. Algebra **41** (1976) 172-174. [370]
688. Smith, S. D., *Nonassociative commutative algebras for triple covers of 3-transposition groups*, Mich. Math. J. **24** (1977) 273-287. [78, 397, 399]
689. Smits, P. R. J. M. and J. C. W. van Vroonhoven, *The polarities of the partial geometry pg(5,5,2)*, Geom. Dedicata **21** (1986) 51-54. [373]
690. Snover, S. L., *The uniqueness of the Nordstrom-Robinson and the Golay binary codes*, Ph. D. Thesis, Dept. of Math., Michigan State Univ. (1973). [55]
691. Solomon, L., *The orders of the finite Chevalley groups*, J. Algebra **3** (1966) 376-393. [308]
692. Solomon, L., *A Mackey formula in the group ring of a Coxeter group*, J. Algebra **41** (1976) 255-268. (With an appendix by J. Tits.) [297]

693. Somma, C., *An infinite family of perfect codes in antipodal graphs*, Rend. Mat. Appl. (7) **3** (1983) 465-474. MR 86f:94025; Zbl 546.94014 [385]

694. Song, S.-Y., *Products of distance regular graphs*, Util. Math. **29** (1986) 173-175. Zbl 591.05046 (576.05040) [26]

695. Sprague, A. P., *Characterization of projective graphs*, J. Combin. Th. (B) **24** (1978) 294-300. [271]

696. Sprague, A. P., *Pasch's axiom and projective spaces*, Discrete Math. **33** (1981) 79-87. [257, 270]

697. Stanton, D., *Some Erdős-Ko-Rado theorems for Chevalley groups*, SIAM J. Alg. Disc. Meth. **1** (1980) 160-163. MR 81h:05027 [279]

698. Stanton, D., *Some q-Krawtchouk polynomials on Chevalley groups*, Amer. J. Math. **102** (1980) 625-662. MR 82a:33015 [Note that the page numbers in issues 3 and 4 overlap; this paper is in issue 4.] [275, 276]

699. Stanton, D., *Generalized n-gons and Chebychev polynomials*, J. Combin. Th. (A) **34** (1983) 15-27. MR 84g:05038 [201]

700. Stanton, D., *t-Designs in classical association schemes*, Graphs Combin. **2** (1986) 283-286. [60]

701. Stellmacher, B., *Einfache Gruppen, die von einer Konjugiertenklasse von Elementen der Ordnung drei erzeugt werden*, J. Algebra **30** (1974) 320-354. [411, 412]

702. Stemple, J. G., *Geodetic graphs of diameter two*, J. Combin. Th. (B) **17** (1974) 266-280. [38]

703. Stoer, J. and R. Bulirsch, *Einführung in die numerische Mathematik*, Springer, Berlin (1979). [129]

704. Suzuki, M., *Group Theory I, II*, Grundlehren d. math. Wiss. 247, 248, Springer, Berlin (1982, 1986). [214]

705. Sylvester, J. J., *Elementary researches in the analysis of combinatorial aggregation*, Philos. Mag. **24** (1844) 285-296. Reprinted in: Coll. Math. Papers 1, Cambridge Univ. Press, 1904, pp. 91-102. [30]

706. Sylvester, J. J., *A demonstration of the theorem that every homogeneous quadratic polynomial is reducible by real orthogonal substitutions to the form of a sum of positive and negative squares*, Philos. Mag. **4** (1852) 138-142. Reprinted in: Coll. Math. Papers 1, Cambridge Univ. Press, 1904, pp. 378-381. [86]

707. Tallini, G., *On a characterization of the Grassmann manifold representing the lines in a projective space,* pp. 354-358 in: *Finite Geometries and Designs*, London Math. Soc. Lecture Note Ser. 49 (P.J. Cameron, J.W.P. Hirschfeld & D. Hughes, eds.), Cambridge University Press, Cambridge, 1981. [270]

708. Tamaschke, O., *Zur Theorie der Permutationsgruppen mit regulärer Untergruppe I, II*, Math. Z. **80** (1963) 328-354, 443-465. [70]

709. Tamaschke, O., *Schur-Ringe*, Hochschulskripten 735/735a, Bibliographisches Institut, Mannheim (1970). [65]

710. Taylor, D. E., *Regular 2-graphs*, Proc. London Math. Soc. (3) **35** (1977) 257-274. MR 57#16147 [14, 33, 95, 96, 103, 228]

711. Taylor, D. E. and R. Levingston, *Distance-regular graphs,* pp. 313-323 in: *Combinatorial Mathematics, Proc. Canberra 1977*, Lecture Notes in Math. 686 (D.A. Holton & J. Seberry, eds.), Springer, Berlin, 1978. MR 80i:05077; Zbl 387.05026 [viii, 7, 14, 15, 133, 167, 168, 175]

712. Tchuda, F. L., Построение автоморфного графа на 280 вершинах ц помощь' конечных геометрий *(Construction of an automorphic graph on 280 vertices using finite geometries),* pp. 169-174 in: Исследования по алгебраической теории комбинаторных объектов *(Investigations in the Algebraic Theory of Combinatorial Objects)* (M.H. Klin & I.A. Faradjev, eds.), Institute for System Studies, Moscow, 1985. (Russian) MR 88i:05093 [viii, 364]

713. Terwilliger, P., *The diameter of bipartite distance-regular graphs*, J. Combin. Th. (B) **32** (1982) 182-188. MR 83f:05063; Zbl 487.05038 [171, 184]

714. Terwilliger, P., *Eigenvalue multiplicities of highly symmetric graphs*, Discrete Math. **41** (1982) 295-302. Zbl 495.05045 [viii, 163]
715. Terwilliger, P., *Distance-regular graphs and (s,c,a,k)-graphs*, J. Combin. Th. (B) **34** (1983) 151-164. MR 84i:05100; Zbl 511.05052 [21, 171]
716. Terwilliger, P., *Counting 4-vertex configurations in P- and Q-polynomial association schemes*, Algebras Groups Geom. **2** (1985) 541-554. Zbl 592.05011 [239]
717. Terwilliger, P., *Distance-regular graphs with girth 3 or 4, I*, J. Combin. Th. (B) **39** (1985) 265-281. Zbl 588.05039 [33, 169]
718. Terwilliger, P., *A class of distance-regular graphs that are Q-polynomial*, J. Combin. Th. (B) **40** (1986) 213-223. Zbl 579.05029 (568.05029) [33, 169]
719. Terwilliger, P., *The Johnson graph $J(d,r)$ is unique if $(d,r) \neq (2,8)$*, Discrete Math. **58** (1986) 175-189. Zbl 587.05038 [258]
720. Terwilliger, P., *A new feasibility condition for distance-regular graphs*, Discrete Math. **61** (1986) 311-315. Zbl 606.05045 [viii, 158]
721. Terwilliger, P., *A characterization of P- and Q-polynomial association schemes*, J. Combin. Th. (A) **45** (1987) 8-26. MR 88g:05038 [73, 75, 76]
722. Terwilliger, P., *Distance regular graphs with girth 3 or 4, II*, Preprint (1987). [178]
723. Terwilliger, P., *P- and Q-polynomial schemes with $q = -1$*, J. Combin. Th. (B) **42** (1987) 64-67. [The preprint was called 'Towards a classification of distance-regular graphs with the Q-polynomiality property' and contains slightly more information.] [164, 241, 244]
724. Terwilliger, P., *Root system graphs*, Lin. Alg. Appl. **94** (1987) 157-163. [113]
725. Terwilliger, P., *Root systems and the Johnson and Hamming graphs*, European J. Combin. **8** (1987) 73-102. Zbl 614.05048 [The preprint was called 'A characterization of the Johnson and Hamming graphs'.] [170, 258, 264]
726. Terwilliger, P., *Balanced sets and Q-polynomial association schemes*, Graphs Combin. **4** (1988) 87-94. MR 88j:05010; Zbl 644.05016 [239]
727. Terwilliger, P., *The classification of distance-regular graphs of type IIB*, Combinatorica **8** (1988) 125-132. [164, 244, 264]
728. Terwilliger, P., *A new inequality for distance-regular graphs*, Discrete Math. (submitted). [239, 244]
729. Thas, J. A., *On 4-gonal configurations with parameters $r = q^2+1$ and $k = q+1$, part I*, Geom. Dedicata **3** (1974) 365-375. [30]
730. Thas, J. A., *Some results concerning $\{(q+1)(n-1);n\}$-arcs and $\{(q+1)(n-1)+1;n\}$-arcs in finite projective planes of order q*, J. Combin. Th. (A) **19** (1975) 228-232. [387]
731. Thas, J. A., *Two infinite classes of perfect codes in metrically regular graphs*, J. Combin. Th. (B) **23** (not 22) (1977) 236-238. Zbl 357.05070 (322.05148) [357, 385]
732. Thas, J. A., *Polar spaces, generalized hexagons and perfect codes*, J. Combin. Th. (A) **29** (1980) 87-93. Zbl 444.05027 [357]
733. Thas, J. A., *A remark on the theorem of Yanushka and Ronan characterizing the 'generalized hexagon' $H(q)$ arising from the group $G_2(q)$*, J. Combin. Th. (A) **29** (1980) 361-362. [204]
734. Thas, J. A. and F. de Clerck, *Partial geometries satisfying the axiom of Pasch*, Simon Stevin **51** (1977) 123-137. [282]
735. Thompson, J. G., *Bounds for orders of maximal subgroups*, J. Algebra **14** (1970) 135-138. [217]
736. Tietäväinen, A., *On the nonexistence of perfect codes over finite fields*, SIAM J. Appl. Math. **24** (1973) 88-96. Zbl 251.94009 (233.94004) [355]
737. Tietäväinen, A., *A short proof for the nonexistence of unknown perfect codes over $GF(q)$, $q > 2$*, Annales Acad. Scient. Fennicae, Ser. A (I. Mathematica) **580** (1974) 1-6. Zbl 289.94004 [355]
738. Tietäväinen, A. and A. Perko, *There are no unknown perfect binary codes*, Ann. Univ. Turku. Ser. A I **148** (1971) 3-10. [355]
739. Tilborg, H. C. A. van, *Uniformly packed codes*, Thesis, Univ. of Techn. Eindhoven (1976). Zbl 333.94004 [355, 356]

740. Timmesfeld, F. G., *Groups generated by root-involutions. I*, J. Algebra **33** (1975) 75-134. [396]
741. Timmesfeld, F. G., *Groups generated by root-involutions. II*, J. Algebra **35** (1975) 367-441. [396]
742. Tits, J., *Les groupes de Lie exceptionnels et leur interprétation géométrique*, Bull. Soc. Math. Belg. **8** (1956) 48-81. [339]
743. Tits, J., *Sur la trialité et certains groupes qui s'en déduisent*, Publ. Math. I.H.E.S. **2** (1959) 14-60. [200]
744. Tits, J., *Structures et groupes de Weyl*, Sém. Bourbaki no. 288 (1964/1965), W.A. Benjamin, New York (1966). [323]
745. Tits, J., *Le problème des mots dans les groupes de Coxeter*, Ist. Naz. Alta Mat., Symposia Math. **1** (1967/1968) 175-185. [297]
746. Tits, J., *Le groupe de Janko d'ordre 604,800*, pp. 91-95 in: *The theory of finite groups* (R. Brauer & C.-H. Sah, eds.), Benjamin, 1969. [408]
747. Tits, J., *Buildings of spherical type and finite BN-pairs*, Lecture Notes in Math. 386, Springer, Berlin (1974). MR 57#9866 [276, 277, 294, 323, 329, 331, 344]
748. Tits, J., *Classification of buildings of spherical type and Moufang polygons: a survey*, pp. 229-246 in: *Coll. Intern. Teorie Combin. Acc. Naz. Lincei, Roma 1973*, Atti dei convegni Lincei 17, Rome, 1976. MR 56#3140 [205]
749. Tits, J., *Nonexistence de certains polygones généralisés, I*, Invent. Math. **36** (1976) 275-284. MR 55#8208 [205]
750. Tits, J., *Endliche Spiegelungsgruppen, die als Weylgruppen auftreten*, Invent. Math. **43** (1977) 283-295. MR 57#478 [326]
751. Tits, J., *Nonexistence de certains polygones généralisés, II*, Invent. Math. **51** (1979) 267-269. [205]
752. Tits, J., *Quaternions over* $\mathbf{Q}(\sqrt{5})$*, Leech's lattice and the sporadic group of Hall-Janko*, J. Algebra **63** (1980) 56-75. [409]
753. Tits, J., *A local approach to buildings*, pp. 519-547 in: *The Geometric Vein (the Coxeter Festschrift)* (C. Davis, B. Grünbaum, F.A. Sherk, ed.), Springer Verlag, Berlin, 1981. [153, 294]
754. Tonchev, V. D., *A characterization of designs related to the Witt system* $S(5,8,24)$, Math. Z. **191** (1986) 225-230. MR 87g:05040 [367, 368]
755. Tůma, J., *A structure theorem for lattices of generalized partitions*, pp. 737-757 in: *Contributions to Lattice Theory*, Colloq. Math. Soc. János Bolyai 33 (A.P. Huhn & E.T. Schmidt, eds.), North-Holland Publ. Co., Amsterdam, 1983. (Proc. of the 2nd colloq. on lattice theory, Szeged, 1980.) [257]
756. Turán, P., *Egy gráfelméleti szélsőérték feladatról (On an extremal problem in graph theory)*, Mat. Fiz. Lapok **48** (1941) 436-452. (Hungarian. German summary) MR 8, p. 284 [181]
757. Tutte, W. T., *A family of cubical graphs*, Proc. Cambridge Philos. Soc. **43** (1947) 459-474. [209, 221]
758. Tutte, W. T., *A non-Hamiltonian graph*, Canad. Math. Bull. **3** (1960) 1-5. [221, 383]
759. Tutte, W. T., *Connectivity in graphs*, Univ. of Toronto Press, Toronto (1966). [206, 209]
760. Ustimenko, V. A., *On some properties of the geometries of the Chevalley groups and their generalizations*, to appear. [279]
761. Vanden Cruyce, P., *A finite graph which is locally a dodecahedron*, Discr. Math. **54** (1985) 343-346. Zbl 571.05047 [5]
762. Vanden Cruyce, P., *Geometries related to* $PSL(2,19)$, European J. Combin. **6** (1985) 163-173. Zbl 575.51013 [403]
763. Varga, R. S., *Matrix iterative analysis*, Prentice-Hall, Englewood Cliffs, New York (1962). [81]
764. Vasil'ev, Ju. L., О негрупповых плотно упакованных кодах *(On nongroup close-packed codes)*, Probl. Kibern. **8** (1962) 337-339. (Russian; translated in Probleme der Kybernetik **8** (1965) 375-378.) MR 29#5661 [355]

765. Verma, D.-N., *Möbius inversion for the Bruhat ordering on a Weyl group*, Ann. Sci. École Norm. Sup. (4) **4** (1971) 393-398. MR 45#139 [298]

766. Vijayan, K.S., *On a class of distance transitive graphs*, J. Combin. Th. (B) **25** (1978) 125-129. MR 80a:05174; Zbl 406.05035 [226]

767. Wallis, J.S., *On the existence of Hadamard matrices*, J. Combin. Th. (A) **21** (1976) 188-195. [20]

768. Wallis, W.D., A.P. Street, and J.S. Wallis, *Combinatorics: Room squares, Sum-Free sets, Hadamard Matrices*, Lecture Notes in Math. 292, Springer Verlag, Berlin (1972). [10, 20]

769. Wan, Zh.-x., *Studies in finite geometries and the construction of incomplete block designs. I. Some 'Anzahl' theorems in symplectic geometry over finite fields*, Acta Math. Sinica **15** (1965) 354-361. MR 32#6313a (Chinese. English transl. in Chinese Math. **7** (1965) 55-62.) [274]

770. Wan, Zh.-x., Z.-d. Dai, X.-n. Feng, and B.-f. Yang, *Some studies on finite geometries and incomplete block designs*, Science Press, Beijing (1966). (Chinese.) [275]

771. Wan, Zh.-x. and B.-f. Yang, *Studies in finite geometries and the construction of incomplete block designs. III. Some 'Anzahl' theorems in unitary geometry over finite fields and their applications*, Acta Math. Sinica **15** (1965) 533-544. MR 32#6313c (Chinese. English transl. in Chinese Math. **7** (1965) 252-264.) [274]

772. Wegner, G., *A smallest graph of girth 5 and valency 5*, J. Combin. Th. (B) **14** (1973) 203-208. [210]

773. Weichsel, P.M., *On distance-regularity in graphs*, J. Combin. Th. (B) **32** (1982) 156-161. Zbl 477.05047 and 487.05043 [139]

774. Weichsel, P.M., *Products of highly regular graphs,* pp. 497-507 in: *Progress in Graph Theory (Waterloo, 1982)*, Acad. Press, Toronto, 1984. Zbl 555.05050 [26]

775. Weiss, R., *Elations of graphs*, Acta Math. Acad. Sci. Hungar. **34** (1979) 101-103. [viii, 217]

776. Weiss, R., *The nonexistence of certain Moufang polygons*, Invent. Math. **51** (1979) 261-266. [205]

777. Weiss, R., *s-Transitive graphs,* pp. 827-847 in: *Algebraic Methods in Graph Theory*, vol. 2 (L. Lovász & V.T. Sós, eds.), North Holland, 1981. [231]

778. Weiss, R., *Distance-transitive graphs and generalized polygons*, Arch. Math. **45** (1985) 186-192. Zbl 568.05030 [221]

779. Weiss, R., *On distance-transitive graphs*, Bull. London Math. Soc. **17** (1985) 253-256. Zbl 572.05031 [214, 218, 220]

780. Wester, M., *Endliche fahnentransitive Tits-Geometrien und ihre universellen Überlagerungen*, Mitt. Math. Sem. Gießen **170** (1985). [394]

781. Wielandt, H., *Zur Theorie der einfachen transitiven Permutationsgruppen*, Math. Z. **40** (1935) 582-587. [67]

782. Wielandt, H., *Finite permutation groups*, Academic Press, New York (1964). [46, 47, 63, 65, 66, 68, 438]

783. Wielandt, H., *Permutation groups through invariant relations and invariant functions*, Ohio State Univ. Lecture Notes (1969). [68]

784. Wielandt, H., *Subnormal subgroups and permutation groups*, Lecture Notes, Ohio State University (1971). [217]

785. Wild, P., *Generalized Hussain graphs and semibiplanes with $k \leqslant 6$*, Ars Combin. **14** (1982) 147-167. [18]

786. Wilkinson, J.H. and C. Reinsch, *Handbook for automatic computation II: Linear Algebra*, Grundlehren der Math. Wiss. 186, Springer, New York (1971). [47, 130]

787. Wilson, R.A., *The complex Leech lattice and maximal subgroups of the Suzuki group*, J. Algebra **84** (1983) 151-188. [411, 412]

788. Wilson, R.A., *The geometry of the Hall-Janko group as a quaternionic reflection group*, Geom. Dedicata **20** (1986) 157-173. [409]

789. Wilson, R. M., *The exact bound in the Erdős-Ko-Rado theorem*, Combinatorica **4** (1984) 247-257. [261]
790. Witt, E., *Spiegelungsgruppen und Aufzählung halbeinfacher Liescher Ringe*, Abh. Math. Sem. Hansischen Univ. **14** (1941) 289-322. [100]
791. Wolfmann, J., *Codes projectifs à deux poids, 'caps' complets et ensembles de différences*, J. Combin. Th. (A) **23** (1977) 208-222. [356]
792. Wong, P. K., *On the uniqueness of the smallest graphs of girth 5 and valency 6*, J. Graph Th. **3** (1978) 407-409. [210]
793. Wong, P. K., *Cages - a survey*, J. Graph Th. **6** (1982) 1-22. [210]
794. Wong, P. K., *On the smallest graphs of girth 10 and valency 3*, Discrete Math. **43** (1983) 119-124. [210]
795. Wong, W. J., *Determination of a class of primitive permutation groups*, Math. Z. **99** (1967) 235-246. [222]
796. Yanushka, A., *Generalized hexagons of order* (t,t), Israel J. Math. **23** (1976) 309-324. [205]
797. Yap, H. P., *Some topics in graph theory*, Cambridge Univ. Press (1986). [231]
798. Yoshizawa, M., *Remarks on distance-regular graphs*, Discrete Math. **34** (1981) 93-95. MR 82g:05082; Zbl 458.05053 [180]
799. Zanten, A. J. van, *The degree of the eigenvalues of generalized Moore geometries*, Discrete Math. **67** (1987) 89-96. [208]
800. Zinov'ev, V. A. and V. K. Leont'ev, **Несуществование совершенных кодов над полями** Галуа *(On non-existence of perfect codes over Galois fields)*, Probl. Control Inform. Theory **2** (1973) 123-132. (Russian, English abstract; English transl. in Appendix 2, nr. 2, pp. 16-24.) Zbl 318.94013 [355]

Symbols and notation

Intersection arrays

(Arranged in lexicographic order of *d*, *v*, *k*; no references to Chapter 14 are given.)

266 {12,11,11; 1,1,12}, 225
280 {27,16,9; 1,3,12}, 170
352 {36,25,8; 1,4,20}, 160
352 {175,102,1; 1,102,175}, 370
352 {175,72,1; 1,72,175}, 370
364 {12,9,9; 1,1,4}, 225
410 {81,64,1; 1,16,81}, 386
506 {15,14,12; 1,1,9}, 156, 177, 367
512 {21,20,16; 1,2,12}, 363
552 {275,112,1; 1,112,275}, 373
552 {275,162,1; 1,162,275}, 373
598 {72,45,16; 1,8,54}, 160
625 {24,18,16; 1,2,9}, 170
729 {24,22,20; 1,2,12}, 206, 359
729 {42,28,5; 1,2,30}, 152
759 {30,28,24; 1,3,15}, 206, 366, 367
819 {18,16,16; 1,1,9}, 204
1024 {22,21,20; 1,2,6}, 206, 362
1024 {231,160,6; 1,48,210}, 151, 160, 211
1122 {50,42,1; 1,2,50}, 36
1285 {256,204,1; 1,51,256}, 386
1296 {35,30,26; 1,3,10}, 169
1296 {35,28,22; 1,2,14}, 170
1326 {50,42,9; 1,2,42}, 36
1491 {496,330,1; 1,165,496}, 212, 390
1521 {38,30,24; 1,2,15}, 170
2002 {125,112,9; 1,8,125}, 148
2002 {126,125,9; 1,9,126}, 148
2048 {23,22,21; 1,2,3}, 206, 361
2048 {253,210,3; 1,30,231}, 151, 211, 362
2401 {50,40,27; 1,2,20}, 152
2457 {24,16,16; 1,1,3}, 204
2500 {49,42,36; 1,3,14}, 169

18 {3,2,2,1; 1,1,2,3}, 221
28 {3,2,2,1; 1,1,1,2}, 211, 221, 382
30 {3,2,2,2; 1,1,1,3}, 221
32 {4,3,3,1; 1,1,3,4}, 223, 266, 267
32 {5,4,1,1; 1,1,4,5}, 212, 223, 266
32 {8,7,4,1; 1,4,7,8}, 20, 224
36 {6,5,4,1; 1,2,5,6}, 223, 398
42 {10,6,3,1; 1,3,6,10}, 165
45 {4,2,2,2; 1,1,1,2}, 223, 230
45 {6,4,2,1; 1,1,4,6}, 212, 223, 398
48 {12,11,6,1; 1,6,11,12}, 20, 225
50 {5,4,4,1; 1,1,4,5}, 223
52 {6,5,4,3; 1,2,3,6}, 154
54 {9,8,6,1; 1,3,8,9}, 224
63 {10,6,4,1; 1,2,6,10}, 37, 166, 212, 224, 399
64 {8,7,6,1; 1,2,7,8}, 224
80 {4,3,3,3; 1,1,1,4}, 225
81 {8,6,4,2; 1,2,3,4}, 224
96 {10,5,4,2; 1,2,2,10}, 152
98 {7,6,6,1; 1,1,6,7}, 223
100 {8,7,6,4; 1,2,4,8}, 157
100 {15,14,10,3; 1,5,12,15}, 212, 393
112 {10,9,1,1; 1,1,9,10}, 372
128 {8,7,7,1; 1,1,7,8}, 224
162 {6,5,5,4; 1,1,2,6}, 212, 373
162 {9,8,8,1; 1,1,8,9}, 224
170 {5,4,4,4; 1,1,1,5}, 223
176 {5,4,3,3; 1,1,1,2}, 177
184 {26,25,5,1; 1,5,25,26}, 134
200 {10,9,8,6; 1,2,4,10}, 157
242 {11,10,10,1; 1,1,10,11}, 224
243 {20,18,4,1; 1,2,18,20}, 212, 365
256 {12,9,6,3; 1,2,3,4}, 225
266 {11,10,6,1; 1,1,5,11}, 148, 211, 224, 406
280 {9,8,6,3; 1,1,3,8}, 156, 177, 211, 224, 364
288 {12,11,10,7; 1,2,5,12}, 212, 371
288 {21,14,9,3; 1,3,6,21}, 170
315 {10,8,8,2; 1,1,4,5}, 206, 224, 408, 409
330 {7,6,4,4; 1,1,1,6}, 156, 211, 223, 368
338 {13,12,12,1; 1,1,12,13}, 225
364 {22,15,9,4; 1,3,6,10}, 165
378 {45,32,12,1; 1,6,32,45}, 160, 212, 399
425 {8,4,4,4; 1,1,1,2}, 224
486 {45,44,36,5; 1,9,40,45}, 360
525 {12,8,6,4; 1,1,2,9}, 410
576 {23,16,14,2; 1,2,8,14}, 152
792 {21,20,18,6; 1,3,6,12}, 183
1134 {117,80,24,1; 1,12,80,117}, 160, 212, 400
1152 {16,15,14,13; 1,2,3,16}, 154
1170 {9,8,8,8; 1,1,1,9}, 224
1300 {21,18,12,4; 1,1,6,21}, 148
1694 {55,54,44,40; 1,11,15,55}, 173
1755 {10,8,8,8; 1,1,1,5}, 224
2835 {26,24,22,2; 1,2,3,22}, 155
2925 {12,8,8,8; 1,1,1,3}, 165, 225
4096 {24,23,22,21; 1,2,3,24}, 212, 360
4096 {63,48,45,3; 1,3,16,45}, 170
4160 {22,21,21,3; 1,1,3,22}, 148
22880 {280,243,144,10; 1,8,90,280}, 211, 410
421500 {120,119,118,116; 1,2,4,120}, 157
920808 {31671,28160,2160,1; 1,1080,28160,31671}, 212

20 {3,2,1,1,1; 1,1,1,2,3}, 1, 212, 221
20 {3,2,2,1,1; 1,1,2,2,3}, 221
50 {4,3,3,2,1; 1,1,2,3,4}, 180
100 {7,6,6,1,1; 1,1,6,6,7}, 212, 223, 393
112 {10,9,8,2,1; 1,2,8,9,10}, 212, 224
112 {10,9,4,2,1; 1,2,4,9,10}, 372, 417
154 {16,15,12,4,1; 1,4,12,15,16}, 212
200 {22,21,16,6,1; 1,6,16,21,22}, 212
243 {10,8,6,4,2; 1,2,3,4,5}, 224
310 {7,6,6,4,4; 1,1,3,3,7}, 223
729 {22,20,18,2,1; 1,2,9,20,22}, 212, 365
2420 {13,12,12,9,9; 1,1,4,4,13}, 225
2912 {105,90,49,7,1; 1,7,49,90,105}, 160

70 {4,3,3,1,1,1; 1,1,1,3,3,4}, 260
126 {3,2,2,2,2,2; 1,1,1,1,1,3}, 204, 221
189 {4,2,2,2,2,2; 1,1,1,1,1,2}, 204, 225

270 $\{12,11,9,8,3,1;1,3,4,9,11,12\}$, 173
728 $\{4,3,3,3,3,3;1,1,1,1,1,4\}$, 223
729 $\{12,10,8,6,4,2;1,2,3,4,5,6\}$, 225
1024 $\{21,20,16,6,2,1;1,2,6,16,20,21\}$, 151, 212, 365
1456 $\{6,3,3,3,3,3;1,1,1,1,1,2\}$, 223
1518 $\{15,14,12,6,1,1;1,1,3,12,14,15\}$, 368
1536 $\{21,20,16,8,2,1;1,2,4,16,20,21\}$, 157, 365
2048 $\{21,20,16,9,2,1;1,2,3,16,20,21\}$, 212, 365
2048 $\{22,21,20,3,2,1;1,2,3,20,21,22\}$, 151, 212, 365
4554 $\{15,14,12,8,1,1;1,1,1,12,14,15\}$, 368
67648 $\{70,69,68,65,2,1;1,2,5,68,69,70\}$, 416
132860 $\{10,9,9,9,9,9;1,1,1,1,1,10\}$, 224

102 $\{3,2,2,2,1,1,1;1,1,1,1,1,1,3\}$, 211, 221, 403
2048 $\{22,21,20,16,6,2,1;1,2,6,16,20,21,22\}$, 212
4096 $\{23,22,21,20,3,2,1;1,2,3,20,21,22,23\}$, 212

90 $\{3,2,2,2,2,1,1,1;1,1,1,1,2,2,2,3\}$, 212, 221, 398
660 $\{7,6,4,4,3,1,1,1;1,1,1,3,4,4,6,7\}$, 156
990 $\{7,6,4,4,4,1,1,1;1,1,1,2,4,4,6,7\}$, 156, 177, 212, 223, 369

$\{k;1\}$, 8, 222

$\{2\mu,\mu;1,\mu\}$, 8
$\{k,k-1-\lambda;1,\mu\}$, 8
$\{k,k-1;1,k\}$, 222
$\{q(q^3+1)\frac{q^8-1}{q-1}, q^7\cdot\frac{q^5-1}{q-1}; 1, (q^3+1)\frac{q^4-1}{q-1}\}$, 335
$\{s(t+1), st; 1, t+1\}$, 386

$\{2k-2, k-1, k-2; 1, 1, 4\}$, 149
$\{3\mu+1, 2\mu, 1; 1, \mu, 3\mu+1\}$, 24
$\{3\mu-3, \mu, 1; 1, \mu, 3\mu-3\}$, 33
$\{b-i, b-i-l+1, i; 1, 1, b-2i+15\}$, 42
$\{k,\lambda,1;1,\lambda,k\}$, 228
$\{k,\mu,1;1,\mu,k\}$, 13, 16, 212, 228, 242
$\{k,\mu(r-1),1;1,\mu,k\}$, 242, 431
$\{k,k-1-\lambda,1;1,1,k\}$, 40
$\{k,k-1,1;1,2,3\}$, 173
$\{k,k-1,1;1,k-1,k\}$, 16, 173, 222, 432
$\{k,k-1,k-\mu;1,\mu,k\}$, 17, 212, 432

$\{\frac{(q^{12}-1)(q^9-1)}{q^4-1}, q^8(q^4+1)(q^5-1), q^{16}(q-1); 1, q^4(q^4+1), q^8\frac{q^{12}-1}{q^4-1}\}$, 341
$\{q^2-1, q^2-q, 1; 1, q, q^2-1\}$, 286, 358
$\{q^2-q, q^2-q-2, q+1; 1, 1, q^2-2q\}$, 42, 211, 383
$\{q^{2j+1}-1, q^{2j+1}-q, q^{2j}(q-1)+1; 1, q, q^{2j}-1\}$, 358
$\{q^3(q+1), q^4, q^4; 1, 1, q+1\}$, 335
$\{q^n-1, q^n-q^{n-1}, 1; 1, q^{n-1}, q^n-1\}$, 385
$\{q^n-1, (r-1)q^n/r, 1; 1, q^n/r, q^n-1\}$, 385
$\{q(q^3+1), q^4, q^4; 1, 1, q^3+1\}$, 335
$\{q(q^8+q^4+1)\frac{q^9-1}{q-1}, q^9(q^4+1)\frac{q^5-1}{q-1}, q^{17}; 1, (q^4+1)\frac{q^5-1}{q-1}, (q^8+q^4+1)\frac{q^9-1}{q-1}\}$, 335
$\{q(q+1), q^2, q^2; 1, 1, q+1\}$, 205, 335
$\{q, q-2, 1; 1, 1, q\}$, 40, 387
$\{q, q-m-1, 1; 1, m, q\}$, 386
$\{q, (r-1)\mu, 1; 1, \mu, q\}$, 389
$\{r\mu+1, (r-1)\mu, 1; 1, \mu, r\mu+1\}$, 23, 212, 388, 431
$\{r\mu-1, (r-1)\mu, 1; 1, \mu, r\mu-1\}$, 212
$\{r^{2j+1}-1, r^{2j+1}-r, r^{2j}(r-1)+1; 1, r, r^{2j}-1\}$, 211
$\{st, s(t-1), 1; 1, t-1, st\}$, 212, 385
$\{v-1, 2(k-\lambda-1), 1; 1, 2(k-\lambda-1), v-1\}$, 16

$\{2\mu, 2\mu-1, \mu, 1; 1, \mu, 2\mu-1, 2\mu\}$, 19, 20, 22, 183
$\{2\mu^2+\mu, 2\mu^2-\mu-1, \mu^2, \mu; 1, \mu, \mu^2-\mu, 2\mu^2+\mu\}$, 248, 419
$\{k, k-\lambda-1, \frac{2}{3}\mu, 1; 1, \frac{1}{3}\mu, k-\lambda-1, k\}$, 397
$\{k, k-\mu, (r-1)\mu, 1; 1, \mu, k-1, k\}$, 421
$\{k, k-1, k-\mu, 1; 1, \mu, k-1, k\}$, 212
$\{m\mu, m\mu-1, (m-1)\mu, 1; 1, \mu, m\mu-1, m\mu\}$, 18, 425
$\{n^2-n+1, n(n-1), n(n-1), n; 1, 1, (n-1)^2, n^2-n+1\}$, 387
$\{q^2, q^2-1, q^2-q, 1; 1, q, q^2-1, q^2\}$, 358
$\{q^2(q^4+1), q^6, q^6, q^6; 1, 1, 1, q^4+1\}$, 335
$\{q^2-q+1, q^2-q, q^2-q, q; 1, 1, (q-1)^2, q^2-q+1\}$, 212
$\{q^{2j+1}, q^{2j+1}-1, q^{2j+1}-q, q^{2j}(q-1)+1; 1, q, q^{2j}-1, q^{2j+1}\}$, 212, 358
$\{q^4(q^2+1), q^6, q^6, q^6; 1, 1, 1, q^2+1\}$, 335
$\{q^d, q^d-1, q^d-q^{d-1}, 1; 1, q^{d-1}, q^d-1, q^d\}$, 273
$\{q, q-1, q-1, 1; 1, 1, q-1, q\}$, 273

$\{2\mu^2+\mu, 2\mu^2+\mu-1, \mu^2, \mu, 1; 1, \mu, \mu^2, 2\mu^2+\mu-1, 2\mu^2+\mu\}$, 417
$\{2\mu^2+\mu, 2\mu^2+\mu-1, 2\mu^2, \mu, 1; 1, \mu, 2\mu^2, 2\mu^2+\mu-1, 2\mu^2+\mu\}$, 418
$\{k, k-1, k-1, \frac{1}{2}k, 1; 1, 1, \frac{1}{2}k, k-1, k\}$, 180

Author index

Subject index

3260000R00277

Printed in Great Britain
by Amazon.co.uk, Ltd.,
Marston Gate.